ORIGINS

Cellular Origin and Life in Extreme Habitats and Astrobiology

Volume 6

Series Editor:

Joseph Seckbach
Hebrew University of Jerusalem, Israel

Origins

Genesis, Evolution and Diversity of Life

Edited by

Joseph Seckbach
Hebrew University of Jerusalem, Israel

KLUWER ACADEMIC PUBLISHERS
DORDRECHT / BOSTON / LONDON

A C.I.P. Catalogue record for this book is available from the Library of Congress.

ISBN 1-4020-1813-4
ISBN 1-4020-2522-X (e-book)

Published by Kluwer Academic Publishers,
P.O. Box 17, 3300 AA Dordrecht, The Netherlands.

Sold and distributed in North, Central and South America
by Kluwer Academic Publishers,
101 Philip Drive, Norwell, MA 02061, U.S.A.

In all other countries, sold and distributed
by Kluwer Academic Publishers,
P.O. Box 322, 3300 AH Dordrecht, The Netherlands.

Printed on acid-free paper

DEDICATION

This volume is dedicated to a friend, colleague and prominent scientist **Professor Julian Chela-Flores**, the organizer of the conferences on ***Chemical Evolution and Origin of Life***, which have been taken place in the International Center for Theoretical Physics, Trieste, Italy. My best blessing and wishes to him and his family.

This volume is also dedicated to our second-generation offspring: **Oz, Aniam, Eyal, Yiska, Ysrael and Kedem.**

TABLE OF CONTENTS

III. Alternatives to the Standard Scenario

IV. The First Steps of Biological Evolution

V. The First Cells

Biodata of **Leslie E. Orgel** author of the ***Foreword*** for this book.

Dr. Leslie Orgel is a Fellow and Research Professor at the Salk Institute for Biological Studies where he directed the Chemical Evolution Laboratory. He is also an Adjunct Professor in the Department of Chemistry at the University of California, San Diego, and one of five Principal Investigators in the NASA-sponsored NSCORT program in Exobiology. His research is concerned with chemical evolution. He and his coworkers have studied non-enzymatic polymerization reactions in aqueous solution that imitate nucleic acid replication and protein synthesis. In particular, they have developed methods for transcribing DNA sequences into RNA with high fidelity. Leslie E. Orgel was born in London, England. He received his Ph.D. in Chemistry at Oxford in 1951. Dr. Orgel is a Fellow of the Royal Society (1962) and of the American Academy of Arts and Sciences (1985), and also a Member of the National Academy of Sciences (1990).

E-mail: **orgel@salk.edu**

FOREWORD

It seems almost certain that there was once an RNA word in which RNA functioned both as a genetic material and a source of essential catalytic activities. It helps to think of the problem of the Origin of Life as made up of three possibly overlapping sub-problems. How did the RNA world come into existence? How did the RNA world "invent" the last common ancestor of life (LCA), a microorganism of the DNA/RNA/Protein world? How did the LCA diversify into the multitude of early life forms that we know from the fossil record or infer from extant life? This book has much to offer those interested in the first and last of these problems. The possibility that life exists elsewhere in the solar system or the Universe is another topic addressed by several authors.

The core of the book is concerned with the first problem. Life must have originated in an aqueous environment, but was it on a snowball earth, in a hydrothermal vent or on the shores of a temperate ocean? The bad news is that two of these hypotheses must presumably be wrong: the good news is that one of them is probably right. This is a frustrating problem that you can't resist thinking about although you know that you are not likely to find a generally accepted solution, at least for the present. Fortunately, several chapters attempt to justify one or other of these different points of view and they make instructive reading.

How did the organic molecules involved in the earliest stages of chemical evolution accumulate on the primitive earth? Were they formed as a prebiotic soup when reactive intermediates that had formed in the atmosphere dissolved in surface waters and reacted together, or did they arrive in meteorites and comets? Maybe there never was a prebiotic soup, just a thin layer of prebiotic paste formed on and firmly stuck to the surface of minerals such as iron sulfides. All of these possibilities are explored. If you pay your money (to buy this book) you can take your choice.

Now we come to the most challenging problem of all. Let us assume that an adequate prebiotic soup or paste was available. How did the transition from chemical chaos to biological order come about? Darwinian selection must have emerged early as a source of complexity, but what was the first system to appear on the primitive earth that was capable of Darwinian selection? Was it RNA, some simpler covalent polymer of perhaps some non-covalent aggregate? Are there complex evolvable metabolisms that do not depend on a genetic material?

You wont find definitive answers to any of these important questions in this or any other book on the Origin of Life, but here you will find them discussed from a variety of different perspectives in a stimulating way. You may even be tempted to do an experiment!

The Salk Institute for Biological Studies, **Leslie E. Orgel**
La Jolla, CA 92037 USA
E-mail: **orgel@salk.edu**

15 June 2003

Biodata of **Joseph Seckbach**, editor all volumes of the "**C**ellular **O**rigins, **L**ife in **E**xtreme Habitats and Astrobiology (COLE)" book series, published by Kluwer. He is the author of ***"Introduction to the Extremophiles"*** (with co-author A. Oren) chapter and of the ***"Introduction"*** to this book.

Dr. Joseph Seckbach is the initiator and chief editor of ***Cellular Origins, Life in Extreme Habitats and Astrobiology*** **(COLE)** book series, see: **http://www.wkap.nl/prod/s/COLE** and author of several chapters in this series. He is the co-author (with R. Ikan) of the ***Chemistry Lexicon*** (1991, 1999). Dr. Seckbach earned his Ph.D. from the University of Chicago, Chicago, IL (1965) and spent his postdoctoral years in the Division of Biology at Caltech (Pasadena, CA). Then he headed at the University of California at Los Angeles (UCLA) a team for searching of extraterrestrial life. Seckbach has been appointed to the faculty of the Hebrew University (Jerusalem, Israel) and performed algal research and taught Biological courses. He spent his sabbatical periods in Tübingen (Germany), UCLA and Harvard University. He served at Louisiana State University (LSU), (1997/1998) as the first selected occupant of the John P. Laborde endowed Chair for the Louisiana Sea Grant and Technology transfer, and as a visiting Professor in the Department of Life Sciences at LSU (Baton Rouge, LA).

Among his publications are books, scientific articles concerning plant ferritin (phytoferritin), cellular evolution, acidothermophilic algae, and life in extreme environments. He also edited and translated several popular books. His recent interest is in the field of enigmatic microorganisms and life in extreme environments.

E-mail: **seckbach@huji.ac.il**

"Good are the luminaries that our God has created. He has fashioned them with wisdom, with insight and discernment. Strength and power has He granted them, to be dominant within the world. Filled with luster and radiating brightness, their luster is beautiful throughout the world."

From the Sabbath Prayer Book

PREFACE

The Origin of Life and the new field of Astrobiology are vibrant topics on which, in recent years, several books have been published and many conferences have taken place. Moreover, of late, journals devoted to these interdisciplinary sciences have begun to appear.

This book is the sixth volume in the Kluwer series "Cellular Origins, Life in Extreme Habitats and Astrobiology (COLE)." A variety of scholars—eminent biologists, chemists, astro-, nucleo- and theoretical physicists— stemming from 15 countries contributed this book's 40 chapters. To introduce the reader to the range of interests and knowledge of each writer, a short biodata appears at the beginning of each contribution. With an eye on maintaining a high standard of scholarship, almost all chapters were reviewed by peer reviewers or external referees.

The book is divided into seven major sections: (1) Origin and evolution of life; (2) The standard scenario; (3) Alternatives to the standard scenario; (4) The first steps of biological evolution; (5) The first cells; (6) Extremophiles and Biodiversity; (7) Distribution and destiny of life in the universe, which encompasses multidisciplinary fields, such as origin of life, extremophiles, research on the extraterrestrial moons and planets as well as the possibilities of finding living forms on their surface or in their subsurface. The section 7 discusses comets sources, bacterial fossils, past and present Martian life, and habitats inside other celestial bodies. Indeed, some microbial extremophiles may present analogs of lower life forms, perhaps existing in other ethereal bodies, such as within Mars, Europa (Jovian moon), and other planets or moons.

Darwin conceived the origin of life arising within "warm little ponds," but since the famous experiment by Stanley Miller (1953), science has moved from the idea of "primordial soup" into the full subject of Astrobiology. In addition, one of chapters deals with life's origin on clay particles. Recently, there are proposals suggesting that in the early Earth UV radiation was intense, temperatures were elevated, and a large number of rocks and meteorites bombarded certain bodies in our solar system. These circumstances did not provide an amenable background for the emergence of living cells. Life could have originated in subsurface areas protected from those harsh environmental conditions during the early Earth. Actually, several proposals suggest that chemical and biological evolution toward the origin of life may have taken place within protected depths under oceanic hydrothermal vents, while others suggest that it took place in the undersurface of earth. Bacteria have indeed been detected in the ocean depths at several kilometers and also at the depths of the earth surface. Such bacteria live under conditions of high pressure and elevated temperatures. New research has shown that some bacteria can survive pressures equivalent to several dozens of kilometers beneath the earth's crust or under water. Thus, the subsurface may be regarded as a potential cradle of life's origin on Earth and on other bodies of the Solar System.

Are the bodies of outer space a factor in the origin of life on earth? Can microbes remain alive after such a long voyage to Earth? Experiments have indicated that bacterial spores that have been exposed to outer space may suggest that small meteorites could carry microbial life from one planet to another. A few years ago, publications appeared about bacteria revival after a long stay on the moon surface. The Martian meteorite picked up in Antarctica in 1984 (AHL84001) suggested some biological structures and biochemical residues. Interpretation of these "nano-worms" inside such extraterrestrial meteorites pointed out (in a Science 1996 article) that those structures are "nano-bacterial" images. If this is the case, these data might support a past history of ancient life on Mars. Now that images of the red planet had been obtained, it has been suspected that Mars was once warmer and wetter—possessing running rivers and, perhaps, even oceans—and once life may have existed on it.

Other ideas (not popular among the scientific community) consider the bombardment notion and the AHL84001 findings as perhaps suggesting that life had been imported from outside planet Earth (for example, the "Panspermia" theory).

The purpose of this volume is to enhance the knowledge of researchers, lecturers, and graduate students as well as any "open-minded" reader regarding the ancient question of "Genesis" and the origin of life as well as the question, "Are we alone in the Universe?" We hope the papers included here will stimulate the readers' curiosities about these puzzles for mankind and shed light on possible answers.

I must thank the numerous colleagues who assisted in the formation of this book. Special mention must be made of Prof. Aharon Oren, Prof. David Chapman, Prof. Chris McKay and Greg Cook, who kindly reviewed, deeply and constructively, several chapters.

The Hebrew University of Jerusalem
E-mail: **seckbach@huji.ac.il**

Joseph Seckbach

10 August 2003

ACKNOWLEDGEMENTS

This volume is dedicated with admiration to all of our "early birds" who submitted their chapters in due time. I thank them for their patience and understanding. I wish to thank so much to so many people who assisted in various ways with this volume. Special appreciation is due to his wife Fern L. Seckbach for constant linguistic advice, proofreading, hints in the word-processing technique and general interest in this project.

I am grateful to the Editorial Board Members:
Professors David J Chapman [UCSB], Julian Chela-Flores [JCTP, Trieste, IT]; Francois Raulin [University of Paris] and last but not least Professor Aharon Oren [The Hebrew University of Jerusalem]. All of them assisted in reviewing papers, advising and cooperation toward the "making" of this volume. Professor Aharon Oren who never refused to be ready and has been our constant adviser by his efficient and pedantic reviewing of several chapters of this volume as well as other chapters in our book series of Cellular Origins, Life in Extreme Habitats and Astrobiology (http://www.wkap.nl/prod/s/COLE).

I send my gratitude to the entire peer reviewers and external referees for performing an efficient task. Our reviewers include the following names (Alphabetical sorted) M. D. Brasier (UK); R. Castenholz (USA); D.J. Chapman (USA) – for reviewing a few papers; J. Chela-Flores (ICTP, Trieste, Italy); D. Cohen (Israel); G. Cook (New Zealand); D. Deamer (USA); Ch. Erard (France); J. P. Ferris (USA); Iris Fry (Israel); Daniel Gautier (France); R. Gupta (Canada); G. Horneck (Germany); R.M. Kellogg (The Netherlands); Noam Lahav (Israel); D. Lancet; (Israel) C. Lineweaver (Australia): R. Lorenz (USA); L. Luigi (Switzerland); M. Madigan (USA); C. Matthew (USA); Chris McKay (NASA, USA) for his reviewing several chapters; H. Morowitz (USA); S.J. Mojzsis; R. Navarro (Mexico); Y. Ne'eman (Israel); M. Nishiguchi (USA), A. Oren (Israel), for his constant help and advice; L. Orgel (USA); Florence Raulin (USA); François Raulin (France) for her recommendations; Francois Robert (France); D. Segre (USA); R. Shapiro (USA); J. Svoboda (Canada); J. Toporski (USA); D. Wharton. (New Zealand).

I am gratefulness to those colleagues who offered illustrations, or granted us the copy right for using their figures, from their collection for the book cover. Among the are: Elsevier publisher; Krumbein; W.E., Lancet; D., Lerman, L., Lineweaver; C., NASA; Nishiguchi; M., Soina V., Sterrenburg, F.A.S., Tuck; A. and Webber; P.

Thanks are due to Professors D. Lancet, Julian Chela-Flores and Louis Lerman for advising in the title of this book. Last but not least, I appreciate the faithful assistance from Kluwer Academic Publishers' workers who were involved in the publication of this book. My main appreciation is aimed to Ms. Claire van Heukelom and to Dr. Frans van Dunne. They have been always ready to give their efficient advice and practical services whenever we approach them.

The Hebrew University of Jerusalem **Joseph Seckbach**
E-mail: **seckbach@huji.ac.il**

15 August 2003

LIST OF AUTHORS FOR "*ORIGINS: GENESIS, EVOLUTION AND THE BIODIVERSITY OF LIFE.*"

All senior authors are underlined

AGUILERA, A.
Chapter 28.
CENTRO DE ASTROBIOLOGÍA (CSIC-INTA), TORREJÓN DE ARDOZ, 28855 MADRID, SPAIN

AMARAL-ZETTLER, L.A.
Chapter 28.
MARINE BIOLOGICAL LABORATORY, WOODS HOLE, MASSACHUSETTS 02543, USA.

AMILS, R.
Chapter 28.
CENTRO DE ASTROBIOLOGÍA (CSIC-INTA), TORREJÓN DE ARDOZ, 28855 MADRID, SPAIN, AND CENTRO DE BIOLOGÍA MOLECULAR (CSIC-UAM), U. AUTÓNOMA DE MADRID, CANTOBLANCO, 28049 MADRID, SPAIN.
E-mail: Ramils@correo.cbm.uam.es

BAKER, VICTOR R.
chapter 37.
DEPARTMENT OF HYDROLOGY AND WATER RESOURCES, THE UNIVERSITY OF ARIZONA, TUCSON, ARIZONA 85721-0011 USA
E-mail: baker@hwr.arizona.edu

BELL, P. J. L.
chapter 22.
MACQUARIE UNIVERSITY, NEW SOUTH WALES, 2109, AUSTRALIA
E-mail: pbell@rna.bio.mq.edu.au

BRACK, A.
chapter 5.
CENTRE DE BIOPHYSIQUE MOLECULAIRE, CNRS. RUE CHARLES SADRON, 45071 ORLEANS CEDEX 2, France
E-mail: brack@cnrs-orleans.fr

CAMPINS, HUMBERTO
chapter 34
PHYSCIS DEPARTMENT, UNIVERSITY OF CENTRAL FLORIDA, ORLANDO, FL 32816, USA
E-mail: campins@phy.physics.ucf.edu

CASTENHOLZ, RICHARD W.
chapter 27
CENTER FOR ECOLOGY AND EVOLUTIONARY BIOLOGY, UNIVERSITY OF OREGON, EUGENE, OR 97403, USA
E-mail: rcasten@darkwing.uoregon.edu

CHELA-FLORES, JULIAN
chapter 40.
THE ABDUS SALAM, INTERNATIONAL CENTRE FOR THEORETICAL PHYSICS. P.O.BOX 586; STRADA COSTIERA 11; 34136 TRIESTE, ITALY AND INSTITUTO DE ESTUDIOS AVANZADOS, CARACAS 1015A, VENEZUELA.
E-mail: chelaf@ictp.trieste.it

COHEN, DAN
chapter 30
DEPT. OF EVOLUTION, SYSTEMATICS AND ECOLOGY.THE SILBERMAN INSTITUTE OF LIFE SCIENCES. AND THE CENTER FOR RATIONALITY AND INTERACTIVE DECISIONS THE HEBREW UNIVERSITY OF JERUSALEM. ISRAEL.
E-mail: dancohen@vms.huji.ac.il

DEAMER, DAVID W.
chapter 6
DEPARTMENT OF CHEMISTRY AND BIOCHEMISTRY,UNIVERSITY OF CALIFORNIA, SANTA CRUZ, SANTA CRUX CA 95064 USA
E-mail: deamer@hydrogen.ucsc.edu

EROL AKCA
chapter 18
INSTITUTE OF MICROBIOLOGY AND WINE RESEARCH, JOHANNES GUTENBERG-UNIVERSITY, 55099 MAINZ, GERMANY.

FERNANDEZ-REMOLAR, D.
Chapter 28.
CENTRO DE ASTROBIOLOGIA (CSIC-INTA), TORREJON DE ARDOZ, 28855 MADRID, SPAIN

GOMEZ, F.
Chapter 28.
CENTRO DE ASTROBIOLOGIA (CSIC-INTA), TORREJON DE ARDOZ, 28855 MADRID, SPAIN

GONZALEZ-TORIL, E.
Chapter 28.
CENTRO DE BIOLOGIA MOLECULAR (CSIC-UAM), U. AUTONOMA DE

MADRID, CANTOBLANCO, 28049 MADRID, SPAIN,

GUPTA, RADHEY S.
chapter 17
DEPARTMENT OF BIOCHEMISTRY, MCMASTER UNIVERSITY, HAMILTON, ON CANADA L8N 3Z5
E-mail: gupta@mcmaster.ca

HARALD CLAUS
chapter 18
INSTITUTE OF MICROBIOLOGY AND WINE RESEARCH, JOHANNES GUTENBERG-UNIVERSITY, 55099 MAINZ, GERMANY.

HOSSEINMARDI, AZARNOUSH
chapter 18
INSTITUTE OF CHEMICAL ENGINEERING FUNDAMENTALS AND PLANT DESIGN, GRAZ UNIVERSITY OF TECHNOLOGY INFELDGASSE 25, A-8010, GRAZ, AUSTRIA

HOWARD-WILLIAMS, CLIVE
chapter 29
NATIONAL INSTITUTE OF WATER AND ATMOSPHERIC RESEARCH, CHRISTCHURCH, NEW ZEALAND
E-mail: c.howard-williams@niwa.co.nz

ITOH, T.
chapter 20
DIVISION OF CHEMISTRY, CENTER FOR NATURAL SCIENCE, KITASATO UNIVERSITY,1-15-1 KITASATO, SAGAMIHARA, KANAGAWA 228-8555, JAPAN.
E-mail: itoht@clas.kitasato-u.ac.jp

ITOH, Y. H.
chapter 20
DEPARTMENT OF BIOENGINEERING, FACULTY OF ENGINEERING, SOKA UNIVERSITY,1-236 TANGI-CHO, HACHIOJI, TOKYO 192-8577, JAPAN.
E-mail: yhitoh@t.soka.ac.jp

JONES, BRYAN W
chapter 32
NEW MEXICO STATE UNIVERSITY, DEPARTMENT OF BIOLOGY, BOX 30001, MSC 3AF, LAS CRUCES, NM, 88003-8001 USA
E-mail: bryjones @nmsu.education

KAŹMIERCZAK, J.
chapter 21

INSTITUTE OF PALEOBIOLOGY, POLISH ACADEMY OF SCIENCES, TWARDA 51/55, PL-00818, WARSZAWA, POLAND
E-mail: jkaz@twarda.pan.pl

KEMPE, S.
chapter INSTITUTE FOR APPLIED GEOSCIENCES, TU-DARMSTADT 22, SCHNITTSPAHNSTR. 9, D-64287 DARMSTADT, GERMANY.
E-mail: kempe@geo.tu-darmstadt.de.

KEVBRIN, V.V.
chapter 24
INSTITUTE OF MICROBIOLOGY RAS, PROSPECT 60-LETIJA OCTIABRIA, 7/2, 117312 MOSCOW, RUSSIA
E-mail: kevbrin@inmi.host.ru

KÖNIG, HELMUT
chapter 18
INSTITUTE OF MICROBIOLOGY AND WINE RESEARCH, JOHANNES GUTENBERG-UNIVERSITY, 55099 MAINZ, GERMANY.
E-mail: hkoenig@uni-mainz.de

KRING, DAVID A.
chapter 34
LUNAR AND PLANETARY LABORATORY, UNIVERSIYT OF ARIZONA, TUCSON, AZ 85721, USA

KRITSKY, M. S.
chapter 14
A.N. BACH INSTITUTE OF BIOCHEMISTRY, RUSSIAN ACADEMY OF SCIENCES, LENINSKY PROSPEKT 33, MOSCOW, 119071 RUSSIAN FEDERATION
E-mail: mkritsky@inbi.ras.ru

LAMMER, HELMUT
chapter 38
DEPARTMENT OF EXTRATERRESTRIAL PHYSICS SPACE RESEARCH INSTITUTE, AUSTRIAN ACADEMY OF SCIENCES, SCHMIEDLSTR. 6, A-8042 GRAZ, AUSTRIA
E-mail: helmut.lammer@oeaw.ac.at

LANCET, D.
chapter 7
DEPARTMENT OF MOLECULAR GENETICS AND THE CROWN GENOME CENTER, THE WEIZMANN INSTITUTE OF SCIENCE, 76100 REHOVOT, ISRAEL.
E-mail: Doron.Lancet@weizmann.ac.il

LERMAN, LOUIS
chapter 4.
DEPARTMENT OF CHEMISTRY, PHILIPPS UNIVERSITY, MARBURG, GERMANY AND CORRESPONDENCE TO: ENGRAM ASSOCIATES, 270 E. FLAMINGO #337, LAS VEGAS, NEVADA 89109 USA.
E-mail: bubbles@dnai.com

LICHTENEGGER, HERBERT I. M.
chapter 18
Department of Experimental Space Physics, Space Research Institute, Austrian Academy of Sciences, A-8042, Graz, Austria

LINEWEAVER, CHARLES H.
chapter 15
SCHOOL OF PHYSICS, UNIVERSITY OF NEW SOUTH WALES, SYDNEY, AUSTRALIA
E-mail: c.lineweaver@unsw.edu.au

LOVLEY, DEREK R.
chapter 19
Department of Microbiology, University of Massachusetts-Amherst, Amherst MA 01003 USA
E-mail: dlovley@mcrobio.umass.edu

MALKI, M.
Chapter 28.
CENTRO DE BIOLOGIA MOLECULAR (CSIC-UAM), U. AUTONOMA DE MADRID, CANTOBLANCO, 28049 MADRID, SPAIN,

MATSUNO, KOICHIRO
chapter 11
DEPARTMENT OF BIOENGINEERING, NAGAOKA UNIVERSITY OF TECHNOLOGY, NAGAOKA 940-2188, JAPAN
E-mail: kmatsuno@vos.nagaokaut.ac.jp

MATTHEWS, CLIFFORD N.
chapter 8
Department of Chemistry, University of Illinois at Chicago, Chicago, Illinois 60607, USA
E-mail: hcnmatthews@cs.com

MORRISON, PHILIP
chapter 2
Room 37-641, Astrophysics, MIT, 77 Mass. Ave.,Cambridge MA 02139 USA
E-mail: philmorr@MIT.edu

MUELLER, DEREK
chapter 29
CENTRE D'ÉTUDES NORDIQUES, UNIVERSITÉ LAVAL, QUÉBEC CITY, QC G1K 7P4, CANADA

NAVARRO-GONZÁLEZ, RAFAEL
chapter 9
LABORATORIO DE QUÍMICA DE PLASMAS Y ESTUDIOS PLANETARIOS, INSTITUTO DE CIENCIAS NUCLEARES, UNIVERSIDAD NACIONAL AUTÓNOMA DE MÉXICO. CIRCUITO EXTERIOR, CIUDAD UNIVERSITARIA. APARTADO POSTAL 70-543, MÉXICO, D.F. 04510 MEXICO.
E-mail: navarro@nuclecu.unam.mx

NE'EMAN, YUVAL
chapter 16
School of Physics and Astronomy, Tel-Aviv University, Tel-Aviv, Israel 69978
E-mail: matildae@tauex.tau.ac.il

NEGRON-MENDOZA, A.
chapter 12
INSTITUTO DE CIENCIAS NUCLEARES, UNAM, *A.P. 70-543, CIUDAD UNIVERSITARIA, D.F. 04510 MÉXICO.*
E-mail: negron@nuclecu.unam.mx

NISHIGUCHI, MICHELE K.
chapter 32
NEW MEXICO STATE UNIVERSITY, DEPARTMENT OF BIOLOGY, BOX 30001, MSC 3AF, LAS CRUCES, NM, 88003-8001 USA
E-mail: nish@nmsu.edu

OREN, AHARON
chapters: 23 & 25
DIVISION OF MICROBIAL AND MOLECULAR ECOLOGY, THE INSTITUTE OF LIFE SCIENCES, AND THE MOSHE SHILO MINERVA CENTER FOR MARINE BIOGEOCHEMISTRY, THE HEBREW UNIVERSITY OF JERUSALEM, 91904 JERUSALEM, ISRAEL
E-mail: orena@cc.huji.ac.il

ORGEL, LESLIE E.
see Foreword
THE SALK INSTITUTE FOR BIOLOGICAL STUDIES LA JOLLA, CA 92037 USA
E-mail: orgel@salk.edu

ORÓ, JOAN
chapter 33
DEPARTMENT OF BIOLOGY & BIOCHEMISTRY, UNIVERSITY OF HOUSTON,

HOUSTON, TX 77204-5934, USA.
E-mail: fjo@fjo.es

RAMOS-BERNAL, S.
chapter 12
INSTITUTO DE CIENCIAS NUCLEARES, UNAM, *A.P. 70-543, CIUDAD UNIVERSITARIA, D.F. 04510 MÉXICO.*
E- mail: ramos@nuclecu.unam.ms

RAULIN-CERCEAU, F.
chapter 3
MUSEUM NATIONAL D'HISTOIRE NATURELLE, CENTRE ALEXANDRE KOYRE (EHESS-CNRS-MNHN-UMR 8560) 57 RUE CUVIER, F - 75005 PARIS, FRANCE.
E-mail: raulin@mnhn.fr

REITNER, JOACHIM
chapter 13
DEPARTMENT OF GEOBIOLOGY, GZG, UNIV. GÖTTINGEN, GOLDSCHMIDTSTR. 3, 37077 GÖTTINGEN. GERMANY
E-mail: jreitne@gwdg.de

RODRIGUEZ, N.
Chapter 28.
CENTRO DE ASTROBIOLOGIA (CSIC-INTA), TORREJON DE ARDOZ, 28855 MADRID, SPAIN

ROMANEK, C. S.
chapter 24
SAVANNAH RIVER ECOLOGY LABORATORY, AIKEN, SC AND DEPARTMENT OF GEOLOGY, UNIVERSITY OF GEORGIA, ATHENS, GA 30602 USA.
E mail: romanek@srel.edu

SCHWARTZMAN, DAVID
chapter 15
DEPARTMENT OF BIOLOGY, HOWARD UNIVERSITY, WASHINGTON DC, USA
E-mail: dws@scs.howard.edu

SECKBACH, JOSEPH
see: Preface, Introduction & chapter # 23
THE HEBREW UNIVERSITY OF JERUSALEM
E-mail: seckbach@huji.ac.il

SEGRÈ, D.
chapter 7
LIPPER CENTER FOR COMPUTATIONAL GENETICS AND DEPARTMENT OF GENETICS, HARVARD MEDICAL SCHOOL, 200 LONGWOOD AVE., BOSTON, MA 02115, USA
E-mail: dsegre@genetics.med.harvard.edu

SEGURA, ANTIGONA
chapter 9
LABORATORIO DE QUÍMICA DE PLASMAS Y ESTUDIOS PLANETARIOS, INSTITUTO DE CIENCIAS NUCLEARES, UNIVERSIDAD NACIONAL AUTÓNOMA DE MÉXICO. CIRCUITO EXTERIOR, CIUDAD UNIVERSITARIA. APARTADO POSTAL 70-543, MÉXICO, D.F. 04510 MEXICO.
E-mail: antigona@essc.psu.edu

SIMAKOV, M.B.
chapter 39
GROUP OF EXOBIOLOGY, INSTITUTE OF CYTOLOGY RAS, TIKHORETSKY AV., 4, ST.PETERSBURG, 194064, RUSSIA
E-mail: exobio@mail.cytspb.rssi.ru

SOINA, V.S.
chapter 26
SOIL SCIENCE DEPARTMENT, MOSCOW LOMONOSOV STATE UNIVERSITY, VOROBYOVY GORY, MOSCOW, 119899, RUSSIA
E-mail: soina@yandex.ru

STEELE, ANDREW
chapter 35
GEOPHYSICAL LABORATORY, CARNEGIE INSTITUTION OF WASHINGTON, 5251 BROAD BRANCH ROAD NW, WASHINGTON DC, 20015 USA
E-mail: j.toporski@gl.ciw.edu

SWINDLE, TIMOTHY D.
chapter 34
LUNAR AND PLANETARY LABORATORY, UNIVERSIYT OF ARIZONA, TUĆSON, AZ 85721, USA

TELEGINA, T. A
chapter 14
A.N. BACH INSTITUTE OF BIOCHEMISTRY, RUSSIAN ACADEMY OF SCIENCES, LENINSKY PROSPEKT 33, MOSCOW, 119071 RUSSIAN FEDERATION

TENG, JACQULINE
chapter 4.
THE INSTITU (BENGEN, WASHINGTON) AND UNIVERSITY OF CALIFORNIA,

BERKELEY, USA
E-mail: chalice@Dragonwarden.com

TERVAHATTU, HEIKKI
chapter 10
DEPARTMENT OF LIMNOLOGY AND ENVIRONMENTAL PROTECTION, UNIVERSITY OF HELSINKI, P. O. BOX 62, FIN-00014 HELSINKI, FINLAND
E-mail: heikki.tervahattu@helsinki.fi

TEWARI, VINOD CHANDRA
chapter 31
WADIA INSTITUTE OF HIMALAYAN GEOLOGY, DEHRA DUN 248 001, UTTARANCHAL, INDIA AND INTERNATIONAL CENTER FOR THEORETICAL PHYSICS, MIRAMARE 34100 TRIESTE, ITALY
E-mail: vctewari@rediffmail.com

TOPORSKI, JAN
chapter 35
GEOPHYSICAL LABORATORY, CARNEGIE INSTITUTION OF WASHINGTON, 5251 BROAD BRANCH ROAD NW, WASHINGTON DC, 20015 USA
E-mail: j.toporski@gl.ciw.edu

TUCK, ADRIAN
chapter 10
NOAA AERONOMY LABORATORY R/AL6, 325 BROADWAY, BOULDER CO 80305-3328, USA
E-mail: tuck@al.noaa.gov

VAIDA, VERONICA
chapter 10
DEPARTMENT OF CHEMISTRY AND BIOCHEMISTRY, UNIVERSITY OF COLORADO, BOULDER CO 80309-0215, USA
E-mail: Vaida@colorado.edu

VAN HOVE, PATRICK
chapter 29
CENTRE D'ÉTUDES NORDIQUES, UNIVERSITÉ LAVAL, QUÉBEC CITY, QC G1K 7P4, CANADA

VINCENT, WARWICK F.
chapter 29
CENTRE D'ÉTUDES NORDIQUES, UNIVERSITÉ LAVAL, QUÉBEC CITY, QC G1K 7P4, CANADA
E-mail: warwick.vincent@bio.ulaval.ca

VOROBYOV, E.A.
chatper 26
SOIL SCIENCE DEPARTMENT, MOSCOW LOMONOSOV STATE UNIVERSITY, VOROBYOVY GORY, MOSCOW, 119899, RUSSIA

WHARTON, DAVID A.
chapter 36
DEPARTMENT OF ZOOLOGY, UNIVERSITY OF OTAGO, P.O. BOX 56, DUNEDIN, NEW ZEALAND.
E-mail: david.wharton@stonebow.otago.ac.nz

WIEGEL, J.
chapter 24
DEPARTMENT OF MICROBIOLOGY, UNIVERSITY OF GEORGIA, ATHENS, GA 30602-2605, USA
E-mail: jwiegel@uga.edu

ZULUAGA, J.
Chapter 28.
DEPARTAMENTO DE QUÍMICA FÍSICA APLICADA, U. AUTÓNOMA DE MADRID, CANTOBLANCO, 28049, SPAIN,

LEGENDS FOR THE COVER PHOTO

Title: "Major Events in the Evolution of the Universe and Life on Earth".

Caption: The left hand side of this picture represents the major events known to have taken place in the evolution of the Universe, from the Big-bang to the formation of planetary systems. Time here flows from the top to the bottom. The right hand side of the picture represents the early evolution of our planet, the formation of chemical structures leading to Life and its evolution from the seas into land. Time flows like a letter "U": from the upper right hand side, down then to the right in the bottom and finally again up. In the upper right-hand side, we see man as another species but capable with its consciousness of asking deep questions about the Universe, its Origin and its Evolution, even creating machines capable to carry us into other worlds. Man however is affecting the planetary environment and many species are being affected by our impact on the planet. The evolution of the Cosmos and Life has taken place through major transitions in which the expansion of the Universe must have been the main driver of its large-scale changes in complexity. At smaller scales the arrow of time combined with other forces are the main drivers in the evolution of complexity. The main objectives of Astrobiology are to understand the nature of many of these punctuated events, specially the ones referring to Life, and to try to scientifically understand the Origin, Evolution and Future of Life on Earth and, possibly, elsewhere in the Universe.

The editor of this volume (J.S) thanks Professor Juan Perez-Mercader (The Director Centro de Astrobiologia (CSIC/INTA) Madrid SPAIN) for contributing this illustration to the cover of the present book.

I. Origin & Evolution of Life

INTRODUCTION

This is volume number six in the series entitled "Cellular Origins, Life in Extreme Habitats and Astrobiology (COLE)" http://www.wkap.nl/prod/s/COLE (edited by J. Seckbach, published by Kluwer Academic Publishers). It covers possible events in the genesis and evolution of living cells and conditions of life in the universe. Some chapters analyze the essential question of chemical evolution and the "Fact of Life." The subjects covered in this book are among the most frequently discussed in the world of science and at the forefront of scientific frontiers. Indeed, lately numerous books have been published, journals initiated, international conferences and societies organized associated with the aspects of science addressed in this new volume. Several of the chapters in this volume complement and expand studies published in previous volumes of "COLE."

The questions of origins of life and age of Earth and the Universe are not new – earlier generations pondered these interesting, seemingly insolvable puzzles. In some cases, religious faiths continue to consult holy documents about the "Genesis" for answers. The fossil record is seen as contradicting religious traditions, regarding topics such as the times when various forms of life where on the Earth. Or, the radioactive decay of elements that tells something about the age of earth. Some thinkers have tried to bridge the differences in these approaches, and this very topic has stimulated a wide range of ideas. There remain proponents of the "Panspermia" concept who see life "seeds" as having been imported from other planets or from further celestial regions. The ancient martian meteorite AHL84001 collected in Antarctica and meteorites from other sites have been claimed by some to carry a payload of biological traces. Such meteorites represent a ready supply of extraterrestrial material delivered to Earth for study.

The open questions, already addressed in the past, are still unanswered: How did life emerge on Earth? When and where did the first forms of life appear? What was the state of the early atmosphere was it reducing, neutral or oxidative? Are there other worlds which harbor life? Is the planet Earth the only place in our solar system, or perhaps even the galaxy where living organisms have emerged? was there a role for exogenous life in the history of the biosphere?

Currently, many scientists consider a plausible scenario the origin of life within hot environments during the early ages of the Earth. Primeval cells had to protect themselves from the intense UV radiation on the surface and against the energetic impacts stemming from the last stages accretion to the early Earth. Some scientists assume that the organics for the first life may have resulted from the organics delivered to the surface of earth by meteorites and comets. Protective zones could have been present in the depths of oceans and subsurface of the Earth. Recent discoveries have uncovered bacterial communities that live within deep marine hydrothermal vents, thriving at high temperatures (113°C) and under elevated pressure. Other microorganisms have been isolated from deep within rocks and sediments from drill cores on land. A minority view holds that the origin of life took place in cryophilic environments of melted ice zones rather than in warm or hot environments.

J. Seckbach (ed.), Origins, 3.

Several mechanisms have been proposed as to how the proto-cells emerged on Earth. The chemical reactions leading to the necessary precursors of life are considered to have occurred within an ill-defined process of "chemical reactions". Arguments persist regarding the chemical composition and abundance of these initial substances, were there molecules that lead to life such as HCN, nucleotides, and other reduced molecules? Did there exist at some point in biological evolution an RNA World, in which this nucleic acid acted also as ribozyme and was superceded by DNA and protein synthesis? The reactions leading to chemical assembly of prebiotic molecules may have taken place either in a "primordial soup" in "little warm ponds", or perhaps on the surface of clay which, e.g., catalyzes the formation of RNA and the formation liposomes from micelles. The beginnings of experimental prebiotic chemistry hearken back to 1953 when Stanley Miller performed his famous experiment at the University of Chicago in which he mixed some abiotic primordial gases, boiling water, and electrical discharges and obtained bioorganic compounds.

The origin of microbial life has been determined (from geological data) to have occurred sometimes before 3.8 bya (billion years ago). Recently the appearance of the LUCA (Last Universal Common Ancestor) has been calculated to 4.3 bya. It is unknown what the metabolic nature of the first true cells was, however among the earliest recognized biological communities are those that formed sedimentary structures termed stromatolites found as far back in the geologic record as ca. 3.5 bya.

The extremophiles represent various microorganisms living in very harsh environments, which most other organisms are not able to survive. The reviews published in this book discuss most of the severe habitats in which these fascinating organisms thrive. These microbes are found at both edges of temperature range for viable biological activity (from –20 to +113°C), from very acidic to alkaline pH limits, under pressure of thousand atmospheres, and in saturated solutions of salt, others are able to tolerate high doses of UV radiation. Extremophiles cope with very severe growth conditions and may serve as candidates for analogous cells in, or under, the surface of some planetary bodies. Among the leading solar system candidates for harboring microorganisms are Mars–Earth's brother planet; Europa–the Jovian satellite; or perhaps even Titan, the big moon of Saturn. Astrobiology is the broad-based discipline which deals with the topic of the origin of life and the search for extraterrestrial life.

Astrobiology is the broad base discipline that deals with the topic of the origin of life extraterrestrial life. This is a new interdisciplinary area of research covering several aspects of origin of life, chemical and biological evolution, possibilities of extraterrestrial life as well as characteristics of space.

Images of Mars suggest that in the past this planet was substantially warmer than at present and was wet with running waters and oceans, perhaps also harboring living forms. Microbial life may still be present at the subsurface where water may be stored in ice and in local pockets of hydrothermal activity. Europa is believed to have liquid water under its icy layers warmed by tidal interactions in a gravitational dance with Jupiter and the other jovian satellites. Titan (even further away from Earth than Europa) is believed to be a paradise of organic compounds (such as methane and higher hydrocarbons). It is interesting to speculate that some living forms might be found under the surfaces of these bodies as yet within our own solar system.

It is hoped that this volume will benefit a wide range of readers, from beginning students through to professional scholars, in divisions of microbiology, astrophysics, chemistry and ecology.

The Hebrew University of Jerusalem **Joseph Seckbach**
E-mail: **seckbach@huji.ac.il**

18 August 2003

Biodata of **Philip Morrison** author of ***"The Fact of Life"***

Dr. Philip Morrison is a professor (emeritus) of the Physics/Astronomy Department, MIT. He was born in 1915 in New Jersey, USA, and obtained his B.S. in Physics at Carnegie Tech in 1936 and his Ph.D. at UC Berkeley in Theoretical Physics in 1940 (his supervisor was J. R. Oppenheimer). His main faculty posts include: Physics at Cornell, Ithaca, NY (1946-1962); MIT, Physics/Astronomy (1962 to present). Dr. Morrison's main research involves the radioactive origin of gas well helium 3; gamma ray astronomy; cosmic ray propagation in solar system and outside; the search for interstellar communication; active galaxy nuclei and cosmic ray origin (1960s). AGN bean composition limits; rotation models of AGN (1970s); high-energy output of stars (1980s); exoplanets; unusual AGN (Her A) from 1990 to 2002.
Dr. Morrison has reviewed a thousand books on science for magazines and counting; Co-author (with Phylis Morrison) of ***Power of Ten, Ring of Truth*** (*TV and book*). He has visited campuses in North America and four other continents to teach and learn. Federation of American Scientists, founding member; Fellow of US National Academy of Science. Books and columns in support of public understanding of science.

E-mail: **philmorr@MIT.edu**

J. Seckbach (ed.), Origins, 7–13.

THE FACT OF LIFE

PHILIP MORRISON
Room 37-641, Astrophysics,
MIT, 77 Mass. Ave.,Cambridge MA 02139 USA

1. Introduction

We know of only one cosmic location that surely exhibits the grand fact of life: our earth itself. As cells define the structure of the copious life we know, so our minute Sun-warmed local domain fulfills the sufficient ecological conditions for life—once we are able to recognize them... We examine here what this view suggests.

2. Our Galaxy

Our Milky Way Galaxy, is a big home spiral that spreads out stars in their clouds and clusters over a distance ten or twenty thousand times the distance between our Sun star and our closest sun-like starry neighbor. The external galaxies beyond ours count crowds of stars, each one in multiple billions, a scale that parts us dauntingly from the close-up astronomy of a mere life search. Our Galaxy is one of a member of a small group of similar nearby galaxies, its age under ten gigayears, double that of our star, the sun, though about as old as anything we can see.

An astronomically based close up shows physical limits by no means atypical, and a few functional properties of importance. A single star of the populous main sequence, our sun is a stable whitish-yellow dwarf star of modest luminosity and age, called a G2 dwarf of luminosity class V. Two-thirds of all such stars are multiple, their starry companions usually within a fraction of a light year away. Our sun's nearest neighbor, Alpha Centauri is a sun-like dwarf star too, but it is a multiple star about four light years away

The sun's mass is a thousand times that of the largest solar planet, Jupiter. That one giant planet Jupiter would balance three hundred earths. The sun is plenty to anchor planets in lasting orbits, a light hour or two out to Jupiter. (The sun-earth distance is 500 light seconds) Our star orbits the far-off Galaxy center in a complex path (discussed below) and carries its bound companions along. The kinetic energy of a planet's orbital motion is typically a thousand-fold the energy of its own spin, a relic of its formation.

3. Substance

The substance of the planets was drawn from the forming sun itself long ago, and partitioned out among the planets during the early years of the solar system's birth. Some diffuse gas and dust is a similar old residue, or perhaps a later feature.. The entire array of planets moons, meteoric matter, and a trillion far-flung icy comets arose out of the disc of matter thrown out of the newly condensing sun, often through several steps of collecting and rejecting. Rather more mass was lost to outer space—planets and even a starry mass or two – than was retained as smaller fry, and much mass fell inward to a fiery end.

Our biocentric system centers still on the sun , a radiant glowing sphere, as does the system of local gravitational force. The major output of energy is the local flow of visible photons, nourishing and warming planets nearby ever since it began. That flow carries far beyond, to enter the galactic domains of the diverse and myriad stars .The local planets are so much nearer to us than are the stars beyond the sun that they seem to glow by mere reflection, their wanderings suggesting their relative proximity even to watchers long ago. The stars were fixed within the steadily turning sky, with no relative motion between star and star that could be noticed during human history, until the coming of the telescope. The intuitive judgment that stars are but distant suns seems to have won the Buddhist scholars, who added no quantitative argument to all for their clever conclusion.

In Holland the brilliant Christiaan Huygens first compared the brightness of the sun seen in a tiny spherical mirror to that of bright star Sirius seen directly, to argue from the optics that our sun would resemble such a star if viewed from a plausible distance of a few light years. The visible stars do range over a deep domain of space beyond our solar system.

4. Simple Orbits

Look at the planetary orbits around the sun in some detail. Nearly all lie roughly in one plane, they are well spaced, and they are nearly circular.(Small Pluto is no typical planet.) The big gaseous planets lie in a far-off cold zone where the abundant light gases (the sun is 98 % hydrogen and helium gas by atomic count) can be held for a long time by gravity. We are unsure of the whole story, but it is easy to accept that the big planets we call gas giants formed where helium and hydrogen abound, and the small, solid ones, like Earth made mostly of refractory rock lie closer in to the hot sun. In short, the motions, places, and substances of the planets seem simple as can be.

Is this not the effect of long evolution, when all excess energy that allows an orbit to stay far from circular, or much out of the average plane of the planetary motions, would have been dissipated by the interactions between the planets and the other matter over time? Our system has been "worn down" over time by slow losses of energy to become simpler.

In 1995 we saw the first exoplanets that resemble our own system, familiar run-of-the mill sun-like stars orbited by planets generically like our gas giants. It is not that the big planets certainly dominate, for unhappily we now cannot detect the smaller ones, if any. Our present technique (just wait till 2007!) cannot pick up distant earths. They are targets a hundredfold too small, although of our sun's four inner planets, not gas balls but rocky ones, Venus and Earth are roughly matched in size, Mars and Mercury are distinctly smaller. A hundred exosystems, most with only single large planets, are now known; none look like the solar system, for none of the exosystems appear worn down to apparent simplicity. The frequent elliptical orbits with super-Jupiters and even super-Saturns, somewhat smaller, are commonplace out there. (Our Jupiter is more than three times the mass of our Saturn).

In our system the four nearest planetary orbits all hold small planets. Among these new systems we simply could not pick up any small planets were they present., and the large ones—gas giants bigger than our Jupiter, none much smaller than our local runner-up, Saturn—show up closest to their sun as well as far away, closer indeed than our own small Mercury, the smallest orbit of our set of nine planets. and some with gas giants farther out than any of ours. Orbits are less nearly circular, and tilts from the average plane are bigger. That air of long evolution to a simpler state is not present out there; those systems have not been combed out to simplicity and minimal energy .

Planets out there are far more common than we knew ten years back. We saw not one good example until 1995, yet now with new means we find them monthly. So far we have studied stars out to about a hundred light years from our Sun.. We have picked so far some 1100 stars that closely match our own Sun in mass, in age, and in brightness. Some six to eight percent of those sun-stars have real planetary systems, but they seem unlike our familiar home.

5. Planets can migrate

It is well known both from general calculations and by stricter numerical simulations that planets can be lost, once made. Two Jupiters in neighboring orbits whose radii differ by only ten percent or so must interact by their own gravity, causing disturbances in the orbits that after very many passages can cause one of them to fall into the sun. Sometimes the more disturbed one will instead leave the entire region, to escape outward into space beyond. Long repetition in the end succeeds. It is a matter of resonance, as in the famous example of breaking a wine glass by bowing the right tone.

The repeated crest and troughs of the sound wave are replaced in the case of planetary expulsion by the repeated closest approaches of the two interacting planets, over and over again as they pass in time. Tens of millions of passages with the right timing can move a planet from its stable orbit, where the fiddle usually takes only thousands of crests to break the glass. But the orbits have endured for billions of repeats

The more massive the planet, the greater its force, the closer the two mutually disturbed orbits approach the greater the effect. The familiar musical one- to- one resonance fit of orbit motion cannot take place, for the orbit period depends on its size. But eventually resonances will build, even less tidy, "syncopated" fits, with such time ratios as 3 to 2, up to rare higher ones, a chord of orbits. Motions such as the to and fro motions of elliptical orbits, or the leveling pulls between tilted orbits, have similar effects. The system we call home has been simplified in time. There may well have been many planets formed and since lost by resonances.

Circular orbits of small well-spaced planets all in the same plane of symmetry is the poster model of gravitational stability, and a genuine mathematical proof has been found. Our own system lies outside the proved limits of stability, though close to them. There are signs that our Mercury, so small and so frequently rounding the sun, lacks a truly stable orbit. Even though it has endured 20 billion rounds of the sun, it may yet leave its routine orbit during the next billion or two years, and big Jupiter may share the blame. It looks as though ours is a simpler example than these hundred exoplanets. We can conjecture why, recognizing that the full complex case is still far from complete. We see many stars within flat discs of gas and dust. These are either early stages of the disc genre, or discs so small in mass as to be unable to form significant planets.

The big Jupiters, their apparent relocation toward the suns they accompany, and the frequent presence of biggish planets, suggest that in these cases there was too much disc mass here for our local planet pattern.. A broken disc fragment , as it begins to orbit, acts as an eccentric, transient, and hugely heavy planet, draws at the outer planets, confuses all simple motions, and makes for many large planets. There may be familiar systems, but perhaps we find mainly the residues from bigger initial discs, discs too big for our local economy model. the systems we miss have no Jupiter-like planets at all, there might be many more. Even if they are few, with only a distant Jupiter taking quite a few tens of years to orbit, we may not yet have caught it in the slower small motions of a star we have studied for less than ten years. More big Jupiters are being found farther from their own sun.

6. A Second Zodiac at Galactic Scale

Suppose we now see most of what systems are there. At one percent of sun-like stars we might expect about five or ten million of the light disc systems to occupy the same galactocentric ring in the great Year of the Galaxy, a region not too dense with sun-stars nor too scant in them. Within the band nearest to our own orbital ring about the Milky Way we might find a new zodiac, a band of millions of sun-systems in the mighty group of galactic orbits, near-circles that all the stars follow very roughly. Each 250 million solar years our own trip around the distant center of the Milky Way more or less repeats. The orbit of the sun in the galaxy is much more complex than the small solar orbit. The sun is the sharp focus of gravitating force hereabouts. In the Galaxy, however, there is a much larger disc of central stars whose overall effect is felt, not only one tiny star. Our sun's big orbit in the Galaxy, mingled with many similar star orbits, winding and unwinding around the Galaxy center with additional small complicated excursions, helps fill up a tubeful, out here some 25,000 light years from that heavy center. It will take a while indeed to search them out, even sampling far away...

A planet is of course far from living; it is only an address for life. Many homes may never have been occupied. Simple microorganisms may be found widely, stabler habitats for larger creatures have more constraints. But among the few million candidates we can expect someday to detect the collisional ruckus of planet formation taking place here and there around the whole Galactic ring. Distances are big, and we will need to learn how to look afar. But the way to proceed is at hand. Space probes capable of searching out earth-like planets will be launched in about five years time.

Ten years back we could not point at any single exoplanet. Now we know a hundred, at most a few of them open to simple forms of life. Ten years hence....or a century? The *fact* of life here with us is the surest clue. Its implication remains ambiguous, for we know too little. But it offers hope that we might find life soon enough, with effort. No sign marks our place as unique among so many enduring stars in this our Milky Way...and other big galaxies are strewn like so much glittering dust throughout cosmic space.

Biodata of **Florence Raulin-Cerceau** author of "***Historical Review of the Origin of Life and Astrobiology***"

Dr. Florence Raulin-Cerceau is a senior Lecturer at the National Museum of Natural history of Paris. She obtained her Ph.D. in 1983 from the University of Paris VI in the field of Astronomy and SpaceTechniques, on the topic of the organic compounds in the atmosphere of Titan. She is also a researcher in History of Astrobiology at the Center Alexander Koyré (History of Science).

E-mail: **raulin@mnhn.fr**

J. Seckbach (ed.), Origins, 15–33.

HISTORICAL REVIEW OF THE ORIGIN OF LIFE AND ASTROBIOLOGY

F. RAULIN-CERCEAU
Muséum national d'Histoire naturelle
Centre Alexandre Koyré (EHESS-CNRS-MNHN-UMR 8560) 57 rue Cuvier, F - 75005 Paris, France.

1. Introduction

Wondering about the place of humanity in the cosmos is probably one of the oldest questions philosophers have raised since ancient times. The problem of the very essence of life or the existence of other worlds possibly inhabited by human-like living beings is still debated. Are humans the only beings in the cosmos trying to find the key to the origins of life and of the universe? Numerous centuries have passed and such a question still remains to be answered. The history and the philosophy of sciences can probably contribute to understanding the structure of this new huge field of research - from the solution to the problem of the origin of life on Earth to the search for life elsewhere in the universe.

Discussing the origin of life only takes on a rational meaning (with a present-day significance) if we consider that life started from a single beginning somewhere (for instance, on Earth) under certain conditions and at a certain time in the past. This point of view is in fact rather a recent one for the following reasons: (1) the concept of biological evolution did not emerge until the middle of the nineteenth century, (2) the concept of chemical evolution was not clearly expressed until the second decade of the twentieth century, (3) very few things were known about the basic molecular constituents of life until the middle of the twentieth century, (4) we had to wait until the middle of the twentieth century to obtain more accuracy on the age of the oldest terrestrial rocks. Therefore, it seems that before the end of the nineteenth century, the problem of a single starting point for the whole living world could not have been implied by the term "origin". And the consequence of this statement is that the question of " origin " couldn't have been, at this time, the central one. Moreover, the theory of spontaneous generation, having its roots in Antiquity, clearly prevailed for many centuries. It particularly caused many controversies from the seventeenth century to the end of the nineteenth century. This heritage of the past traveled down the ages until appropriate experiments proved that to appear somewhere, life needed life: there was no birth of living beings without a lineage, even for microbes. The theory of descent (Lamarck, Darwin) positively changed our idea of the living world, as well as the perspective proving the non-existence of spontaneous generation (Pasteur). But if life has always derived from living organisms and has reached complexity through evolutionary mechanisms, where did very primitive forms of life come from?

The search for life elsewhere in the universe seems to be a separate problem, historically speaking. The hypothetical idea of beings (as humans) living in other worlds has for many centuries been known as " pluralism ". Possibly for the reasons mentioned in this introduction, biological considerations came very late in the history of pluralism. We had to wait until the second part of the twentieth century to see a link between the research focused on the quest for extraterrestrial life and that concerning the origin of life. Problems of biological contamination during the manned space flights to the Moon linked biology to astronomy. Since 1961 this official link has been known as of Exobiology (this term obviously reflects the influence of biology). Further developments in astronomical methods enabling the detection of extraterrestrial signals shed new light on this subject leading to the birth of Bioastronomy. Astrobiology (in fact this term dates back to the 1950's), as well as Exobiology or Bioastronomy, became in a few decades the official research fields for clarifying the enigma of the origins and distribution of life in the universe.

2. Spontaneous Generation Through the Ages and the First Controversies

The generation (the term generation takes here its primeval signification: action of generating something) of living beings is a very old question. During Antiquity, Aristotle (384-322 BC) saw a continuous generation coming from matter in a state of decay: animals are brought to life not only by coupling but also from the decaying soil and dung. In the Middle Ages the opinion prevailed that *generatio aequivoca* (equivocal generation as opposed to univocal generation) was responsible for the generation of life because of the uncertain nature of this process (Farley, 1997). This term had been used throughout the centuries up to the middle of the nineteenth century when Pasteur and Pouchet began their famous contest to demonstrate (or disprove) the part played by preexisting germs in the birth of microorganisms. This well-known debate bears the name of the spontaneous generation controversy (to differentiate from " parental " generation). To clarify the statement of facts of this extremely complex subject, only the term " spontaneous generation " will be used in the rest of this text.

Up to the end of the seventeenth century, there was no reason to question the occurrence of spontaneous generation: it was an observational fact (Farley, 1997). Spontaneous generation was viewed in terms of putrefaction, an explanation deriving from Antiquity. This view was also strengthened by the mechanical philosophy provided by the French mathematician René Descartes (1596-1650). Descartes reduced natural science to the consideration of matter and motion alone: when decomposed matter was agitated, it became possible for the more dense particles to create an organism. But, by the middle of this century, it became more and more difficult to accept that the complex form of living beings could ever have been produced solely by the general laws of motion (Farley, 1997). The first experimental criticism opposing the theory of spontaneous generation was from the Italian naturalist Francesco Redi (1626-1697). He described in 1668 a whole series of experiments proving that maggots were not generated by meat in a state of putrefaction (as it was currently believed at this time), but rather and solely by the flies coming to lay their eggs (Raulin *et al*, 1997).

In the latter part of the seventeenth century, spontaneous generation was much debated. The impact of the new anatomical and microscopic findings -the Dutch microbiologist Antoine Van Leeuwenhoek (1632-1723) started in 1674 observing microorganisms with microscopic techniques – revealing the complex organization of living beings, even for the simplest forms. The problem of sexual generation was confronted by spontaneous generation, and likewise, medical fields, such as the development of pathological methods, influenced considerably the debate. Contagious diseases and the existence of parasitic worms provided support for spontaneous generation. Conversely, the doctrine of the preexistence of germs led to a denial of any form of spontaneous generation: this doctrine upheld that the germ of preformed parts was created by God at the beginning and was conserved in that state until the moment of its development (Farley, 1997).

The concept of organic molecules released from the decomposition of decaying organisms and later incorporated into other organisms has been called *heterogenesis* (a form of spontaneous generation). The French naturalist Buffon (1707-1788) believed in *heterogenesis*, as well as the British naturalist Turberville Needham (1713-1781) who attributed a vegetative force rather than a combination of organic molecules in all substances to explain spontaneous generation. The experimental findings and theoretical assumptions of Needham and Buffon were challenged by the Italian naturalist Abbe Lazzaro Spallanzani (1729-1799) who in 1765 severely criticized spontaneous generation. To Spallanzani, the generation of living forms was only possible through the preexistence of germs. Spallanzani was persuaded that nature was static, with an infinite series of preformed individuals.

During the first three decades of the nineteenth century, most of the naturalists came to believe in the spontaneous generation of infusorians and parasitic worms (Farley, 1997). In 1809, the French naturalist Jean-Baptiste Lamarck (1744-1829) postulated the theory of transformism in his work *Philosophie zoologique (Zoological Philosophy)*. In this theory, the simple forms of life would have resulted from a perpetual spontaneous generation, and, through progressive perfection, would have formed a more and more complex species. In Germany, spontaneous generation obtained strong support with the help of the doctrine of *Naturphilosophie* born among the " romantic " philosophers during the early nineteenth century. *Naturphilosophie* assumed that there was no distinction between living and nonliving entities and that life was defined in terms of the unity and totality of all processes occurring in a living being. This concept of the unity of nature had a profound influence on the spontaneous generation debate among German biologists: since life was a constant manifestation of unity through multiplicity, spontaneous generation for simple organisms was a process logically integrated in the processes of the organization of matter. These organisms could have emerged from inorganic matter or from matter coming from living or dead organisms (Farley, 1997). The *Naturphilosophen*, such as Lorenz Oken (1779-1851), accepted the continuous occurrence of spontaneous generation.

Then, following two independent concepts, one coming from the French naturalists and the other from the German philosophers, the idea of spontaneous generation for simple forms of life was widely accepted up to 1830.

3. How to Build Primitive Life: *abiogenesis* or *heterogenesis*?

The French microbiologist Louis Pasteur (1822-1895) was famous for his pertinent experiment rejecting spontaneous generation. This experiment had been realized in a rather special context that requires explanation. The French director of the Museum of Natural History of Rouen, Felix Pouchet (1800-1972), published in 1859 a provocative writing entitled *Hétérogénie ou Traité de la génération spontanée* (*Heterogeny or Spontaneous generation Treatise*) in which the author criticized materialism and transformism and strongly supported spontaneous generation for microorganisms. According to Pouchet, who was against Darwin's theory, only God was able to furnish enough vital force to create new eggs of microorganisms from organic matter. In his opinion, it was the only way to generate various biological organizations that would explain the different species living on Earth ((Latour, 1989). Pouchet was a bench scientist and supported his thesis with appropriate experiments. He wrote to Pasteur asking for advice regarding his observations of microorganisms by spontaneous generation in laboratory testing. Pasteur politely replied to Pouchet but tried to make him realize that his experiments were probably unreliable. Pasteur particularly reproached Pouchet for asserting spontaneous generation. It was in this way that Pasteur was brought into the controversy of spontaneous generation. The French Academy of Sciences wanted to resolve once and for all the problem of spontaneous generation and decided to give the Alhumbert Prize to the scientist able to provide the best experiment to prove this. In fact, in France, Pasteur was quite popular and the scientific community was *a priori* in favor of renouncing the theory of spontaneous generation. On April 7, 1864, in front of the Academy of Sciences, Pasteur produced evidence of the presence of germs in the atmosphere and their action to generate microorganisms, and he won the Alhumbert Prize. With the help of this pertinent experiment, it seemed clear that the appearance of microorganisms was not coming from any sort of spontaneous generation.

After Pasteur's work, the question of the emergence of life brought forth new inquiries. However, the conclusion of this experiment did not completely destroy the doctrine of spontaneous generation - which continued to be debated until about 1900. Between 1860 and the first years of the nineteenth century, the question of the emergence of life was indeed a very complex one. Very few scientists and philosophers dismissed the theory of spontaneous generation for microorganisms: this very strong dogma remained on certain levels a valid theory to explain the emergence of very primitive life or its prime material. Living matter could emerge *via* contemporary spontaneous generation from another living (or in putrefaction) matter (*heterogenesis*), or *via* transition from mineral matter to biogenic matter (*abiogenesis*): but both were relevant to spontaneous generation (Raulin-Cerceau, 2001).

This idea prevailed for some specific reasons. First, accepting that life emerged only in the past, a long time ago on the primitive Earth by the action of purely physical causes, required the integration of a very large time-scale. Therefore, it could not have been obvious to imagine, at that time, a transformation of inorganic matter into organic matter using a very long time-span. Secondly, the fact that the emergence of primitive organisms occurred only once during the history of life meant a real paradigm shift. Thirdly, very little was known at that time concerning the physical and chemical conditions of the primitive Earth; the age of the Earth itself was very uncertain. It is probably the reason why, during this period, various suggestions were made to elucidate the problem of the production of primitive organisms. Even after Darwin and Pasteur's conclusions, some general considerations included the theory of evolution and integrated *abiogenesis* as a part of the process of evolution, without excluding some forms of spontaneous generation.

In *The Modes of Origin of Lowest Organisms* published in 1871, Henry Charles Bastian (1837-1915), professor of pathological anatomy at the University College in London, considered four different ways by which bacteria could have come into being: direct *homogenesis* (generation coming by fission from pre-existing bacteria); indirect *homogenesis* (deriving from more complex organisms); *heterogenesis* (reorganization of particles coming from living matter) and *archeobiosis* (generation of life in solutions that only contained mineral ingredients - a kind of *abiogenesis*) (Kamminga, 1980). Bastian argued in his writings *The Beginnings of Life* (1872) and *Evolution and the Origin of Life* (1874) that evolutionists must accept the abiogenic origin of life (Farley, 1997): the manifestations of life could not be independent of physical and chemical laws ultimately derived from the mineral world (Kamminga, 1980). An abiogenic origin of life was then accepted, but it was clear that this type of consideration did not exclude a form of spontaneous generation. Bastian's arguments in favor of *archeobiosis* were based on the invariable association of microorganisms with fermentation processes. However, conversely to Pasteur, Bastian thought that fermentation was a purely chemical process in which living organisms played no causative role (Farley, 1997). This view had been supported by a large number of experiments carried out by Bastian himself. With the help of these arguments, he was able to give an explanation for both the origin of life on Earth and for the persistence of lower organisms. Nevertheless, he didn't call on any sort of vitalism to explain the living world; only combinations and rearrangements of certain kinds of molecules would be necessary for the formation of living matter (*heterogenesis* was thought to be mediated by a special force called vital force or vitalism).

In Germany, the mechanist reaction against *Naturphilosophie* gave rise to scientific materialism around the 1850's, exemplified by the work of Luwig Büchner (1824-1899), Jacob Moleschott (1822-1893) and Karl Vogt (1817-1895) (Kamminga, 1980). During the second part of the nineteenth century, the naturalist Ernst Haeckel (1834-1919) was probably the most prolific German writer on the subject of primitive organisms and their emergence. Adopting the materialist movement, and inspired by Kant's mechanist principle (1755), he explained the natural phenomena with only mechanical causes. According to Haeckel, matter - organic as well as inorganic - was a necessary product of natural forces. All the phenomena related to the organization of matter were explainable without any supernatural intervention, any miracle or any vital force (Haeckel, 1874). In his books *Generelle Morphologie der Organismen* (*General morphology of organisms*) (1866) and *Natürliche Schopfungsgeschichte* (*Natural history of creation*) (1868), he supported the idea that studying living matter could have been reduced to the study of the properties of the carbon element (carbon theory) and to those of proteins, in particular the protoplasm (resulting from the combination of proteins), seen as the basic substance of life (Haeckel, 1874).

Considering the problem of the origin of life, Haeckel could be considered as a forerunner: for him, life had begun on Earth at a very precise moment in the past and he supported the thesis, in *The Wonders of Life* (1904), of a gradual development of matter from inorganic elements to life. His position was inspired by the writings of famous philosophers (such as Goethe), naturalists (Lamarck, Oken, Darwin), and chemists (Pflüger). However, Haeckel accepted even so a form of spontaneous generation (by *autogony* - a kind of *archebiosis*, or *plasmogony* - a kind of *heterogeny*) for extremely simple biological substances (such as the *Monera*).

In 1857, Thomas H. Huxley (1825-1895), a strong supporter of Darwin's theory, identified a mysterious gelatinous substance that was named *Bathybius haeckelii* (to pay tribute to Haeckel). This discovery seemed to consolidate the possibility of a present-day life emergence from chemical elements by *autogony* in very deep oceans. The story was as follows. Huxley examined some mud dredged from the sea bottom during the voyage of the ship H.M.S. Cyclops. After examination under the microscope, he reported that the mud contained " innumerable lumps of a transparent, gelatinous substance " in which were embedded " granules, coccoliths and foreign bodies " (Girard, 1874). Huxley interpreted the *Bathybius haeckelii* as one of the *Monera* described by Haeckel and thought he had discovered a sample of an exceedingly simple organism from which all evolution had started. However, in 1876 John Buchanan, a chemist analyzing *Bathybius*, found it was an inorganic calcium sulphate deposit. Despite this demonstration, the acceptance of a present-day formation of simple living beings from primitive matter was not refuted, even if the distinction between a past process and a supposed every-day process was becoming clearer. For instance, the French biologist Felix Isnard (in *Spiritualisme et matérialisme - Spiritualism and Materialism)* (1879, Paris) used the term " spontaneous origin of life " to distinguish the every-day processes of spontaneous generation from those of the beginning of transformism in the past (Farley, 1997).

4. Birth of a New Paradigm: A Single Origin of Life (on Earth)

The necessity of a primeval *abiogenesis* appeared more clearly when arguments assuming a very distant time and a very different environment for the primitive Earth were strengthened. This assumption led some researchers to reconstruct the first steps of chemical processes on the primitive Earth. The first attempts to formulate these hypothetical stages, preceding biological evolution, arose from the chemical research field. The German physiologist Eduard Pflüger (1829-1910) realized that the key to understanding the origin of life in the past would necessarily tie with organic chemical syntheses (Raulin-Cerceau, 2001). He argued in 1875 that this key could have been the synthesis of radical cyanogen, added to the synthesis of other organic molecules when very warm conditions existed on the early Earth. Proteins could have been formed later from these chemical reactions when the Earth was cooler (Tirard, 1996).

At the beginning of the twentieth century, the French naturalist Edmond Perrier (1844-1921) supported the idea that the origin of life should be explained in terms of organic chemistry. According to Perrier, the birth of living beings was not the result of a special substance that would have been the physical basis of life (such as the *Bathybius*) but rather the result of processes of chemical synthesis putting together ordinary inert substances. Perrier stressed the following important fact in his book *La Terre avant l'histoire: les origines de la vie et de l'homme* (*The Earth before History: The Origins of Life and Mankind*) (1920): the science of the origin of life faces a major problem because it is a historical science. Experimentally speaking, there was no possibility of going back to the initial conditions (Perrier, 1920). With the help of this conclusion, the origin of life could take on its modern meaning, that is to say a single starting point for the whole living world on Earth.

The following episode is more widely-known. The Russian biochemist A.I.Oparin (1894-1980), one of the pioneers of biochemistry in the Soviet Union, proposed in 1936 (Russian edition, 1938 for the English edition) a well-argued scientific scenario leading to bioorganic molecules on primitive Earth (his first publication –in Russian- on the origin of life dated 1924). Oparin was particularly interested in the basic question of whether there was a fundamental difference between living and dead things. He felt that to discover the conditions under which the properties of life were conjoined, would be to explain the origin of life (Dick, 1996). He also used recent astronomical data to confirm his scenario on chemical evolution. Independently, in 1929, the British biologist J.B.S. Haldane (1892-1964) also suggested a scenario with a gradual process of evolution. It must be noticed that this pertinent concept of chemical evolution, integrating a very large time-scale in a very different environment from present-day Earth, inherited progressive ideas previously enunciated by a few scientists such as the Italian Ermano Giglio Tos (Tirard, 1996).

With these new concepts, the mechanism of *archeobiosis*, promulgated during the nineteenth century, took on a practical meaning. The suggested stages to transform small atmospheric organic molecules into living matter *via* chemical synthesis created a turning point in the field of organic chemistry and initiated new laboratory experiments.

The American biochemist M. Calvin and his colleagues undertook in 1951 the first experiments designed to synthesize organic compounds under conditions consistent with the primitive Earth's environment (Raulin-Cerceau *et al*, 1998). However, Calvin's experiments did not follow Oparin's proposal (carbon dioxide had been chosen as an atmospheric component instead of methane). The subsequent experimental work carried out by the American chemists S. Miller and H. Urey (1953) was the first to put into practice the hypothesis of the chemically reduced atmosphere suggested by Oparin. These famous results led the way to a large and new scientific field devoted to prebiotic chemistry.

5. Sowing the Planets by Panspermia

During the second part of the nineteenth century, one of the consequences of the denial of spontaneous generation was the idea that life was necessarily antecedent to life. With this point of view, the following question resurfaced: was life an eternal process, as been suggested before by the philosophers of Antiquity? If so, life would have been born with the universe (eternal itself) and would have developed anywhere in the universe where life's spores could have germinated. This theory has been called *Panspermia*, and was first proposed in 1865 by the German physician Hermann E. Richter (1808-1876), a defender of Darwin's theory. He advocated for a meteoritic origin of life on Earth in which the meteorites carried the germ cells (*lithopanspermia*). In 1871, William Thomson (1824-1907) (widely-known as Lord Kelvin) and the German physicist Hermann von Helmholtz (1821-1894) supported the same theory of a sowing of microbes carried by meteorites falling onto the Earth. The French botanist Philip Van Tieghem (1839-1914) defended this hypothesis in his *Traité de Botanique* (*Botanical Treatise*) in 1884. In 1907, the Swedish chemist Svante Arrhenius (1859-1927) proposed the *radiopanspermia*, another way of carrying microbes. He suggested that these microbes would have been ejected from planets and would have been scattered in the galaxy, where, carried by the radiation pressure of stars, they would have encountered and seeded our planet (Arrhenius, 1910).

In the beginning of the twentieth century, the French agronomist Paul Becquerel (a nephew of the physicist Henri Becquerel) undertook experiments to test Arrhenius' hypothesis. He studied UV interactions with spores and bacteria at very low temperatures and in a high vacuum. The results showed that the spores didn't resist the UV conditions he created. Becquerel concluded that the spores must have been destroyed during their journey between the stars and the interplanetary medium would have been completely sterilizing (Becquerel, 1910).

The experiments testing Panspermia have considerably evolved since the first attempts carried out by Becquerel. Since the 1970s, the survival of bacterial spores in space can be tested by experiments undertaken on board spacecrafts and artificial satellites. Sophisticated experiments performed on spacelabs and satellites (LDEF, EURECA, FOTON) have given some insight into the responses of microorganisms to the parameters of space (Horneck *et al*, 2002). Some conclusions have been drawn from this modern data. In particular, it seems that solar UV radiation would be the most destructive parameter reducing the survival of spores if exposed without any shielding (Horneck, 1994). On the other hand, ejection from other planets of microbes inside rocks and their transport through the solar system seems to be a feasible process. If protected against solar UV and galactic cosmic radiation, spores could survive inside meteorites over extended periods of time (Horneck *et al*, 2002).

6. Plurality of Worlds and the Different Cosmologies

Debating about the plurality of worlds, also named " pluralism ", is a very old question. Through the ages, the different ways to answer this problem have been closely linked to the various concepts of "universe". We must stress that for the period of Antiquity, it would be better to use the term cosmos or world (ether filled with stars and planets) instead of " universe ", which is a later astronomical concept. With the meaning coming from Antiquity, the entire visible universe composed one cosmos or one world. This cosmos or world designated an independent and ordered system, composed of fixed stars and planets, with one "Earth" and one "Sun". Each of these worlds could have simultaneously existed or followed one another.

During Antiquity, Greek philosophers used cosmological sketches to make their position on the existence of other worlds clear. Adopting an idea coming from Leucippe (460-370 BC), Democritus (460-370 BC) believed in infinite worlds, each of them offering a beginning, a development and an end, followed by a new world. Aristotle (384-322 BC), student of Plato (427-348/347 BC), believed in a single world that was not infinite but on the contrary presented a center and limits: enclosed with fixed stars, it was spherical and revolved around a fixed axis where the motionless Earth was located. The Earth was then the center of this world. But the Earth was more than the physical center; it was also the center of motion: everything moved with respect to that single center. Following his doctrine of natural motion and place, Aristotle concluded that if there were more than one world, the elements of earth and fire would have more than one natural place toward which to move (Dick, 1996).

Epicurus (342/341-271/270 B.C.) took up Democritus' philosophy about an infinite number of atoms evolving in an infinite vacuum (also named *atomism*). He supported the idea of the plurality of worlds since an infinite number of atoms (likely to collide with each other) could have led to an infinite number of worlds. In his *Letter to Herodote*, he clearly proposed arguments in favor of the existence of other worlds inhabited by living beings (Raulin *et al*, 1997). Atomism was also adopted by Lucretius (99-55 BC): other worlds must exist " since there is unlimited space in every direction, and since seeds innumerable in number and unfathomable in sum are flying about in many ways driven in everlasting movement " (Lucretius, quoted *in* Dick, 1996).

But it was Aristotle's system that was transmitted to the Latin West, where it was repeatedly commented on in the context of the Christian system (Dick, 1996): Aristotle's system, rejecting the possibility of other worlds was in accordance with Christianity since a plurality of worlds directly confronted its omnipotent God. The influence of Aristotle has been significant throughout the ages and the position of a plurality of worlds did not find many supporters as long as the Earth was considered the center of a single world created by God. During the Middle Ages, the German philosopher and theologian, Albertus Magnus (1193-1280), wondered about pluralism, but, as many people of that time, was influenced by Aristotle's thought and chose to reject pluralism. However, from the end of the thirteenth century, it became conceivable to believe in an omnipotent God, creator of numerous worlds. The French philosopher Nicolas Oresme (1325-1382) examined Aristotle's writings and presented his personal critique on Aristotle's cosmology in *Le Livre du ciel et du monde* (*The book on the Heavens and the Cosmos*) (1377).

The Polish astronomer Nicolas Copernicus (1473-1543) orchestrated a decisive turning point in cosmology by decentralizing the position of the Earth in the cosmos in *De revolutionibus* (1543). Even if Copernicus did not directly discuss the problem of pluralism, he opened up new ideas. The new paradigm of a heliocentric system demonstrated the existence of possible other " Earths " revolving around other " Suns ". The Italian philosopher Giordano Bruno (1548-1600) picked up on this new paradigm and imagined unlimited inhabited worlds in an infinite universe (*De l'infinito universo e mondi*, 1584). According to Bruno, the greatness of Divine power and the metaphysical concept of unity in Nature led to the existence of infinite worlds. This view was in accordance with his philosophical position, labeled as heretic by the Catholic Church during the Inquisition. Besides, he was probably condemned in 1600 to be burnt at the stake more for his entire philosophy than for his opinion on the plurality of worlds (Raulin-Cerceau, 2002).

Another upheaval in astronomy was the invention of the telescope in 1610. In *Siderius nuncius*, Galileo noted that the surface of the Moon was " not unlike the face of the Earth " (Dick, 1996). Galileo's observations (surface of the Moon, moons of Jupiter) opened up the way to the comparison between the different bodies of the solar system. Such was the case with the German astronomer Johannes Kepler (1571-1630) who wondered about the similarities between the Earth, the Moon and Mars. At the end of his life, Kepler became well-known for his relevant experimental method announcing that the planet revolutions around the Sun were ellipses rather than circles. One of his publications (a posthumous one, published in 1634) entitled *Somnium* (*The Dream*) was most likely a forgotten masterpiece among the many works Kepler focused on in the field of astronomy. Often considered as a book of fiction, *Somnium* described the universe in the eyes of observers living on the Moon. It was not however fiction. This changing of reference (the Moon instead of the Earth) was considered by Kepler as an easy way to offer some answers to significant questions such as: why some astronomical phenomena occur both on the Earth and on the Moon, and why some show variations and offer specific parameters? To Kepler, *The Dream* was a didactic exercise to demonstrate the motion of the Earth around the Sun and its axis. In this way, Kepler provided revolutionary arguments to prove that laws of physics were appropriate on the Earth as well as on other celestial bodies of the solar system (such as the Moon) (Lear, 1965). In fact, if *The Dream* looked like fiction in the narrative part of the book, numerous and rich footnotes (longer than the main text) provided a lot of accurate information on astronomy and on data related to pluralism. For instance, these footnotes furnished pertinent remarks on the possibility of inhabiting the Moon and on the question of the origin of life (Kepler, *in* Lear, 1965).

The French philosopher and mathematician René Descartes (1596-1650) formulated in his *Principia Philosophiae* (1644) a rational cosmology in which every star was seen as a " Sun " potentially surrounded by planets. The universe was seen as a plenum filled with innumerable atoms. By putting forward a physical system in which each star was considered a " Sun " and, as such, possibly surrounded by planets, Descartes had opened a door through which many would pass to pluralism (Crowe, 1986). But Descartes did not confirm or deny the possibility that these planetary systems could be inhabited.

Cosmological worldviews adopted during the seventeenth century greatly influenced the popularity of pluralism. By releasing the Earth from its singular position in the universe, both the Copernican and Cartesian theories gave vent to many notions of life elsewhere. The French philosopher and poet Bernard le Bovier de Fontenelle (1657-1757) proposed in his *Entretiens sur la pluralité des mondes* (*Discussions on the Plurality of Worlds),* (1686) an amusing and lively description of the inhabitants of every planet of the solar system. Widely distributed in Europe, this book became very famous and popularized the idea of the existence of human-type beings on the other planets of our solar system.

The first attempts at permitting the detection of other planetary systems, were due to the Dutch astronomer Christiaan Huygens (1629-1695). Huygens, who had already shown his belief in pluralism in *Systema Saturnium* (1659), expounded his scientific methods supposed to provide evidence of stars surrounded by planets in a (posthumous) book named *Cosmotheros* (1698). Using astronomical data, such as the brilliance of the stars and their assessed distance from the sun, he was able to reach some conclusions, unfortunately announcing the difficulties in detecting other planetary systems (Raulin-Cerceau, 2002). A few years later, the British physicist Isaac Newton (1642-1727), following his " natural theology " guided by God, voiced, in *Optiks (1706)*, his conviction of the existence of superior beings on other planetary systems (Crowe, 1986).

In the middle of the eighteenth century, pluralism gained popularity with the help of the position of physicists, astronomers and writers. For this reason and thanks to the development of the techniques of astronomical observation, many astronomers started studying intensively the surface of the Moon. The Earth's satellite was becoming one of the favorite planetary bodies where extraterrestrial life was imagined. This second part of the eighteenth century probably displayed the more prolific comments related to the habitation of the planets of our solar system. For instance, Johann Elert Bode (1747-1826) suggested that life was possible everywhere in the solar system, provided the atmospheric conditions and temperatures were favorable. His extreme position on the question of pluralism was labeled as " cosmic panpopulationism " (Crowe, 1986). The French mathematician Pierre Simon de Laplace (1749-1827) elaborated in his *Exposition du système du monde* (*Exposition of the World's System*) (1796), a theory on the formation of the solar system involving the rotation, the contraction and the condensation of a pre-solar nebula. In this hypothesis, the formation of the planets was an integral part of the one of the stars, and then in the majority of cases the stars would have been surrounded by planets.

7. Pluralism Entering Astronomy as a New Research Field

The nineteenth century saw the amplification of all sorts of popularizing works using pluralism to make the writings more attractive. For instance, the Director of the Observatory of Paris, François Arago (1786-1853) pointed out in his *Astronomie populaire* (*Popular Astronomy*) several possibilities indicating that the Sun and the comets were inhabited. At the same time, the French physicist François Edouard Plisson raised some basic questions to solve the question of pluralism in *Les Mondes* (*The Worlds*) published in 1847. In this book, he gave details of every planetary body of the solar system, for instance the presence of an atmosphere, the nature of the surface, the range of temperatures... finally, all the parameters likely to be necessary for living beings -those answering to the laws of physics and biology.

During the second part of the nineteenth century, the topic of Extraterrestrial Life became integrated into the research fields of many astronomers and was often discussed in the scientific community. Pluralism changed from the position of an ideological trend to the position of a scientific field directly influenced by the rapid progress of the sciences. The scientific context of that time, marked by the idea of biological evolution (mainly coming from Darwin in 1859), opened up a larger implication of this idea, that is to say on the evolution of the matter in the universe and the origin of the celestial bodies. At the same time, the astronomical field was enriched with new techniques, in particular with spectroscopy, a means of delving deeper into the composition of the radiation emitted by the celestial bodies; spectroscopic methods were used to establish the presence (or the absence) of an atmosphere, and to give details on the chemical composition of such an atmosphere. In this way, and maybe for the first time in the history of pluralism, there was a concrete link between scientific data (conditions of a planetary environment) and the search for extraterrestrial life.

The results of spectroscopy applied to astronomy also strengthened the idea of the universality of chemical processes - those which were of biological interest. Therefore spectroscopy could predetermine, if the case arose, the universality of life. This view was strongly supported by the French astronomer Jules Janssen (1824-1907) who thought he had discovered water in the spectrum of Mars during his observations. According to Janssen, if the planets had the presence of water in common, there was a good chance that life arose somewhere else in the solar system (Janssen, 1929).

Pluralism continued to arouse interest among writers in the astronomical field. The British Richard Proctor (1837-1888), although he was not an astronomer, gained the confidence of the astronomical community with his writings intended for the general public (for instance: *Other Worlds than Ours*, 1870- *Our Place Among Infinities*, 1876- *The Orbs Around Us*, 1899) (Crowe, 1986). In France, the astronomer Camille Flammarion (1842-1925) published his *Pluralité des mondes habités* (*The Plurality of Inhabited Worlds*) in 1862, a book translated in many languages and followed by numerous other writings.

8. Mars: the Best Place to Search for Extraterrestrial Beings

At the end of the nineteenth century, the interest in knowing whether or not Mars was inhabited was at its peak. Since the detection of the *canali* by Giovanni Schiaparelli (1835-1910) in 1877, the existence of artificial canals (and more generally of living beings) on the surface of Mars was much debated. The American astronomer Percival Lowell (1855-1916) invested in the construction of the Observatory of Flagstaff (Arizona) in order to study the habitability of the other planets, especially of Mars. In his view, the canals observed by Schiaparelli were the work of intelligent beings. Lowell developed his arguments in the following books: *Mars* (1895), *Mars and its Canals* (1906) and *Mars as the Abode of Life* (1908) (Dick, 1996). According to

Lowell, the canals could have been built to carry the water during the summer coming from melted ice of the Martian polar caps, and therefore, irrigate the land cultivated by some superior form of living beings. The controversy about the canals ended in 1909 with the observations of the Greek-born astronomer Eugene Antoniadi (1870-1944). He gave a different interpretation as to the nature of the canals: the geometrical canal network was only an optical illusion (Coliolo, 2002).

If the possibility of intelligent life on Mars was ruled out at the very beginning of the twentieth century, nevertheless there remained a strong hypothesis concerning Martian vegetation. The changing of colors on the Martian surface encouraged an interpretation connected with the changing of the seasons on the planet. The climatic conditions on Mars would have controlled the growth of a primitive form of vegetation. From the 1920's to the end of the 1950's, many astronomers (Coblentz, Slipher, Dollfus, Tikhov, Sinton, Kuiper...) from various observatories (Mt Wilson, Lowell, Pic du Midi, Harvard...) were involved in such observations, attempting to solve the question about the Martian vegetation (Dick, 1996).

At the dawn of the Space Age, the canal controversy had waned and much was known about the physical conditions of the planet Mars (Dick, 1998). Vegetation of some sort was still a real possibility and the search for life *via* spacecrafts appeared to be the best way to prove or disprove this hypothesis. But, up to now, experiments on board spacecrafts have failed to detect any sort of life on Mars.

In the space of a few decades, it has become more and more clear that Mars is an extremely diverse planet in geological terms. This geological diversity has become a fact of paramount importance in the search for life on its surface. The modern approach is matched by finding forms of life in a huge variety of geological settings, environments and ecosystems on Earth (Brack, Fitton, Raulin, 1999). With terrestrial analogues of potential relics of extraterrestrial life furnished by the Earth's oldest sediments, it is now feasible to have a standard from which it is possible to reasonably evaluate remnants of analagous fossil life - if ever found on other planets, particularly on Mars (Schidlowski, 2002). It seems to be a reasonable supposition to resort to a paleontological inventory of the oldest terrestrial sediments (3.5-3.85 Gyr old) for guidance concerning potential fossil evidence from identically aged Martian rocks (Brack, Fitton, Raulin, 1999).

Paleontological evidences stored in the oldest recorded sediments were quite recently discovered. In 1953, new methods of dating (Pb isotopic composition) applied to the Earth's rocks revealed an age of our planet of 4.5 Gyr. The oldest sedimentary rocks dated back to 3.8 Gyr. Since 1983, the microscopic or cellular evidence (chert-embedded microflora from the Warrawoona Group of Western Australia) (Awramik *et al*) is supposed to go back to at least 3.5 Gyr. After an initial controversy about the authenticity of these fossil communities on grounds of imprecisely constrained petrographic background parameters for the host lithology, it has been accepted that these observed morphotypes were microfossils (Brack, Fitton, Raulin, 1999). However, reports of chemical fossils, microfossils and stromatolites dating from 3.4-3.8 Gyr have recently all been questioned (Riding, 2003). Reinterpretation of the structures previously identified as the oldest bacterial and cyanobacterial microfossils from 3.465 Gyr old Apex cherts of the Warrawoona Group shows that these structures could be secondary artefacts formed from amorphous graphite within multiple generations of metalliferous hydrothermal vein chert (Brasier, 2003). Graphite in metamorphic rocks at Isua (Greenland), identified as chemical fossils (3.8 Gyr old) has been reinterpreted as inorganically derived from non-sedimentary iron carbonates formed by high temperature solution alteration. Finally, doubts have also been raised about the biogenicity of Warrawoona stromatolites (Riding, 2003).

Controversies are recurrent in history of science. For instance, the analysis of the Martian meteorite ALH 84001 and the observations correlated with the possible past presence of Martian bacteria (McKay *et al.,* 1996) represent an interesting example of the significance of scientific criticism and doubt, in the context of the history of scientific concepts. Such an example could be picked out in the following questioning: in view of these uncertainties regarding chemical fossils - cell-like objects and stromatolites - how should the search for life in old rocks, on Earth and elsewhere, such as on Mars, be advanced? (Riding, 2003).

9. New Methods of Investigations: Beyond the Solar System

The very late years of the nineteenth century saw the arrival of some constraints which severely restricted the research field of pluralism: the theory of the gas escape velocity in planetary atmospheres, and even spectroscopy, furnished much data in contradiction with the requisite conditions for the possible development of living beings. At the same time, the idea of directly contacting extraterrestrial intelligent beings germinated in the minds of several people. It was the case for the inventor of the phonograph, the French inventor Charles Cros (1842-1888) who published in 1869 the *Etudes sur les moyens de communication avec les planètes (Studies on communication techniques with planets*) in the Journal *Cosmos* (Raulin-Cerceau, 2002).

Moreover, at the end of the nineteenth century a new method arose which could be developed into a form of communication of much greater power: the use of radio waves discovered by the German physicist Heinrich Hertz (1857-1894) in 1887 (Dick, 1996). Communication with other planets became conceivable and some scientists entered into this branch, working on techniques necessary for sending signals towards the celestial bodies (planets and stars). The Serbian-born American physicist Nikola Tesla (1856-1943) supported the idea on feasible radio communications with other planets. He even claimed to have observed in 1899 unusual electrical disturbances that he interpreted as possible extraterrestrial signals from intelligent beings. As early as 1909, the American astronomer David P. Todd (1855-1939) had suggested that Martians might have communicated with the Earth using Hertzian waves. A few years later, in 1919 in the *New York Times*, Guglielmo Marconi (1874-1937) demonstrated that radio waves were a really significant way to communicate with intelligent beings living around other stars. He also claimed he had received unexplained signals that might have originated from the stars (Dick, 1996). In France, the astronomer Alain Mercier elaborated a plan to send signals of light towards the planet Mars (Coliolo, 2002).

Confronted with the great difficulty of observing planets around other stars, the astronomers realized very early that they had to develop indirect techniques to detect extrasolar-planets. As soon as 1844, the German astronomer Friedrich Wilhelm Bessel (1784-1846) noticed tiny disturbances in some stars due to the possible presence of invisible companions. At the end of the nineteenth century indeed, astrometry, photometry and spectroscopy were already used as indirect methods to detect massive companions of stars. In fact, detection was limited to other stars (double stars-system), not planets, because of the limitation of the techniques.

Since 1937, the Dutch-born astronomer Peter Van de Kamp, a student of Ejnar Hertzsprung (1873-1967) who put the astrometric method into practice for double stars, has embarked on a research program devoted to the detection of extrasolar-planets with the help of the parallax method. While looking for perturbations in the motions of stars, he identified in 1963 what he thought was a Jupiter-size body revolving around Barnard's star (Van de Kamp, 1969). However, this discovery was later invalidated due to misinterpretation because of the limitation of instrumentation. Ever since the 1930's, the increased interest in detecting low-mass companions has resulted in many discoveries (such as the ones obtained in the 1930s-1940s, respectively by Holmberg and Strand, using the photometric astrometry method). But the astronomers were aware of the difficulty in interpreting their observations regarding the unseen companions: were they dwarf stars or planets? Whatever the problems that were found, the field of research for the detection of low-mass star companions was launched. However, following these disappointments, several years of doubt hampered the scientific community - when planetary systems were considered to be extremely rare in the universe (Raulin-Cerceau *et al*, 1998).

Interstellar communications found their origin in the development of radioastronomy. The first strategy using microwaves to detect hypothetical extraterrestrial intelligent beings came in 1959 from the American physicists Giuseppe Cocconi and Philip Morrison. The first application of this strategy was carried out by the astronomer Franck Drake in 1960 and a few years later (in 1965), the committee CETI (Communication with Extra Terrestrial Intelligence) was established. The term CETI changed to SETI (Search for Extra Terrestrial Intelligence) in 1986 in order to express its growing interest in every possible kind of search for extraterrestrial civilizations (Almar, 1988).

10. Conclusion

By changing from an every-day process of spontaneous generation to a single process of chemical evolution having occurred once in the past, the problem of the origin of life switched from the status of an observational science to the one of a historical reconstruction. That is the reason why in a few decades (around the middle of the twentieth century), the methods of approaching the solution to this problem have considerably evolved. An experimental chemistry field (in-laboratory) quickly grew, attempting to reconstruct the first stages of the chemical evolution. Faced with the difficulty of the task, the laboratories became specialized, integrating the new astronomical and geological data and that of the expanding space exploration. Since the 1960s, the space missions towards the other celestial bodies of the solar system have led the way to more and more refined and precious insights on what could have been the environment of the primitive Earth. Moreover, since it was obvious that it was impossible to re-create life experimentally, celestial bodies presenting a complex organic chemistry have become probably the best laboratories to obtain a better understanding of some of the main stages that could have led to the beginning of life. From this moment, the link between the search for the origin of life and the search for extraterrestrial beings definitely existed.

Around the 1940s biochemistry had significantly developed and had brought a detailed analysis of the basic constituents of living beings and of their functioning, stressing both their unity and their differences. The field of the origin of life became aware of new experimental processes focused on the reconstruction of the molecules of life. With the integration of astronomy, biology, chemistry, geology, micropaleontology, molecular biology, the branch of exobiology widened to become a very broad pluridisciplinary science, while the " old " field of pluralism provided in 1982 the foundations for the construction of bioastronomy, originally organized by the community of astronomers. Today, exobiology and bioastronomy seem to have merged into one, since astrobiology provides another dimension: the study of life's future on the Earth and beyond.

11. References

Almar, I. (1988) The History of IAA SETI Committee, In: G. Marx. (ed.) *Bioastronomy-The Next Steps*, Kluwer Academic Publishers, Dordrecht, The Netherlands, pp. XV-XVII.

Arrhenius, S. (1910) *L'évolution des Mondes*, Béranger, Paris.

Awramik, S.M., Schopf, J.W. and Walter, M.R. (1983) Filamentous fossil bacteria from the Archean of Western Australia, *Precambrian Research* **20** : 357-374.

Becquerel, P. (1910) L'action abiotique de l'ultraviolet et l'hypothèse de l'origine cosmique de la vie, *Comptes Rendus de l'Académie des Sciences*, séance du 10 mai, Paris, pp. 86-88.

Brack, A., Fitton, B. and Raulin, F. (1999) *Exobiology in the Solar System & The Search for Life on Mars, Report from the ESA Exobiology Team Study 1997-1998*, Edited by Andrew Wilson, ESA Publications Division.

Brasier, M.D., Green, O.R., Jephcoat, A.P., Kleppe, A., Grassineau, N., Lindsay, J.F. and Steele, A. (2003) Questioning the evidence for Earth's oldest fossils, In: *Astrobiology 2003, Abstracts and Symposium Schedule*, March 27 and 28, Girton College, Cambridge, UK, p. 13.

Coliolo, F. (2002) *L'histoire et la recherche de la vie extraterrrestre en France entre 1850-1950*, Rapport de Recherches interne, Paris.

Crowe, M. J. (1986) *The Extraterrestrial Life Debate 1750-1900*, Cambridge University Press, Cambridge.

Dick, S.J. (1996) *The Biological Universe, The twentieth-century extraterrestrial life debate and the limits of science*, Cambridge University Press, Cambridge.

Dick, S.J. (1998) *Life on other worlds, The 20th-century extraterrestrial life debate*, Cambridge University Press, Cambridge.

Farley, J. (1997) *Spontaneous Generation Controversy from Descartes to Oparin*, The Johns Hopkins University Press, Baltimore and London.

Girard, J. (1874) *Les explorations sous-marines*, Savy, Paris, p. 89.

Haeckel, E. (1874) *Histoire naturelle de la Création*, Reinwald, Paris.

Horneck, G. (1994) Long-term survival of bacterial spores in space, *Adv. Space Res.* Vol.14, No 10, pp. (10)41-(10)45.

Horneck, G., Mileikowsky, C., Melosh, H.J., Wilson, J.W., Cucinotta, F.A., and Gladman, B. (2002) Viable Transfer of Microorganisms in the Solar System and Beyond, In: G. Horneck and Ch. Baumstark-Khan (eds), *Astrobiology, The Quest for the Conditions of Life*, Springer, pp. 57-76.

Janssen, J. (1929) *Œuvres Scientifiques recueillies et publiées par Henri Dehérain*, Société d'Editions géographique, maritimes et coloniales, Paris, p. 119.

Kamminga, H. (1980) *Studies in the History of ideas on the Origin of Life from 1860*, Thesis of Philosophy, University of London, London.

Latour, B. (1989) Pasteur et Pouchet : hétérogenèse de l'histoire des sciences, In: M. Serres (ed.) *Eléments d'Histoire des Sciences,* Bordas, Paris, pp. 423-445.

Lear J. (1965) *Kepler's Dream, with the full text and notes of Somnium, Sive Astronomia Lunaris*, University of California Press, Berkeley and Los Angeles.

McKay, D.S., Gibson, E.K.Jr., Thomas-Keprta, K.L., Vali, H., Romanek, C.S., Clemett, S.J., Chillier, X.D.F., Maechling, C.R. and Zare, R.N. (1996) Search for past life on Mars; possible relic biogenic activity in Martian meteorite ALH84001, *Science* **273**: 924-930.

Perrier, E. (1920*) La Terre avant l'histoire : les origines de la vie et de l'homme*, La Renaissance du Livre, Paris, p. 81.

Raulin, F. Raulin-Cerceau F. and J. Schneider (1997) *La Bioastronomie*, Presses Universitaires de France, Que Sais-je ?, Paris.

Raulin-Cerceau, F., Maurel, M.C. and Schneider, J. (1998) From Panspermia to Bioastronomy, The Evolution of the Hypothesis of Universal Life, *Origins of Life and Evolution of the Biosphere*, **28:** 597-612.

Raulin-Cerceau, F. (2001) Theories on Origins of Life between 1860 and 1900, In: J. Chela-Flores, T. Owen and F. Raulin (eds.) *First Steps in the Origin of life in the Universe*, Kluwer Academic Publishers, Dordrecht, The Netherlands, pp. 39-42.

Raulin-Cerceau, F. (2002) Histoire de la notion de vie ailleurs, In: F. Raulin-Cerceau, P. Lena and J. Schneider (eds.), *Sur les Traces du Vivant*, Le Pommier, Paris, France.

Riding, R. (2003) Cells, filaments, biofilms: physical evidence of ancient microbial life, In: *Astrobioloy 2003, Abstracts and Symposium Schedule,* March 27 and 28, Girton College, Cambridge, UK., pp. 56-57.

Schidlowski, M. (2002) Search for Morphological and Biogeochemical Vestiges of Fossil Life in Extraterrestrial Settings: Utility of Terrestrial Evidence, In: G. Horneck and Ch. Baumstark-Khan (eds.) *Astrobiology, The Quest for the Conditions of Life*, Springer, pp. 373-386.

Tirard, S. (1996) *Les travaux sur l'origine de la vie de la fin du XIXème siècle jusqu'aux années 1970*, Thèse de l'Université Paris VII (Epistémologie et Histoire des Sciences), Paris.

Van de Kamp, P. (1969) Parallax, Proper Motion, Acceleration, and Orbital Motion of Barnard's Star, *Astron. J.* **74**, 2, pp. 238-240.

Biodata of **Louis Lerman** author (with coauthor **Jacqueline Teng**) of the chapter: ***"In the Beginning: A Functional First Principles Approach to Chemical Evolution"***

Dr. Louis Lerman's interest in how complex systems arise from simple ones seems to have "spontaneously arisen" from his background in astrophysics, geophysics, and high-energy nuclear physics. Louis received his Ph.D. from Philipps University (Marburg, FRG) in 2002, his research exploring a novel form of symmetry-breaking in relativistic heavy ion collisions. Started while a visiting scientist at Lawrence Berkeley National Laboratory, it eventually grew into a 100 man-year international research project under his direction. His M.S. from Stanford University (1992) was on self-organizing planetary processes and the origin of life. While at Stanford, as a Hertz Fellow in the Applied Physical Sciences, Louis also developed the Flotilla concept for low-earth orbiting unmanned space systems. His non-scientific activities include fine-art photography and studying classical voice.

E-mail: **Bubbles@DNAi.com**

Biodata of **Jacqueline Teng** coauthor (with author **Louis Lerman**) of the chapter: ***"In the Beginning: A Functional First Principles Approach to Chemical Evolution"***

Jacqueline Teng is a senior at the University of California, Berkeley; majoring in Electrical Engineering, Computer Science, and English (with a minor in Creative Writing). She is also currently employed by the Insitu Group, helping to develop long-duration UAV's for scientific, government, and commercial projects. A software expert and specialist in high-performance computer systems, Jacqueline has also worked as a professional in computer graphics and digital art. Along with her writing of science-fiction and fantasy, she has directed storylines and developed tools for interactive game sites. Jacqueline is an accomplished pianist, specializing in Chopin, who invariably needs more time for practice.

E-mail: **Chalice@Dragonwarden.com**

J. Seckbach (ed.), Origins, 35–55.

IN THE BEGINNING
A Functional First Principles Approach to Chemical Evolution

LOUIS LERMAN[1] and JACQUELINE TENG[2]
[1]*Department of Chemistry, Philipps University (Marburg, FRG)*[*] *and Pteranodon Ventures (Las Vegas, NV., USA)* [2]*University of California, Berkeley (Berkeley, CA., USA)*

1. Prologue: Games Prebioticists Play

Just as the biochemistry of contemporary organisms can be viewed as a "fossil" record of biogenesis, the geochemical physics of the contemporary earth can be used to delineate the self-organizing processes underlying prebiotic chemistry. Pre-biotic chemical evolution presupposes successive generations of increasingly complex organic molecules combinatorially synthesized from previous generations. Not so obvious is how this combinatorial chemistry occurred or how the overall process of chemical self-organization was functionally supported at each stage of its occurrence. This latter point, the functional support of prebiotic chemical evolution within a realistic geophysical/geochemical environment, is what this work is about.

2. Chemical Evolution as a Functional *Fermi Question*

The physicist Enrico Fermi delighted in posing fundamental questions without initial assumptions or boundary conditions. *"How far can a bird fly?"* was one, or *"How many piano tuners are there in New York?"* His delight has become widely shared, for these *Fermi Questions* embody a powerful methodological approach to problem solving. It allows one to construct a qualitatively meaningful quantitative answer to a complex problem where one is ignorant of its boundary conditions and details. More than the search for a specific numerical answer, the fundamental problem is itself elucidated through a series of questions and justified order-of-magnitude assumptions. Indeed, the field of astrobiology can be argued to have come from what is perhaps the most famous of Fermi Questions: *Where are They?* "They" refers to other intelligent life in the universe, and the attempt to answer this open-ended question led directly to the Drake Equation (Drake 1961).

In this paper, we apply a similar approach to the question of how chemical evolution came about. We look at functional requirements from first principles and utilize only the most basic and transparent of necessary assumptions and imposed boundary conditions. Nonetheless, because the origin of life is a unique event with no other examples to study, compare, and interpret a few axiomatic assumptions are necessary.

---

[*] Work performed at Stanford University and Lawrence Berkeley Laboratory (University of California, Berkeley) *Correspondence should be sent to*: Louis Lerman c/o Pteranodon Ventures 270 E. Flamingo, Suite 305 Las Vegas, NV 89109 USA

Keeping Occam's Razor well sharpened, a "scientific" study of prebiotic evolution can be based on the following minimalist set of axioms:

I. The laws of physics and chemistry have been constant over the last 5 billion years
II. Life evolved on the early earth from the simple to the complex
III. Preceding biological evolution was the chemical evolution of organic molecules
IV. Chemical evolution occurred within the boundary conditions of the geophysics and geochemistry of the early earth

3. Chemical Evolution as Process

Continuing on in a Fermi Question way, we (necessarily!) assume ignorance of the totality of reactions involved in the creation of successive generations of increasingly complex prebiotic organic molecules. The stochastic nature of physics (from Axiom 1 above) applied to the simple-to-complex nature of chemical self-organization (Axiom 3) means that each level of self-organization must have evolved from broadly ranging sequences of chemical trial and errors. Some pathways will work, leading to products useful for the construction of the next level of organization...but most will not. Hence, in the abstract, our postulated process of chemical evolution requires the existence of a *global chemical engineering system* which at the system level amplified the probabilities of stochastic self-organization at the molecular level. For this amplification to be effective at the system level our still-to-be discovered processes must be:

- *First-order ("had-to-have-been" mechanisms)*
- *Robust (highly efficient and rapid processes)*
- *Diverse (exploring many different possible chemical routes and mechanisms)*
- *Selective (for the good stuff!)*
- *Semi-closed (able to retain "useful" materials in the total system)*

Meanwhile at the molecular level the race between the self-organization of increasingly complex organic structure and its entropy-driven dissipation requires the following functional operations:

- *Selective concentration of the desired organics and metal ions needed as reactants*
- *Stabilization and coordination of these reactants*
- *Controlled energy and "directed" synthesis*
- *Cycle continuity (products become the reactants for the next stage of the cycle)*
- *Most importantly, all must occur in a likely geophysical/geochemical environment*

Hence, a global chemical engineering system is required which does all of the above; and which had to have been in existence on the early Earth (or an early Mars with surface water). Such a first-order-system exists on the contemporary earth in the Bubble-Aerosol-Droplet Cycle; or as it will be abbreviated here...the *bubblesol* cycle. It is the most fundamental, robust, and far-reaching of geophysical/chemical super-cycles involving the ocean and atmosphere. Outlined in Figure 1, the *bubblesol* cycle embodies a nested hierarchy of microenvironments which is certain to have existed on the early earth; and which was capable of supporting each of the above functional requirements.

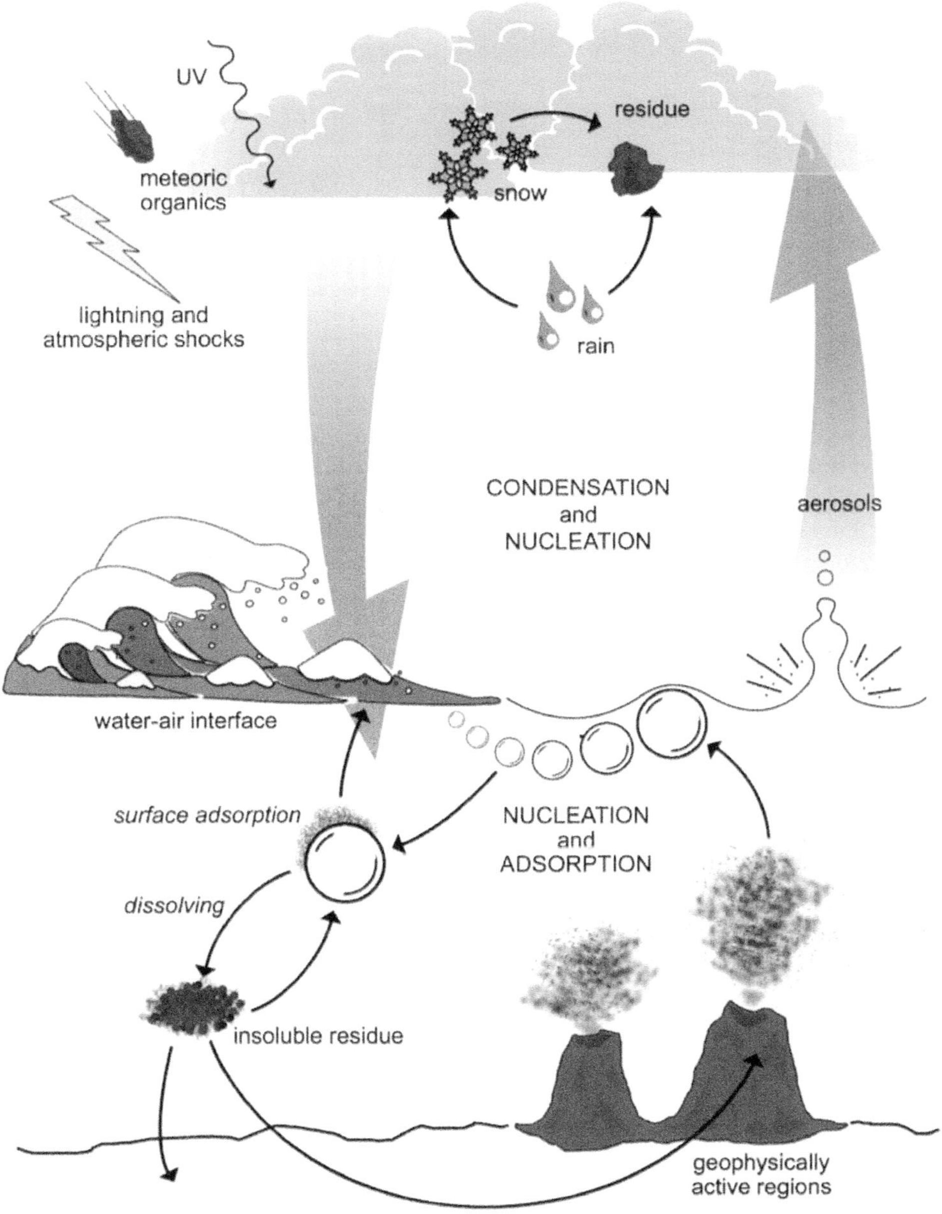

Figure 1. ***The Bubble-Aerosol-Droplet Cycle***
Each arrow represents the principal mode of mass transfer between the contemporary oceanic and atmospheric reservoirs, and each node the principal site of heterogeneous chemistry in its regime. (after Lerman 1992)

4. The Bubble-Aerosol-Droplet Cycle

Referring to Figure 1, a single stage of this cycle includes bubble formation, the consequent adsorption of surface active materials, bubble dissolution, and the non-equilibrium energetics of bubble bursting. With regard to organic molecules...

- *Organic materials, selected metals, and clay particles are preferentially adsorbed onto the surface of the bubble.*
- *This stabilizes the bubble, and when the organically "dirty" bubble finally dissolves or bursts, leads to a highly concentrated particle rich in organics.*
- *The organic and metal rich residue of a dissolving bubble nucleates other bubble formation or will be adsorbed by other bubbles.*
- *The bursting of bubbles injects into the atmosphere particulate matter highly concentrated in organics, metal ions, and minerals.*
- *These injected materials are then coupled to aerosol formation and the subsequent nucleation of atmospheric condensation, leading to the complex chemistry and physics of rain and snow.*

Throughout this super-cycle, made up of many intricate sub-cycles, hydration-dehydration cycles are in abundance. These are of potentially unique importance to the polymerization reactions upon which chemical evolution was necessarily based. Additionally, the bubblesol cycle promotes the existence of temporary membrane-like phase boundaries, critical in the transition from organic chemistry to biochemistry.

Strong phenomenological support for this cycle comes from the fact that it incorporates the principal modes of mass transfer between the contemporary oceanic and atmospheric reservoirs, as well as the principal sites of heterogeneous chemistry (Duce and Hoffman 1976). Each of the arrows in Figure 1 represent the *primary* physical transfer mechanism between the nodes and their respective regimes as a whole (*i.e.* the atmosphere and oceans). The nodes represent the principle sites of heterogeneous chemistry and, not coincidentally, are just those with the highest surface area to volume ratio (bubbles, aggregates, aerosols, and precipitation). From this comes the main mechanisms for the concentration, fractionation, and transport of dissolved organics and metal ions in the upper ocean, bubbles (MacIntyre 1974); as well as the scavenging of organics in the atmosphere by aerosols and precipitation (Gill et al 1983).

The common element throughout is the air-water interface, being relatively independent of the specific chemistry of the atmosphere and ocean. Indeed, one can in the abstract look at bubbles and aerosols as complements: one is a metastabilized fluctuation of air in water, the other of water in air. On the terrestrial scale, the continuous layers of bubble clouds formed beneath breaking waves are the analogues of low stratus or stratocumulus clouds in the atmosphere (Kraus and Businger 1994). Whether floating under water, bobbing at the sea-surface, or drifting in the atmosphere, it is the adsorption of amphiphiles that drives these phenomena by decreasing the local surface energy, thus metastabilizing the microenvironment and its structure.

5. The Bubble-Aerosol-Droplet Cycle on the Early Earth (and elsewhere...)

This cycle is based only on physico-chemical phenomena – bubble formation by the Rayleigh-Taylor instability (Sharp 1984) without regard to specific details of the ocean and atmosphere chemistry. It is thus impossible to imagine a similar cycle not being active in the early Archean. Its initiation would require only the existence of an air-water interface disrupted by mechanical turbulent energy (from waves, meteorites, or geophysically active regions). Its maintenance requires only a metastabilization by simple amphiphilic compounds which at the very least would have come from meteorites. This latter statement is strongly supported by the fact that 70% of the organic compounds in the Murchison meteorite are polar (Cronin and Chang 1993).

Of critical importance to prebiotic chemistry, natural cycles of hydration-dehydration of organics on salt and other mineral surfaces are coupled to sites of chemical concentration. This is the case at each phase of the bubblesol cycle (bubble bursting, aerosol formation, precipitation scavenging, and precipitation-induced bubble formation). Because polymerization of biological and prebiotic molecules is a dehydration reaction this is an almost unique advantage of the bubblesol cycle.

In proposing the functional cycle represented in Figure 1, we have made the least number of assumptions on the conditions of the early earth; with each assumption based on the strongest principles. One can with greater confidence assume the existence of bubbles on the early earth than, say, the widespread availability of a particular montmorillonite clay (which should have been around as well). Because the existence of these bubbles, cavities, and droplets are such common phenomena in nature, they are functionally relevant to chemical evolution on the early earth or elsewhere in the solar system. In analogy to contemporary processes, such a prebiotic geophysical cycle provides the potential for consecutive cycles of selective concentration, stabilization, and synthesis of increasingly complex organic molecules. *In particular, the dynamics of the water-air interface formed in bubbles and droplets may have specific concentrating and catalytic properties leading to condensation or polymerization reactions.* And as discussed in later sections, the nested hierarchy of microenvironments naturally embodies and functionally supports in a real-world way the majority of specialized environments and reaction possibilities postulated by other workers in the field of prebiotic chemistry.

6. Bubbles.... Adsorption

Central to the function and partially closed nature of this cycle is the ubiquitous bubble. Isolated bubbles come into existence as regimes of a gas in a liquid in which the surface of the bubble-water interface exists due to a local increase of surface free energy (supported by the vapor pressure within the bubble). Hence, a potential energy difference exists between the surface and interior of a bubble, allowing well-matched surfactants the opportunity to build more complex and stable structures through the lowering of this energy (Adamson 1997). This adsorption is driven by a reduction of the surface free energy as material is added. Lacking the stabilization due to adsorbed materials, a pure water bubble without any sort of a "skin" has a surface lifetime of less than a second. Most of this is a function of the curved air-water interface and is therefore also applicable to the aerosol (inverted bubble) phase of the bubblesol super-cycle. The robustness of the selective concentration abilities of bubbles is underlined by the large number of industrial processes such as mineral flotation and secondary and tertiary petroleum extraction based on the use of bubbles and foams (Lemlich 1972).

7. The Concentration of Organics, Phosphates, and Metals

Bubbles are not only primarily responsible for the concentration and transport of organics to the sea-surface microlayer, but they are also the principal source of sea-salt aerosols in the atmosphere (Duce and Hoffman, 1976). As seen in Table 1, bubble-generated processes are therefore the greatest source of matter in the atmosphere by both total mass and number of particles (Chester 1986).

Table 1. *Estimates of the Global Emissions of Particulate Matter into the Atmosphere (Units = 10^{12} g/yr)*

Source			Global Production 1	2	3
Man-made:					
	Direct Particle Production		30		
	Particles Formed from Gases				
		Converted Sulphates	200		
		Others	50		
	Total Man-made		280		200
Natural:					
	Direct Particle Production				
		Forest Fires	5	36	
		Volcanic Emissions	25	10	
		Vegetation		75	
		Crustal Weathering (mineral dust)	250	500	
		Sea Salt Aerosol (from bubbles)	**500**	**1000**	
	Particles Formed from Gases				
		Converted Sulphates	335		
		Others	135		
	Total Natural		1250		

In the contemporary oceans, this organic concentrating ability of bubbles is fundamental to the support of a number of critical ecological niches, from plankton to whales. So one asks...could such bubblesol phenomena be the active mechanism whereby the earliest organic molecules (previously synthesized in low yields by other means) were concentrated, and then condensed into the larger organics required for biological chemistry? A highly relevant observation to this point on the contemporary earth is summarized by Duce and Hoffman (1976):

"Based on the available data, the only substances having positive enrichments of more than a few percent in the atmosphere that can be clearly ascribed to chemical fractionation occurring during sea salt particle production (i.e. through the bubble transport and bursting process) are I, PO_4^-, probably some organic nitrogen, and organic carbon, and possibly some heavy metals and K in biologically productive waters."

The heavy metals enriched in these bubble generated processes include the biologically necessary elements Fe, Zn, Mn, Mg, Cu, and Ni, as well as Pb, Hg, Cd (Szekielda 1972). The concentrations of these metals can be greater than 10^4 compared with those in the bulk solution (ibid, as well as Duce et al. 1972). Also as seen in Figure 2a, phosphates – which are often the limiting factor in ecosystems – can be enriched by factors of a 1000 (MacIntyre 1969). Indeed, the bubble scavenging process is so efficient that it can selectively scavenge out of solution organic materials of less than a 10^{-9} molar concentration. (This is why so much of the chemical engineering industry is based around these separation and selection processes.) As with other components, organic enhancements can be up to a factor of 10^4. Figure 2b shows such an enhancement for n-alkanes and phthalates. Based on contemporary biochemistry, many of these chemical species are just those needed for chemical evolution.

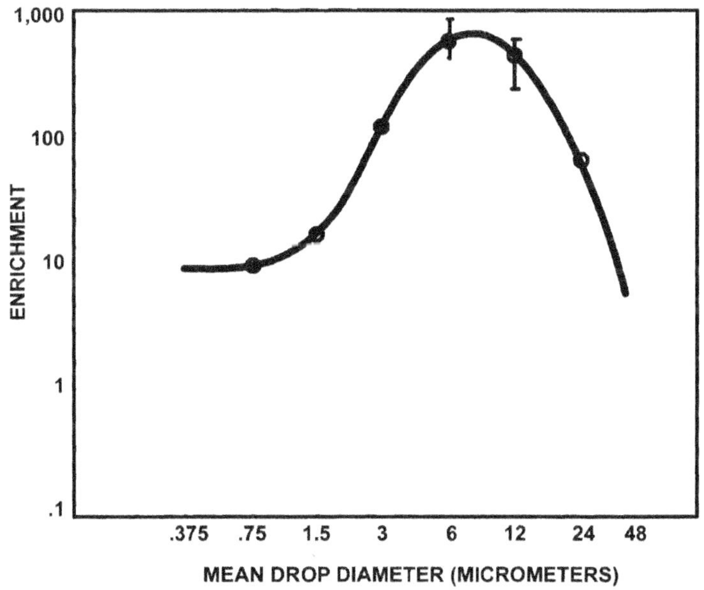

Figure 2a. ***Phosphate Enrichment due to Bubble Concentrating and Bursting*** (MacIntyre 1974c)

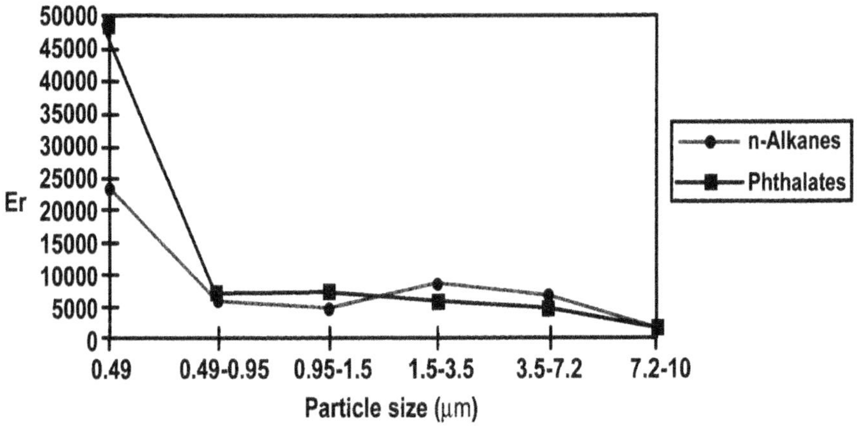

Figure 2b. ***Enrichment Ratio (E_r) of n-alkanes and Phthalates (relative to Na^+) vs. Aerosol Particle Size*** (Cincenelli 1999)

8. The Sea-Surface Microlayer

Because of the bubble mass transfer from below, and both dry and wet deposition from the atmosphere, in the sea-surface microlayer the concentration of select organics, metals, ions, and particulate matter is 10^4 (or more) greater than the concentration of these materials in the bulk water. Upon inertial bubble formation, it is this surface layer which becomes the inner surface of the bubble. And a bubble being a micro-version of the air-sea interface, it will tend to further adsorb those materials that congregate at the air-sea interface. This includes fatty acids and alcohols, proteins, polysaccharides, metal ions, colloidal silica, and clays (kaolin and montmorillonite). Some observations have indicated that foams made from proteins and sugars (thought by some to be the majority of contemporary sea surface films) are stabilized by silica and clays. The congruence of this confluence of organics and minerals to contemporary prebiotic chemistry experiments is satisfying to say the least.

Support for a possible direct connection of these microlayer processes to the origin of life comes from the work of Banin and Navrot (1975), illustrated in Figure 3. They observe that plots of elemental enrichment factors (ratio of concentration of an element in an organism to its concentration in the earth's crust) vs. ionic potentials for the four major groups of organisms (bacteria, fungi, plants, and land animals) are fascinatingly congruent to that of seawater. What is most intriguing with respect to the bubblesol cycle is the fact that for ionic potentials greater than 10 (sulfur, carbon, nitrogen, and other of the primary biochemical elements), organisms have enrichments in the range of 10 – 10,000 over the nominal ocean value. This is the primary range of enrichments due to bubble-processes at the sea-surface! Hence, there is the strong hint that the enhanced elemental enrichments of living systems may be correlated with the sea surface microlayer enrichments resulting from bubblesol processes.

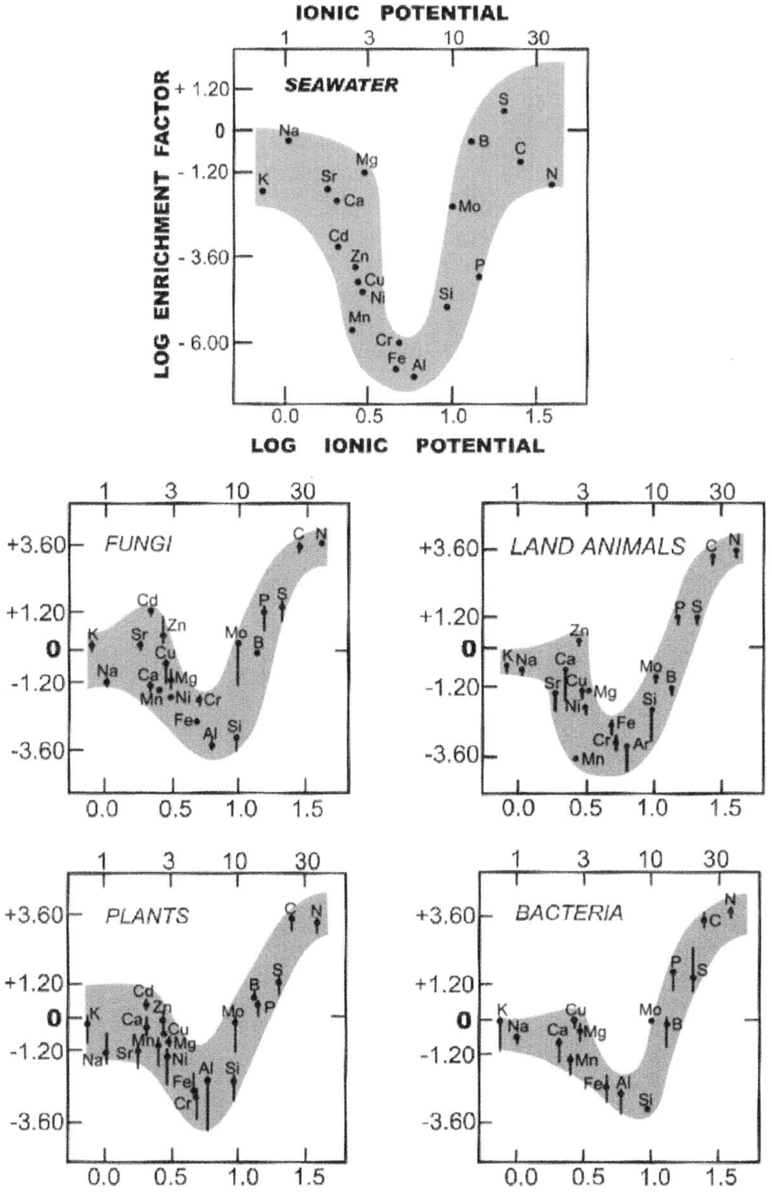

Figure 3. ***Elemental Enrichment Factors vs. Ionic Potentials: A Congruence between Seawater and Life*** Elemental Enrichment Factors (ratio of concentration of an element in an organism to its concentration in the earth's crust) vs. Ionic Potentials for the four major groups of organisms (bacteria, fungi, plants, and land animals) [Banin & Navrot 1975]

9. Bubble Life Cycles

Below the water surface, bubbles may be formed by cavitation or by outgassing (such as from a discharging magma, hydrothermal vent microenvironments, or thermally induced pyrolysis of previously collected organic materials). In this latter case, the interior gas will naturally reflect the composition of the outgassing source. At the surface of the ocean, bubbles can be formed by infalling raindrops, snowflakes, and whitecaps (Blanchard and Woodcock 1957). A direct mode of bubble formation is also the bursting of other bubbles (MacIntyre 1972). As previously discussed, the inner lining of the newly formed bubble was formerly the surface of the water. This is important to our prebiotic scenario since as we have seen the chemical composition of the surface microlayer is highly differentiated from that of the bulk ocean.

Once a bubble exists, it tends to collect surface-active surfactants, the first of several different concentrating phenomena associated with bubbles. These initially concentrated materials, predominantly organic (Blanchard 1975), stabilize the bubble allowing it to follow pathways which in turn offer greater concentrating opportunities.

1) The bubble can follow potentially complex trajectories in the water, resulting from the sum of the hydrodynamic currents and buoyant forces upon it. Under these circumstances, a single bubble could pass through a variety of environmental conditions. It can also cycle through regions of both high and low pressure, deflating and inflating respectively (Johnson and Cooke 1981).

2) If the bubble remains below the surface, it will eventually dissolve, leaving behind a concentrated organic rich residue (Johnson and Cooke 1980). This residue can in turn be adsorbed by other bubbles, or act as a nucleation site for the formation of more bubbles.

3) Bubbles with adsorbed surfactants can deposit these materials on the microlayer surface of the ocean.

4) The bubble may arrive (or remain after formation) at the water-air interface and subsequently burst (MacIntyre 1972), leading to the formation of sea-salt aerosols and condensation nuclei (Cipriano and Blanchard 1981).

10. Aerosols

There are two bursting mechanisms possible for an isolated bubble, each leading to the formation of a different class of particulate matter injected into the atmosphere (Figure 4). For small bubbles (<0.5 mm), surface tension driven "jet drops" are formed, which in turn are the principal source of sea-salt aerosols; and are by some estimates the largest source (by mass) of particulate matter injected into the atmosphere. For larger bubbles (> 0.5 mm) an instability mechanism due to the gravity-driven draining of the bubble cap leads to the formation of large numbers of "film cap drops", particles much smaller than jet drops.

Figure 4. ***Formation of Jet Drops from a Bursting Bubble***
The material making up successive jet drops from a single bursting bubble comes from successively microtoned layers of its air-water interface. (Figure courtesy of Ferren MacIntyre)

The latter mechanism is possibly the largest source (by number) of particles injected into the atmosphere; and may be the major source of cloud condensation nuclei (CCN) as well. The particles injected into the atmosphere by bursting bubbles have different airborne lifetimes depending on the mass and composition of the particle. The heaviest particles injected immediately fall back to the water, forming more bubbles upon their impact. Particles swept up into the atmosphere will enter into the complex tropospheric aerosol cycle (Figure 5). Some particles will act as condensation nuclei for precipitation, others will be scavenged by one of a variety of precipitation bodies…themselves nucleated by bubble generated particles.

A condensation nucleus of an aerosol can undergo a number of hydration-dehydration cycles, typically ~10 over a week's time (Graedel, private communication), before eventually falling to the ocean surface. Precipitation will further concentrate organics and minerals from the atmosphere, regardless of source (meteoritic, terrestrial, or oceanic). During this scavenging process, there is some evidence for the formation of an organic skin around the precipitation object whether rain, snow, or hail (Gill, et al. 1983). Fog particles have been found to contain mostly organic carbonates, esters, and proteins; while the fatty acids in aerosols have been shown to have an oceanic source. Cycle continuity follows from the fact that the impact of the precipitation object upon the water surface forms bubbles, returning us to the bubble phase of the cycle (Blanchard and Woodcock 1957). While in the atmosphere these organic-metal rich objects will be exposed to a variety of energy sources, including solar radiation and the plasma and shock effects associated with lightning.

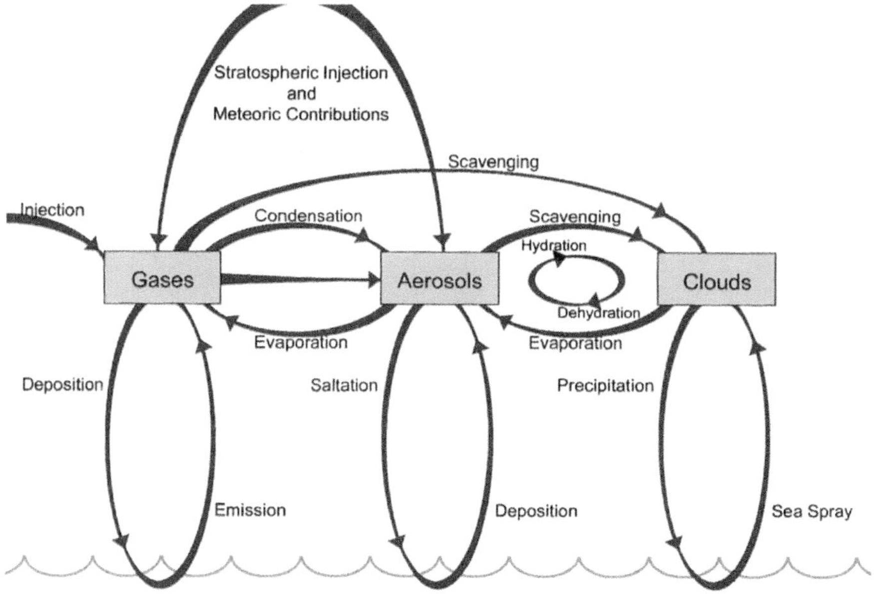

Figure 5. ***Atmospheric Cycles of Condensation, Accretion, and Evaporation over Multiple Length-Scales*** (After Turco et al 1982)

11. Potential Mechanisms of Bubble and Droplet Chemistries

In the laboratory modeling of prebiotic chemistry, much attention has been paid to reactions in homogeneous aqueous solutions (Chang et al. 1983). In addition, many studies have concentrated on the catalytic processes on the surfaces of solids, especially clays (Rao et al. 1980). However, very little is known in the prebiotic context of the reactions that occur at the gas-liquid interface, especially those at the surface of bubbles and droplets. An interface determined by phase boundaries can act as a region where the chemistry may take quite different pathways than in the bulk homogeneous environment. Not only are organics and metals selectively concentrated, but there is the potential for relatively anhydrous and short-lived high temperature non-equilibrium regimes involving the transduction of free energy from the mechanical into the chemical. Taken together, these geophysical/geochemical microenvironments allow new classes of reactions to occur under what are very likely prebiotic conditions. (Further details can be found in Lerman (1992) as well as papers under preparation.)

As in the case of bubbles, substantial reasons exist to explore the special chemical environments associated with these "micro-Miller bottles". A different chemistry than that of bubble phenomena may be associated with aerosols (Figure 6), making them of additional interest to us in their further explorations of the phase space of combinatorial chemical possibilities. Lerman (1992) presents a more detailed overview of the energetics and heterogeneous chemistry available in this phase of the bubblesol cycle.

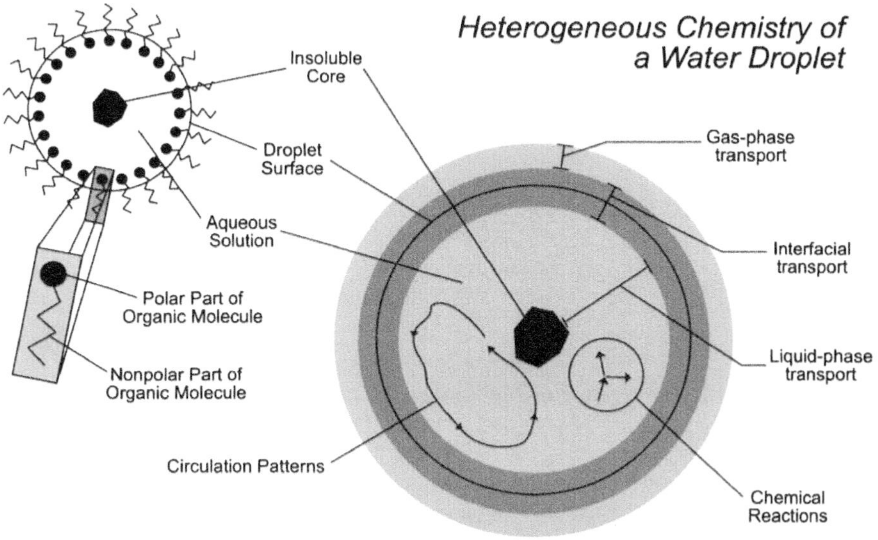

Figure 6. ***Heterogeneous Regimes of an Atmospheric Aerosol: Organic Skin & Transport Regimes*** [transport components, Turco et al (1982); chemical components, Graedel and Wechsler (1981)]

12. On Amphiphilic Bilayers and Things to Go in Them

The production of topologically closed amphiphilic bilayers, surrounding a solute rich in other organics and metal ions in non-equilibrium proportions, may result from the rehydration of previously dehydrated films of the involved matter (Deamer 1989). Taking the simplest single hydrocarbon chain amphiphilcs as a sample case, this requires relatively stringent chemical conditions:

1) An admixture of amphiphiles > C_{10}
2) pH of 8.5
3) Hydration time scales sustained from hours to days
4) Constant high concentrations of amphiphilic compounds, hence requiring…
5) Continual source of fresh amphiphilic material (the decay constant for anything but the simplest molecules is days to weeks).

This is remarkably close to the aerosol-droplet environments described above:

- Aerosols (derived from oceanic bubbles and atmospheric scavenging) would be rich in fatty acids, phospholipids, proteins, and any other organic compounds available.

- Such atmospheric bodies would likely have a pH in the range 7.8-8.3 from a variety of metallic-amphiphilic soaps.

- Contemporary aerosols have a lifetime of about one week, and in the process go through perhaps 10 cycles of hydration/dehydration.

Hence, we have a possibility of not just making a ladder of increasingly complex organic molecules, but of "wrapping" them up as well

13. From Chemical Evolution to Biogenesis?

This leads directly to the following hypothesis: In spite of considerable effort, no synthesis of RNA has yet been found "in the wild". Hence it would seem natural to suggest then that RNA synthesis, based on the precursors previously created, might have occurred inside the microenvironments of these just described protocells. This hypothesis is supported by the fact that these microenvironments seem to produce conditions fascinatingly close to the optimum for RNA activity: a pH of 8-9, high salinity, divalent ions such as Mg, and monovalent ones such as Na and K (private communication Laura Landweber). RNA synthesis involving hydration-dehydration induced condensation reactions (Zaug and Cech 1985) could then be supported by these dynamic microenvironments. Another possibility that these microenvironments might help explain are those hints of hydrophobic/hydrophilic codon-anticodon correlations (Lacey and Mullins 1983). Lerman (1992) addresses these questions in a broader context as well as describing a class of early experiments, designed by Anastasia Kanavaroti, to test the possibilities of nucleotide polymerization.

Going one step further, given such a situation it should be easier to make (and make use of) nucleic acids in the more protected confines of a metabolizing, ATP utilizing, bilayered protocell. For example, the polymerization of ATP cannot occur in water due to electrostatic hindrances, but can on an anhydrous surface. This would be a broad-based enough mechanism to support not just the conventional hypothesis of the creation of a RNA world, but the more generally stated two-origin theory of life put forward by Freeman Dyson (1999) and Robert Shapiro (1999): metabolism first, replication second. Additionally, the Bubble-Aerosol-Droplet phenomena would seem to provide a realistic geophysical/chemical environment to support some of the necessary (concentrating) conditions for Carl Woese's new ideas on an evolutionary analogue to annealing, i.e. the consolidation of molecular and metabolic functions (Woese 1998). Finally, the bubblesol cycle provides a natural starting point for the considerations of Vaida and Tuck (in this volume) who have begun to further explore the physical organic chemistry of the aerosol environment with respect to possible implications for prebiotic processes. But in all of these "more advanced" applications, caution must be used in extrapolating from simple morphological congruences to those of dynamic functionality.

A further speculation by one of the authors (LL) is that of the potential relationship between bubbles, bacteria, and Lynn Margulis' theory of the endosymbiotic origins of eukaryotes (Margulis 1981). The extraordinary efficiency by which bacteria are collected and concentrated by bubbles invites the question: could bubblesol phenomena have helped catalyze the endosymbiotic relationship between prokaryotes that resulted in the development of eukaryotes? In particular, did the hydrophobic/philic properties of bubbles concentrate the otherwise disparate bacteria leading to unusual populations at the sea-surface, in bubble-created marine snow, and in the sea-salt aerosol? Analogous to the contemporary world, an Archean version of marine snow would have provided a superior microenvironment for collections of bacteria to survive, and disparate species of bacteria collected onto a jet drop would undergo a variety of environmental assaults ranging from enhanced UV radiation, to repetitive hydration-dehydration and freezing cycles. Early cell membranes could hardly have remained completely intact during all of this, quite possibly leading to a mixing of bacterial components. Mutations as well would more likely have been induced, to the occasional advantage of "interesting" combinations of cellular components.

14. The Bubble-Aerosol-Droplet Model – Historical Perspectives

Although the suggestion to apply this contemporary geophysical/chemical cycle as a whole to prebiotic problems (i.e. as a global chemical reactor) is original to Lerman (Lerman – 1986,1992,1994,1996,2001,2002), it provides a geophysical/chemical basis that ties together suggestions made by a host of other prebiotic studies. Almost forty years ago, Peter Wangersky (1965), in an elegant and far-reaching paper on seawater chemistry, commented in passing that the scavenging ability of bubbles for organics was superior to the similar role played by clays. Goldacre (1958) discussed the potential prebiotic role of surface films. Michael Anbar (1968), among others, has suggested that bubbles undergoing sonolysis could fix nitrogen from the atmosphere. Polymerization is always a fundamental concern, and the soluble salt substrate model of Noam Lahav and Sherwood Chang (1982) using hydration-dehydration cycles to efficiently drive polymerization is naturally supported in the bubblesol super-cycle. And Peter Raven with George Johnson (2001) offer insightful historical perspective on the more general concept of "bubbles" and prebiotic processes. In regard to the atmospheric component of the supercycle presented here, Carl Woese first suggested that atmospheric droplets might be useful in prebiotic synthesis (Woese 1979), and following the work of Lerman, this was expanded upon by Oberbeck et al (1991). More recently, Dobson, Ellison, Tuck, & Vaida (2000) have begun to explore the potential role of the aerosol phase.

15. Conclusions

Because of its strong phenomenological basis in contemporary geophysics and geochemistry the hypothesis of the fundamental role played by the Bubble-Aerosol-Droplet Cycle seems to overcome a number of stumbling blocks in the current field of prebiotic chemistry. In particular it addresses the problems of selectivity, concentration, and stabilization of organic products in prebiotic environments. The Bubble-Aerosol-Droplet Cycle offers the possibility of non-equilibrium heterogeneous chemical processes different from conventional solution chemistry, as well as the possibility of coupling a supply of mechanical free energy to chemical processes and reactant concentration mechanisms. Especially important is the creation of relatively non-aqueous chemical environments within an aqueous medium. Condensation reactions are but one of the more intriguing consequences. Cycling of hydration-dehydration conditions is easily produced, supported by a high surface area to volume ratio for the reactant substrates. A wide range of initial conditions of "chemical phase space" are sampled, and just as importantly, once organics enter into the bubblesol cycle they tend to remain within it.

All of the above processes are spatially localized and temporally coincident, with the bubblesol cycle providing a natural macroscopic framework for the microscopic realization of the majority of specific chemical model environments developed by others. Indeed, Sherwood Chang (1994) has called the sea-surface microlayer/Bubble-Aerosol-Droplet Cycle one of the two most important new ideas when considering the geological setting of potential prebiotic processes (the other, naturally, being the underwater microenvironments of black smoker communities).

On the contemporary earth the Bubble-Aerosol-Droplet Cycle is the primary way in which organics are concentrated and transported throughout the ocean and atmosphere. Because its initiation depends only on the Rayleigh-Taylor instability, its application to liquid water environments elsewhere in the solar system's history seems certain. Applied to the early Earth, the fundamental mechanisms of this super-cycle, requiring only a liquid water-gas interface, are *independent* of the less reliable assumptions of specific atmospheric, oceanic, or geological chemistry. Components of this model are similarly applicable to any other planetary body having had liquid water at some point in its history. Whether Earth, Mars, Europa or Titan ... the bubblesol cycle is a global-scale geochemical reactor that must have existed within the framework of their respective planetary environments. Therefore, in the search for life and its origins, the Bubble-Aerosol-Droplet Cycle is of critical importance.

16. Acknowledgements

To owe so much to so many, and have so little space to thank them all! First and foremost, the Fannie and John Hertz Foundation is thanked for their remarkably generous support of my early work on self-organizing systems. Jim Lawless and Sherwood Chang of NASA-Ames have, since the first germinating ideas of the work presented here encouraged and challenged my interest in the origin of life as a scientific problem. The infectious enthusiasm of Anastassia Kanavarotti, also of NASA-Ames, has been much appreciated and the section on physical chemical reaction mechanisms owes much to discussions with Michael Anbar of SUNY-Buffalo. Indeed parts of this document were born in a NASA Director's Fund Proposal I wrote for experiments to be done with Chang, Anbar, and Kanavarotti (Lerman et al 1985). Opportunities to present and discuss this work at a number of Gordon Conferences, ISSOL, NASA Astrobiology Institute, and LPSC meetings have tightened the logic immeasurably. The gracious hospitality of Duncan Blanchard, Ferren MacIntyre, and Tom Graedel at their respective institutions introduced me, first hand, to some of the more amazing properties of bubbles and aerosols. Laura Landweber and David Deamer are both thanked for their scientific insights and David for his editorial contributions as well. The diligent library research of Sarah Kuwabara is most appreciated. And warmest thanks for the constancy of their support go to Jerry van Andel of Cambridge University, Norm Sleep of Stanford, James Symons of Lawrence Berkeley Laboratory, Paul Bratterman of the University of North Texas, and Jen Blank of Lawrence Livermore National Labs.

17. References

Adamson, A. (1997) *The Physical Chemistry of Surfaces 6th edition,* Wiley-Interscience, New York.

Anbar, M. (1968) Cavitation during Impact of Liquid Water on Water: Geochemical Implications, *Science* **161**, 1343-1344.

Banin, A. and Navrot, J. (1975) Origin of Life: Clues from Relations between Chemical Compositions of Living Organisms and Natural Environments, *Science* **189**, 550-551.

Blanchard, D.C. and Woodcock, A.H. (1957) Bubble Formation and Modification in the Sea and its Meteorological Significance, *Tellus* **9,** 145-158. (n.b. This is a seminal paper!)

Blanchard, D.C. (1975) Bubble Scavenging and the Water to Air Transfer of Organic Material in the Sea, In R. Baier (ed) *Applied Chemistry at Protein Interfaces, Advances in Chemistry* **145,** 360-387.

Chang, S. (1994) The Planetary Setting of Prebiotic Evolution, In S. Bengtson (ed) *Early Life on Earth* Nobel Symposium No. 84, Columbia University Press, New York, 10-23.

Chang, S., DesMarais, D., Mack, R., Miller, SL, and Strathearn, GE (1983) Prebiotic Organic Synthesis and the Origin of Life, In J. Schopf (ed) *Earth's Earliest Biosphere: It's Origin and Early Evolution*, 53-92.

Cincinelli, A., Lepri L., Checchini L., Perugini M., (1999) The Role of the Surface Microlayer in Contaminant Concentration and Pollutant Transport via Marine Aerosols, In F. Briand (ed) *The Role of Sea Surface Microlayer Processes in the Biogeochemistry of the Mediterranean Sea*, CIESM Workshop Series **9**, 23-25.

Cipriano, R.J. & Blanchard, D.C. (1981) Bubble and Aerosol Spectra Produced by a Laboratory 'Breaking Wave', *Journal of Geophysical Research* **86**, 8085-8092.

Cronin, J. R. and Chang, S. (1993) Prebiotic synthesis in planetary environments, In J. M. Greenberg, C.X. Mendoza-Gómez, and V. Pirronello (eds) The Chemistry of Life's Origin, Kluwer, Netherlands, 209-258.

Deamer, D. and R.M. Pashley. 1989. Amphiphilic Components of the Murchison Carbonaceous Chondrite: Surface Properties and Membrane Formation. *Origins of Life and Evolution of the Biosphere* **19,** 21-38.

Dobson, C., Ellison, G., Tuck, A., and Vaida, V. (2000), Atmospheric Aerosols as Prebiotic Chemical Reactors, *Proc. Nat. Acad. Sci.* **97**, 11864-11868.

Drake, F. (1961) First SETI Meeting (see http://www.seti.org/science/drake-bg.html).

Duce, R.A. and Hoffman, E.J. (1976) Chemical Fractionation at the Air/Sea Interface, *Ann. Rev. Earth Planet. Sci.* **4**, 187-228.

Duce, R.A., Quinn, J.G., Olney, C.E., Piotrowicz, S.R., and Ray, B.J. (1972) Enrichment of Heavy Metals and Organic Compounds in the Surface Microlayer of Narranganset Bay, R.I., *Science* **176,** 161-163.

Dyson, F. (1999) *Origins of Life*, Cambridge University Press, New York.

Gill, P., Graedel, T., and Weschler, C. (1983) Organic Films on Atmospheric Aerosol Particles, Fog Droplets, Cloud Droplets, Raindrops, and Snowflakes, *Reviews of Geophysics and Space Sciences.* **21**, 903-920.

Goldacre, R.J. (1958) Surface films: Their collapse on compression, the shapes and sizes of cells, and the origin of life. In J.F. Danielli, K.G.A. Pankhurst, and A.C. Riddiford (eds.), *Surface Phenomena in Biology and Chemistry*. Pergamon Press, New York., 12-27.

Graedel, T.E. and Weschler, C.J. (1981) Chemistry Within Aqueous Atmospheric Aerosols and Raindrops, *Review of Geophysics and Space Sciences* **19,** 505-539.

Johnson, B.D. and Cooke, R.C. (1980) Organic Particle and Aggregate Formation Resulting from the Dissolution of Bubbles in Seawater, *Limnology & Oceanography* **25,** 653-661.

Johnson, B. and Cooke, R. (1981) Generation of Stabilized Microbubbles in Seawater,*Science* **213**, 209-211.

Kraus, E. and Businger, J. (1994) *Atmosphere-Ocean Interaction 2d edition*, Oxford University Press, NY.

Lahav, N and Chang, S. (1982) The Possible Role of Soluble Salts in Chemical Evolution, *Journal of Molecular Evolution* **19,** 36.

Lemlich, R., (ed.) (1972) *Adsorptive Bubble Separation Techniques*, Academic Press, New York.

Lerman, L. (1986) Potential Role of Bubbles and Droplets in Primordial and Planetary Chemistry: Exploration of the Liquid-Gas Interface as a Reaction Zone for Condensation Processes, *Origins of Life* **16**, 201-202.

Lerman, L. (1992) *Exploration of the Liquid-Gas Interface as a Reaction Zone for Condensation Processes: The Potential Role of the Air-Water Interface in Prebiotic Chemistry* Thesis, Stanford University.

Lerman, L. (1994a) The Bubble-Aerosol-Droplet Cycle as Natural Reactor for Prebiotic Chemistry (I), *Origins of Life and Evolution of the Biosphere* **24**, 111-112.

Lerman, L. (1994b) The Bubble-Aerosol-Droplet Cycle as Natural Reactor for Prebiotic Chemistry (II), *Origins of Life and Evolution of the Biosphere* **24**, 138-139.

Lerman, L. (1996) The Bubble-Aerosol-Droplet Cycle: A Prebiotic Geochemical Reactor *(That Must Have Been), Origins of Life and Evolution of the Biosphere* **26**, 369-370.

Lerman, L. (2001) Consequences and Artifacts: Terrestrial Findings and Martian Analogues of an Air-Water Interface, in NASA Astrobiology Institute General Meeting 2001, *Astrobiology* **1**, 365-366.

Lerman, L. (2002a) Consequences and Artifacts: Terrestrial Findings and Martian Analogues of an Air-Water Interface, *33rd Lunar and Planetary Science Conference*, March 11-15,2002, NASA Johnson Space Center (http://www.lpi.usra.edu/meetings/lpsc2002/pdf/2062.pdf).

Lerman, L. (2002b) How the Chemical Physics of the Microscale Drives Prebiotic Evolution, presented at The 10th ISSOL Meeting and the 13th International Conference on the Origin of Life. Oaxaca, Mexico *Origins of Life and Evolution of the Biosphere* **32**, 419-420.

Lerman, L. (2002c) (Intermittent) Water on Mars: Chemical Evolution & the Air-Water Interface, presented at The 10th ISSOL Meeting and the 13th International Conference on the Origin of Life. Oaxaca, Mexico *Origins of Life and Evolution of the Biosphere* **32**, 538-539.

Lerman, L., Kanavarotti, A., Anbar, M., and Chang, S. (1985) Exploration of the Liquid-Gas Interface as a Reaction Zone for Condensation Processes: The Potential Role of Bubbles and Droplets in Primordial Chemistry, A Proposal to the Director, NASA-Ames Nov. 5, 1985.

Macintyre, F. (1970) Geochemical Fractionation During Mass Transfer from Sea to Air by Breaking Bubbles, *Tellus* **22**, 451-462.

MacIntyre, F. (1972) Flow Patterns in Breaking Bubbles, *Journal of Geophysical Research* **77**, 5211-5228.

MacIntyre, F. (1974a) Chemical Fractionation and Sea-Surface Microlayer Processes, In E.D. Goldberg (ed) *The Sea (Marine Chemistry)* **5**, 245-299.

MacIntyre, F. (1974b) Non-Lipid related Possibilities for Chemical Fractionation in Bubble Film Caps, Journal de Recherche Atmospherique, **7**, 515.

MacIntyre, F. (1974c) The Top Millimeter of the Ocean, *Scientific American, May,* **230**, 62-77.

MacIntyre, F. and Winchester, J. W. (1969) Phosphate Ion Enrichment in Drops from Breaking Bubbles, *J. Phys. Chem.* **73**, 2163-2169.

Margulis, L. (1981) *Symbiosis in Cell Evolution: Life & Its Environment on the Early Earth,* W.H. Freeman.

Oberbeck, V. et al. (1991) Prebiotic Chemistry in Clouds, *Journal of Molecular Evolution,* **32**, 296.

Rao, M., et al. (1980) Clays in Prebiotic Chemistry, *Journal of Molecular Evolution* **15**, 317.

Raven, P.H. and Johnson, G. B. (1999) *Biology* 5th edition, McGraw-Hill, Boston.

Shapiro, R. (1999) *Planetary Dreams: The Quest to Discover Life Beyond Earth*, J. Wiley, New York.

Sharp, D.H. (1984) An Overview of Rayleigh-Taylor Instability, *Physica D* **12D**, 3-18.

Szekielda, K.H., Kupferman, S.L., Kllemas, V., Polis, D.F. (1972) Element Enrichment in Organic Films and Foam Associated with Aquatic Frontal Systems, *Journal of Geophysical Research* **77**, 5278-5282.

Turco, R., Toon, O.B., Whitten, R.C., Keesee, R.G., and Hamill, P. (1982) Importance of Heterogeneous Processes to Tropospheric Chemistry: Studies with a One-Dimensional Model, In D. Schryer (ed*) Heterogeneous Atmospheric Chemistry* Geophysics Monograph Series **26**, 231-240.

Wangersky, P. (1965) The Organic Chemistry of Sea Water, *American Scientist* **53**, 358-374.

Wangersky, P. (1976) The Surface Film as Physical Environment, *Ann. Rev. Ecol Sys.* **7**, 161-176.

Weschler, C. and Graedel, T. (1982) Theoretical Limitations on Heterogeneous Catalysis by Transition Metals in Aqueous Atmospheric Aerosols, In D. Schryer (ed*) Heterogeneous Atmospheric Chemistry* Geophysics Monograph Series **26**, 196.

Wilford, J.N. and Schidlovsky, D. (1993) Bubbles May Have Speeded Life's Origins, *New York Times* 7/6/93.

Woese, C.R. (1979) A Proposal Concerning the Origin of Life on the Planet Earth, *Journal of Molecular Evolution* **13**, 95-101.

Woese, C.R. (1998) The Universal Ancestor, *Proc. Natl. Acad. Sci.* US **95**, 6854- 6859.

Zaug, A.J. and Cech, T.R. (1985) Oligomerization of Intervening Sequence RNA Molecules in the Absence of Proteins, *Science* **229**, 1060-1064.

18. Figure Credits

Table 1 appears with the kind permission of Dr. Roy Chester and Kluwer Academic Publishers. The table originally appeared in Chester, R. (1986) "The Marine Mineral Aerosol", In P. Buat-Menard (ed) *The Role of Air-Sea Exchange in Geochemical Cycling*, Reidel, 443-476. Copyright 1986 Reidel (Kluwer Academic Publishers)

Figure 2a was reprinted with permission from MacIntyre, F. and Winchester, J.W. (1969) “Phosphate Ion Enrichment in Drops from Breaking Bubbles”, *Journal of Physical Chemistry* **73**, 2163-2169. Copyright 1969. American Chemical Society.

Figure 2b appears with the kind permission of Dr. A. Cincinelli and CIESM. The figure was reprinted from Cincinelli, A., Lepri L., Checchini L., Perugini M., (1999) "The Role of the Surface Microlayer in Contaminant Concentration and Pollutant Transport via Marine Aerosols", In F. Briand (ed) *"The Role of Sea Surface Microlayer Processes in the Biogeochemistry of the Mediterranean Sea"* CIESM Workshop Series **9**, 23-25. Monaco <www.ciesm.org/publications/Paris.pdf> .

Figure 3 was reprinted with permission from Banin, A. and Navrot, J. (1975) “Origin of Life: Clues from Relations between Chemical Compositions of Living Organisms and Natural Environments”, *Science* **189**, 550-551. Copyright 1975 AAAS.

Figure 4 appears with the generous permission of Prof. Ferren MacIntyre.

Figures 5 and 6 were reproduced/modified by permission of the American Geophysical Union. They originally appeared in Turco, R., Toon, O.B., Whitten, R.C., Keesee, R.G., and Hamill, P. (1982) “Importance of Heterogeneous Processes to Tropospheric Chemistry: Studies with a One-Dimensional Model”, In D. Schryer (ed) *Heterogeneous Atmospheric Chemistry* Geophysics Monograph Series **26**, 231-240. Copyright 1982 American Geophysical Union; and Graedel, T.E. and Weschler, C.J. (1981) “Chemistry Within Aqueous Atmospheric Aerosols and Raindrops”, *Review of Geophysics and Space Sciences* **19**, 505-539. Copyright 1981 American Geophysical Union.

The photograph on the biodata page is from archives of the Family Bach and appears with the generous permission and kind coonsideration of Dr. Cheshiru Bach. Copyright 2003 Louis Lerman.

II. The Standard Scenario

Biodata of **André Brack** author of "***The Chemistry Of Life's Origins***"

Professor André Brack is Directeur de Recherche at the Centre National de la Recherche Scientifique, Centre de Biophysique Moléculaire, Orléans. He was trained in polymer chemistry at the University of Strasbourg. Dr. Brack received his M.S. in chemistry in 1962 and did his dissertation on polypeptide chemistry in 1970. He started his career at the Centre de Recherche sur les Macromolecules in Strasbourg (1963-1968) and moved to the Centre de Biophysique Moléculaire in Orléans in 1968. He spent one year (1975) at the Salk Institute for Biological Studies in San Diego. His field of expertise concerns the chemistry of the origins of life, as well as the search for life in the Solar System. He is a former president of the International Society for the Study of the Origin of Life (1996-1999) and is president of the European Exo/Astrobiology Network since 2001. He is Principal Investigator of two ESA space experiments: AMINO (on ISS EXPOSE) and STONE (artificial Martian meteorite experiments on FOTON). Professor Brack has published over 150 articles in peer-reviewed journals, 4 books and 15 science-popularizing articles in magazines.

E-mail: **brack@cnrs-orleans.fr**

J. Seckbach (ed.), Origins, 59–73.

THE CHEMISTRY OF LIFE'S ORIGINS

A. BRACK
Centre de Biophysique Moléculaire, CNRS
Rue Charles Sadron, 45071 Orléans cedex 2, France

1. Introduction

Humans in every civilization have always been intrigued by the question of the origin of life. Over thousands of years, the comforting theory of spontaneous generation seemed to provide an answer to this enduring enigma. Pasteur gave the finishing blow to spontaneous generation in 1865 when he designed a rigorous experimental setup for sterilization. The demonstration of Pasteur opened the question of the historical origin of life. Charles Darwin first formulated the modern approach to the chemical origin of life. In February 1871, he wrote in a private letter to Hooker: "If (and oh, what a big if) we could conceive in some warm little pond, with all sorts of ammonia and phosphoric salts, light, heat, electricity, etc., present that a protein compound was chemically formed, ready to undergo still more complex changes, at the present day such matter would be instantly devoured or adsorbed, which would not have been the case before living creatures were formed". For 50 years, the idea laid dormant. In 1924, the young Russian biochemist Aleksander Oparin pointed out that life must have arisen in the process of the evolution of matter thanks to the reducing nature of the atmosphere. In 1928, the British biologist J.B.S. Haldane, independently of Oparin, speculated on the early conditions suitable for the emergence of life. Subjecting a mixture of water, carbon dioxide, and ammonia to UV light should produce a variety of organic substances, including sugars and some of the materials from which proteins are built up. Before the emergence of life they must have accumulated in water to form a hot, dilute "primordial soup". Almost 20 years after Haldane's publication, J.D. Bernal (1949) conjectured that clay mineral surfaces were involved in the origin of life. Stanley Miller (1953) reported the formation of the amino acids, glycine, alanine, aspartic acid and glutamic acid, when subjecting a mixture of methane, ammonia and water to electric discharges. Miller's publication really opened the field of experimental prebiotic chemistry.

2. The Quest For Geological Records

The oldest sediments have been found in Greenland. The isotopic signatures of the organic carbon in these metasediments provide indirect evidence that life may be >3.7 billion years old (Schidlowski, 1988; Mojzsis et al., 1996; Appel and Moorbath, 1999; Rosing, 1999). Taking the age of the Earth as 4.5 billion years, this means that life began as a quite early event in the Earth's history. This isotopic evidence stems from the fact that the carbon atom has two stable isotopes, carbon 12 and carbon 13. The $^{12}C/^{13}C$ ratio in abiotic mineral compounds is 89. In biological material, the process of photosynthesis gives a preference to the lighter carbon isotope and raises the ratio to about 92. Consequently, the carbon residues of previously living matter may be identified by this enrichment in ^{12}C. A compilation has been made of the carbon isotopic composition of over 1,600 samples of fossil kerogen (a complex organic macromolecule produced from the debris of biological matter) and compared with that from carbonates in the same sedimentary rocks. This showed that the biosynthesis by photosynthetic organisms was involved in all the sediments studied. In fact, this offset is now taken to be one of the most powerful indications that life on Earth was active nearly 3.9 Ga ago because the sample suite encompasses specimens right across the geological time scale. This conclusion is consistent with the diversity of the 3.5 b.y.-old fossilized microflora found in the Barberton and Pilbara greenstone belts in South Africa and Australia, respectively (Schopf, 1993; Westall et al., 2001). Unfortunately, the direct clues that could help chemists to identify the molecules, which participated in the emergence of life on Earth about 4 b.y. ago, have been erased by the combined action of plate tectonics, the permanent presence of running water, the unshielded solar ultraviolet radiation and atmospheric oxygen.

3. Primitive Life In A Test Tube

It is difficult to define what is meant by the word "life". One generally considers as living any chemical system able, *a minima*, to transfer its molecular information via self-reproduction and also capable of evolution. The concept of evolution implies that the system normally transfers its information fairly faithfully but makes a few random errors, leading potentially to a higher efficiency and possibly to a better adaptation to environmental changes. Schematically, the constituents of primitive life can be compared to parts of "chemical robots". By chance, some parts self-assembled to generate robots capable of assembling other parts to form identical robots. Sometimes, a minor error in the building generated more efficient robots, which became the dominant species. By analogy with contemporary life, it is generally believed that the parts were made of organic matter - carbon skeletons flanked by H, O, N, S, atoms - migrating into liquid water. The number of parts required for the first robots is still unknown. This number is expected to have been rather small because the chemical robots emerged when the Earth was heavily bombarded. A simple self-reproducing system would have been more robust and offered better chances to resist the cataclysmic impacts during the heavy bombardment. If the number of parts was small, chemists have reasonable chances to reproduce primitive chemical robots in a test tube. If the number of parts was very large, the mission of the chemists will be almost impossible.

3.1. WATER

Liquid water played a major role in the appearance and evolution of life by favoring the diffusion and exchange of organic molecules. Liquid water has many peculiarities. Water molecules establish hydrogen bonds with molecules containing hydrophilic groups. In water, organic molecules containing both hydrophilic and hydrophobic groups self-organize in response to their hydrophobic and hydrophilic properties. This duality generates interesting prebiotic situations such as the stereoselective aggregation of short peptide sequences of alternating hydrophobic - hydrophilic residues into thermostable β-sheet structures endowed with chemical activity, as shown below. In addition to the H-bonding capability, water exhibits a large dipole moment (1.85 debye) as compared to alcohols (<1.70 debye). This large dipole moment favors the dissociation of ionizable groups such as $-NH_2$ and -COOH leading to ionic groups, which can form additional H-bonds with water molecules, thus improving their solubility. With a high dielectric constant ε of 80, water is also an outstanding dielectric compound. When organic groups with opposite charges Q and Q' separated by a distance r are formed, their recombination is unfavored, because the force of attraction for reassociation is given by $F=QQ'/\varepsilon r^2$. This is also true for metal ions, which have probably been associated with organic molecules since the beginning of life. Liquid water was probably active in prebiotic chemistry as clay producer and heat dissipator. Liquid water is also a powerful hydrolytic chemical agent. As such, it allows pathways, which would have few chances to occur in an organic solvent. Liquid water is therefore generally considered as a prerequisite for the emergence of life on Earth (Brack, 2001).

3.2. THE QUEST FOR ORGANIC MOLECULES

Originally, the carbon needed to construct the organic building blocks of life was available as simple volatile compounds, either reduced (methane) or oxidized (carbon monoxide or carbon dioxide).

3.2.1. Endogenous production of organic molecules

The Russian biochemist Aleksander Oparin suggested that the small reduced organic molecules needed for primitive life were formed in a primitive atmosphere dominated by methane. Stanley Miller (1953) identified four of the twenty naturally occurring amino acids when exposing a mixture of methane, ammonia, hydrogen and water to electric discharges. Since this historic experiment, seventeen natural amino acids have been obtained *via* the intermediate formation of simple precursors, such as hydrogen cyanide and formaldehyde. It has been shown that spark discharge synthesis of amino acids occurs efficiently when a reducing gas mixture is used. However, the true composition of the primitive Earth's atmosphere remains unknown. Today, geochemists favor a non-reducing atmosphere dominated by carbon dioxide. Under such conditions, the production of amino acids appears to be very limited.

Deep-sea hydrothermal systems may also be likely environments for the synthesis of prebiotic organic molecules (Baross and Hoffman, 1985; Holm and Andersson, 1995; 1998). Hydrocarbons containing 16 to 29 carbon atoms have been detected in the Rainbow ultramafic hydrothermal system, Mid-Atlantic Ridge (Holm and Charlou, 2001). Amino acids have been obtained, although in low yields, under conditions simulating these hydrothermal vents (Yanagawa and Kobayashi, 1992). Hydrothermal vents are often disqualified as efficient reactors for the synthesis of bioorganic molecules, because of the high temperature. However, the products that are synthesized in hot vents are rapidly quenched in the surrounding cold water, which may preserve those organics formed. If the carbon source for life was carbon dioxide, the energy source required to reduce the carbon dioxide might have been provided by the oxidative formation of pyrite from iron sulfide and hydrogen sulfide. Pyrite has positive surface charges and bonds the products of carbon dioxide reduction, giving rise to a two-dimensional reaction system, a "surface metabolism" (Wächtershäuser, 1994 and 1998). Laboratory work has provided some support for this new promising hypothesis (Huber and Wächtershäuser, 1997 and 1998).

3.2.2. Extraterrestrial delivery of organic molecules

Besides abundant hydrogen and helium, 114 interstellar and circumstellar gaseous molecules are currently identified in the interstellar medium. Ultraviolet irradiation of dust grains may result in the formation of complex organic molecules. The interstellar dust particles are assumed to be composed of silicate grains, surrounded by ices including carbon containing molecules. In a simulation experiment, ices of H_2O, CO_2, CO, CH_3OH, and NH_3 were deposited at 12 K and a pressure of 10^{-7} mbar and irradiated with electromagnetic radiation representative of the interstellar medium. The solid layer that developed on the solid surface was analyzed by enantioselective gas chromatography and mass spectrometry GC-MS. In order to exclude contamination, parallel experiments were performed with ^{13}C-containing chemicals. After the analytical steps of extraction, hydrolysis, and derivatization, 16 amino acids were identified in the simulated ice mantle of interstellar dust particles (Muñoz Caro et al., 2002). The results were confirmed by the ^{13}C-experiments, which definitely excluded contamination. The chiral amino acids were identified as being totally racemic. The results strongly suggest that amino acids are readily formed in interstellar space.

The incorporation of interstellar matter in meteorites and comets in the pre-solar nebula provides the basis of the cosmic dust connection. A comparison of interstellar and cometary ices using recent data from ESA Infrared Space Observatory ISO launched in November 1995 has revealed important similarities between interstellar ices and volatiles measured in the coma of some comets. Comets show substantial amounts of organic material, as was nicely demonstrated by ESA cometary mission Giotto in 1986. On average, dust particles ejected from the Comet Halley nucleus contained 14% organic carbon by mass (reviewed by Delsemme, 1998). About 30% of cometary grains are dominated by the light elements C, H, O, and N, and 35% are close in composition to carbonaceous chondrites. Among the molecules identified in comets are hydrogen cyanide and formaldehyde. The presence of purines, pyrimidines, and formaldehyde polymers has also been inferred from the fragments analyzed by Giotto Picca and Vega Puma mass spectrometers. However, there is no direct identification of the complex organic molecules present in the dust grains and in the cometary nucleus. Many chemical species of interest for exobiology have been detected in Comet Hyakutake in 1996, including ammonia, methane, acetylene, acetonitrile and hydrogen isocyanide. In addition to the hydrogen cyanide and formaldehyde seen in several earlier comets, the Comet Hale-Bopp was also shown to contain methane, acetylene, formic acid, acetonitrile, hydrogen isocyanide, isocyanic acid, cyanoacetylene, and thioformaldehyde. Cometary grains might, therefore, have been an important source of organic molecules delivered to the primitive Earth (reviewed in Ehrenfreund and Charnley, 2000; Ehrenfreund et al., 2002).

The study of meteorites, particularly those called the carbonaceous chondrites that contain up to 5% by weight of organic matter, has allowed close examination of the extraterrestrial organic material that has been delivered to the Earth. Nucleic acid bases, purines and pyrimidines, have been found in the Murchison meteorite (Stoks and Schwartz, 1982). One sugar (dihydroxyacetone), sugar-alcohols (erythritol, ribitol) sugar-acids (ribonic acid, gluconic acid) have been detected in the Murchison meteorite but no ribose, the sugar moiety which links together the nucleic acid building blocks (Cooper et al., 2001). Vesicle-forming fatty acids have been extracted from different carbonaceous meteorites (Deamer, 1985).

Eight proteinaceous amino acids have been identified in the Murchison meteorite, among more than 70 amino acids found therein (Cronin et al., 1988). These amino acids are asymmetric. The two enantiomers, L and D, (mirror image configurations of the same amino acid) are generally found in equal proportions. However, small (1.0-9.2%) L-enantiomeric excesses (%L-%D) were found in six α-methyl-α-amino acids from the Murchison (2.8-9.2%) and Murray (1.0-6.0%) carbonaceous chondrites (Pizzarello and Cronin, 2000). The presence of L-enantiomeric excesses in these meteorites points towards an extraterrestrial process of asymmetric synthesis of amino acids, asymmetry that is preserved inside the meteorite. These excesses may help to understand the emergence of a one-handed life. Pasteur was probably the first to realize that the biological asymmetry (one-handedness) could best distinguish between inanimate matter and life. For example, proteins are built up with twenty different amino acids. Each amino acid, with the exception of glycine, exists in two mirror-image forms, the L and the D form. But proteins actually use only the L-form amino acids. Life that would simultaneously use both the right- and left-handed forms of the same biological molecules appears, in the first place, very unlikely for geometrical reasons. Proteins adopt asymmetrical rigid geometries, right-handed α-helices and β-sheets, which play a key rôle in the catalytic activity. Enzyme β-pleated sheets cannot form when both L- and D-amino acids are present in the same chain (Brack and Spach, 1979). Since the catalytic activity of an enzyme is intimately dependent upon the geometry of the chain, the absence of β-pleated sheets would impede, or at least considerably reduce, the activity spectrum of the enzymes. The use of one-handed biomonomers also sharpens the sequence information of the biopolymers. For a polymer made of n units, the number of sequence combinations will be divided by 2^n when the system uses only homochiral (one-handed) monomers. Taking into account the fact that enzyme chains are generally made up of hundreds of monomers, and that nucleic acids contain several million nucleotides, the tremendous gain in simplicity offered by the use of monomers restricted to one handedness is self evident. Thus, homochirality can be a crucial signature for life.

The excess of the one-handed amino acids, as found in the meteorites, may result from the processing of the organic mantles of the interstellar grains from which the meteorite was originally formed. That processing could occur, for example, by the effects of circularly polarized synchrotron radiation from a neutron star, a remnant of a supernova. Strong infrared circular polarization, resulting from dust scattering in reflection nebulae in the

Orion OMC-1 star-formation region, has been observed (Bailey et al., 1998). Circular polarization at shorter wavelengths might have been important in inducing this chiral asymmetry in interstellar organic molecules that could be subsequently delivered to the early Earth (Bailey, 2001).

Dust collection in the Greenland and Antarctic ice sheets and its analysis show that the Earth captures interplanetary dust as micrometeorites at a rate of about 50-100 tons per day. About 99% of this mass is carried by micrometeorites in the 50-500 μm size range. This value is much higher than the most reliable estimate of the meteorite flux, i.e. about 0.03 tons per day. In the Antarctic micrometeorites, a high percentage of unmelted chondritic micrometeorites from 50 to 100 μm in size has been observed, indicating that a large fraction entered the terrestrial atmosphere without drastic thermal treatment. In this size range, the carbonaceous micrometeorites represent 80% of the samples and contain 2% of carbon, on average. This flux of incoming micrometeorites might have brought to the Earth about 10^{23}g of carbon over a period of 300 million years, corresponding to the late terrestrial bombardment phase (Maurette, 1998; Maurette et al., 2000). For comparison, this delivery represents more carbon than that engaged in the surficial biomass, i.e. about 10^{18}g. These grains also contain a high proportion of metallic sulfides, oxides, and clay minerals that belong to various classes of catalysts. In addition to the carbonaceous matter, micrometeorites might also have delivered a rich variety of catalysts, having perhaps acquired specific crystallographic properties during their synthesis in the microgravity environment of the early solar nebula. They may have also functioned as tiny autonomous chemical reactors when reaching oceanic water, catalyzing the processing of their organic constituents.

Before reaching the Earth, organic molecules are exposed to UV radiation in interstellar space and in the solar system. Amino acids have been exposed in Earth orbit to study their survival in space. The UV flux (λ <206 nm) in diffuse interstellar medium is about 10^8 photons cm^{-2} s^{-1}. In Earth orbit, the corresponding solar flux is in the range of 10^{16} photons cm^{-2} s^{-1}. This means that one week irradiation corresponds to 275,000 years in the interstellar medium. In comparison with ground experiments, space allows the exposure of samples to all space parameters simultaneously and to irradiate a great number of samples under strictly identical conditions. Amino acids like those detected in the Murchison meteorite have been exposed to space conditions in Earth orbit on board unmanned Russian satellites FOTON 8 and 11, free and associated with clay minerals. Free exposed aspartic acid and glutamic acid were partially destroyed during exposure to solar UV. However, decomposition was prevented when the amino acids were embedded in clays (Barbier et al, 1998; Barbier et al., 2002). Amino acids and peptides have also been subjected to solar radiation outside the MIR station for 97 days as solid films as well as embedded into mineral material (montmorillonite clay, basalt powder and Allende meteorite powder). Different thicknesses of meteorite powder films were used to estimate the shielding threshold. After three months exposure, about 50% of the amino acids were destroyed in the absence of mineral shielding. Peptides exhibited a noticeable sensitivity to space vacuum and sublimation effects were detected. Decarboxylation was found to be the main effect of photolysis. No polymerization occurred and no racemization (the conversion of L-amino acids into the D-enantiomers) was observed. Among the different minerals used as 5μm films, meteoritic powder offered the best protection whereas montmorillonite was the less efficient. Significant protection from solar radiation was observed when the thickness of the meteorite mineral was 5 μm or greater.

3.3. RECONSTRUCTING A PRIMITIVE CELLULAR LIFE

By analogy with contemporary living systems, it was long considered that primitive life emerged as a cellular object, requiring boundary molecules able to isolate the system from the aqueous environment (membrane). Also needed would be catalytic molecules to provide the basic chemical work of the cell (enzymes) and information retaining molecules that allow the storage and the transfer of the information needed for replication (RNA).

Fatty acids are known to form vesicles when the hydrocarbon chains contain more than ten carbon atoms. Such vesicle forming fatty acids have been identified in the Murchison meteorite (Deamer, 1995 and 1998). However, the membranes obtained with these simple amphiphiles are not stable over a broad range of conditions. Stable neutral lipids can be obtained by condensing fatty acids with glycerol or with glycerol phosphate, thus mimicking the stable contemporary phospholipid. Primitive membranes could initially have also been formed by simple isoprene derivatives (Ourisson and Nakatani, 1994).

Most of the catalytic chemical reactions in a living cell are achieved by proteinaceous enzymes made of 20 different homochiral L-amino acids. Amino acids were most likely available on the primitive Earth as complex mixtures. The number of condensing agents capable of assembling amino acids into peptides in water is limited, especially when looking for prebiotically plausible compounds. Some carbodiimides, R-N=C=N-R, can be used in water. The simplest carbodiimide, H-N=C=N-H, can be considered as a transposed form of cyanamide NH_2-CN which is present in interstellar medium. In water, cyanamide forms a dimer, dicyandiamide, which is as active as carbodiimides in forming peptides. However, the reactions are very slow and do not proceed beyond the tetrapeptide. Clays and salts can also be used to condense amino acids in water. Amino acid adenylate anhydrides have been reported to condense readily in the presence of montmorillonite (Paecht-Horowitz et al., 1970). When subjecting mixtures of glycine and kaolinite to wet-dry and 25°C-94°C temperature fluctuations, the formation of oligopeptides up to five glycines long has been observed. Efficient mineral-catalyzed (hydroxylapatite, illite clay) condensation of amino acids into long peptides have been recently reported (Ferris et al., 1996). According to De Duve (1998), the first peptides may have appeared *via* thioesters. Thioesters lead to short peptides in the presence of mineral surfaces (Bertrand et al., 2001).

Thermal condensation of amino acids has been described by Fox. He has shown that dry mixtures of amino acids polymerize when heated at 130°C to give "proteinoids" (Fox and Dose, 1977). In the presence of polyphosphates, the temperature can be decreased to 60°C. High molecular weights were obtained when an excess of acidic or basic amino acids were present. In aqueous solutions, the proteinoids aggregated spontaneously to form microspheres of 1-2 μm, presenting an interface resembling the lipid bilayers of living cells. The microspheres increased slowly in size from dissolved proteinoids and are sometimes able to bud and to divide. These microspheres were described as catalyzing the decomposition of glucose and were able to work as esterases and peroxydases. The main advantage of proteinoids is their organization into particles but they represent a dramatic increase in complexity.

Chemical reactions capable of selectively condensing protein amino acids more readily than non protein ones have been described. Helical- and sheet-structures can be modeled with the aid of only two different amino acids, one hydrophobic, the second hydrophilic. Polypeptides with alternating hydrophobic and hydrophilic residues adopt a water soluble β-sheet geometry because of hydrophobic side-chain clustering. Due to the formation of β-sheets, alternating sequences display a good resistance toward chemical degradation. Aggregation of alternating sequences into β-sheets is possible only with homochiral (all-L or all-D) polypeptides (Brack, 1993). Short peptides have also been shown to exhibit catalytic properties (Barbier and Brack, 1992).

In contemporary living systems, the hereditary memory is stored in nucleic acids built up with bases (purine and pyrimidine), sugars (ribose for RNA, deoxyribose for DNA) and phosphate groups. The accumulation of significant quantities of natural RNA nucleotides does nor appear as a plausible chemical event on the primitive Earth (Schwartz, 1998). Purines are easily obtained from hydrogen cyanide or by subjecting reduced gas mixtures to electric discharges. No pyrimidine synthesis from electric discharges has been published so far whereas hydrogen cyanide affords only very small amounts of these bases. Condensation of formaldehyde leads to ribose among a large number of other sugars. Although the synthesis of purine nucleosides (the combination of purine and ribose) and of nucleotides have been achieved by heating the components in the solid state, the yields are very low and the reactions are not regioselective. Interestingly, efficient clay-catalyzed condensation of nucleotides into oligomers up to 55 monomers long have been reported (Ferris et al., 1996).

The synthesis of oligonucleotides is much more efficient in the presence of a preformed pyrimidine-rich polynucleotide acting as a template. Non-enzymatic replication has been demonstrated by Orgel and his coworkers (Inoue and Orgel, 1982). The preformed chains align the nucleotides by base-pairing to form helical structures which bring the reacting groups into close proximity. However, the prebiotic synthesis of the first oligonucleotide chains remains an unsolved challenge.

3.4. PRIMITIVE LIFE BASED ON RNA? ON SIMPLER MOLECULES?

Cech found that some RNAs, the ribozymes, have catalytic properties (Zaug and Cech, 1986). For example, they increase the rate of hydrolysis of oligoribonucleotides. They also act as polymerization templates since chains up to 30 monomers long can be obtained starting from a pentanucleotide. Since this primary discovery, the catalytic spectrum of these ribozymes has been considerably enlarged by directed test tube molecular evolution experiments initiated in Joyce's and Szostak's laboratories (reviewed by James and Ellington, 1995 and 1998). Since RNA was shown to be able to act simultaneously as the genetic material and the catalysts, RNA has been considered the first living system on the primitive Earth (the "RNA world"). This is because it can simultaneously be the genotype and phenotype and can fulfill the basic cycle of life consisting of self-replication/mutation/selection. Strong evidence for this proposal has been obtained from the discovery that modern protein synthesis in the ribosome is catalyzed by RNA (Ban et al., 2000). One should, however, remember that RNAs synthesis under prebiotic conditions remains an unsolved challenge. It seems unlikely that life started with RNA molecules because these molecules are not simple enough. The RNA world appears to have been an episode in the evolution of life before the appearance of cellular microbes rather than as the spontaneous birth of life.

RNA analogs and surrogates have been studied. The initial proposal was that the first RNA was a flexible, achiral derivative in which ribose was replaced by glycerol (Spach, 1984; Joyce et al., 1987) but these derivatives did not polymerize under prebiotic conditions. The ease of forming pyrophosphate (double phosphate) bonds prompted investigation of linking nucleotides by pyrophosphate groups. This proposal was tested using the reactions of the diphosphorimidazolides of deoxynucleotides (Schwartz and Orgel, 1985; Visscher and Schwartz, 1989; Visscher and Schwartz, 1990). Their reaction in the presence of magnesium or manganese ion resulted in the formation of 10-20 mers of the oligomers.

Considering the ease of formation of hexose-2,4,6-triphosphates from glycolaldehyde phosphate in a process analogous to the formose reaction, Eschenmoser and coworkers (Eschenmoser, 1999) chemically synthesized polynucleotides containing hexopyranose ribose (pyranosyl-RNA or p-RNA) in place of the usual "natural" pentofuranose ribose found in RNA. p-RNAs form Watson-Crick-paired double helices that are more stable than RNA. Furthermore, the helices have only a weak twist which should make it easier to separate strands during replication. Replication experiments have had marked success in terms of sequence copying but have failed to demonstrate template-catalysis turnover numbers greater than one. p-RNA seems to be an excellent choice as a genetic system if it can be demonstrated that the prebiotic synthesis of pyranosyl nucleotides is much easier than synthesis of the standard isomers. The chemical synthesis of threo furanosyl nucleic acid, TNA, an RNA analog built on the furanosyl form of the tetrose sugar threose, was also reported by the Eschenmoser group (Schöning et al. 2000). TNA strands are much more stable than RNA to hydrolysis in aqueous solution. They form complementary duplexes between complementary strands but, of even greater potential importance, they form complementary strands with RNA. This raises the possibility that TNAs could have served as templates for the formation of complementary RNAs by template-directed synthesis. TNA is a more promising precursor to RNA than pRNA because tetroses have the potential to be synthesized from glycolaldehyde phosphate and other two carbon precursors which may have been present in quantities greater than those of ribose on the primitive Earth.

Peptide nucleic acids (PNA) first synthesized by Nielsen and co-workers (Nielsen et al., 1991), consist of a peptide-like backbone to which nucleic acid bases are attached. PNAs form very stable double helical structures with complementary strands of PNA (PNA-PNA), with DNA (PNA-DNA) and with RNA (PNA-RNA) and even stable PNA_2-DNA triple helices. Information can be transferred from PNA to RNA, and vice-versa, in template-directed reactions (Schmidt et al., 1997). Although PNA hydrolyzes rather rapidly, thus restricting considerably the chances of PNA to have ever accumulated in the primitive terrestrial oceans, the PNA-PNA double helix illustrates that genetic information can be stored in a broad range of double helical structures.

Chemists are also tempted to consider that primitive self-replicating systems must have used simpler informational molecules than biological nucleic acids or their analogs. Since self-replication is, by definition, autocatalysis, they are searching for simple autocatalytic molecules capable of mutation and selection. Different templates have been tested including peptides (Severin et al., 1997) and other molecules (Terfort and von Kiedrowski, 1992; Wintner et al., 1994) reviewed by Burmeister (1998). In most cases, the rate of the autocatalytic growth was not linear. The initial rate of autocatalytic synthesis was found to be proportional to the square root of the template concentration, i.e. the reaction order in these autocatalytic self-replicating systems was found to be 1/2 rather than 1, a limiting factor as compared to most autocatalytic reactions known so far. Autocatalytic reactions are particularly attractive since they might amplify small enantiomeric excesses, eventually extraterrestrial, to homochirality (Shibata et al., 1998).

4. Conclusion

On Earth, life probably appeared about 4 billion years ago, when some assemblages of organic molecules processed by liquid water began to transfer their chemical information and to evolve by making a few accidental transfer errors. The number of molecules required for those first assemblages is still unknown. The problem is that on Earth, those molecules have been erased. The discovery of a second independent genesis of life, on a body presenting environmental conditions similar to those which prevailed on the primitive Earth, would strongly support the idea of a rather simple genesis of terrestrial life. More than just a societal wish, the discovery of a second genesis of life is a scientific need for the study of the origin of life. It will demonstrate that life is not a magic one-shot process but a rather common phenomenon.

Recent discoveries have allowed a better estimate of the chances to discover an extraterrestrial life form. Biologists have shown that bacterial life can survive under extreme conditions. Life has continued to develop very well in water which is very acidic, alkaline, or is a strong brine solution. It has also survived and flourished in water at high pressure and at temperatures up to 113°C. A flourishing biosphere has been discovered a kilometer deep below the surface of the Earth.

There is clear evidence that water existed in substantial amounts on the surface of Mars at some earlier epoch (Boynton et al, 2002, Mitrofanov, 2002). Therefore, primitive life might also have developed on Mars. How long water was present at the surface is unknown. Nor is it yet certain if water still exists in subsurface aquifers, although there are clear indications of the existence of large permafrost regions and of old water flows out of associated areas. Given the discovery of a flourishing biosphere even a kilometer deep below the surface of the Earth, it would seem possible that a similar microbial community might still be present below the surface of Mars, having long ago retreated into that ecological niche following the disappearance of a surface water environment. The possibility that life may have evolved on Mars during an early period when water existed on the surface and that life may still exist deep below the surface, marks it as a prime candidate in a search for life beyond the Earth. It warrants serious attention as a potential objective of a future European space mission. Part of the organic molecules which participated in the emergence of life might also have been made in hydrothermal oceanic vents. Europa may have an ocean of liquid water beneath its icy crust, as suggested by data and theory. If submarine volcanism exists on Europa, the question arises whether such activity could support life, as takes place on volcano-hydrothermal sites on the Earth's seafloor. Organic chemistry has been shown to be universal, since over sixty different organic molecules have been identified in the interstellar medium by radioastronomers. As of November 2002, one hundred extrasolar planets (exoplanets) have also been discovered, which begin to raise the future possibility of detecting water harboring planets beyond the solar system.

The search for life elsewhere in the solar system and beyond remains as one of the great scientific endeavors of our age. The missions planned to land on Mars are milestones on the road towards that objective. But it must surely be followed by other more ambitious attempts to probe deep below the surface of Mars, to find the relics of an earlier life form or, just possibly, a novel living world.

5. References

Appel, P.W.U. and Moorbath, S. (1999) Exploring Earth's oldest geological record in Greenland. Eos **80**, 257-264.

Bailey, J., Chrysostomou, A., Hough, J.H., Gledhill, T.M., McCall, A., Clark, S., Ménard, F. and Tamura M. (1998) Circular polarization in star formation regions: Implications for biomolecular homochirality. Science **281**, 672-674.

Bailey, J. (2001) Astronomical sources of circularly polarized light and the origin of homochirality. Origins Life Evol. Biosphere **31**, 167-183.

Ban, N., Nissen, P., Hansen, J., Moore, P.B. and Steitz, T.A. (2000) The complete atomic structure of the large ribosomal subunit at 2.4 A resolution. Science **289**, 905-920.

Barbier, B. and Brack, A. (1992) Conformation-controlled hydrolysis of polyribonucleotides by sequential basic polypeptides. J. Am. Chem. Soc. **114**, 3511-3515.

Barbier, B., Chabin A., Chaput D. and Brack A. (1998) Photochemical processing of amino acids in Earth orbit, Planet. Space Sci. **46**, 391-398.

Barbier, B., Henin, O., Boillot, F., Chabin, A., Chaput, D. and Brack, A. (2002) Exposure of amino acids and derivatives in the Earth orbit. Planet. Space Sci. **50**, 353-359.

Baross, J.A. and Hoffman, S.E. (1985) Submarine hydrothermal vents and associated gradient environment as sites for the origin and evolution of life. Orig. Evol. Life Evol. Biosphere **15**, 327-345.

Bernal, J. D. (1949) The Physical Basis of Life. Proc. Roy. Soc. London **357A**, 537-558.

Bertrand, M., Bure, C., Fleury, F. and Brack, A. (2001) Prebiotic polymerisation of amino acid thioesters on mineral surfaces, In: S. Nakashima, S. Maruyama, A. Brack and B.F. Windley (eds.) *Geochemistry and the Origin of Life*, Universal Academy Press, Inc., Tokyo, Japon, pp. 51-60.

Boynton, W.V. et al. (2002) Distribution of hydrogen in the near surface of Mars: evidence for subsurface ice deposits. Science, **297**, 81-85.

Brack, A. and Spach, G. (1979) β-Structures of polypeptides with L-and D-residues. Part I. - Synthesis and conformational studies. J. Molec. Evolution **13** , 35-46.

Brack, A. (1993) From amino acids to prebiotic active peptides: A chemical restitution. Pure &Appl. Chem **65**(6), 1143-1151.

Brack, A. (2001) Water, the spring of life, In: G. Horneck and C. Baumstark-Khan (eds.) *Astrobiology. The Quest for the conditions of life*, Springer, Berlin..pp. 79-88.

Burmeister, J. (1998) Self-replication and autocatalysis, In: A. Brack (ed.) *The molecular origins of life assembling pieces of the puzzle*, Cambridge University Press, Cambridge. pp. 295-312.

Cooper, G., Kimmich, N., Belisle, W., Sarinana, J., Brabham, K. and Garrel, L. (2001) Carbonaceous meteorites as a source of sugar-related organic compounds for the early Earth. Nature **414**, 879-883.

Cronin, J.R., Pizarello, S. and Cruikshank, D.P. (1988) Organic matter in carbonaceous chondrites, planetary satellites, asteroids, and comets, In: J.F. Kerridge and M.S. Matthews (eds.) *Meteorites and the Early Solar System.*, University of Arizona Press, Tucson . pp. 819-857

De Duve, C. (1998) Possible starts for primitive life. Clues from present-day biology: the thioester world, In: A. Brack: (ed.) *The molecular origins of life: assembling pieces of the puzzle*,. Cambridge University

Press, Cambridge. pp. 219-236.

Deamer, D.W. (1985) Boundary structures are formed by organic components of the Murchison carbonaceous chondrite, Nature **317**, 792-794.

Deamer, D.W. (1998) Membrane compartments in prebiotic evolution, In: A. Brack (ed.) *The molecular origins of life: assembling pieces of the puzzle*, Cambridge University Press, Cambridge. pp. 189-205.

Delsemme, A. (1998) Cosmic origin of the biosphere, In: A. Brack (ed.) *The Molecular Origins of Life*, Cambridge Univ. Press, Cambridge, pp. 100-118.

Ehrenfreund, P. and Charnley, S.B. (2000) Organic molecules in the interstellar medium, comets and meteorites. *Ann. Review Astron. Astrophys.* **38**, 427-483.

Ehrenfreund, P. et al. (2002) Astrophysical and astrochemical insights into the origin of life. Rep. Prog. Phys. **65**, 1427-1487.

Eschenmoser, A. (1999) Chemical etiology of nucleic acid structure. Science **284**, 2118-2124.

Ferris, J. P., Hill, A. R., Jr., Liu, R. and Orgel, L. E. (1996) Synthesis of long prebiotic oligomers on mineral surfaces. Nature **381**, 59-61.

Fox, S. W. and Dose, K. (eds.) (1977) *Molecular Evolution and the Origin of Life* (Revised Edition), Marcel Dekker, New York.

Holm, N.G. and Andersson, E.M. (1995) Abiotic synthesis of organic compounds under the conditions of submarine hydrothermal systems : a perspective. Planet. Space Sci. **43**, 153-159.

Holm, N.G. and Andersson, E.M. (1998) Organic molecules on the early Earth: hydrothermal systems, In: A. Brack (ed.) *The molecular origins of life: assembling pieces of the puzzle*, Cambridge University Press, Cambridge. pp. 86-99.

Holm, N.G. and Charlou, J.-L. (2001) Initial indications of abiotic formation of hydrocarbons in the Rainbow ultramafic hydrothermal system, Mid-Atlantic ridge. Earth Planet. Sci. Lett. **191**, 1-8.

Huber, C. and Wächtershäuser, G. (1997) Activated acetic acid by carbon fixation on (Fe, Ni)S under primordial conditions. Science **276**, 245-247.

Huber, C. and Wächtershäuser, G. (1998) Peptides by activation of amino acids with CO on (Ni, Fe) surfaces: implications for the origin of life. Science **281**, 670-672.

Inoue, T. and Orgel, L.E. (1982) Oligomerization of guanosine 5'-phosphor 2-methylimidazolide on poly(C). An RNA polymerase model. J. Mol. Biol. **162**, 204-217.

James, K.D. and Ellington, A.D. (1995) The search for missing links between self-replicating nucleic acids and the RNA world. Origins Life Evol. Biosphere **25**, 515-530.

James, K.D. and Ellington, A.D. (1998) Catalysis in the RNA world, In: A. Brack (ed.) *The molecular origins of life: assembling pieces of the puzzle*, Cambridge University Press, Cambridge. pp. 269-294.

Joyce, G.F., Schwartz, A.W., Miller, S.L. and Orgel, L.E. (1987) The case for an ancestral genetic system involving simple analogs of the nucleotides. Proc. Nat. Acad. Sci. USA **84**, 4398-4402.

Maurette, M. (1998) Carbonaceous micrometeorites and the origin of life. Origins Life Evol. Biospher, **28**, 385-412.

Maurette, M., Duprat, J., Engrand, C., Gounelle, M., Kurat, G., Matrajt, G. and Toppani, A. (2000) Accretion of neon, organics, CO_2, nitrogen and water rom large interplanetary dust particles on the early Earth. Planet. Space Sci. **48**, 1117-1137.

Miller, S.L. (1953) The production of amino acids under possible primitive Earth conditions. Science **117**, 528-529.

Mitrofanov, I. et al. (2002) Maps of subsurface hydrogen from the High Energy Neutron Detector, Mars Odyssey. Science **297**, 78-81.

Mojzsis, S.J., Arrhenius, G., Keegan, K.D., Harrison, T.H., Nutman, A.P., Friend, C.L.R. (1996) Evidence for life on Earth before 3,800 million years ago. Nature **384**, 55-59.

Muñoz Caro, G.M., Meierhenrich, U.J., Schutte, W.A., Barbier, B., Arcones Segovia, A., Rosenbauer, H., Thiemann, W.H.-P., Brack, A. and Greenberg, J.M. (2002) Amino acids from ultraviolet irradiation of interstellar ice analogues. Nature **416**, 403-406.

Nielsen, P.E., Egholm, M., Berg, R.H. and Buchardt, O. (1991) Sequence-selective recognition of DNA by strand displacement with a thymine-substituted polyamide. Science **254**, 1497-1500.

Ourisson, G. and Nakatani, Y. (1994) The terpenoid theory of the origin of cellular life : the evolution of terpenoids to cholesterol. Chemistry and Biology **1**, 11-23.

Paecht-Horowitz, M., Berger, J. and Katchalsky, A. (1970) Prebiotic synthesis of polypeptides by heterogeneous polycondensation of amino-acid adenylates. Nature **228**, 636-639.

Pizzarello, S. and Cronin, J. R. (2000) Non-racemic amino-acids in the Murray and Murchison meteorites. Geochim. Cosmochim. Acta **64**(2), 329-338.

Rosing, M.T. (1999) ^{13}C depleted carbon microparticles in >3700 Ma seafloor sedimentary rocks from West Greenland. Science **283**, 674-676.

Schidlowski, M. (1988) A 3,800-million-year isotopic record of life from carbon in sedimentary rocks. Nature **333**, 313-318.

Schmidt, J.G., Nielsen, P.E. and Orgel, L.E. (1997) Information transfer from peptide nucleic acids to RNA by template-directed syntheses. Nucleic Acids Res. **25**, 4797-4802.

Schöning, K.-U., Scholz, P., Guntha, S., Wu, X., Krishnamurthy, R. and Eschenmoser, A. (2000) Chemical etiology of nucleic acid structure: the α-threofuranosyl-(3'$\rightarrow$2') oligonucleotide system. Science **290**, 1347-1351.

Schopf, J.W. (1993) Microfossils of the Early Archean Apex Chert: new evidence of the antiquity of life. Science **260**, 640-646.

Schwartz, A. W. and Orgel, L. E. (1985) Template-directed synthesis of novel, nucleic acid-like structures. Science **228**, 585-587.

Severin, K.S., Lee, D.H., Martinez, J.A. and Ghadiri, M.R. (1997) Peptide self-replication via template-directed ligation. Chem. Eur. J. **3**, 1017-1024.

Shibata, T., Yamamoto, J., Matsumoto, N., Yonekubo, S., Osanai, S. and Soai, K. (1998) Amplification of a slight enantiomeric imbalance in molecules based on asymmetry autocatalysis. J. Amer. Chem. Soc. **120**, 12157-12158.

Spach, G. (1984) Chiral versus chemical evolutions and the appearance of life. Origins Life Evol. Biosphere **14**, 433-437.

Stoks, P.G. and Schwartz, A.W. (1982) Basic nitrogen-heterocyclic compounds in the Murchison meteorite. Geochim. Cosmochim. Acta **46**, 309-315.

Szostak, J. W. and Ellington, A. D., (eds.) (1993) In vitro selection of functional RNA sequences. First edit. *The RNA World.* Edited by R. F. Gesteland, and J.F. Atkins, Cold Spring Harbor, New York: Cold Spring Harbor Laboratory Press.

Terfort, A. and von Kiedrowski, G. (1992) Self-replication by condensation of 3-aminobenzamidines and 2-formyl-phenoxyacetic acids. Angew. Chem. Int. Ed. Engl. **31**, 654-656.

Visscher, J. and Schwartz, A. W. (1989) Manganese-catalyzed oligomerizations of nucleotide analogs. J. Mol. Evol. **29**, 284-287.

Visscher, J. and Schwartz, A. W. (1990) Oligomerization of cytosine-containing nucleotide analogs in aqueous solution. J. Mol. Evol. **30**, 3-6.

Wächtershäuser, G. (1994) Life in a ligand sphere. Proc. Natl. Acad. Sci. USA **91**, 4283-4287.

Wächtershäuser, G. (1998) Origin of life in an iron-sulfur world, In: A. Brack (ed.) *The molecular origins of life: assembling pieces of the puzzle*, Cambridge University Press, Cambridge, pp. 206-218.

Westall, F., De Wit, M.J., Dann, J., Van Der Gaast, S., De Ronde, C., Gerneke, D. (2001) Early Archaean fossil bacteria and biofilms in hydrothermally-influenced, shallow water sediments, Barberton greenstone belt, South Africa. Precambrian Res. **106**, 93-116.

Wintner, E.A., Conn, M.M. and Rebek, J. (1994) Studies in molecular replication. Acc. Chem. Res. **27**, 198-203.

Yanagawa, H. and Kobayashi, K. (1992) An experimental approach to chemical evolution in submarine hydrothermal systems. Origins. Life Evol. Biosphere **22**, 147-160.

Zaug, A.J. and Cech, T.R. (1986) The intervening sequence RNA of Tetrahymena is an enzyme. Science **231**, 470-475.

Biodata of **David W. Deamer** author of ***"Prebiotic Amphiphilic Compounds: Self-Assembly and Properties of Early Membrane Structures"***

Dr. David W. Deamer is a Professor at the Department of Chemistry and Biochemistry, University of California at Santa Cruz, California, and currently serves as the acting chair of Biomolecular Engineering. He obtained his Ph.D. in 1965 from the Department of Physiological Chemistry, Ohio State University, School of Medicine, Columbus OH. He has been a visiting scientist at the University of Bristol, Cambridge University, Australian National University (Canberra), and the Weizmann Institute, Rehovot, Israel. He received honors as a Guggenheim Fellow (1985), Redman Lecturer, (McMaster University, Ontario, Canada 1998), and is a Fellow of ISSOL (International Society for the study to Origin of Life.) Dr. Deamer served as the organizing chair of numerous conferences including Evolution of Light-transducing membranes, Protons and Membrane Reactions, Polymers in Confined Spaces, Intracellular pH, and Single Molecule Physics. His public service includes the NRC Panel on the Origin and Evolution of Life, the NASA review panel for Exobiology, and currently the NASA Space Science Advisory Committee. Dr. Deamer is a member of the editorial boards of several journals (Membrane Biology, Bioenergetics and Biomembranes, Astrobiology, Origins of Life and Evolution of the Biosphere) and has served as treasurer of the Biophysical Society and ISSOL.

E-mail: **deamer@hydrogen.ucsc.edu**

J. Seckbach (ed.), Origins, 75–89.

PREBIOTIC AMPHIPHILIC COMPOUNDS
Self-assembly and properties of early membrane structures

DAVID W. DEAMER
Department of Chemistry and Biochemistry
University of California, Santa Cruz, Santa Crux CA 95064 USA

1. Introduction

This review focuses on the origin and self-assembly of complex organic molecules in the prebiotic environment. Membranous boundary structures define all life today, and a source of membrane-bounded compartments on the early Earth was essential for cellular life to begin. Cellular life arose from countless natural experiments in which assemblages of organic molecules were subjected to energy sources as such light, heat, and electric discharge. At some point, membrane-bounded systems of catalytic-replicating molecules appeared that could grow and reproduce by using energy and nutrients from the environment.

Here we will discuss possible sources of amphiphilic molecules, self-assembly processes leading to membrane structures, encapsulation mechanisms by which large molecules can be captured in membrane-bounded microenvironments, and finally describe recent experimental model systems that incorporate certain properties related to the origin of cellular life.

2. Sources of Amphiphilic Compounds on the Early Earth

There are only two possible sources of organic compounds on a primitive planetary surface: delivery during late accretion, followed by chemical evolution, or synthesis by geochemical processes in the early atmosphere or hydrosphere. Earlier investigations focused on chemical synthesis of monomers common to the primary macromolecules involved in living systems, with the goal of determining whether it was possible that compounds such as amino acids, sugars and purines/pyrimidines were available on the primitive Earth. The classic experiments of Miller and Urey (1953, 1959) showed that impressive yields of amino acids could be obtained when a mixture of reduced gases was exposed to an electrical discharge. The mixture was assumed to be a simulation of the original terrestrial atmosphere, which, by analogy with the outer planets, would have contained hydrogen, methane, ammonia and water vapor. At sufficiently high energy fluxes, such mixtures of reduced of gases generate hydrogen cyanide and formaldehyde, which in turn react to produce amino acids, purines and a variety of simple sugars.

The possibility that organic compounds could be synthesized under prebiotic conditions was given additional weight when it was convincingly shown that carbonaceous meteorites contained amino acids, hydrocarbons, and even traces of purines (Kvenholden 1970; Lawless and Yuen, 1979). If such meteorites represent samples of the primitive solar system that underwent synthetic chemical reactions, it was reasonable to assume that similar reactions may have occurred on the Earth's surface. This view was challenged in the late 1970s when it became increasingly clear that the early atmosphere was composed of carbon dioxide and nitrogen rather than the mixture of reducing gases assumed by the Miller-Urey model (Holland 1984; Kasting 1998). Carbon dioxide does not support synthetic pathways leading to chemical monomers, so interest was drawn to the second potential source of organic material, extraterrestrial infall in the form of micrometeorites and comets. This was first proposed by Oro (1961) and Delsemme (1984) and more recently extended by Anders (1989) and Chyba and Sagan (1992. The total organic carbon added by extraterrestrial infall over $\sim10^8$ years of late accretion can be estimated to be in the rage of 10^{16} - 10^{18} kgs. which is several orders of magnitude greater than the total organic carbon now circulating in the biosphere. From such calculations it seems reasonable that extraterrestrial infall was a significant source of organic carbon in the prebiotic environment. Even today a surprising amount of infall can be detected, which increases our confidence in such estimates. For instance, by counting interplanetary dust particles (IDP) captured in the upper atmosphere, Love and Brownlee (1993) estimated the annual amount of micrometeoritic dust collected by the Earth to be in the range of 40,000 metric tons. Since IDP have several percent of their mass as carbon compounds, it follows that thousands of tons of organic carbon are still entering the terrestrial environment every year. This rate would have been several orders of magnitude greater during the late accretion phase of the early Earth (Chyba and Sagan, 1992).

The discovery of biologically relevant compounds in meteorites immediately raised the question of sources and synthetic pathways. Clues to a possible source have been provided by radio astronomy. Spectral features obtained from dense molecular clouds of the interstellar medium indicate the presence of a hundred or more carbon-containing compounds (Ehrenfreund and Charnley, 2000; Sandford 1996). Since molecular clouds are the birthplace of stars and solar systems, it seems reasonable that the organic substances present in comets and the parent bodies of meteorites were derived from carbon compounds in the original molecular cloud that gave rise to the solar system. Possible synthetic pathways have been established by interstellar ice simulations in the laboratory, which show that a variety of organic compounds can be synthesized by photochemical processes driven by UV photons (Greenberg and Mendoza-Gomez, 1993; Bernstein et al. 1995, 2002). It follows that a major fraction of the organic compounds available for delivery to a primitive planetary surface were synthesized on interstellar grain mantles, then swept up in the planet forming process and delivered to planetary surfaces such as the Earth.

3. Mineral-Water Interfaces: Organic Films as Precursors to Life

Given that organic compounds are made available on a planetary surface as described above, they must then be organized into mixtures that are sufficiently concentrated to undergo chemical reactions. Most of the compounds would accumulate in extensive early oceans, rather than the relatively small area of available Archaean volcanic terrain. However, this means that global concentrations of organic material in the early ocean would have been too dilute to permit synthetic chemical reactions to occur (Stribling and Miller, 1987). For this reason, it has been suggested that mineral/aqueous interfaces were involved in concentrating and organizing organic solutes, and perhaps catalyzing specific reactions related to life processes [Hazen 2001. See Nakashima et al. 2001, for review]. Corliss et al (1981) and Baross and Hoffman (1985) first proposed that life began as an organic film on mineral surfaces in subsurface geothermally active sites. Such films would provide a microenvironment of low water activity so that hydrolytic back reactions would not continuously degrade more complex polymeric molecules formed by condensation reactions. As an energy source, either dissolved hydrogen gas or the mineral surface itself would provide a source of reducing power ((Pace, 1991; Wächtershäuser, 1988a, 1988b). According to this scheme, membrane encapsulation and a system of information transfer would evolve at a later time.

As a specific chemical example of a mineral-dependent reaction pathway, Huber and Wächtershäuser (1997) described an experimental model in which a slurry of nickel and iron sulfide was found to promote the formation of acetic acid from carbon monoxide and methyl mercaptan (CH_3SH). Peptide bond formation was also demonstrated in subsequent experiments (Huber and Wächtershäuser, 1998). These conditions were considered to represent a simulation of a primordial geothermal system in which metal sulfides at high temperatures (~100° C) provide a reaction pathway for the initial steps of an autotrophic metabolism. This is an interesting result that is pertinent to the synthesis of organic material, and confirms earlier observations that a variety of free energy sources can drive the formation of simple organic molecules.

Martin and Russell (2002) have taken this concept a step further. They note that certain iron sulfide minerals contain microscopic pores in the size range of cells (~10 - 100 micrometers) and propose that such cavities could provide a mineral version of a membranous boundary structure. The authors suggest that the cavities can concentrate nutrient organic solutes that would then serve as reactants in primitive metabolic pathways. They also propose that the iron sulfide membranes could provide a source of chemical energy, perhaps even chemiosmotic energy to drive early metabolism. This idea has merit, and deserves further testing. In particular, it should be determined whether mineral membranes are able to act as true permeability barriers to the free diffusion of solutes. So far, permeability barriers have only been demonstrated in lipid bilayer membranes having a hydrophobic phase that has the capacity to maintain concentration gradients of polar and ionic solutes.

4. Assembly of Amphiphilic Compounds Into Membrane Structures

Self-assembly of certain kinds of organic compounds occurs through hydrogen bonding and nonpolar forces that stabilize orderly arrangements of small and large molecules. Examples include the self-assembly of DNA strands into a double helix, and the folding of newly synthesized protein molecules into structures having catalytic functions as enzymes. Self-assembling systems related to the origin of life were first investigated in the form of microscopic gel structures called coacervates (Oparin et al 1976) and proteinoid microspheres (Fox, 1973). However, neither coacervates nor proteinoid microspheres have a true boundary membrane that can act as a selective permeability barrier.

A more useful model is based on liposomes, defined as microscopic vesicles composed of one or more lipid bilayers that encapsulate a volume (see Deamer 1997 for review). If liquid water is added to dried films of amphiphilic compounds, vesicular structure are readily produced and provide a variety of cell-sized environments. Self-assembly of such vesicles occurs when small amphiphilic molecules spontaneously associate by hydrophobic interactions into more complex structures having defined compositions and organization. Examples include the assembly of amphiphilic molecules into micelles, monolayers, and bilayers in the form of vesicles (Figure 1).

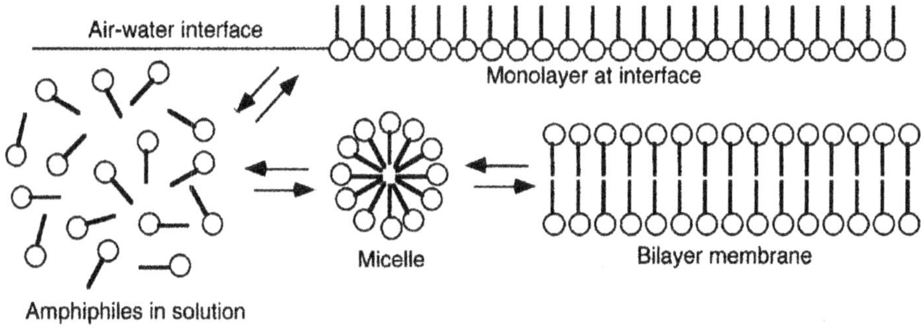

Figure 1. Self-assembled structures produced by amphiphilic molecules. (See text for discussion.)

Such structures are highly concentrated relative to the bulk phase, thereby overcoming the dilution that occurs with water-soluble organic species. Vesicular structures are able to keep specific groups of macromolecular species in a single compartment, which facilitates their interaction and provides a form of speciation that is lacking in bulk phase environments. The membranes of self-assembled compartments also have the potential to maintain concentration gradients of ions, thus providing a source of free energy that can drive energetically uphill processes. Finally, if certain components of the prebiotic organic inventory happened to be non-polar pigments, they would partition into the hydrophobic phase of a membrane and potentially capture light energy (Deamer, 1991). This set of biophysical properties can only arise from amphiphilic molecules that assemble into vesicular structures, and it is clear that such structures must have been present on the early Earth for cellular life to begin.

What physical properties are required if a molecule is to become incorporated into a stable bilayer? All bilayer-forming molecules are amphiphiles, with a hydrophilic "head" and a hydrophobic "tail" on the same molecule (Figure 2). If amphiphilic molecules were present in the mixture of organic compounds available on the early Earth, it is not difficult to imagine that their self-assembly into molecular aggregates was a common process.

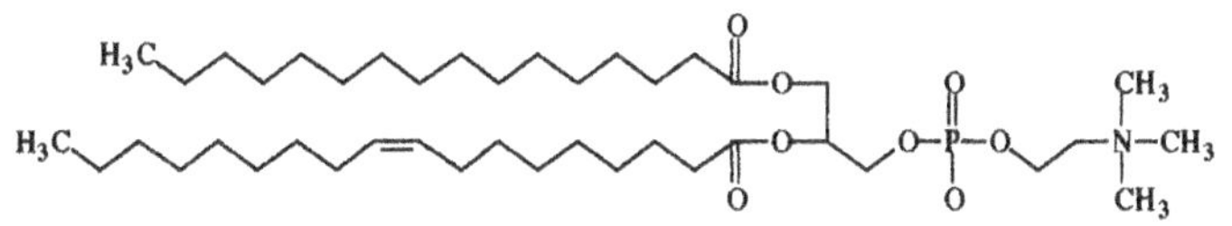

A.

B. $H_3C\text{-}(CH_2)n\text{-}COOH$

Figure 2. Comparison of a modern phospholipid (A) with an amphiphilic monocarboxylic acid present in carbonaceous meteorites. The phospholipid forms robust lipid bilayers that are relatively impermeable to ionic solutes. The monocarboxylic acid, with n = 8 – 12 (B) forms bilayers that are less stable and relatively permeable to ionic solutes (Apel et al. 2002).

Is this a plausible premise? In order to approach this question, we can assume that the mixture of organic compounds in carbonaceous meteorites such as the Murchison meteorite resembles components available on the early Earth through extraterrestrial infall. A series of organic acids represents the most abundant water-soluble fraction in carbonaceous meteorites (Cronin et al. 1988). Samples of the Murchison meteorite were extracted in an organic solvent commonly used to extract membrane lipids from biological sources (Deamer, 1985; Deamer and Pashley, 1989). When this material was allowed to interact with aqueous phases, one class of compounds with acidic properties was clearly capable of forming membrane-bounded vesicles.

Figure 3 shows micrographs of such vesicles, as well as vesicles formed by products of interstellar ice simulations and fatty acid dispersions. The vesicle-forming behavior of all of these amphiphiles is similar. The organic compounds present in the meteoritic extract and those synthesized in the simulation of grain mantle photochemistry both contain amphiphilic compounds capable of self-assembly into membranous boundary structures. The vesicles produced from the interstellar simulations, like those of the meteoritic compounds, can also capture and maintain a gradient of ionic dye molecules (Dworkin et al. 2001).

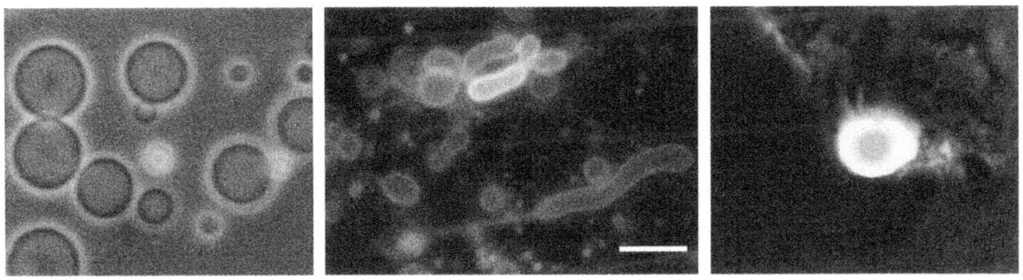

Figure 3. Membranous vesicles self-assemble from a variety of organic amphiphiles. Left: Murchison meteorite extract. Center: Decanoic acid. Right: Photoproduct from laboratory simulation of interstellar grain mantle ices. (Dworkin et al. 2001.) The photoproduct vesicle has captured pyranine, a fluorescent dye, which indicates the presence of a boundary membrane. Bar shows 10 microns.

From these results, it is reasonable to conclude that amphiphilic molecules were present on the early Earth which could participate in the formation of primitive membrane structures. The long chain acids and alcohols that contribute the amphiphilic property of contemporary membrane lipids are one possible component of prebiotic membrane structures. These compounds are present in carbonaceous meteorites (Lawless and Yuen, 1979; Shimoyama et al., 1989) and are synthesized under simulated geochemical conditions (McCollom et al. 1999; Rushdi and Simoneit, 2001). Significantly, such simple amphiphiles can also form vesicles, as shown in Figure 3B (Hargreaves and Deamer, 1978; Apel et al. 2002). Stability of the vesicles is strongly dependent on chain length, concentration, amphiphile composition, temperature, and head group characteristics. For example, even a 9-carbon monocarboxylic acid—nonanoic acid—can form vesicles at concentrations of 85 mM and pH 7.0, which is the pK of the acid in bilayers (Apel et al. 2002). Addition of small amounts of an alcohol (nonanol) further stabilizes the bilayers due to hydrogen bonding between the alcohol and acidhead groups, and vesicles can form at lower concentrations (~20 mM) at pH ranging from 6 to 11. The vesicles provide a selective permeability barrier, as indicated by osmotic activity and ionic dye capture. As chain length increases, stability also increases and vesicles form at lower concentrations.

5. Environmental Constraints on the First Cell Membranes

Although self-assembly of amphiphilic molecules promotes the formation of complex molecular systems, the physical and chemical properties of an aqueous phase can significantly inhibit such processes, possibly constraining the environments in which cellular life first appeared. One such constraint is that temperature strongly influences the stability of vesicle membranes. It has been proposed that the last common ancestor, and even the first forms of life, were hyperthermophiles that developed in geothermal regions such as hydrothermal vents (Baross and Hoffman, 1985) or deep subterranean hot aquifers (Pace, 1991). Such environments have the advantage of providing chemical energy in the form of redox potentials as well as abundant mineral surfaces to act as potential catalysts and adsorbants. However, because the intermolecular forces that stabilize self-assembled molecular systems are relatively weak, it is difficult to imagine how lipid bilayer membranes assembling from plausible prebiotic constituents would be stable under these conditions. All hyperthermophiles today have highly specialized lipid components, and it seems likely that these are the result of more recent adaptation than a molecular fossil of early life.

A second concern is related to the ionic composition of a marine environment. The high salt concentration of the present ocean (near 0.5 M NaCl) has the potential to exert significant osmotic pressure on any closed membrane system. All marine organisms today have evolved highly developed membrane transport systems that allow them to maintain osmotic equilibrium against substantial salt gradients across their membranes. Furthermore, the concentrations of divalent cations, in particular Mg^{2+} and Ca^{2+}, were likely to exceed 10 mM in the early oceans. In the absence of oxygen, Fe^{2+} would also be present at similar concentrations. All such divalent cations have a strong tendency to bind to the anionic head groups of amphiphilic molecules, strongly inhibiting their ability to form stable membranes (Monnard et al. 2002).

These considerations suggest that, from the perspective of membrane biophysics, a plausible planetary environment for the origin of life would be at moderate temperature ranges (<60 C), with the ionic content at low ionic strength, pH values near neutrality (pH 5 - 8) and divalent cations at submillimolar concentrations. This suggestion is in marked contrast to the view that life most likely began in a marine environment, perhaps even the extreme environment of a hydrothermal vent. A marine site for life's beginning seems plausible because fresh water would be rare on the early Earth. Even with today's extensive continental crust, fresh water only represents ~1% of the contemporary Earth's reservoir of liquid water. Another concern about a fresh water origin of life is that the lifetime of fresh water bodies tends to be short on a geological time scale. On the other hand, if seawater, with its high content of sodium chloride and divalent ions, markedly inhibits self-assembly processes and reactions that are essential to the emergence of cellular life, we may need to reconsider the assumption that life inevitably began in a marine environment. A more plausible site for the origin of cellular life may be a low ionic strength lacustrine environment such as a pond or lake. After the first form of cellular life was able to establish itself in a relatively benign environment, it would rapidly begin to adapt through Darwinian selection to more rigorous environments, including the extreme temperatures, salt concentrations and pH ranges that we associate with the limits of life on the Earth today.

6. Membrane Permeability and Early Bioenergetics

Although membranes define all living cells, the membrane also reduces ready access to nutrients and energy sources. It seems unlikely that the first living cellular systems had time to evolve highly specialized membrane transport systems, which brings up the question of how early cells overcame the membrane permeability barrier. To give a perspective on permeability and transport rates, we can compare the fluxes of relatively permeable and relatively impermeable solutes across contemporary lipid bilayers. The measured permeability of lipid bilayers to small, uncharged molecules such as water, oxygen and carbon dioxide are ~10^9 greater than the permeability to ions. For instance, the permeability coefficient of water is approximately 10^{-3} cm s^{-1}, and the permeability coefficient of potassium ions is 10^{-12} cm s^{-1}. These values mean little by themselves, but make more sense when put in the context of time required for exchange across a bilayer. Measurements show that half the water in a liposome exchanges in milliseconds, while potassium ions have half-times of exchange measured in days.

We can now consider some typical nutrient solutes like amino acids and phosphate. Such molecules are ionized, which means that they would not readily cross the permeability barrier of a lipid bilayer. Permeability coefficients of liposome membranes to phosphate and amino acids have been determined (Chakrabarti and Deamer 1994) and were found to be in the range of 10^{-11} – 10^{-12} cm s^{-1}, in the same range as other solutes such as sodium and chloride ions. From these figures one can estimate that if a primitive microorganism depended on passive transport of phosphate across a lipid bilayer composed of a typical phospholipid, it would require several years to accumulate phosphate sufficient to double its DNA content, or pass through one cell cycle. In contrast, a modern bacterial cell can reproduce in as short a time as 20 minutes.

If solutes like amino acids and phosphate are so impermeable, how could primitive cells have had access to these essential nutrients? One clue may be that modern lipids are highly evolved products of several billion years of evolution, and typically contain hydrocarbon chains 16 to 18 carbons in length. These chains provide an interior "oily" portion of the lipid bilayer that represents a nearly impermeable barrier to the free diffusion of ions such as sodium, potassium and protons. If an ion dissolved in water attempts to leave the water and dissolve in the oil phase, it faces a very high energy barrier called Born energy, which is associated with the difference in energy for an ion solvated in a high dielectric medium (water with dielectric constant of 80) compared to the same ion in a low dielectric medium (hydrocarbon with a dielectric constant of 2). This energy barrier is immense, up to 40 kcal per mole (Parsegian, 1969).

However, recent studies have shown that permeability is strongly dependent on chain length (Paula et al 1996). For instance, shortening phospholipid chains from 18 to 14 carbons increases permeability to ions by several orders of magnitude. The reason is that thinner membranes have increasing numbers of transient defects that open and close on nanosecond time scales, so that ionic solutes can get from one side of the membrane to the other without dissolving in the oily interior phase of the bilayer. Ionic solutes even as large as ATP can diffuse across a bilayer composed of dimyristoylphosphatidylcholine, a 14-carbon phospholipid (Monnard and Deamer, 2001). On the early Earth, shorter hydrocarbon chains would have been much more common than longer chain amphiphiles, suggesting that the first cell membranes were sufficiently leaky so that ionic and polar nutrients could enter, while still maintaining larger polymeric molecules in the encapsulated volume.

7. Encapsulation of Macromolecules by Bilayer vesicles

The origin of cellular life presumably took place in a concentrated mixture of chemical components that had already established simple metabolic pathways and catalyzed polymerization of a primitive genetic material. In order for lipid bilayer vesicles to capture large molecules and thereby produce the first cells, a robust encapsulation mechanism would be required. If encapsulation is to occur, a reversible process must be possible by which the bilayer barrier first is broken, allowing entry of large molecules, then resealed. One encapsulation mechanism that has the potential to function in the prebiotic environment depends on the fact that when lipid vesicles are dried in the presence of macromolecules, they tend to fuse into multilayered structures that "sandwich" the solutes (Deamer and Barchfeld, 1982). The macromolecules are then captured upon rehydration when the lipid layers reseal into vesicles. It is not difficult to imagine such cycles occurring in fluctuating environments such as ponds or intertidal zones, and it follows that encapsulated systems of macromolecules may have been common in the prebiotic environment.

8. Model Systems of Primitive Cells

A central event in the origin of life was the self-assembly of a molecular system in which catalytic polymers could interact with a second class of polymers having the capacity to store information in a sequence of monomers. That sequence in turn would in some manner determine the sequence of monomers in the catalyst, so that the resulting catalytic-information cycle was able to undergo directed growth. In contemporary cells, the cycle is represented by protein catalysts (enzymes) and nucleic acids that store genetic information and have the potential to transmit that information to a second molecule by replication or transcription. However, in a protocell, both catalytic and information-containing sites could be present in the same molecule, as suggested by recent studies of RNA ribozymes (Johnson et al. 2001). Several approaches to artificial cells have been proposed to test various scenarios for the origin of cellular life (Koch, 1985; Morowitz, 1992; Cavalier-Smith, 1995; Luisi, 1999; Segre et al. 2001; Szostak et al. 2001; Pohorille and Deamer, 2002). An ideal model cell would incorporate an encapsulated polymerase activity together with a template of some sort, so that sequence information in the template can be transcribed to a second molecule. The membrane must be sufficiently permeable to allow the polymerase to have access to externally added substrates. Furthermore, the membrane itself should be able to grow in order to accommodate the growth of the encapsulated polymers. Finally, in an ideal cell model, the polymerase itself would be reproduced from information in the template, so that the entire system is able to grow and evolve.

Substantial progress has been made in laboratory investigations of such systems. For instance, Walde et al. (1994) have shown that vesicles composed of oleic acid can grow and "reproduce" as oleoyl anhydride spontaneously hydrolyzes in the reaction mixture, thereby adding additional amphiphilic components (oleic acid) to the vesicle membranes. To demonstrate polymerase activity in a model cell, Chakrabarti et al. (1994) encapsulated polynucleotide phosphorylase in vesicles composed of dimyristoylphosphatidylcholine (DMPC). This enzyme can produce RNA from nucleoside diphosphates such as adenosine diphosphate (ADP) and does not require a template, so it has proven useful for initial studies of encapsulated polymerase activity. Furthermore, DMPC liposomes are sufficiently permeable so that 5 - 10 ADP molecules per second enter each vesicle. Under these conditions, measurable amounts of RNA in the form of polyadenylic acid were synthesized and accumulated in the vesicles after several days incubation. The enzyme-catalyzed reaction could be carried out in the presence of a protease external to the membrane, demonstrating that the vesicle membrane protected the encapsulated enzyme from hydrolytic degradation. Similar behavior has been observed with monocarboxylic acid vesicles [Walde et al. 1994], so complex phospholipids are not required for an encapsulated polymerase system to function.

In other work, the Q-beta replicase (Oberholzer et al 1995a) and the components of the polymerase chain reaction (Oberholzer et al 1995b) have also been encapsulated, together with templates and substrates in the form of nucleoside triphosphates. and are functional in liposomes. Both of these enzyme systems use templates, so it is clear that template-dependent polymer synthesis can occur in an encapsulated environment. The phospholipids used in these studies were relatively impermeable, so that substrates were necessarily encapsulated along with enzyme and template. This limited the amount of nucleic acid replication that could occur to a few molecules per vesicle. More recently, a template directed reaction was established in DMPC liposomes in which external substrate was used to supply the enzyme (Monnard and Deamer, 2002). In this study, T-7 RNA polymerase and a circular 4000 bp plasmid template were encapsulated, and substrates were provided by addition of the ribonucleotides ATP, GTP, CTP and UTP. RNA synthesis was monitored by incorporation of radiolabeled UTP, and transcription was confirmed by reverse PCR (polymerase chain reaction). Figure 4 shows a micrograph of the resulting structures containing RNA synthesized within the vesicle volume.

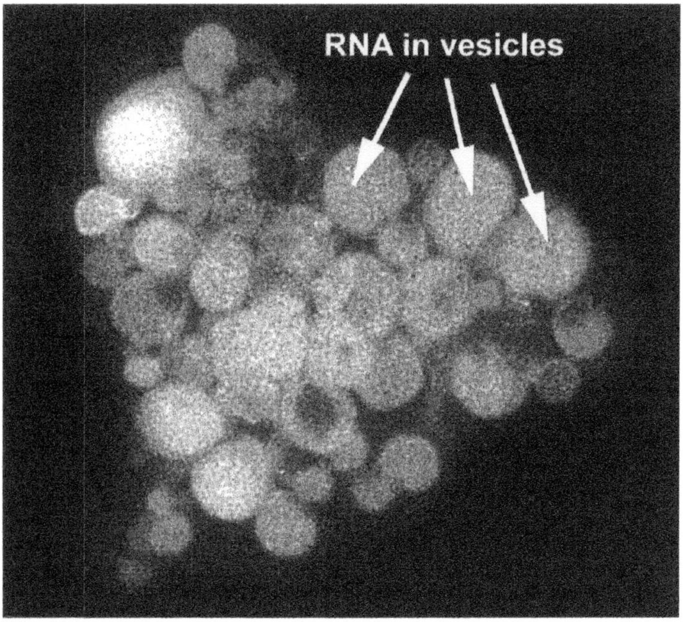

Figure 4. Lipid vesicles with encapsulated T7 RNA polymerase and DNA template. A mixture of four nucleoside triphohsphates was added, and these diffused into the vesicles and were used by the polymerase to synthesize RNA from the DNA template. The RNA was stained with ethidium bromide and appears as fluorescent material within the vesicles. Note that some of the vesicles do not contain fluorescent RNA, presumably because they lacked sufficient enzyme or template.

An important next step in modeling such systems will be to encapsulate an evolving ribozyme system (Beaudry and Joyce, 1992; Wilson and Szostak 1994; Johnson et al. 2001) within vesicles formed from amphiphilic mixtures that are optimized for stability and permeability. It seems likely that one such mixture will have an optimal set of properties that permit it to encapsulate a catalytic polymerase system and template, with sufficient permeability to allow substrate access to the enzyme at reasonable rates. Replication and ribozyme evolution would then occur in immensely large numbers of microscopic volumes represented by the liposome interiors, rather than in the macroscopic volume of a test tube. Under these conditions, the rare ribozyme that happens to undergo a favorable mutation would be readily selected, whereas in a test tube it is lost among trillions of other similar molecules.

9. Summary

The results described here provide a useful perspective on the most primitive forms of cellular life. In the early Earth environment, a variety of amphiphilic hydrocarbon derivatives could self-assemble into bilayer boundary structures and encapsulate polymers that were being synthesized by a separate process (Ferris et al. 1996, 2002). The vesicle membranes would have been sufficiently permeable to allow passage of smaller ionic substrates required for metabolism and biosynthesis, yet maintain larger molecules within a boundary. Encapsulated catalysts and information-bearing molecules would thus have access to nutrients required for growth. Furthermore, specific groupings of macromolecules would be maintained in a compartment, rather than diffusing indefinitely. This would allow true selection of such groupings to occur, a process that could not as easily take place in mixtures of molecules free in solution. A small number of the encapsulated molecular systems were likely to have the specific set of properties that allowed them to capture free energy and nutrients from their environment and undergo growth by polymerization. At some point, polymerization reactions would first become catalyzed by the encapsulated polymers, and then directed by an emerging genetic process. Such structures would be on the evolutionary path to the first forms of cellular life.

10. Acknowledgements

Portions of this review were adapted in part from Deamer (1997) and Deamer et al. (2002). Research on primitive amphiphiles in our laboratory is supported by a grant from the NASA Exobiology program.

11. References

Apel, C.L., Deamer, D. W. and Mautner, M. (2002) Self-assembled vesicles of monocarboxylic acids and alcohols: Conditions for stability and for the encapsulation of biopolymers. *Biochim Biophys Acta* **1559**, 1.

Baross, J.A. and Hoffman, S. E. (1985) Submarine hydrothermal vents and associated gradient environments as sites for the origin and evolution of life. *Origins of Life* **15**, 327.

Beaudry, A. A. and Joyce, G. F. (1992) Directed evolution of an RNA enzyme. *Science* **342**, 255.

Bernstein, M. P., Sandford, S. A., Allamandola, L. J., Chang, S. and Scharberg, M. A. (1995) Organic compounds produced by photolysis of realistic interstellar and cometary ice analogs containing methanol. *Astrophys. J.* **454**, 327-344.

Bernstein, M. P., Dworkin J. P., Sandford S. A., Cooper G. W. and Allamandola L. J.. (2002) The formation of racemic amino acids by ultraviolet photolysis of Interstellar Ice Analogs. *Nature* **416**, 401.

Cavalier-Smith, T. (1987) The origin of cells: A symbiosis between genes, catalysts and membranes. Cold Spring Harbor Symposia on Quantitative Biology, Volume **LII**,. 805- 824.

Cairns-Smith, G. (1982) Genetic Takeover and the Mineral Origins of Life. Cambridge University Press, Cambridge UK

Chakrabarti, A. and Deamer, D.W. (1994) Permeation of membranes by the neutral form of amino acids and peptides: Relevance to the origin of peptide translocation. J. Mol. Evol. **39,** 1-5.

Chakrabarti, A.,. Breaker, R.R,. Joyce, G.F. and Deamer, D.W. (1994) Production of RNA by a polymerase protein encapsulated within phospholipid vesicles. J. Mol. Evol. **39**, 555-559.

Chyba, C.F. and Sagan, C. (1992) Endogenous production, exogenous delivery and impact-shock synthesis of organic molecules: An inventory for the origin of life. Nature **355,**125-13.

Corliss, J.B.,. Baross, J.A and Hoffman, S.E. (1981) An hypothesis concerning the relationship between submarine hot springs and the origin of life on Earth Acta. Proc. 26th Intl. Geolog. Congress, Geology of the Oceans Symposium, Paris. p 59-69..

Cronin, J.R., Pizzarello, S. and Cruikshank, D.P. (1988) Organic matter in carbonaceous chondrites, planetary satellites, asteroids and comets. In: J.F. Kerridge and M.S. Matthews (eds.) *Meteorites and the Early Solar System*, University of Arizona Press, Tucson AZ, pp. 819-857.

Deamer, D. W. (1985) Boundary structures are formed by organic components of the Murchison carbonaceous chondrite. Nature **317**, 792-794.

Deamer, D.W. (1991) Polycyclic aromatic hydrocarbons as potential pigments in primitive membrane structures. Adv. Spac Sci. **12,**183-189.

Deamer D. W. (1997) The first living systems: A bioenergetic perspective. *Microbiol. Mol. Biol. Rev.* **61**, 230-261.

Deamer, D.W. and. Barchfeld, G.L. (1982) Encapsulation of macromolecules by lipid vesicles under simulated prebiotic conditions. J. Mol. Evol. **18**, 203-206.

Deamer, D.W. and Pashley, R.M. (1989) Amphiphilic components of carbonaceous meteorites. Orig, Life Evol. Biosphere. **19**,:21-33.

Delsemme, A. (1984) The cometary connection with prebiotic chemistry. Origins of Life **14,** 51-60.

Dworkin, J. P., Deamer, D. W., Sandford, S. A. and Allamandola, L. J. (2001). Self-assembling amphiphilic molecules: Synthesis in simulated interstellar/precometary ices. *Proc. Nat. Acad. Sci. USA* **98,** 815-819.

Ehrenfreund, P. and Charnley, S. B. (2000). Organic molecules in the interstellar medium, comets, and meteorites: A voyage from dark clouds to the early Earth. Ann. Rev. Astron. Astrophys. **38,** 427-483.

Ferris, J.P., Hill, A. R., Liu, R. and Orgel, L. E. (1996) Synthesis of long prebiotic oligomers on mineral surfaces. *Nature* **381**, 59.

Ferris, J.P. (2002) Montmorillonite catalysis of 30–50mer oligonucleotides: Laboratory demonstration of potential steps in the origin of the RNA world. Orig. Life Evol Biosphere **32**, 311-332.

Fox, S.W. (1973) Origin of the cell: experiments and premises. Naturwissenschaften. **60**, 359-368.

Greenberg, M. and Mendoza-Gomez, C.X. (1993) Interstellar dust evolution: A resevoir of prebiotic molecules. In: M. Greenberg, C.X. Mendoza-Gomez and V. Pironella, (eds.) *The Chemistry of Life's Origins.* Kluwer Academic Publishers, Dordrecht. pp. 1 - 32.

Hargreaves, W.R. and Deamer, D.W. (1978) Liposomes from ionic, single-chain amphiphiles. Biochemistry **17,** 3759-3768.

Hazen, R.M., Filley, T.R. and Goodfriend, G.M.. (2001) Selective adsorption of L- and D-amino acids on calcite: Implications for biochemical homochirality. Proc. Natl. Acad. Sci. USA **98,** 5487-5490.

Holland, H.D. (1984) The Chemical Evolution of the Atmosphere and Oceans. Princeton University Press, Princeton, NJ.

Huber, C. and Wächtershäuser, G. (1997) Activated acetic acid by carbon fixation on (Fe,Ni)S under primordial conditions. *Science* **276**,: 245

Huber, C. and Wächtershäuser, G. (1998) Peptides by activation of amino acids with CO on Ni,Fe surfaces: Implication for the origin of life. *Science* **281,** 670-672.

Johnston, W.K., Unrau, P.J., Lawrence, M.S., Glasner, M.E. and Bartel D.L. (2001) RNA catalyzed RNA polymerization: accurate and general RNA-templated primer extension. Science **292,** 1319-1325.

Kasting, J.F., and Brown, L.L. (1998). The early atmosphere as a source of biogenic compounds. In: A. Brack (ed.) *The Molecular Origins of Life*. Cambridge UK: Cambridge University Press. pp 35-56..

Koch, A.L. (1985) Primeval cells: Possible energy-generating and cell division mechanisms. J. Mol. Evol. **21**,270-277.

Krishnamurthy R. V., Epstein S., Cronin J. R., Pizzarello S. and Yuen G. U. (1992) Isotopic and molecular analyses of hydrocarbons and monocarboxylic acids of the Murchison meteorite. *Geochim. Cosmochim. Acta* **56**, 4045-4058.

Kvenvolden, K.A., Lawless, J.G., Pering, K., Peterson, E., Flores, J., Ponnamperuma, C., Kaplan, I.R. and Moore, C. (1970) Evidence for extraterrestrial amino acids and hydrocarbons in the Murchison meteorite. Nature **28**: 923.

Koch, A.L. and T.M. Schmidt (1991) The first cellular bioenergetic process: Primitive generation of a proton motive force. J. Mol. Evol. **33,** 297-304.

Lawless, J.G. and Yuen G.U. (1979) Quantitation of monocarboxylic acids in the Murchison carbonaceous meteorite. Nature **282,** :396-398.

Love, S. G., and Brownlee, D. E. (1993). A direct measurement of the terrestrial mass accretion rate of cosmic dust. *Science* **262**, 550-553.

Luisi, P.L. 1996. Self-reproduction of micelles and vesicles: Models for the mechanisms of life from the perspective of compartmented chemistry. Adv. Chem. Phys. **92,** 425-438.

Martin, W. and Russell, M.J. (2002) On the origins of cells: A hypothesis for the evolutionary transitions from abiotic geochemistry to chemoautotrophic prokaryotes, and from prokaryotes to nucleated cells. Phil. Trans. R. Soc. London B.

McCollom, T.M., Ritter, G. and Simoneit, B.R.T. (1999) Lipid synthesis under hydrothermal conditions by Fischer-Tropsch-type reactions. Orig. Life Evol. Biosphere **29,** 153-166.

Miller, S. L. (1953) Production of amino acids under possible primitive Earth conditions. Science **117,** 528-529.

Miller, S. L. and Urey, H. C. (1959) Organic compound synthesis on the primitive Earth. *Science* **130,** 245-251.

Monnard, P.-A. and Deamer D. W. (2001) Loading of DMPC based liposomes with nucleotide triphosphates by passive diffusion: A plausible model for nutrient uptake by the protocell. Orig. Life Evol. Biosphere. **31,**147-155.

Monnard, P.-A. and Deamer, D. W. (2002) Models of primitive cellular life: Polymerases and templates in liposomes. Orig. Life Evol. Biosph. (Submitted).

Monnard, P.-A., Apel, C. L., Kanavarioti, A. and. Deamer, D. W. (2002) Influence of ionic solutes on self-assembly and polymerization processes related to early forms of life: Implications for a prebiotic aqueous medium. Astrobiology (In press).

Morowitz, H.J. (1992) Beginnings of Cellular Life.. New Haven CT: Yale University Press.

Nakashima, S., Maruyama, S., Brack, A. and Windley, B.F. (eds.) *Geochemistry and the Origin of Life*. Tokyo: Universal Academy Press. 2001.

Oberholzer, T.,. Wick R.,. Luisi, P. L., and Biebricker C. K. (1995) *Biochem. Biophys. Res. Commun.* **207**, 250.

Oberholzer, T., Albrizio, M. and Luisi, P.L.(1995) Polymerase chain reaction in liposomes. Current Biology **2**, :677-682.

Oparin, A.I., Orlovskii, A.F., Ya. Bukhlaeva, V., and Gladilin, K.L. (1976) Influence of the enzymatic synthesis of polyadenylic acid on a coacervate system. Dokl. Akad. Nauk SSSR. **226**, 972-974.

Pace, N. R. (1991) Origin of life - Facing up to the physical setting. Cell **65**, 531-533.

Parsegian, A. (1969) Energy of an ion crossing a low dielectric membrane: solutions to four relevant electrostatic problems, Nature, **221**, 844-846.

Paula, S., Volkov, A.G., Van Hoek, A.N., Haines, T.H. and Deamer, D.W. (1996) Permeation of protons, potassium ions, and small polar molecules through phospholipid bilayers as a function of membrane thickness. Biophys. J. **70**, 339-348.

Pohorille, A. and Deamer, D. W.. (2002) Artificial cells: Prospects for biotechnology. *Trends Biotech.* **20**, 123.

Rushdi, A.I. and Simoneit, B. (2001) Lipid formation by aqueous Fischer-Tropsch type synthesis over a temperature range of 100 to 400° C. Orig. Life Evol. Biosphere **31,** 103 -0 118.

Sandford, S. A. (1996). The Inventory of Interstellar Materials Available for the Formation of the Solar System. *Meteoritics and Planetary Science* **31**, 449-476.

Segré, S., Deamer, D. W. and Lancet D. (2001) The Lipid World. Orig Life Evol. Biosphere **31**, 119-145.

Shew, R. and Deamer, D.W. (1983) A novel method for encapsulating macromolecules in liposomes.Biochim. Biophys. Acta. **816,** 1-8.

Shimoyama,-A., Naroka, H.,. Komiya, M. and Harada, K. (1989) *Geochem. J.* **23**, 181.

Sowerby, S.J., Petersen, G.B. and Holm, N. G. (2002). Primordial coding of amino acids by adsorbed purine bases. Orig. Life Evol. Biosphere **32,** 35-46.

Stribling, R. and Miller, S.L. (1987) Energy yields for hydrogen cyanide and formaldehyde synthesis: The HCN and amino acid concentration in the primitive ocean. Orig. Life Evol. Biosphere **17**, 261-73.

Szostak, J.W., Bartel, D.P. and Luisi, P.L. (2001) Synthesizing life. *Nature* **409**, 387-390.

Wächtershäuser, G. (1988a) Before enzymes and templates: Theory of surface metabolism. Microbiol. Rev. **52**, 452-484.

Wächtershäuser, G. (1988b) Pyrite formation, the first energy source for life: A hypothesis. Syst. Appl. Microbiol. **10**, 207-210.

Walde, P., Wick, R., Fresta, M., Mangone, A. and Luisi, P.L. (1994a) Autopoietic self-reproduction of fatty acid vesicles. J. Am. Chem. Soc. **116**, 11649-11654.

Walde, P., Goto, A., Monnard, P.-A., Wessicken, M. and Luisi, P.L.(1994b) Oparin's reactions revisited: Enzymatic synthesis of poly (adenylic acid) in micelles and self-reproducing vesicles. J. Am. Chem Soc. **116,** 7541-7547.

Wilson, C. and Szostak, J.W. (1994) In vitro evolution of a self-alkylating ribozyme. Nature **374**, 777-782.

Biodata of **Daniel Segrè** author (with co-author D. Lancet) of ***"Theoretical and computational approaches to the study of the origin of Life"***

Dr. Daniel Segrè has completed his studies towards the Laurea (M.Sc.) in theoretical physics at the University of Trieste, Italy, in 1994. He received his Ph.D. in biology at the Department of Molecular Genetics of the Weizmann Institute of Science, Israel in 2001. His Ph.D. thesis work, proposing new paradigms and mathematical models for the Origin of Life was published in journals, highlighted in the news and presented at several International conferences. Dr. Segrè has received training in Complex Systems at the Santa Fe Institute summer school, and has been actively involved in introducing courses on Origin of Life at the Weizmann Institute. Since 2001 he is a postdoctoral research fellow at the Department of Genetics, Harvard Medical School, working on whole-cell models of metabolic networks. Dr. Segrè's research interests include theoretical population genetics and the evolution of human language.

E-mail: **dsegre@genetics.med.harvard.edu**

Dr. Doron Lancet is a Professor at the Weizmann Institute of Science, Israel, and currently is the head of the Crown Human Genome Center at the Weizmann Institute of Science and of Israel's National Laboratory for Genome Infrastructure. He obtained his Ph.D. degree in 1978 at the Weizmann Institute of Science. Dr. Lancet spent his postdoctoral training at Harvard University (on transplantation proteins) and at Yale University (on the biology of the sense of smell). He headed Weizmann's Department of Membrane Research and Biophysics. He pioneered genome research in Israel; His research interests span human genome research, the genetic basis sense of smell, bioinformatics (the merger of biology and computing) and the evolution of life on earth. Professor Lancet team has pioneered the studies of the molecular basis of olfaction, discovering how smell signals are amplified and terminated. They also developed GeneCards, a world-known compendium of human genes. Professor Lancet teaches at Weizmann Institute graduate courses on Bioinformatics and computational Genomics, as well as on molecular recognition. He received the Hestrin Prize of the Israel Biochemical Society, the first Takasago Award of the American Association for Chemoreception Science, and the USA R.H. Wright Award in Olfactory Research.

E-mail: **Doron.Lancet@weizmann.ac.il**

Daniel Segrè

Doron Lancet

J. Seckbach (ed.), Origins, 91–120.

THEORETICAL AND COMPUTATIONAL APPROACHES TO THE STUDY OF THE ORIGIN OF LIFE

D. SEGRÈ[1] AND D. LANCET[2]
[1]*Lipper Center for Computational Genetics and Department of Genetics Harvard Medical School, 200 Longwood Ave., Boston, MA 02115.*
[2]*Department of Molecular Genetics and The Crown Genome Center The Weizmann Institute of Science, 76100 Rehovot, Israel.*

1. Prebiotic synthesis and molecular diversity

1.1. THE SUBSTRATE OF EARLY EVOLUTION

Investigating the evolutionary path that led to present-day cells is a little like tracing a maze, where the entrance and exit are known, but much work remains to find the paths that connect them. For the origin of life, the "entrance" is the multifaceted environment of a water- and carbon-rich planet and the "exit" is a protocellular structure subject to Darwinian evolution.

Whether terrestrial life started on our planet or elsewhere in the Universe is a recurrent subject of debate, often nourished by proposals of Panspermia scenarios (Crick, 1981; McKay, 1997), and recently inspired by the finding of candidate relics of martian life forms (Kerr, 1996; McKay et al., 1996). An extraterrestrial origin cannot be absolutely discarded using the present knowledge, but there are two good reasons for setting aside this debate: (i) there is currently no solid evidence of extraterrestrial life, and (ii) moving the problem elsewhere still requires solving it. While exciting recent analyses have certified the presence of planetary systems other than the solar one (Lissauer et al., 2000), and future explorations might one day give us the opportunity to analyze frozen relics of extraterrestrial life, most origin of life scenarios assume that the cradle of life-as we-know-it has been planet Earth (de Duve, 1995; Lahav, 1999; Oparin, 1957).

Although the details of prebiotic Earth conditions and of the nature of the earliest forms of life are probably lost forever, a lot is known about the general features primordial Earth. For example, a time window for the transition from inanimate to animate matter has been estimated using geophysical and astronomical methods. The lower bound of this window, 3.9 billion years ago, corresponds to the end of a period of intense meteorite bombardment (Chyba and Sagan, 1992), as testified, for example, by the analysis of the Moon's craters (Maher and Stevenson, 1988). For the upper bound we can rely on geological records, such as stromatolites and microscopic fossils, which provide evidence that 3.8 billion years ago life was already present on Earth (Mojzsis et al., 1996; Mojzsis et al., 1998; Schopf et al., 1965).

Since Oparin's first detailed description of how the emergence of life could be in principle explained on a molecular basis (Oparin, 1957), several experiments have been performed for evaluating potential paths towards the spontaneous emergence of a primordial cell. This has been mostly guided by the chemistry of present-day organisms, and in particular by the fundamental role of nucleic acids (Gesteland et al., 1999; Gilbert, 1986; Orgel, 1998) and proteins (Fox, 1991; Lee et al., 1996). This concept is inspired by an extreme of the Continuity Principle, stating that every stage in evolution is connected continuously to the previous ones (Morowitz, 1992; Orgel, 1968). It is assumed that very early life contained the same biopolymers as present day life (cf. Orgel, 1998).

At the other extreme, inorganic clay minerals were proposed to be the first substrate for life-like processes (Cairns-Smith, 1982). In the model proposed by Graham Cairns-Smith, imperfections in a growing crystal are the first form of inheritable information, and organic life is proposed to have arisen later, through a genetic takeover (Cairns-Smith, 1982). Clays are minerals made of micron-size particles, composed of complex crystals based on silicon, aluminum, iron and other inorganic compounds. The fluidity and the complexity of collections of these particles in water inspired the analogy with organic macromolecules. The clay mineral model, however, is considered a less likely predecessor of the cell, because it constitutes an extreme deviation form the Continuity Principle (Fry, 2000; Morowitz, 1992).

An intermediate approach may be envisaged, in which continuity is used to indicate that life started with carbon-based molecules, but without any assumption about the detailed structure of the primordial organic compounds. As opposed to the view invoking the initial abundant presence of present-day building blocks, this scenario envisages a selection process that gradually reduces the diversity of molecular building blocks, in analogy to the screening of large random libraries in random chemistry experiments (Segrè and Lancet, 1999).

1.2. SOURCES OF MOLECULAR DIVERSITY

The formation of the chemical elements necessary for the emergence of life started with the Big Bang (Macià et al., 1997; Orò, 2000; Sciama, 1971). Light nuclei, such as H, D, ^{3}He and ^{4}He, are believed to have formed during the early dense stages of the expanding Universe (Arnett, 1996). Additional synthesis of He and of heavier elements by thermonuclear burning followed the formation of stars and galaxies. Newly synthesized elements, from C to U, were ejected by exploding supernovae, and eventually incorporated into other stars or into planets (Arnett and Bazan, 1997; Trimble, 1997). The nuclear reaction responsible for the formation of the nuclide ^{12}C is the condensation of three alpha particles (Oberhummer et al., 2000). This reaction, as well as the formation of O in large amount, S, and small fractions of P (Macià et al., 1997) is thought to happen especially in carbon stars (Doty and Leung, 1998; Orò, 2000). Upon migrating to the circumstellar and cooler regions of the star, at temperatures of a few thousands degrees, the biogenic elements (H, C, O, N, S, P) produce more complex molecules (Doty and Leung, 1998), through ordinary chemical reactions. The newly formed compounds include diatomic and triatomic species such as C_2, CN, CO, CH, NH, OH and H_2O, as observed also in the atmosphere of the Sun (Orò, 2000).

Anders (Anders, 1989) and Chyba and Sagan (Chyba and Sagan, 1992) have found evidence that interplanetary dust particles (IDP) were the most abundant source of extraterrestrial organic carbon in the late accretion phase of the early Earth. Comets and carbonaceous meteorites would contribute smaller amounts. The total delivery over a period of 100 million years of the late Hadean - early Archean eras is estimated to be on the order of 10^{16}-10^{18} kg (Chyba and Sagan, 1992). For comparison, the total organic carbon in the present biosphere is $6 \cdot 10^{14}$ kg, i.e. 10 to 1000 times smaller. Hence, organic material delivered as extraterrestrial infall was likely to have been a significant contribution to the organic inventory of the early Earth environment. In addition to extratrerrestrial delivery, several potential prebiotic reactions that could produce simple organic compounds on early Earth have been proposed (de Graaf et al., 1995; Huber and Wachtershauser, 1997; Miller and Urey, 1959; Schlesinger and Miller, 1983; Wachtershauser, 1988). Energy sources available to drive such reactions range from volcanoes and hydrothermal vents to lightning discharges, solar photochemistry and pyrite-dependent reduction. It should be stressed that in many cases, in addition to water-soluble organic compounds that are easily analyzed, insoluble residue (e.g. "tar" or "tholins" (Bernstein et al., 1999; McDonald et al., 1996) are reported in prebiotic synthesis simulations (Miller and Urey, 1959; Schlesinger and Miller, 1983). These are likely to contain sparingly soluble hydrophobic and amphiphilic compounds of relatively large molecular mass.

Why should one be interested in this large diversity of molecular shapes formed abiotically on early Earth or in outer space? Many scholars have been screening potential prebiotic mixtures in search for amino acids (Miller, 1953; Miyakawa et al., 2002), nucleosides (Ponnamperuma et al., 1963a; Robertson and Miller, 1995), and other precursors of present-day cellular life (McCollom et al., 1999; Ponnamperuma et al., 1963b; Reid and Orgel, 1967; Sagan, 1963). The success of such searches is sometimes mistakenly appraised as the solution to the problem of the origin of life. Indeed, the conclusion that a prebiotic mixture contains some of the building blocks of a present-day living cell constitutes a fundamental link between the inanimate and the animate worlds (Morowitz, 1992). Yet, a prominent alternative view is that one should take into account the whole primordial repertoire of organic compounds, and ask how its physical and chemical properties may lead to life-like characteristics (Fraser and Folsome, 1975; Lifson, 1996; Morowitz, 1992). This question, of how a "wild" prebiotic mixture (primordial soup) could self-organize to give rise to a self-reproducing and evolving system remains largely unanswered (Dyson, 1999).

Mathematical and computational models aimed at understanding these early stages of biological organization do not generally depend on a detailed knowledge of the composition of a primordial soup (Dyson, 1999; Kauffman, 1993). Rather, they make use of global measures, like the diversity of molecular species, or statistical parameters, such as the probability that a given molecules catalyzes the synthesis of any other molecule in a set. The question of prebiotic molecular diversity has been addressed explicitely by Morowitz (Morowitz, 1992), who reported the number of isomeric alkanes as a function of carbon number. The numbers fit an exponential law

$$N_G = Ke^{\alpha C} \qquad (1)$$

where N_G indicates the global number of different chemical species possible, C is the number of carbon atoms, and $K=6.7 \cdot 10^{-3}$ and $\alpha=0.915$ are parameters characteristic of the specific class of compounds. For 10 carbons the number of isomers is 75. A double number of carbons yields more than 360,000 compounds. It should be stressed that this calculation is relative to carbon and hydrogen atoms only. Hence, the above formula could be considered as a possible lower bound for molecular diversity. This kind of calculation, in addition to being useful for explicit mathematical modeling of prebiotic phenomena, should remind us of how narrow the chemical spectrum of present-day cellular life is with respect to the variegated universe of all possible chemical structures. Living cells typically contain less than thousand different low molecular weight "monomeric" compounds.

2. Chemical evolution

2.1. DARWINIAN EVOLUTION AT THE MOLECULAR LEVEL

The three classical elements of Darwin's theory, self-reproduction, mutation and selection, are widely accepted as the fundamental ingredients of evolution at the level of organisms. As stated by Li and Graur (Li and Graur, 1991), natural selection is the "differential reproduction of genetically distinct individuals or genotypes within a population". This definition could be extrapolated to early forms of life with primitive genomes, or even to freely floating lone genes capable of copying themselves (Eigen, 1971). But such definitions appear inapplicable to systems that do not possess genetic material. Any model that assumes early protocells that constitute ensembles of molecules without genetic machinery, would appear unable to explain how an early Darwinian process could take place. This puzzle is one facet of the Chicken-Egg paradox, summarized as follows: (i) Biopolymer-based self-reproducing systems must have evolved from simpler non-self-reproducing systems, (ii) Evolution can act only upon systems based on self-reproducing bio-polymers. Apparently, when dealing with primordial life, we should consider relinquishing one of these propositions. Relaxing proposition (i) implies that the first self-reproducing system emerged without evolving, i.e. that life originated in a one-time "lucky" jump. Such a view is sometimes accepted as an axiom of the inexplicability of life's origin (Yockey, 1992). Alternatively, self-organization processes have been suggested to potentially happen at the molecular level in absence of selective pressure (Horowitz, 1945), perhaps in analogy to random genetic drift (Dyson, 1999). Often the argument is presented that despite the extremely low probability for the chance sudden emergence of a self-reproducing system, the long available time and the huge set of possible geophysical settings could make such an event reasonably likely. An alternative view involves relaxing proposition (ii). This implies that a kind of natural selection may act on systems that do not contain self-reproducing biopolymers, perhaps on systems totally lacking genes and enzymes (Woolfson, 2000; Segrè et al., 2000). Is this possible? Could there be self-reproduction before genes? What would be the meaning of natural selection in a population of geneless protocellular systems?

2.2. INHERITANCE OF WHAT?

In modern living organisms, as the central dogma teaches us, biological information is contained in the genome (Crick, 1981). The flow of information is basically unidirectional, starting from DNA and ending with proteins. It is generally assumed that inheritance has to do with the copying of the message contained in the genome (Dawkins, 1996; Muller, 1966; Orgel, 1998). Indeed, this mechanism is so important, that the mutation of a single letter in this message might critically determine the fate of the cell receiving the modified sequence. When we talk about biological inheritance, however, we are not considering abstract systems that send and receive symbols like in Shannon's theory (Yockey, 2000). A DNA sequence alone has no meaning. Biological inheritance involves the self-reproduction of a complex molecular structure - the cell. The fact that one of its components, DNA, possesses an alphabet based string-like structure does not imply that information and inheritance should be ascribed only to sequences embodied by nucleic acid polymers. During cell division, lipids, proteins, whole organelles, cytosolic molecules and many other components are being inherited together with DNA. What is then biological information? What is inherited during a self-reproduction process?

Along the history of life, from an inanimate mixture of organic compounds to a sophisticated reproducing cell, there must have been intermediate stages involving proto-organisms capable of undergoing evolution without relying on the complex genetic machinery present today (Morowitz et al., 1988; Shapiro, 2000). It is conceivable that biopolymers were initially much shorter, and yet, an evolutionary process took place. It is necessary to envisage a simpler form of inheritance, which does not involve genes, coding and translation. This form of inheritance is the transmission of a peculiar chemical composition, as pioneered by Oparin (Oparin, 1957), and elaborated by many others (Dyson, 1982; Kauffman, 1986; Morowitz, 1967; Morowitz et al., 1988; New and Pohorille, 2000; Woolfson, 2000; Segrè et al., 2000). This is embodied in the "garbage bags" scenario (Dyson, 1999) where assortments of diverse simple molecules are capable of growing and splitting inaccurately. And interestingly, it is essentially true even today, where molecular compositions inherited during mitosis contain highly evolved, specialized and complex biopolymers, and the fate of each molecule upon cell division is carefully programmed. This compositional inheritance view is, however, still debated by many investigators. One of the reasons for this is the relatively small number of proposed formal analyses of such a process.

2.3. MOLECULAR SELF-REPLICATION

One of the fundamental characteristics of a living system is the capacity to make more of itself. The simplest chemical system bearing such attribute is an autocatalytic molecule, i.e. a molecule capable of accelerating the rate of its own production from precursors. (Lifson, 1996; Orgel, 1992; Szathmary and Maynard Smith, 1997). Autocatalytic molecules were experimentally designed in the laboratory, and it was shown that molecular self-replication in vitro is indeed possible. Examples involve modified nucleotides (Li and Nicolaou, 1994; Rebek, 1994; Tjivikua et al., 1990), as well as short peptides (Lee et al., 1996). One of the most attractive forms of autocatalyisis is the one involving biopolymers, as explored by Orgel's through the template-directed synthesis of oligonucleotide analogues (Orgel, 1992; Schwartz and Orgel, 1985; Sievers and von-Kiedrowski, 1994). Despite the increasing success in the design of autocatalytic molecules, it should be mentioned that no relevant example has been reported within any life form. Neither DNA nor RNA is truly autocatalytic, as they require the aid of specialized enzymes. How likely is it that nature could randomly hit upon a contrived self-replicating molecule similar to molecules whose synthesis presently requires the careful engineering of experienced organic chemists? This is a question that would require considerable future pursuit.

2.4. FORMAL ANALYSIS OF SELF-REPLICATING MOLECULES

Setting aside the plausibility of self-replicating molecules, it is legitimate to assume their existence and investigate their kinetic behavior. This approach brought to the initiation of a formal description of molecular systems that may undergo evolutionary processes. Such formalism is analogous to the evolutionary dynamics of more highly organized living systems. The most prominent mathematical model was developed by Manfred Eigen and Peter Schuster in the 70's (Eigen, 1971; Eigen and Schuster, 1979). This is analogous to the classical formalisms used in population biology for describing the increase and regulation of animal populations (Wilson and Bossert, 1971), based on a Malthusian growth law (Szathmary and Maynard Smith, 1997), as summarized by the logistic equation

$$\mathrm{d}N/\mathrm{d}t = aN - bN^2 \qquad (2)$$

In Eigen's scenario, a set of RNA polymers is enclosed in a flow reactor. Each such polymer is assumed to catalyze its own replication from constantly supplied activated monomers, and can therefore be referred to as a self-replicating information carrier, or replicator (Szathmary, 1999; Szathmary and Maynard Smith, 1997). The fidelity of a replicator is not always absolute: errors during copying may occur (Eigen and Schuster, 1979; Swetina and Schuster, 1982), whereby a polymer may catalyze the production of a variant polymer, rather than its own. Such replication errors are analogous to DNA sequence mutations.

In one of the possible formulations (Constant Population (CP)) the equation for the concentration of replicator i can be written as follows (Kuppers, 1983)

$$\frac{dx_i}{dt} = (W_i - E(t))x_i + \sum_{j \neq i} \psi_{ij} x_j \qquad (3)$$

where W_i is the autocatalytic rate enhancement (or selection value) of replicator i, $E(t)$ is the average excess productivity, which keeps the overall concentration constant (in fulfillment of the CP constraint), and ψ_{ij} is the probability that replicator i will erroneously give rise to replicator j. An overall matrix of kinetic rates may be visualized as composed of the autocatalysis W_i terms (diagonal values) and the error rates ψ_{ij} (off-diagonal values). Eigen shows that a steady state is reached (Eigen, 1971; Eigen and Schuster, 1979; Kuppers, 1983), for which only a specific linear combinations of replicators, called a quasi-species, persists. In a transformed linear system of equations, the quasi-species corresponds to the eigenvector associated with the largest eigenvalue of the linear operator (Jain and Krishna, 1998). The quasi-species is hence associated with a distribution of RNA sequences with maximal replication rate. The larger the ψ_{ij} values with respect to W_i, the broader the distribution of different sequences contributing to the quasi-species.

2.5. ERROR CATASTROPHES

An important aspect of Eigen's analysis is the characterization of a threshold value for the rate of replication mistakes, which critically determines the information transfer property of the quasi-species (Alves and Fontanari, 1998; Eigen, 2000). Small error rates do not affect the replicating polymers in a dramatic way: the distribution of sequences at steady state is centered around a master polymer (i.e. the species with highest autocatalytic rate), whose information is faithfully propagated through the generations. When a critical value (the error threshold) is reached, however, the information of the master sequence is lost, and the mixture of self-replicating polymers becomes randomized (Swetina and Schuster, 1982). This occurs in a way analogous to a phase transition, referred to as the error catastrophe (Eigen, 2000). It should be mentioned that while a slow mutation rate prevents the loss of information, a fast mutation rate can increase the capacity of a system to explore the sequence space and optimize its fitness. A balance between these two opposite tendencies is what determines, according to Eigen, an optimal mutation rate (Eigen and Schuster, 1979). The error threshold analysis imposes a limit to the length of sequences that can be replicated with sufficient accuracy. Structures with a higher level of organization, the hypercycles, involving protein enzymes coded by the nucleic acid polymers, are shown to be necessary in order to afford longer polymers with the same accuracy of replication (Eigen, 2000).

2.6. RNA WORLD

An important origin of life scenario involving molecular replication is the RNA-world (Gesteland et al., 1999; Gilbert, 1986; Joyce, 1991). Following the discovery that RNA can perform catalyitc activities (Cech, 1986), it was proposed that the early evolution of life may have undergone through the following stages: (i) RNA polymers could act as primitive genes and enzymes (ribozymes, cf. Ellington et al., 1997; Lohse and Szostak, 1996), and propagate information through self-replication (Cech, 1986; Szostak, 1992), then (ii) RNA acquired the capability of synthesizing coded polypeptides (Lohse and Szostak, 1996), which would later specialize into becoming the major biocatalyst, and (iii) DNA took over for the role of genetic information carrier (Robertson and Miller, 1995). Ribozymes were invoked as possible indications of RNA's prebiotic role, since they persist today in tRNAs and within many cofactors, as well as in the active portion of the ribosome (Wimberly et al., 2000).

The interest in the potential role of RNA as an omnipotent biological molecule, generated major efforts in the last fifteen years (Gesteland et al., 1999), towards demonstrating whether indeed all the major cellular catalytic functions could be in principle performed by RNA. It was recently discovered, for example, that RNA may catalyze the synthesis of further nucleic acids building blocks (Unrau and Bartel, 1998). A comprehensive list of catalytic activities displayed by ribozymes, generated through *in vitro* RNA evolution (Ellington et al., 1997; Wright and Joyce, 1997), has been reported (Bartel and Unrau, 1999). In parallel to the interest in the historical path that has led to life, the RNA-world body of knowledge is often directed towards the attempt of engineering an artificial self-sustaining form of cellular life (Szostak et al., 2001). An engineered experimental realization of an RNA-based cell, however, would be unlikely to resemble the real protocells that once populated our planet. The spontaneous prebiotic emergence of a set of autocatalytic RNA polymers, in addition, encounters serious difficulties, mostly due to the implausibility of RNA synthesis and stability under prebiotic conditions (Bartel and Unrau, 1999; Niesert et al., 1981; Shapiro, 1984; Shapiro, 2000; Yarus, 1999).

2.7. ENSEMBLE REPRODUCTION

Despite the fact that the model of Eigen was constructed specifically for self-replicating nucleic acid polymers, the idea of describing molecular evolution with kinetic equations is quite general, and can be applied to cases of mutual catalysis rather than autocatalysis (Bagley and Farmer, 1991; Farmer et al., 1986; Segrè et al, 1998a; Segrè et al., 2000). The interpretation of variables and parameters in the two cases, though, is radically different. In Eigen's model the emphasis is on the autocatalytic replicating biopolymers, i.e. on the diagonal term of the matrix of interactions. The molecular interactions utilized in the compositional inheritance model (Bagley and Farmer, 1991; Segrè et al., 2000), on the other hand, do not require the ad hoc assumption of a self-replication capacity for individual molecules. The kinetic equations in this case are for catalyzed growth of entire molecular ensembles, due to the catalyzed aggregation and/or the formation of small organic compounds within non-covalent assemblies. The individual molecular species are not replicators, nor do they carry information. Both self-reproduction and inheritance are obtained as the emergent property of a collection of such molecules (Kauffman, 1986; Dyson, 1985). Interestingly, phenomena that resemble the quasi-species error threshold behavior can also observed for collective replication (Kauffman, 1993; Segrè et al., 2001a).

2.8. KAUFFMAN'S MODEL

This model (Kauffman, 1986) was one of the first attempts to formalize the idea that mutually catalytic metabolism-like networks of interacting molecules. It may lead to self-reproduction even if none of the individual molecules is a replicator (cf. McMullin, 1995; Oparin and Gladilin, 1980). Kauffman and colleagues demonstrated that a set of random polymers might undergo collective self-reproduction, provided that it entails sufficient diversity of molecular species (Farmer et al., 1986; Kauffman, 1986; Stadler et al., 1993). In Kauffman's words, this is an attempt to show that "self-reproduction and homeostasis, basic features of organisms, are natural collective expressions of polymer chemistry". According to this view, self-reproduction of a mixture of molecular species, exemplified by, but not restricted to oligomers, is caused by growth of the concentration of all the molecular species included. This is due to the mutual catalysis, or mutual rate acceleration, that each polymer exerts on the synthesis of others. Based on a graph theory derivation, Kauffman argues that any sufficiently complex set of molecules may attain the emergent collective property of catalytic closure, whereby every member of the set has at least one of the possible last steps in its formation catalyzed by some member of the set.

In the mathematical formulation of this model, monomers of different kinds can form polymers with length up to a maximal value M. A constant probability parameter P describes the chance that a molecular species in the set will catalyze any reaction leading to the formation of another species. As one considers increasing values of M, the number of possible polymer types goes up as an exponent of M, but the number of synthesis and cleavage reactions grows even faster. The ratio of reaction count to polymer types is shown then to increase linearly with M. When M is large enough, catalytic closure is realized and therefore each polymer in the set is efficiently synthesized. The outcome is that the whole set produces more of itself, despite the fact that none of the components of the set is assumed to be an autocatalyst. This concept of self-organizing metabolism-like network has been explored in different forms, including the study of systems of differential equations for the kinetics of polymers formation (Bagley and Farmer, 1991; Bagley et al., 1991) and the construction of an abstract artificial chemistry based on λ calculus (Fontana and Buss, 1994).

Kauffman's complexity-dependent closure reflects a widely diffused view that the emergence of life-like properties should resemble a phase transition, i.e. that the tuning of a fundamental parameter may determine in a critical way whether a system is in a state of randomness, or has undergone a sudden self-organization (Dyson, 1999). In the formalism of Eigen, this phase transition corresponds to the error threshold behavior described above. In Kauffman's model, the tuning parameter is the number of different mutually catalytic polymers, and the ensuing self-organization property is represented by catalyitc closure (Kauffman, 1986; Kauffman, 1993).

The self-reproduction behavior ensuing from a network of mutually catalytic interactions has been experimentally demonstrated using complementary nucleotide-based oligomers (Sievers and von-Kiedrowski, 1994), as well as an enhanced version of Ghadiri's self-replicating peptides (Kauffman, 1996; Lee et al., 1997; Lee et al., 1997). Both these experiments involve at most dual catalysis. More complex networks of mutually catalytic molecules are still missing experimental verification.

2.9. DYSON'S MODEL

While Kauffman champions the role of increasing diversity in leading to self-replication, Dyson's approach expresses the importance of a time-dependent process that leads to homeostasis. Both Dyson's and Kauffman's models describe possible transitions between a poorly organized catalytic network and a well connected system of catalytic interactions, capable of increasing the global rate of synthesis of the constituent molecules from externally available "food" material.

Dyson assumes that molecules enclosed in a bounded microenvironment are composed of monomers that exist in two states: catalytically active and catalytically inactive. An autocatalytic phenomenon is assumed, whereby active molecules will convert others into a similarly privileged state. Since Dyson's model includes a backward reaction, in which an active species may be inactivated, the exponential explosion typical of autocatalysis is prevented, and steady states are reached under certain conditions. Dyson shows that given the probability 1/a of specific catalytic activation (a is roughly the number of monomer types) and if one assumes a single mean-field catalytic enhancement factor b, then the efficacy of an existing molecular population in promoting the formation of new catalysts within it may be computed as a function of x, the fraction of active catalysts. For some combinations of the parameters a and b, there are three steady states. An assembly can move through single substitution mutations from the disordered steady state (α) to the ordered one (γ), by passing through the unstable steady state (β). This process leads to the emergence of an organized metabolism. The average time for this transition to occur can be expressed as a function of a, b, and of the total number of monomers.

The simplifications implied by Dyson's mean field approach are limiting the applicability of such model to reality. Nevertheless, the increasing available computational power may allow the implementation of simulations where explicit interactions between chemical species could be reproduced, as foreseen by Dyson himself. Dyson's point γ would then be substituted by a whole landscape of possible organized states, with different levels of complexity, stability and effectiveness in metabolizing external resources. The presence of multiple steady states may open also the way to a competition between different mutant assemblies, allowing evolutionary processes to occur. Interestingly, an experimental example of non-equilibrium bistability resembling Dyson's transition has been reported for mixed surfactant systems (Buhse et al., 1997)

It should be stressed that while in the first version of his model (Dyson, 1982; Dyson, 1985) Dyson depicted the monomers as linked into polymers that float in a vesicle-enclosed aqueous solution, in the 1999 edition (Dyson, 1999) the same membrane is assumed to enclose unbound monomers. Moreover, in the novel scenario, the catalyzed chemical reactions of activation and inactivation are identified with adsorption and desorption at sites of the vesicle rather than with changes in their position within polymers. This change of view reflects the concept of catalyzed amphiphilic monomers aggregation, which characterizes the Amphiphile-GARD model (Segrè and Lancet, 1998; Segrè and Lancet, 2000; Segrè et al., 2000).

2.10. THE MOROWITZ BOUNDARY

In models of collective self-reproduction, like the ones proposed by Dyson and Kauffman, the production of similar progeny is not evident as in the case of single molecule replication. If self-reproduction of a noncovalent assembly is to be obtained through a process of ongoing accretion or synthesis of molecular species through mutual catalysis, the generation of two or more entities out of a single one requires an additional step of physical separation. This has been classically described as a process of budding or splitting (Fox, 1976; Koch, 1985; Luisi et al., 1999; Morowitz, 1992; Morowitz et al., 1988). Under the assumption of randomly disposed molecules within an assembly, splitting may critically determine the fate of the progeny, i.e. whether the mixture of chemicals inherited by each daughter assembly may keep the catalytic network intact and guarantee continuing homeostasis. Statistical calculations may ascertain whether enough of each component is transferred to the progeny upon division. Morowitz shows how constraints on the transmissibility of information through direct inheritance of a molecular composition are related to the size of the assembly and the diversity of its molecular species (Morowitz, 1967). It is demonstrated that if an assembly contains m molecular types (the complexity of the assembly, or its intricacy (Smith and Morowitz, 1982)), and if each is present with an average copy number 2r per assembly, then P_d, the probability that all molecular types are represented in the progeny in at least one copy, is given by $P_d = (1 - e^{-r})\cdot m$ Assuming that all the components are catalytically essential, P_d represents the likelyhood of a successful functional division. If, for example, m=4 and 2r=32, then the probability of faithful division P_d is very near to 1 (only one in a million chance for a "wrong" division). However, if the assembly was much smaller, say with 2r=4, then the splitting success probability would go down to about $P_d = 0.55$. For a given average copy number, it is thus possible to define a "Morowitz boundary", corresponding to the assembly size that yields $P_d = 0.5$.

This concept may be crucial for determining specific ranges of parameters allowing the spontaneous emergence of protocellular structures capable of faithful collective self-reproduction (Segrè et al. 2000; Segrè et al., 2001a). In a description of membrane-enclosed assortments of genes, Szathmary proposed the "stochastic corrector model" for inheritance through protocell division, which is based on the same principle as above, but refers to a later evolutionary stage (Szathmary and Maynard Smith, 1997).

2.11. AUTOPOIESIS

An important demonstration that self-reproduction can be the property of a molecular assembly derives from the experiments done by Pier Luigi Luisi (Bachmann et al., 1992; Luisi et al., 1999). Inspired by the concept of autopoiesis (from the Greek 'auto' (self) and 'poiesis' (formation)), first proposed by Maturana and Varela (Luisi, 1997), Luisi demonstrated that micelles and vesicles can catalyze the production of their own constituents, and hence display a growth behavior akin to self-reproduction. Micelles and vesicles are self-assembled systems containing 10^3-10^6 amphiphilic monomers and forming thermodynamically stable physical structures through hydrophobic forces (Deamer, 1986; Morowitz et al., 1988; Tanford, 1978). The universal presence of structures like these in present day cells may indicate that lipid-like molecules played a major role during early evolution (Luisi et al., 1999).

In some of Luisi's experiments, long fatty acid chains were used as an amphiphile. A precursor of the surfactant (e.g. fatty acids anhydrides or esters) was initially deposited on a water solution, forming oil droplets floating on the surface. In a classical example, the hydrolysis of the ethylcaprylate precursors is responsible for the formation of caprylic acid vesicles (Bachmann et al., 1992). The hydrolysis is initially very slow, but its rate increases as vesicles are being formed. In other words, formed vesicles catalyze the production of further vesicle-forming components, in an autocatalytic reaction. Interestingly, it is the whole vesicle that acts as a catalyst rather than a single molecule, similarly to the cases of collective replication proposed by Kauffman and Dyson. Here too, single monomers are not replicators.

One of the drawbacks of Luisi's paradigm is the lack of diversity and therefore of evolution. Self-replicating micelles or vesicles would always look the same, unless different kinds of lipid monomers were introduced. Even if examples of mixed autocatalytic vesicles were designed (Bonaccio et al., 1994), the concept of compositional inheritance was not seriously taken into account, and the incorporation of self-replicating bio-polymers within self-reproducing vesicles was pursued as a paradigm for a minimal protocell (Oberholzer et al., 1995). In contrast, the recently proposed Lipid World hypothesis (Segrè et al., 2001) expands the autopoiesis concept to the realm of complex prebiotic mixtures, and envisages lipid vesicles and micelles as possible carriers of compositional information.

2.12. A WORKING DEFINITION OF LIFE

Several formal definitions of the living state have been proposed (for a broad collection, see Lahav, 1999; Rizzotti, 1996), but no general agreement exists. This may be related to the fact that phenomenological descriptions of life's characteristics often refer to debated scenarios for the emergence of life. A good definition of life should describe a set that contains all the living systems we know, as well as any form of life possibly different from the one we know. This implies trying to distill, from our understanding of life, all those features that seem to be characteristic of the processes observed, without restrictions to specific molecular substrates. The latter may reflect "historical accidents" that happened along evolution. Without further reasoning on this matter, which has complex philosophical implications, one can formulate a working definition of life that is related to the problems under study. This will be helpful as a reference, since descriptions of life-like characteristics appear frequently in the models for prebiotic evolution presented here.

Life may be defined as an open chemical system far from thermodynamic equilibrium, whose linked reactions are organized in such a way that homeostasis and self-reproduction ensue. In more detail, the following features could be listed:

a) A living system is an open chemical system far from thermodynamic equilibrium. This involves the capacity of exchanging matter and energy with the surrounding environment. For a living system to emerge and perpetuate itself, an external energy source and an external energy sink must exist (Morowitz, 1979).

b) Each molecular species in the system (possibly an organic compound) is linked to other molecular species, forming a network of chemical transformations. The same molecular species may be involved in multiple functions within the reaction network and may be at the same time a substrate, a product and a catalyst for different reactions (Kauffman, 1986).

c) The network of chemical reactions and catalytic interactions displays a certain degree of organization. This may involve the presence of cycles or other ordered topologies (Morowitz, 1979).

d) From a functional perspective this organization implies two fundamental properties: the first is Homeostasis (Dyson, 1985), i.e. the capacity of the system to sustain itself, and maintain its internal order in spite of moderate fluctuations of environmental conditions. The second, intimately connected to the first one, is self-reproduction, i.e. the possibility to replenish molecular species that are diluted due to a growth in the total size and mass of the system. Upon a scission process, this maintenance of molecular concentrations upon growth ultimately leads to the duplication of the original system.

e) The self-reproduction process described above is, importantly, an imperfect one. The system is therefore prone to acquire mutation-like changes and may therefore evolve according to the Darwinian paradigm.

3. Random chemistry

What can be said about the interactions among organic molecules in a primordial mixture, without entering into details about their structures? Models for ensemble collective self-reproduction often utilize statistical assumptions, e.g. Kauffman's fixed probability P of catalysis between to molecules in a random set, or Dyson's mean field approximation for catalytic activity within a monomers' assembly. A deeper insight about the statistical properties of noncovalent interactions in large molecular ensembles re-quires a detour into the field of random and combinatorial chemistry.

3.1. PROBABILISTIC RECOGNITION

Many biological phenomena involve probabilistic recognition between randomly encountered molecules. This is especially manifest in biological repertoires, which have evolved to contain the molecular diversity necessary for binding any randomly encountered ligand with a functionally sufficient affinity. The immune repertoires, immunoglobulins and T-cell receptors, provide the most well known examples for systems displaying probability-based interactions. Other examples include the Multi Drug Resistance (MDR) proteins that underlie the cellular efflux of a large set of compounds (Bolhuis et al., 1997), the olfactory receptor repertoire, which recognizes diverse odorant molecules (Buck and Axel, 1991; Lancet, 1986; Lancet and Ben-Arie, 1993; Glusman et al., 2001), and bio-transformation enzymes like cytochrome P450, which provide examples of the phenomenon of "probabilistic catalysis" (Cupp and Tracy, 1998). Probabilistic interactions are also found in catalytic antibodies (Janda et al., 1997; Schultz and Lerner, 1995).

Probability-based recognition is also at the core of the field of combinatorial chemistry. Here, ligand repertoires are used to find new binders for specific molecular targets (Collins, 1997; Hoogenboom, 1997; Lohse and Szostak, 1996; Lorsch and Szostak, 1994; Plunkett and Ellman, 1997; Scott and Smith, 1990). Within the realm of probabilistic recognition, the need for a formal depiction of the statistics that govern the interactions within ligand and receptor repertoires has brought to the concept of Affinity Distribution. This constitutes a frequency histogram for the affinities obtained when a single target is tested against numerous members within a repertoire. The existence of such an affinity distribution is widely recognized (Burnet, 1963; Inman, 1978; Kauvar et al., 1995; Lancet et al., 1993; Levitan, 1997; Levitan, 1998; Macken and Perelson, 1991; Richards, 1975; Vant-Hull et al., 1998). However, no general agreement exists on the specific functional shape of such distribution.

An appealing suggestion is that biological recognition between receptors and ligands obeys a simple, and very general, statistical law (Inman, 1978; Lancet et al., 1994a; Lancet et al., 1993; Rosenwald, 1998; Rosenwald et al., 2002). In other words, it is possible that a simple mathematical model could describe the affinity distribution for many different repertoire types, including receptor multi-gene families and combinatorial ligands. One of these statistical models, the Receptor Affinity Distribution (RAD) model (Lancet et al., 1993) suggests that the number of interactions L contributing to the energy of binding between a receptor and an arbitrary set of ligands is distributed binomially,

$$P(L) = \frac{B!}{L!(B-L)!}\left(\frac{1}{S}\right)^{L}\left(1-\frac{1}{S}\right)^{B-L} \qquad (4)$$

where B is the size of the binding site and S is the subsite diversity. The affinity of binding K is related to L through the following relation: $L = (RT/\alpha)log(K)$ where α is the average energy contribution for a single subsite interaction, R is the gas constant and T is the absolute temperature. This distribution, in a Poisson approximation has recently been verified in reference to experimental data (Rosenwald et al., 2002).

3.2 QUANTITATIVE KINETICS OF MUTUAL CATALYSIS IN RANDOM ENSEMBLES

Besides their pharmaceutical importance for drug design, combinatorial chemistry tools have been used for addressing important questions about biocatalysis and self-replication in proteins (Moore et al., 1997) and nucleic acids (Gold et al., 1997). In the realm of origin of life, a statistical chemistry approach would view the random emergence of diverse organic molecules, and their assemblies, as a natural consequence of undirected prebiotic organosynthesis (Schwartz, 1996; Ehrenfreund and Charnley, 2000). It would then ask how life-like processes could emerge within such primordial random assortments (Cousins et al., 2000), rather than study their specific molecular content (e.g. for α-amino acids or nucleotide bases). One should not forget, however, that while in the traditional combinatorial chemistry studies one of the counterparts is usually a highly evolved biological macromolecule, introduced by the experimenter, prebiotic scenarios assume that both reactants are derived from the same combinatorial collection (Lancet et al., 1994b).

The importance of a probabilistic recognition approach to systems involving large numbers of random chemicals has been shown to extend beyond its initial scope. In previous models for prebiotic catalytic interactions (Chou et al., 1994; Dyson, 1982; Kauffman, 1986) the assumptions about unevolved molecular recognition processes was not based on explicit, biochemically rigorous results. This is where the RAD model can serve as a tool for exploring in a more accurate and quantitative way mutually catalytic sets (Segrè et al., 2001a; Shenhav et al., 2002).

The connection between RAD-governed binding affinity and catalytic activity in random collections of oligomers was explored previously (Lancet et al., 1994b). This expresses the fact that, since a fundamental step in catalytic reactions is the binding of the catalyst to the transition state, the extent of the catalytic enhancement can be assumed to distribute similarly to the binding affinity in a random repertoire.

The power of the statistical approach to mutually catalytic sets is greatly enhanced when combined with a quantitative kinetic analysis (Bagley and Farmer, 1991; Bagley et al., 1991; Wills 1997; Lancet 1994; Segrè et al., 2001a). This is essential for a rigorous demonstration that such sets may undergo self-replication. A computational approach realistic enough for faithfully simulating interactions among molecules in a prebiotic environment should thus best involve equations of chemical kinetics, as well as a statistical account of catalytic events.

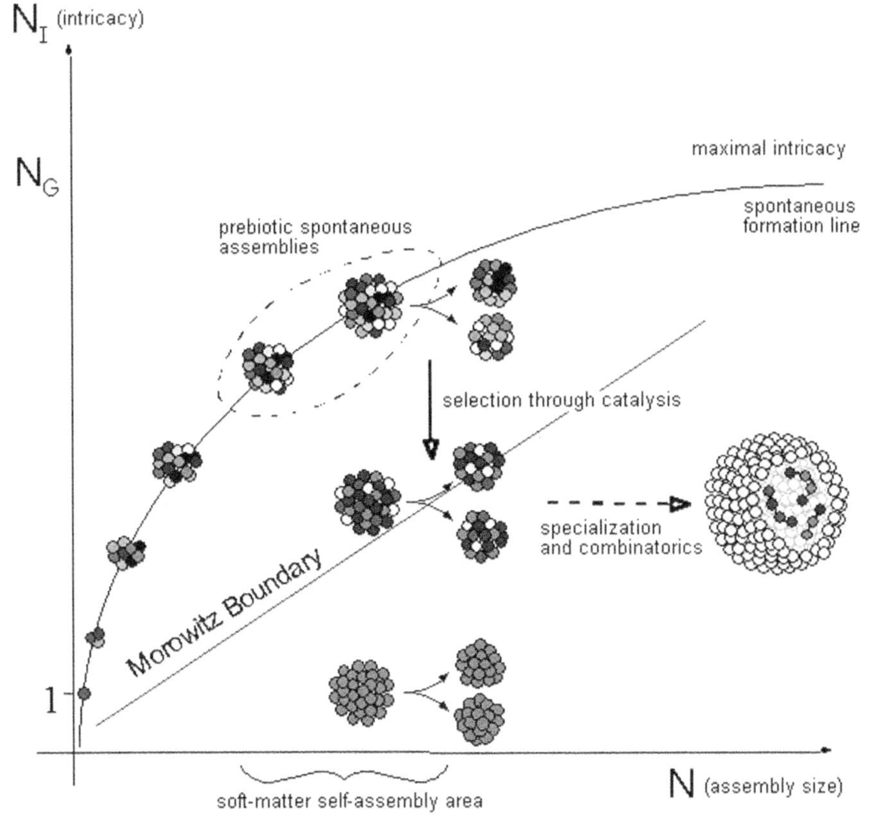

Figure 1 A schematic depiction of how assemblies with different Intricacy (number of different types of molecular species inside an assembly) and N (size, in total number of molecular components), can display different physical and self-organization properties relevant to the emergence of inheritance and self-replication (for more details, see Segrè and Lancet, 1999; Segrè and Lancet, 2000).

4. Accepting the twilight zone: the GARD model and the Lipid World

4.1. THE GARD "MANIFESTO"

Based on the aforementioned theoretical and experimental notions it is possible to delineate a set of guidelines for exploring theoretically the emergence of a primitive form of inheritance in the absence of biopolymers:

• Origin of life research can be viewed as the study of physico-chemical principles that were involved in primordial self-organization (Dyson, 1999; Kauffman, 1993; Oparin, 1957). The Graded Autocatalysis replication Domain (GARD) model (Segrè and Lancet, 1998; Segrè et al., 1998a) aims at distilling such principles and showing how life-like features could emerge based on testable assumptions. The exact details of the early history of life-as-we-know-it on Earth (Smith and Morowitz, 1982) are out of the scope of such approach.

• The central players of the model are molecules. At a minimal level of detail, the kinetic behaviors of ensembles of molecules are translated into differential equations based on the laws of mass action and average statistical properties. However, in some cases, the fluctuations related to stochastic effects of single molecule transformations become significant, as in the realm of mesoscopic physics, defining a high level of detail. The phenomena studied here - similarly to present-day cellular life - lie at the border between these two levels (Bartholomay, 1962; Bolhuis and Frenkel, 1997; Shenhav et al., 2002).

• A viable way of coping with the ignorance about the detailed composition of a primordial mixture is to use the statistical knowledge available about large repertoires of molecules, and their mutual interactions. In this respect, the emergence of life could be thought of as a planet-scale random chemistry experiment. The Receptor Affinity Distribution (RAD) model provides a statistical basis for modeling the probability of graded catalytic efficiencies among random molecules (Lancet et al., 1993, Rosenwald et al. 2002).

• Continuity of functions, rather than continuity of molecular structures, has been used as a criterion for depicting a self-organization scenario (Morowitz, 1992; Morowitz et al., 1988). Some simple features of modern cell biochemistry, such as the widespread prevalence of hydrophobic forces (Tanford, 1978), or the ubiquity of networks of weak catalytic effects (Altreuter and Clark, 1999; Cousins et al., 2000) could be readily obtained in an unevolved prebiotic scenario. Therefore such features should be included in a model for early evolution.

• A model aimed at explaining the emergence of self-reproduction need not include molecular autocatalysis and self-reproduction in its foundations. Rather, self-reproduction should be pursued as an emergent phenomenon (Dyson, 1999; Fontana and Buss, 1994; Kauffman, 1993; Morowitz et al., 1988; New and Pohorille, 2000; Oparin, 1957; Segrè and Lancet, 2000). This behavior may be derived from a large number of mutually catalytic interactions among different molecules, none of which is necessarily self-replicating. This view of self-reproduction requires a thorough quantitative description, similar to classical molecular autocatalysis (Eigen, 1971; Eigen and Schuster, 1979).

• Good candidate molecules for a chemical embodiment of early self-reproduction should (i) be among the possible products of abiotic synthesis, (ii) entail self-aggregation properties and soft-matter attributes, and (iii) possess the capacity to form spatially confined domains potentially capable of growing and dividing. Lipid-like amphiphilic molecules seem the best organic candidates. However, these could be rather different from present-day lipids, and could have diverse chemical properties, particularly in their polar head group (Segrè et al., 2001).

4.2. THE GARD SCENARIO

The GARD model was introduced in four fundamental forms: a) the Monomer-GARD, in which N_G kinds of monomeric species A_j can inter-convert through a common precursor molecule A_0. Molecular species are enclosed in a compartment, or vesicle; b) the Dimer-GARD, in which $N_G=F^2$ kinds of dimeric species D_{ij} are formed from F different monomeric precursors M_i; c) the Amphiphile-GARD, in which N_G kinds of amphiphilic monomers participate in reactions of spontaneous aggregation into noncovalent assemblies, such as micelles and vesicles. In this case the enclosure and the set of mutually catalytic molecules are the same entity; and d) the Chiral-GARD, for which the N_G Monomer-GARD molecular species have a chiral center, and can appear in two enantiomeric forms (A_{jL} and A_{jD}) (Kafri, 2002).

4.2.1. Monomer GARD

The Monomer-GARD model provides a framework for a thermodynamic and kinetic analysis of mutually catalytic assemblies combined with statistical tools. Similarly to other scenarios (Dyson, 1999; Eigen and Schuster, 1979; Kauffman, 1993) it assumes a finite enclosure (e.g. an amphiphilic vesicle), containing the catalytic set members, and absorbing energy-rich chemical precursors ("foodstuff") from the external environment, represented by a single species A_0. GARD's chemical and physical rules are computationally implemented by numerical solution of differential equations or by Monte Carlo simulation procedures. The flow of free energy that maintains the system far from thermodynamic equilibrium involves a volume expansion, combined with the supply of high free energy precursor molecules. The dynamics of the molecular species is dominated by the β matrix of kinetic parameters for mutual catalysis, rather than by the thermodynamic equilibrium constants. Despite the simplicity of its reaction topology, the Monomer-GARD model allows one to quantify the homeostatic capacity of molecular ensembles and relate it to the microscopic kinetic properties of intermolecular interactions. Simply formulated, if an enclosed, externally-fed catalytic set undergoes expansion, and if its catalytic network is effective, then its idiosyncratic internal composition will tend to be homeostatically preserved, despite the continuous growth of volume. Such preservation is a basic hallmark of life forms, and a prerequisite for more elaborate characteristics, such as reproduction and selection. The constant supply of free energy-rich precursors, together with the process of vesicle expansion, maintain the GARD system far from thermodynamic equilibrium. This is somewhat analogous to Eigen's assumption in the quasi-species scenario, involving a combination of a monomers inflow and a polymers outflow. This property is embodied in a measure of homeostatic preservation, λ_C, defined as the highest expansion rate that still allows an above-threshold maintenance of the original composition. The λ_C measure introduces for the first time a quantitative, graded scale for the efficiency of a catalytic network. The strength of this criterion for measuring homeostatic efficiency is demonstrated by its capacity to lead to complex network self-organization, when used as a fitness parameter for stochastic simulations of GARD evolution. In such simulations, based on a Metropolis algorithm for mutation and selection, stable compositions are obtained, which are resistant throughout millions of mutation steps, and possess highly organized topologies of catalytic interconnections. The evolutionary time curve of the fitness parameter λ_C displays long plateaus of invariant states, alternated to burst of changes and avalanches of reorganization steps. The distribution of the lengths of the plateaus is shown to obey a power law with exponent -1, possibly revealing a self-organized critical behavior (Bak and Sneppen, 1993; Bak et al., 1987), or other kinds of known self-organization dynamics (Newman and Sneppen, 1996).

The GARD formalism does not describe a specific molecular scenario, and may therefore apply equally well to different kinds of chemistries (e.g. peptides, nucleotides or amphiphiles) and reaction schemes (isomerizations, covalent dimerization or noncovalent associations, as formulated in the different versions of GARD). In the context of Kauffman's definition of catalytic closure (Kauffman, 1986), GARD is formally always catalytically closed, but it is possible to define within it realms, through the λ_C parameter, the *degree* of catalytic closure for every conceivable assembly. Scenarios for prebiotic mutually catalytic sets (Kauffman, 1993, Dyson, 1982) had only included a qualitative discussion of the link between the catalyzed replenishment of a molecular ensemble subject to dilution, and the self-reproduction potentially ensuing. Our work has provided for the first time a detailed mathematical and computational description of this principle.

4.2.2. Dimer-GARD

The generation of diversity through combinatorics is one of the basic features of the chemistry of life. In the Monomer-GARD model (as well as in the Amphiphile-GARD), the diversity of monomer kinds is an assumption rather than a result, and further diversity is obtained at the level of molecular ensembles, through the combinatorics of different compositions. The fact that these compositions can be inherited constitutes one of the main implications of this part of my work. The Dimer-GARD goes one step further, by incorporating the covalent combinatorial diversity of dimers within the setup of a composition-based assembly.

One of the exciting prospects of the GARD view is the possibility of exploring a stepwise transition from such primordial form of homeostasis and implicit inheritance, to more advanced systems, in which the initially available monomers can bind covalently to each other to form longer and longer polymers. The Dimer-GARD constitutes a first step in this direction. Dimer-GARD was shown to be formally equivalent to Monomer-GARD, and therefore to share with it the property of homeostatic preservation during growth. It is expected that collections of even longer oligomers would behave similarly, and that more powerful computer simulations will allow to track in a graded fashion the transition from assemblies of monomers to complex mixtures containing longer polymers with specialized catalytic roles. It should be stressed that in an alternative interpretation, the transition from Monomeric to Dimeric GARD may be viewed as a progression from an absolutely heterotrophic to a somehow autotrophic organism (Maden, 1995), namely from a system which only recruits available components from its surrounding to one that performs covalent reactions for generating some of its constituents (Ben-Eli, 2000; Shenhav et al., 2003).

4.2.3. Amphiphile-GARD

Through the Amphiphile-GARD (A-GARD, Segrè and Lancet, 1998; Segrè et al. 1998; Segrè et al. 2000) a self-consistent and comprehensive scenario for the emergence of primordial self-reproduction is proposed. This brings together for the first time the concept of mutually catalytic networks (Dyson, 1982; Kauffman, 1986; Morowitz et al., 1988), the fundamental role of self-assembling molecules (Bachmann et al., 1992; Deamer, 1986), a mathematical formulation of chemical evolution through chemical kinetics equations (Eigen, 1971), and the stochastic nature of mesoscopic processes (Bartholomay, 1962; Gillespie, 1977). This synthesis is mainly due to the following innovative features with respect to the Monomer- and Dimer- GARD: (i) A natural process of assembly growth is introduced. In the first two GARD embodiments, as well as in other models, the coupling between the potential growth of a vesicle-like enclosure and the internal catalytic activity is not clearly defined. In Amphiphile-GARD it is importantly assumed that the same molecules that form the self-enclosed assembly, a micelle-like structure, are also responsible for the mutually catalytic functions. In this way, growth of assemblies becomes a kinetic consequence of successful mutual catalysis. (ii) Amphiphile assemblies have a facile capacity to undergo spontaneous, physically-dictated splitting, generating primitive progeny. The homeostasis ensuing from steady concentration of mutually catalytic molecular species within an expanding boundary, nicely devised by Dyson and Kauffman, is subject, in A-GARD, to the splitting events, which are chance-regulated sampling processes. This allows a natural definition of inheritance, which is only implicit in the previous GARD versions. One may ask whether a given assembly is "robust" enough to survive and keep propagating its composition despite the fluctuations due to such splitting events. The Morowitz boundary formula (Morowitz, 1967) (see 2.10 above) provides an approximate answer to this question. In addition, the newly defined Heritability measure η (Segrè et al. 2001a) represents a more advanced tool, which allows one to evaluate the probability of reaching different degrees of fidelity in compositional inheritance fidelity during the sampling process. This allows one also to express in a rigorous way the fact that an assembly, whose composition vector is highly biased, has higher probability of a faithful transmission of its molecular components. (iii) In A-GARD equations and simulations, an assembly is described through the counts of molecules of different types, rather than through their concentrations. In this setup, the kinetics of the catalyzed self-assembly process is governed by a stochastic single-molecule discrete approximation of the Monomer-GARD differential equations. Therefore the compositional mutations, which in other settings (including the evolutionary process for Monomer-GARD) are ad hoc externally imposed discontinuous events, become natural consequences of the dynamics of the system. This results in a situation where the dynamic model for primordial evolution may begin in a condition of near-randomness, with compositional mutations being the rule rather than the exception, and end up in a stationary state characterized by a relatively small mutation rate. (iv) A-GARD simulations provide a pragmatic way of addressing the propagation of information along generations of growing and splitting assemblies. A new definition for assembly similarity H (Segrè et al. 2000) is employed to follow both the changes in an assembly's composition with respect to its past configurations (cross correlation), and the resemblance to reference compositions, such as the eigenvectors of the linear operator used for the linearized GARD equations. In the observed quasi-stationary states (QSSs) assemblies behave homeostatically despite the on-going growth and splitting processes. In most cases these states have a finite lifetime, and decay spontaneously into other states, as a consequence of large compositional fluctuations. In other cases the stationary states seem extremely stable. When homeostasis prevails for numerous A-GARD cycles we say that a rudimentary form of inheritance is taking place. What is inherited is the specific composition of the assembly, which may be regarded as a *Compositional Genome*. The different QSSs observed during A-GARD simulations correspond to different assortments that have the capacity of being propagated faithfully for a detectable number of generations. Such different states are called Composomes.

These features of A-GARD allow one to tackle profound questions about the properties that characterize prebiotic self-reproducing assemblies. Populations of assemblies may be simulated, so that different composomes can compete and give rise to a compositional genome-based evolutionary process. We envisage that in simulations with larger N_G a process of Darwinian evolution could take place under these conditions.

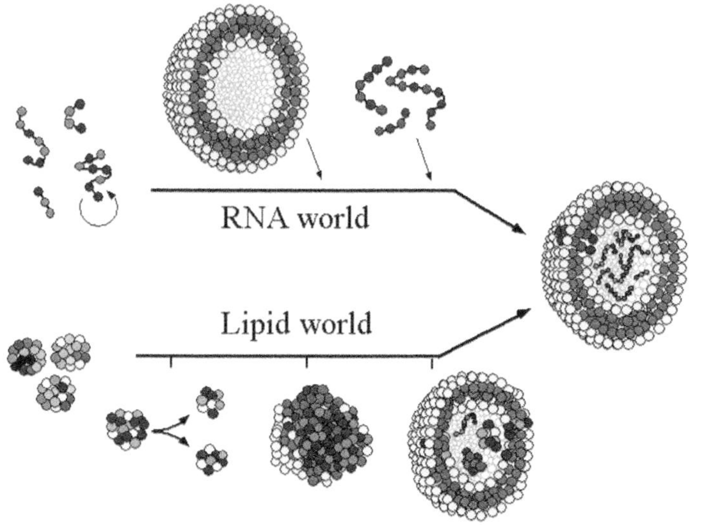

Figure 2 Comparison of two possible views for the emergence of a reproducing protocell: The RNA world (see chapter 2.6) and the Lipid World (chapter 4.3).

4.3. THE LIPID WORLD

The experimental evidence that lipid assemblies may entail many of the properties required by the A-GARD model led to the formulation of a novel scenario for the origin of life, "The Lipid World" (Segrè et al. 2001). As part of the Lipid World view, several novel concepts were proposed as fundamental unexplored aspects of early prebiotic evolution. One notion relates to the possible combinatorial nature of lipid-like chemicals. Thousands of different configurations can be achieved with minimal variations on phospholipid-like structures. The Lipid Word View is compatible with the finding that molecular species found in the Murchison meteorite display aggregation properties in water (Deamer, 1985), as well as with experimental proofs that amphiphiles could form under likely pre-biotic conditions (McCollom et al., 1999).

In analogy to the concept of ribozyme (Lohse and Szostak, 1996), emphasizing the catalytic role of RNA, we coined the term lipozyme (Segrè et al. 2001), by which we indicate any noncovalent assembly, whose components exert catalytic action on any chemical reaction. An autocatalytic lipozymes represents a specific case that can synthesize more of its constituents. It should be noted that while this concept is compatible with the idea of mutually catalytic network, it does not involve necessarily autocatalytic molecules. This is supported by statistical analysis of a large amount of data for micelle and liposome catalysis, extracted from a list compiled by Janos Fendler's (Fendler and Fendler, 1975; Fendler, 1982; Segrè et al. 2001). It was found that the distribution of catalytic rate enhancement factors obeys the lognormal shape predicted by the RAD model (Lancet et al., 1993, Rosenwald et al. 2002) and used throughout GARD simulations.

The fluid nature of lipozymes may be responsible for the formation of transient efficient catalytic sites on a micelle or membrane, due to random encounter of molecules freely diffusing on the interface (Fendler, 1987; Schneider-Henriquez et al., 1992). This might be an intermediate step towards the formation of covalent bonds, which would entail more specific and efficient, but less flexible catalytic functions. Tests of some of these ideas could derive from extensions of Luisi's experiments with autocatalytic micelles and vesicles (autopoietic units (Walde et al., 1994)), whereby many different molecular kinds could be used. Technical problems, especially related to the detection of single assemblies compositions, need to be overcome in order for such an experiment to be useful.

4.4. FUTURE PROSPECTS

An understanding of the evolutionary significance of the idea of a primordial inheritance of a compositional genome requires a thorough description of large heterogeneous populations or assemblies of interacting molecules. GARD and other simulations mostly embodied up to N_G=100 different types of molecular species. Realistic prebiotic mixtures, in contrast, may contain a much larger number of different molecules, in the range of 10^6-10^9 (cf. Morowitz, 1992). Larger N_G values are expected to generate a remarkable assortment of different meta-stable compositions, many of which might arise only under certain sets of initial conditions, or as a consequence of very rare compositional mutations. Large N_G values, however, impose serious computational limitations. For example, the β matrix for catalytic enhancement factors grows quadratically with N_G.

Two main possible approaches can be envisaged for these simulations. The first, and so far most productive, has been the one based on a fixed number (N_G) of chemical kinetics differential equations. This includes GARD kinetics, stochastic Amphiphile-GARD simulations, and many other kinetic models of mutually catalytic networks. In a second possible approach the repertoire of molecular species available is potentially unlimited. This is because the composition of an assembly is represented by a complete list of its molecular components, and no constraints exist on the number of possible kinds. Such an implementation requires however a consistent description of catalytic rate acceleration factors for any possible molecular interaction. This could be obtained by defining a set of fundamental recognition rules for a given set of N_G monomers, and implementing a universal catalytic interaction function, based on string matching rules analogous to the RAD formalism (Bagley and Farmer, 1991; Shenhav et al., 2003), for computing the expected catalytic rate exerted between oligomers.

A new generation of computer simulations, based on stochastic dynamics rather than on rigidly defined differential equations, should be designed in order to cope with such an unlimited number of different possible molecular species. One problem with such an embodiment is that small molecular assemblies might wander indefinitely in the unlimited compositional space without ever falling into a stationary state, and without displaying any visible sign of organization. Only relatively large assemblies may be expected to contain a sufficient number of molecular components, in accordance with the optimal N/N_G ratio (Segrè et al. 2001a).

Extended computations should also include more detailed physico-chemical features, and conform accurately to standard physical chemistry rules for self-assembly, as previously explored in different contexts (Goetz and Lipowsky, 1998; Nekovee et al., 2000). An ultimate GARD setting would be represented by a mesoscopic-scale molecular dynamics simulation (Shenhav et al., 2002), in which membrane and polymer formation would be the consequences of elementary interaction rules. A richer description of the dynamics of GARD populations, involving splitting and fusion of assemblies of different sizes, may pave the way to a novel interpretation of early prebiotic evolution phenomena, whereby single molecule compositional mutations could be extended to more elaborate methods for the generation of compositional diversity, in analogy to the role of genetic recombination.

Under these circumstances, it is conceivable that large-scale supercomputer simulations, analogous to the ones currently utilized for astrophysics and particle physics realistic simulations (Norman, 1998) may in the future open new horizons in our understanding of early molecular evolution. One should bear in mind that such large-scale simulations would produce an enormous amount of complex data, whose interpretation would require a major theoretical effort, and perhaps the invention of new computational techniques, a counterpart of the experimental ones used by biologists for deciphering the real biochemical world. It is becoming also increasingly apparent that interesting connections exist between the study of present-day biochemical networks (Segrè et al., 2002) and the mutually catalytic networks explored in origin of life research (Shenhav et al., 2002). The study of prebiotic evolution and the rising discipline of Systems Biology, both involved in an effort to uncover general principles of biological organization, may end up displaying deep and perhaps yet unknown links.

5. References

Allamandola, L.J., Hudgins, D.M., Bauschlicher Jr., C.W. and Langhoff, S.R. (1999) Carbon chain abundance in the diffuse interstellar medium. Astron. Astrophys., 352, 659-664.

Altreuter, D.H. and Clark, D.S. (1999) Combinatorial biocatalysis: taking the lead from nature. Curr. Opin. Biotechnol., 10, 130-136.

Alves, D. and Fontanari, J.F. (1998) Error threshold in finite populations. Phys. Rev. E, 57, 7008-7013.

Anders, E. (1989) Prebiotic organic matter from comets and asteroids. Nature, 342, 255-257.

Arnett, D. (1996) Supernovae and nucleosynthesis. Princeton Univ. Press, Princeton, NJ.

Arnett, D. and Bazan, G. (1997) Nucleosynthesis in stars: recent developments. Science, 276, 1359-1362.

Bachmann, P., Luisi, P. and Lang, J. (1992) Autocatalytic self-replicating micelles as models for prebiotic structures. Nature, 357, 57-59.

Bagley, R.J. and Farmer, J.D. (1991) Spontaneous emergence of a metabolism. In: C.G. Langton, C. Taylor, J.D. Farmer and S. Rasmussen (eds.), Artificial Life II. Addison-Wesley, Vol. X, pp. 93-140.

Bagley, R.J., Farmer, J.D. and Fontana, W. (1991) Evolution of a metabolism. In: C.G. Langton, C. Taylor, J.D. Farmer and S. Rasmussen (eds.), Artificial Life II. Addison-Wesley, Vol. X, pp. 141-158.

Bak, P. and Sneppen, K. (1993) Punctuated equilibrium and criticality in a simple model of evolution, Phys. Rev. Lett., 71, 4083-4086.

Bak, P., Tang, C. and Wiesenfeld, K. (1987) Self-Organized Criticality: An explanation of 1/f Noise, Phys. Rev. Lett., 59, 381-384.

Bartel, D.P. and Unrau, P.J. (1999) Constructing an RNA world. TIBS (Millennium Issue), 24, M9-M13.

Bartholomay, A.F. (1962) A stochastic approach to statistical kinetics with application to enzyme kinetics. Biochemistry, 1, 223-230.

Ben-Eli, D. (2000) M.Sc. Thesis. The Weizmann Institute of Science, Rehovot, Israel.

Bernstein, M.P., Sandford, S.A., Allamandola, L.J., Gilette, J.S., Clemett, S.J. and Zare, R.N. (1999) UV irradiation of polycyclic aromatic hydrocarbons in ices: production of alcohols, quinones, and ethers. Science, 283, 1135-1138.

Bolhuis, H., van, V.H., Poolman, B., Driessen, A.J. and Konings, W.N. (1997) Mechanisms of multidrug transporters. Fems Microbiol. Rev., 21, 55-84.

Bolhuis, P.G. and Frenkel, D. (1997) Microscopic and mesoscopic simulation of entropic micelles. Physica A, 244, 45-58.

Bolli, M., Micura, R. and Eschenmoser, A. (1997) Pyranosyl-RNA: chiroselective self-assembly of base sequences by ligative oligomerization of tetranucleotide-2',3'-cyclophosphates (with a commentary concerning the origin of biomolecular homochirality). Chemistry & Biology, 4, 309-320.

Bonaccio, S., Walde, P. and Luisi, P.L. (1994) Liposomes containing purine and pyrimidine bases: stable unilamellar liposomes from phosphatidyl nucleosides. J. Phys. Chem., 98, 6661-6663.

Buck, L. and Axel, R. (1991) A novel multigene family may encode odorant receptors: a molecular basis for odor recognition. Cell, 65, 175-187.

Buhse, T., Pimienta, V., Lavabre, D. and Micheau, J.-C. (1997) Experimental evidence of kinetic bistability in a biphasic surfactant system. J. Phys. Chem. A, 101, 5215-5217.

Burnet, F.M. (1963) Cold Spring Harbor Symp. Quant. Biol., 32, 1.

Cairns-Smith, G. (1982) Genetic takeover and the mineral origins of life. Cambridge University Press, Cambridge, UK.

Cech, T.R. (1986) A model for the RNA-catalyzed replication of RNA. Proc. Natl. Acad. Sci. U S A, 83, 4360-4363.

Chou, H.-H., Reggia, J.A., Navarro-Gonzalez, R. and Wu, J. (1994) An extended cellular space method for simulating autocatalytic oligonucleotides. Computers Chem., 18, 33-43.

Chyba, C.F. and Sagan, C. (1992) Endogenous production, exogenous delivery and impact-shock synthesis of organic molecules: An inventory for the origin of life. Nature, 355, 125-132.

Collins, J. (1997) Phage display. In: A.D. Ellington, W.H. Moos, M.R. Pavia and B.K. Kay (eds.), Annual Reports in Combinatorial Chemistry and Molecular Diversity. ESCOM, The Netherlands, Vol. 1, pp. 210-262.

Cousins, G.R., Poulsen, S.A. and Sanders, J.K. (2000) Molecular evolution: dynamic combinatorial libraries, autocatalytic networks and the quest for molecular function. Curr. Opin. Chem. Biol., 4, 270-279.

Crick, F. (1981) Life itself. Simon and Schuster, New York.

Cupp, M.J. and Tracy, T.S. (1998) Cytochrome P450: new nomenclature and clinical implications. Am. Fam. Physician, 57, 107-16.

Dawkins, R. (1996) The River out of Eden. Phoenix Press.

de Duve, C. (1995) Vital Dust. Harper Collins Publishers.

de Graaf, R.M., Visscher, J. and Schwartz, A.W. (1995) A plausibly prebiotic synthesis of phosphonic acids. Nature, 378, 474-477.

Deamer, D.W. (1985) Boundary structures are formed by organic components of the Murchison carbonaceous chondrite. Nature, 317, 792-794.

Deamer, D.W. (1986) Role of amphiphilic compounds in the evolution of membrane structure on the early earth. Origins of Life, 17, 3-25.

Doty, S.D. and Leung, C.M. (1998) Detailed chemical modeling of the circumstellar envelopes of carbon stars: application to IRC +10216. The Astrophys. J., 502, 898-908.

Dyson, F. (1999) Origins of Life. Cambridge University Press, Cambridge.

Dyson, F.J. (1982) A model for the origin of life. J. Mol. Evol., 18, 344-350.

Dyson, F.J. (1985) Origins of life. Cambridge University Press, Cambridge.

Ehrenfreund, P. and Charnley, B. (2000) Organic molecules in the interstellar medium, comets and meteorites: a voyage from dark clouds to the early Earth. Annu. Rev. Astron. Astrophys., 38, 427-483.

Eigen, M. (1971) Selforganization of matter and the evolution of biological macromolecules. Naturwissenschaften, 58, 465-523.

Eigen, M. (2000) Natural selection: a phase transition? Biophys. Chem., 25, 101-123.

Eigen, M. and Schuster, P. (1979) The Hypercycle. Springer Verlag, Berlin.

Ellington, A.D., Robertson, M.P. and Bull, J. (1997) Ribozymes in Wonderland. Science, 276, 546-547.

Farmer, J.D., Kauffman, S.A. and Packard, N.H. (1986) Autocatalytic replication of polymers. Physica, 22D, 50-67.

Fendler, H.J. and Fendler, E.J. (1975) Catalysis in Micellar and Macromolecular Systems. Academic Press, New York.

Fendler, J.H. (1982) Membrane Mimetic chemistry. Wiley, New York.

Fendler, J.H. (1987) Atomic and Molecular Clusters in Membrane Mimetic Chemistry. Chem. Rev., 87, 877-899.

Fontana, W. and Buss, L.W. (1994) What would be conserved if "the tape were played twice"? Proc. Natl. Acad. Sci. USA., 91, 757-761.

Fox, S.W. (1976) The evolutionary significance of phase-separated microsystems. Origin Life Evol. Biosphere, 7, 49-68.

Fox, S.W. (1991) Synthesis of life in the lab? Defining a protoliving system. Q. Rev. Biol., 66, 181-5.

Fraser, C.L. and Folsome, C.E. (1975) Exponential kinetics of formation or organic microstructures. Orig. Life, 6, 429-33.

Fry, I. (2000) The Emergence of Life on Earth : A Historical and Scientific Overview. Rutgers Univ. Press.

Gesteland, R.F., Cech, T.R. and Atkins, J.F. (1999) The RNA world, Second Edition. Cold Spring Harbor Laboratory Press, Cold Spring Harbor, New York.

Gilbert, W. (1986) The RNA world. Nature, 319, 618-618.

Gillespie, D. (1977) Exact stochastic simulation of coupled chemical reactions. J. Phys. Chem., 81, 2340-2360.

Glusman G., Yanai I., Rubin I., Lancet D. (2001) The complete human olfactory subgenome. Genome Res. 2001, 11, 685-702.

Goetz, R. and Lipowsky, R. (1998) Computer simulations of bilayer membranes: Self-assembly and interfacial tension. J. Chem. Phys., 108, 7397-7409.

Gold, L., Brown, D., He, Y.-Y., Shtatland, T., Singer, B.S. and Wu, Y. (1997) From oligonucleotide shapes to genomic SELEX: Novel biological regulatory loops. Proc. Natl. Acad. Sci. USA, 94, 59-64.

Hoogenboom, H.R. (1997) Designing and optimizing library selection strategies for generating high-affinity antibodies. Trends Biotechnol., 15, 62-70.

Horowitz, N.H. (1945) On the evolution of biochemical syntheses. Proc. Natl. Acad. Sci. USA, 31, 153-157.

Huber, C. and Wachtershauser, G. (1997) Activated acetic acid by carbon fixation on (Fe, Ni)S under primordial conditions. Science, 276, 245-247.

Inman, J.K. (1978) The antibody combining region: Speculations on the hypothesis of general multispecificity. In: G.I. Bell, A.S. Perelson and G.H.Jr. Pimbly (eds.), Theoretical Immunology. Marcell Dekker, New York, pp. 243-278.

Irvine, W.M. (1998) Extraterrestrial organic matter: a review. Origins Life Evol. Biosphere, 28, 365-383.

Jain, S. and Krishna, S. (1998) Autocatalytic sets and the growth of complexity in an evolutionary model. Phys. Rev. Lett., 81, 5684-5687.

Janda, K.D., Lo, L.C., Lo, C., Sim, M.M., Wang, R., Wong, C.H. and Lerner, R.A. (1997) Chemical selection for catalysis in combinatorial antibody libraries. Science, 275, 945-8.

Joyce, G.F. (1991) The rise and fall of the RNA world. The New Biologist, 3, 399-407.

Kafri, R. (2002) M.Sc. Thesis, Weizmann Institute of Science, Israel.

Kauffman, S. (1996) Even peptides do it. Nature, 382, 496-497.

Kauffman, S.A. (1986) Autocatalytic sets of proteins. J. Theor. Biol., 119, 1-24.

Kauffman, S.A. (1993) The origins of order, Oxford University Press, Oxford.

Kauvar, L.M., Higgins, D.L., Villar, H.O., Sportsman, J.R., Engqvist-Goldstein, A., Bukar, R., Bauer, K.E., Dilley, H. and Rocke, D.M. (1995) Predicting ligand binding to proteins by affinity fingerprinting. Chem. Biol., 2, 107-18.

Kerr, R.A. (1996) Ancient life on Mars? Science, 273, 864-866.

Koch, A.L. (1985) Primeval cells: possible energy-generating and cell-division mechanisms. J. Mol. Evol., 21, 270-277.

Kondepudi, D.K. and Nelson, G.W. (1985) Weak neutral currents and the origin of bio-molecular chirality. Nature, 314, 438-441.

Kuppers, B.-O. (1983) Molecular theory of evolution. Springer-Verlag, Berlin-Heidelberg.

Lahav, N. (1999) Biogenesis: Theories of Life's origin. Oxford University Press, NY.

Lancet, D. (1986) Vertebrate olfactory reception. Annu. Rev. Neurosci., 9, 329-55.

Lancet, D. and Ben-Arie, N. (1993) Olfactory receptors. Current Biology, 3, 668-674.

Lancet, D., Horovitz, A. and Katchalski-Katzir, E. (1994a) Molecular recognition in biology: Models for analysis of protein/ligand interactions. In: J.-P. Behr (ed.), Perspectives in supramolecular chemistry. J. Wiley New-York, pp. 25-71.

Lancet, D., Kedem, O. and Pilpel, Y. (1994b) Emergence of order in small autocatalytic sets maintained far from equilibrium:application of receptor affinity distribution (RAD) model. Ber. Bunsenges. Phys. Chem., 98, 1166-1169.

Lancet, D., Sadovsky, E. and Seidemann, E. (1993) Probability model for molecular recognition in biological receptor repertoires: Significance to the olfactory system. Proc. Natl. Acad. Sci. USA., 90, 3715-3719.

Lee, D.H., Granja, J.R., Martinez, J.A., Severin, K. and Ghadiri, M.R. (1996) A self-replicating peptide. Nature, 382, 525-528.

Lee, D.H., Severin, K., Yokobayashi, Y. and Ghadiri, M.R. (1997) Emergence of symbiosis in peptide self-replication through a hypercyclic network. Nature, 390, 591-594.

Levitan, B. (1997) Models and search strategies for applied molecular evolution. In: A.D. Ellington, W.H. Moos, M.R. Pavia and B.K. Kay (eds.), Annual Reports in Combinatorial Chemistry and Molecular Diversity. ESCOM, The Netherlands, Vol. 1, pp. 95-152.

Levitan, B. (1998) Stochastic Modeling and Optimization of Phage Display. J. Mol. Biol., 277, 893-916.

Li, T. and Nicolaou, K.C. (1994) Chemical self-replication of palindromic duplex DNA. Nature, 369, 218-221.

Li, W.-H. and Graur, D. (1991) Fundamentals of Molecular Evolution, Sinauer Associates, Sunderland, MA.

Lifson, S. (1996) On the crucial stages in the origin of animate matter. J. Mol. Evol., 44, 1-8.

Lissauer, J.J., Marcy, G.W. and Ida, S. (2000) Extrasolar planets. Proc. Natl. Acad. Sci. USA, 97, 12405-12406.

Lohse, P.A. and Szostak, J.W. (1996) Ribozyme-catalysed amino-acid transfer reactions. Nature, 381, 442-444.

Lorsch, J.R. and Szostak, J.W. (1994) In vitro selection of RNA aptamers specific for cyanocobalamin. Biochemistry, 33, 973-982.

Luisi, P.L. (1997) Self-reproduction of chemical structures and the question of the transition to life. In: C.B. Cosmovici, S. Bowyer and D. Werthimer (eds.), Astronomical and Biochemical Origins and the Search for Life in the Universe, Editrice Compositori, Bologna, pp. 461-468.

Luisi, P.L., Walde, P. and Oberholzer, T. (1999) Lipid vesicles as possible intermediates in the origin of life. Curr. Opinions in Colloid & Interface Sci., 4, 33-39.

Macià, E., Hernandez, M.V. and Oró, J. (1997) Primary sources of phosphorus and phosphates in chemical evolution. Origins Life Evol. Biosphere, 27, 459-480.

Macken, C.A. and Perelson, A.S. (1991) Affinity maturation on rugged landscapes. In: A.S. Perelson and S.A. Kauffman (eds.), Molecular Evolution on Rugged Landscapes: Santa Fe Institute Studies in the Sciences of Complexity. Addison-Wesley, Reading, MA, Vol. 9, pp. 93-118.

Maden, B.E. (1995) No soup for starters? Autotrophy and the origins of metabolism. TIBS, 20, 337-341.

Maher, K.A. and Stevenson, J.D. (1988) Impact frustration of the origin of life. Nature, 331, 612-614.

Mason, S.F. (1984) Origins of biomolecular handedness. Nature, 311, 19-23.

McCollom, T.W., Ritter, G. and Simoneit, B.R.T. (1999) Lipid synthesis under hydrothermal conditions by Fisher-Tropsch-type reactions. Origins Life Evol. Biosphere, 29, 153-166.

McDonald, G.D., Whited, L.J., DeRuiter, C., Khare, B.N., Patnaik, A. and Sagan, C. (1996) Production and chemical analysis of cometary ice tholins. Icarus, 122, 107-117.

McKay, C.P. (1997) Life in comets. In: P.J. Thomas, C.F. Chyba and C.P. McKay (eds.), Comets and the origin and evolution of life. Springer Verlag, New York, pp. 273-282.

McKay, D.S., Gibson Jr., E.K., Thomas-Keprta, K.L., Vali, H., Romanek, C.S., Clemett, S.J., Chillier, X.D.F., Maechling, C.R. and Zare, R.N. (1996) Search for past life on Mars: possible relic biogenic activity in Martian meteorite ALH84001. Science, 273, 924-930.

McMullin, B. (1995) Replicators Don't! In: F. Moran, A. Moreno, J.J. Merelo and P. Cachon (eds.), Advances in Artificial Life. Springer, Granada, Spain, pp. 158-169.

Miller, S.L. (1953) A production of amino acids under possible primitive earth conditions. Science, 117, 528-529.

Miller, S.L. and Urey, H.C. (1959) Organic compound synthesis on the primitive Earth. Science, 130, 245-251.

Miyakawa S., Yamanashi H., Kobayashi K., Cleaves H.J. and Miller SL. (2002) Prebiotic synthesis from CO atmospheres: implications for the origins of life. Proc. Natl. Acad. Sci. U S A, 99, 14628-14631.

Mojzsis, S.J., Arrhenius, G., McKeegan, K.D., Harrison, T.M., Nutman, A.P. and Friend, C.R. (1996) Evidence for life on Earth before 3,800 million years ago. Nature, 384, 55-59.

Mojzsis, S.J., Krishnamurthy, R. and Arrhenius, G. (1998) Before RNA and After: Geophysical and Geochemical Constraints on Molecular Evolution. In: R.F. Gesteland, T.R. Cech and J.F. Atkins (eds.), The RNA world, Second Edition. Cold Spring Harbor Laboratory Press, Cold Spring Harbor, New York, pp. 1-47.

Moore, J.C., Jin, H.-M., Kuchner, O. and Arnold, F.H. (1997) Strategies for the in vitro evolution of protein function: enzyme evolution by random recombination of improved sequences. J. Mol. Biol., 272, 336-347.
Morowitz, H.J. (1967) Biological self-replicating systems. In: F.M. Snell (ed.), Progress in Theoretical Biology. Academic Press, Vol. 1, pp. 35-58.
Morowitz, H.J. (1979) Energy flow in biology. Academic Press, New York.
Morowitz, H.J. (1992) Beginnings of cellular life. Yale University Press, London.
Morowitz, H.J., Heinz, B. and Deamer, D.W. (1988) The chemical logic of a minimum protocell. Origins Life Evol. Biosphere, 18, 281-287.
Muller, H.J. (1966) The gene material as the initiator and the organizing basis of life. Am. Nat., 100, 493-517.
Nekovee, M., Coveney, P.V., Chen, H. and Boghosian, B.M. (2000) Lattice-Boltzmann model for interacting amphiphilic fluids. Phys. Rev. E, 62, 8282-8294.
New, M.H. and Pohorille, A. (2000) An inherited efficiencies model of non-genomic evolution. Simulation Practice and Theory, 8, 99-108.
Newman, M.E.J. and Sneppen, K. (1996) Avalanches, scaling and coherent noise. Phys. Rev. E, 54, 6226-6231.
Nicolis, G. and Prigogine, I. (1977) Self-organization in nonequilibrium systems. Wiley & Sons, Toronto.
Niesert, U., Harnasch, D. and Bresch, C. (1981) Origin of life between scylla and charybdis. J. Mol. Evol., 17, 348-353.
Norman, M. (1998) Cosmos in a Computer. Science, 280, 1522-1523.
Oberholzer, T., Wick, R., Luisi, P.L. and Biebricher, C.K. (1995) Enzymatic RNA replication in self-reproducing vesicles: an approach to a minimal cell. Biochem. Biophys. Res. Commun., 207, 250-257.
Oberhummer, H., Csoto, A. and Schlattl, H. (2000) Stellar production rates of carbon and its abundance in the universe. Science, 289, 88-90.
Oparin, A.I., and Gladilin, K., L.,. (1980) Evolution of self-assembly of probionts. BioSystems, 12, 133-145.
Oparin, A.I. (1957) The origin of life on the earth. Oliver and Boyd, London.
Orgel, L. (1998) The origin of life - a review of facts and speculations. TIBS, 23, 491-495.
Orgel, L.E. (1968) Evolution of the genetic apparatus. J. Mol. Biol., 38, 381-393.
Orgel, L.E. (1992) Molecular replication. Nature, 358, 203-209.
Orò, J. (2000) Organic matter and the origin of life in the solar system. Sheridan Books, Michigan.
Plunkett, M.J. and Ellman, J.A. (1997) Combinatorial chemistry and new drugs. Sci. Am., 276, 68-73.
Ponnamperuma, C., Mariner, R. and Sagan, C. (1963a) Formation of adenosine by ultraviolet irradiation of a solution of adenine and ribose. Nature, 198, 1199-1200.
Ponnamperuma, C., Sagan, C. and Mariner, R. (1963b) Synthesis of adenosine triphosphate under possible primitive earth conditions. Nature, 199, 222-226.
Rebek, J.J. (1994) Synthetic self-replicating molecules. Sci. Am., July, 34-40.
Reid, C. and Orgel, L.E. (1967) Synthesis of sugars in potentially prebiotic conditions. Nature, 216, 455.
Richards, F.F., Konigsberg, W. H., Rosenstein, R. W., and Varga, J. M. (1975) On the specificity of antibodies. Science, 187, 130-137.
Rizzotti M. (1996) Defining Life: the central problem in theoretical biology, University of Padova.
Robertson, M.P. and Miller, S.L. (1995) Prebiotic Synthesis of 5-Substituted Uracils: A Bridge Between the RNA World and the DNA-Protein World, Science, 268, 702-705.
Rosenwald, S. (1998) M.Sc. Thesis. The Weizmann Institute of Science, Rehovot, Israel.
Rosenwald, S., Kafri, R. and Lancet, D. (2002) Test of a statistical model for molecular recognition in biological repertoires, J. Theor. Biol., 216, 327-36.
Sagan, C. (1963) Primordial ultraviolet synthesis of nucleoside phosphates. In: S.W. Fox (ed.), The origin of prebiological systems, pp. 207-219.
Salam, A. (1991) The Role of Chirality in the Origin of Life. Journal of Molecular Evolution, 33, 105-113.
Schlesinger, G. and Miller, S.L. (1983) Prebiotic synthesis in atmospheres containing CH4, CO, and CO2. I. Amino acids. J. Mol. Evol., 19, 376-382.
Schneider-Henriquez, J.E., Denicourt, N. and Fendler, J.H. (1992) Surface Forces in Supported Spherical Lipid Membranes. Langmuir, 8, 3063-3072.
Schopf, J.W., Barghoorn, E.S., Maser, M.D. and Gordon, R.O. (1965) Electron microscopy of fossil bacteria two billion years old. Science, 149, 1365-1357.
Schultz, P.G. and Lerner, R.A. (1995) From molecular diversity to catalysis: lessons from the immune system. Science, 269, 1835-42.
Schwartz, A.W. (1996) Did minerals perform prebiotic combinatorial chemistry? Chemistry and Biology, 3, 515-518.

Schwartz, A.W. and Orgel, L.E. (1985) Template-directed synthesis of novel, nucleic acid-like structures. Science, 228, 585-587.
Sciama, D.W. (1971) Modern Cosmology. Cambridge University Press, Cambridge, UK.
Scott, J.K. and Smith, G.P. (1990) Searching for peptide ligands with an epitope library. Science, 249, 386-390.
Segrè, D., Pilpel, Y., Glusman G. and Lancet, D. (1997) Self-Replication and Evolution in Primordial Mutually Catalytic Sets. In: C.B. Cosmovici, S. Bowyer and D. Werthimer (eds.), Astronomical and Biochemical Origins, Editrice Compositori, Bologna, pp. 469-476.
Segrè, D. and Lancet, D. (1998) Mutually catalytic amphiphiles: simulated chemical evolution and implications to exobiology, In: J. Chela-Flores and F. Raulin (eds.), Exobiology: Matter, Energy and Information in the Origin and Evolution of Life in the Universe, Kluwer Ac. Pub., 123-131.
Segrè, D., Pilpel, Y. and Lancet, D. (1998) Mutual catalysis in sets of prebiotic organic molecules: evolution through computer simulated chemical kinetics, Physica A, 249, Nos. 1-4, 558-564.
Segrè, D., Kedem, O., Lancet, D. and Pilpel, Y. (1998a) Graded Autocatalysis Replication Domain (GARD): kinetic analysis of self-replication in mutually catalytic sets, Origins Life Evol. Biosphere, 28, 501-514.
Segrè, D. and Lancet, D. (1999) A statistical chemistry approach to the origin of life, Chemtracts - Biochemistry and Molecular Biology, 12 (6), 382-397.
Segrè, D. and Lancet, D. (2000) Composing Life, EMBO Reports, 1(3), 217-222.
Segrè, D., Ben-Eli, D. and Lancet, D. (2000) Compositional genomes: prebiotic information transfer in mutually catalytic non-covalent assemblies, Proc. Natl. Acad. Sci. USA, 97(8), 4112-4117.
Segrè, D., Ben-Eli , D. and Lancet, D. (2000a) Prebiotic evolution of amphiphilic assemblies far from equilibrium: from compositional information to sequence-based biopolymers. In: C.B. Cosmovici, S. Bowyer and D. Werthimer (eds.), Astronomical Society of the Pacific Conference Series Vol.213, Sheridan Books, Michigan, pp. 373-378.
Segrè, D., Ben-Eli, D., Deamer, D.W. and Lancet, D. (2001) The Lipid World, Origins Life Evol. Biosphere, 31, 119-145.
Segrè, D., Shenhav, B., Kafri, R. and Lancet, D. (2001a) The molecular roots of compositional inheritance, J. Theor. Biol., 213, 481-491.
Segrè, D., Vitkup, D. and Church, G.M. (2002) Analysis of optimality in natural and perturbed metabolic networks. Proc. Natl. Acad. Sci. U S A, 99, 15112-15117.
Shapiro, R. (1984) The improbability of prebiotic nucleic acid synthesis. Origins Life Evol. Biosphere, 14, 565-570.
Shapiro, R. (2000) A replicator was not involved in the origin of life. IUBMB Life, 49, 173-176.
Shenhav, B., Segrè, D., and Lancet, D. (2002) Mesobiotic emergence: molecular assemblies that self-replicate without biopolymers. Advances in Complex Systems, submitted.
Shenhav, B., Kafri, R. and Lancet, D. (2003) The transition from compositional to polymer based replication. Origins Life Evol. Biosphere, in press.
Sievers, D. and von-Kiedrowski, G. (1994) Self-replication of complementary nucleotide-based oligomers. Nature, 369, 221-224.
Smith, T. and Morowitz, H. (1982) Between history and physics. J. Mol. Evol., 18, 265-282.
Soai, K., Shibata, T., Morioka, H. and Choji, K. (1995) Asymmetric autocatalysis and amplification of enantiomeric excess of a chiral molecule. Nature, 378, 767-768.
Stadler, P.F., Fontana, W. and Miller, J.H. (1993) Random catalytic reaction networks. Physica D, 63, 378-392.
Stadler, P.F., Schnabl, W., Forst, C.V. and Schuster, P. (1995) Dynamics of Small Autocatalytic Reaction Networks - II. Replication, Mutation and Catalysis. Bulletin of Mathematical Biology, 57, 21-61.
Swetina, J. and Schuster, P. (1982) Self-replication with errors - A model for polynucleotide replication. Biophys. Chem., 16, 329-345.
Szathmary, E. (1999) Chemes, Genes, Memes: a revised classification of replicators. Lectures on mathematics in the life sciences, 26, 1-10.
Szathmary, E. and Maynard Smith, J. (1997) From Replicators to Reproducers: the First Major Transitions Leading to Life. J. Theor. Biol., 187, 555-571.
Szostak, J.W. (1992) In vitro genetics. TIBS, 17, 89-93.
Szostak J.W., Bartel D.P. and Luisi P.L. (2001) Synthesizing life. Nature, 409 (Suppl.), 387-90.
Tanford, C. (1978) The hydrophobic effect and the organization of living matter. Science, 200, 1012-1018.
Tjivikua, T., Ballester, P. and Rebek, J. (1990) A self-replicating system. J. Am. Chem. Soc., 112, 1249-1250.
Trimble, V. (1997) Origin of the biologically important elements. Origins Life Evol. Biosphere, 27, 3-21.
Unrau, P.J. and Bartel, D.P. (1998) RNA-catalysed nucleotide synthesis. Nature, 395, 260-263.

Vant-Hull, B., Payano-Baez, A., Davis, R.H. and Gold, L. (1998) The mathematics of SELEX against complex targets. J Mol Biol, 278, 579-97.
Wachtershauser, G. (1988) Before Enzymes and Templates: Theory of Surface Metabolism. Microbiol. Rev., 52, 452-484.
Walde, P., Wick, R., Fresta, M., Mangone, A. and Luisi, P.L. (1994) Autopoietic Self-Reproduction of Fatty Acid Vesicles. J. Am. Chem. Soc., 16, 11649-11654.
Wills, P.R., Kauffman, S.A., Stadler, B.M.R. and Stadler, P.F. (1998) Selection Dynamics in Autocatalytic Systems: Templates Replicating Through Binary Ligation. Bull. Math. Biol., 60, 1073-1098.
Wilson, E.O. and Bossert, W.H. (1971) A primer of population biology. Sinauer Assoc., Sunderland, Mass.
Wimberly, B.T., Brodersen, D.E., Clemons, W.M.J., Morgan-Warren, R.J., Carter, A.P., Vonrhein, C., Hartsch, T. and Ramakrishnan, V. (2000) Structure of the 30S ribosomal subunit. Nature, 407, 327-339.
Woolfson, A. (2000) Life without genes. Harper Collins, London.
Wright, M.C. and Joyce, G.F. (1997) Continuous in vitro evolution of catalytic function. Science, 276, 614-617.
Yarus, M. (1999) Boundaries for an RNA world. Curr. Opin. Chem. Biol., 3, 260-267.
Yockey, H.P. (1992) Information theory and molecular biology. Cambridge University Press.
Yockey, H.P. (2000) Origin of Life on Earth and Shannon's Theory of Communication. Computers and Chemistry, 24, 105-123.

Biodata of **Clifford Matthews** author of ***"The HCN World: Establishing Protein Nucleic Acid Life via Hydrogen Cyanide Polymers"***

Professor Clifford Matthews was born in Hong Kong in 1921 and received his early education there and in England at London University (B.Sc., 1950). He obtained his Ph.D. degree in 1955 at Yale University, New Haven, Connecticut, USA. Following several years in industry he became a Professor of Chemistry at the University of Illinois at Chicago in 1969, becoming emeritus in 1992. His research covered cosmochemistry and the origin of life, with particular emphasis on the role of hydrogen cyanide polymers in prebiotic and extraterrestrial chemistry. Among the courses he taught regularly were *Chemistry and Life*, *Environmental Chemistry* and *Chemical Evolution*. Interdisciplinary seminar subjects included *Science and Society*, *Models of the Future*, *Symmetry in Art and Science*. He organized symposia and minicourses with titles like *Milky Way to DNA*, *Cosmic Beginnings and Human Ends*, and *At Home in the Universe.*

E-mail: **hcnmatthews@cs.com**

J. Seckbach (ed.), Origins, 121–135.

THE HCN WORLD

Establishing Protein – Nucleic Acid Life via Hydrogen Cyanide Polymers

CLIFFORD N. MATTHEWS
Department of Chemistry, University of Illinois at Chicago
Chicago, Illinois 60607, USA

1. Introduction

Hydrogen cyanide polymers may have been among the earliest naturally occurring macromolecules on Earth, as foreseen by Pflüger over a century ago (1875) when he surmised that on the newborn planet 'cyanogen and its compounds had plenty of time and opportunity to follow their great tendency to transformation and polymerization and, by the addition of oxygen and later of water and salts, to change to a labile protein, which constitutes living matter.' This long suspected connection between cyanide, proteins and the origin of life is explored in the following pages through a discussion of HCN polymerization and its significance for prebiotic and extraterrestrial chemistry.

2. Hydrogen Cyanide Polymers: Synthesis, Structure and Reactions

Liquid HCN (bp 25° C) polymerizes spontaneously to a dark brown or black solid at ambient temperatures in the presence of a base such as an amine or ammonia (Matthews and Moser, 1967). Polymerization also occurs readily in non-aqueous solvents or in water (Völker, 1960; Matthews and Moser, 1966, 1967). Two types of structural units appear to be present in these solid materials. Most stable are the ladder polymers (Völker, 1960) shown in Figure 1, formally derived from the olefinic tetramer of HCN, diaminomaleonitrile (A), which is usually found among the products. It seems probable, however, that polymerization to the substituted polymethylene B proceeds by way of an HCN dimer (Völker, 1960; Kliss and Matthews, 1962) for which several structures have been proposed (see Evans *et al.*, 1991). Cyclization to C and D then leads to ladder structures possessing conjugated >C=N- bonds, as proposed by Völker (1960) on the basis of extensive physical and chemical investigations further supplemented by the work of Umemoto *et al.* (1987).

More controversial is the existence of the polyamidine structures shown in Figure 1. Polyaminomalononitrile (F) can be considered an addition polymer of the reactive trimer aminomalononitrile (E), though again it is possible that polymerization occurs

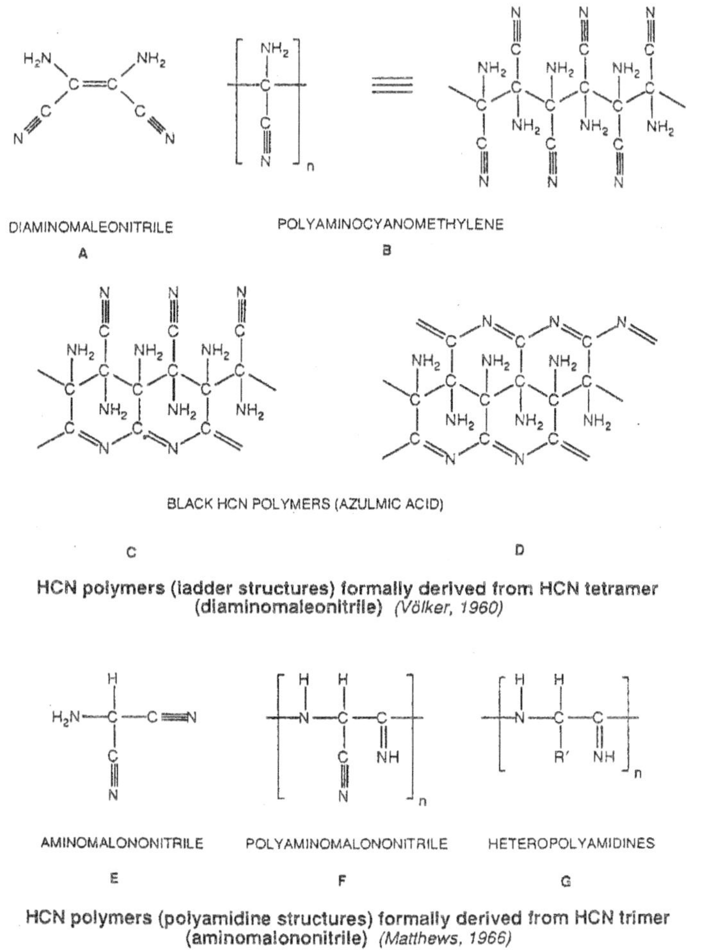

Figure 1 Structures of hydrogen cyanide polymers. A sample of HCN polymer may possess any or all of these structures including hybrids (multimers).

through an HCN dimer (Kliss and Matthews, 1962; Matthews and Moser, 1967; Jameson and Yang, 1972; Yang *et al.*, 1976). Cumulative reactions of HCN on the highly activated nitrile groups of F then yield the heteropolyamidines G which are readily converted by water to heteropolypeptides (Matthews and Moser, 1966, 1967) with release of ammonia and CO_2. Overall, this series of reactions constitutes a route for the direct synthesis of heteropolypeptides without the intervening formation of α-amino acids (Matthews, 1992) (Figure 2).

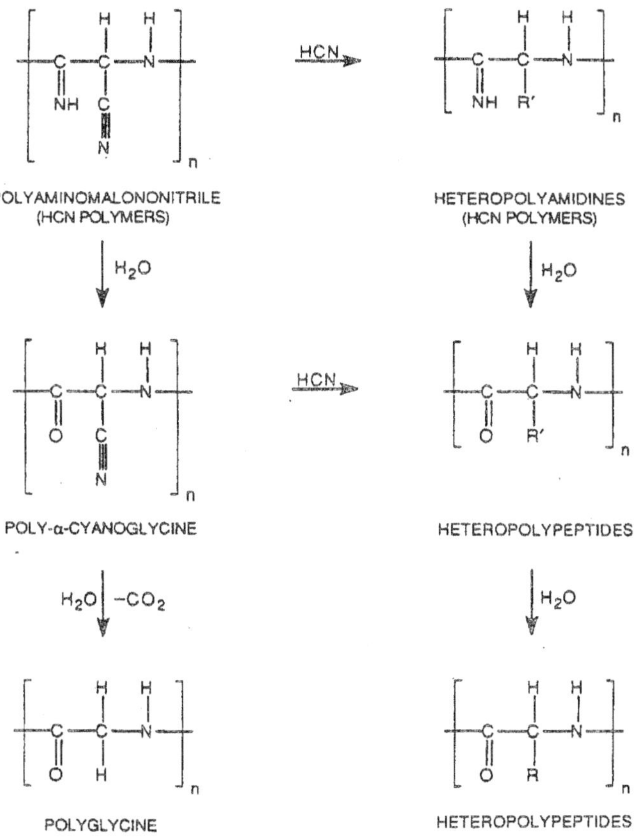

Figure 2 Polypeptides from polyamidines. Cumulative reactions of HCN on polyaminomalononitrile yield heteropolyamidines (with side chains R') which are converted stepwise by water to heteropolypeptides (with side chains R).

Several kinds of experiments have provided results consistent with this polyamidine model. In general, water-soluble, yellow-brown solids can be extracted from the products of each of the following types of reactions:

1. base-catalyzed polymerization of liquid HCN, alone, in water or in non-aqueous solvents (Matthews and Moser, 1967);
2. electric discharge experiments producing HCN from methane-ammonia mixtures (Matthews and Moser, 1966);
3. alkaline hydrolysis of aminoacetonitrile (Moser and Matthews, 1968), aminomalononitrile (E, HCN trimer) (Moser *et al.*, 1968), and diaminomaleonitrile (A, HCN tetramer) (Moser *et al.*, 1968), all of which are ready sources of HCN at high pH;
4. HCN modification of the reactive nitrile side chains of poly-α-cyanoglycine, a polyamide analog of polyaminomalononitrile (F) synthesized from the N-carboxyanhydride related to α-cyanoglycine (Warren et al., 1974; Minard et al., 1975).

Acid hydrolysis of these yellow brown polymers yields not just glycine, the major product, but other α-amino acids as well, such as alanine, aspartic acid, glutamic acid, serine and threonine, together with some α-amino acids not found in proteins. Further GC-MS studies show that the glycine is perdeuterated when D_2O/DCl is used for hydrolysis instead of H_2O/HCl (Matthews et al., 1977), an indication of the highly acidic nature of the methine carbons of F and of the great reactivity of its nitrile groups brought about by its delocalized structure.

Non-destructive analysis of the total solid product obtained from HCN, as well as of separate components, became possible with the advent of cross-polarization magic-angle spinning solid-state NMR spectroscopy (^{13}C and ^{15}N). In particular, the unambiguous present of secondary amide groups, as in peptides, has been established by double-cross-polarization studies on polymers synthesized from equimolar amounts of $H^{13}CN$ and $HC^{15}N$ (Schaefer *et al.*, 1982; Matthews *et al.*, 1984; McKay *et al.*, 1984; Garbow *et al.*, 1987).

Several state-of-the-art methods for the separation and identification of these intriguing polymeric materials, including photoacoustic Fourier transform infrared spectroscopy, supercritical fluid extraction chromatography and pyrolysis mass spectrometry have also revealed fragmentation patterns and chemical functionalities consistent with the presence of polymeric peptide precursors both in HCN polymers and in the Murchison meteorite (Liebman *et al.*, 1993; Pesce-Rodriguez *et al.*, 1993). The presence of long-lived free radicals, primarily carbon-based, in the polymer was shown by multifrequency electron spin resonance studies (Budil *et al.*, 2003).

Especially effective have been thermochemolysis/GC-MS investigations of the polymer with tetramethylammonium hydroxide. This brings about bond cleavage and *in situ* methylation, producing a suite of stable products including derivatives of α-amino acids and nitrogen heterocycles, purines and pyrimidines in particular (Minard *et al.*, 1998). Another major product is diketopiperazine indicating the presence of polyglycine structures (Minard *et al.*, 2002) in water-treated HCN polymers.

3. Hydrogen Cyanide Polymers in the Solar System

Comparable polymers are synthesized when methane and ammonia mixtures are converted to hydrogen cyanide by electric discharges (Matthews and Moser, 1967; Woeller and Ponnamperuma, 1969). The original presence on cometary nuclei of frozen volatiles such as methane, ammonia, and water subjected to high energy sources (Whipple, 1974) thus makes them possible sites for the formation and condensed-phase polymerization of HCN (Matthews and Ludicky, 1986, 1992). Dust emanating from the nucleus, contributing to the coma and tail, would also arise partly from the polymer. Results from the 1986 Halley missions support this view: following Vega observations of a dark surface largely masked by clouds (Sagdeev et al., 1986), the Giotto multicolor camera showed the presence of a potato-shaped core ($\approx$ 15 x 10 km) with a very low albedo (2-4%) (Reinhard, 1986; Keller et al., 1986). We have suggested that this non-volatile black crust is largely composed of hydrogen cyanide polymers and related compounds (Matthews and Ludicky, 1986, 1992), a conclusion supported by the detection in its coma of free hydrogen cyanide, lots of cyanide radicals, and solid particles consisting only of H, C, and N or only of H, C, N, O. Cruikshank *et al.* (1990, 1991) have detected molecules containing cyano groups in the very dark solids in the dust of "new" comets. The infrared reflectance spectra of these bodies resemble the comparable spectra of HCN polymers in the same 0-3 μm region.

Continuing internal polymerization of HCN—a strongly exothermic process—may even be the cause of the sudden cometary outbursts that have been noted by observers for many years (Rettig *et al.*, 1992). Most dramatically, the dark brown patches arising from the impacts of fragments of comet P/Shoemaker-Levy on Jupiter in 1994 could have been caused mainly by the deposition and synthesis of HCN polymers (Matthews, 1997). Most intriguing will be the analysis of cometary material brought to Earth around 2006 from Comet Wild by the ongoing Stardust mission. I predict that HCN polymers and related compounds will be major components (Matthews, 2000).

The recognition that cyanide chemistry could be proceeding on Halley and other comets suggests that HCN polymerization might also take place on carbonaceous chondrites from the asteroid belt. Studies of the Murchison meteorite have shown that free α-amino acids are present (Kvenvolden *et al.*, 1973) together with acid-labile amino acid precursors of undefined structure (Cronin, 1976). Deuterolysis of these water-soluble yellow-brown meteorite extracts yields perdeuterated glycine, as would be expected from peptide segments in polymers derived from HCN (Matthews *et al.*, 1977, 1980). Hydrogen cyanide polymerization could account, too, for much of the yellow-brown-orange coloration of Jupiter and Saturn (Matthews and Moser, 1966; Woeller and Ponnamperuma, 1969; Matthews, 1982), especially since HCN has been found in Jupiter's reducing atmosphere and in the atmosphere of Titan, the largest moon of Saturn. Most intriguing is an orange haze high in Titan's stratosphere that may consist of organic polymers (Owen, 1982). These could be polycyanides formed directly from HCN, some of which would be converted by water to heteropolypeptides after settling on the frozen surface of the satellite (Matthews, 1982).

It seems possible, then, that many dark bodies in the outer solar system are covered by organic solids consisting largely of HCN polymers. Laboratory simulations of Jupiter and Titan chemistry—reactions of methane and ammonia (or nitrogen) subjected to high energy sources—have yielded tholins (Sagan and Khare, 1979; Khare *et al.*, 1984; McDonald *et al.*, 1991), dark materials of undefined structure whose infrared reflectance spectra are remarkably similar to the spectra of HCN polymers (Matthews and Ludicky, 1986; Cruikshank *et al.*, 1991). In view of the results of our original spark experiments with methane and ammonia (Matthews and Moser, 1966, 1967) we would expect polymers of hydrogen cyanide to be significant components of these tholins, especially since they yield α-amino acids on hydrolysis. Sagan and Khare, however, question any meaningful presence of HCN polymers. In their studies of organic aerosols in the atmosphere of Titan (Khare *et al.*, 1994) they report that the optical properties of several tholins differ markedly from those of HCN polymer. The particular sample they used, however, cannot be regarded as representative, since it was formed accidentally rather than intentionally in an industrial process for the production of HCN and must have passed through many different sets of conditions before being deposited on pipes or walls of the reactor. Not surprisingly, we have found the material to be quite different from the HCN polymers we have studied in the past. In any case, we expect differences between HCN polymers and tholins since the latter must also possess hydrocarbon components from methane by way of acetylene intermediates. A direct comparison of an HCN polymer with a Titan tholin prepared by charged particle bombardment of an N_2/CH_4 mixture (Khare *et al.*, 1984) was made by Minard *et al* (2000) through TMAH thermochemolysis/GC-MS analysis. Similar breakdown products from each sample included glycine, alanine, aspartic acid, glutamic acid, adenine, uracil, urea and sym-triazine derivatives, indicating that hydrogen cyanide polymers were major components of the tholin (kindly supplied by Bishun Khare). Parallel ESR studies (Budil *et al.*, 2003) showed, too, the presence of long-lived free radicals in HCN polymers and the tholin. A search for HCN polymers in Titan's atmosphere should clearly be attempted by the Cassini-Huyghens mission expected to reach Saturn/Titan by the year 2004.

Beyond the solar system, the existence of HCN polymers would be expected in dense interstellar clouds where cyanide infrared signatures have been detected (Pendleton and Cruikshank, 1994), and in planetary nebulae where HCN molecules are present (Sun Kwok, 2002). Large amounts of HCN exist in a massive protostar GL 2591 (Boonman *et al.*, 2001) suggesting that HCN is crucial in starbirth.

4. Hydrogen Cyanide Polymers and the Origin of Life

The pioneering demonstration by Miller and Urey (Miller, 1953; Miller and Urey, 1959) that α-amino acids are readily formed from methane, ammonia and water subjected to electric discharges has naturally strengthened the widespread assumption that the original formation of primitive proteins occurred through the polycondensation of these monomers. We emphasize here the opposite conclusion (Matthews and Moser, 1966, 1967) that the amino acids are actually secondary products arising from polypeptides formed by way of HCN polymers. This polymeric pathway seems more plausible than the Strecker route proposed by Miller (Peltzer, *et al.*, 1984) for the synthesis of α-amino acids, requiring the prior formation of aminoacetonitriles from intermediate aldehydes and HCN:

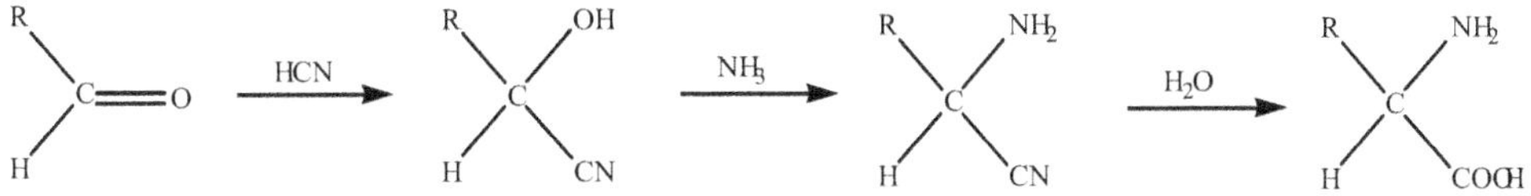

Carrying out spark reactions with methane and ammonia in the absence of water (Matthews and Moser, 1966, 1967), demonstrated that amino acid precursors—HCN polymers—could be synthesized without any aldehydes being involved. It should be noted too that the appearance of a-hydroxy acids together with the a-amino acids occurs not only in spark reactions, as predicted by the Strecker mechanism (Pelzer *et al.*, 1984), but also by hydrolysis of HCN polymers (Cronin *et al.*, 1993). In addition to achieving the direct synthesis of polypeptides from hydrogen cyanide and water (Matthews and Moser, 1966, 1967) it was further shown (Moser and Matthews, 1968) that aminoacetonitrile is unstable in alkali, giving off HCN which rapidly polymerizes. This explains the otherwise surprising result that alkaline hydrolysis of aminoacetonitrile yields not only glycine but also other α-amino acids such as alanine, aspartic acid, glutamic acid, serine and threonine. These unexpected products (and most of the glycine) clearly arise from the HCN polymer. It is only under acidic conditions not expected on the primitive Earth that aminoacetonitrile is stable enough to persist and give rise to high yields of glycine (but no other α-amino acids). Quite revealing, too, are the results of deuterolysis experiments (D_2O/DCl) (Matthews *et al.*, 1977). The glycine obtained from HCN polymer became perdeuterated i.e., all six hydrogen atoms were replaced by deuterium, as predicted from the delocalized structure of the polymer. In contrast, the glycine from aminoacetonitrile or polyglycine incorporates little carbon-bound deuterium.

Another route to α-amino acids was opened up by the pioneering experiments of Oró *et al* (Oró and Kamat, 1961; Oró and Kimball, 1962; Oró and Lazcano-Araujo, 1980) leading to the synthesis both of α-amino acids and adenine from HCN in ammonium hydroxide solution. This drew attention to the important role in such reactions of HCN oligomers such as the trimer (Figure 1, E) and the tetramer (Figure 1, A). While these oligomers readily yield glycine following acid hydrolysis, we have shown that under prebiotic conditions—in alkali—they decompose like aminoacetonitrile to HCN which then forms polymers which can be further hydrolyzed to give glycine and several additional α-amino acids (Moser *et al.*, 1968). These oligomers are probably key intermediates for the laboratory synthesis of nitrogen heterocycles (Oró and Lazcano-Araujo, 1980; Ferris and Hagan, 1984; Ponnamperuma, 1989). HCN polymers, however, would be most involved in the abiogenesis of proteins, purines and pyrimidines (Minard *et al.*, 1998).

The ubiquitous presence of HCN in reducing environments invites the reexamination and possible reinterpretation of almost all previous research concerned with the origin of α-amino acids, including simulations of the chemistry of primitive atmospheres (Miller and Orgel, 1976), studies of aqueous cyanide chemistry (Oró and Lazcano-

Araujo, 1980; Ferris and Hagan, 1984), and meteorite analysis (Kerridge, 1991). These investigations of reactions ostensibly yielding α-amino acids actually supply evidence for the abundant prebiotic and extraterrestrial existence of polymeric protein ancestors—heteropolypeptides synthesized directly from hydrogen cyanide and water (Matthews and Moser, 1966, 1967; Matthews, 1984, 1985). The detection of amino acid polymers in some of these experiments (Lowe *et al.*, 1963; Matthews and Moser, 1966, 1967; Woeller and Ponnamperuma, 1969; Su *et al.*, 1989; Khare *et al.*, 1989; McDonald *et al.*, 1991) lends credence to this view.

The ready conversion by water of polyamidines to polypeptides demonstrated by our investigations suggests that the polyamidines—HCN polymers—might have played a further essential role in chemical evolution. In the absence of water—on land—they could have been the original condensing agents of prebiotic chemistry giving rise to essential polymeric structures. Their reactive amidine groups, eager to become amides, would have brought about the stepwise formation of nucleosides, nucleotides, and polynucleotides from available sugars, phosphates, and nitrogen bases by a series of dehydration reactions. Most significant would have been the parallel synthesis of polypeptides and polynucleotides arising from the dehydrating action of polyamidines on nucleotides (see Figure 3):

$$\text{polyamidines} + \text{nucleotides} \longrightarrow \text{polypeptides} + \text{polynucleotides}$$

Optimum conditions might well have existed on a primitive Earth where photochemical synthesis of organic molecules proceeded in the atmosphere in three overlapping stages defined by the relative volatility of methane, ammonia, and water. As I see it, hydrocarbon chains formed first from methane by way of acetylenes. Then, as ammonia became more involved, hydrogen cyanide and cyanoacetylene became major reactants leading to the synthesis of nitriles and carboxylic acids. Polymeric peptide precursors were formed as described above together with nitrogen heterocycles possessing the basic skeletons of purines, pyrimidines, and pyridines, presumably through the ladder structures shown in Figure 1. When most of the ammonia had been used up, photolysis of water vapor that had been confined to lower levels became possible, leading to the third stage when oxygen-containing molecules such as formaldehyde and sugars were synthesized, as well as phosphates from phosphine. Unlike the prevailing dilute soup picture of chemical evolution, this atmospheric model supplies the necessary prebiotic compounds in sequence, in the right place at the right time. As Earth's surface became covered with this organic shower, potential membrane material—carboxylic acids, carbohydrates, polypeptides—accumulated in lakes and oceans, rapidly becoming cellular through surface tension forces, while on land the simultaneous synthesis of polypeptides and polynucleotides, potential genetic material, was promoted by cyanide polymers, perhaps assisted by clays. Interaction on beaches of this potential "software" with the metabolic "hardware" in protocells produced elementary replicating systems and, eventually, the genetic code. On our dynamic planet this polypeptide-polynucleotide symbiosis mediated by polyamidines may have set the pattern for the evolution of protein-nucleic acid systems, controlled by enzymes, so characteristic of life today (Figure 4) (Matthews, 2000).

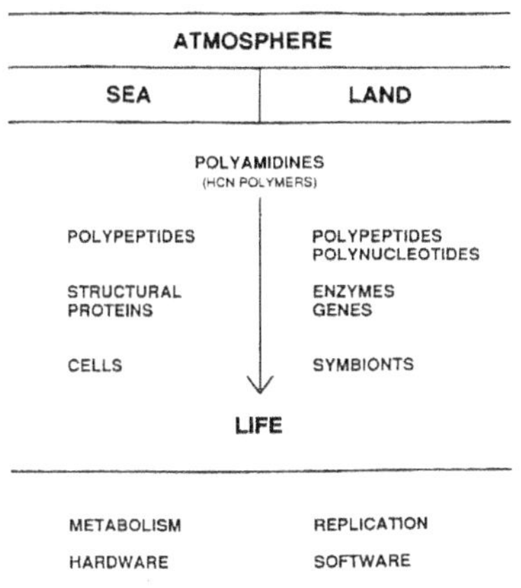

Figure 3 HCN polymers on land and sea, giving rise to life on Earth.

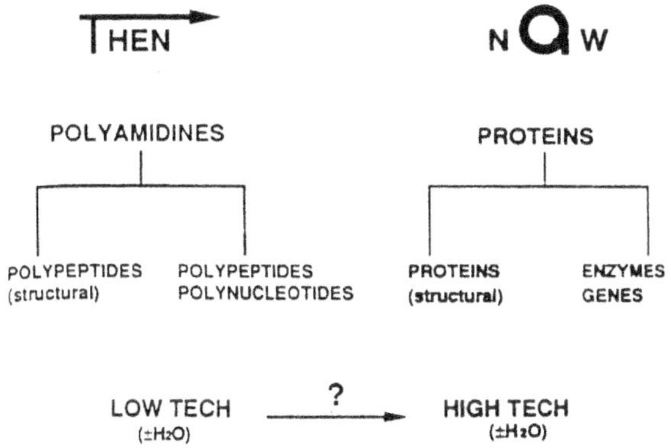

Figure 4 On our dynamic planet, this polypeptide/polynucleotide symbiosis mediated by polyamidines may have set the pattern of protein/nucleic acid systems controlled by enzymes, the mode characteristic of life today.

5. Conclusions

Underlying much of past thinking about the origin of life has been the question: Which came first, proteins (enzymes) or nucleic acids (genes)? A popular view today is that RNA was in at the beginning, before DNA and proteins, since it has been shown that some RNA can act both as a catalyst and as a carrier of information. This RNA world would eventually have evolved into today's world with its genetic code connecting nucleic acids to proteins (Joyce and Orgel, 1993). By contrast, in the HCN world described above, HCN polymer came first, giving rise simultaneously to polypeptides and polynucleotides, precursors of proteins (enzymes) and nucleic acids (both DNA and RNA). This is an argument from abundance, given the ease of formation in our Milky

Way galaxy both of HCN and its polymers. It seems, then, that primitive Earth, perhaps a billion years old, became covered with HCN polymers, as well as other organic compounds, through bolide bombardment or by photochemical reactions in the atmosphere. In an aqueous reducing environment, life emerged from this vital dust, woven out of air by light.

Why do these ideas, based on a reasonable balance of fact and speculation, remain so controversial? Certainly they touch on and challenge much of today's research on chemical evolution. So there is a technical challenge—Show me!—which of course is very welcome since it leads to further research and additional results, pro and con. There is also a broader objection that the overall hypothesis is much too complicated (see Matthews, 2000). I would reply that, if anything, I believe simplicity is its main strength, simplicity of a type that generates complexity. According to this model, polypeptide structures arose essentially from HCN and water. Only later were their breakdown products, amino acids, used to reconstitute the parent polymer, making further use of HCN polymers—polyamidines—as dehydrating agents. In the beginning—at all beginnings—things were necessarily different from today.

To conclude, laboratory and extraterrestrial studies suggest that hydrogen cyanide polymerization is a truly universal process that accounts not only for the past synthesis of protein ancestors on Earth but also for chemistry proceeding today elsewhere in our solar system, on planetary bodies and satellites around other stars, in planetary nebulae, and in the dusty molecular clouds of spiral galaxies. This preferred pathway surely points to the widespread existence of protein-dominated life on Earth-like planets in our hydrogen-rich universe.

6. Acknowledgements

I thank my continuing collaborators, Shirley Liebman and Bob Minard, for their imaginative state-of-the-art investigations aimed at elucidating the structure and function of these ubiquitous HCN polymers.

7. References

Boonman, A.M.S., Stark, R., Van der Taak, F.F.S., van Dischoek, E.F., van der Waal, P.B., Schäfer, F., de Lange, G., and Lauwen, W.M. (2001) Highly abundant HCN in the inner hot envelope of GL 2591: Probing the birth of a hot core?, *Astrophysical J.* **553**, L63-L67.

Budil, D.E., Roebber, J.L., Liebman, S.A., and Matthews, C.N. (2003) Multifrequency electron spin resonance detection of solid state organic free radicals in HCN polymers and a Titan tholin, *Astrobiology*, in press.

Cronin, J.R. (1976) Acid-labile amino acid precursors in the Murchison meteorite, *Origins of Life* **7**, 337-348.

Cronin, J.R., Pizzarello, S., and Matthews, C.N. (1993) unpublished results.

Cruikshank, D.P., Hartmann, W.K., Tholen, D.J., Allamandola, L.J., Brown, R.H., Matthews, C.N., and Bell, J.F. (1991) Solid CN bearing material on outer solar system bodies, *Icarus* **94**, 345-353.

Ehrenfreund, P., Boon, J.J., Commandeur, J., Sagan, C., Thompson, W.R., and Khare, B. (1994) Analytical pyrolysis experiments of Titan aerosol analogues in preparation for the Cassini-Huyghens mission, *Adv. Space Res.* **15** (3) 335-342.

Evans, R.A., Lorencak, P., Ha, T.-K., and Wentrup, C. (1991) HCN dimers: iminoacetonitrile and N-cyanomethanimine, *J. Am. Chem. Soc.* **113**, 7261.

Ferris, J.P., and Hagan, W.J. (1984) HCN and chemical evolution: the possible role of cyano compounds in prebiotic synthesis, *Tetrahedron* **40**, 1093-1120.

Garbow, J.R., Schaefer, J., Ludicky, R.A., and Matthews, C.N. (1987) Detection of secondary amides in HCN polymers by dipolar rotational spin-echo ^{15}N NMR, *Macromolecules* **20**, 305-309.

Jameson, C.J., and Yang, W. (1972) Quantum-mechanical treatment of the electronic structures and geometry of the hydrogen cyanide dimer, *J. Theor. Biol.* **35**, 247-257.

Joyce, G.F., and Orgel, L.E. (1993) Prospects for understanding the origin of the RNA world in Gesteland, R.F., and Atkins, J.F. (eds) *The RNA World*, Cold Spring Harbor Lab Press, Ch. 1.

Keller, H.U., Arpigny, C., Barbieri, C., Bonnet, R.M., Cazes, S., Coradini, M., Cosmovici, C.B., Delamere, W.A., Huebner, W.F., Hughes, D.W., Jamar, C., Malaise, D., Reitsema, H.J., Schmidt, W.K.H., Seige, P., Whipple, F.L., and Wilhelm, K. (1986) First Halley multicolour imaging results from Giotto, *Nature* **321**, 320-326.

Kerridge, J.F. (1991) A note on the prebiotic synthesis of organic acids in carbonaceous meteorites, *Origins of Life* **21**, 19-21.

Khare, B.N., Sagan, C., Thompson, W.R., Arakawa, E.T., Suits, F., Calcott, T.A., Williams, M.W., Shrader, S., Ogino, H., Willingham, T.O., and Nagy, B. (1984) The organic aerosols of Titan, *Adv. Space Res* **4**, 59-68.

Khare, B.N., Sagan, C., Thompson, W.R., Flynn, L., and Morrison, M.A. (1989) Amino acids and their polymers in the lower clouds of Jupiter? Preliminary findings, *Origins of Life* **19**, 495-496.

Khare, B.N., Thompson, W.R., Sagan, C., and Arakawa, F.T. (1993) Organic aerosols in the atmosphere of Titan, *Origins of Life* **24**, 95-96.

Khare, B.N., Sagan, C., Thompson, W.R., Arakawa, E.T., Meiss, C., Tuminello, P.S. (1994) Optical properties of poly-HCN and their astronomical applications, *Can. J. Chem.* **72**, 678-694.

Kliss, R.M., and Matthews, C.N. (1962) Hydrogen cyanide dimer and chemical evolution, *Proc. Natl. Acad. Sci. U.S.* **48**, 1300-1306.

Kvenvolden, K.A., Lawless, J.G., and Folsome, C.E. (1973) Organic matter in meteorites, *Sci. Amer.* **227**, June, 38-54.

Liebman, S.A., Pesce-Rodriguez, R.A., and Matthews, C.N. (1995) Organic analysis of hydrogen cyanide polymers: prebiotic and extraterrestrial chemistry, *Adv. Space Res.* **15**, (3) 71-80.

Lowe, C.U., Rees, M.W., and Markham, R. (1963) Synthesis of complex organic compounds from simple precursors: formation of amino acids, amino acid polymers, fatty acids and purines from ammonium cyanide, *Nature* **199**, 219-222.

Matthews, C.N., and Moser, R.E. (1966) Prebiological protein synthesis, *Proc. Natl. Acad. Sci. U.S.* **56**, 1087-1094.

Matthews, C.N., and Moser, R.E. (1967) Peptide synthesis from hydrogen cyanide and water, *Nature* **215**, 1230-1234.

Matthews, C.N., Nelson, J., Varma, P., and Minard, R.D. (1977) Deuterolysis of amino acid precursors: evidence for hydrogen cyanide polymers as protein ancestors, *Science* **198**, 622-624.

Matthews, C.N., Nelson, J.E., and Minard, R.D. (1980) Deuterolysis of amino acid precursors: evidence for hydrogen cyanide polymers as protein ancestors, *Abstracts, 6th Int. Conf. on the Origins of Life*, Jerusalem, June, 100.

Matthews, C.N. (1982) Heteropolypeptides on Titan?, *Origins of Life* **12**, 281-286.

Matthews, C.N. (1984) Chemical evolution: protons to proteins, *Proc. Roy. Inst. Gt. Britain* **55**, 199-206.

Matthews, C.N., Ludicky, R.A., Schaefer, J., Stejskal, E.O., and McKay, R.A. (1984) Heteropolypeptides from hydrogen cyanide and water: solid state ^{15}N NMR investigations, *Origins of Life* **14**, 243-249.

Matthews, C.N., and Ludicky, R.A. (1986) The dark nucleus of comet Halley: hydrogen cyanide polymers, *Proc. 20th ESLAB Symposium on the Exploration of Halley's Comet*, **ESA SP-250**, 273-277.

Matthews, C.N. (1986) Simultaneous synthesis of polypeptides and polynucleotides? Hydrogen cyanide polymers as prebiotic condensing agents, *Origins of Life* **16**, 500.

Matthews, C.N. (1988) Cosmic metabolism: the origin of macromolecules, in Marx, G. (ed.), *Bioastronomy—The Next Steps*, Kluwer, Dordrecht, Holland, 167-178.

Matthews, C.N. (1990) Simultaneous synthesis of polypeptides and polynucleotides? Hydrogen cyanide polymers as prebiotic condensing agents, in Ponnamperuma, C., and Eirich, F.R. (eds.), *Prebiological Self Organization of Matter*, Deepak, Hampton, VA, 171-178.

Matthews, C.N. (1992) Dark matter in the solar system: hydrogen cyanide polymers, *Origins of Life* **21**, 421-434.

Matthews, C.N., and Ludicky, R.A. (1992) Hydrogen cyanide polymers on comets, *Adv. Space Res* **12**, (4) 21-24.

Matthews, C.N. (1997) Hydrogen cyanide polymers from the impact of Comet P/Shoemaker/Levy 9 on Jupiter, *Adv. Space Res.* **19**, (7) 1087-1091.

Matthews, C.N. (2000) Chemical evolution in a Hydrogen Cyanide World, in Margulis, L., Haselton, A., and Matthews, C.N. (eds.), *Environmental Evolution, 2nd Edition*, MIT Press, Cambridge, MA, Ch. 3.

McDonald, G.D., Khare, B.N., Thompson, W.R., and Sagan, C. (1991) $CH_4/NH_3/H_2O$ spark tholin: chemical analysis and interaction with Jovian aqueous clouds, *Icarus* **94**, 354-367.

McKay, R.A., Schaefer, J., Stejskal, E.O., Ludicky, R.A., and Matthews, C.N. (1984) Double-cross-polarization detection of labeled chemical bonds in HCN polymerization, *Macromolecules* **17**, 1124-1130.

Miller, S.L. (1953) A production of amino acids under possible primitive earth conditions, *Science* **117**, 528.

Miller, S.L., and Urey, H.C. (1959) Organic compounds synthesis on the primitive earth, *Science* **130**, 245-250.

Miller, S.L., and Orgel, L.E. (1974) *The Origins of Life on the Earth*, Prentice-Hall, Englewood Cliffs, NJ.

Minard, R.D., Yang, W., Varma, P., Nelson, J., and Matthews, C.N. (1975) Heteropolypeptides from poly-α-cyanoglycine and hydrogen cyanide: a model for the origin of proteins, *Science* **190**, 387-389.

Minard, R.D., Hatcher, P.G., Gormley, R.C., and Matthews, C.N. (1998) Structural investigations of hydrogen cyanide polymers. New insights using TMAH Thermochemolysis/GC-MS, *Origins of Life* **28**, 461-473.

Minard, R.D., Stauffer, E., Kahle, K., and Matthews, C.N. (2000) Titan N_2/CH_4 tholin contains HCN polymers: evidence from TMAH Thermochemolysis/GC-MS analysis, *Abstracts, Astrobiology Sci. Conf.*, April, 2000.

Minard, R.D., Parsons, K., Johnson, A., Shields, E., Fiddler, M., Jones, A.D., and Matthews, C.N. (2002) Surface autocatalytic HCN polymerization. Evidence for glycine oligomers in $(HCN)_x$ and a water-soluble HCN hexamer, *Abstracts, Astrobiology Sci. Conf.*, April, 2002.

Moser, R.E., and Matthews, C.N. (1968) Hydrolysis of aminoacetonitrile: peptide formation, *Experientia* **24**, 658-659.

Moser, R.E., Claggett, A.R., and Matthews, C.N. (1968) Peptide formation from diaminomaleonitrile (HCN tetramer) and from aminomalononitrile (HCN trimer), *Tetrahedron Lett.* 1599-1608.

Oró, J., and Kamat, (1961) Amino acid synthesis from hydrogen cyanide under possible primitive earth conditions, *Nature* **190**, 442-443.

Oró, J., and Kimball, A.P. (1962) Synthesis of purines under possible primitive earth conditions. I. Adenine from hydrogen cyanide, *Arch. Biochem. Biophys.* **96**, 293-295.

Oró, J., and Lazcano-Araujo, A. (1980) The role of HCN and its derivatives in prebiotic evolution, in Vennesland, B., Conn., E.E., Knowles, C.J., Westley, J., and Wissing, F., (eds.), *Cyanide in Biology*, Academic Press, New York, NY, 517-541.

Owen, T. (1982) Titan, *Sci. Amer.* **246**, February, 98-109.

Peltzer, .E.T., Bada, J.L., Schlesinger, G., and Miller, S.L. (1984) Strecker synthesis of amino acids, *Adv. Space Res.* **4**, 69.

Pendleton, Y.J., and Cruikshank, D.P. (1994) Life from the Stars?, *Sky and Telescope*, March, 36-42.

Pesce-Rodriguez, R.A., Liebman, S.A., and Matthews, C.N. (1993) Characterization of hydrogen cyanide polymers and the Murchison meteorite: implications for prebiotic and extraterrestrial chemistry, *Origins of Life* **24**, 123-124.

Pflüger, E. (1875) *Arch. ges. Phisiol.* **10**, 641 (see Oró and Lazcano-Araujo, 1980).

Ponnamperuma, C. (1989) Chemical Evolution, *J. Brit. Interplanetary Soc.* **161**, 1005-1010.

Reinhard, R. (1986) The Giotto encounter with comet Halley, *Nature* **321**, 313-318.

Rettig, T.W., Tegler, S.C., Pasto, D.J., and Mumma, M.J. (1992) Comet outbursts and polymers of HCN, *Ap. J.* **398**, 293-298.

Sagan, C., and Khare, B.N. (1979) Tholins: organic chemistry of grains and gas, *Nature* **277**, 102-107.

Sagdeev, R.Z., Blamont, J., Galeev, A.A., Moroz, V.I., Shapiro, V.D., Shevchenko, V.I., and Szegö, K. (1986) Vega spacecraft encounters with comet Halley, *Nature* **321**, 259-262.

Schaefer, J., Stejskal, E.O., Jacob, G.S., and McKay, R.M. (1982) Natural abundance N_{15} NMR of the solids from the reaction of HCN and ammonia, *Appl. Spectr.* **36**, 179-182.

Su, Y.-L., Honda, Y., Hare, P.E., and Ponnamperuma, C. (1982) Search of peptide-like materials in electric discharge experiments, *Origins of Life* **19**, 237-238.

Sun Kwok (2002) *Astronomy*, June.

Umemoto, K., Takahashi, M., and Yokota, K. (1987) Studies on the structure of HCN oligomers, *Origins of Life* **17**, 283-293.

Völker, T. (1960) Polymeric hydrogen cyanide, *Angew. Chem.* **72**, 379-384.

Warren, C.B., Minard, R.D., and Matthews, C.N. (1974) Synthesis of α-cyanoglycine N-carboxyanhydride and α cyanoglycine, *J. Org. Chem.* **39**, 3375-3378.

Whipple, F.L. (1974) The nature of comets, *Sci. Amer.* **230**, February, 48-57.

Woeller, F., and Ponnamperuma, C. (1969) Organic synthesis in a simulated Jovian atmosphere, *Icarus* **10**, 386-392.

Yang, W., Minard, R.D., and Matthews, C.N. (1976) Azacyclopropenylidenimine: a low energy dimer of hydrogen cyanide, *J. Theor. Biol.* **56**, 111-123.

Biodata of **Rafael Navarro-González** author (with co-author Antigona Segura) of ***"The Possible Role of Volcanic Lightning in Chemical Evolution"***

Dr. Rafael Navarro-González is a distinguished research professor at the Laboratory of Plasma Chemistry and Planetary Studies of the Institute of Nuclear Science of the National Autonomous University of Mexico. He obtained his B. Sc. in biology in 1983 at this University, and his Ph.D. in chemistry in 1989 from the University of Maryland at College Park working under the supervision of Dr. Cyril Ponnamperuma. His current research interest focuses on the origin and evolution of life on Earth and other planetary environments. Dr. Navarro-González is also active in future space missions to Mars and Europa, and in terraforming Mars. His studies on Mars brought him recently to Mars-like environments such as the Atacama Desert in Chile, one of the driest places on Earth, and Pico de Orizaba, Mexico, a volcano with the highest treeline in the world. Dr. Navarro-González has been a visiting professor at the University of Maryland, The Massachusetts Institute of Technology, the Universties of Paris 7 and 12.

E-mail: **navarro@nuclecu.unam.mx**

Dr. Antigona Segura is now a postdoctoral student at the Pennsylvania State University working with Dr. James Kasting in theoretical modeling of planetary atmospheres. She obtained her B. Sc. in physics in 1994 at the Universidad Autonóma de San Luis Potosí, in Mexico, graduating with honors as the best student in the country. During her undergraduate studies she started to do research in the detection of binaries in the Hyades cluster by statistical methods under the direction of Dr. Arcadio Poveda and Dr. Miguel Angel Herrera. She continued to do this research towards her master's degree, graduating in astronomy in 1997. Looking for a research subject related to extraterrestrial life she joined Dr. Rafael Navarro-Gonzalez' research group and changed to the Earth Science graduate program, where she got her Ph.D. in 2001. Her research during her PhD was on the synthesis of reactive nitrogen compounds relevant for prebiotic chemistry in the volcanic plumes of early Mars.

E-mail: **antigona@essc.psu.edu**

R. Navarro-González

A. Segura

J. Seckbach (ed.), Origins, 137–152.

THE POSSIBLE ROLE OF VOLCANIC LIGHTNING IN CHEMICAL EVOLUTION

RAFAEL NAVARRO-GONZÁLEZ AND ANTIGONA SEGURA
Laboratorio de Química de Plasmas y Estudios Planetarios, Instituto de Ciencias Nucleares, Universidad Nacional Autónoma de México. Circuito Exterior, Ciudad Universitaria. Apartado Postal 70-543, México, D.F. 04510 MEXICO.

1. Introduction

Nitrogen and phosphorus are fundamental elements for life. Nitrogen is present in structural (*e.g.*, proteins), catalytic (*e.g.*, enzymes and ribozymes), energy transfer (*e.g.*, ATP) and information storage (RNA and DNA) biomolecules. Atmospheric and planetary models suggest that nitrogen was abundant in the early atmospheres of Earth and Mars as dinitrogen (N_2), an inert gas under normal atmospheric conditions. To be available for prebiotic synthesis it must be converted into hydrogen cyanide, ammonia and/or nitrate, in a process referred to as nitrogen fixation. Due to the strength of the triple bond in N_2, nitrogen fixation, while thermodynamically favored is kinetically restricted (Navarro-González *et al.*, 2001). In a reducing atmosphere dominated by CH_4-N_2, thunderstorm lightning efficiently produces HCN and NH_3 (Toupance *et al.*,1975; Stribling and Miller, 1987; Chameides and Walker, 1981). Nevertheless, photochemical and geochemical constraints strongly suggest that the early atmosphere was weakly reducing, dominated by CO_2 and N_2 with traces of CH_4, CO, and H_2 (Kasting, 1993). Under these conditions, HCN is no longer synthesized in the lightning channel and instead NO is formed (Navarro-González, *et al.*, 2001 and references therein). NO has not yet been implicated in the syntheses of amino acids, purines or pyrimidines under prebiotic conditions. The lack of formation of HCN by thunderstorm lightning introduces serious barriers to the process of chemical evolution in the early Earth.

Phosphorus, on the other hand, is present in ribozymes, ATP, RNA and DNA molecules. This element occurs naturally as apatite ($Ca_5(PO_4)_3(F, Cl)$), an insoluble mineral (Glindemann *et al.*, 1999). In order to be available for prebiotic process, phosphorus must be reduced to hydrophosphites or phosphites.

A promising environment for the synthesis of organic molecules and their rapid removal from the reaction zone is volcanic ash-gas clouds (Markhinin and Podkletnov, 1977; Podkletnov and Markhinin, 1981; Lavrentiev *et al.*, 1984; Hill, 1992), where all the gaseous (or vaporous) components necessary for the formation of prebiotic compounds are present under sharp pressure gradients. At the same time, ash particles are represented by minerals possessing large surface area and well-expressed catalytic properties. Ash-gas clouds are also characterized by the presence of various temperatures as well as powerful electric discharges, which can serve as an efficient energy source (Schwartz and Henderson-Sellers, 1983; Hill, 1992). Recent experimental work conducted in this laboratory and elsewhere shows that the lightning formed in volcanic clouds during highly explosive eruptions may be an important source of reactive nitrogen and phosphorus. This paper reviews the available knowledge on the role of volcanic lightning in chemical evolution, the available data on this environment, and the most important chemical and physical parameters relevant to the Archean.

2. Volcanic Activity

As the collapse of the presolar nebula began, planetesimals coalesced into small planetary embryos. The heating produced by accretion, radioactive decay and electromagnetic induction, from a putative T Tauri phase of the early Sun, led to planetary differentiation and magmatic activity on several planetary embryos in the inner Solar System. Magmatic activity on small planetary objects has been inferred for approximately 70 parent bodies represented by igneous meteorites, such as the Angra dos Reis, the aubrites, the ureilites and the howardite-eucrite-diogenites (Hewins and Newsom, 1988), and some carbonaceous chondrites (Kurat and Kracher, 1980). Remote sensing has indicated the presence of basaltic rocks on the surface of a large asteroid, Vesta (Gaffey, 1983). As these planetary embryos continued to grow into terrestrial planets, their interiors became very hot, with temperatures essentially at the solidus.

Thermal history models of the Earth suggest that the planet cooled rapidly in the first several 100 Myr by a vigorous mantle convection system (Schubert *et al.*, 1989). During this period volcanism should have been very intense, characterized by extraordinarily explosive eruptions on account of the higher content of volatiles in the magma and the greater temperatures in the mantle.

In addition to the above classical mantle-derived volcanism, there was an additional type of volcanism during the late heavy bombardment process (4.5 to 3.8 Gyr) induced by impacts from space bodies. During this period, a large number of projectiles (20 km $\leq$ diameter $\leq$ 300 km) collided with the Earth accreting a substantial mass, estimated in the range of 1×10^{25} to 4×10^{25} g (Chyba *et al.*, 1994). Such catastrophic events were likely to destroy the incipient crust and trigger volcanism. Evidence for impact-triggered volcanism has been found on the Moon at the crater Tsiolkovski and on the Earth at the Sudbury Structure in Ontario, Canada, dated at about 2 Gyr old (French, 1970).

Subsequent to a few 100 Myr after the end of the heavy bombardment process, the Earth underwent a slow and gradual cooling at a rate of the order of 100 K Gyr^{-1} controlled by deep convective heat transport and near-surface conductive heat transport through a mobile lithosphere (Schubert *et al.*, 1989). At this stage, volcanic activity gradually declined to its present rate. The contemporaneous global volcanic activity on Earth has been estimated at about 4 km^3 of lava and pyroclasts emitted yearly (Decker and Decker, 1982); this figure includes material vented by subduction zone (about 1 km^3 yr^{-1}), rift (2.5 km^3 yr^{-1}) and hot-spot (0.5 km^3 yr^{-1}) volcanoes.

3. Explosive Volcanism

Volcanoes emit three types of physical products: gases, liquids, and solids; their relative importance is determined by the degree of explosivity of the eruption, which depends on (1) the viscosity of the magma, (2) its gas content, (3) the rate of emission, and (4) the environment of the vent (Decker and Decker, 1982). Gentle or effusive volcanoes discharge small amounts of fumes and airfall fragments, and large volumes of lava, which may boil out in spectacular fire fountains or simply flow from cracks or vents into streams of lava. Explosive volcanoes generate large volumes of gases and hot solid fragments that can either billows upward as a huge ash cloud or avalanche rapidly down slopes as fluidized flows. The volume of ejecta in explosive eruptions ranges from 10^6 m^3 to 10^{12} m^3 and the height of the ash cloud column can vary from 1-5 km up to 25 km depending on the rate of emission of pyroclasts (Settle, 1978; Simkin *et al.*, 1981). Additionally, explosive volcanoes exhibit intense lightning activities, which makes them particularly relevant to prebiotic chemistry. There are two dominant mechanisms leading to explosive volcanism; these are: (1) the magmatic process, which involves exsolution of dissolved volatiles from the melt during the rise and decompression of magma; and (2) the hydromagmatic process, which operates during contact of melt with external water (liquid or solid) at or near the surface of the Earth. These two processes may operate simultaneously during an eruption if the magma composition and environmental factors permit. It seems likely that these mechanisms also operated in the Archean.

4. Characteristics of Volcanic Clouds

Volcanic clouds have been proposed as a viable environment for prebiotic synthesis because volcanism was a very common phenomenon in the Archean (Fox and Harada, 1961; Harada and Fox, 1964). During these eruptive episodes, the magma and part of the volcanic system are fragmented in particles known as pyroclasts. A volcanic plume is formed by the discharge of a mixture of pyroclasts and magmatic gases into the atmosphere. Such plumes typically exhibit strong electric fields and generate copious lightning discharges of hundreds of meters in length at rates of $10\text{-}10^2$ flash min^{-1} in and around the plume near the eruption site (Navarro-González *et al.,* 1996). The compounds formed in the volcanic lightning channel can escape the high temperature zone of the eruption site at sonic or supersonic speeds and be injected into the troposphere and stratosphere where they would undergo photochemical processing. These characteristics allow volcanic plumes to serve as natural chemical reactors. This environment has been recently reviewed by Basiuk and Navarro-González (1996), and Navarro-González *et al.,* 1996, and an overview is given below.

4.1. GAS PHASE

The composition of volcanic gases can vary considerable from volcano to volcano, and even within the same volcano that undergoes different eruptive phases. A large variety of volatiles are emitted, being the most abundant water steam, CO_2, SO_2, N_2, H_2S, H_2, CO,

HCl, HF, Ar, CH_4, NH_3, COS (Basiuk and Navarro-González, 1996). Large variations, within several (up to 8) orders of magnitude, are observed in the concentrations of these volatiles. Steam is on the average the most abundant and has a relatively constant content. These fluctuations are attributed to a heterogeneous mantle from which the volatiles are released, namely from the deep (less altered) and the upper (more recycled) mantles. Analyses of isotopic composition of noble gases in a suite of volcanic glasses from different sources by Allègre *et al.* (1983) indicate that the volatiles emitted by Hawaiian volcanoes are richer in nonradiogenic noble gases. This has been interpreted to imply the existence of a primordial, undegassed reservoir deep in the Earth's mantle (Allègre *et al.*, 1983, 1993). Therefore, the volatiles emitted by Hawaiian volcanoes could, perhaps, exemplify more closely the nature of gases emitted by Archean volcanoes. The typical composition of volatiles emitted by Kilauea, one of the Hawaiian volcanoes is (Gerlach, 1993): H_2O: 52% (±14%); CO_2: 31% (±13%); SO_2: 15 (±4%); CO: 1% (±0.4%); H_2: 0.8 (±0.2%); and H_2S: 0.2 (±0.1%).

4.2. ASH

Volcanic ash is a fine-grained magmatic material (Kearey *et al.*, 1993), composed of a combination of oxides of the major rock-forming elements. It is formed by disordered Si-O tetrahedra with inclusions of the ions of Mg, Fe, Ca, Na, and K. Depending on the composition, magmatic materials may be classified under the following types: (1) dacite; (2) basalt; (3) andesite; (4) rhyolite; and (5) komatiite. For the early phase of the evolution of the Earth, komatiite was the dominant type of volcanic material erupted. The typical composition of ashes from this type of volcanism is: SiO_2: 48.6%; Al_2O_3: 6.3%; FeO: 11.2%; CaO: 5.7%; Na_2O: 0.1%; MgO: 30.8%; K_2O: 0.02%; and TiO_2: 0.3%.

4.3. TEMPERATURE AND PRESSURE

The initial temperatures at which the magmatic materials reaches the planetary surface depends on the melt composition. For instance, basaltic andesite magma of the 1979 eruption of Soufriere volcano was at 1000°C (Brazier *et al.*, 1982); basaltic lava of the 1980 eruption of Large Tolbachik volcano was at 980°C to 1070°C (Fedotov *et al.*, 1980); for Hawaiian basalts, 980°C to 1200°C (Peck *et al.*, 1979); lava of the Niragongo 1977 eruption, 1100°C (Tazieff, 1977). The temperature of the Mount St. Helens dacitic/andesitic melt varied from 920°C to 990°C (Casadevall *et al.*, 1983). Nisbet (1985) has pointed out that in the Archean the temperatures of primary igneous melts rising up from the mantle were probably much higher than today (1700°C or more). However, the temperature of erupted materials rapidly drops with altitude; for instance for Mount St. Helens, the plume temperature 100 m above the crater was only slightly higher than the ambient temperature (Lawrence *et al.*, 1980).

Magma and gases are compressed at very high pressures prior to eruption, *e.g.* 2200±500 bar for Mt. Pinatubo's eruption on June 15, 1991 (Rutherford, 1991). However, it is very difficult to estimate the pressures at which volcanic matter is erupted from the crater, although it is thought that they cannot exceed a few hundred atmospheres. But after the eruption, the pressure equilibrates with the atmospheric one even faster than the temperature does. Therefore, this plume parameter is generally defined by the altitude. The densest part of the Mount St. Helens plume on May 18, 1980, was reported to reach the upper tropospheric and lower stratospheric levels, which are characterized by the pressure of about 300 and 200 mbar, respectively (Carey and Sigurdsson, 1982). Thus one may accept the pressure range for present-day volcanic ash-gas clouds to be from 200 to 1000 mbar.

4.4. ELECTRIC FIELD

Although there are few reports aimed at studying electric field potentials in volcanic plumes, these seem to lead to the conclusion that injection of volcanic fumes and/or ash into the atmosphere results in a cloud positively charged, both horizontally and vertically, with an abnormal electrostatic field, *i.e.* it differs from the regular fine-weather field of +100-130 V m^{-1} by several orders of magnitude (Anderson *et al.*, 1965; Brook *et al.*, 1974; Cobb, 1980). Anderson *et al.* (1965) recorded electric field potentials during the formation of Surtsey volcano. The cloud had an electric potential as high as 30 kV m^{-1} in the upper regions. The net charge decreased rapidly as the cloud was carried away from the crater by wind. At sea level the electric potential was recorded to be about 8 kV m^{-1} and point discharge currents were measured in the microampere magnitude. Brook *et al.* (1974) detected potential gradients sometimes exceeding 7 kV m^{-1} in the Westmann island of Heimaey's eruption. Kikuchi and Endoh (1982) measured 15 kV m^{-1} about 5 km away from the Mount Usu volcano. Cobb (1980) measured 20 kV m^{-1} at the ground whereas Hobbs and Lyons (1983) registered in an aircraft 10 kV m^{-1} at 175 km from Mount St. Helens on May 18, 1980.

During the explosive phase of a volcano, the potential gradient is oscillating drastically between high and low values (Anderson *et al.*, 1965; Brook *et al.*, 1974). The drop in the potential toward the fine weather values is caused by neutralization of charges through lightning activity whereas its growth is related to new injection of positively charged particles into the atmosphere by the volcano.

The electrification process responsible for the positive charge in the high-velocity tephra eruptions of marine volcanoes might be due to interaction of lava with ocean waters (Anderson *et al.*, 1965). This is supported by the experiments of Blanchard (1964) and Pounder (1980), who found that contact of seawater with molten lava results in the production of a positively charged cloud of particles. Woodcock and Spencer (1961) have shown that these particles are composed of sea salt. Undoubtedly, other charge-separation mechanisms must play a role since land volcanoes are also electrified even though sea water is not present, such as in the eruptions of Vesuvius, Paricutin and Redoubt volcanoes, for example. Possible mechanisms include rock fracture (Fujinawa *et al.*, 1992; Takahashi, 1993), fragmentation of pyroclasts (Cheng, 1982), and grazing collisions between ash particles (Hatakeyama and Uchikawa, 1952).

4.5. VOLCANIC LIGHTNING

Despite several reports of volcanic lightning in the past, few researchers have been interested in studying volcanic lightning. Some possible explanations for such an apparent lack of interest are the unpredictability of the event (though a volcano's violent eruption phase can extend to several months with a sustained lightning activity), the hazards associated to perform field studies, and the inapplicability of typical lightning remote sensing techniques (*e.g.*, visible detection by satellites of thunderstorm lightning at nighttime) to monitor volcanic lightning.

Lightning discharges of hundreds of meters in length are frequently generated during volcanic eruptions, in which both gases and tephra are emitted simultaneously into the atmosphere. They have been observed in explosive eruptions caused by magmatic (Green, 1944), hydromagmatic (Anderson *et al.*, 1965; Brook *et al.*, 1974) and glacier-pyroclastic processes (Hoblitt, 1994). Several types of lightning are produced, such as intracloud, cloud-to-ground, ground-to-cloud, and air-discharges (Anderson *et al.*, 1965; Brook *et al.*, 1974; Salanave, 1980; Hoblitt, 1994); however, there are no available statistics about their relative frequencies. Maximum flashing rates have been reported to be 10 flash min^{-1} for the February 4, 1964 eruption of Surtsey (Anderson *et al.*, 1965), 11 flash min^{-1} for March 18, 1980 eruption of Mt. St. Helens (Cobb, 1980), 16 flash min^{-1} for the February 15, 1990 eruption of the Redoubt Volcano (Hoblitt, 1994), and >60 flash min^{-1} for the April-June, 1979 eruption of the Soufriere Volcano (Shepherd *et al.*, 1979).

5. Energy Dissipation Rate

Volcanism is an expression of heat transfer from the mantle to the surface. The planet loses internal heat (i) by conduction from the interior to the surface and then into space, (ii) by advection where melts are generated in the interior and migrate to the surface or near the surface, and (iii) by convection involving gravity-induced turn-over by plastic flow of the mantle. The current mean surface heat flow due to conduction has been estimated to be about 56.6 mW m^{-2} (Schubert, 1997; Schubert *et al.*, 1980 and 1989). Volcanism is the surface manifestation of advection and/or convection (Park, 1997). At present the volcanic flux has been estimated at about 4 km^3 yr^{-1} (Decker and Decker, 1982); this includes material vented by subduction zone (~ 1 km^3 yr^{-1}), rift (~ 2.5 km^3 yr^{-1}) and hot-spot volcanoes (~ 0.5 km^3 yr^{-1}). Taking into account a density of ~ 2900 kg m^{-3}, and an enthalpy of melting ~418 J g^{-1} for basalts (Yoder, 1976), the surface heat flow due to volcanism is estimated to be ~ 0.3 mW m^{-2}. Approximately 30% of this value is dissipated explosively by subaerial eruptions in the form of tephra (Pyle, 1995) where volcanic lightning is generated. Based on thermal evolution models, it is predicted that 4 billion years ago the surface heat flow of the Earth was ~ 400 mW m^{-2} (BVSP, 1981; Schubert *et al.*, 1980; 1989). If we assume that the surface heat loss due to conduction through the lithosphere and hydrosphere has remained constant over time, the surface heat flux due to volcanic activity is derived to be about 340 mW m^{-2}. This value is somewhat of an upper limit since the heat flow by conduction may have varied with time. This implies that approximately 4.6×10^3 km^3 of basalt erupted annually. Eruption rates of this magnitude are consistent with major flood basalt episodes of the recent geologic past (250 to 14 Myr ago) (Richards *et al.*, 1989; Longhi, 1997). Observations suggest that about 10 wt% of the basaltic magma is fragmented into fine ash and entrained into volcanic plumes above flood eruptions (Stothers *et al.*,

1986; Woods, 1993a,b). The mass flux of tephra (*MF*) injected to the atmosphere is therefore calculated to be ~ 1.3×10^{15} kg yr^{-1}. Measurements of the electrical charge on ash particles from explosive volcanic eruptions indicate that the particles are nearly saturated with charge. The charge-to-mass ratio (Q / m) averages ~ 8×10^{-4} C kg^{-1} for both positive and negative charges (Gilbert and Lane, 1994). Injection of these electrically charged tephra particles into the atmosphere leads to the generation of strong electric fields within the volcanic plume. The plume typically exhibits a dipole structure with negatively charged particles in its lower region close to the vent site and positively charged particles in its upper portion. The electric field potential (V) between these two regions has been recorded in a number of volcanic plumes and ranges from 10 to 30 kV m^{-1} (Navarro-González *et al.*, 1996). Electric breakdown within volcanic plumes leads to the generation of lightning discharges with a typical length (L) of about 500. The maximum electric power (P) available in explosive volcanic eruptions may be readily estimated according to the following equation: $P = MF\ V\ L\ Q / m$. Considering a mean value of 20 kV m^{-1} for V, this results in about ~ 1×10^{19} J yr^{-1} accessible for nitrogen fixation (Navarro-González *et al.*, 1998).

6. Nitrogen Fixation

The principal reservoir of nitrogen is molecular dinitrogen in the atmosphere. However to be available for chemical evolution it must be in the form of ammonia or nitrate, forms known as fixed nitrogen. Due to the strength of the triple bond in N_2, nitrogen fixation, while thermodynamically favored is kinetically restricted. Navarro-González *et al.* (1998) examined the effect of volcanic lightning as a mechanism to fix nitrogen on early Earth. For their experiment, it was assumed that the chemical composition of Archean volcanic gases was similar to that of Hawaiian volcanoes. Volcanic gases generally undergo rapid mixing with underground or ground water or with the surrounding atmosphere present outside the volcanic vent environment. Consequently, the effects of dilution of volcanic gases with water and the atmosphere were investigated, assuming that early Earth's atmosphere was composed by 80% carbon dioxide and 20% molecular nitrogen (Kasting, 1990).

The magmatic gases gas mixture was excited by flowing into a microwave discharge cavity, and the products were analyzed in a flow system coupled to chemical ionization mass spectroscopy. Figure 1 shows the mass spectrograms for the various gas mixtures examined. A quantitative study of the experimental production efficiency of nitric oxide (NO) is given in Figure 2 as a function of dilution of the volcanic gas mixture with water and atmospheric gases. In addition, this figure also includes the predicted trends for NO and other minor species expected to form in the cooling channel of volcanic lightning. The experimental and predicted trends for NO agree quite well under most conditions except at high dilutions of water vapor (≥ 60%). This discrepancy is attributed to the injection of excess water in the form of aerosols when a high helium flow passes through the water bubbler. The predicted NO energy yield is estimated to be ~ 1×10^{16} molecule

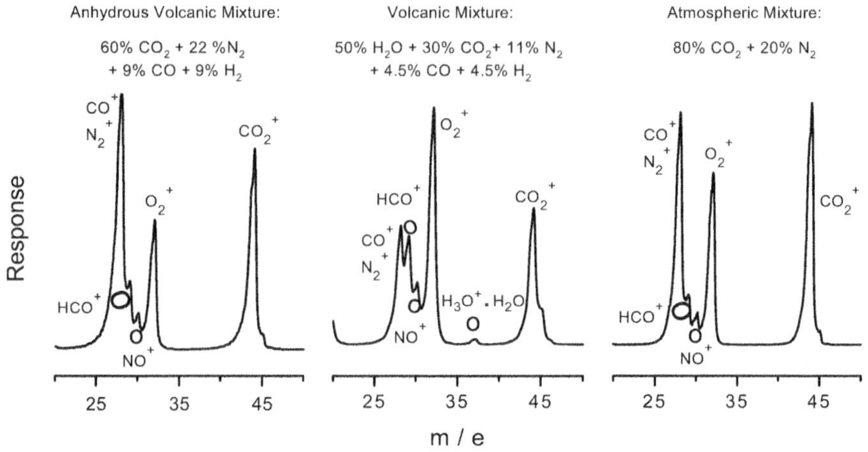

Figure 1. Chemical ionization mass spectrograms of effluents from different types of volcanic gas mixtures subjected to a microwave discharge.

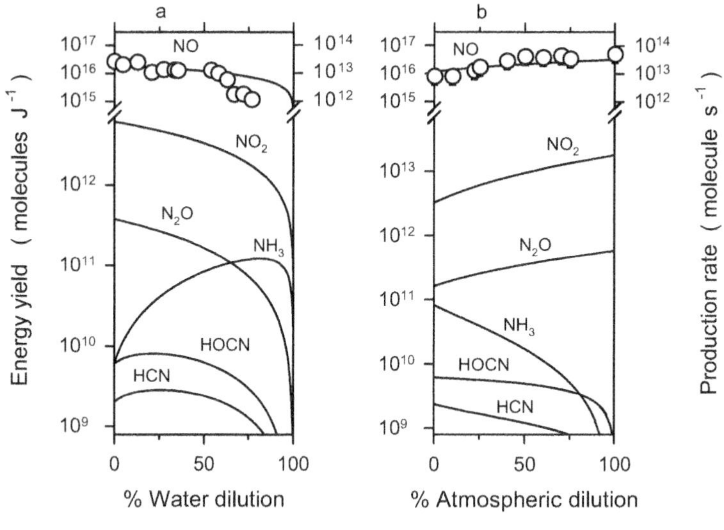

Figure 2. Production yields for several volcanic lightning products as a function of dilution of volcanic gases with (a) water and (b) the surrounding atmosphere. Lines are predicted trends from the thermochemical-hydrodynamic model whereas symbols are the experimental values derived in a microwave discharge.

J^{-1} for volcanic lightning produced in a Hawaiian style volcanic mixture. This value can increase to ~2×10^{16} molecule J^{-1} in an anhydrous volcanic mixture or to ~3×10^{16} molecule J^{-1} at 97.5% dilution of the magmatic gases with the surrounding atmosphere.

If the magmatic gases are significantly diluted with water, this value can decrease down to ~2×10^{15} molecule J^{-1} at 97.5% dilution. Other minor products predicted (see Fig. 2) were not detected experimentally on account of their extremely low product energy yields. A detailed analysis of volcanic lightning compared with other endogenous and exogenous sources showed that this source was the most important to fix nitrogen in early Earth (Navarro-González *et al.*, 2000).

Segura and Navarro-González (2000) examined the effect of volcanic lightning in a more reducing atmosphere from a simulated Martian volcanic plume. The chemical composition of the gas mixture used in these experiments was derived from the accretion model developed by Kuramoto and Matsui (1996) that was applied to Mars by Kuramoto (1997). The composition of volcanic gases was 64% CH_4, 24% H_2, 10% H_2O, 2% N_2. This was irradiated by focusing a high-energy infrared laser beam to simulate lightning, and the resulting products were analyzed by gas chromatography coupled to infrared and mass spectrometry. Figure 3 shows a gas chromatogram of the irradiation products. It has found that HCN was the most important form of fixed nitrogen with an energy yield of ~6×10^{14} molecules J^{-1}. HCN is considered a key intermediate in chemical evolution since it can lead to the formation of amino acids, purines and pyrimidines (Ferris and Hagan, 1984).

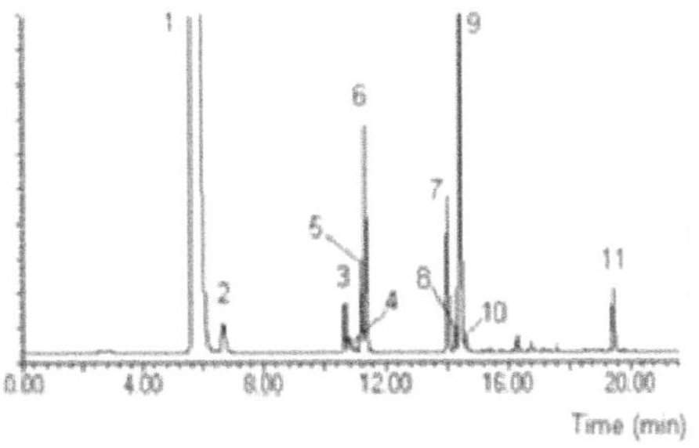

Figure 3. Gas chromatogram of compounds produced by lightning in a simulated Martian volcanic plume. Peak identification: 1. Acetylene + ethylene; 2. Ethane; 3. Propene; 4. Hydrogen cyanide; 5. 1,2-propadiene; 6. Propyne; 7. 1-buten-3-yne; 8. 1-butyne; 9. 1,3-butadiyne; 10. 2-butyne; 11. Benzene.

7. Phosphate Reduction

Glindemann *et al.* (1999) and de Graaf and Schwartz (2000) studied the reduction of phosphate by volcanic lightning using a spark discharge which was generated by applying a microwave field between the ends of a quartz tube containing the sample. In a first set of experiments, Glindemann *et al.* (1999) used samples as solution of Na_2HPO_4 or as pastes (mixtures of fluorapatite with montmorillonite) in water-saturated with a nitrogen atmosphere containing 1-10% CH_4. A spark discharge was generated in the device using a domestic microwave oven. The products were analyzed by gas chromatography-mass spectrometry and the results are summarized in Table 1. In 10%

TABLE 1. Formation of phosphite by reduction of orthophosphate in $CH_4 + N_2$ mixtures.

Phosphate source	Matrix	%CH_4 (in N_2)	%Yield
Na_2HPO_4	-	10	11
Na_2HPO_4	-	1	4
Na_2HPO_4	-	0	0.1
Hydroxyapatite	Montmorillonite	10	6
Fluorapatite	Montmorillonite	10	7
Fluorapatite	Montmorillonite	1	4
Fluorapatite	Montmorillonite	0	0.5
-	Montmorillonite	10	-

methane an average yield of 11% phosphite was obtained using Na_2HPO_4 as the source of phosphorus. As the content of methane in the atmospheric mixture decreased, the production efficiency of phosphite dropped to 4% in 1% CH_4 and to 0.1% in pure N_2. In order to extend these results to a geophysical more plausible model, Glindemann *et al.* (1999) examined the reduction of fluorapatite, the most important phosphate mineral in igneous rocks. In these experiments, a yield of 7% phosphite was obtained in the presence of 10% methane. As with the experiments of Na_2HPO_4, the yield of phosphite decreased but was still substantial. Small amounts of methylphosphonic acid could also be detected in small quantities in these experiments.

As was proposed by Segura and Navarro-González (2000, 2001), these reduced compositions may be geologically relevant for Mars, where phosphorus is an abundant element. Accretion models predicted concentrations exceeding those on Earth for volatile and moderatlely volatile elements (Dreibus and Wänke, 1985). Martian meteorites analyses (Banin *et al.*, 1992) and Mars Pathfinder mission results (Dreibus *et al.*, 2000) have verified the high phosphorus content on Mars. Apatite grains have been detected in the Martian meteorites (Leshin, *et al.*, 1996, Watson, *et al.*, 1994); therefore this was an available source of phosphates on Mars.

De Graaf and Schwartz (2000) extended the research using gas mixtures contained 60% CO_2, 22-40% N_2 and variable concentrations of H_2 and CO. These mixtures are based on a volcanic outgassing model studied by Navarro-González *et al.* (1998, 2000). For these experiments, fluorapatite was mixed in a clay mineral matrix in order to have a more representative composition of igneous rocks. Samples undergo the same experimental and analysis processes that in the experiment developed by Glindemann *et al.* (1999). Their results demonstrated that several percent reduction of apatite occurs even in the presence of as little as 1% H_2 and CO_2 in the gas mixture. In addition, they demonstrated the formation of polyphosphate production (see Table 2) as a result of heating the mineral apatite in the presence of other minerals by the lightning discharge under various atmospheric chemical compositions.

TABLE 2. Analysis for orthophosphate (Pi), pyrophosphate (PPi) and tripolyphosphate (PPPi) in water extracts after exposure to electric discharge.

Matrix	Conditions	Pi	PPi	PPPi	% Recovery
Montmorillonite	60% CO_2 + 40% N_2 + H_2O_{vapor}	36±8	10±3	1.7±0.8	48
Montmorillonite	Same as above	39±1	13±2	1.7±0.5	54
Bentonite	Same as above	30±2	10±1	1.7±0.8	42
Montmorillonite	Same as above	12±3	4±1	-	16
Montmorillonite	Same as above but - water vapor	5±1	5±1	-	10
Montmorillonite	100% N_2 + H_2O $_{vapor}$	35±5	12±3	-	47

8. Concluding Remarks

The experiments presented above demonstrate that nitrogen fixation is an efficient process in volcanic lightning. Nitric oxide and hydrogen cyanide are the major volcanic lightning products expected for early Earth and Mars, respectively. In the case of phosphite, it is produced in high yields even in neutral atmospheres. Therefore electrical discharges associated with volcanic eruptions supplied a pathway by which nitrogen and phosphorus atoms were incorporated into prebiotic molecules needed for the emergence and sustainability of life. We hope that the work described in this review may motivate additional research on volcanic lightning. For instance, it is essential a thorough characterization of intracloud, cloud-to-ground, ground-to-cloud and air discharges in terms of frequency, mode of propagation, number of return strokes, discharge length, peak current, energy dissipation and lightning spectra. In addition, good laboratory simulations have to be designed to evaluate the possible contributions of volcanic lightning to the origins of life on Earth; particularly exploring the effects of high water vapor content and the presence of catalytically active surfaces from tephra. These studies should be complemented with field work, analyzing the contemporaneous chemical effects induced by volcanic lightning in the atmosphere.

9. Acknowledgements

This work was supported by grants from the National autonomous University of Mexico (DGAPA-UNAM IN119999) and the National Council of Science and technology of Mexico (CONACYT No. 32531-T and F323-M9211.

10. References

Allègre, C.J., Sarda, P. and Staudacher, T. (1993) Speculations about the cosmic origin of He and Ne in the interior of the Earth. *Earth Planet. Sci. Let.* **117**, 229-233.

Allègre, C.J., Staudacher, T, Sarda, P. and Kurz, M. (1983) Constraints on evolution of earth's mantle from rare gas systematics. *Nature* **303**, 762-766.

Anderson, R., Bjornsson, S., Blanchard, D.C., Gathman, S., Hughes, J., Jonasson, S., Moore, C.B., Survilas, H.J. and Vonnegut, B. (1965) Electricity in volcanic clouds. *Science* **148**, 1179-1189.

Banin, A., Clark, B.C., and Wänke H. (1992) Surface Chemistry and Mineralogy, In: H. H. Kieffer, B.M. Yakosky, C.W. Snyder and M.S. Matthews (eds.) *Mars*, The University Arizona Press, pp. 594-625.

Basaltic Volcanism Study Project (1981) *Basaltic Volcanism on the Terrestrial Planets*, Pergamon Press, Inc, New York.

Basiuk, V.A. and Navarro-González, R. (1996) Possible role of volcanic ash-gas clouds in the Earth's prebiotic chemistry. *Origins Life Evol. Biosph.* **26**, 173-194.

Blanchard, D.C. (1964) Charge separation from saline drops on hot surfaces. *Nature* **201**, 1164-1166.

Brazier, S., Davis, A.N., Sigurdsson, H. and Sparks, R.S.J. (1982) Fall-out and deposition of volcanic ash during the 1979 explosive eruption of the soufriere of St. Vincent. *J. Volcanol. Geotherm. Res.* **14**, 335-359.

Brook, M., Moore, C.B. and Sigurgeirsson, T. (1974) Lightning in volcanic clouds. *J. Geophys. Res.* **79**, 472-475.

Carey, S.N. and Sigurdsson, H. (1982) Influence of particle aggregation on deposition of distal tephra from the May 18, 1980, eruption of Mount St. Helens volcano. *J. Geophys. Res.* **87**, 7061-7072.

Casadevall, T., Rose, W., Gerlach, T., Greenland, L.P., Ewert, J., Wunderman, R., and Symonds, R. (1983) Gas emissions and the eruptions of Mount S. Helens through 1982. *Science* **221**, 1383-1385.

Chameides, W.L., and. Walker, J.C.G (1981) Rates of Fixation by Lightning of Carbon and Nitrogen in Possible Primitive Atmospheres. *Origins Life Evol Biosph.* **11**, 291-302.

Cheng, R.J. (1982) The mechanisms of fine particle generation and electrification during Mount St. Helens volcanic eruption in Atmospheric effects and potential climatic impact of the 1980 eruptions of Mount St. Helens. In: A. Deepak (ed.) *Atmospheric Effects and Potential Climatic Impact of the 1980 Eruptions of Mount St. Helens*, NASA CP-2240, Washington, D.C., pp.211-217.

Chyba, C., Owen, T., and Ip, W.H. (1994) Impact Delivery of Volatiles and Organic Molecules to Earth. In: T. Gehrels (ed.) *Hazards Due to Comets and Asteroids*, University of Arizona Press, Tucson, 9-58.

Cobb, W.E.: 1980, Electric fields and lightning in the Mt. St. Helens volcanic clouds. *Eos. Trans. Am. Geophys Union* **61**, 978.

De Graaf, R.M., and Schwartz, A.W. (2000) Reduction and activation of phosphate on the primitive Earth. *Origin Life Evol. Biosph.* **30**, 405-410.

Decker, R. and Decker, B. (1982) *Volcanoes and the Earth's Interior*, W.H. Freeman and Co., Sn Francisco.

Dreibus, G., and Wänke, H. (1985) Mars, a volatile-rich planet. *Meteoritics* **20**, 367-381.

Dreibus, G., Brückner, J., and Wänke, H. (2000) Phosphorus in Martian Rocks and Soils and the Global Surface Chemistry of Mars as Derived from APXS on Pathfinder. 31st Lunar and Planetary Science Conference, abstract 1127.

Fedotov, S.A., Chirkov, A.M., Gusev, N.A., Kovalev, G.N., and Slezin, Yu.B. (1980) The large fissure eruption in the region of Plosky Tolbachik volcano in Kamchatka 1975-1976. *Bull. Volcanol.* **43**, 47-60.

Ferris, J.P. and Hagan, W.J.Jr (1984) HCN and chemical evolution: The possible role of cyano compounds in prebiotic synthesis. *Tetrahedron* **40**, 1093-1120.

Fox, S.W. and Harada, K. (1961) Synthesis of uracil under conditions of a thermal model of prebiological chemistry. *Science* **133**, 1923-1924.

French, B.M. (1970) Possible relations between meteorite impact and igneous petrogenesis, as indicated by the Sudbury structure, Ontario, Canada. *Bull. Volcanol.* **34**, 466-517.

Fujinawa, Y., Kumagai, T. and Takahashi, K. (1992) A study of anomalous underground electric field variations associated with a volcanic eruption. *Geophys. Res. Lett.* **19**, 9-12.

Gaffey, M.J. (1983) The Asteroid (4) Vesta: Rotational Spectral Variations, Surface Material Heterogeneity, and Implications for the Origin of the Basaltic Achondrites. *Lunar Planet. Sci.* **XIV**, 231-232.

Gerlach, T.M. (1993) Oxygen buffering of Kilauea volcanic gases and the oxygen fugacity of Kilauea basalt. *Geochim. Cosmochim. Acta* **57**, 795-814.

Gilbert, J.S. and Lane, S.J. (1994) Electrical phenomena in volcanic plumes In: T.J. Casadevall, (ed.) *Volcanic Ash and Aviation Safety: Proceedings of the First International Symposium on Volcanic Ash and Aviation Safety*, U.S. Geological Survey Bulletin 2047, Washington, D.C., 31-38.

Glindemann, D., De Gaaf, R.M., and Schwartz, A.W. (1999) Chemical reduction of phosphate on the primitive Earth. *Origins Life Evol. Biosph.* **29**, 555-561.

Green, J.A. (1944) Paricutín, the cornfield that grew a volcano. *Natl Geograph. Mag.* **85 (2),** 129-156.

Harada, K. and Fox, S.W. (1964) Thermal synthesis of natural amino-acids from a postulated primitive terrestrials atmosphere. *Nature* **201**, 335-336.

Hatakeyama, H. and Uchikawa, K. (1951) On the disturbance of atmospheric potential gradient caused by eruption-smoke of the volcano Aso. *Pap. Met. Geophys.* **2**, 85-89.

Hewins, R.H. and Newsom, H.E. (1988) Igneous activity in the early Solar System, In: J.F. Kerridge and M.S. Mattews (eds.), *Meteorites and the Early Solar System*, University of Arizona Press, Tucson, pp.73-101.

Hill, R.D. (1992) An efficient lightning energy source on the early Earth. *Origins Life Evol. Biosphere* **22**, 277-285.

Hobbs, P.V. and Lyons, J.H. (1983) *Final Report to IRT Corporation* under P.O. 31433, 24p.

Hoblitt, R.P. (1994) An experiment to detect and locate lightning associated with eruptions of Redoubt volcano. *Volcanol. Geotherm Res.* **62,** 499-517.

Kasting J.F. (1993) Earth's early atmosphere. *Science* **259**, 920-926.

Kasting, J.F (1990) Impacts and the oxidation state of carbon in the Earth's early atmosphere. *Origins Life Evol. Biosph.* **20,** 199-231.

Kearey P. *et al.* (eds.): 1993, *The Encyclopedia of the Solid Earth Sciences*, Blackwell Scientific Publications, Oxford.

Kikuchi, K. and Endoh, T. (1982) Atmospheric electrical properties of volcanic ash particles in the eruption of Mt. Usu Volcano, 1977. *J. Met. Soc. Jap.* **60 (1),** 548-561.

Kuramoto K. and Matsui T. (1996) Partitioning of H and C between the mantle and core during the core formation in the Earth: Its implications for the atmospheric evolution a redox state of early mantle. *J. Geophys. Res.* **101 (E6),** 14,909-14,932.

Kuramoto, K. (1997) Accretion, core formation, H and C evolution of the Earth and Mars. *Phys. Earth Planet. Int.* **100,** 3-20.

Kurat, G. and Kracher, A.: (1980) Basalts in the Lance carbonaceous chondrite. *Z. Naturforsch.* **35a**, 180-190.

Lavrentiev, G.A., Strigunkova, T.F. and Egorov, I.A. (1984) Abiological synthesis of amino acids, purines and pyrimidines under conditions simulating the volcanic ash-gas clouds. *Origins of Life* 14, 205-212.

Lawrence, W.S., Qamar, A., Moore, J., and Kendrick, G. (1980) A comparison of thermal observations of Mount St. Helens before and during the first week of the initial 1980 eruption. *Science* **209**, 1526-1527.

Leshin, L.A., Harvey, R.P., McCoy, T.J., and McKeegan K.D. (1996) Water in Apatite from Shergottite QUE94201: Abundance and D/H. *Meteoritics Planet. Sci.* **31,** A79.

Longhi, J. (1997) Flood Basalt, In: J. H. Shirley and R. W. Fairbridg (eds.), *Encyclopedia of Planetary Sciences*, Chapman & Hall, London, pp. 249-251.

Markhinin, E.K. and Podkletnov, N.E. (1977a) The phenomenon of formation of prebiological compounds in volcanic processes. *Origins Life Evol. Biosphere* **8**, 225-235.

Navarro-González, R., Basiuk, V.A., and Rosembaum, M. (1996) Lightning associated to Archean volcanic ash clouds, In: J. Chela-Flores and F. Raulin (eds.*), Chemical Evolution: Physics of the Origin and Evolution of Life*, Kluwer Academic Publishers, Netherlands, pp 123-142.

Navarro-González, R., McKay, C.P. and Nna Mvondo, D. (2001) A possible nitrogen crisis for Archean life due to reduced nitrogen fixation by lightning. *Nature* **412**, 61-64.

Navarro-González, R., Molina, M.J. and. Molina, L.T. (1998) Nitrogen fixation by volcanic lightning in the early Earth. *Geophys. Res. Lett.* **25**, 3123-3126.

Navarro-González, R., Molina, M.J., and Molina, L.T. (2000) The chemistry of Archean volcanic lightning, In: M Akobayashi, N. Fujii and R. Navarro-González (eds.), *The Role of Radiation in the Origins of Life and Evolution of Life*. Kyoto University Press, 121-141.

Nisbet, E.G. (1985) The geological setting of the erliest life forms. *J. Mol. Evol.* **21,** 289-298.

Park, A.F. (1997) Volcanism in the Solar System, In: J. H. Shirley and R. W. Fairbridge (eds.), *Encyclopedia of Planetary Sciences*, Chapman and Hall, London, pp. 915-922.

Peck, D.L., Wright, T.L., and Decker, R.W. (1979) The lava lakes of Kilauea. *Sci. Am.* **241(4),** 114-128.

Podkletnov, N.E. and Markhinin, E.K. (1981) New data on abiogenic synthesis of prebiological compounds in volcanic processes. *Origins Life* **11**, 303-315.

Pounder, C. (1980) Volcanic lightning. *Weather* 35, 357-360.

Pyle, D.M. (1995) New data on abiogenic synthesis of prebiological compounds in volcanic processes. *Geophys. Res. Lett.* **22**, 563-566.

Richards, M.A., Duncan, R.A. and Courtillot, V. E. (1989) Flood basalts and hot spots tracks: Plume heads and tails. *Science* **246,** 103-107.

Rutherford, M.J.: 1991. Pre-eruption conditions and volatiles in 1991 Pinatubo magma. *Eos, Trans. Am. Geophys. Union,* **72**, 62.

Salanave, L.E. (1980) *Lightning and Its Spectrum. An Atlas of Photographs*, Univ. of Arizona Press, Tucson.

Schubert, G. (1997) J. H. Shirley and R. W. Fairbridge (eds.), Thermal evolution of planets and satellites, In: *Encyclopedia of Planetary Sciences*, Chapman and Hall, London, pp. 808-814.

Schubert, G., Stevenson, D., and Cassen, P. (1980) Whole planet cooling and the radiogenic heat source contents of the Earth and Moon. *J. Geophys. Res.* **85**, 2531-2538.

Schubert, G., Turcotte, D.L., Solomon, S.C., and. Sleep, N.H (1989) Coupled evolution of the atmospheres and Interiors of Planets and Satellites, In: S. K. Atreya, J. B. Pollack and M. S. Matthews (eds.), *Origin and Evolution of Planetary and Satellite Atmospheres*, Univ. Arizona Press, Tucson, pp. 450-483.

Schwartz, A.W. and Henderson-Sellers, A. (1983) Glaciers, volcanic islands and the origins of life. *Precambrian Res.* **22**, 167-174.

Segura, A., and Navarro-González, R. (2000) Experimental simulation of volcanic lightning in early Mars, In J. Chela-Flores, G.A. Lemarchand and J. Oró (eds.), *Astrobiology: Origins from the Big bang to Civilisation*, Kluwer Academic Publishers, Netherlands, pp 293-296.

Segura, A., and Navarro-González, R. (2001) Experimental simulation of early martian volcanic lightning. *Adv. Space. Res.* **27,** 201-206.

Settle, M. (1978) Volcanic eruption clouds and the thermal power output of explosive eruptions *J. Volcanol. Geotherm. Res.* **3,** 309-324.

Simkin, T., Siebert, L., McClelland, Bridge, D., Newhall, C. and Latter, J.H. (1981) *Volcanoes of the World,* Hutchinson Ross Pub. Co., Stroudsburg, Pennsylvania.

Stothers, R. S., Wolff, J. A. Self, S., and Rampino, M. R. (1986) Basaltic fissure eruptions, plume heights, and atmospheric aerosols. *Geophys. Res. Lett.* **13,** 725-728.

Stribling, R., and Miller, S.L. (1987) Energy yields for the hydrogen cyanide and formaldehyde synthesis: the HCN and amino acid concentrations in the primitive ocean. *Origins Life* **17,** 261-273.

Takahashi, K. (1993) *J. Commun. Res. Lab.* 40, 3.

Tazieff, H. (1977) An exceptional eruption: Mt. Niragongo, Jan. 10th, 1977. *Bull. Volcanol.* **40**, 189-200.

Toupance, G., Raulin, F. and Buvet, R. (1975) Formation of preobiological compounds in models of primitive Earth's atmosphere. I: CH_4-NH_3 and CH_4-N_2 atmospheres. *Origins Life Evol. Biosph.* **6,** 83-90.

Watson, L.L., Hutcheon, I.D., Epstein, S., and Stolper, E. (1994) Water on Mars - Clues from Deuterium / Hydrogen and Water Contents of Hydrous Phases in SNC Meteorites. *Science* **265**, 86-90.

Woodcock, A.H. and Spencer, A.T. (1961) Lava-sea-air contact areas as sources of sea salt particles in the atmosphere. *J. Geophys. Res.* **66 (9),** 2873-2887.

Woods, A.W. (1993a) Moist convection and the injection of volcanic ash into the atmosphere *J. Geophys. Res.* **98,** 17627-17636.

Woods, A.W. (1993b) A model of the plumes above basaltic fissure eruptions. *Geophys. Res. Lett.* **20,** 1115-1118.

Yoder, H. S. (1976) *Generation of Basaltic Magma* National Academy of Science, Washington, D.C.

Biodata of **Heikki Tervahattu** author (with **A. Tuck** and **V. Vaida**) of ***"Chemistry in Prebiotic Aerosols: A Mechanism for the Origin of Life"***

Dr. Heikki Tervahattu is a Researcher at the University of Helsinki, Department of Limnology and Environmental Protection. He obtained his Ph.D. in 1969 from the University of Oulu, Finland. Dr. Tervahattu is a Leader in the Urban Dust Project (UDP). The overall objective of the Project is to produce new knowledge about the composition and sources of urban particulate matter in order to improve air quality. His group is specialized to use electron microscopy and related methods in aerosol particle research. UDP was first focused to road dust, which cause health impacts every spring in Nordic countries. More recently, long-range transport (LRT) of aerosol particles has become the main object of UDP. In the LRT study of sea-salt particles, the group found particles, which had an organic surface layer composed mainly of fatty acids. This discovery was the reason why Dr. Tervahattu became interested about the origin of life and the theory of prebiotic aerosols as potential reactors in the chemical development. E- mail: **heikki.tervahattu@helsinki.fi**

Dr. A.F. Tuck is currently Chief of the Meteorological Program at the NOAA Aeronomy Laboratory. He obtained his Ph.D. in Physical Chemistry (1969) from Cambridge University (UK) and a Certificate from the Meteorological Office College in 1976. His previous scientific activities include leadership of the NASA/NOAA/NSF Airborne Antarctic Ozone Experiment (AAOE), the Airborne Arctic Stratospheric Expedition (AASE), the Airborne Southern Hemisphere Ozone Experiment (ASHOE) and the WB57F Aerosol Mission (WAM). Dr. Tuck has also received several awards, such as the Medal of Honour (Helsinki University), the Buchan Prize (Royal Meteorological Society), L.G. Groves Memorial Prize for Meteorology and a Fulbright Scholarship (University of California, San Diego) E-mail: **tuck@al.noaa.gov**

Dr. Veronica Vaida is a Professor of Chemistry and Biochemistry and a Fellow of the Cooperative Institute for Research in Environmental Sciences at the University of Colorado. She obtained her Ph.D. in 1977 from Yale University, New Haven, CT and was an Assistant and Associate Professor of Chemistry at Harvard University before coming to Colorado in 1984. Her current research at the interface of physical chemistry and atmospheric science involves the study of spectroscopy and photoreactivity of atmospheric radicals, their complexes and aerosols. She obtained several awards from various sources. E-mail: **Vaida@colorado.edu**

Heikki Tervahattu

A.F. Tuck

Veronica Vaida

J. Seckbach (ed.), Origins, 153–165.

CHEMISTRY IN PREBIOTIC AEROSOLS: A MECHANISM FOR THE ORIGIN OF LIFE

HEIKKI TERVAHATTU[1], ADRIAN TUCK[2] and VERONICA VAIDA[3]*

[1]*Department of Limnology and Environmental Protection, University of Helsinki, P. O. Box 62, FIN-00014 Helsinki, Finland;* [2] *NOAA Aeronomy Laboratory R/AL6, 325 Broadway, Boulder CO 80305-3328, USA;* [3]*Department of Chemistry and Biochemistry, University of Colorado, Boulder CO 80309-0215, USA*

1. Introduction

The appearance of life in the universe must have involved a sequence of events (Oparin, 1924) starting with the synthesis from inorganic precursors of simple organic compounds such as hydrocarbons, amino acids, cyanides, purines, pyrimidines and other chemical building blocks of life (Miller, 1953; Fox and Dose, 1972; Miller 1998; Rode, 1999). These carbon-containing compounds must have formed more and more complex substances and polymers with high-energy bonds, which would then have had the ability to self-assemble into structures capable of growth and multiplication. Such supramolecular structures would have had to acquire more elaborate metabolism and, through prebiological selection, would have led to the formation of primordial organisms.

It is generally accepted that on Earth and elsewhere, several plausible mechanisms can be contemplated for the generation of organic compounds from inorganic precursors under prebiotic conditions. The prebiotic synthesis of lipids, amino acids and short chain peptides, nucleic acids and short chain polynucleotides, polyamines, polyphosphates etc. has been suggested by many researchers, first by Oparin in 1924 (Oparin,1924; Norris and Raine,1998; Aylward and Bofinger, 2001; Sowerby et al., 2002; Wills and Bada, 2001). Miller showed in 1953 that important biomolecules such as amino acids could be synthesized under simulated conditions of the early Earth, with carboxylic acids, purines and pyrimidines all having been so synthesized subsequently (Miller 1953; Ourisson and Nakatani, 1994; Bada and Lazcano, 2002). The proposals include the use of discharges and UV radiation to form amino acids, sugars, purines and pyrimidines. Recently, the high pressure and high temperature conditions characteristic of hydrothermal vents were shown to be environments for the synthesis of organic compounds (Huber and Wächtershäuser, 1998; Amend and Shock, 1998). Many such organic compounds have been discovered in interstellar space or extracted from meteorites, suggesting that chemical compounds can be and are formed with some likelihood in extraterrestrial

**author for correspondence*

environments and brought to Earth by meteoritic and cometary infall (Anders, 1989; Chyba and Sagan, 1992). From the Murchison meteorite one can estimate that complex aromatic hydrocarbon polymers, soluble organic acids, aliphatic and aromatic hydrocarbons, amino acids, urea, ketones, alcohols, aldehydes and purines were delivered to Earth (Cronin et al., 1988). In what follows, we discuss scenarios for the chemical evolution of life-like properties assuming that diverse monomeric organic molecules were present in sufficient concentration.

The main question addressed by our proposal is how simple organic compounds (amino acids, purines, pyrimidines, fatty acids and sugars), can be concentrated sufficiently and ordered to undergo polymerization reactions to produce the high energy biopolymers. Several approaches have been considered to overcome the bottleneck in the sequence of events described above. Some models for the emergence of life focus on self replicating molecules, "RNA World", while others describe replicating molecular assemblies, “metabolism first” and "the lipid world". Interesting alternatives are hydrothermal vents, panspermia and various templates for biopolymer formation such as minerals and clays. The ideas developed in this chapter make use of atmospheric aerosols and of the ocean-atmosphere interactions to explore the role that organic films on the surface of atmospheric aerosols and of the ocean could have played in concentrating organic compounds and in promoting the synthesis of high-energy bonds used in biochemistry. The nature of the aerosols and the chemical heterogeneity of the planetary surface probably implies co-evolution of the common classes of biopolymer – lipids, proteins and nucleic acids - from the start.

The present sea surface microlayer has generally been found to be enriched with various chemical and microbiological components. This enrichment is caused by physico-chemical and biological effects both from the underlying water and from the atmosphere, and is of both natural and anthropogenic origin (e.g., Kuznetsova and Lee, 2001). In the prebiotic ocean, enrichment in the surface microlayer probably took place by air bubbles rising to the surface and collecting any material, dissolved or particulate, that was surface active and which made contact with the bubbles (Blanchard,1989). Tervahattu et al. (2002a, 2002b) have recently given direct observational evidence for the presence of an outer film of surfactant molecules (palmitic, oleic and stearic acids) on marine aerosols, as predicted by Ellison et al. (1999).

In this chapter, we outline a model for the emergence of life based on atmospheric aerosols (Shah, 1970; Lerman, 1986; Dobson et al., 2000). We describe the underlying geophysical conditions for aerosol formation and discuss the unique properties of organic films on aqueous solutions in controlling the morphology and chemistry at the interfaces (Goldacre, 1958). We stress the chemical advantages of interfaces in connecting geochemistry to biochemistry in the emergence of life on Earth or extraterrestrial environments where atmospheric aerosols could have existed.

2. Implications of the Prebiotic Geophysical Environment

Both the age of the Earth and the geometry of its orbit around the sun have direct implications for the geophysical processes which set the scene upon which geochemistry was transformed into biochemistry. The spectral output of the sun, approximately characterized now by the Planck function for a black body at 5800 K, has always been central since the planet acquired a liquid ocean some 4.4 x 10^9 years ago (Wilde et al., 2001), which may have been evaporated one or more times during the late heavy meteor bombardment (Sleep and Zahnle, 1998; Nisbet and Sleep, 2001). Depending upon whether this bombardment completely sterilized the Earth or not, the starting point for the organic aerosols as prebiotic chemical reactors was either 4.4 x 10^9 or 4 x 10^9 years ago. If Schopf et al. (2002) are correct about the age of the earliest microbes in the fossil record, life was present 3.5 x 10^9 years ago and had had 500 to 900 million years in which to originate. If Brasier et al. (2002) are correct, the earliest unequivocally identified fossil bacteria occur at 2.7 x 10^9 years ago (Buick, 1988), implying a range of 1300 to 1700 million years for life to originate. Donaldson et al. (2002), using results from Dyson (1999), estimated that the organic aerosol mechanism would have provided an upper limit of $\sim 10^{32}$ opportunities for the transition from a geochemical aerosol population to a biochemical bacterial and viral population. The number could have been larger if allowance is made for the non-interaction of the evolving containers in Dyson's model; in reality the aerosols could coagulate and, asymmetrically, divide.

The flux of solar visible and ultraviolet radiation to the Earth, approximately a beam of photons characteristic of a black body at 5800 K, represents a much lower entropy state than the outgoing infrared flux approximately characteristic of a black body at 245 K, emitted through a solid angle of 4π steradians. Since the planet is in approximate enthalpy (heat) balance, there is a large source of Gibbs free energy to maintain the system far from equilibrium at the average surface temperature. The biosphere absorbs approximately 1% of this incident solar radiation during photosynthesis (Nebel and Wright, 1993).

The tilt and rotation of the Earth's axis results in a diurnal cycle, a seasonal cycle, and a systematic circulation of the atmosphere and ocean, primarily in response to entropy production at low latitudes by evaporation of oceanic water and entropy disposal by infrared radiation to space at higher altitudes and latitudes. This circulation can result in residence times ranging from years in the upper stratosphere, with exposure to the full solar spectrum, to days in the troposphere. The temperatures in the atmosphere can range from –100 deg C at the tropical tropopause to +40 deg C over the surface of subtropical deserts, with 6 months darkness in the polar night and 6 months of visible light at the summer pole. Thunderstorms produce lightning discharges and phase changes of water are frequent. The tilted axis results in equator-to-pole solar heating gradients, which in combination with the Coriolis force associated with the planetary rotation, produces winds. These winds interact with the sea surface to produce an oceanic circulation through its entire depth, and also result in waves which when they break make air bubbles in the ocean (whitecaps). The air bubbles are buoyant, and as they reach the air-sea interface, they burst and produce aerosols by the cap-and-jet mechanism (Mason, 1954; Blanchard, 1964). These aerosols, frequently recycled, would have sampled the whole range of temperatures, humidities and radiation fields in the atmosphere. There is every reason to suppose that the land and sea were as heterogeneous and chemically varied as they are now, both in the bulk and at their surfaces.

The salting out effect (Demou and Donaldson, 2002) means that even small organic molecules, in addition to their larger companions and polymers, tend to concentrate at the ocean surface; both there and on aerosols, the amphiphilic film provides a naturally hydrophobic environment, an advantage for condensation reactions in which water is eliminated from the monomers which form peptides, proteins and nucleic acids. The sea surface would act as an integrator as well as a concentrator; fluxes of aerosols containing meteoritic elements (Murphy et al., 1998) would ultimately impinge upon it, as would the products from thermal sea floor vents (Simoneit, 1995).

The solar flux incident upon a fluid medium will eventually "naturally select" a layer of molecules capable of self-shielding from the shorter, photodissociative wavelengths; it is ozone in the current atmosphere (Chapman, 1930; Goody and Walker, 1972) but could have been a wide variety of functionalized organic molecules and their oligomers, either in the atmosphere or at the ocean surface (Tuck, 2002).

We note that the balance between gravity and atmospheric density (via aerodynamic drag) determines the size of particles with appreciable residence times, and so this latter property is in principle predictable for other planets and moons. If the asymmetric fission of Donaldson et al (2001, 2002) is at work, the virally-sized daughter particle has mass ~3 orders of magnitude less than that of the bacterially-sized daughter, and subject to coagulative behavior, will have a longer residence time.

Finally, we note that there are major implications for the prebiotic chemistry occurring in fluid media such as the atmosphere, ocean and their interface, arising from the fat-tailed (i.e., non-gaussian) probability distributions and power law scaling of the reactants (Schertzer and Lovejoy, 1985; Seuront et al., 1999; Tuck et al., 2003). The fractal dimension of the atmosphere for example is neither 3 as it would be for isotropic turbulent mixing, or 2 as it would be for horizontally stratified transport, but has an intermediate dimension. The emergence of order from randomness is possible in such fluids in a way that it is not for régimes in which molecular diffusion is the sole transport mechanism (Tél et al., 2000). Much as is possible *in vivo* (Adam and Delbrück, 1968) and *in vitro* (Dewey, 1997), in the atmosphere such fractal turbulence can accelerate reaction rates (Tuck et al., 2003). There are obvious implications for *in silico* modelling of the origins of life, not least of which is the ability of fluid flow to enable counter-gradient transport of reactants and products, unlike diffusion.

3. Properties of organic films on aqueous solutions

Organic films at the interface of aqueous solutions provide unique reaction environments. Even soluble small organic molecules have recently been shown to partition preferentially to the surface of aqueous solutions (Demou and Donaldson, 2002). Lipids and other amphiphiles will be the major constituents of organic surface films; their role in prebiotic evolution has been addressed (Ourisson and Nakatani, 1994; Norris and Raine, 1998; Luisi et al., 1999). Some prebiotic syntheses have been shown to include the formation of lipid-like amphiphilic molecules, long chain hydrocarbons and their derivatives (Hargreaves et al., 1977; Leach et al., 1978; Rao et al., 1982). Another source of prebiotic organic material is delivery to the Earth by meteoritic and cometary infall (Anders, 1989; Chyba and Sagan, 1992). Amphiphilic compounds present in carbonaceous meteorites could self-assemble into membranous vesicles or form atmospheric aerosols with an organic film coating ("inverted micelle"). This possibility has been demonstrated using compounds extracted from meteorites (Deamer, 1985; Deamer and Pashley, 1989; Deamer, 1997).

Conditions found in hydrothermal vents (high pressure, high temperatures) have been shown to be auspicious for the generation of organics under prebiotic conditions (Huber and Wächtershäuser, 1998; Amend and Shock, 1998). When organic acids are formed, the maximum of the carbon chain length distribution is found at 5 and 6 carbons. Making use of selective evaporation, dissolution and ability to form stable films, the aerosol surface can select from such a distribution the acids with longer chains of 12 to 24 carbons, compounds which are known to make stable self-assembled films and membranes. The net result of the selection of organics at the surface of aqueous solutions is the self-assembly and orientation of organics on atmospheric aerosols and the sea surface (Ellison et al., 1999; Dobson et al, 2000; Kuznetsova and Lee, 2001).

Probably the vesicle itself would not contain all the compounds needed for the chemical development. Therefore, it is necessary to get molecules and ions from the environment, which was supposed to be enriched with both inorganic and organic molecules. We note that gas phase production of peptides from amino acids has been reported recently (Wincel et al., 2000; Remko and Rode, 2001). Membranes form selective permeability barriers to the transport of ions, nutrients and waste products across protocellular walls (Pohorille and Wilson 1995, Apel et al. 2002). However, transport can proceed without the aid of complex molecules, which mediate this process in contemporary cells. Ions are transported across membranes through deep thinning defects, which make the membranes by 14 to 17 orders of magnitude more permeable (Pohorille and Wilson, 1995; Fry, 2000). The experiments of Monnard and Deamer (2001) demonstrate that passive diffusion could be used by the earliest forms of cellular life for transport of important nutrients such as amino acids, phosphate, and phosphorylated organic solutes.

The organic films generated should form a semipermeable enclosure, whose stability and permeability depend on the specific conditions. The two dimensional mobility of surface active compounds leads to formation by self assembly of "rafts" and "islands" with boundaries between them. Domain boundaries are expected to provide interesting sites for reaction as well as permeability and morphological changes.

The aqueous subphase can be an important factor in controlling the packing and therefore the permeability of the organic film. The salinity and pH of the aqueous phase are just two examples of parameters expected to play an important role in this regard.

It is expected that basic solutions such as the ocean with a pH of 8 will lead to ionization of the organic acid groups. Under these conditions, ion-ion repulsion produces a more distended film, which is also more permeable. In the present atmosphere, aerosols become acidified as gases like CO_2 and SO_2 reach the aqueous core and form the corresponding acids. The solution pH is lowered, the organic acids are no longer ionized and pack much more tightly. At this point, the organic film is less permeable and may delay or even limit the uptake of small molecules. The opposite situation arises when gases such as NH_3 reach the aqueous core as they will lead to an increase in the solution pH. Atmospheric aerosols, unlike their solution counterparts, will undergo such pH and salinity "switching" in the course of their atmospheric journey.

The presence of an organic film on the surface of an aerosol allows the particles to divide as well as coagulate. This is a new result (Donaldson et al., 2001, 2002); fission is denied to a homogeneous liquid droplet because its spherical shape is already that which minimizes the surface free energy. However, there is enough free energy available in a compressible amphiphilic film that for a typical molecule, stearic acid, on a typical

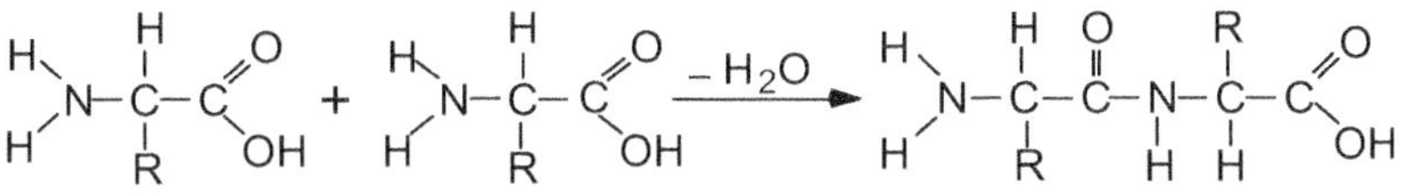

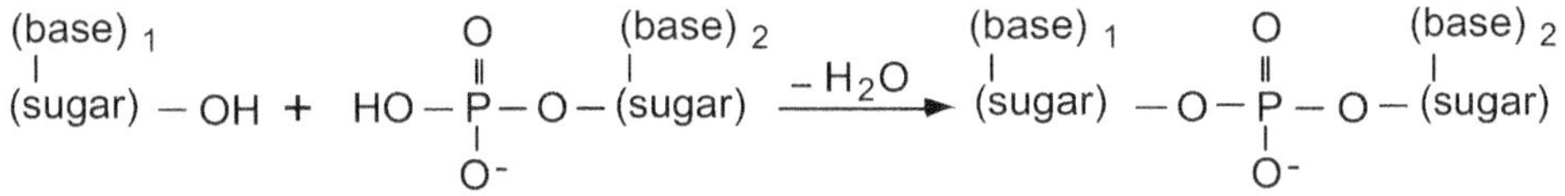

aerosols and in the ocean. In such solutions, the reactions between amino acids to form peptides, and between nucleotides to form nucleic acids are extremely unfavourable as water has to be eliminated in an aqueous environment. The thermodynamic barrier is surmountable if water can be removed (Fox et al., 1956). The equilibrium lies on the side of the reagents. Accordingly, attempts to synthesize peptides under prebiotic conditions (in the absence of an enzyme) have only been successful when the reaction is conducted in anhydrous environments (Fox et al., 1956). The thermodynamic constraints could be alleviated if reaction occurs at the surface where a water-poor environment is available (Ranganathan et al., 1989; Oliver and Singh, 1999; Kumar and Oliver, 2002). Energy for this process could be supplied by ultraviolet radiation, present in abundance in the prebiotic atmosphere, before significant amounts of oxygen were present and before ozone was present in large quantities.

In aqueous solutions, amino acids are ionized to form zwitterions, a process very much affected by the solution pH. Depending on the pH, organic acids and bases will be selectively ionized so that at low pH, $^{+}H_3N$-HCR-COOH is most likely, while at high pH, H_2N-HCR-COO^- will predominantly form. The kinetic problem facing peptide bond formation stems from the fact that -NH_3^+ is not a good electron donor and -COO^- is not a good electron acceptor, each of these properties precluding the formation of the peptide bond. In the case of the ocean of today, which is a slightly basic solution (pH 8), most of the amine groups remain unionized (-NH_2) and therefore are good electron donors. Under these circumstances, the acid group on the amino acid as well as on the organic acids, which form the enclosure for the organic, "inverted micelle" aerosols, will be largely ionized. These ions will need to be activated by a catalyst in the presence of a salt (Rode and Tauler, 1990; Eden and Rode, 1994; Schwendinger et al., 1995; Rode and Schwendinger,1990; Suwannochot and Rode, 1999; Plankensteiner et al., 2002). Plankensteiner et al. (2002), Rode (1999) and Suwannachot and Rode (1999) have studied the reactions involved in the mechanism called Salt-Induced Peptide Formation (SIPF). Under the proposed conditions of the primordial earth SIPF is a facile reaction pathway to explain the formation of the first peptides. Prebiotically, some of these catalysts could have been provided as transition metal ions. Such ions would have been delivered by meteorites, in active forms and large amounts. In the presence of such a catalyst, the negative charge is tied up allowing the organic acid group to accept electrons from the amine. The net result of this interfacial mechanism is that the kinetic problems are overcome and peptide bonds generated with a higher propensity.

The spatial structure of the polypeptide chain attached to the lipid membrane might play an important role in the further chemical evolution. According to Nilson (2002), molecules attached to an oil-water interface can interact with each other in two dimensions (diffusion along a plane), instead of in three dimensions (a volume). The resulting high interfacial concentration would be the most important for molecular interactions, in contrast to other systems where some components are attached to a solid surface (low mobility) and to systems completely in the bulk solution (high dilution). At an oil-water interface, peptides and proteins could interact with metal-ions, possibly catalyzing stereospecific reactions. Peptides and proteins could also act as ion-exchangers for concentrating phosphate ions. Lateral proton conduction has been found to be 20 times faster at phospholipid-water interfaces compared with that of bulk water.

Pohorille and Wilson (1995) reported that membrane-water interfaces have the ability to organize short peptides into ordered structures, providing that they have a proper sequence of hydrophobic and hydrophilic residues. Computer simulation techniques have been used to investigate peptide bond formation (Martin, 1998; Chalmet et al., 2001; Chaudhuri and Canuto, 2002; Texler et al., 1998).

4. Summary

The sea surface would have automatically provided a location in which amphiphilic molecules would assemble into anhydrous films. Because of its location, it would also automatically act as a global integrator of fluxes of monomers and amphiphiles from meteorites, sea floor production and bulk oceanic production, and cycle them repeatedly through atmospheric aerosols, which would have been present in very large numbers, via wind driven whitecap formation. The films on both the sea surface and on the aerosols would have provided potential sites for the condensation reactions necessary to polymerize amino acids and nucleotides. We have suggested how such reactions, particularly for peptide synthesis from amino acids, could work at these interfacial sites. The large, geophysical inevitable populations of chemically heterogeneous aerosols would have provided many opportunities for natural selection to transform geochemistry into biochemistry. Aerodynamics and atmospheric density determine the size distribution, which is of bacterial dimensions. The surface films on the aerosols also allow the thermodynamic possibility of the co-emergence and co-evolution of bacterially and virally sized populations upon asymmetric aerosol fission. Chemical differentiation during such fission is possible.

We have discussed a model for the origin of life based on the structural, optical and chemical properties of organic films at the surface of atmospheric aerosols. Organic atmospheric aerosols with an interfacial lipid membrane comprise supramolecular assemblies with the ability to concentrate organic molecules at their surface and to act as a template for the synthesis of biomolecular building blocks of life. Analogies in size, structure, morphology and function (metabolism and replication) between atmospheric aerosols and primitive cells can be drawn to explore the unique chemical properties of interfaces characteristic of atmospheric aerosols. The possible roles played by the air-lipid and lipid-saline solution interfaces-are considered to be important in the conversion of geochemistry to biochemistry.

5. References

Adam, G. and Delbrück, M. (1968) Reduction of dimensionality in biological diffusion processes, In: A. Rich and N. Davidson (eds.) *Structural Chemistry and Molecular Biology,* W. H. Freeman, San Francisco, pp.198-215.

Amend, J. P., and Shock, E. L. (1998) Energetics of amino acid synthesis in hydrothermal ecosystems, *Science*, **281**,1653-1662.

Anders, E. (1989) Pre-biotic organic matter from comets and asteroids, *Nature*, **342**, 255-257.

Apel, C. L., Deamer, D. W., Mautner, M. N.(2002) Self-assembled vesicles of monocarboxylic acids and alcohols: conditions for stability and for the encapsulation of biopolymers, *Biochim. Biophys. Acta-Biomembranes*, **1559**, 1-9.

Aylward, N. and Bofinger, N. (2001) The reactions of methanimine and cyanogen with carbon monoxide in prebiotic molecular evolution on earth, *Orig. Life Evol.. Biosph.*, **31**, 481-500.

Bada, J. L. and Lazcano, A. (2002) Origin of life - some like it hot, but not the first biomolecules, *Science,* **296**, 1982-1983.

Blanchard, D. C. (1964) Sea-to-air transport of surface active material, *Science,* **146,** 396-397.

Blanchard, D. C. (1989) The ejection of drops from the sea and their enrichment with bacteria and other materials: a review, *Estuaries,* **12**, 127-137.

Brasier, M. D., Green, O. R., Jephcoat, A. P., Kleppe, A. K., Van Kranendonk, M. J., Lindsay, J. F., Steele, A. and Grassineau, N. V. (2002) Questioning the evidence for Earth's oldest fossils, *Nature,* **416,** 76-81.

Buick, R. (1988) Carbonaceous filaments from North Pole, Western Australia: are they fossil bacteria in Archaean stromatolites? A reply, *Precambrian Res.*, **39,** 157-172.

Carothers, W. H. (1936) Polymers and polyfunctionality, *Trans.Farad. Soc.*, **32**, 39-53.

Chalmet, S., Harb, W., and Ruiz-Lopez, M. F. (2001) Computer simulation of amide bond formation in aqueous solution, *J. Phys. Chem. A*, **105**, 11574-11581.

Chapman, S. (1930) A theory of upper atmospheric ozone, *Mem. Roy. Met. Soc.*, **3,** 103-125.

Chaudhuri, P., and Canuto, S. (2002) An ab inito study of the peptide bond formation between alanine and glycine: electron correlation effects on the structure and binding energy, *J. Mol. Struct. (Theoschem)*, **577**, 267-273.

Chyba, C. F., and Sagan, C. (1992) Endogenous production, exogenous delivery and impact-schock synthesis of organic molecules: an inventory for the origin of life, *Nature*, **355**, 125-132.

Cronin, J. R., Pizzarello, S., and Cruickshank, D. P. (1988) Organic matter in carbonaceous chondrites, planetary satellites, asteroids and comets, In: J.F. Kerridge and M.S. Matthews (eds). *Meteorites and the Early Solar System, Tucson, AZ*, Univ. of Arizona Press. pp. 819-857.

Deamer, D.W. (1997) The first living systems: a bioenergetic perspective, *Microbiol. Molecular Biology Reviews*, **61**, 239-261.

Deamer, D. W. and Pashley, R. M. (1989) Amphiphilic Components of Carbonaceous meteorites, *Origin of Life and Evolution of the Biosphere*, **19**, 21-33.

Deamer, D. W. (1985) Boundary structures are formed by organic components of the Murchison carbonaceous chondrite, *Nature*, **317**, 792-794.

Demou, E., and Donaldson, D. J. (2002) Adsorption of atmospheric gases at the air-water interface. 4: the influence of salts, *J. Phys. Chem. A*, **106,** 982-987.

Dewey, T. G. (1997) *Fractals in Molecular Biophysics,* Oxford University Press.

Dobson, C. M., Ellison, G. B., Tuck, A. F. and Vaida, V. (2000) Atmospheric aerosols as prebiotic chemical reactors, *Proc. Nat. Acad. Sci.*, **97**, 11864-11868.

Donaldson, D. J., Tuck, A. F. and Vaida, V. (2001) Spontaneous fission of atmospheric aerosol particles, *Phys. Chem. Chem. Phys.*, **3**, 5270-5273.

Donaldson, D. J., Tuck, A. F. and Vaida, V. (2002) The asymmetry of organic aerosol fission and prebiotic chemistry, *Orig. Life Evol. Biosph.*, **32**, 237-245.

Dyson, F. J. (1999) *Origins of Life*, 2nd edition, Cambridge University Press.

Eden, A. H. and Rode, B. M. (1994) Influence of alkali and alkaline-earth-cations on the "salt-induced peptide formation" reaction, *J. Chem. Soc. Dalton. Trans.*, 1125-1130.

Ellison, G. B., Tuck, A. F. and Vaida, V. (1999) Atmospheric processing of organic aerosols, *J. Geophys. Res.*, **104**, 11633-11641.

Fox, S. W. and Dose, K. (1972) Molecular Evolution and the Origin of Life, Chapter 5, *Macromolecules*, W. H. Freeman and Company, 135-195.

Fox, S. W., Johnson, J. E. and Vegotsky, A. (1956) On biochemical origins and optical activity, *Science*, **124,** 923-925.

Fry, I. (2000) *The Emergence of Life on Earth*, Rutgers University Press.

Gennis, R.B. (1989) *Biomembranes,* Springer-Verlag.

Goldacre, R. J. (1958) Surface films, their collapse on compression, the shapes and sizes of cells, and the origin of life, In: J. F. Danielli, K. G. A. Parkhurst and A. C. Riddiford (eds.) *Surface Phenomena in Chemistry and Biology,* Pergamon Press, pp. 12-27.

Goody, R.M. and Walker, J.C.G. (1972) *Atmospheres,* Prentice Hall.

Hargreaves, W. R., Mulvihill, S., and Deamer, D. W. (1977) Synthesis of phospholipids and membranes in prebiotic conditions, *Nature*, **266**, 205-207.

Huber, C. and Wachtershauser, G. (1998) Peptides by activation of amino acids with CO on (Ni, Fe)S surfaces: implications for the origin of life, *Science*, **281**, 670-672.

Kumar, J. K. and Oliver, J. S. (2002) Proximity effects in monolayer films: kinetic analysis of amide bond formation at the air-water interface using H NMR spectroscopy, *J. Am. Chem. Soc.*, **124**, 11307-11314.

Kuznetsova, M. and Lee, C. (2001) Enhanced extracellular enzymatic peptide hydrolysis in the sea-surface microlayer, *Mar. Chem.*, **73**, 319-332.

Lerman, L. (1986) Potential role of bubbles and droplets in primordial and planetary chemistry: exploration of the liquid-gas interface as a reaction zone for condensation, *Orig. Life Evol. Biosph.,* **16,** 201-202.

Leach, W. W., Mooner, D. W., and Oró J. (1978) Abiotic synthesis of fatty acids, *Orig. Life Evol. Biosph.,* **9**, 113-122.

Luisi, P. L., Walde, P., and Oberholzer, T. (1999) Lipid vesicles as possible intermediates in the origin of life, *Curr. Opin. Colloid Interface Sci.*, **4**, 33-39.

Martin, R. B. (1998) Free energies and equilibria of peptide bond hydrolysis and formation, *Biopolymers*, **45**, 351-353.

Mason, B.J. (1954) Bursting of air bubbles at the surface of sea water, *Nature*, **174**, 470-471.

Miller, S. L. (1953) A production of amino acids under possible primitive earth conditions, *Science* **117**, 528-529.

Miller, S. L. (1998) The endogenous synthesis of organic compounds, Chapter 3, *The Molecular Origins of Life,* A. Brack, ed., Cambridge University Press.

Monnard, P. A. and Deamer, D. W.(2001) Nutrient uptake by protocells: a liposome model system, *Orig. Life Evol. Biosph*, **31**, 147-155.

Murphy, D.M., Thomson, D.S. and Mahoney, M.J. (1998) In situ measurements of organics, meteoritic material, mercury and other elements in aerosols at 5 to 19 kilometers, *Science*, **282**, 1664-1669.

Nebel, B. J. and Wright, R. T. (1993) *Environmental Science,* 4th Edition, Prentice Hall.

Nilson, F.P.R (2002) Possible impact of a primordial oil slick on atmospheric and chemical evolution, *Orig. Life Evol. Biosph.*, **32**, 247-253.

Nisbet, E. G. and Sleep, N. H. (2001) The habitat and nature of early life, *Nature*, **409**, 1083-1091.

Norris, V. and Raine, D. J. (1998) A fission-fusion origin for life, *Orig. Life Evol. Biosph.*, **28**, 523-537.

Oliver, J. S. and Singh, J. (1997) Unambiguous demonstrations of compression promoted dipeptide synthesis in a monolayer film, *J. Org. Chem.*, **62**, 6436-6438.

Oparin, A. I. (1924) *The Origin of Life on Earth*, Pabochii.

Ourisson, G. and Nakatani, Y. (1994) The terpenoid theory of the origin of cellular life: the evolution of terpenoids to cholesterol, *Chem. & Biol.*, **1**, 11-23.

Plankensteiner, K., Righi, A., and Rode, B. M. (2002) Glycine and diglycine as possible catalytic factors in the prebiotic evolution of peptides, *Orig. Life Evol. Biosph.*, **32**, 225-236.

Pohorille, A. W. and Wilson, M. A.: (1995) Molecular dynamics studies of simple membrane-water interfaces: structure and functions in the beginnings of cellular life, *Orig. Life Evol. Biosph.* **25**, 21-46.

Ranganathan, D., Singh, G. P., and Ranganathan, S. (1989) Peptide bond formation at the micellar interface, *J. Am. Chem. Soc.*, **111**, 1144-1145.

Rao, M., Eichenberg, J., and Oró, J. (1982) Synthesis of phosphatidylcholine under possible primitive earth conditions, *J. Mol. Evol.*, **18**, 196-202.

Remko, M. and Rode, B.M. (2001) Catalyzed peptide bond formation in the gas phase, *Phys. Chem. Chem. Phys.*, **3**, 4667-4673.

Rode, B. M. and Schwendinger, M. G. (1990) Copper-catalyzed amino acid condensation in water - a simple possible way of prebiotic peptide formation, *Orig. Life Evol. Biosph.,* **20**, 401-410.

Rode, B. M. and Tauler, R. (1990) Reaction of Copper (II) with glycine and glycylglycine in aqueous solution at high concentrations of sodium chloride, *Inorg. Chim. Acta*, **173**, 93-98.

Rode, B. M. (1999) Peptides and the origin of life, *Peptides*, **20**, 773-786.

Schertzer, D. and Lovejoy, S. (1985) The dimension and intermittency of atmospheric dynamics, In: L. J. S. Bradbury, F. Durst, B. E. Launder, F. W. Schmidt and J. H. Whitelaw, *Turbulent Shear Flows,* Springer-Verlag, **4**, pp. 7-33.

Schopf, J. W., Kudryavtsev, A. B., Agresti, D. G., Wdowiak, T. J. and Czaja, A. D. (2002) Laser-Raman imagery of Earth's oldest fossils, *Nature,* **416,** 73-76.

Schwendinger, M. G., Tauler, R., Saetia, S., Liedl, K. R., Kroemer, R. T., and Rode, B. M. (1995) Salt induced peptide formation: on the selectivity of the copper induced peptide formation under possible prebiotic conditions, *Inorg. Chim. Acta,* **228**, 207-214.

Seuront, L., Schmitt, F., Lagadeuc, Y., Schertzer, D. and Lovejoy, S. (1999) Universal multifractal analysis as a tool to characterize multiscale intermittent patterns: example of phytoplankton distribution in turbulent coastal waters, *J. Plankton Res.,* **21,** 877-922.

Shah, D. O. (1972) The Origin of Membranes and Related Surface Phenomena, Chapter 7, *Exobiology,* C. Ponnamperuma, ed., North Holland. pp. 235-265.

Simoneit, B. R. T. (1995) Evidence for organic synthesis in high temperature aqueous media – facts and prognosis, *Orig. Life Evol. Biosph.,* **25,** 119-140.

Sleep, N.H. and Zahnle, K. (1998) Refugia from asteroid impacts on early Mars and the early Earth, *J. Geophys. Res*., **103**, 28529-28544.

Sowerby, S. J., Petersen, G. B., and Holm, N. G. (2002) Primordial coding of amino acids by adsorbed purine bases, *Orig. Life Evol. Biosph.* **32**, 35-46.

Suwannachot, Y. and Rode, B. M. (1999) Mutual amino acid catalysis in salt-induced peptide formation, *Orig. Life Evol. Biosph.*, **29**, 463-471.

Tél, T., Karolyi, G., Péntek, A., Scheuring, J., Toroczkai, Z., Grebogi, C. and Kadtke, J. (2000) Chaotic advection, diffusion and reaction in open flows, *Chaos,* **10,** 89-98.

Tervahattu, H., Hartonen, K., Kerminen, V.-M., Kupiainen, K., Aarnio, P., Koskentalo, T., Tuck, A., Vaida, V. (2002a) New evidence of an organic layer on marine aerosols, *J. Geophys. Res.*, **107**, 10.1029/2000JD000282.

Tervahattu, H., Juhanoja, J., Kupiainen, K. (2002b) Identification of an organic coating on marine aerosol particles by TOF-SIMS, *J. Geophys. Res.,* **107,** 10.1029/2001JD001403.

Texler, N. R., Holdway, S., Neilson, G. W., and Rode, B. M. (1998) Monte Carlo simulations and neutron diffraction studies of the peptide forming system 0.5 mol kg^{-1} $CuCl_2$-5 mol kg^{-1} NaCl-H_2O at 293 and 353 K, *J. Chem. Soc. Farad. Trans.*, **94**, 59-65.

Tuck, A. (2002) The role of atmospheric aerosols in the origin of life, *Surv. Geophys.,* **23**, 379-409.

Tuck, A. F., Hovde, S. J., Gao, R-S. and Richard, E. C. (2003) The law of mass action in the Arctic lower stratospheric polar vortex January-March 2000: ClO scaling and the calculation of ozone loss rates in a turbulent fractal medium, *J. Geophys. Res.,* **108,** 10.1029/2002JD002832, in press.

Wilde, S.A., Valley, J.W., Peck, W.H. and Graham, C.M. (2001) Evidence from detrital zircons for the existence of continental crust and oceans on the Earth 4.4 Gyr ago, *Nature*, **409**, 175-178.

Wills, C. and Bada, J. (2001) *The Spark of Life,* Perseus Publishing, Cambridge, MA., pp. 291.

Wincel, H., Fokkens, R. H., and Nibbering, N. M. M. (2000) Peptide bond formation in gas-phase ion/molecule reactions of amino acids: a novel proposal for the synthesis of prebiotic oligopeptides, *Rapid Commun. Mass Spectrom*, **14**, 135-140.

III. Alternatives to the Standard Scenario

Biodata of **Koichiro Matsuno** author of ***"Prebiotic Interplay Between Fatty Acids and Amino Acids in Hydrothermal Environments"***

Dr. Koichiro Matsuno is a Professor of Biophysics in the Department of Bioengineering at the Nagaoka University of Technology, Nagaoka, Japan. He received his Ph.D. in physics from the Massachusetts Institute of Technology in 1971. Dr. Matsuno was a visiting Professor (1982-1984) at the Institute for Molecular and Cellular Evolution of the University of Miami, Florida, USA. His specialized studies are in theoretical evolution processes, experimental chemical evolution and muscle contraction. He is the author of the monographs ***Protobiology: Physical Basis of biology*** (CRC Press, Boca Raton Florida, 1989) and ***What is Internal Measurement*** (Seido-sha, Tokyo, 2001). Dr. Matsuno edited the volumes of: ***Molecular Evolution and Protobiology*** with K. Dose, D. Rohlfing, and K. Harada. (Plenum, New York, 1984), ***The Origin and Evolution of the Cell,*** with H. Hartman (World Scientific, Singapore, 1992) and ***Uroboros: Biology between Mythology and Philosophy***, with W. Lugowski (Arbortum, Wroclaw, Poland, 1999).

E-mail: **kmatsuno@vos.nagaokaut.ac.jp**

J. Seckbach (ed.), Origins, 169–179.

PREBIOTIC INTERPLAY BETWEEN FATTY ACIDS AND AMINO ACIDS IN HYDROTHERMAL ENVIRONMENTS

KOICHIRO MATSUNO
Department of Bioengineering
Nagaoka University of Technology
Nagaoka 940-2188, Japan

1. Introduction

In our laboratory we are already quite experienced in synthesizing organic molecules from precursors in our simulations of hydrothermal chemistry. The next major step within the context of evolution is to use these synthesized products as precursors for even more complex subsequent reactions. This necessitates the creation of a simulated prebiotic environment with its own evolutionary capabilities. In such a system there must be ongoing energy transactions. That is, the system must have the ability to let usable energy in and disposable energy out.

On the primitive Earth, there were at least two candidates for implementing such a system. One obvious choice would be utilizing solar energy and dissipating the disposable energy in the form of heat. The other would utilize higher temperatures and dissipate unused energy at a lower temperature. The latter process manifests in hydrothermal environments in the primitive ocean. Capturing solar photons requires a molecular complex that would already have evolved to a significant extent. In contrast, transaction or utilization of heat energy in the interface between the hot and the cold water was already in place. We created a simulation of such an environment in our laboratory.

There has been a considerable amount of experimental effort focused on prebiotic oligomerization of monomers under conditions that could have been available on the primitive earth. Mineral surfaces on rocks can certainly provide electrostatic forces to attract charged monomeric reactants and form oligomers thereupon (Orgel, 1998). Even some of the reducing mineral surfaces such as pyrite can supply reaction energy to drive oligomeric synthesis during the oxidization (Wächterhäuser, 1992). This observation naturally begs the further question: How does the energy from primary sources come to be locally tailored so as to facilitate elongation of oligomers that are quite localized in three-dimensional space?

Oligomerization of activated monomers is a demonstration of enhancing the energy concentration locally in the form of chemical bonds that appear in the process of energy dissipation on a much larger scale. Mineral surfaces undoubtedly serve as such an agency of dissipating the excess energy available to each monomer attached on them in an extremely anisotropic manner so as to form chemical bonds in sequence. Nonetheless, mineral surfaces are not the only means to actualize anisotropic energy dissipation to facilitate Oligomerization of monomers. Oligomerization on mineral surfaces is in fact one particular instance of actualizing energy dissipation accompanying adiabatic quenching, in which many dissipation processes proceed towards their respective thermal equilibria independent of each other and with different rates. Here we explore a possible evolutionary significance of energy dissipation accompanying adiabatic quenching, with a special emphasis on its occurrence at submarine hydrothermal vents in the context of prebiotic oligomerization (Corliss et al, 1979; Edmond et al, 1982).

Adiabatic quenching is ubiquitous in many material processes. For instance, nucleosynthesis of a carbon from a helium and a beryllium atom is certainly a consequence of adiabatic quenching in the sense that when the ambient temperature decreases, the carbon atom formed through the synthesis can decrease its temperature while holding the structure without disintegrating into the constituent helium and beryllium atoms. This is due to the difference of dissipation rates of energy between a carbon atom and a mere aggregate of one helium and one beryllium atom. The energy dissipation while maintaining the form of a carbon atom happens to proceed at a much faster rate. Anisotropic energy structuring occurring in the process of energy dissipation can be protected against isotropic energy homogenization if the adiabatic walling is sufficiently high enough to prevent the disintegration. A similar idea of adiabatic quenching and walling can also apply to prebiotic oligomerization of monomers in the vicinity of submarine hydrothermal vents since both energy sources and sinks are available there. More specifically, what is specific to hydrothermal environments is its unique facilitation of adiabatic quenching.

2. Adiabatic Quenching

For simplicity, let us consider a peptide bond formation between an amino acid and an oligopeptide in the vicinity of submarine hydrothermal vent. A peptide synthesis requires crossing of the free energy barrier $\Delta G \cong 2\text{\textasciitilde}4 kcal/mol$ (Fox and Dose, 1977). When the temperature is T, the corresponding Boltzmann factor

$$B(T) = exp\left(-\frac{\Delta G}{\kappa T}\right) \qquad (1)$$

determines the fraction of the quantum states occupied by a peptide at temperature T in thermal equilibrium, where κ is Boltzmann's constant. Once a peptide bond is formed inside hot hydrothermal vents at temperature T_2, rapid and adiabatic transference of a small portion of the fluid into the surrounding cold water at temperature T_1 ($T_2 > T_1$) could keep the Boltzmann factor for the peptide bond almost intact on the same level as at the high temperature T_2, of the form

$$B(T_2) = exp\left(-\frac{\Delta G}{\kappa T_2}\right). \qquad (2)$$

This is due to the fact that if thermal communication between the two states, a peptide bond formed and no such bond formed, is blocked up by an adiabatic wall associated with the adiabatic quenching, the occupation number of the quantum states while maintaining the peptide bond formed could remain as it was in the hot vents. The peptide bond surviving in the cold region could then be characterized by the Boltzmann factor

$$B(T_2) = exp\left(-\frac{\Delta G}{\kappa T_1} \frac{T_1}{T_2} \right) \tag{3}$$

instead of $B(T_1) = exp\left(-\Delta G / \kappa T_1 \right)$.

The adiabatic quenching equipped with the associated adiabatic walling sufficiently high enough to prevent thermal communication between the two states of a peptide-bond present and absent, can effectively enhance the Boltzmann factor for the case of a peptide bond formed, when it is rapidly transferred into the low temperature environment. When the temperature of the hot hydrothermal vents is $523K$ ($250^{o}C$) and the surrounding cold water is at $273K(0^{o}C)$, the Boltzmann factor is enhanced about 11.2 times more when the free energy barrier ΔG is chosen to be $3kcal/mol$. Enhancement of the Boltzmann factor could be multiplicative in increasing the population of peptides if the fluid containing reactants of both amino acids and oligopeptides is allowed to circulate through the hot vents repeatedly.

Repeated cycles of the sequence of heating and adiabatic quenching accompanied by the adiabatic walling can increase the number of peptide bonds available in the fluid, with a further consequence of elongation of oligopeptides thus formed. The resulting elongation of oligopeptides is eventually counterbalanced by their inevitable disintegration due to, for instance, hydrolysis. However, once the elongated oligomers are disintegrated into smaller pieces that could again participate in the elongation process (Orgel, 1986), the combined effect could become multiplicative in yielding an exponential growth of oligomers at least initially (Matsuno, 1982; Kauffman, 1986).

Further enhancement of oligomerization could be expected from microscopic encapsulation of reacting molecules within spherical vesicles endogenously formed in the reaction solution (Deamer, 1997; Luisi et al, 1999; Rushdi and Simoneit, 2001, Apel, et al, 2002; Tsukahara et al, 2002) Hydrothermal environments could also be capable of synthesizing fatty acids as component molecules for making lipid vesicles (McCollom and Simoneit, 1999). These observations come to suggest a possibility of testing the experimental likelihood of enhancing oligomerization of amino acids in the presence of fatty acids in hydrothermal environments.

3. Hydrothermal vents and their experimental models

In our flow reactor simulating a hydrothermal environment (Matsuno, 1997; Imai et al, 1999a, b), the reaction solution was injected at a flow rate of 7.5 mL per min from a 15 mL heated chamber at 200^{0} C and 22 MPa. The reactants passed through a nozzle (diameter 0.8 mm) into a larger, cooler chamber at 25^{0} C and approximately the same pressure. Cycle time for the total volume of fluid (500 mL) was 62.5 min, but with stirring the reactants recycled once per minute. Each run lasted 2 hours and was started at room temperature, requiring approximately 10 min to reach 200^0 C. The circulated reaction solution consisted of 100mM glycine, 10mM oleic acid and 1% methanol. Oleic acid was chosen simply as a model fatty acid. Ionic salts were not added, but the solution contained trace amounts of iron due to the stainless steel chamber (SUS-316). The reaction products were profiled by HPLC analysis (A195 nm). Oleic acid vesicles were prepared by first forming oleic acid micelles as adding NaOH into the reaction solution up to pH 10. Then, the micelles were transformed into oleic acid vesicles as decreasing the pH down to 9 by adding HCl.

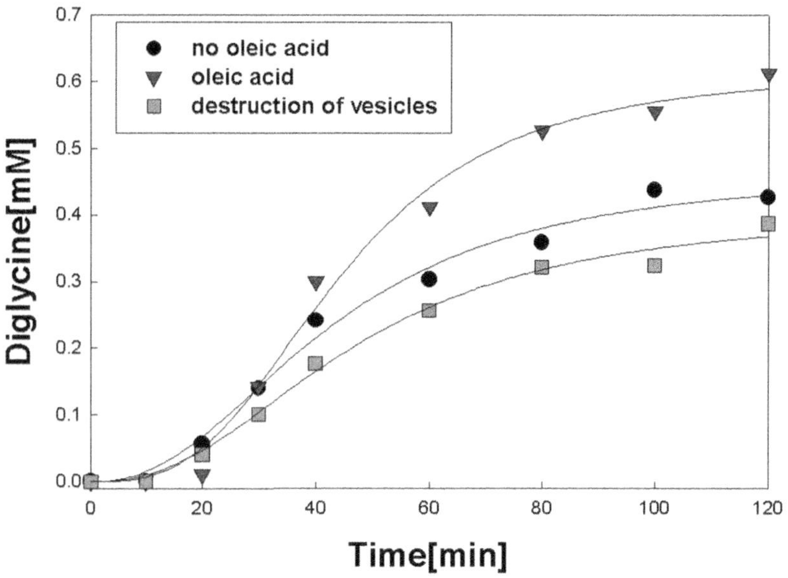

Figure 1. Yields of diglycine after 2 hours operation of the flow reactor simulating a hydrothermal environment. The reaction solution consisted of 100mM glycine and 10 mM oleic acid. Dissolution of oleic acid vesicles was completed by adding 10% tritonX-100 after the reactor operation.

4. Results

The yields of diglycine and triglycine were compared in the presence of oleic acids vesicles and also in the absence of the vesicles due to the further addition of 10% tritonX-100 as a surfactant (Figures 1 and 2). The effect of the dissolution of the oleic acid vesicles was examined by comparing it to control experiment carrying no oleic acid in the reaction solution from the start. The presence of oleic acid vesicles certainly helped synthesizing oligoglycine in the simulated hydrothermal environment.

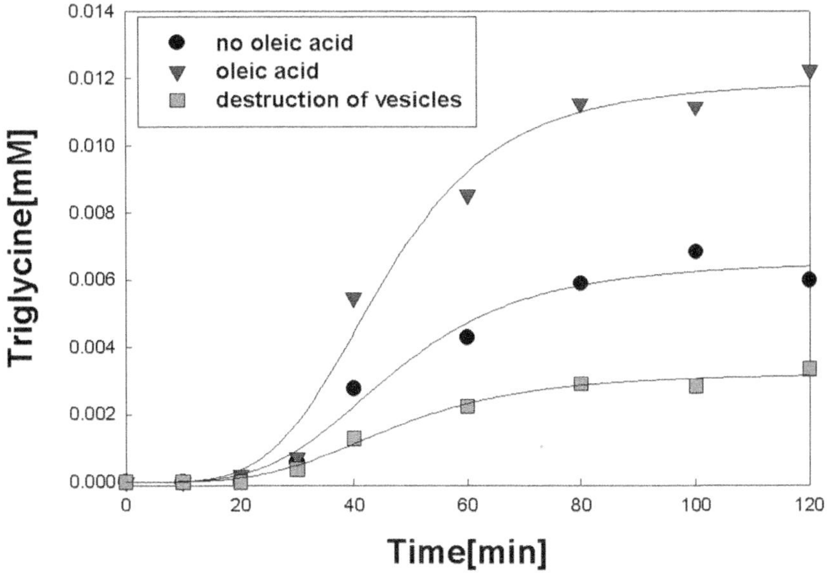

Figure 2 Yields of triglycine after 2 hours operation of the flow reactor simulating a hydrothermal environment. The reaction solution consisted of 100mM glycine and 10 mM oleic acid. Dissolution of oleic acid vesicles was completed by adding 10% tritonX-100 after the reactor operation.

5. Implications

Oligomers of prebiotic significance could easily have been synthesized from their monomeric constituents in a multiplicative manner in the vicinity of marine hydrothermal vents on the primitive earth. Theoretical justifications of this likelihood are twofold. One is that a covalent bond formed between a monomer and an oligomer in the hot water inside the vents could be fixed through adiabatic quenching when the product is suddenly injected into the surrounding cold water. One more justification is within the geological structures of hydrothermal vents themselves enabling seawater carrying the products from the adiabatic quenching to circulate through the vents repeatedly as rendering the preceding products constantly to be the subsequent reactants. Hydrothermal vents on the Earth had the intrinsic evolutionary capacity to convert prior reaction products into subsequent reactants even before the emergence of reaction cycles of biochemical origin.

Adiabatic quenching of synthesized monomers and oligomers could have been a major stepping-stone for subsequent synthetic reactions. Chemical evolution leading to the origin of life could have been a material process of harnessing the flow of energy dissipation from its primary sources. Synthesis of monomers and their oligomerization in the form of anisotropic energy structuring could be a dominant mode of harnessing the energy flow of degradation. Submarine hydrothermal vents could most probably serve as one of the primary energy sources available on the primitive Earth that could generate a rich catalogue of chemical synthesis (Ferris, 1992).

Anisotropic energy structuring in the form of monomers and oligomers of prebiotic significance that could appear in the middle of the flow of energy degradation is sandwiched between two interfaces. One is towards the primary energy sources, and the other towards heat sinks. A most likely candidate of anisotropic energy structuring is the one that can minimize the rate of disintegration with regard to the downward flow of energy dissipation (Matsuno, 1974, 1977). The winning structure is the least corrosive, though this sounds almost tautological. On the other hand, when it comes to energy uptake, a most likely energy structuring is the one that is fastest in uptake. Henceforth, a principle for the most likely candidate of anisotropic energy structuring in chemical evolution leading to the origin of life would be "First come, first served individually, while the winner takes them all collectively". Even Darwinian evolution can be taken in the context of anisotropic energy structuring, focusing on the energy consumers that are fastest in the uptake of usable energy (Matsuno and Swenson, 1999).

Submarine hydrothermal vents on the Earth should be considered as a candidate in raising and maintaining such an anisotropic energy structuring of prebiotic significance. Adiabatic quenching and walling of chemical species synthesized inside the hot vents can give rise to a least corrosive energy structuring in the form of long-lived monomers and oligomers. At the same time, elongation of oligomers thus formed and necessarily associated with the inevitable disintegration processes could be dominated by the one that is fastest in the energy uptake. Prebiotic oligomerization expected in the vicinity of submarine hydrothermal vents could thus serve as a concrete instance demonstrating the principle on the fastest energy uptake and the least corrosive structuring in the context of prebiotic evolution.

6. From Thermodynamic Perspective

Chemical evolution leading to the origin and evolution of biological organization must have been of material origin and ubiquitous in the cosmological context. A common thread is consumer-dominating thermodynamics focusing on energy consumers in general and on various biological organisms in particular. Ice grains in diffuse clouds in interstellar medium and hydrothermal vents on the Earth, both demonstrate how ubiquitous consumer-dominating thermodynamics could be on material grounds. What has been difficult for us, on the other hand, is our long-held perception towards supplier-domination while associating ourselves with the supplying agency when thermodynamics is focused. Once it is duly recognized that there is more than one agency in the common framework of thermodynamics as in the form of supplying and consuming heat energy, no material processes can be seen free from the interplay of multiple agencies (Matsuno, 1989). The origin and evolution of life in the cosmological context just manifests a likelihood of the material dynamics or thermodynamics of multiple agencies, especially with the occurrence of consumers.

When it is applied to the origin and evolution of biological organizations as a nonequilibrium extension, thermodynamics comes to be supplemented by an agential capacity. Even in its original equilibrium framework, thermodynamics remains under-complete in itself and requires to be supplemented by something else. Take, for instance, an ideal gas as a representative prototype of all of the thermodynamic systems conceivable. The set of three thermodynamic variables such as temperature, internal energy and volume of the gas do not specify their values explicitly by themselves because of their intrinsic incompleteness. They have to be supplemented by the ideal gas law as a derivative of mechanics. This does not change even if thermodynamics is extended to the nonequilibrium case. In retrospect, however, all of these thermodynamic variables have been identified as measurable quantities. The retrospective determination of the thermodynamic variables makes any thermodynamic system agential in specifying the values of the variables on its own.

Agential activities latent in thermodynamic systems manifest in the present progressive mode. Concrete determination of thermodynamic variables upon an agential activity sometime serves as a cause for succeeding determination of other variables in the neighborhood with the aide of thermodynamic equalities such as the First Law of Thermodynamics. Although such an agential capacity cannot be appreciated if one restricts oneself only to the description in the present mode, the present progressive mode addresses itself to the agential interplay of thermodynamic variables for their own internal determination.

A most conspicuous case of the agential determination of thermodynamic variables is seen in the agential interplay between temperature and internal energy. As constantly facing the issue of which is the first to change, either temperature or internal energy, for fulfilling the First Law as being perturbed by similar activities in the neighborhood, a thermodynamic system can maintain in itself a certain extent of intrinsic freedom. Indefiniteness associated with the capacity of agency can allow both plastic integration in one extreme case and random disintegration in the other. Taming the agential capacity of thermodynamic variables should now be considered to be a crucial matter for the issue of the origin and evolution of biological organizations put in a thermodynamic perspective. What we have observed in this regard is that temperature adjustment following changes in the amount of internal energy must be a typical pathway of harnessing the thermodynamic agential capacity. Consumer-dominated thermodynamics seems to be just one particular instance of taming the thermodynamic agential capacity so that it could account for and accommodate the origin and evolution of biological organizations.

7. Summary

Hydrothermal environments in the primitive ocean on our Earth were capable of synthesizing various molecules including fatty acids and amino acids. In addition, the complex of fatty acids and amino acids could have further exhibited the capability of synthesizing various oligomers or oligopeptides on or inside the lipid vesicles made of fatty acids. We constructed in the laboratory a simulated hydrothermal environment, and examined the molecular organization that could be synthesized from oleic acid and glycine. The simulated environment was maintained between hot and cold water, and the reaction fluid was circulated between the hot and the cold region repeatedly. Glycine was found to adhere on or to be encapsulated into the lipid vesicles made of oleic acid even though the vesicles were formed and dissolved as traversing the high temperature regions. Synthesis of oligoglycine was enhanced in the presence of the lipid vesicles of oleic acid, compared to the cases otherwise. Furthermore, the temperature of the hot water required for the proper synthesis of oligoglycine could be lowered in the presence of oleic acid vesicles. These observations, when combined together, suggest that lipid vesicles of fatty acids could have been functional in facilitating prebiotic oligomerization even prior to the evolutionary emergence of phospholipids and the vesicles thereof.

8. Acknowledgments

Experimental works referred to in this article were done in collaboration with Ryo Furuuchi, Ei-ichi Imai, Hajime Honda and Kuniyuki Hatori. Thanks are also due to Charles L. Apel for helpful comments.

9. References

Apel, C. L., Deamer, D. W. and Mauntner, M. N. (2002). Self-assembled vesicles of monocarboxylic acids and alchohols: conditions for stability and for the encapsulation of biopolymers. Biochim. Biophys. Acta. **1559**, 1-9.

Corliss, J. B., Dymond, J., Gordon, L. I., Edmond, J. M., von Herzen, R. P., Ballard, R. Green, K. K., Williams, D., Bainbridge, A., Crane, K. and van Andel, T. H. (1979) Submarine thermal springs on the Galapagos rift, Science, **203**, 1073-1083.

Deamer, D. W. (1997) The first living systems: a bioenergetic perspective. Microbiol. Mol. Biol. Rev. **61**, 239-261.

Edmond, J. M., Von Damm, K. L., McDuff, R. E. and Measures, C. I. (1982) Chemistry of hot springs on the East Pacific rise and their effluent dispersal, Nature **297**, 187-191.

Ferris, J. P. (1992) Chemical markers of prebiotic chemistry in hydrothermal systems. Origins Life Evol. Biosphere **22**, 109-134.

Fox, S. W. and Dose, K. (1977) *Molecular Evolution and the Origin of Life*, Revised Edition, Marcel Dekker, New York, p. 142.

Imai, E., Honda, H., Hatori, K., Brack, A., and Matsuno, K. (1999a) Elongation of oligopeptides in a simulated submarine hydrothermal system. Science **283**, 831-833

Imai, E., Honda, H., Hatori, K., and Matsuno, K (1999b) Autocatalytic synthesis of oligoglycine in a simulated submarine hydrothermal system. Origins Life Evol. Biosphere **29**, 249-259.

Kauffman, S. A. (1986) Autocatalytic sets of proteins. J. Theor. Biol. **119**, 1-24.

Luisi, P. L., Walde, P. and Oberholzer, T. (1999) Lipid vesicles as possible intermediates in the origin of life. Curr. Opinion Coll. Interface Sci. **4**, 33-39.

Matsuno, K. (1974) Probabilistic kinetics of macroprocesses in broken microscopic reversibility. J. Stat. Phys. **11**, 87-132.

Matsuno, K. (1977) Prebiological molecular evolution as physicochemical processes. J. Phys. Soc. Japan **42**, 1691-1693.

Matsuno, K. (1982) Natural self-organization of polynucleotides and polypeptides in protobiogenesis: appearance of a prtohypercycle. BioSystems **15**, 1-11.

Matsuno, K. (1989) *Protobiology: Physical Basis of Biology*. CRC Press, Boca Raton Florida.

Matsuno, K. (1997) A design principle of a flow reactor simulating prebiotic evolution. Viva Origino **25**, 191-204.

Matsuno, K. and Swenson, R. (1999) Thermodynamics in the present progressive mode and its role in the context of the origin of life. BioSystems **51**, 53-61.

McCollom, T. M. and Simonet, B. R. T. (1999) Abiotic formation of hydrocarbons and oxygenated compounds during thermal decomposition of iron oxalate. Origins Life Evol. Biosphere **29**, 167-186.

Ogata, Y., Imai, E., Honda, H., Hatori, K. and Matsuno, K. (2000) Hydrothermal circulation of seawater through hot vents and contribution of interface chemistry to prebiotic synthesis. Origins Life Evol. Biosphere **30**, 527-537.

Orgel, L.E. (1986) RNA catalysis and the origin of life. J. Theor. Biol. **127**, 127-149.
Orgel, L. E. (1998) Polymerization on the rocks: theoretical introduction. Origins Life Evol. Biosphere **28**, 227-234.
Rushdi, A. I. and Simonet, B. R. T. (2001) Lipid formation by aqueous Fischer-Tropsch-type synthesis over a temerature range of 100 to 400°C. Origins Life Evol. Biosphere **31**, 103-118.
Wächterhäuser, G. (1992) Groundworks for an evolutionary biochemistry - the iron sulfur world. Prog. Biophys. Mol. Biol. **58**, 85-201.
Tsukahara, H., Imai, E., Honda, H., Hatori, K. and Matsuno, K. (2002) Prebiotic oligomerization on or inside lipid vesicles in hydrothermal environments. Origins Life Evol. Biosphere **32**, 13-21.

Biodata of **Alicia Negron-Mendoza** author (with co-author S. Ramos-Bernal) of "***The Role of Clays in the Origin of Life***"

Dr. Negron-Mendoza works in radiation chemistry applied to problems in chemical evolution at the Institute of Nuclear Sciences (Universidad Nacional Autonoma de Mexico) in Mexico City. She obtained her Ph.D. (in 1980) under the direction of Professor C. Ponnamperuma at the University of Mariland, USA.

E-mail: **negron@nuclecu.unam.mx**

Dr. Sergio Ramos-Bernal is a researcher at the Institute of Nuclear Sciences Universidad Nacional Autonoma de Mexico) in Mexico City, Mx. He obtained his Ph.D. (in 1974) at the University of Manchester, England. Dr. Ramos-Bernal carries out research in the field of solid-state physics connected with problems of prebiotic processes.

E- mail: **ramos@nuclecu.unam.ms**

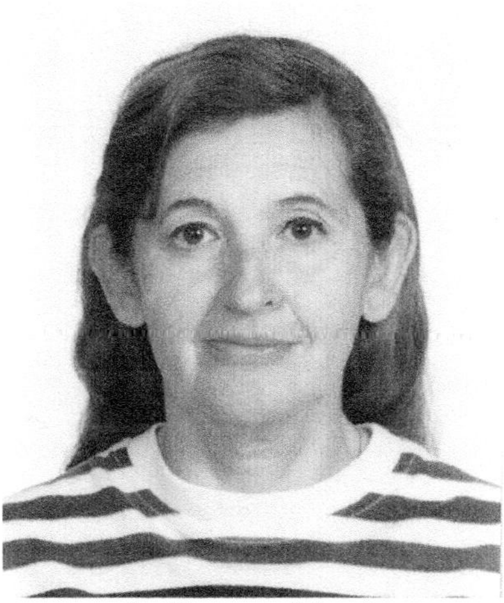

A. Negron-Mendoza

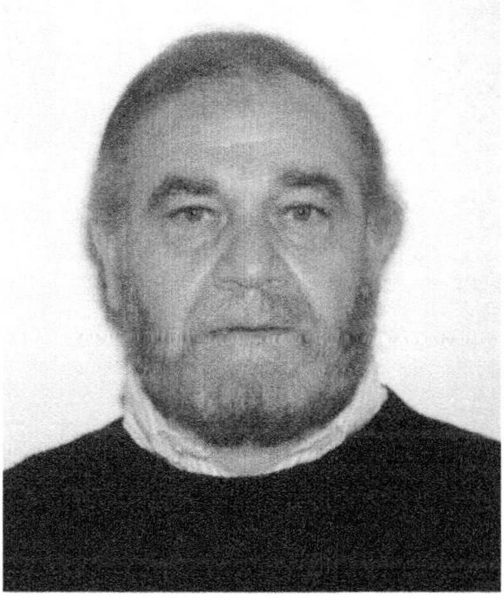

S. Ramos-Bernal

J. Seckbach (ed.), Origins, 181–194.

THE ROLE OF CLAYS IN THE ORIGIN OF LIFE

A. NEGRON-MENDOZA AND S. RAMOS-BERNAL
Instituto de Ciencias Nucleares, UNAM
A.P. 70-543, Ciudad Universitaria, D.F. 04510 México.

1. Introduction

One premise in chemical evolution is that the chemical composition of modern organisms might be a reflection of the chemical conditions that were present in the origin of life. Early organisms needed to create a pathway for synthesis and breakdown processes that we know in the most complex form as metabolism (Oparin, 1965). Oparin and Haldane proposed that important molecules could be a result of many prebiotic processes (Oparin, 1965). The experimental simulations of the possible processes on the primitive Earth and the comparisons with simple organisms help us to understand the early steps in the origin of life.

An important aspect in simulated experiments related to chemical evolution on Earth, is their relation with a geologically relevant scenario. Multiphase systems represent more realistic and plausible simulated environments of the primitive Earth (Lahav and Chang, 1976, Lahav, 1999). In this geological framework it is important the contribution of solid surfaces. Among the most important surfaces are silicates, carbonates, sulfides, and in particular, clays (Lahav and Chang, 1982). In solids, the adsorption of organic molecules produces concentration of the adsorbate. Without such a concentration, most prebiotic scenarios come to a halt, and highly concentrated prebiotic environments may be implausible. In particular, some significant fraction of monomers would have formed on the surface of solid minerals. On the other hand, it is necessary to have both matter and energy to have chemical reactions. An important characteristic of the type of energy source required for the interaction with solids is that it needs to be very penetrating.

Clay minerals are capable of selective concentrating and internally binding compounds. They were probably the most geologically relevant and abundant surfaces on the primitive Earth surface. It is very likely that the role of clay minerals in primordial organic chemistry was very important. Solid surfaces adsorbed organic molecules, protected them against ultraviolet radiation, and induced ordered arrangements that probably favored the polymerization of those molecules.

Clays are ubiquitous minerals on the Earth; and it is very likely that they appeared in the early steps of the formation of our planet. The Isua sediment's analysis in which metamorphic clays may be present at least 3.8 x 10^9 years ago support this premise (Moorbath, 1995). While various clays probably formed on the continents of the Earth in Archaean times, the largest area of clay production was the seabed. The genesis of continental smectites can be traced back 3 or 4 billion years ago (Odin, 1988).

Several research groups with different purposes (Lazlo, 1987; Theng, 1974) have studied the adsorption process of organic molecules in clays. Besides adsorption, clays are important in catalysis. Clay particles catalyze reactions in several ways: by energy transfer processes, redox reactions, stabilization of intermediates, and as Bronsted or Lewis acids. Data in the literature on this subject is so extensive that we will restrict our introductory remarks to the role of clay minerals in chemical evolution.

2. Some Properties and Structural Characteristics of Clay Minerals.

Clay is a natural, earthy, fine-grained material that develops plasticity when mixed with a limited amount of water. This property is its most prominent macroscopic characteristic that is related to its microscopic structure. The components of clays are silica, alumina, water, and appreciable quantities of alkaline and alkaline earth elements. The individual structural units are measured in angstroms, but their aggregates are of microns in diameter. From the microscopic standpoint clays have the following structure: the individual units are in parallel plates stacked one above the other. Kaolinite is two-layered clay; each layer consists of tetrahedrically-coordinated silica, combined through shared oxygen atoms in a layer of octahedrally coordinated magnesia or alumina. Montmorillonite is three-layered clay where each laminae has an octahedral alumina sandwiched between two tetrahedral silica layers. Interlamellar channel is the name of the space between the silica layers (Swartzen-Allen and Matijevic, 1974). The tetrahedral sheet results from the sharing of corners occupied by oxygen ions. This arrangement is optimal for screening the positive coulomb field produced by the Si^{4+} ions at the centers of the tetrahedral. The octahedral sheets are building up through the sharing of edges, which bring neighboring metal cations nearer to one another (Sposito, 1984). The relative number of tetrahedral and octahedral sheets in the parallel plates varies from clay to clay. For example in kaolinite, its proportion is 1:1, and in smectite is 2:1. This composed unit continuous in two directions of a plane in the c direction to form packets of two to fifteen elementary units.

According to their variations of atomic structure and chemical composition the classification of clays is: (1) allophane, (2) kaolinite, (3) halloysite, (4) smectite[1], (5) illite, (6) chlorite, (7) vermiculite, (8) attapulgite, (9) sepiolite, (10) palygorskite, and (11) mixed-layer clay minerals. In prebiotic chemistry, montmorillonite is the prototype clay more extensively used for adsorption processes. Other clay minerals differ in structural detail, but they possess similar characteristics.

To these microstructural aspects of clay particles, we should now superimpose the electrostatic phenomena, which arise on the clay surfaces exposed to a watery environment. It develops due to exposed structural hydroxyls and broken bonds at the edges of the particles. However, it is more important the metallic ions isomorphous substitution within the lattice. For example, Al^{3+} or Fe^{2+} can replace Si^{4+} in the tetrahedral layer, and Mg^{2+}, Fe^{2+}, Zn^{2+}, Ni^{2+}, Li^{+}, and others may substitute Al^{3+} in the octahedral sheet. These substitutions create an electrical unbalance with a net negative charge on the structural units.

---

[1] Included in the smectites are montmorillonite, nontronite, and beidellite, which are dioctahedral (i.e., 2/3 of the aluminum octahedral sites are occupied by other cations), and saponite, hectorite, and sauconite, which are trioctahedral.

The negative charge prevails at all pH values above two or three. Counterions or interstitial cations (such as Na^+, K^+, Ca^{2+}, etc.) in the interlamellar space compensate the negative charge.

It is important to remark that in neutral and acid pH ranges, the edges of the clay present a positive charge. The central atom of the octahedral layer is the origin of this charge (Swartzen-Allen and Matijevic, 1974, Nicol and Hunter, 1970). Prominent phenomena related to the counterions located in the interlamellar space are their exchangeability with respect to other metallic cations. For alkali ions the order for ion exchange to montmorillonite, vermiculite, and kaolinite is $Li^+ < Na^+ < K^+ < Rb^+ < Cs^+$ (i.e. each ion in these series can displace the one to its left in competition for adsorption centers). The particular distinction of the montmorillonite structure from the other types is that water, free inorganic cations, and polar molecules can occupy the interlamellar space. These cause the lattice to expand in the direction of stacking. X-ray diffraction methods show this expansion. The basal distance may vary from 10 Å in totally collapsed montmorillonite to 20-50 Å, depending upon the exchangeable cation (Anderson and Banin, 1975). Figure 1 shows a proposal structure of montmorillonite according to Edelman and Favejee (1940).

The microscopic characteristic of intercalation of layers of liquid water between the clay lattice structures imparts the macroscopic plasticity and expansible properties common to clays. Furthermore, from this structural property it also derives the astonishing large surface area available to adsorption processes. Total clay dimension is below 2 μm, but clay particles show specific areas that vary from clay to clay. For kaolinite, this area is 15-20 m^2/g, for illite is 80-90 m^2/g, and for montmorillonite is 500-760 m^2/g (Laszlo, 1987). It is this large specific area of montmorillonite, which has attracted so much attention in adsorption studies for prebiotic systems. However, internal surfaces are not common to all clays. Illite and kaolinite have no internal surface, whereas in vermiculite a large fraction (0.8-0.9) of its total specific area corresponds to internal surfaces. In montmorillonite, it ranges between 0.1-0.9 (Anderson and Banin, 1975).

3. Possible Role of Clays in Prebiotic Chemistry

Bernal proposed in a lecture in 1947, six years before the Miller-Urey experiment, that clay minerals might have an important role in prebiotic synthesis. Clay minerals were necessary: (1) to concentrate the organic molecules present in a diluted ocean by adsorption on clay deposits; (2) catalyze the polymerization of adsorbed organic compound. (3) Protect these organic molecules from destruction by ultraviolet light. (4) Induce the chirality in contemporary molecules. The investigation of the role of clays in prebiotic organic synthesis ranges from mere adsorbents and catalysts to the controversial claim that clays were the first functional templates (Cairns-Smith, 1966).

Prebiotic scenarios in nature can be found in mud beds, shores, hydrothermal vents at the sea bottom, and regions alternatively wet and dry as tidal estuaries, etc. These environments stress the significance of multiphase systems in prebiotic processes. The reproductions of these scenarios are a difficult task to achieve in the laboratory. However, there are good approximations in the literature that simulated these environments. We describe some experiments made with this purpose.

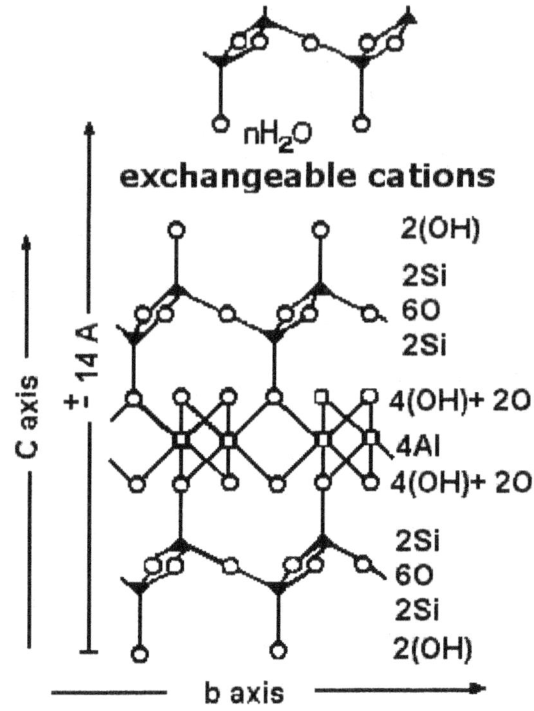

Figure 1. Structure of Montmorillonite.

3.1 ADSORPTION PROCESS IN CLAY MINERALS.

Adsorption is the process through which a net accumulation of a substance occurs at the common boundary of two contiguous phases. Adsorption mechanism of organic molecules with clays is a complex phenomenon where different processes can take place: H-bonding, ion-dipole and interactions of the van der Waals type. Organic compound also may complex with counterions of the clay or it may undergo ionic exchange (Swartzen-Allen and Matijevic, 1974). There are experiments in which one ton of montmorillonite incorporated up to 550 kg of protein, 150 kg of fatty acids or 200 kg of carbohydrates (Weiss, 1969).

Clays have excess surface charge, and they act as natural ion exchangers. The adsorption depends on the type of clay, initial equilibrium concentration of adsorbate, nature of adsorbate molecule and its shape, pH of bulk solution, metal in the interchange position, and the quantity of water molecules. Other important factor in the adsorption is the co-adsorption phenomena. A review of these phenomena shows that indeed the presence of other molecules, besides the adsorbate, leads to a change in the extent of the adsorption (Mosqueira et al., 1996). The Langmuir equation is by far the most common description of adsorption processes. However, large organic molecules and polymers do not follow this equation.

The adsorption data for bioorganic compounds showed that the adsorption is larger in acidic pH. Sometimes the binding of these types of compounds drops almost to null adsorption at pH 8. Since the pH of the primitive ocean probably was around pH 8, this may disqualify some results. Nevertheless, there are natural microenvironments that attain higher acidities, improving adsorption binding, which supports the possible role of clays in the primitive Earth. There are many studies of clay minerals related to the adsorption of biological compounds (amino acids, peptides, nucleic acid bases and their derivatives, carboxylic acids and sugars) (Theng, 1974, Mortland, 1970; Weiss, 1969). This process is the first step for the variety of reactions that can occur inside the clay. Rao et al., 1980 and Ponnamperuma et al., 1982, made a review of these studies.

3.1.1. Amino acids.

Some general trends for amino acids in the adsorption properties are (1) the adsorption increases with decreasing pH and increasing the concentration. (2) The major adsorption process is by cation exchange, but ion-dipole coordination interactions and hydrogen bonds also take place in the process. (3) The dependence of these mechanisms is mainly due to the isoelectric point, molecular size, shape of the molecule, and pH of the solution. (4) The amino acid molecules are oriented so that the main chain of the molecule, (from one functional group to other) lies parallel to the basal surface of the clay. These may affect further polymerization (Hsu, 1977).

Friebele et al., (1981) studied the adsorption of protein and non-protein amino acids in montmorillonite, and they did not find preferential adsorption for protein amino acids over non-protein amino acids. To look for a preferential adsorption of enantiomers in active surfaces, some experiments were made with D and L- amino acids or with a racemic mixture. Degens studied the asymmetric polymerization of amino acids in clays (Degenes et al., 1970). There has not been conclusive evidence of asymmetric adsorption or polymerization of isomers by clay (Shimoyama and Ponnamperuma, 1980, Friebele et al., 1981).

For oligomers such as tetraglycine, the adsorption increases at neutral pH and increasing molecular weight (Greenland, 1962). The fixation characteristics of protein to mineral not only varies with the type of mineral, but also with the kind of protein involved (Weiss, 1969). For proteins, the highest absorption was at pH around the isoelectric point of the protein (Albert and Harter, 1973).

3.1.2 *Nucleic Acid Bases, nucleosides, nucleotides and poly-nucleotides*

The binding of nucleic acid bases, nucleosides, nucleotides and poly-nucleotides in clay minerals has been extensively studied (Lailach et al., 1968a 1968b; Odom, et al., 1979; Graf et al., 1980; Perezgasga et al., 2003). Some general observations are (1) the adsorption is pH dependent. (2) The adsorption is more extensive in acidic pH, and then it decreases steadily to a very low adsorption level at pH >6. (3) The extent of adsorption is larger in purines than in pyrimidines. (4) The adsorption of nucleotides and nucleosides is in less extent than their corresponding bases.

There are two sites of binding, namely the interlamellar channel and the edges of the clay. Nucleic bases are adsorbed mainly in the interlamellar channel. Experiments blocking selectively the interlamellar channel with alkylamines or poisoning the edges with phosphates show this (Ertem and Ferris, 1998; Franchi et al., 1999; Perezgasga, et al., 2003).

3.1.3. Sugars

Experimental data on the adsorption of sugars in clay minerals are scarce, but they all show very low adsorption (Greenland, 1956; Mitra et al., 1957; Jepson and Williams, 1972). The edges of the clay platelets adsorb anionic polymers (Schott, 1968; Perezgasga et al., 2003).

3.2 SYNTHESIS OF BIOORGANIC COMPOUNDS IN THE PRESENCE OF CLAY MINERALS

Because catalysts might have been important in chemical evolution, clays in combination with metal cations and simple organic species may have performed such functions. Factors influence the reactivity, selectivity and specificity of clays. These factors are transition state, size, surface chemical equilibrium, and reactant pair proximity (Pinnavaia et al., 1979; Weiss, 1981).

Clays can enhance chemical reactions on their surfaces via Lewis or Bronsted acid sites. The influence of water content on the activity of these sites is profound. To make Lewis sites available, it is necessary to remove water. Structural metal cations at the mineral edges form Lewis acid sites (Solomon, 1968). In addition, exchange metal cations can function as Lewis sites if they are electron acceptors. This way, exchangeable transition metal cations that have unfilled "d" orbitals are specially suited for this function, and solvate water molecules are coordinated to potential electron acceptors. On the other hand, the hydrogen ions occupying exchange sites on the interlamellar channel are responsible of Bronsted acidity. The larger the charge to radius ratio of the cation, the stronger the Bronsted acid will be. In addition, the water content markedly affects its Bronsted acidity. Clay surfaces show acidities considerably greater than those of clay suspensions.

The theory of a "RNA world" is part of the current thinking concerning the origin of life (Gilbert, 1986). This was after the discovery that certain ribonucleic acids have catalytic properties. These discoveries led to a new horizon of prebiotic experiments. The "RNA world" means that life could exist before the appearance of DNA and proteins. RNA preceded DNA as a genetic molecule. Orgel and his group developed the template-directed oligomerzation of mononucleotides with very promising results (Gibbs et al., 1980; Inoue and Orgel, 1982a; 1982b; Schwartz and Orgel, 1985).

The condensation of activated RNA monomers has been achieved by clay catalysis. The purpose of this research was on the primer elongation reactions that may have occurred in the primitive Earth. Another objective was to establish the sequence and selectivity of phosphodiester bond formation (Ferris et al., 1989; Ferris, 1991; Ferris and Ertem, 1992a and 1992b; Miyakawa and Ferris, 2002). The self-condensation of ImpA on montmorillonite and other elongation process may have resulted in the RNA formation.

The activity of the silicate structure in the catalytic process of clays suggests other important reactions might have been feasible such as transamination and decarboxylation. The idea of simple organic molecules combining with inorganic materials to perform some biochemical function in chemical evolution requires further research.

The diversity of organic reactions catalyzed by clay minerals is impressive (Theng, 1974). Table I presents a summary of studies related to chemical evolution.

TABLE 1. Abiotic synthesis of amino acids and polymeric material in the presence of a clay mineral

Reactants	Products	Clay	References
H_2 + CO +NH_3	Asp, glu, ser, gly, ala, leu, lyc, arg	Montmorillonite	Yoshino et al., 1971.
CO + NH_3	Thr, ser, ala, phe, lys	Zeolites	Fripiat et al., 1972.
CO + NH_3	Asp, thr, ser, glu, gly, ala, leu, val	Zeolites	Poncelet et al., 1975.
HCHO + CH_3CHO + Polyglycine	Thr, ser	Kaolinite	Akabori, 1956.
HCHO + $HONH_2$	Amino acids	Montmorillonite, Kaolinite	Hatanaka and Egami, 1977.
HCHO + NH_3 + HCN	Polyglycine	Kaolinite	Akabori, 1955
Amino acyl adenylates	Polypeptides	Montmorillonite	Paecht-Horowitz et al.,1970
Glycine	Oligopeptides up to gly_5	Kaolinite, bentonite Wetting-drying cycles	Greenland, 1962
Aspartic acid	Polymers	Kaolinite	Degens et al.,1970
Diglycine	Gly_3, gly_4, gly_5, gly_6	Ca-montmorillonite or hectoliter	Rode et al., 1999
Asp, or L-glu or O-phospho-L-ser,glu_{10}, carbonyldiimidazole	Oligopeptides	Illite	Hill et al., 1997
β-aa, poly-glu, CDI	Oligomers	Illite	Liu et al., 1998
TMP + imidazole	Oligodeoxyribonucleo-tides up to 4 units	Montmorillonite	Ibañez et al., 1971
TMP	Oligonucleotides	Kaolinite	Odom et al., 1979
PA + EDAC	2',5'-$(pA)_2$; ,3'5'$(pA)_2$, AppA	Na-Montmorillonite	Ferris et al., 1989
5'UMP/Poly A	U_5	Na-Montmorillonite	Ostrochenko and Vasilyeva, 2002
PApC+ ImpA/ImpG	Trimers	Na-Montmorillonite	Miyakawa and Ferris, 2002
DAMN+3'AMP, ImpA	5mers	Montmorillonite	Ferris and Ertem, 1992

3.3 ROLE OF CLAY S AS PROTECTOR AGENTS

Clay minerals might have provided protection against high-energy radiation to adsorbed molecules. The references regarding this particular role of clay are scarce. Guzman et al., 2000 studied the stability of adenine adsorbed in a clay mineral and exposed to a high radiation field. The recovery of adenine after a gamma irradiation was higher in the system containing clay in relation to a system without clay. These series of experiments showed the protective role of clays.

3.4 HETEROGENEOUS CATALYSIS BY RADIATION. RELEVANCE IN CHEMICAL EVOLUTION PROCESSES.

In the gamma irradiation of an organic compound adsorbed in solids, much of the energy adsorbed by the solid decomposes the organic compound. This energy transfer is not well understood. During such energy transfer process, the adsorbent participates in the chemical reaction.

Pre-irradiated catalysts constitute a special case. The long-lived defects generated by radiation at the catalyst surface are active, i.e. there is ordinary heterogeneous catalysis on radiation-modified catalysts.

Experimental data on problems related to the interaction of solids with radiation, involving the origin of life are scarce. However, an outline of the main features of these radiation catalytic effects is possible.

To consider an energy source as a potential source for prebiotic synthesis, it needs to be available, abundant and efficient to induce chemical reactions. Natural energy sources such as radioactive decay and triboelectric energy (i.e., energy of mechanical stress) may have a great importance in those scenarios for being such penetrating sources of energy. Mosqueira et al., 1996 evaluated the radiation doses on the primitive Earth arising from long-lived radionuclides as ^{40}K, ^{238}U and ^{232}Th. These isotopes have a half-life of 10^9 years or more. Their present concentrations make possible to calculate, by backward extrapolation, their concentration and associated radiation doses on the primeval Earth. These calculations involved different physical environments of interest to prebiotic chemistry (clays, sedimentary rocks, igneous rocks and plain seawater. In clays, 3.8 x10^8 years ago, the dose rate considering only the contribution of ^{40}K was 1.75 x 10^{-2} Gy/year (Gy= Gray, unit for absorbed dose, and equivalent to 1x10^{-3} Joule/g). This amount of energy is sufficient to induce chemical reactions.

Under irradiation, additional imperfections will appear on the surface of the crystal. However, these may be active in adsorption and catalytic processes. Energy may be absorbed in a solid by the following processes: (1) electron displacement (ionization, excitation), (2) atom displacement, (3) electron trapping.

Radiation effects in solids produce crystallographic structural defects, and those that are chemical in nature. The predominant type of damage will depend largely on the material (metals, ionic solids, semiconductors, etc.) and the source of radiation. It is important to discriminate how much is due to defects produced in the solid by the radiation, and if they have some responsibility in the reaction of the organic adsorbed.

Several questions are arisen in these types of studies: (1) the first is the simple question of whether the presence of a solid surface would alter the nature of the products of radiolysis. (2) The next question is in relation to the site of binding in the clay.

Montmorillonite has two different sites of binding. These sites can be either at the edges or in the interlamellar channel. (3) Energy transfer mechanisms from the clay to the organic compound adsorbed in it.

An example of studies in heterogeneous catalysis by irradiation has been presented by Ramos-Bernal and Negrón-Mendoza. They studied the radiation-induced decomposition of a carboxylic acid adsorbed in a clay mineral (Ramos-Bernal and Negrón-Mendoza, 1992; Negrón-Mendoza, et al., 1993; Negrón-Mendoza, et al., 1995, and Negrón-Mendoza and Ramos-Bernal, 1998). The main objective of the study was to discriminate if the presence of a solid surface alters the formation and distribution of radiolytic products, and compare it to the radiolysis of the carboxylic acid without the surface solid (clay).

TABLE II. Products formed from the irradiation of carboxylic acids in aqueous media, with and without a clay mineral.

Acid	Principal reaction without clay	Main products	Principal reaction with clay	Main products
Acetic	Dimerization	Succinic acid	Decarboxylation	Methane and CO_2
Aconitic	Addition	Tricarballylic and citric acids	Decarboxylation	Itaconic acid and CO_2
α-ketoglutaric	Reduction	Hydroxy-glutaric acids	Decarboxylation	Succinic acid and CO_2
Malonic	Addition	Succinic and carboxysuccinic acids	Decarboxylation	Acetic acid and CO_2
Pyruvic	Dimerization	Dimethyl tartaric acid	Decarboxylation	CO_2

The results showed that the radiolysis of the clay-acid system goes along a defined path rather than showing various pathways of decomposition. The main pathway was the decarboxylation of the target compound rather than condensation/dimerization reactions. Table II summarizes the results from some carboxylic acids irradiated with clay Na-Montmorillonite. The decomposition was mainly due to an interaction of the adsorbed molecules with non-equilibrium charge carriers formed by the ionizing radiation in the clay. However, more investigations of the solid state reactions are needed to extent our understanding of prebiotic process.

4. Concluding Remarks

Clay minerals are important to the process of chemical evolution due to their properties, ancient origin, and wide distribution. To extend the knowledge of their role in the prebiotic epoch, many authors have studied the adsorption, and synthesis of bioorganic compounds in clays. The molecule order going from simple to complex in structure on the surfaces of clays is very important in prebiotic chemistry. In the clay, these processes take place either through direct interaction with surface atoms or by complexation of exchangeable cations at the interlamellar space. There are many promising results, especially with the formation of

RNA-oligomers and with the preferential radiation-induced decomposition of organic compounds adsorbed in clay minerals. Further research is needed in this area, mainly in the synthesis of bio-polymeric molecules, because some of experiments were done in controlled conditions of pH, type of clay, presence of buffers etc., that do not correspond the geological environment of the primitive Earth.

Energy and charge transfer in clays is of particular interest: (1) clays can store energy from the environment and release it in various forms. A consequence of the energy transfer is that solids serve as moderators of energy, (2) there are modes of energy transfer across interfaces that do not exist in homogeneous phases, (3) organic compounds absorbed in clays are subject to energy, as well as, charge transfer from the "activated solid". Energy transfer from clays to adsorbate is possible. Therefore, it is likely that energy deposited in the bulk solid by penetrating energy sources is moderated to a minor energy form before transfer to a distant interfacially adsorbed reactant.

5. Acknowledgment

This work was partially supported by DGAPA-UNAM grant ES116601.

6. References

Akabori, S. (1955) Asymmetric synthesis of amino acids and formation of fore proteins, *Kagaku* **25**, 54-59.

Akabori, S., Okawa, K., and Sato, M. (1956) Introduction of side chains into polyglycine dispersed on solid surface I, Bull. *Chem. Soc. Japan* **29**, 608-611.

Albert, J.T. and Harter, R.D. (1973) Adsorption of lysozyme and ovalbumin by clay. Effect of clay suspension pH, and clay mineral type, *Soil Sci.* **115**, 130-136.

Anderson, D.M. and Banin, A. (1975) Soils and water and its relationship to the origin of life, *Origins of Life* **6**, 23-36.

Bernal, J.D. (1949) *The Physical Basis of Life*, Proc. Phys. Soc. Conference November 1947, **62**, 537-558. (London),

Cairns-Smith, A.G. (1966) The Origin of life and the nature of the primitive gene, *J. Theor. Biol.* **10**, 53-88.

Degenes, E.T., Matheja, J., and Jackson, T.A. (1970) Template catalysis: Asymmetric polymerization of amino acids on clay minerals, *Nature* **227**, 492-493.

Edelman, C.H. and Favejee, J.C.L. (1940) On the crystal structure of montmorillonite and halloysite, *Z. Kris.* **102**, 417-431.

Ertem, G. and Ferris, J.P. (1998) Formation of RNA oligomers on montmorillonite. Site of the reaction, *Origins of Life Evol.Biosphere* **28**, 485-499.

Franchi, M., Bramanti, E., Morassi Bonzi, L., Orioli, P. L., Vettori, C., and Gallori, E. (1999) Clay-nucleic acids complexes: Characteristics and implications for the preservation of genetic material in primeval habitats, *Origins Life Evol. Biosphere* **29**, 297-315.

Ferris, J.P., Ertem, G., and Agarwal, V.K. (1989) The adsorption of nucleotides and polynucleotides on montmorillonite clay, *Origins Life Evol. Biosphere* **19**, 153-164.

Ferris, J.P. (1991) Prebiotic synthesis on minerals: RNA oligomer formation, In: J.M. Greenberg, C.X. Mendoza-Gomez, and J. Pirronello (eds.), *The Chemistry of Life's Origins*, Kluwer Academic Publishers, Dordrecht, The Netherlands. pp.301-322.

Ferris, J.P. and Ertem, G. (1992a) Oligomerization reactions of nucleotides on montmorillonite: Reaction of the 5' -phosphorimidazolide of adenosine, *Science*, **257**, 1387-1389.

Ferris, J.P. and Ertem, G. (1992b) Oligomerization reactions of ribonucleotides: The reaction of the 5'phosphorimidazolides of nucleosides on montmorillonite and other minerals, *Origins of Life Evol. Biosphere* **22**, 369-381.

Friebele, E., Shimoyama, A., and Ponnamperuma, C. (1981) Possible selective adsorption of enantiomers by Na-montmorillonite, In: Y. Wolman (ed.), *Origin life, Proceedings of the Third ISSOL Meeting and the Sixth ICOL Meeting,D. Reidel Publishing Company*, Dordrecht, The Netherlands. *pp. 337-346.*

Fripiat, J.J., Poncelet, G., van Assche, A.T., and Mayandon, J. (1972) Zeolite as catalysts for the synthesis of amino acids and purines, *Clays and Clay Miner* **20**, 331-339.

Gibbs, D., Lohrmann, R., and Orgel, L.E. (1980) Template-directed synthesis and selective adsorption of oligonucleotides on hydroxyapatite, *J. Mol. Evol.* **15**, 347-354.

Gilbert, W. (1986) The RNA world, *Nature* **319**, 618.

Graf, G. and Lagaly, G. (1980) Interaction of clay minerals with adenosine-5-phosphates, *Clay and Clays Minerals* **28**, 12-18.

Greenland, D.J. (1956) The adsorption of sugars by montmorillonite I. X-ray studies, *J. Sil Sci.***7**, 319-328.

Greenland, D.J., Laby, R.H., and Quirk, J.P. (1962) Adsorption of glycine and its di-, tri- and tetrapeptides by montmorillonite, *Trans.Faraday Soc.* **58**, 829-841.

Guzmán, A., Ramos-Bernal, S., and Negrón-Mendoza, A. (2000) Irradiation of adenine adsorbed in Na-montmorillonite. Implication to chemical evolution studies, In: J. Chela-Flores, G A. Lemarchand, and J. Oró (eds.), *Astrobiology: Origins from the Big Bang to Civilization*, Kluwer Academic Publisher, Dordrecht, The Netherlands. pp. 271-273.

Hatanaka, H., Egami, F. (1977) The formation of amino acids and related oligomers from formaldehyde and hydroxylamine in modified sea mediums related to prebiotic conditions, *Bull. Chem. Soc. Japon* **50**, 1147-1156.

Hill, A.R., Bohler, C., and Orgel, L. (1998) Polymerization on the rocks: Negatively charged α-amino acids, *Origins of Life Evol. Biosphere* **28**, 235-243.

Hsu, Shuei-Chi, (1977) Ph.D. The adsorption of peptides on montmorillonite, Dissertation, Polytechnic Institute of New York, Brooklyn, N.Y.

Ibañez, J.D., Kimball, A.P., and Oró, J. (1971) Possible prebiotic condensation of mononucleotides by cyanamide, Science **173**, 444-446.

Inoue, T., Orgel. L.E. (1982a) Oligomerization of (guanosine 5'-phosphor)-2 methylimidazolide on Poly(C). A RNA polymerase model, *J. Mol. Evol.* **162**, 201-217.

Inoue, T., Orgel. L.E. (1982b) A non-enzymatic RNA polymerase model. Science **219**, 859-862.

Jepson, W.B., Williams, J.F. (1972) Adsorption of water by clays, *Clay Miner.* **9**, 275-279.

Lahav, N. and Chang, S. (1976) The possible role of solid surface area in condensation reactions during chemical evolution: Re-evaluation, *J. Mol. Evol.* **8**, 357-380.

Lahav, N., and Chang, S (1982) The possible role of soluble salts in chemical evolution, J. Mol. Evol. **19**, 36-46.

Lahav, N (1999) *Biogenesis*, Oxford University Press, Oxford.

Lailach, G.E., Thompson, T.D., and Brindley, G.W. (1968a) Absorption of pyrimidines, purines, and nucleosides by Li-, Na-, Mg-, and Ca-montmorillonite (clay-organic studies XII), *Clays and Clay Mineral* **16**, 285-293.

Lailach, G.E., Thompson, T.D., and Brindley, G.W. (1968b) Absorption of pyrimidines, purines, and nucleosides by Co-, Ni-, Cu-, and Fe(III)-montmorillonite (clay-organic studies XIII), *Clays and Clay Mineral* **16**, 295-301.

Laszlo, P. (1987) Chemical reactions on clays, *Science* **235**, 1473-1477.

Liu, R. and Orgel, L. (1998) Polymerization on the rocks: β-amino acids and arginine, *Origins of Life Evol.Biosphere* **28**, 245-257.

Miyakawa, S. and Ferris, J. (2002) Sequence selectivity in the trimers formed in the montmorillonite clay catalyzed reactions of RNA dimers (pNpN) with the 5'phosphorimidazolides of nucleosides (ImpN), In: A. Negron-Mendoza, L. Delaye, E. Chacón, and A. Becerra (eds), *Abstracts of the 10th ISSOL Meeting,* Oaxaca, Mexico, pp. 78.

Mitra, S.P., Misra, S.G., and Panda, N. (1957) Adsorption of glucose by calcium bentonite, *Proc. Natl.Acad. Sci. India* **26A**, Pt 1, 72-74.

Moonrbath, S. (1995) Private communication.

Mortland, M. M.. (1970) Clay-organic complexes and interactions, Adv. *Agron.* **22**, 75-117.

Mosqueira, F.G., Albarrán, G., and Negrón-Mendoza, A. (1996) A review of conditions effecting the radiolysis due to ^{40}K for nucleic acid bases and their derivatives adsorbed on clay minerals, *Origins Life Evol. Biosphere* ***26***, 75-94.

Negrón-Mendoza, A., Albarrán, G., and Ramos-Bernal S. (1993) Transformation of malonic acid on a clay mineral by gamma radiation, *Radiat. Phys. Chem.* **42**, 4-6.

Negrón-Mendoza, A., Ramos-Bernal, S., and. Albarrán, G. (1995) Enhance decarboxylation reaction of carboxylic acids in clay minerals, *Radiat. Phys. Chem.* **46**, 565-568.

Negrón-Mendoza, A. and Ramos-Bernal, S. (1998) Radiolysis of carboxylic acids adsorbed in clay minerals, Radiat. Phys. Chem. **52**, 395-399.

Nicol, S.K.and Hunter, R.J. (1970) Rheological and electrokinetic properties of kaolinite suspensions, *Aust.J. Chem.* **23**, 2177-2186.

Odom, D., Lahav, N., and Chang, S. (1979) Association of nucleotides with homoionic clays, *J. Mol. Evol.* **12**, 365-367.

Odin, G.S. (1988) The origin of clays on Earth, In: A.G. Cairns-Smith and H. Hartman (eds.), *Clay Minerals and the Origin of Life*, Cambridge University Press, Cambridge, Great Britain. pp. 81-89.

Oparin, I. A., (1965) The pathways of primary development of metabolism and artificial model of this development, In: S. Fox (ed.) *The Origin of Prebiological Systems and Their Molecular Matrix*, Academic Press, New York. U.S.A. pp. 331.

Otroschenko, V. and Vasilyeva, N.V. (2000) The oligonucleotide synthesis from non-activated mononucleotides, directed by polynucleotide template adsorbed on the mineral surfaces, *Origins Life Evol. Biosphere* **30**, *256.*

Paecht-Horowitz, M., Berger, J., and Katchalsky, A. (1970) Prebiotic synthesis of polypeptides by heterogeneous condensation of amino-acid adenylates, *Nature 228*, 636-639.

Perezgaga, L. Negrón-Mendoza, A., Mosqueira, G., Serratos, A., and De Pablo, L. (2003) Site of adsorption of purines, pyrimidines and its corresponding derivatives on sodium montmorillonite. To be published.

Pinnavaia, T.J., Raythatha, R., Lee, J.G., Halloran, L.J., and Hoffman, J.F. (1979) Intercalation of catalytically active metal complexes in mica-type silicates: rhodium hydrogenation catalysis, *J. Am. Chem. Soc.* **101**, 6891-6897.

Ponnamperuma, C., Shimoyama, A., and Friebele, E. (1982) Clay and the origin of life, *Origins of life* **12**, 9-39.

Shimoyama, A and Ponnamperuma, C. (1980) In: P.E. Hare, T.C. Hoering and K. King, Jr. (eds.), *Biogeochemistry of Amino Acids*, Wiley, New York, U.S.A. pp. 145-151.

Solomon, D.H. (1968) Clay minerals as electron acceptors and or electron donors in organic reactions, *Clays and Clay Miner.* **16**, 31-39.

Sposito, G. (1984) *The Surface Chemistry of Soils*, Clarendon Press, Oxford.

Schwartz, A. and Orgel L. (1991) Biology and theory: RNA and the origin of life, In: J.M. Greenberg, C.X. Mendoza-Gomez, and V. Pirronello (eds.), *The Chemistry of Life's Origins*, Kluwer Academic Publishers, Dordrecht, The Netherlands. pp. 323-344.

Swartzen-Allen, S.L., and Matijevic, E. (1974) Surface and colloid chemistry, Chem. *Rev.* **74**, 385-400.

Rao, M., Odom, D.G. and Oro, J. (1980) Clays in Prebiological Chemistry, *J. Mol. Evol.* **15**, 317-331.

Theng, B.K.G. (1974) *The Chemistry of clay-organic reactions*, John Wiley & Sons, New York.

Weiss, A. (1969) Organic derivatives of clay minerals, zeolites and related minerals, In: G. Eglinton and M. T.J. Murphy, (eds.), Organic *Geochemistry*, Springer-Verlag, New York, pp. 737-781.

Weiss, A. (1981) Replication and evolution in inorganic systems, Angew. *Chem. Int. Ed. Engl.*, **20**, 850-860.

Yoshino, D., Hayatsu, R., and Anders, E. (1971) Origin of organic matter in early solar system III. Amino acids: Catalytic synthesis, *Geochim. Cosmochim. Acta* **35**, 927-938.

Biodata of **Joachim Reitner** author of ***"Organomineralization: A Clue to the Understanding of Meteorite-related "Bacteria-Shaped" Carbonate Particles"***

Professor Joachim Reitner is currently the Dean of the Faculty of Geoscience and Geography in the University of Göttingen, Germany. He obtained his Ph.D. from the University of Tübingen in 1984 and continued his studies and research at the Free University of Berlin. Professor Reitner's scientific interests are in the areas of: evolution of Metazoans, taxonomy and phylogeny of Porifera, biology and geobiology of biofilms, geomicrobiology and biomineralization, and evolution of reef systems.

E-mail: **jreitne@gwdg.de**

J. Seckbach (ed.), Origins, 195–212.

ORGANOMINERALIZATION

A Clue to the Understanding of Meteorite-related "Bacteria-Shaped" Carbonate Particles

JOACHIM REITNER
Department of Geobiology, GZG, Univ. Göttingen
Goldschmidtstr. 3, 37077 Göttingen. Germany

1. Introduction

There is an intense ongoing discussion on the origin of calcium carbonate particles in various meteorites (e.g. Martian meteorites, further achondrites and certain carbonaceous chondrites) whether they are formed biologically on their parent body and, therefore, true microbial fossils, or whether they are mediated by non-biological organic matrices or formed by various terrestrial processes. It is the purpose of the presented paper to demonstrate that large carbonaceous but potentially abiogenic molecules and/or organic matrices may form nano scale, bacteria-shaped calcium carbonate particles comparable to those found in meteorites.

It is common use to distinguish formally between inorganic, biologically induced, and biologically controlled mineralization (Lowenstam, 1981, 1986). Trichet and Defarge (1995) introduced the term "organomineralization" in order to describe a mineral formation linked to non-living organic substances in contrast to biologically-induced mineralization which results from ionic pumping and metabolic processes, and biomineralization which is completely biologically controlled. The presence and chemical properties of acidic macromolecules seem to be crucial factors in $CaCO_3$ precipitation, ranging from non-skeletal particles such as ooids to sophisticated skeletons like those of mollusks. Organomineralization could explain distinctive carbonate particles commonly described as "Nannobacteria" (Folk 1993) and structures known from certain Martian meteorites and from various carbonaceous chondrites (e.g. McKay et al.1996, Barrat et al. 1998, Benzerara et al. 2003).

The purpose of this paper is to demonstrate that organic (i.e. carbonaceous) macromolecules are able to nucleate calcium carbonate minerals (fig.8, 16, 17) in the absence of microbes.

2. Concept

The significance of non-enzymatically controlled $CaCO_3$ precipitation, commonly attributed to bacterial activity or "microbes" in general, is increasingly claims in Phanerozoic reef formation (e.g., Webb, 1996). Following Lowenstam (1981) and

Lowenstam & Weiner (1989) biomineralization is defined to be biologically controlled by the organisms. Consequently skeletons are formed that are integral functional parts of the organisms. Organomineralization and automicrite formation (Defarge and Trichet, 1995; Trichet and Defarge, 1995, Reitner 1993, Reitner et al. 1995) are used as terms to describe mineralization processes that involve organic molecules or particles, either linked to living organisms or not. Thus, organomineralization is not necessarily linked to living organisms, but may also happen anywhere within non-living, bioinorganic, re-organized macromolecular films or aggregates (e.g. Reitner et al., 1995, 1997).

The transfer of the organic matrix-concept to problems of biofilm mineralization resulted in a consequence a new view on the role of free organic macromolecules in microbialite and ooid formation (Reitner, 1993; Reitner et al., 1997, Arp et al. 1998, 1999a,b, 2001).

Organic macromolecules that are involved in biogenic mineralization are polyanionic polymeres of dividing chains composed of monomers such as amino acids and sugars (e.g., Marsh, 1994). Their substantial participation in biomineralization is indicated by the fact that organic macromolecules can be up to 5% of the total mass of a biomineral (Addadi and Weiner 1985; Berman et al. 1988). As a consequence of molecular size and abundance of polar groups, the organic macromolecules of biominerals comprise "soluble organic matrices (SOM)" that are water soluble after dissolving the mineral by EDTA (ethylenediamine-tetraacetate) and "insoluble organic matrices (IOM)". The latter are more or less neutrally charged and polymerized to a high degree. These substances are important as frame-building matrices like collagen, cellulose, chitin, or silk proteins. Rigid mineralized surfaces serving as adsorption/binding sites for SOM may be considered analogous to insoluble matrices. SOM molecules are often highly acidic, due to abundant free negatively carboxylate groups that result from inserted amino acids like aspartic acid (asp) or glutamic acid (glu). With their two carboxylate groups these amino acids remain negatively charged after peptide bonding, in contrast to amino acids with aliphatic, basic, or aromatic residues. SOM usually exhibit relatively low mean molecular weights of 10-30 kd only. Common are saccharide groups which are covalently (glycosidic) bonded with the acidic proteins forming glycoproteins. They have free carboxylate- and/or sulfate groups which are also proton donors. Other acidic macromolecules are acidic proteoglycanes (mucopolysaccharides), glycoseamino-glycans and proteins, (glycoseaminoglycan = Uronic acid + amino sugars), and polysaccharides only. Non-biologically formed macromolecules like humic acids and kerogens (geopolymers) may also nucleate calcium carbonate minerals, however also based on reactive side chains (Neuweiler et al. 1999). The crucial process of mineral precipitation is the formation of seed crystals via heterogenic nucleation (Sigg and Stumm, 1989). Heterogenic nucleation happens spontaneously, when very small extraneous particles like molecules, ions, and foreign atoms are present in an oversaturated solution. The extraneous particles have a catalytic effect by decreasing the activation energy of crystal forming ions.

In organo- and biomineralization heterogeneous crystal nucleation is provided by acidic organic macromolecules that attract and bind divalent cations such as Ca^{2+}, Mg^{2+}, Sr^{2+} by their free carboxylate and/or sulfate groups. Acidic mollusc glycoproteins (SOM), that are adsorbed to a insoluble organic matrix, have been shown to induce $CaCO_3$ mineral formation, whereas they strongly inhibit precipitation when they are in solution (Addadi and Weiner, 1985, 1989, 1992). This is explained by Addadi and Weiner (1985, 1989, 1992) in a model that postulates the formation of a flat molecular monolayer of the acidic macromolecules in polypeptide ligature chains. The chains are linked by hydrogen bonding and are arranged on a neutral non-reactive surface in a faulted secondary structure to form a so-called "β-sheet structure" (Worms and Weiner, 1986; Addadi and Weiner, 1989; Addadi et al., 1990; Boskey, 1996). The negatively charged COO^- groups of the β-sheets are responsible for the binding of Ca^{2+} from the liquid phase. In certain cases the carboxylate groups of asp and glu in the β-sheets are arranged distally (far from the insoluble/frame-building matrix) in one plane. Extremely acidic macromolecules, predominantly glycoproteins, inhibit precipitation upon Ca^{2+} binding until saturation of their acidic groups (fig.8, 9). These very acidic species are inhibitors of any mineralization or they allow the crystal to grow only in selected directions (further discussion in Addadi et al., 1990, Borbas et al., 1991; George et al., 1996; Westbroek et al., 1994).

3. Methods and Experiments

Samples were fixed with 4% buffered formol and subsamples were stained in bulk with fluorochrome dyes (calcein, calcein-Na_2, aureomycine, tetracycline-HCl, acridine orange). Thin sections were cut with a Leica hardpart microtome. Image stacks with a Z-spacing of 0.5 or 0.25 μm were obtained by using a piezo-mover (Physik Instrumente GmbH and Co, Waldbronn) attached to a "Plan-Apochromat" 63x-objective (Zeiss, NA=1.4) of a Zeiss Axiplan. Image processing was carried out by using the Metamorph® Imaging software (Universal Imaging Corporation, West Chester, PA) and the EPR™ deconvolution software (Scanalytics, Billerica, MA). Paraffin sections of decalcified samples were stained with various histochemicals, dapi for DNA detection and various oligonucleotides probes for fluorescence in situ hybridization (FISH). Samples for TEM and SEM/FEM studies were fixed with 4% buffered glutardialdehyde and postfixed with 2% OsO_4.. Glutardialdehyde fixed samples were dried with Peldri II (Pelco) avoiding drying artifacts. Uncoated samples were investigated with a Field Emission Microscope (FEM) "LEO Gemini" at less than 1 kV. Some samples were investigated fully hydrated by use of an Oxford-Cryosystem combined with the FEM in order to avoid dehydration artifacts or collapse of the structures of the organic matter.

The samples for biochemical analyses (High Pressure Liquid Chromatography-HPLC, gas chromatography + mass spectrometer and electrophoresis) were dried and/or frozen. For decalcification specimens were put in pH 4 controlled acetic acid for 24 hours or EDTA. The insoluble fraction was removed by centrifugation. The remaining organic matter was desalted by low-pressure gel filtration chromatography on PHARMACIA-G25C with UV detection (280 nm). Amino acids: 24 hrs, 110°C 6N HCl hydrolysis; PITC derivatization; HPLC Reverse phase chromatography (BECKMAN system) with Hypersil 100 column (C18 5μm 250x4.6mm); UV detection 245 nm. Monosaccharides: 6 hrs 2N TFA hydrolysis; HPLC Ion exchange chromatography (DIONEX system) with CarboPac PA1(250x4mm); pulsed amperometric detection. Molecular weights were also determined with HPLC gel filtration using aTSK G2000SWXL column eluted by sodium citrate adjusted to pH 4.5 (0.33M NaCl added); UV scan detection (LDC

Analytical-Spectro-Monitor 5000 photodiode array detector); LKB differential refractometer. Amino acid and monosaccharide analyses were carried out on the total Soluble Organic Matrix (SOM) after desalting, ultrafiltration against MilliQ water (FILTRON ultrafiltration cells, 3K), and lyophilization. Amino acid composition was tested again on separated fractions collected during HPLC gel filtration used as a preparative fractionating system. Before the usual hydrolysis and derivatization procedures, fractions were vacuum concentrated, ultrafiltered against MilliQ water, and lyophilized. For electrophoresis all samples were decalcified with EDTA. Protein concentrations were determined by the nanoOrange protein quantification assay (company Molecular Probes). Samples were electrophoresed on a 10 % SDS gel and silver stained. Isoelectrical focusing of proteins was done with precast gels (pH-range 3-10, SERVA). Total protein content was determined by the BCA method (Smith et al., 1985). Molecular weights were determind using SDS-Page electrophoresis (SDS-polyacrylamid).

4. *In vitro* Mineralization - "Inhibiton" Assay

This experiment is crucial to understand the role of organic molecules during the calcium carbonate crystal formation. *In vitro* mineralization inhibition experiments were carried out based on the procedure described by Wheeler et al. (1981), Gunthorpe et al. (1990), and Lange et al. (2001) using bulk intracrystalline native organic matter not treated with SDS. No living material is used for this experiments strictly under sterile conditions. It is possible to control the molecular weights using SDS-page electrophoresis, cutting the stained bands and extraction of the organic matter from the gel.. Inhibition experiments give hints on the Ca^{2+} binding ability of the organic matter and consequently the importance for skeletal and organomineral formation. Mineralization was studied in a reaction tube containing 400 µl $NaHCO_3$ (1 M), 19 ml doubly distilled waterdoubly distilled water, and 200 µl sample or 200 µl doubly distilled waterdoubly distilled water in the control assay, respectively. The solution contains total proteins or Ca^{2+}-binding proteins. Protein concentration was 0.5 µg x ml^{-1} if not indicated otherwise. To start the reaction, 400 µl $CaCl_2$ (1 M) were added. Inhibition of mineralization was determined by measuring the pH over a certain period. As standard reaction molecules the extremely acidic artificial peptide polyasparagine 30 KD and aqua doubly distilled water (purely demineralized) were used. The mineralization event could be observed by the occurrence of the turbidity of the solution and the drop of the pH to 7. During this process protons were released from $NaHCO_3$ according to the following reaction:

$$NaHCO_3 + CaCl_2 \rightarrow CaCO_3 + H^+ + 2\,Cl^- + Na^+$$ (fig.5, 18)

The calcified products have the characteristics of calcareous, abiotically formed organominerals, as discussed below.

5. Case Studies

5.1. GREAT SALT LAKE OOIDS

Ooids are ball- and egg-shaped, small (0,1mm-1cm in average) calcareous, non-skeletal carbonates (Fig.1). They mainly occur in warm shallow marine carbonate platforms (e.g. Bahamas Islands) or in terrestrial, highly saline and alkaline environments like the Great Salt Lake in Utah.

The hypersaline Great Salt Lake (GSL) is a classical locality of Recent radial-fibrous ooids (Eardley, 1938; Kahle, 1974; Sandberg, 1975; Gwynn and Murphy, 1980, Mitterer, 1968, 1972, Mitterer and Cunningham, 1985). Arguments suggesting that ooids may still form today are a hydrochemical setting favorable for $CaCO_3$ precipitation and findings of recently formed microbialitic crusts (Pedone and Folk, 1996). Our observations with regard to structure and composition are briefly summarized in Reitner et al., 1997. The aragonitic GSL ooids are principally composed of three different structural elements: Dominant radial-fibrous layers with or without depressions/pits, smooth tangentially structured layers, and intercalated corrosion surfaces (fig.1). Nuclei are quartz grains, calcified fecal pellets of the brine shrimp *Artemia*, and ooid clasts (Eardley, 1938). The ooids exhibit nano-sized calcified particles growing in organic mucus (fig.2-4). Ooids decalcified with EDTA exhibit a water-insoluble organic framework that reflects intercrystalline spaces, microcrystalline patches and laminae, corrosion interlayers, and surface films. The organic substances are only slightly osmiophilic (low contrast in TEM) and only weakly to moderately acidic (histochemical staining, HPLC analyses). The organic content of ooids has been analysed with respect to amino acid and sugar composition. The total organic extracts of purified ooids comprises up to 520 µg per g carbonate. The analyses of surface cleaned ooids include fluid inclusions, intra- and intercrystalline organic matter of the ooids. Up to 450 µg/g carbonate form the EDTA-soluble fraction that is composed of 25% proteinaceous material and 75% sugars. The soluble matrices are characterized by an extremely high content of acidic amino acids. Glutamic acid forms 50% and aspartic acid 11-15% of the AS fraction. The glucidic compounds of soluble matrices are dominantly composed of mannose, glucose, galactose, fucose, glucosamine, and galactosamine. In contrast hereto, the insoluble matrices show high amounts of Pro (17-22%), Leu (16.1-16.7%), Gly (11.8-16%), Val (12.3-12.7%), and Ala (9.2-11.4%). Acidic amino acids are only weakly present (Glu: 7.4-8.2%; Asp: 2.1-3%). The high contents of glutamic- and aspartic acid might reflect a high concentration in the solution from which the carbonate was precipitated. Inhibition experiments proof this assumption (fig.5) The extreme salinity of the Great Salt Lake is known to slow down certain physiological reactions, so that intermediate products such as acetate (in pore water at salinities above 15-20%) are accumulated, that are otherwise consumed rapidly. We speculate that the considerable reduction of some key reactions in the carbon cycle might also be responsible for the unusual AS spectra. The GSL ooids probably result from organomineralization processes, that are not linked to bacterial biofilms or direct physiological influence of living microorganisms.

It is important to note that the organomineralization from our experiments discussed above, comprise very small ball-shaped (20-200nm) carbonate particles (fig.2-4) which exhibit certain similarities with the calcified structures of the ALH84001 SNC meteorite (fig.8-9).

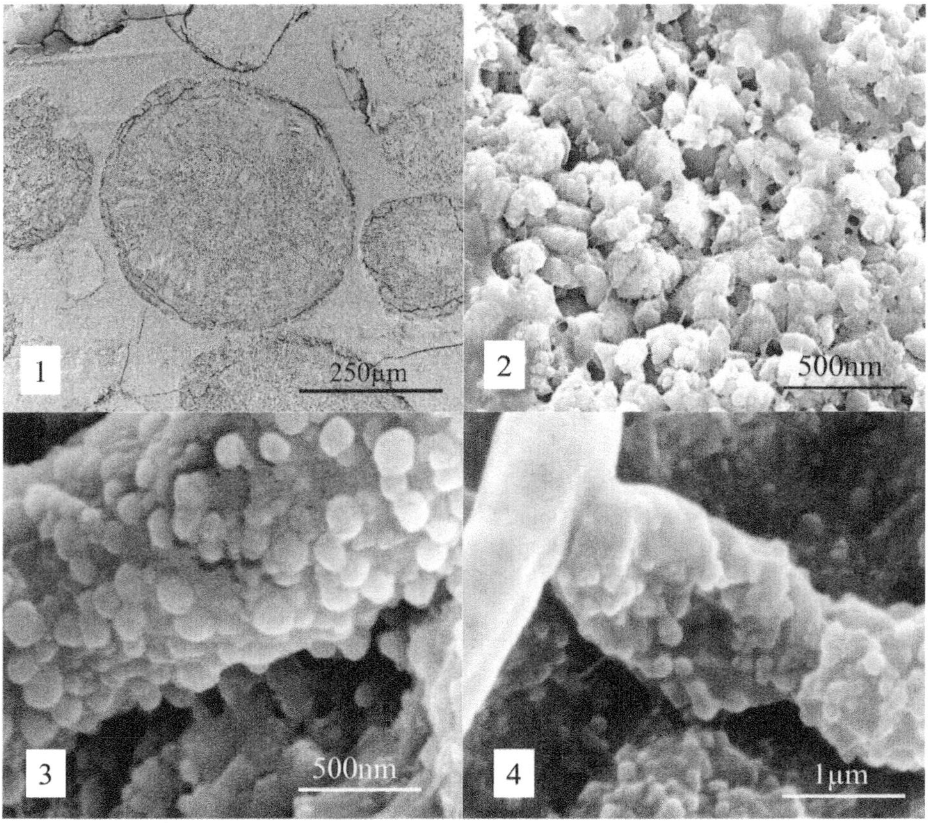

Figure 1. Median sections of ooids. The ooids exhibiting radial fibrous aragonitic crystals. (Fig. 1-4: Great Salt Lake ooids, Bridger Point, Great Salt Lake Utah. Field emission scanning electron microscope Leo "Gemini; samples are carbon coated). **Figure 2**. Peldri II dried calcifying surface. Samples fixed with glutardialdehyde (Fig.2-4). Nano-aragonite anhedral crystals are growing in an extremely acidic organic mucus. **Figure 3**. Larger carbonate particle coated with an acidic macromolecular mucus with aragonitic nanobacteria-like structures (20-100nm). **Figure 4.** Collapsed bacterial cell coated also with aragonitic nanobacteria-like structures

5.2. BIOFILM EXOPOLYMER-MEDIATED (EPS) CALCIFICATION

Almost any water-substrate interface is covered by a biofilm, that consists of microorganisms attached to the solid substratum and embedded in an organic polymer matrix (Characklis and Marshall, 1990). Extracellular polymeric substances (EPS) play a central role in biofilm development and structure (for review see Decho 1990). The highly hydrated polymer matrix comprises mainly carbohydrates (polysaccharides), but also proteinaceous macromolecules. With regard to sedimentary rock formation, inclusive of reef carbonates, the considerable interference of their functional/reactive groups with pure physicochemical precipitation in nature is of crucial importance. EPS calcification is a key process in the formation of stromatolites and thrombolites (e.g. Reitner 1993). The calcification products via microbial EPS are highly variable in shape

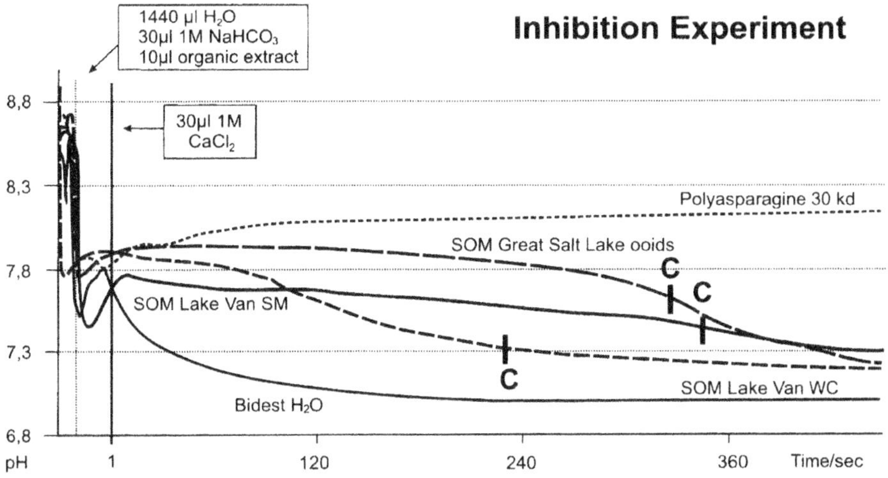

Figure 5 Inhibition curves of polyasparagine (30kd), demineralized water, acidic soluble organic matter from the Great Salt Lake ooids, and two carbonate types from Lake Van microbialites. One type is extracted SOM from an anorganic white carbonate precipitate which is immediately formed in the mixinig zone of HCO_3^- -rich lake water and Ca^{2+} -rich spring water (SOM Lake Van WC). Second type is extracted from cyanobacterial microbialite from top of the carbonate towers (SOM Lake Van SM). The latter one exhibits a strong inhibition. Symbol C represents the calcification point.

and structure. The most intrinsic of these carbonate structures were observed in highly alkaline environments (Arp et al. 1998, 1999a,b) and in dark reef caves with normal marine water from the Great Barrier Reef and St. Croix in the Caribbean (Reitner, 1993, Reitner et al. 1995, 2000). Reef cave microbialites grow at the interface between slightly increased HCO_3^- (4 - 5 meq/l) reef porewater and Ca^{2+}-rich seawater with normal marine alkalinity (2 meq/l). The pore water alkalinity is controlled by ammonification and sulfate reduction in deep cryptic reef pore space. The reef cave biofilms are free of cyanobacteria and are normally formed by various types of alpha, gamma, and delta proteobacteria including heterotrophic and chemoautotrophic types (Fe- and Mn oxidizers) and often associated with thin biofilm-type sponges (Reitner, 1993, Reitner et al. 2000). Generally the biofilms are thin (50-100 µm) and extremely slow growing, e.g. 5 cm thick reef cave microbialites of the Great Barrier Reef need ca. 3000 years to grow. Reef cave microbialites are formed by a micritic high Mg-calcite (automicrite) which is precipitated within the thin EPS-layers (fig.6-7). The automicrite does not exhibit special calcification fabrics due to the thin EPS-layers. Inhibition experiments of the low molecular weight fraction (ca. 25 kd) show a strong inhibition of calcium carbonate nucleation. The ratio between glucidic and proteinaceous soluble organic matter (4:1) is the same as observed in other microbialites from various environments (for details see Reitner et al. 1995).

In highly alkaline soda lakes the calcification in mixing zones is often much faster and happens in days or weeks due to increased fluid flow and thicker biofilms and microbial mats. Mixing of Ca^{2+}-supplying spring water and HCO_3^--rich lake water in a hydrochemically forced precipitation, often resulted in thrombolitic spring mounds. Autotrophic CO_2 removal (e.g. by photosynthesis), HCO_3^--production by sulfate reduction, and ammonification are unlikely to substantially affect the calcium carbonate system in biofilms of a highly alkaline, highly supersaturated macroenvironment (Arp et al. 2001). Precipitation in and upon biofilms is inhibited at first due to acidic mucus substances, mainly carboxylated and sulfated polysaccharides. They adsorb Ca^{2+} from the liquid phase, thus preventing $CaCO_3$ precipitation at first (Addadi and Weiner, 1989; Arp et al. 1998, 1999a,b, Reitner, 1993). Diffusion is generally slowed down by these highly hydrated polymers of the biofilms (Decho, 1990; Costerton et al., 1995). The EPS is therefore considered as a Ca^{2+}-buffer, the capacity of which has to be surpassed before nucleation is possible.

The writer and his group have made detailed studies of soda lakes in China, USA, Indonesia, and Turkey (Arp et al. 1998, 1999a,b, 2001, 2003, Landmann and Reimer 1996). The microbialites from Lake Van in eastern Anatolia are of particular interest, because some calcified products exhibit some similarities with the ALH84001 (Kazmierczak and Kempe 2003). In mixing zones between the highly alkaline lake water and Ca^{2+}-rich groundwater, special types of tower-shaped large microbialites formed (for details see Kempe et al. 1991, and Landmann and Reimer 1996). We have studied in detail the calcified EPS of coccoid cyanobacteria on top of the microbialite towers (fig., 10-12). The calcified products exhibit very small (50nm) globular bodies arranged in rows and forming highly organized aggregates (fig.11-12). These aragonitic aggregates show high comparability with ALH84001 carbonate particles (fig.13). This special fabric is controlled by the internal fabric of thick polysaccharide-rich EPS which exhibits a strong birefringence using crossed Nicols. We have extracted 310 µg total intracrystalline organic matter from 1 g EPS-related carbonate, of which 110 µg were EDTA soluble. The soluble macromolecular organic matter is composed of ca. 80% of glucidic and ca. 20% of proteinaceous material. We have not yet analyzed the saccharidic components in detail. Histochemical staining with specially adapted histochemical Alciane Blue demonstrates high amounts of carboxylated polysaccharides (fig.10). The proteinaceous material is weakly acidic (15 mol% asp and 10 mol% glu). However, the inhibition experiment of the soluble bulk organic matter exhibits an inhibition pattern (fig.5).

5.3. MURCHISON METEORITE

The two presented terrestrial examples have shown that organic films and large organic molecules are able to nucleate carbonate minerals, sometimes with characteristic crystal

Figure 6. Lithified aphotic biofilm with a rough stromatolitic fabric (fig.6-9: Great Barrier Reef reef cave microbialite (Lizard Island, Australia). **Figure 7**. Peloidal organomicrites formed in a acidic mucus-rich open pore. These stellate-shaped spheroids are comparable with the in vitro formed ones from the Murchison meteorite (fig. 17). **Figure 8**. Mg-calcite crystal growing in a carbohydrate-rich mucus extracted from the Great Barrier Reef microbialite. In this particular case the interface between the large organic molecules and the crystal is visible (001 plane). This plane has a very characteristic rough surface, typical for organomineral base planes. Visible is the ß-sheet monolayer. **Figure 9**. High magnification of the 001 plane of fig.11. The calcifying domains of the monolayer are visible.

shapes depending on the composition of the organic films. Cementation processes of terrestrial carbonate rocks happen under certain conditions in the same way. Could this process explain the carbonate minerals in meteorites? It is not the purpose of this paper to speculate about a biological origin for all these carbonate particles. However, fluids enriched in divalent cations and reactive organic films and components may well cause formation of these carbonate particles.

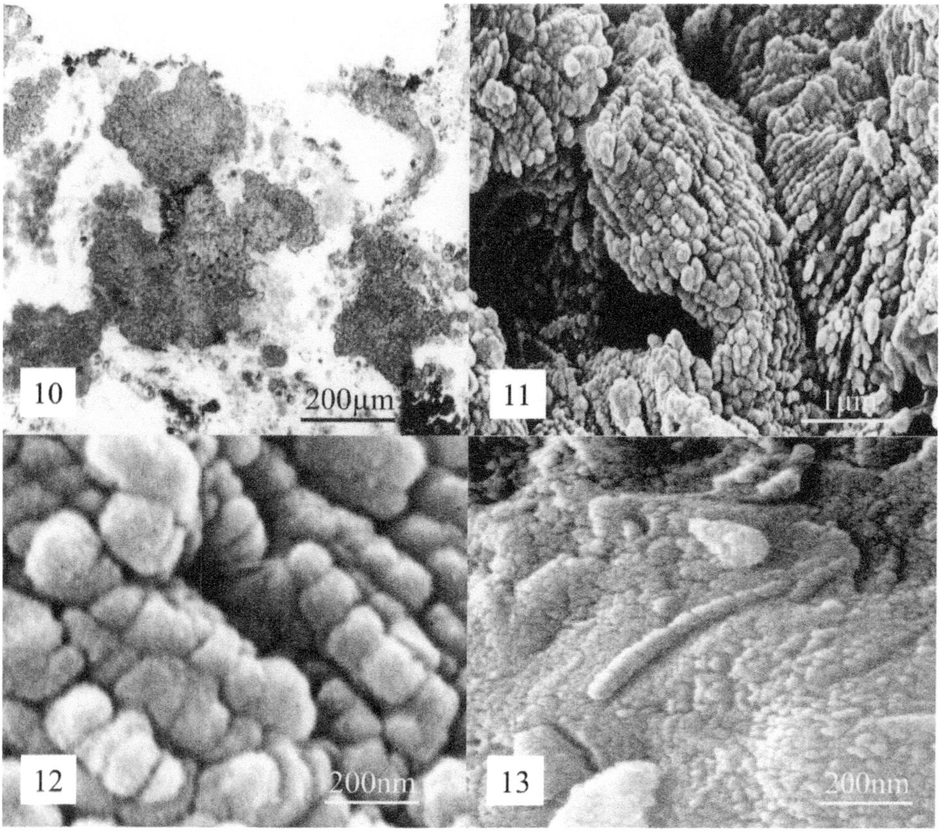

Figure 10. Hardpart microtome section of the active growing surface of a Lake Van microbialite. Section stained with alcian blue, a special dye to detect acidic polysaccharides, here adapted to check carboxylated functional groups (COO^-). The biofilm is mainly constructed of coccoid cyanobacteria. The calcified EPS is showing a thrombolitic fabric. (fig.10-12: Lake Van microbialite from Turkey – East Anatolia. Samples were provided by Dr. A. Reimer, Univ. of Göttingen). **Figure 11**. FEM micrograph of the calcified EPS. The hollows are remains of the coccoid cyanobacteria (*Pleurocapsa*?). The calcified EPS is exhibiting a distinct structure based on the orientation of thick polysaccharide bundles. The EPS is constructed of larger (10-30 μm) spindle-like bodies which itself constructed rows of round shaped may be amorphic aragonitic nano-crystals. **Figure 12**. Detail magnification of the nano-crystal rows. A single nano-crystal has a size of 20-50 nm. These structures are matrix-mediated organominerals. **Figure 13**. The ALH84001 nano-crystal aggregate exhibits a comparable shape and size (modified from McKay et al. 1996). It is assumed that the ALH84001 nano-crystal aggregate is a abiotic organomineral.

All meteorites on earth are contaminated with microbes and terrestrial organic matter (fig.14). Their open pore space filled with water fluids are occupied by microorganisms as we know from our deep biosphere investigations in crystalline rocks and contamination tests with various DNA sensitive fluorochromes such as dapi, plus fluorescence *in situ* hybridization (FISH). These data fit well with investigations by others groups noted in the Carnegie Institution Year book 2001-2002. Andrew Steele reported fungal and bacterial contaminations in the Murchison Meteorite.

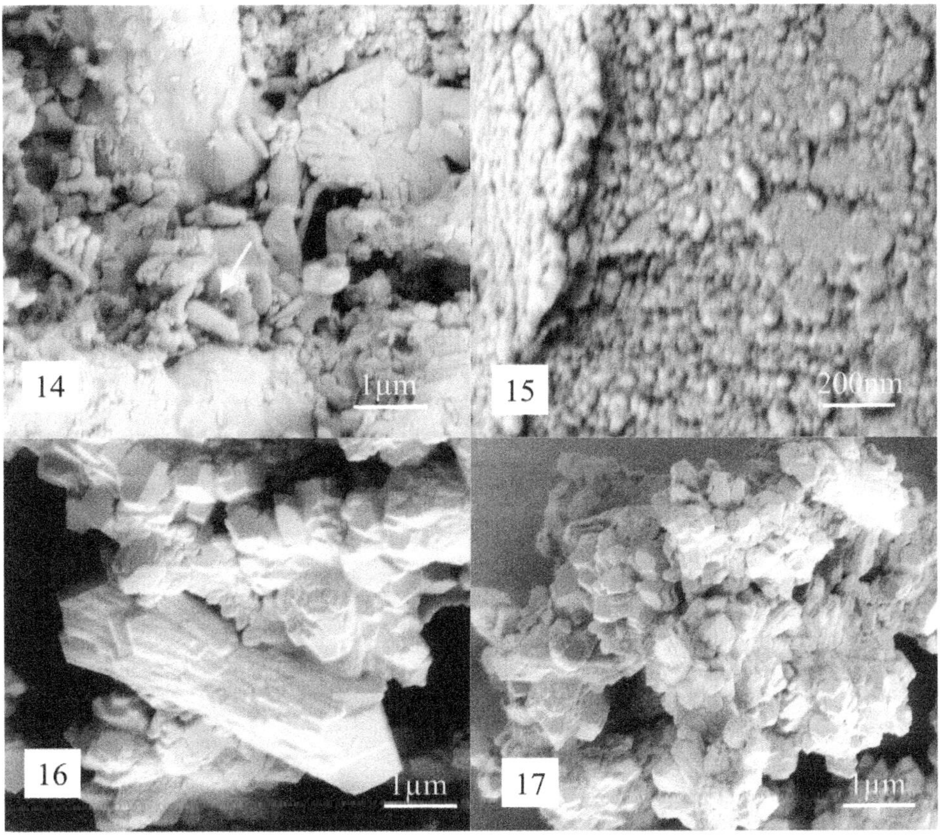

Figure 14. Small fracture, filled and coated with rod-shape carbonate bodies (arrow). These structures are may be remains of terrestrial microbes (fig. 14-17 Murchison CM2 meteorite). **Figure 15**. Topography of an olivine surface with very small carbonate seed crystals (20-50 nm). These features are probably also a result of an organomineralization (see figure 18). **Figure 16**. Dumbbell-shaped calcitic crystal formed during an inhibition experiment. The inhibition time of the soluble organic matter is comparable with 30 kd polyasparagine. **Figure 17.** Spherulitic calcitic crystal aggregates also formed during an inhibition experiment. These spherulitic aggregates are "peloids" in carbonate petrology. Peloids are formed in degraded organic matter and common carbonate components in Phanerozoic carbonate platforms and reefs.

Nevertheless the Murchison CM2 is enriched with organic matter and is therefore an excellent target for mineralization experiments. The organic compounds from the Murchison CM2 have been intensively studied (e.g. Engel and Macko, 1997, Engel and Macko 2001, Engel and Nagy 1982, Cronin and Pizzarello, 1983, a.m.o.). Of particular interest are the amino acid composition and generally the soluble organic matter. More than seventy different amino acids are known from this meteorite. Some of them show

L- and D-enantiomers and extremely heavy $\partial^{13}C$ values of +20 ‰ (Engel and Macko 2001) which clearly indicate an extraterrestrial origin.

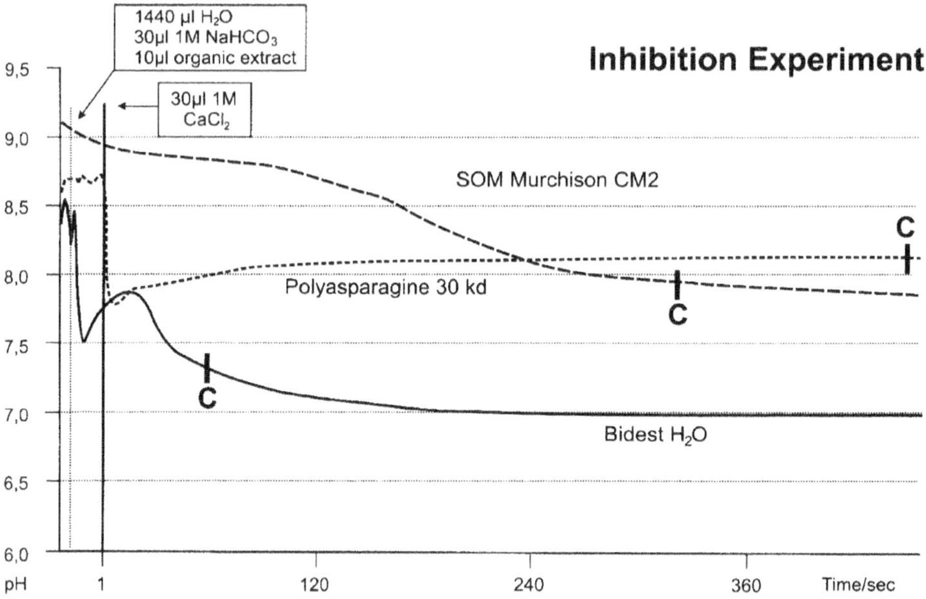

Figure 18. Inhibition curves of polyasparagine (30kd), demineralised water, acidic soluble organic matter from Murchison CM2. Symbol C represents the calcification point. The inhibition curve of the Murchison SOM is extremely acidic and comparable with polyasparagine!

The preliminary investigation presented here is based on a 60 g stone of the Murchison meteorite housed in the collection of the museum of the geoscience department of the University of Goettingen. From 1 g of the meteorite material the total organic material was extracted for microbiological, SDS electrophoresis, and HPLC investigations. We have extracted 105µg, of which 30% were soluble. In a first step we determined 30 amino acids from the soluble organic matter with special respect to the acidic amino acids, with concentrations of 20.2 mol%/l glu and 7.5 mol/l asp. With the bulk soluble organic material we carried out mineralization experiments (fig.16, 17, 18). The Murchison SOM exhibits a strong inhibition comparable with artificial polyaspargine (fig.18). The calcified products are spherulitic and dumbbell shaped calcitic aggregates (fig. 16, 17). We do not yet know the chemical character of the remaining organic compounds.

It has proved possible, in the present experiments, to produce *in vitro* small carbonate bodies (fig.16, 17) which exhibit similarities to carbonate peloids, calcitic dumbbells known from bacterial EPS (e.g. Chafetz, 1986, Buczynski and Chafetz, 1991, Chafetz and Buczynski 1992, Reitner et al. 2000) as a well as small coccoid- and rod-shaped bodies like those reported from other meteorites (e.g., McKay et al. 1996).

6. Conclusion

Nanocrystalline carbonates are observed in some achondrites (e.g. ALH84001 and Tatahouine) and these have been interpreted as calcified "nanobacteria" and sometimes as calcified extraterrestrial life forms. The carbonates from the Tatahouine achondrite are contamination from recent microbes on the surface and in open pore spaces of the meteorite and cannot be related to any extraterrestrial origin (Benzerara et al. 2003, Gillet et al. 2000). The famous ALH84001 achondrite seems to be contaminated with terrestrial, possibly fungal microbiota (Steele et al. 2000) and terrestrial amino acids, except for some d-alanine (Bada et al. 1998). Questionable are the recently reported rod shaped, calcareous nanostructures (fig.13) which were interpreted as "fossilized Martian microorganisms" by some astrobiologists (McKay et al. 1996). ALH84001 bears some traces of PAH – polycyclic aromatic hydrocarbons, common organic molecules in cosmic clouds and everywhere in terrestrial environments. Whilst it might be claimed that the analysis of the stable carbon isotopes of these PAHs could solve the problem of origin, it not longer looks safe to accept negative carbon isotopic values as biomarkers that are unique to life! However, under certain circumstances they may form special types of geopolymers together with terrestrial microbial EPS-remains in rock fractures. Due to degradation process by terrestrial microbes negatively charged functional groups may be formed. These abiotic organic films may nucleate nanocrystalline carbonates in a way as shown in certain terrestrial carbonates like oolites and microbialites.

The abiotic amino acid-rich organic matter from the Murchison CM2 has nucleated *in vitro* nanocrystalline carbonate particles (peloids, dumbbells, rods) comparable to those known from degrading microbial mats (fig.16,17).

As an assumption and conclusion, meteorite-linked calcium carbonates could be abiotic organomineral products.

7. Acknowledgements

The "Deutsche Forschungsgemeinschaft" is gratefully acknowledged for financing the investigations on biomineralization (Re 665/18, Re665/12 Leibniz award – Evolution of Multicellular Systems and Organomineralization EMSO). Dr. Pascale Gautret (Orsay/Paris, France) provided some amino acid spectra of the Great Salt Lake ooids, and Dr. Andreas Reimer (Göttingen, Germany) microbialites from the Lake Van. The Geobiology group of Goettingen is also greatly acknowledged. Last but not least many thanks to Prof. Dr. Joseph Seckbach and the two anonymous reviewers for help to improve the manuscript and patience!

8. References

Addadi, L. and Weiner, S. (1985) Interactions between acidic proteins and crystals: Stereochemical requirements in biomineralization. Proc. Natl. Acad. Sci. USA **82**, 4110-4114.

Addadi, L. and Weiner, S. (1989) Stereochemical and Structural Relations between Macromolecules and Crystals in Biomineralization. In: S. Mann, J. Webb. and R.J.P. Williams (eds.): *Biomineralization.* - Weinheim (VCH) pp. 133-156,.

Addadi, L. and Weiner, S. (1992) Kontroll- und Designprinzipien bei der Biomineralization. Angew. Chemie **104**, 159-176.

Addadi, L.; Berman, A.; Moradian-Oldak, J.and Weiner, S. (1990) Tuning of Crystal Nucleation and Growth by Proteins: Molecular Interactions at Solid-Liquid Interfaces in Biomineralization. Croatica Chem. Acta **63**, 539-544.

Arp, G., Hofmann, J. and Reitner, J. (1998) Microbial Fabric Formation in Spring Mounds ("Microbialites") of Alkaline Salt Lakes in the Badain Jaran Sand Sea, PR China. Palaios **13**, 581-592.

Arp, G., Reimer, A. and Reitner, J. (1999) Calcification in cyanobacterial biofilms of alkaline salt lakes. Eur. J. Phycol.. **34**, : 393-403

Arp, G., Reimer, A. and Reitner J. (2001) Photosynthesis-induced biofilm calcification and Calcium concentrations in Phanerozoic Ocean. Science **292**, 1701-1704.

Arp, G., Reimer, A., and Reitner, J. (2003) Microbialite formation in seawater of increased alkalinity, Satonda Crater Lake, Indonesia. Jour. Sed. Res. **73**, 105-127.

Arp, G., Thiel, V., Reimer, A., Michaelis, W. and Reitner, J. (1999) Biofilm exopolymers control microbialite formation at hot springs discharging into alkaline Pyramid Lake, Nevada, USA. Sedimentary Geology **126**, 159-176.

Bada, J.L., Glavin, D.P., McDonald, G.D. and Becker, L. (1998) A search for endogenous amino acids in Martian meteorite ALH84001. Science **279**, 362-365.

Barrat, J.A., Gillet, P., Lécuyer, C., Sheppard, S.M.F. and Lesourd, M. (1998) : Formation of carbonates in the Tatahouine meteorite.- Science, **280**, 412-414.

Benzerara, K., Menguy, N., Guyot, F., Dominaci, Chr. and Gilet, P. (2003) Nanobacteria-like calcite single crystals at the surface of the Tatahouline meteorite. PNAS **100**, 7438-7442

Berman, A., Addadi, L. and Weiner, S. (1988) Interactions of sea-urchin skeleton macromolecules with growing calcite crystals - a study of intracrystalline proteins. Nature **331**, 546-548.

Borbas, J. E.; Wheeler, A. P. and Sikes, C. S. (1991) Molluscan Shell Matrix Phosphoproteins: Cirrelation of Degree of Phosphorylation to Shell Mineral Microstructure and to In Vitro Regulation of Mineralization. J. Exp. Zool. **258**, 1-13.

Boskey, A.L. (1996) Matrix Proteins and Mineralization: An Overview. Connec. Tiss. Research., Amsterdam. **35**, 357-363;.

Buczynski, C. and Chafetz, H. S. (1991) Habit of Bacterially induced Precipitates of Calcium Carbonate and the Influence of medium Viscosity on Minerology. J. Sed. Petrol. Tulsa. **61**, 226-233

Carnegie Institution Yearbook 2001-2002, 26-29

Chafetz, H. S. (1986) Marine Peloids: A Product of Bacterillay induced Precipitation of Calcite. J. Sed. Petrol. **56**, 812-817.

Chafetz, H. S. and Buczynski, C. (1992) Bacterially induced Lithification of Microbial Mats. Palaios **7**, 277-293.

Characklis, W.G. and Marshall, K.C. (eds.) (1990) Biofilms. New York (Wiley). p. 796 .

Collins, M.J., Bishop, A.N.and Farrimond, P. (1995) Sorption by mineral surfaces: Rebirth of the classical condensation pathway for kerogen formation. Geochim. Cosmochim. Acta **59**, 2387-2391.

Costerton, J.W., Lewandowski, Z., Caldwell, D.E., Korber, D.R., and Lappin-Scott, H.M., (1995) Microbial Biofilms. - Ann. Rev. Microbiol. **49**, 711-746.

Cronin, J.R. and Pizzarello, S. (1983) Amino acids in meteorites. Adv.Space Res. **3**, 5-18.

Decho, A.W. (1990) Microbial exopolymer secretions in ocean environments: Their role(s) in food webs and marine processes. Oceanography and Marine Biology Annual Review **28,** 73-153.

Defarge, C. and Trichet, J. (1995) From biominerals to "organominerals": The example of the modern lacustrine calcareous stromatolites from Polynesian atolls. Bulletin de l'Institut Oceanographique Monaco, no. spec., **14 (2)**, 265-271.

Eardley, A.J. (1938) Sediments of Great Salt Lake, Utah. Bull.AAPG, Tulsa. **22**, 1305-1411,.

Engel, M.H. and Macko, S.A. (1997) Isotopic evidence for extraterrestrial non-racemic amino acids in the Murchison meteorite. Nature **389**, 265-268.

Engel, M.H. and Macko, S.A. (2001) the stereochemistry of amino acids in the Murchison meteorite.- Precambrian Research, 106, 35-45.

Engel, M.H. and Nagy, B. (1982): Distribution and enantiomeric composition of amino acids in the Murchison meteorite. Nature **296**, 837-840.

George, A., Bannon, L., Sabsay, B., Dillom, J.W., Malone, J., Veis., A., Jenkins, N.A., Gilbert, D.J. and Copeland, N.G. (1996) The carboxyl-terminal domain of phosphophoryn contains unique extended triplet amino acid repeat sequences forming ordered carboxyl-phosphate interactions ridges that may be essential in biomineralization process. The Journal of Biological Chemistry **272 (51)**, 32869-32873;

Gillet, Ph., Barrat, J.A., Heulin, T., Achouak, W., Lesourd, M., Guyot, F. and Benzerera, K. (2000) Bacteria in the Tatahouine meteorite: nanometric-scale life in rocks. Earth and Planetary Sc. .Letters **175**, 161-167.

Gunthorpe, M. E., Sikes, C. S. and Wheeler, A. P. (1990) Promotion and Inhibition of Calcium Carbonate Crystallization in Vitro by Matrix Protein from Blue Crab Exoskeleton. Biol. Bull. **179**, 191-200.

Gwynn, J.W. and Murphy, P.J. (1980) Recent sediments of the Great Salt Lake basin. In: J.W. Gwynn (ed.): *Great Salt Lake. A scientific, historical and economic overview*. Utah Geol. Min. Surv., Bull. Salt Lake City. **116**, 83-96,

Kahle, C.F. (1974) Ooids from Great Salt Lake, Utah, as an analogue for the genesis and diagenesis of ooids in marine limestones. J.sedim.Petrol. , Tulsa **44**, 30-39.

Kazmierczak, J. and Kempe, St. (2003) Modern terrestrial analogues for the carbonate globules in Martian ALH84001. Naturwissenschaften, **90**, 167-172

Kempe, St., Kazmierczak, J., Landmann, G., Konuk, T., Reimer, A. and Lipp, A. (1991) Largest known microbialites discoverd in Lake Van, Turkey. Nature **394**, 605-608

Landmann, G. and Reimer, A. (1996): Climatically induced lake level changes at Lake Van, Turkey, during Pleistocene/Holocene transition.- Global Biochemical Cycles **10**, 797-808

Lange, R., Bergbauer, M., Szewzyk, U. and Reitner, J. (2001) Soluble Proteins Control Growth of Skeleton Crystals in three Coralline Demosponges. Facies **45**, 195-202

Lowenstam, H.A. (1981) Minerals formed by organisms. Science **211**, 1126-1131.

Lowenstam, H.A. (1986) Mineralization processes in monerans and protocists. In: B.S.C. Leadbeater, and R. Riding (eds.) *Biomineralization in lower plants and animals*. Oxford (Clarendon) pp. .1-17.

Lowenstam, H. A. and Weiner, S. (1989) On Biomineralization. Oxford (Oxford Univ.Press)..324 pp.

Mann, S. (1989) Crystallochemical strategies in biomineralization. In: S. Mann, J. Webb and R.J.P. Williams (eds.) *Biomineralization*, , Weinheim (VCH Verlagsgesellschaft) pp. 35-62.

Marsh, M.E. (1994) Polyanions and biomineralization. Bulletin de l'Institut Oceanographique Monaco, no. spec., Monaco **14 (1)**, 121-128.

McKay, D.S. and Gibson Jr, E.K., Thomas-Keprta, K.L., Vali, H., Romanek, Chr.S., Clemett, S.J., Chillier, X.D.F., Maechling, C.R. and Zare, R.N. (1996) Search for past life on Mars: Possible recil biogenic activity in Martian meteorite ALH84001. Science **273**, 924-930.

Mitterer, R.M. (1968) Amino acid composition of organic matrix in calcareous oolites. Science **162**, 1498-1499.

Mitterer, R.M. (1972) Biogeochemistry of aragonite mud and oolites. Geochimica et Cosmochimica Acta **36**, 1407-1422.

Mitterer, R.M. and Cunningham, R.Jr. (1985) The interaction of natural organic matter with grain surfaces: implications for calcium carbonate precipitation. Soc. Econ. Paleont. Mineral., Spec. Publ. **36**, 17-31.

Neuweiler, F., Gautret, P., Thiel, V., Lange, R., Michaelis, W. and Reitner, J. (1999) Petrology of Lower Cretaceous carbonate mud mounds (Albian, N.Spain): insights into organomineralic deposits of the geological record. Sedimentology **46**, 837-859.

Pedone, V. and Folk, R.L. (1996) Formation of aragonite cement by nannobacteria in the Great Salt Lake, Utah. Geology **24**, 763-765.

Reitner, J. (1993) Modern cryptic microbialite/metazoan facies from Lizard Island (Great Barrier Reef, Australia). Formation and Concepts. Facies **29**, 3-39.

Reitner, J., Arp, G., Thiel, V., Gautret, P., Galling, U. and Michaelis, W. (1997) Organic matter in Great Salt Lake Ooids (Utah, USA) First approach to a formation via organic matrices. Facies **36**, 210-219.

Reitner, J, Gautret, P, Marin, F, and Neuweiler, F. (1995) Automicrites in a modern marine microbialite. Formation model via organic matrices (Lizard Island, Great Barrier Reef, Australia). Bull Inst océanograph Monaco no spec **14 (2)**, 237-263.

Reitner, J., Thiel, V., Zankl, H., Michaelis, W., Wörheide, G., and Gautret, P. (2000) Organic and Biochemical Pattern in Cryptic Microbialites. In: R.E. Riding and S.M. Awramik, *Microbial Sediments*, Springer . pp. 149-160.

Sandberg, P.A. (1975) New interpretations of Great Salt Lake ooids and of ancient non-skeletal carbonate mineralogy. Sedimentology **22**, 497-537.
Sigg, L.and Stumm, W. (1989) Aquatische Chemie. Zürich (VDF). p. 388.
Steele, A., Goddard, D.T., Stapleton, D., Toporski, J.,K.W., Peters, V., Bassinger, V., Sharpples, G., Wynn-Williams, D.D.and McKay, D.S. (2000) Investigatinos into an unknown organism on the Martian meteorite Allan Hills 84001. Meteoritics and Planetary Science **35**, 237-241.
Trichet, J. and Defarge, C. (1995) Non-biologically supported organomineralization. Bulletin de l'Institut Oceanographique Monaco, no. spec. **14 (2),** 203-236.
Webb, G.E. (1996) Was Phanerozoic reef history controlled by the distribution of non-enzymatically-secreted carbonates (microbial carbonate and biologically-induced cement)? Sedimentology **43**, 947-971.
Westbroek, P., Buddemeier, B., Coleman, M., Kok, D.J., Fautin, D. and Stal, L. (1994) Strategies for the study of climate forcing by calcification. Bulletin de l'Institut Oceanographique Monaco, no. spec. Monaco. **13**, 37-60,
Wheeler, A. P., George, J. W. and Evans, C. A. (1981) Control of calcium carbonate nucleation and crystal growth by soluble matrix of oyster shell. Science **212**, 1397-1398.
Worms, D. and Weiner, S. (1986) Mollusk shell organic matrix: Fourier transform infared of the acidic macromolecules. J. Exp. Zool. **237**, 11-20.

IV. The First Steps of Biological Evolution

Biodata of **Mikhail S. Kritsky** author (with co-author T. A. Telegina) of ***"Role Of Nucleotide-Like Coenzymes In Primitive Evolution"***

Dr. Mikhail Sergeevich Kritsky is the Head of laboratory of Evolutionary Biochemistry in the A.N. Bach Institute of Biochemistry, Russian Academy of Sciences, Moscow. He graduated from Faculty of Biology of M.V. Lomonosov State University of Moscow in 1959. Dr. Kritsky has been Candidate of Biological Sciences (Biochemistry) since 1966 [which is a degree equivalent to Ph.D.]. He holds a Doctor of Biological Sciences (Biochemistry) degree [which is equivalent to German Dr. Habil] since 1986. His main research interests are in photobiochemistry of coenzymes and photoregulation of microbial metabolism.

E-mail: **mkritsky@inbi.ras.ru**

Taisiya Aleksandrovna Telegina is a senior scientist in Laboratory of Evolutionary Biochemistry of the A.N. Bach Institute of Biochemistry, Russian Academy of Sciences, Moscow. She graduated from the faculty of Chemistry of M.V. Lomonosov State University of Moscow in 1967. She is Candidate of Biological Sciences (Biochemistry) since 1978. Main research interest is a role of photochemically active molecules in evolution.

M.S. Kritsky

T.A. Telegina

J. Seckbach (ed.), Origins, 215–231.

ROLE OF NUCLEOTIDE-LIKE COENZYMES IN PRIMITIVE EVOLUTION

M. S. KRITSKY and T. A. TELEGINA
A.N. Bach Institute of Biochemistry, Russian Academy of Sciences
Leninsky Prospekt 33, Moscow, 119071 - Russian Federation
E-mail: mkritsky@inbi.ras.ru

1. Catalytic Inferiority of the RNA World

Two principles are fundamental for life: (1) Organisms, to maintain their homeostasis, capture energy and building materials from the environment by using catalytic reactions of metabolism. (2) The information about the structure of metabolic catalysts is stored in organism's genome, and after the genome's replication this information can be transferred to the descendants. Since the middle of the 20th century, molecular biologists have rigorously separated the roles of nucleic acids as genetic material on the one hand, and of proteins as catalysts, on the other. Since no reasonable explanation could be given to the question on how a mechanism decoding information from nucleic acids to form proteins had emerged, the origin of primitive organisms possessing both catalytic and genetic systems has turned into a seemingly insoluble problem.

One solution was to assume that at the onset of biological evolution the catalytic and informational functions were not separated between different classes of polymers, but the same molecule could play both roles. Although amino acids abiotically form polymers displaying catalytic and other functions inherent in modern proteins [Fox and Dose, 1972], the inability of polypeptides (unlike polynucleotides) to form complementary replicas did not permit to ascribe them a function of genetic information store and transfer. The idea had been proposed that RNA might possess catalytic activity and the primitive biosphere was an RNA world [Woese, 1967; Crick, 1968; Orgel, 1968]. The RNA-world hypothesis [Gilbert, 1986] hinges on the assumption that early life was based on polyribonucleotides. That is, RNAs, which could have served as their own genes (DNA is a later sophistication of the genome), also performed catalytic functions in the absence of genetically ordered proteins.

The discovery of ribozymes, *i.e.* catalytically active RNAs, has given strong support to the hypothesis almost speculative at its birth. Ribozymes from modern organisms displayed, however, a restricted menu of catalytic activities. It seemed at first that these catalysts mainly promoted phosphodiester transfer and phosphodiester hydrolysis at RNA and, sometimes, DNA linkages [Pyle, 1993]. The development of *in vitro* randomization, selection and amplification methods dramatically increased the repertoire of RNA catalysis. Basically, this new approach for selection of new ribozymes reproduces the principles of chemical evolution, though the technology of the experiment does not have much in common with the conditions of prebiotic Earth. The selected ribozymes are able to catalyze formation of complementary RNA strands [Doudna and Szostak, 1989; Ekland and Bartel, 1996; Paul and Joyce, 2000; Johnston *et al.*, 2002]. They also catalyze reactions relevant to protein synthesis such as aminoacyl and peptidyl transfer and the peptide bond formation [Picirilli *et al.*, 1992; Illangasekare *et al.*, 1995; Lohse and Szostak, 1996; Zhang and Cech, 1996, 1998; Illangasekare and Yarus, 1999; Sun *et al*, 2002]. Moreover, ribosomal 23S RNA was shown to play a catalytic role in protein synthesis, and it may be that the peptidyl transferase in modern ribosomes comprises only RNA [Noller *et al.*, 1992]. Thus, there is reason to believe, that the catalytic properties of the RNA world were sufficient for the development of replication and translation mechanisms.

Although the listed-above examples do not exhaust catalytic activities revealed by now in natural and selected polyribonucleotides [Bartel and Unrau, 1999], some important reaction types which protein enzymes do catalyze, in particular, redox reactions, are yet to be demonstrated in ribozyme catalysis. The failure of organismic and selected nucleic acids to maintain these and other metabolic reactions in the absence of protein enzymes poses a question: what could expand a poor assortment of catalytic activities up to the level sufficient to maintain primitive metabolism? Could the activities necessary to drive metabolism develop within the RNA world before primitive gene-encoded protein enzymes have arisen? The answer to this question can be, possibly, given after the inspection of the chemical boundaries of the RNA world in search of reactive molecules and groups, which could, along with standard nucleotides, participate in the proto-RNAs formation.

The goal of this article is to substantiate a hypothesis, according to which the catalytic potential of primitive RNAs was boosted by the presence in their molecules of heterocycles similar to modern coenzymes. When excited by light, these heterocycles could reveal a high activity as catalysts of redox reactions, thus sufficiently expanding the catalytic repertoire of primitive RNA organisms in the absence of specific apoproteins [Kritsky *et al.*, 2001a]. Among coenzymes, flavins, nicotinamides and pterins attract the notice by their chemical closeness to ribonucleotides and high reactivity of the excited molecules. The chromophore function of flavins and pterins in modern photoreceptor proteins adds interest in the role they had played in evolution.

2. Nucleotides, Nucleotide-Like Coenzymes and the RNA World

The nucleotide structure of RNAs isolated from modern organisms is invariable. All known RNAs' monomers contain a heterocyclic base, which is purine (adenine and guanine) or pyrimidine (uracil and cytosine). The varying structural element in minor bases are side groups, not the heterocycle itself. The base is linked by glycosidic bond to a sugar (D-ribose in furanose conformation), and the phosphodiester bond links 5′ and 3′ hydroxyl groups of neighboring monomers. Because of such invariability and a seemingly perfect fitness of RNA molecules to their functions, this polymer structure is accepted today as an alternative-less version. However, in the light of evolution it makes more sense to regard modern RNAs as just a molecule type which has succeeded to pass through the sieve of natural selection. In other words, we cannot ignore a possible existence in primitive RNAs of other monomers and other linkages.

The view of modern RNAs as just one example among a variety of chemical alternatives finds support from chemical syntheses, demonstrating that 5′, 3′ ribose phosphate linkages are not obligatory for backbone formation in nucleic acid strands and by some properties the alternative polynucleotides are even "better" than organisms' RNAs. For example, synthetic ribopyranosyl RNA with 4′, 2′ phosphodiester bonds was found to have a stronger and more selective base pairing system than 5′, 3′ ribofuranosyl molecules. Ribose itself can be replaced with another sugar such as L-threofuranose linked by 3′, 2′ phosphodiester bonds [Eschenmoser, 1994; 1999]. Moreover, a nucleic acid which a backbone completely lacks the phosphodiester bonds-linked sugar phosphate was synthesized under various conditions including those relevant to prebiotic chemistry [Nielsen *et al.*,1991 Nelson *et al.*, 2000]. The backbone of this polymer, named as peptide nucleic acid (PNA), consists of poly-2-aminoethyl glycine (*syn.* polyethylenediamine monoacetate) linked by amide bonds. The standard RNA bases are attached to this backbone through an acetate linker. Interestingly, the PNA-DNA double strand demonstrated a higher stability than the DNA-DNA helix. Thus, the nucleic backbone is merely one version among the variety of chemically permitted structures, and we may regard the canonical RNA as a combination of monomers selected from a longer slate for their stereochemical fitness and the presence of amino and keto-enol groups in the heterocycle positions suitable to establish complementary base pairing.

Besides their function as RNA monomers, all ribonucleotides or their nucleoside di- and triphosphates are involved into metabolic reactions as coenzymes and the carriers of aminoacyl, phosphoryl and glycosyl residues and the phospholipids' precursors [Lehninger *et al.*, 1993]. Some other coenzymes mimic the ribonucleotide module, consisting of a positively charged heterocyclic "head", a neutral linker, and the negatively charged phosphorylated "tail". Such are flavin mononucleotide (FMN), molybdopterin, pyridoxal phosphate and thiamine diphosphate (TDP) and the coenzymes' precursors, 6,7-dihydroneopterin triphosphates (a precursor of biopterinт) and nicotinamide mononucleotide (NMN) (Fig.1).

In nicotinamide and pyridoxal coenzymes a positive 'head' is a substituted pyridine ring, and in thiamine diphosphate coenzymes, the 'head' is a thiazole-conjugated pyrimidine. The heterocyclic base in FMN and other flavins is 1,3-diketo-7,8-dimethyl izoalloxazine, and a similar heterocycle, in which carbon replaces the nitrogen atom in 5 position is present in deazaflavins. Pterin coenzymes biopterin and molybdopterin as well as other biological pterins and folates contain pterin (2-amino-4-keto-pteridine) as heterocyclic base [Lehninger *et al.*, 1993; Rajagopalan and Johnson, 1992].

The linker varies from ribose and sugar alcohol ribitol in NMN^+ and, correspondingly FMN, to a short aliphatic chain in thiamine diphosphate and pyridoxal phosphate. In pterins, such as biopterin and its precursor 7,8-dihydroneopterin triphosphate, the linker is a hydroxylated aliphatic structure. In another coenzyme of this group, molybdopterin, the "head" and phosphoryl "tail" are connected by a thiolated and hydroxylated butane chain.

The most complex pterin coenzymes, folates also fit in the above module. In their molecules pteridine base is connected with a negatively charged glutamyl or oligoglutamyl 'tail' with a *p*-aminobenzoate (PAB) linker. Since the PAB's amino group is situated closely to the 'head', it can be regarded as its functional part.

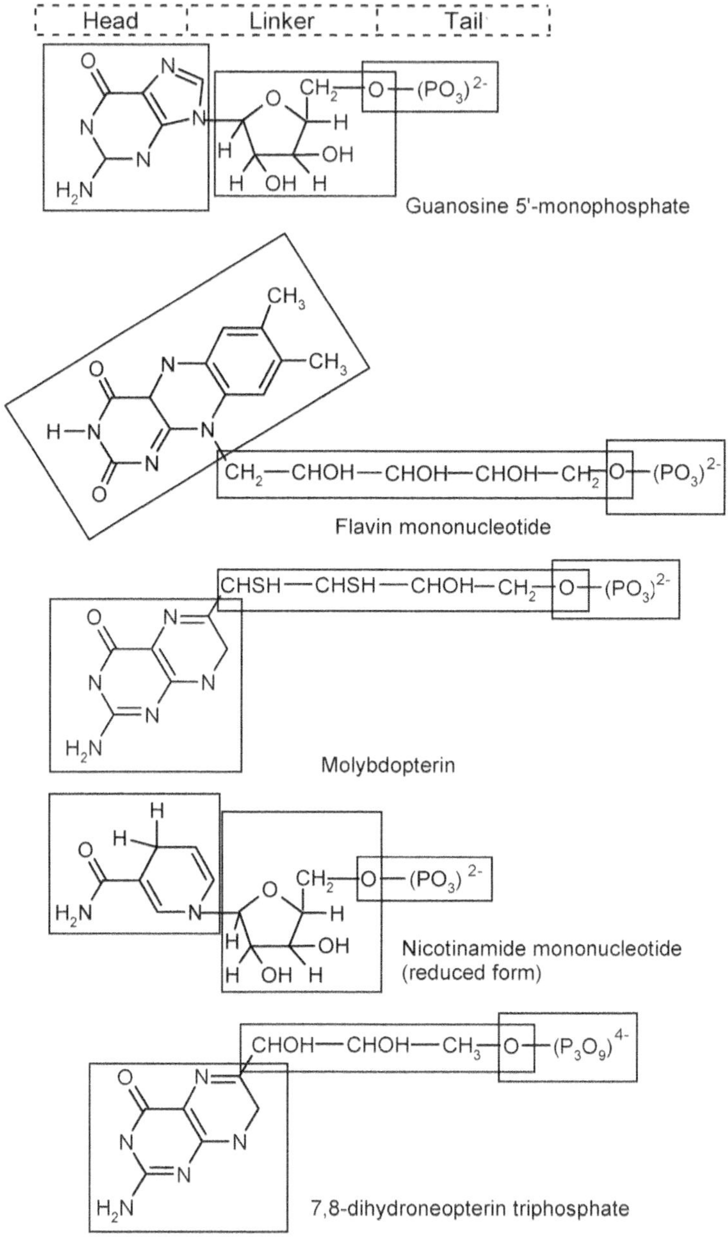

Figure 1. Guanosine 5'-monophosphate is an example of nucleotide module consisting of a "head", which is a heterocyclic base with exposed amino and keto(enol) groups, ribose "linker" and the phosphoryl "tail". Some coenzymes, for instance, flavin mononucleotide and molybdopterin as well as the coenzymes' precursors nicotinamide mononucleotide and 7,8-dihydroneopterin triphosphate mimic the nucleotide module. Their molecules consist of a heterocyclic "head", a linker, which is ribose, ribitol or the hydroxylated or thiolated-aliphatic chain, and the phosphorylated "tail".

Nicotinamide adenine dinucleotide (NAD^+), nicotinamide adenine dinucleotide 3′-phosphate ($NADP^+$), flavin adenine dinucleotide (FAD) and molybdopterin-containing dinucleotides which are products of the coenzyme monomer condensation with common ribonucleotides are quasi-dinucleotides, in which molecules a 5′,5′- pyrophosphate bridge, not the 5′,3′ phosphodiester bond links the units (Fig.2).

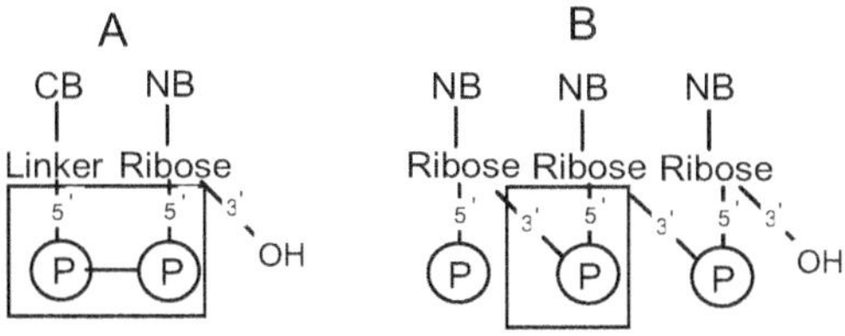

Figure 2. Some coenzymes such as nicotinamide adenine dinucleotide, nicotinamide adenine dinucleotide 3′-phosphate ($NADP^+$), flavin adenine dinucleotide and molybdopterin-containing dinucleotides are products of coenzyme monomer condensation with a common ribonucleotide, for instance, adenosine 5-monophosphate. Their molecules are, in fact, quasi-dinucleotides, where 5′,5′- pyrophosphate bridge (A), not the 5′,3′ phosphodiester bond (B) links the monomer units. CB is the coenzyme's heterocyclic base and NB is adenine or other purine or pyrimidine ribonucleic base.

It has been suggested, that the coenzymes' attachment could expand the catalytic repertoire of the RNA molecules [Visser, 1984; Connel and Christian, 1993]. The coenzyme bases were never reported to be natural components of nucleic acids in modern organisms. However, the covalent attachment to nucleic polymer of some coenzymes and their analogs can be performed both chemically and enzymatically. Synthetic pteridine analogs of guanosine and adenosine were incorporated from phosphorimidate-activated nucleosides into non-terminal positions of nucleic covalent linkage by automated DNA synthesis. Due to a closeness of the pteridine structure to purine (a pyrazine ring replaces the imidazole part of the purine molecule), the presence of pteridine bases was minimally disruptive for DNA structure as evidenced by melting temperatures of analog-containing and control polymers as well as by their digestion by P1 nuclease [Hawkins *et al.*, 1995, 2001; Driscoll *et al.*, 1997; Hawkins. 2001]. Folic acid was conjugated to a ribozyme by using a chemical approach, true enough, hardly relevant to primitive Earth conditions [Matulic-Adamic *et al.*, 2002].

In other experiments coenzymes were covalently attached at the ends of RNA molecules in reactions catalyzed by protein and RNA catalysts. NAD^+ and FAD substituted ATP in the initiation of the template-dependent RNA synthesis catalyzed by *Escherichia coli* RNA-polymerase [Malygin and Shemyakin, 1979]. A shortened version of a *Tetrahymena thermophila* ribozyme was shown to catalyze the self-incorporation of NAD^+ and dephosphorylated coenzyme A (CoA) [Breaker and Joyce, 1995]. Catalytic RNA was shown to create 5′, 5′-pyrophosphate linkages with all nucleotides and coenzymes including FMN, $NADP^+$, TDP and CoA [Huang and Yarus, 1997]. Terminal transferase from calf thymus accepted nicotinamide mononucleoside 5′-triphosphate (NTP) and efficiently added the NMN^+ residue to the 3′-end of the oligodeoxyribonucleotide, and the T4 polynucleotide kinase accepted NMN as substrate Moreover, the same study showed that reduced nicotinamide adenine dinucleotide 3′-

phosphate (NADPH) was an excellent substrate for polynucleotide phosphorylase of *Micrococcus luteus* in polymerization and in primer extension reactions. However, the T3 and T7 RNA polymerases failed to use NTP as substrate in template-directed RNA synthesis [Liu and Orgel, 2000]. (Fig.3).

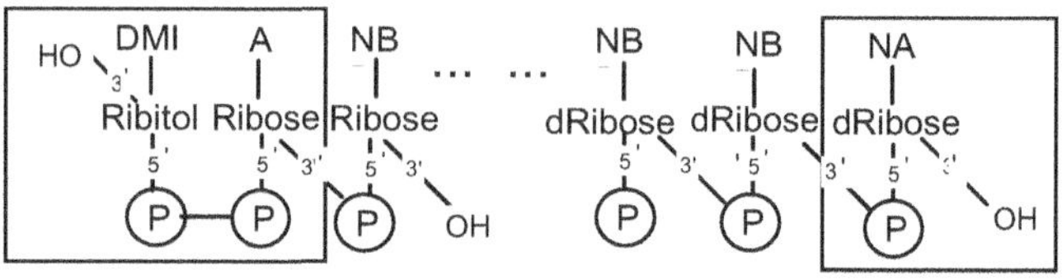

Figure 3. Examples of coenzyme covalent attachment to nucleic molecules. A. FAD can substitute ATP in the 5'-end initiation of template-dependent RNA synthesis catalyzed by *Escherichia coli* RNA-polymerase. DMI is dimethyl-izoalloxazine, A- adenine, NB – nucleic base. NAD^+ is active in this reaction as well [Malygin and Shemyakin, 1979].]. B. Terminal transferase from calf thymus can add the NMN^+ residue to 3′-end of oligodeoxyribonucleotide. NA is nicotinamide, NB –nucleic base [Liu and Orgel, 2000].

Polynucleotides attach coenzymes and other catalytically active ligands by non-covalent bonding as well. The development of *in vitro* selection methods has allowed the construction of specific aptamer complexes, a type of synthetic oligonucleotide, that bind to a particular target molecule [Patel and Suri, 2000; Jaysena, 1999]. By using this approach a number of RNA aptamers were synthesized and selected for their affinity to coenzymes and related molecules such as riboflavin, FMN, FAD, NMN^+ , NAD^+ and reduced forms of latter two molecules [Lauhon and Szostak, 1995; Fan *et al.*, 1996; Roychowdhury-Saha *et al.*, 2002]. The high affinity of the ligand-aptamer interaction caused by a stereochemical conformity of ligand molecule to a cavity in specifically folded polynucleotide strands is additionally enhanced by hydrogen bonding of the coenzyme's heterocycle (*i.e.* 1,3-diketo-7,8-dimethyl-izoalloxazine) with RNA bases [Fan *et al.*, 1996]. The above examples of attachment of coenzymes to polynucleotides do not exhaust all the ways how a nucleic acid molecule could acquire catalytic functions. For instance, chemical modification of the RNA's own purine bases leading to exposition of reactive imidazole radicals [Maurel and Decourt, 1997].

Thus, covalent and non-covalent intervention of coenzymes into the RNA world is not a chemically forbidden process and these substances themselves can be regarded as a part of this world. However, the demonstrated attachment of coenzymes does not explain how their presence could have boosted catalytic activity in the absence of a specific apoprotein.

3. Coenzymes as Photocatalysts

Free coenzymes are less reactive than when combined into a complex with their apoprotein. Photoexcitation enhances their reactivity besides specific apoprotein bonding. When excited, their molecules gain a capacity to participate in numerous reactions forbidden in the dark (Fig.4). The aromatic amino acid residues present in apoprotein may prevent a photochemical process by quenching the excited coenzyme molecule and dissipating its energy.

Photophysical Processes

$Fl+h\lambda \rightarrow {}^1Fl \rightarrow {}^3Fl$ $\quad$ $Pt+h\lambda \rightarrow {}^1Pt \rightarrow {}^3Pt$

$^1O_2$ generation: $\quad {}^3Fl+O_2 \rightarrow Fl+{}^1O_2$ $\quad {}^3Pt+O_2 \rightarrow Pt+{}^1O_2$

Energy transfer between excited

pterin and flavin components

in the DNA-photolyase active center: $\quad {}^1PtH_4+FlH_2 \rightarrow PtH_4+{}^1FlH_2$

Photoreduction and Donor-to-Acceptor Electron Transfer

Photoreduction: $\quad {}^3Fl+D \rightarrow FlH^{\bullet}+D_{ox}$ $\quad {}^3Pt+D \rightarrow PtH^{\bullet}+D_{ox}$

$^3Fl+D \rightarrow FlH_2+D_{ox}$ $\quad {}^3Pt+D \rightarrow PtH_2+D_{ox}$

$^3PtH_2+D \rightarrow PtH_4+D_{ox}$

Donor-to-acceptor

transfer: $\quad {}^3Fl+D+A \rightarrow Fl+D_{ox}+A_{red}$ $\quad {}^3Pt+D+A \rightarrow Pt+D_{ox}+A_{red}$

Cyclic electron transfer

from $FADH_2$ to pyrimidine (Pyr)

dimer in DNA-photolyase: $\quad {}^1FlH_2+Pyr<>Pyr \rightarrow FlH+Pyr<>Pyr^{\bullet} \rightarrow FlH_2+2Pyr$

Photoadduction

$^*Fl+M \rightarrow FL\text{-}M$ $\quad {}^*Fol+C_1\text{-}+D_{red} \rightarrow FolH_4\text{-}C_1+D_{ox}$

Figure 4. Schematic overview of the reaction types of excited flavins (Fl) and pterins (Pt). D stands for electron donor and, correspondingly, A for the acceptor. The redox reactions proceed in solution, but membrane-embedded flavin sensitizes this reaction as well. In photoadduction reactions excited flavin establishes a covalent bond with another molecule (M), and this reaction type functions in the flavin-containing photoreceptor phototropin,. The excited pterin derivatives, folates (Fol), transit to reduced H_2- and H_4-forms and then can add monocarbon (*i.e.* formyl) group to its molecule, thus mimicking enzymatic reactions [After Heelis, 1982; Presti, 1983; Sankar, 1994; Ledbetter *et al.*, 1995; Kritsky *et al.*, 1997,. 2001a, 2001b; Egorov *et al.*, 1999; Telegina *et al.*, 2001].

The photochemical properties of flavins are nowadays the most extensively studied among coenzymes and related molecules [see reviews of Heelis, 1982; Presti, 1983]. The other groups, which today attract the attention of photochemists and photobiologists are nicotinamides and pterins [Nikandrov *et al.*, 1978; Chahidi *et al.*, 1981; Galland and Senger, 1988; Ledbetter *et al.*, 1995; Neverov *et al.*, 1996; Kritsky *et al.*, 1997, 2001b; Egorov *et al.*, 1999; Suárez, *et al.*, 2000; Thomas., *et al.*, 2000]. In addition to UV-radiation with a wavelength maximum at 250÷280 nm absorbed by nucleic bases, flavins, pterins and reduced nicotinamides absorb UV-light of the area between 300 and 400 nm. Flavns are excited by short-wave visible light up to about 500 nm as well. The deazaflavins absorbance area is UV- shifted as compared to common flavins [Nikandrov *et al.*, 1986].

The excited singlet coenzymes are efficiently converted into triplet states. Their heterocycles play a major role in excitation, but side groups strongly influence parameters of the process [Presti, 1983; Kritsky *et al.*, 2001b]. From the standpoint of prebiotic chemistry, the latter are of more interest, because they can interact with other molecules in solution, whereas the excited singlet states are chemically quenched only in ordered systems such as photosynthetic reaction centers. The inherent lifetime of flavin and pterin triplets reaches $10^{-1}\div10^{0}$ seconds [Chahidi *et al.*, 1981; Heelis, 1982; Presti, 1983; Neverov *et al.*, 1996]. The midpoint redox potential value of flavins (E'_0) is - 0.22 V, i.e. reduced flavins behave as rather strong reductans. After transition to triplet state the E'_0 of the pertinent redox pair is shifted to +1,.85 V because of energization of the oxidized form of the pair. Strongly electronophilic, it subtracts electrons from various donors including those with highly positive E'_0 values. The reaction involves a free radical mechanism [Heelis, 1982; Presti, 1983; Ledbetter *et al.*, 1995; Kritsky *et al.*, 2001b] and leads to a formation of active metabolic reductants: dihydroflavins and the dihydro- and tetrahydropterins [Heelis, 1982; Presti, 1983, Kritsky *et al.*, 1997] (Fig.4). Among the efficient electron donors are some believed reagents of prebiotic chemistry such as carbonic and amino acids, alcohols and, interestingly, ethylenediamine monoacetate (the PNA backbone repeating unit) and its more heavily acetylated derivatives (E'_0 for ethylenediamine tetraacetate is *ca.* +0.40 V) [Heelis, 1982]. Pterins demonstrate similar behavior. Their reduced forms, especially tetrahydropterins are strong reductants ((E'_{0}. –0.50 V), and the photoexcited triplet state gets an additional electronic energy of about 2.50 eV [Chahidhi, *et al.*, Neverov *et al.*, 1996; Kritsky *et al.*, 1997].

In the presence of acceptors such as Fe^{3+}-cytochrome *b*, excited flavins and pterins catalyze donor-to-acceptor electron transfer under anaerobic and aerobic condition [Heelis, 1982; Presti, 1983, Lyudnikova *et al.*, 2001]. Oxygen sharply decreases the viability of reduced coenzymes, and the mechanism of photoreactions changes due to the generation of singlet oxygen ($^1O_2$) and other reactive oxygen species [Heelis, 1982; Presti, 1983; Lyudnikova *et al.*, 2001; Krasnovskii *et al.*, 1987]. Nevertheless, like anaerobic photoprocesses, the reactions of flavins and pterins in oxidized atmosphere can lead to accumulation of free energy in products.

Besides the reactions proceeding in aqueous solution, excited lipophilic riboflavin derivative embedded in an artificial lipid membrane was shown to sensitize translocation of redox equivalents across this membrane [Presti, 1983]. At last, due to a difference of proton dissociation constants of the ground state and excited molecule, the excitation and relaxation cycle of flavins was suggested to act as transmembrane proton translocator. Any experiments to prove this hypothesis are not known [Presti, 1983].

The photochemistry of reduced molecules differs from that of oxidized coenzymes. Excited dihydropterin molecule, an intermediate step in pterin two-electron reduction, can act as electron acceptor and transit to tetrahydroforms [Kritsky et al., 1997]. The excited singlet state of reduced dihydroflavin molecule was reported to perform an electron donor function in enzyme DNA-photolyase [Sancar, 1994]. The excited NADH and NADPH exhibit activity as strong reductants. In the absence of oxygen they reduce ferredoxins and methylviologen. The E'_0 value of non-excited nicotinamides is -0.32 V, whereas for ferredoxins it is –0,43 V, and for dihydropyridyl product of methylviologen reduction this value is –0,79 V. It means that reduced nicotinamides sensitize the uphill electron transfer [Nikandrov *et al.*, 1978].

4. Flavin and Pterin Chromophores in Protein Photosensors

Flavin and pterin coenzymes function as chromophores for widespread photoreceptors mediating physiological and developmental responses of organisms to ultraviolet (UV-A and UV-B) and short-wave visible radiation. The excited molecules of reduced nicotinamide coenzymes have been shown to modulate catalytic activity of some enzymes such as hydrogenase [Zhukova *et al.*, 1987]. There is, however, no evidence on the chromophore function of nicotinamides in specific protein photoreceptors. The excitation by UV-radiation of pyridoxal phosphate, which acts as coenzyme for enzymatic decarboxylation of hydroxytryptophan to serotonin has been shown to stimulate the reaction [Fraikin *et al.*, 1989].

The coenzyme-binding photosensors belong to several structurally distant protein families The first group comprises DNA photolyase, the enzyme repairing the UV-damaged nucleic acid molecule by catalyzing a photochemical splitting of cyclobutane dimers formed by two neighboring (mostly pyrimidine) bases [Heelis *et al.*, 1993; Sankar, 1994]. Its structural homologues, cryptochromes, are pigment mediating plant and animal development [Ahmad and Cashmore, 1993; Lin *et al.*, 1995; Hsu *et al.*, 1996]. These proteins' molecules contain two chromophores. In photolyase the first chromophore is reduced FAD ($FADH_2$) situated in the enzyme's active center and the second, a light-harvester, is 5,10-N-methenyl-tetrahydrofolate (MTHF) or in some organisms, 7,8-didemethyl-10-hydroxy-5-deazariboflavin. Cryptochrome, being structurally similar to photolyases, but functionally distant, bind FAD and MTHF.

The polypeptides of second family such as the White Collar 1 (WC-1) photoreceptor governing development of the fungus *Neurospora crassa* and another protein, the plant growth photoregulator phototropin, contain the so-called PAS and LOV regulator domains capable to bind FMN [Linden *et al.*, 1997; Christie and Briggs, 1999; Christie *et al.*, 1999]. It was shown for at least one protein of this family, phototropin, that photoreception act involves photoaddition of izoalloxazine's C4 atom of excited FMN chromophore to one of the cysteinyl residues in LOV domain causing a change of the domain's conformation [Salomon *et al.*, 2000]. These polypeptides, whose conformation and regulatory function are under control of reactions driven by excited flavin, have likely descended from redox-sensing proteins [Taylor and Zhulin, 1999]. Like in LOV (PAS) domain-containing proteins, the activity of the third group, eukaryotic nitrate reductases, depends on the polypeptide's redox state, which is controlled by excitation and photochemical activity of the FAD chromophore [De la Rosa *et al.*, 1989]. The recently discovered fourth group is represented by light activated FMN-binding adenylyl cyclase from *Euglena gracilis* [Iseki *et al.*, 2002].

Considering the striking diversity of primary structure, it can hardly be expected that all these proteins have a single protein ancestor. It is more reasonable to suggest a multiple origin in evolution of coenzyme-binding proteins with photoreceptor functions. The coenzyme photocatalysis in protein-free systems played, probably, an important role in primitive evolution and is regarded today as a metabolic relic [Krasnovskii, 1959; Krasnovsky, 1981; Kritsky *et al.*, 1998]. The presumed function of coenzyme-sensitized photoprocesses as primitive solar energy converters has been totally ousted in evolution by a more efficient chlorophyll-driven photosynthesis. Nevertheless, these compounds remain significant as sensitizers for regulation processes.

5. Coenzymes in the Evolution of the RNA World

Coenzymes are an unalienable component of the metabolism of all organisms. The profound connection with basic segments of metabolism suggests the antiquity of these compounds as primitive organic catalysts and metabolic relics [White, 1976]. According to the results of chemical modeling, the main structural components of coenzyme molecules and, probably, coenzymes themselves were available in prebiotic world. Izoalloxazine, pteridine and nicotinamide could arise abiotically as well as nucleic acid bases [Ferris *et al.*, 1969; Ponnamperuma, 1971; Gladilin *et al.*, 1978; Heinz *et al.*, 1979; Heinz and Reid, 1981; Oro, 2001]. In the context of this article it is important, that flavins abiotically formed after thermolysis of amino acid mixtures exhibited photochemical activity as sensitizers of redox reactions and of the ADP to ATP phosphorylation [Kolesnikov, 1991; Kolesnikov and Kritsky, 2001]. More complex molecules such as riboflavin, ATP, pyridine nucleotides and nucleoside diphosphate coenzymes (UDP-glucose, CDP-ethanolamine and CDP-choline) have been obtained under the conditions mimicking prebiotic processes [Oro, 1994, Cleaves and Miller, 2001; Oró, 2001]. RNA catalyzed processes could, possibly, lead to their formation as well as abiotic condensation reactions [Huang et al., 2000]. The presumed antiquity of at least one group of coenzymes, the flavins, is supported by comparative analysis of the secondary structure pattern in proteins [Rossman *et al.*, 1974]. According to the estimate given, the age of the two oldest protein groups, flavin-binding flavodoxins and FeS cluster-binding ferredoxins is about 3.2÷4.5 Gy, which is comparable with the age of the biosphere. Thus, we may believe that coenzymes such as modern flavins, pterins and nicotinamides, or their chemical analogs were available in the prebiotic era.

We believe that ribonucleotide monomers and heterocyclic bases present in modern RNA are the result of natural selection, which has tested a vast set of heterocycles, monomers and linkages to eliminate most of them and finally shape the existing RNA structure. The bases exhibiting coenzyme activity such as pteridine, izoalloxazine and nicotinamide as well as their nucleotide-like derivatives were probable members of the list of contenders. When covalently attached to proto-RNA structures, they could sufficiently expand the catalytic spectrum of the RNA world. According to stereochemical considerations the presence of such odd bases at the ends of polynucleotide strands seems more plausible, though a non-terminal incorporation cannot be excluded.

Obviously, the ability to establish hydrogen bonds with the bases in complementary strand had to be a necessary prerequisite for the base or monomer could fit the RNA polymer structure. At first glance coenzyme heterocycles show a much worse spatial conformity to RNA structure than the common bases. However, pteridine bases, whose spatial parameters are close to those of purines, demonstrated a hydrogen bond interaction with the complementary base when introduced to a nucleic acid strand [Hawkins *et al.*, 2001] (Fig. 5). The ketone groups of flavins and amido group of nicotinamide have been reported to participate in hydrogen base pairing with nucleic bases, too [Liu and Orgel, 2000; Fan *et al.*, 1996; Hawkins *et al.*, 2001]. Despite the reported capability of these heterocycles to establish hydrogen bonds with nucleic acid bases, their spatial conformity to nucleic linkage, in particular, the significance of the "linker" and "tail' structure, still deserves detailed consideration.

Photosensitizing activity, in particular the ability to sensitize redox reactions, can be regarded as a positive selection character which could favor the system to capture energy from outside and drive metabolism. It should be noted, however, that coenzyme bases neighboring other heterocycles partially lose their excitation energy in stacking interactions. The value of this loss depends on the combination of stacking bases [Heelis, 1982; Presti, 1983; Hawkins *et al.*, 1995, 2001; Driscoll *et al.*, 1997]. Thus, the primary structure of protoRNA molecule might influence the catalytic activity of the attached coenzyme.

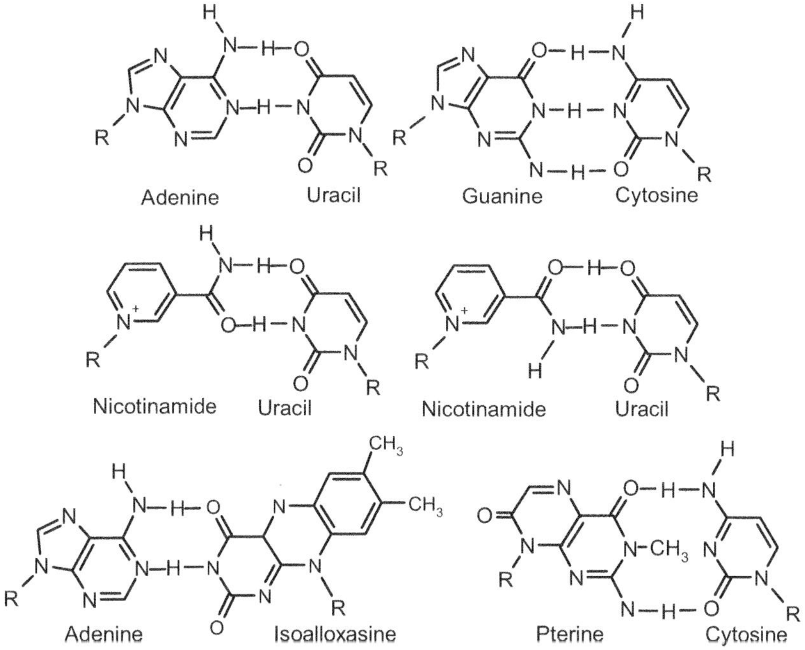

Figure 5. The hydrogen bond interactions of nucleic bases and coenzyme heterocycles. [After Liu and Orgel, 2000; Fan *et al.*, 1996; Hawkins *et al.*, 200].

A negative consequence of photoexcitation was a risk of photodegradation of the bases situated in vicinity of the coenzyme's heterocycle [Hubler *et al.*, 1997]. Both standard and coenzyme bases are more or less equal before the direct deteriorating action of solar UV-radiation below 300 nm. The development of the ozone shield restricted the penetration of radiation with wavelength above 300 nm and minimized the development of degradation risk for standard nucleic bases. At the same time, the ozone shield could not totally eliminate the photodegradation processes sensitized by pterins and, especially, by flavins, which absorb light of longer wavelengths. It is noteworthy to recall that flavins and pterins, but not nucleic bases are active $^1O_2$ generators [Neverov *et al.*, 1996; Egorov *et al.*, 1999]. For this reason pteridine and izoalloxazine had to be more hostile partners in nucleic acid strands than standard bases.

It is probable that after finally having been rejected from polynucleotide carriers, coenzyme molecules have managed to survive in evolution by changing their host polymer. The development of genetically ordered polypeptides had to lead to the appearance of amino acid sequences displaying affinity to coenzymes. In its turn, this affinity could have become a strongly favorable characteristic encouraging the survival of those primitive organisms, whose genotypes were capable to encode such polypeptides, since their presence provided a wider set of catalytic activities.

6. Summary

Certain types of catalytic activities, *e.g.* oxidoreductases, are absent in ribozymes. A possible way the RNA world could compensate for their absence to maintain a primitive metabolism was the attachment of nucleotide-like coenzymes to polynucleotides. Molecules such as flavin, pterin and nicotinamide coenzymes are structurally similar to canonic ribonucleotides and can attach nucleic molecules both covalently and by non-covalent high-affinity binding. Light dramatically elevates the reactivity of coenzymes, especially as redox catalysts. It is suggested that the presence in proto-RNAs of nucleotide-like coenzymes capable to perform photocatalytic functions could expand the repertoire of catalytic activities in the RNA world.

7. Acknowledgement

The support of this work by Russian Foundation for Basic Research Grant 01-04-48268a is appreciated.

8. References

Ahmad, M. and Cahsmore, A.R. (1993) HY-4 genes of *A. thaliana* encodes a protein with characteristics of a blue-light photoreceptor, *Nature*, **366**, 162-166.

Bartel, D.P. and Unrau, P.J. (1999) Constructing an RNA world, *Trends Biochem. Sci.*, **241**, 9-13.

Breaker, R.R. and Joyce, G.F. (1995) Self-replication of coenzymes by ribozymes, *J. Mol. Evol.*, **42**, 551-558.

Chahidi, C., Aubaily, M., Momzikoff, A., Bazin, M., and Santus, R. (1981) Photophysical and photosensitizing properties of 2-amino-4-pteridinone: a natural pigment, *Photochem. Photobiol.*, **33**, 641-649.

Christie, J.M. and Briggs, W.R.(1999) Blue light sensing in higher plants, *J. Biol. Chem,* **276**, 11457-11460.

Christie, J.M., Salomon, M., Nozue, K., Wada, M., and Briggs, W.R. (1999) LOV (light, oxygen, or voltage) domains of the blue light photoreceptor phototropin (*nph1*): Binding sites for chromophore flavin mononucleotide, *Proc. Natl. Acad. Sci. USA USA USA,* **96**, 8779-8783.

Cleaves, H.J. and Miller, S.L. (2001) The nicotinamide biosynthesis pathway is a by-product of the RNA world, *J. Mol. Evol.*, **52**, 73-77.

Connel, G.J. and Christian, E.L. (1993) Utilization of cofactors metabolism in a new RNA world, *Origins of Life Evol. Biosph.*, **23**, 291-297.

Crick, F.T. (1968) The Origin of Genetic Code, *J. Mol. Biol.*, **38**, 367-379.

De la Rosa, M.A., Roncel, M., and Navarro, J.A. (1989) Flavin-mediated photoregulation of nitrate reductase. A key point of control in inorganic nitrogen photosynthetic metabolism, *Bioelectrochem. Bioenerg.*, **27**, 355-364.

Doudna, J.A. and Szostak, J.W. (1989) RNA-catalyzed synthesis of complementary strand RNA, *Nature* , **339**, 618.

Driscoll, S.L., Hawkins, M., Balis, F.M., Pfleiderer, W., and Laws, W.R. (1997) Fluorescence properties of a new guanosine analog incorporated into small oligonucleotides, *Biophys. J.*, **73**, 3277-3286.

Echenmoser, A. (1994) Chemistry of potentially probiological products, *Origins of Life Evol. Biosph.*, **24**, 389-423.

Echenmoser, A. (1999): Chemical etiology of nucleic acids, *Science*, **284**, 2118-2124.

Egorov, S.Yu., Krasnovsky, A.A., Jr., Bashtanov. M.E., Mironov, E.A., Lyudnikova, T.A., and Kritsky, M.S. (1999) Photosensitization of singlet oxygen formation in pterins and flavins. Time-resolved studies of oxygen phosphorescence under laser excitation, *Biochemistry (Moscow)* **64**, 1117-1121.

Ekland, E.H. and Bartel, D.P. (1996) RNA-catalyzed RNA polymerization using nucleoside triphosphates, *Nature*, **382**, 373-376.

Fan, P., Suri, A,K., Fiala, R., Live, D., and Patel, J.D. (1996) Flavin recognition by an aptamer targeted toward FAD, *J. Mol. Biol.*, **258**, 480-500.

Ferris, J.F., Kuder, J.E., and Catalano, A.W. (1969) Photochemical reactions and the chemical evolution of purines and nicotinamide derivatives, *Science*, **166**, 765.

Fox, S.W. and Dose, K. (1972) *Molecular Evolution and the Origin of Life*, S.A. Freeman Co, San Francisco.

Fraikin, G.A., Strakhovskaya, M.G., Ivanova, E.V., and Rubin, A.B. (1989) Near UV-activation ofenzymatic conversion of 4-hydroxytryptophan to serotonin, *Photochem. Photobiol.*, **49**, 425-477.

Gilbert, W. (1986) The RNA World, *Nature*, **319**, 618

Gladilin, K.L., Andreeva, N.A., Tsygankov, A.Iu., Gadzhieva, A.K., and Kirpotin, D.B. (1978) Abiogenous synthesis of pteridine-like compounds, *Doklady Akademii nauk SSSR (Doklady Biochemistry)*, **240**, 1461-1463.

Hawkins, M.E. (2001) Fluorescent pteridine analogs – a window on DNA interaction, *Cell. Biochem. Biophys.*, **34**, 257-281.

Hawkins, M., Pfleiderer, W., Mazumder, A., Pommier, Y.G., and Balis, F.M. (1995) Incorporation of a fluorescent guanosine analog into oligonucleotides and its application to a real time assay for the HTV-1 integrase 3′- processing reaction, *Nucl. Acid. Res.*, **11**, 2872-2880.

Hawkins, M., Pfleiderer, W., Jungmann O., and Balis, F.M. (2001) Synthesis and fluorescence characterization of pteridine adenosine nucleoside analogs for DNA incorporation, *Analyt. Biochem.*, **298**, 231-240.

Heelis, P.F. (1982) The photophysical and photochemical properties opf flavins (izoalloxazines), *Chem. Soc. Rev.*, **11**, 15-39.

Heelis, P.F., Kim, S.T., Okamura, T., and Sancar, A. (1993) The photorepair of pyrimidine dimers by DNA photolyase and model systems, *J. Photochem. Photobiol., B.: Biol.*, **17**, 219-228.

Heinz, B. and Ried, W. (1981) The formation of chromophores through amino acid thermolysis and their possible role as prebiotic photoreceptors, *BioSystems*, **14**, 33-40.

Heinz, B., Ried, W., and Dose, K. (1979) Thermische Erzeugung von Pteridinen and Flavinen aus Aminosäueregemischen, *Angew. Chem.*, **91**, 510-511.

Hsu, D.S., Zhao, X., Kazantsev, A, Wang, R.-P., Todo, T., Wei, Y.-F., and Sancar, A. (1996) Putative human blue light photoreceptors hCRY1 and hCRY2 are flavoproteins, *Biochemistry*, **35**, 13871-13877.

Huang, F.Q, and Yarus, M. (1997) Versatile 5′ phosphoryl coupling of small and large molecules to RNA, *Proc. Natl. Acad. Sci. USA*, **94**, 8965-8969.

Huang, F., Bugg, C.W., and Yarus, M. (2000) RNA catalyzed CoA, NAD, and FAD synthesis from phosphopantetheine, NMN, and FMN, *Biochemistry*, **50**, 15548-15555.

Hubler, C., Burgstaller, P., and Famulok, M. (1997) *In vitro* selection, structural characteristics and photocleavage of an izoalloxazine bonding RNA aptamer. *Nucleosides & Nucleotides*, **16**, 717-720.

Illangasekare, M., Sanchez, D., Nickles,T., and Yarus, M. (1995) Aminoacyl-RNA synthesis catalyzed by an RNA, *Science*, **267**, 643-647.

Illangasekare, M., Yarus, M.: (1999) A tiny RNA that catalyzes both aminoacyl-RNA and peptidyl-RNA synthesis, *RNA*, **5**, 1482-1489.

Iseki, M., Matsunaga, S., Murakami, A., Ohno, K., Shiga, K., Yoshida, K., Sugai, M., Takahashi, T., Hori, T., and Watanabe, M. (2002) A blue-light-activated adenylyl cyclase mediates photoavoidance of *Euglena gracilis, Nature* , **415**, 1047-1051.

Jaysena, S.D. (1999) Aptamers: an emerging class of molecules that rival antibodies in diagnostics, *Clin. Chem.*, **45**, 1628-1650.

Johnston, W.K. and Unrau, P.J., Lawrence, M.S., and Bartel, D.P. (2002) RNA-catalyzed RNA polymerization: Accurate and general RNA template extension, *Science*, **292**, 1319-1325.

Galland, H. and Senger. H. (1988) The role of pterins in the photoreception and metabolism of plants, *Photochem. Photobiol.*, **48**, 811-820.

Kolesnikov, M.P. (1991) Proteinoid microspheres and the process of prebiological photophosphorylation, *Origins of Life Evol. Biosph.*, **21**, 31-37.

Kolesnikov, M.P. and Kritsky, M.S. (2001) Investigation of chemical structure and the photochemical activity of abiogenic flavin pigment, *J. Evol. Bioch. Physiol.*, **37**, 385-390.

Krasnovskii, A.A. (1959) Development of the mode of action of the photocatalytic system in organisms, In A.I. Oparin et al (eds) *The Origin of Life on the Earth*, Pergamon Press, London, *etc.*, pp. 606-618.

Krasnovskii, A.A., Jr., Chernysheva, E.K., and Kritskii, M.S. (1987) Investigation of the role of active forms of oxygen in flavin-photosensitized oxidation of NADH, *Biochemistry (Moscow)*, **52**, 1273-1282.

Krasnovsky, A.A. (1981) Evolution of uphill electron tranfer, *BioSystems*, **14**, 81-87.

Kritsky, M.S., Lyudnikova, T.A., Mironov, E.A., and Moskaleva, I.V. (1997) The UV radiation-driven reduction of pterins in aqueous solution, *J. Photochem. Photobiol., **B**: Biol.*, **48**, 43-48.

Kritsky, M.S., Lyudnikova, T.A., Mironov, E.A., and Neverov, K.V. (1998) The evolutionary roots of coenzyme photobiochemistry, In: J. Chela-Flores and F. Raulin (eds.), *Exobiology: Matter, Energy and Information in the Origin and Evolution of Life in the Universe*. Kluwer Acad. Publshers, Dordrecht, pp. 187-190.

Kritsky, M.S., Kolesnikov, M.P., Lyudnikova, T,A., Telegina, T.A., Otroshchenko, V,A., and Malygin, A.G. (2001a) The nucleotide- and nucleotide-like coenzymes in primitive metabolism photobiology and evolution, in: J. Chela-Flores, *et al.*,(eds.) *First Steps in the Origin of Life in the Universe*. Kluwer Academic Publishers, Dordrecht, pp 237-240.

Kritsky, M.S., Telegina, T.A., Lyudnikova, T.A., Umrikhina, A.V., and Zemskova, Yu.L. (2001b) Participation of free radicals in photoreduction of pterins and folic acid. *Doklady Akademii Nauk (Doklady Biochemistry)* **380**, 408-410.

Lauhon, C.T. and Szostak, J.W (1995) RNA aptamers that bind flavin and nicotinamide refox cofactors, *J. Am. Chem. Soc.*, **117**, 1246-1257.

Ledbetter, J.W., Pfleiderer, W., and Freisheim, J.H. (1995) Photosensitezed reduction of L-biopterin in the active complex of dihydrofolate reductase, *Photochem. Photobiol.*, **27**, 355-364.

Lehninger, A.L., Nelson, D.L., and Cox, M.M.(1993) *Principles of Biochemistry*, Worth Publishers, NY.

Lin, C., Robertson, D.E., Ahmad, M., Raibekas, A.A,, Shuman Jorns, M., Dutton, P.L., and Cashmore, A.R. (1995) Association of flavin adenine dinucleotide with the *Arabidopsis* blue light receptor CRY1. *Science*, **269**, 968-970.

Linden, H., Ballario, P. and Macino, G. (1997) Blue light regulation in *Neurospora crassa, Fungal Genet. Biol.*, **22**, 141-150.

Liu, R. and Orgel, L.E. (2000) Enzymatic synthesis of polymers containing nicotinamide mononucleotide, *Nucl. Acid. Res.*, **23**, 3742-3749.

Lohse, P.A., and Szostak, J.W. (1996) Ribozyme catalysed amino acid transfer reactions, *Nature*, **381**, 442-444.

Lyudnikova, T.A., Telegina, T.A., and Kritsky, M.S. (2001) Issledovanie fotohimicheskikh svoistv pterinov v svjazi s funkcijami v fotobiologicheskih sistemah, in: *III S'ezd Fotobiologov Rossii. Materialy s'ezda. Voronezh, 2001*, ss.126-127, in Russian (Investigation of photochemical properties of pterins in relation to their photobiological functions, in: *Proc. 3rd Congr. Russian Photobiologists, Voronezh*, 2001, pp. 126-127).

Malygin, A.G. and Shemyakin, M.F. (1979) Adenosine, NAD and FAD can initiate template dependent RNA synthesis catalyzed by *Escherichia coli* RNA polymerase, *FEBS Letts.*, **102**, 51-54.

Matulic-Adamic, J., Sanseverino, M., and Beigelman, L. (2002) An efficient synthesis of the ribozyme-folate conjugate, *Tetrahedron Letts.*, **43**, 4439-4441.

Maurel, M.C. and Decout, J.L. (1997) Origins of life. Molecular foundations and new approaches, *Tetrahedron*, **55**, 3141-3182.

Nelson, K.E., Levy, M., and Miller, S.L. (2000) Peptide nucleic acids rather than RNA may have been the first genetic molecule, *Proc. Natl. Acad. Sci. USA*, **97**, 3868-3871.

Neverov, K.V., Mironov, E.A., Lyudnikova, T.A., Krasnovsky, A.A., Jr., and Kritsky, M.S. (1996) Phosphorescence analysis of the triplet states of pterins in connection with their photoreceptor functions in biological systems, *Biochemistry (Moscow)*, **61**, 1149-1155.

Nielsen, P.E., Egholm, M., Berg, R.H., and Buchardt, O. (1991) Sequence-selective recognition of DNA by strand displacement with a thymine substituted polyamide, *Science*, **254**, 1497-1500.

Nikandrov, V.V., Brin, G,P., and Krasnovskii, A.A. (1978) Light-induced activation of NADH and NADPH, Biochemistry (Moscow) **43**, 636-645.

Nikandrov, V.V., Pantskhava, E.S., and Krasnovsky, A.A. (1986) Photobiochemical properties of deazaflavine coenzyme F_{420}: sensitization of NAD^+ photoreduction and hydrogen photoevolution. Photobiochem, Photobiophys., **13**, 105-114.

Noller, H.F., Hoffrath, V., and Zimniak, L. (1992) Unusual resistance of peptidyl transferase to protein extraction procedure, *Science*, **256**, 1416-1419.

Orgel, L.T. (1968) Evolution of genetic apparatus, *J. Mol. Biol.*, **38**, 381-393.

Oró, J. (1994) Chemical synthesis of lipids and the origin of life, *J. Biol. Phys.*, **20**, 135-147.

Oró, J. (2001) Cometary molecules & life origins, In: J. Chela-Flores, *et.al.*(eds.), *First Steps in the Origin of Life in the Universe*. Kluver Academic Publishers, Dordrecht, pp 113-120.

Patel, D.T. and Suri, A.K. (2000) Structure, recognition and discrimination in RNA complexes with substrates, amino acids, drugs and aminoglycosides, *J. Biotechnol.*, **74**, 39-60.

Paul, N. and Joyce, G.F. (2000) A self-replicating ligase ribozyme, *Proc. Natl. Acad. Sci. USA,* **90**, 12733-12740.

Picirilli, J.A., Mc Connel, T.S., Zaug, A.J., Noller, H.I., and Cech, T.R. (1992) Aminoacyl esterase activity of the *Tetrahymena* ribozyme, *Science*, **256**, 1420-1424.

Ponnamperuma, C. and Chang, S. (1971) The role of phosphates in chemical evolution of purines and nicotinamide derivatives, In: R. Buvet and C. Ponnamperuma (eds.), *Chemical Evolution and Origin of Life*, North Holland Publishing Company, NY, pp 216-223.

Presti, D.E. (1983) The photobiology of carotenes and flavins, In D.E. Cosens and D. Vince-Price (eds.), *The Biology of Photoreception,* Soc. Exp. Biol., pp. 133-180.

Pyle, A.M. (1993) Ribozymes – a distinct class of metalloenzymes, *Science*, **261**, 709-714.

Rajagopalan, K.V. and Johnson, J.I. (1992) The pterin molybdenum cofactor, *J. Biol. Chem.*, **267**, 10199-10202.

Rossman, M.G., Moras, D., and Olsen, K.W. (1974) Chemical and biological evolution of a nucleotide-binding protein, *Nature,* **250**, 194-199.

Roychowdhury-Saha, M., Lato, S.M., Shank, E.D., and Burke, D.H. (2002) Flavin recognition by an aptamer targeted toward FAD, *Biochemistry*, **41**, 2492-2499.

Salomon, M., Christie, J.M., Knieb,. E., Lempert, U., and Briggs, W.R., and Briggs, W.R. (2000) Photochemical and mutational analysis of the FMN-binding domains of the plant blue light receptor, phototropin, *Biochemistry*, **39**, 9401-9410.

Sancar, A. (1994): Structure and function of DNA photolyase, *Biochemistry*, **33**, 2-9.

Suárez, G., Cabrerizo, F.M., Lorente, C., Thomas, A.H., and Capparelli, A.L. (2000) Study of the photolysis of 5-carboxypterin in acid and alkaline solutions, *J. Photochem. Photobiol. A: Chem.*, **132,** 53-57.

Sun, L., Cui, Z., Gottlieb R.L., and Zhang, B. (2002) A selected ribozyme catalyzing diverse dipeptidyl synthesis, *Chem. Biol.*, **9**, 619-628.

Telegina, T.A., Lyudnikova, T.A., Zemskova, Yu.L., and Kritsky, M.S. (2001) Fotohimicheskie aspekty fotobiologicheskoj aktivacii folatov, in: *III S'ezd Fotobiologov Rossii. Materialy s'ezda. Voronezh, 2001*, ss. 217-218, in Russian *(Photochemical aspects of photobiological activity of folates*, in: *Proc. 3rd Congr. Russian Photobiologists, Voronezh, 2001)*, pp. 217-218).

Taylor, B.L and Zhulin, I,B. (1999) PAS domains: Internal sensors of oxygen, redox potential and light, *Microbiol. Mol, Biol, Rev.*, **63**, 479-506.

Thomas, A.H., Suárez, G., Cabrerizo, F.M., Martino. R., and Capparelli, A.L. (2000) Study of photolysis of folic acid and 6-formylpterin in acid aqueous solution, *J. Photochem. Photobiol. A: Chem.*, **135,** 147-154.

Visser, C.M. (1984) Evolution of biocatalysis. 2. Nicotinamide and/or flavin-containing RNA molecules as possible pre-genetic-code replicating oxido-reductases, *Origins of Life*, **14**, 301-305.

White, H.B., III (1976) Coenzymes as metabolic fossils, *J. Mol. Evol.*, **7**, 101-104.

Woese, C.R. (1967) *Genetic Code, the Molecular Basis for Genetic Expression*, Harper & Row, NY

Zhang, B.L. and Cech, T.R.(1996) Peptide bond formation by *in vitro* selected ribozymes, *Nature*, **390**, 96-100.

Zhang, B.L. and Cech, T.R. (1998) Peptidyl-transferase ribozymes: trans reactions, structural characterization and ribosomal RNA-like features, *Chem. Biol.*, **5**, 539-553.

Zhukova, L.V., Nikandrov, V.V., and Krasnovsky, A.A. (1987) The activating effect of near ultraviolet light on hydrogen metabolism in *Clostridium butyricum* cells, *Biochemistry (Moscow)*, **52, 655-659.**

Biodata of **Charles H. Lineweaver** author (with co-author D. Schwartzman) of ***"Cosmic Thermobiology: Thermal Constraints on the Origin and Evolution of Life in the Universe"***

Dr. Charles Lineweaver is a research astronomer and senior Lecturer in the School of Physics, University of New South Wales, Sydney, Australia. He received his PhD (1994) in Physics from The University of California at Berkeley. His fields of research are Cosmology and Astrobiology. He teaches the course entitled "Are We Alone?" and published articles in relevant areas (e.g., on Planets and Life in the Universe). He lived and traveled through 62 countries, speaks four languages and was a semi-professional soccer player in Germany.

E-mail: **c.lineweaver@unsw.edu.au**

Dr. David W. Schwartzman is a professor in the Department of Biology at Howard University, Washington, D.C. He obtained his PhD degree from Brown University, Providence, RI. in 1971. His research interests are Biogeochemistry, geochemistry (isotope, environmental) exobiology, philosophy of science, environmental studies. He published (1999 hardcover, 2002 paperback edition) *Life, Temperature and the Earth: The Self-organizing Biosphere*, Columbia University Press.

E-mail: **dws@scs.howard.edu**

Charles Lineweaver

David W. Schwartzman

J. Seckbach (ed.), Origins, 233–248.

COSMIC THERMOBIOLOGY
Thermal Constraints on the Origin and Evolution of Life in the Universe

CHARLES H. LINEWEAVER[1] AND DAVID SCHWARTZMAN[2]
[1]*School of Physics, University of New South Wales, Sydney, Australia*
[2]*Department of Biology, Howard University, Washington DC, USA*

1. Thermal Determinism and the Narrow Roads of Prebiotic Chemistry

Modern cosmology gives us the largest context within which to understand the emergence of life in the universe. Within this context, the transition from molecular to biological evolution is the most recent in a series of transitions that can be easily described as the result of decreasing temperature. In the big bang model for the evolution of the universe, as the universe expanded and cooled from arbitrarily high temperatures, a universal deterministic cascade of structure appeared. Here we describe how the expansion and decreasing temperature of the universe allowed atoms, molecules, stars, planets and life to form. We discuss the thermal constraints on terrestrial biogenesis and suggest that they apply to the earliest forms of life, not only on Earth, but anywhere in the universe.

Cosmology, physics and chemistry are deterministic in the sense that what we learn on Earth will be valid in the most distant parts of the universe. Biological evolution on the other hand seems to be dominated by quirky tinkering and historical contingency. We do not expect elephants or tuataras to have evolved anywhere else in the universe. If life emerged from non-life through a process of molecular evolution, there must be a continuum or a transition between these two paradigms. The first steps in the transition will be inevitable and temperature-induced. The next steps will be less deterministic and the final steps will be as highly contingent and unpredictable as life is today. Here we describe the first thermal constraints of an abiotic deterministic universe and the transition to a less-predictable biotic one. This thermal determinism is similar to the biochemical determinism of "Vital Dust" (deDuve, 1995). deDuve argues that biogenesis and much of what life is, is biochemically inevitable. Here we argue that biogenesis and much of what life is, is thermally inevitable.

2. When Does the Temperature of the Universe Permit the Ingredients of Life to Exist?

The very early universe was filled with radiation at arbitrarily high temperature (Fig. 1). At ~ 10^{-33} seconds after the big bang, the universe cooled enough to allow matter and anti-matter to annihilate, converting rest mass into radiation while leaving a small excess (one part in one billion) of matter. This transition is known as baryogenesis and is the formation of all the matter in the universe. At ~10^{-4} seconds after the big bang the

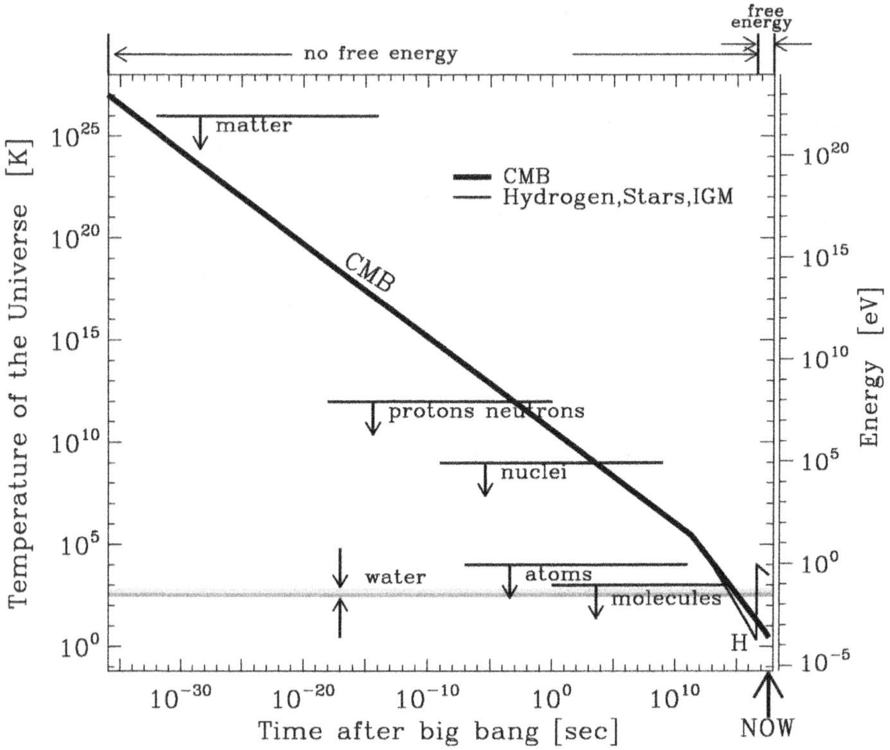

Figure 1. Transitions in the universe as the temperature decreases. Structures which freeze out as the universe cools include, matter, protons and neutrons, nuclei, atoms and molecules. See text and Fig. 2 for details.

thermal energy of the universe was low enough to allow the hot soup of quarks to condense –quarks fell together in triplets under the influence of the strong nuclear force and have remained stable ever since, as protons and neutrons. This epoch is known as the quark-hadron transition. At ~ 100 seconds after the big bang, nucleosynthesis began. The thermal energy of the universe decreased allowing protons and neutrons to bind together under the residual strong nuclear force to form the nuclei of the lightest elements: ^{1}H, ^{2}H, ^{3}He, ^{4}He, ^{7}Li.

For the first 400,000 years after the big bang all the matter was in thermal equilibrium with the sea of photons (cosmic microwave background: CMB) that filled the universe. Then the thermal energy of the CMB fell beneath the ionization energy of hydrogen. This allowed the electromagnetic force between electrons and protons to pull them together to form hydrogen atoms. Without charged particles to maintain thermal equilibrium, the temperature of the matter and the photons went their separate ways.

The temperature of the photons cooled inversely proportional to the size of the universe, while the matter

TABLE 1. Thermal History of the Universe (see Fig. 1)

Structure formed	Temperature [K]	Energy [eV]	Time after Big Bang [sec]
matter	10^{15}	10^{11}	10^{-33}
protons/neutron	10^{12}	10^{8}	10^{-4}
atomic nuclei	10^{9}	10^{5}	100
atoms	10^{4}	~1	400,000 years
molecules	10^{3}	~0.5	few million years
aqueous macromolecules	10^{2}	~0.1	2 billion years

(Kolb and Turner, 1990; Lewis, 1997)

cooled faster, inversely proportional to the square of the size of the universe. For the first time in the history of the universe two temperatures were needed to describe the universe. As the temperature of the hydrogen dropped below the dissociation energy of molecular hydrogen, hydrogen atoms in the densest parts of the universe paired up to form the first molecules in the universe: H_2. There were still no stars or planets. There was no H_2O in the universe. Stars had not yet made oxygen, carbon or nitrogen. There was no life.

The universe continued to expand and cool. When clouds of hydrogen cooled beneath 100 K, the thermal energy decreased enough to allow the weakest force in the universe, gravity, to make the densest regions gravitationally collapse. Gravitational potential energy was converted to heat and lost to outer space. The clouds collapsed further, heated further and finally massive stars were formed. Their UV photons re-ionized the universe. Their supernova explosions induced the collapse of other dense regions. In Figs. 1 and 2 the thin line labeled "H" shows the temperature of the hydrogen. It shot up to 5,000 - 10,000 K as the first stars in the universe broke the equilibrium and provided free energy for the first time in the history of the universe (see region labeled "free energy" above the upper x axis).

About 1% of the normal material in the universe today is not hydrogen and helium. These heavy elements were not produced in the big bang. They were produced gradually by generations of stars. It is the presence of these heavy elements which allows small rocky planets to form when clouds of molecular hydrogen collapse to form stars and gaseous planets.

The Earth formed from a molten ball at ~2000 K about 4.56 Gyr ago (Allègre et al., 1995). The majority of the Earth's mass accreted from planetesimals within the first 100 million years of the Earth's formation (Halliday, 2000). With an initially molten surface, life could not have appeared. The transition from accretion to heavy bombardment included the formation of the Moon by the collision with a Mars-sized object ~4.5 Gyr ago (Hartmann and Davis, 1975; Halliday, 2001; Canup and Asphaug, 2001). We can infer from the dates and sizes of lunar impact craters, whose record goes back to when the Moon formed a solid crust (~4.44 Gyr ago, Sleep et al., 1989) that the surface of the Earth was periodically vaporized and covered with a 2000 K rock vapor atmosphere which lasted for several thousand years (Hartmann et al., 2000; Sleep et al., 2001). These conditions were probably an effective and recurring autoclave for sterilizing the earliest life forms or more generally frustrating the evolution of life. A steadily decreasing heavy bombardment continued until ~ 3.8 Gyr ago.

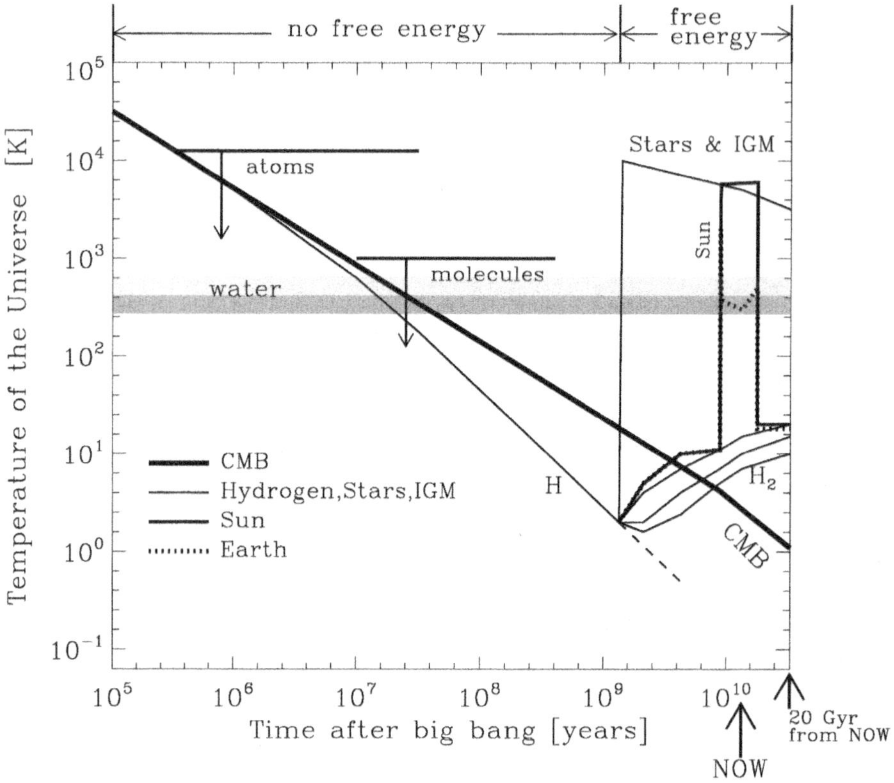

Figure 2. A zoom-in on the transitions in the lower right of Fig. 1. The thin line beneath the CMB line shows how hydrogen cooled more rapidly than the CMB. The dashed line shows how this cooling would have continued if it had not been for the fact that a billion years after the big bang, the thermal energy of the hydrogen sank low enough to allow the weak gravitational binding energy to contract the densest clouds of hydrogen. These clouds then became denser and hotter and eventually formed the first stars in the universe. These massive stars emitted UV photons that heated and ionized the more rarified hydrogen (intergalactic medium : IGM). The formation of these first massive stars and the re-ionization of the IGM are represented by the vertical line at 10^9 years. The other lines originating at the same point represent the temperatures of hydrogen clouds that were dense enough to self-shield and avoid UV ionization. These clouds were gradually enriched with oxygen, carbon, nitrogen, iron and the other waste products of the supernovae explosions of the first, massive, short-lived stars. About 4.6 billion years ago one of these enriched clouds was shocked by a nearby supernova. This initiated collapse and star formation. One of the stars was the Sun. Planetary formation and formation of the Earth was part of this collapse (dotted line). Other, less dense clouds of H_2 (represented by the three other thin lines) collapsed a bit but stayed at 10 or 20 K. The upper x axis shows that free energy is available once the temperature of the hydrogen is low enough to initiate gravitational collapse and star formation. Two adjacent grey strips are labeled "water". The lower darker one is 0-100 C. The higher lighter one is 100-650 C; the highest temperatures at which water, under pressure, can exist.

Events on Mercury, Venus and Mars were similar: an initial period of heavy bombardment abated as planetesimals were swept up. The process of rocky planet formation in our Solar System and elsewhere involves many thermal constraints. The radially dependent temperature of the proto-stellar nebula segregates gases and solids. Stellar winds push the volatiles away, leave the refractories and fractionate elements and isotopes. Rocky planets start off hot, and cool with time. Smaller ones cool off more quickly. On larger ones, the gravitational energy of accretion and the decay of radioactive elements keep the planet hot and power convective currents in the mantle, widespread volcanism, active plate tectonics and large hydrothermal systems. These are probably the temperature constraints that any life in the universe has to begin with if it is to come into existence – these are universal features of hydrogen (enriched with ~1% of other elements) cooling in an expanding universe.

The hyperthermophilic monopoly on the deepest roots of the phylogenetic tree of terrestrial life (Fig. 3) suggests that temperature played a dominant role in biogenesis and/or dominated the selection pressure on the earliest forms of life. This suggests the more general idea that wherever biogenesis occurs in the universe, temperature will play a dominant role not only in setting the stage with stars, planets and the building blocks of life but also in determining, restricting and constraining the earliest and simplest forms of life. Hyperthermophiles may dominate not only the terrestrial tree of life but the trees of life of all planets in the universe. What we find out about terrestrial biogenesis may be representative of biogenesis throughout the universe.

Terrestrial life is made of H, O, C and N: the first, second, third and fourth most abundant elements in the universe (ignoring the chemically inert helium). If life exists elsewhere, pundits expect it to be based on carbon and water. Even if silicon were a possibility, carbon-based life would still be 10 times more abundant than silicon-based life simply because carbon is ten times more abundant in the universe than is silicon. H_2O is the combination of the two most abundant chemically active elements in the universe. It is the most abundant tri-atomic molecule in the universe. H_2O is everywhere. However liquid H_2O is much less common and can only be found within a narrow range of temperatures (Fig. 2).

3. When Do the Sources of Free Energy in the Universe Permit Life to Exist?

In the beginning, 13.5 billion years ago, the universe was very hot (Lineweaver, 1999). There was no life and there were no structures in the universe. The universe was a thermal heat bath of photons. Life is not possible in a heat bath. In thermal equilibrium no free energy is available. Life will not emerge simply because the universe has become the right temperature for H_2O to be a liquid. Life needs a source of free energy. The origin of all sources of free energy can be traced back to the expansion of the universe which led to the cooling of hydrogen, which led (one billion years after the big bang) to the formation of the first stars and planets.

The chemical composition also constrains life. Even with abundant free energy one billion years after the big bang, life could not have formed since water, life and Earth-like planets all require elements that did not exist in any abundance. In Lineweaver (2001) we showed that another billion years was required before Earth-like planets could form.

4. Phylogenetic Thermometry: Hot Ancestors and their Cool Descendants

By identifying the oldest features in extant life we can get an idea of what the thermal constraints on the earliest life were. Molecular biologists have been able to construct phylogenies based on immune responses, molecular weights, protein similarities and DNA and RNA sequences. To examine the deepest roots of the tree of life the most highly conserved homologous sequences are compared. The phylogenetic tree of life in Fig. 3 is based on 16S rRNA (Pace, 1997). Life can be usefully classified into three domains: Bacteria, Archaea and Eukarya (Woese et al., 1990). Trees based on other highly conserved DNA sequences are in good but not complete agreement with this tree. The position of the root of the tree is estimated from ancient gene duplications (Iwabe et al., 1989; Gogarten - Boekels et al., 1995; Grimaldo and Cammarano, 1998). Only

TABLE 2. Maximum Growth Termperatures (see Fig. 3)

Organism	Temperature [degrees C]	References
Archaea		
Methanopyrus	110	Stetter, 1996
Pyrodictium	110	Stetter, 1996
Thermoproteus	97	Stetter, 1996
Methanothermus	97	Stetter, 1996
Archaeoglobus	95	Stetter, 1999; Brock et al., 1994
Thermofilum	95	Stetter, 1996
Thermococcus	93	Stetter, 1998
Methanococcus	91	Stetter, 1999
Sulfolobus	87	Brock et al., 1994
Methanobacterium	75	Brock, 1978
Thermoplasma	67	Kristjansson and Stetter, 1992
Haloferax (Halophiles)	55	Kristjansson and Stetter,1992
Bacteria		
Aquifex	95	Stetter, 1996
Thermotoga	90	Stetter, 1996
Thermus (*T. thermophilus*)	85	Kristjansson and Stetter, 1992
Bacillus	82	Brock, 1978
Clostridium	78	Wiegel, 1992
Synechoccoccus	74	Brock, 1978
Chloroflexus	73	Brock, 1978
Mitochondrion/*Agrobacterium*	60	Brock et al., 1994
Desulfovibrio	< 60	Kristjansson, 1992
Heliobacterium (*H. modesticaldum*)	56	Kimble et al., 1995
Chlorobium	55	Sirevag, 1992
Flavobacterium (*F. autothermophilum*)	55	Aragno, 1992
Plantomyces (*Isophaera pallida*)	55	Schmidt and Starr, 1989
Eukarya	< 60	Madigan et al., 1997

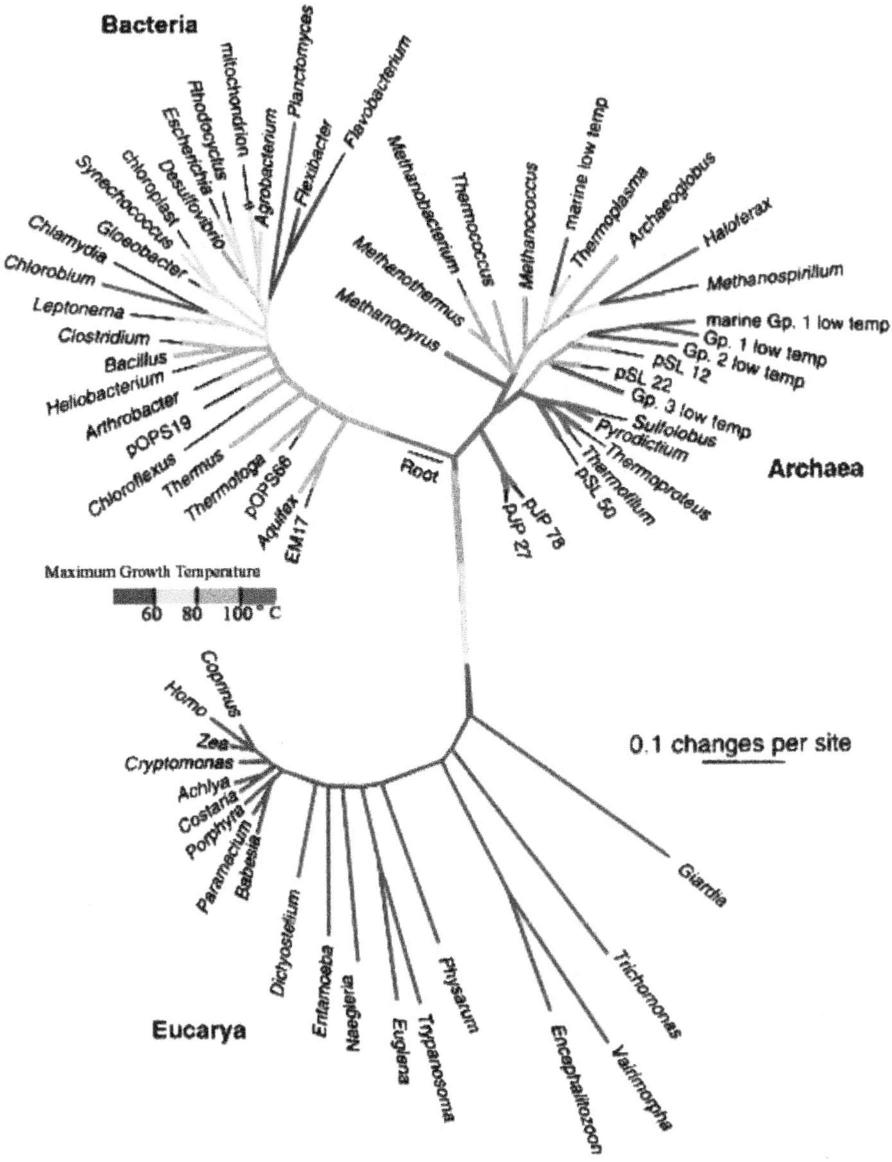

Figure 3. Phylogenetic tree of life based on 16S rRNA sequences. Maximal growth temperatures (Table 2) have been used to assign a grey scale to the branches. See text and Pace (1997) for details.

autotrophic hyperthermophiles from hot springs and hydrothermal vents are found near the root. This suggests that the last common ancestors (LCA) lived in similarly hot environments (Shock et al., 2000; DiGiulio, 2000). The environment of the LCA is our best estimate for thermal constraints on the emergence of life.

Here we attempt to quantify the gradient between hot ancestors and cool descendants. The distance from the root to the tip of any branch is proportional to the number of changes per site (see scale bar). Hyperthermophiles are living fossils in the sense that they occupy the shortest branches – their 16S rRNA has undergone the fewest changes over the past ~ 4 billion years. The fact that hyperthermophiles have a monopoly on the root and that organisms that cannot tolerate heat are only found far from the root suggests that we can use this phylogenetic tree as a crude thermometer. In Schwartzman and Lineweaver (2003) and in Table 2 we assemble from the literature maximum growth temperatures (T_{max}) for as large a subsample of the species shown in Fig. 3, as possible. We have used these temperatures in Fig. 3 to assign shades of grey to the tree (see temperature scale). Since values of T_{max} are derived from extant organisms, the first step was to assign shades of grey only to the tips of the branches for which T_{max} values were available. The tips of the branches for which we have no information are left as thin black lines.

Maximum growth temperature is determined by the fundamental biochemistry of the organism and is therefore a highly conserved feature. This motivates the plausible assumption that one could extend the shade of grey of the tip back through time at least to the closest node (a node is where two or more branches merge). When the branches from a node have the same shade, we extend the same shade one node closer to the root and so on. When the branches from a node do not share the same shade, we extend the shading towards the root with the shade of the branch that matches the next node. The end result is a phylogenetic thermometer. The self-consistency of this thermometer is not built in to the procedure to make it. For example, the phylogenetic position of *Archaeoglobus* among the Archaea is an inconsistency. Its T_{max} is higher than the T_{max} of two ancestral nodes closer to the root. If the T_{max} of an organism were not a highly conserved trait, instead of yielding a highly consistent picture of high T_{max} near the root and low T_{max} further away, we would have a mosaic of greys with no identifiable pattern. There would be many more examples of *Archaeoglobus*-like inconsistency.

The simple picture that emerges from this analysis is of mesophiles evolving from hyperthermophiles and gradually adapting to cool environments. It is easily tested since it makes the simple prediction that branches ending in black tips (for which we have no information) will have T_{max} values consistent with the shade of grey closest to their tips (or slightly cooler). Exceptions like *Archaeoglobus*, for which this is not the case, will be rare or be indications of misplacement in the tree or an indication that *Archaeoglobus*' ability to tolerate heat is a recent adaptation, possibly due to substantial lateral transfer of genes responsible for heat tolerance. The most deeply rooted mesophile, "Gp. 3 low temp" (in the Archaea) is also a candidate for being misplaced in the tree. The few discrepancies between this 16S tree and other trees may correlate with these exceptions.

Figure 3 shows the Eukarya connected to the hyperthermophilic root with a transition from hyperthermophile to thermophile to mesophile. However, the distances from the root at which the ancestors of Eukarya became thermophiles and then mesophiles is unknown. Thus the positions of these transitions are guesses. An endosymbiotic origin for Eukarya even calls into question the idea that these transitions occurred along this branch (Martin and Russell, 2002). This phylogenetic thermometer would be completely undermined if it could be established that the ancestors of hyperthermophiles were mesophiles who evolved a heat tolerance as they learned to adapt to the hottest aqueous environments on Earth (Valley et al., 2002; Matte-Tailliez et al., 2002). A comparison of the antiquity of heat-shock proteins and 'cold-shock' proteins could test this idea; so might further comparative analyses of G + C content.

If life lived in cool environments 3 - 4 billion years ago, then we could expect mesophiles near the root. Either there were no such environments, or life did not live in them or they have left no survivors. A close examination of this idea's most plausible representative, "Gp. 3 low temp", is called for. Until more definitive tests are made, the simple interpretation of this phylogenetic tree, as a remarkably self-consistent thermometer based on a crude but quantifiable gradient in T_{max}, gives us one of the few methods we have for estimating the thermal history of life and the Earth over the past few billion years.

In Fig. 2 of Schwartzman and Lineweaver (2003) we plotted T_{max} values as a function of branch length from the root to the nearest node of the four main trunks. A strong correlation was found. The self-consistent gradient of grey in Fig. 3 from hot root to cold branches is an alternative way to represent the same correlation.

5. Time Calibration of the Phylogenetic Thermometer

Since thermometers on billion year time scales are so rare and approximate, it is important to compare the ones we do have. Our confidence in these thermometers will depend on their consistency. To turn this phylogenetic thermometer into a thermal record of the Earth's history we need to establish a connection between absolute time and the nodes in the main branches of the phylogenetic tree. We will consider the nodes along four main trunks each starting from the root and extending respectively to *Homo* in the Eukarya, *Planctomyces* in the Bacteria, *Methanospirillum* in the Euryarchaeota and "marine Gp. 1 low temp" in Korarchaeota/Crenarchaeota. Along each of these main trunks are nodes. The closer the nodes are to the root, the earlier the time associated with that node. This is a widely accepted interpretation of the relative chronology of these nodes. To convert these relative chronologies into absolute chronologies we will make several plausible assumptions. The end result will be a plausible, approximate time calibration of these nodes with the virtue of being explicit and therefore open for improvement.

First we assume that the LCA lived 4 billion years ago. This is based on the phenotypic similarity of 3.5 Gyr old fossils to extant organisms (e.g. Schopf, 1994) with a small correction for incompleteness (for an analysis of correction factors applied to an incomplete fossil record see Martin, 1993). Thus, we are assuming that the distance from the root to any tip corresponds to the same duration: 4 Gyr. This gives us a measure of how variable the molecular clocks in the various branches are since the dispersion in the distances from the root to the tips of the tree is a measure of the dispersion in the speeds of the molecular clocks.

We also adopt several absolute calibration points. For example, much work has gone into establishing the time of the most recent common ancestor of animals, plants and

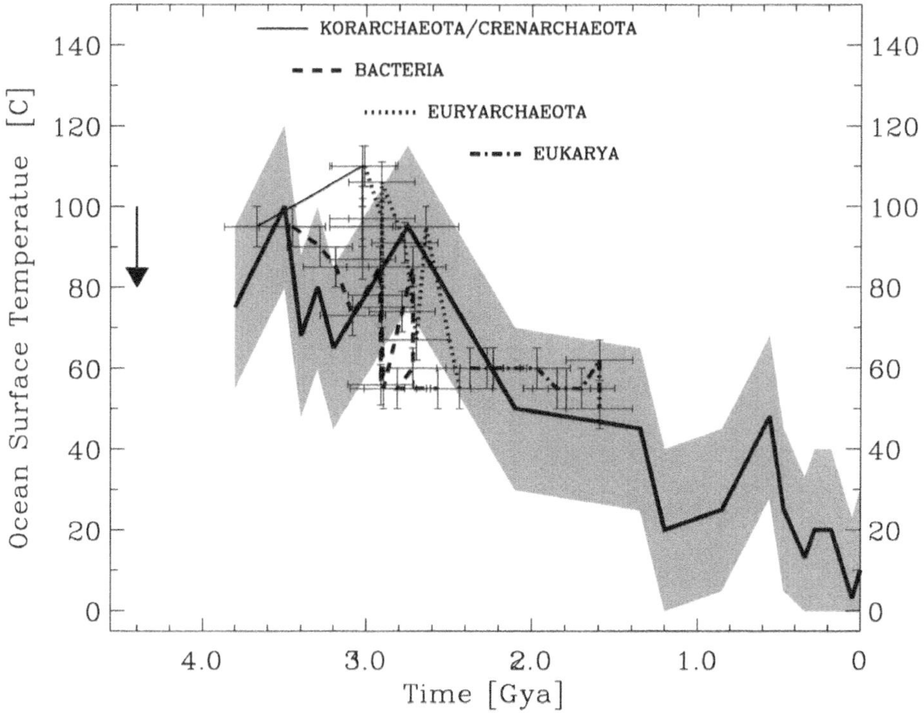

Figure 4. Temperature history of the surface of the Earth based on ^{18}O chert values. The thick solid line is a slightly smoothed version of data from (Knauth, 1992 and Knauth and Lowe, 2003). The grey band is an error estimate. Superimposed on this are the estimated times and T_{max} values of the nodes along the four main trunks of the tree in Fig. 3 (see text).

fungi (node in Eukarya of Fig. 3 where *Homo*, *Zea* and *Coprinus* merge). We adopt 1.6 billion years ago as the time of this node (Hedges, 2002). We then have the constraint that the distance from this node to the root corresponds to 2.4 Gyr (= 4.0 – 1.6). We then make a simple linear mapping between the nodes along this distance and the 2.4 Gyr time interval. Although this mapping has the virtue of simplicity, Woese and others have pointed out a general trend for molecular clocks to slow down with age. We have considered several non-linear scalings to correct for this trend. These corrections compress the horizontal distances between the points on the left side of Fig. 4 but do not change the main result. We have also used values between 2.4 and 2.8 Gyr for the emergence of cyanobacteria (*Synechococcus*/chloroplast node). In this way we obtain a plausible but rough time calibration of Fig. 3 using branch lengths from the root to various nodes along the four main trunks mentioned above. The results from these four trunks are plotted in Fig. 4.

6. Comparison of the Phylogenetic and ^{18}O Chert Thermometers: Implications for the Origin of Life.

In Fig. 4 we compare the phylogenetic thermometer of the temperature of the Earth over the past three to four billion years to the ^{18}O chert thermometer (Knauth, 1992; Knauth and Lowe, 2003). According to the chert data, over the last four billion years ocean surface water has cooled from ~90 C to ~10 C. From 3.5 Gyr ago to 1.5 Gyr ago it decreased from 80 C to 50 C. Over this same time interval, the T_{max} of newly evolved organisms decreased in approximately the same way.

Fig. 3 shows clearly that the roots of the tree of life are hyperthermophilic. Fig. 4 shows an easily identifiable correlation between decreasing temperatures based on ^{18}O in chert and the decreasing T_{max} values of newly evolving organisms at the nodes of the four main trunks in Fig. 3. These two thermometers seem to be consistent. Their similarity helps explain why mesophiles seem to have thermophile ancestors but not vice versa. If we take the ^{18}O chert thermal history seriously, 4 billion years ago there were no environments for mesophiles to live in. If this is true then deep molecular phylogenies of thermophiles, mesophiles and psychrophiles will continue to show a pattern of ancestral thermophily with mesophily and psychrophily as derived adaptation. The evolution from hot to cold could be interpreted as hot environments being more difficult to adapt to than cold ones; an asymmetry in adaptive ability. However, the thermal history from ^{18}O chert suggests that there were only hot environments and these were already occupied. The new challenge for life was the adaptation to cooler places as they emerged. This seems to be the most straight-forward reading of the good but rough correlation between the phylogenetic and ^{18}O chert thermometers. However, although the absolute dates of the ^{18}O thermometer are well established, their interpretation as a measure of ocean surface water temperature is controversial (Des Marais, Kasting, private communication).

We have strong evidence on Earth that life started out simple and that the simplest, most basic pieces of our metabolic pathways are the most ancient. We assume that if there is life on other planets that it too started out simple. Therefore the best candidates for universal features of life are the most ancient features of life that we are able to identify here on Earth. We should expect the most fundamental pathways close to life's origin to have been explored by life elsewhere.

The deepest rooted divergence in the tree of life in Fig. 3 is between the Archaea and the Bacteria. Although we have no examples of extraterrestrial life, we believe it is meaningful to speculate about whether a similar divergence occurs universally in extraterrestrial trees of life. Psychrophiles (cold loving bacteria) have membrane lipids rich in unsaturated fatty acids making the membranes more fluid and flexible at low temperatures. Thermophiles have membrane lipids rich in saturated fatty acids giving them stability at high temperatures since saturated fatty acids form much stronger hydrophobic bonds (Madigan et al., 1997). Hyperthermophiles, virtually all of which are Archaea, do not contain fatty acids in the lipids of their membranes but instead have hydrocarbons of various lengths; phytane with ether linkages (rather than ester linkages in Bacteria). Are these the only two possibilities for membranes to emerge in high temperature environments? This ether/ester, temperature-related structural difference between Archaea and Bacteria is a good candidate for a universal divergence – a divergence that may be quasi-deterministic – a step in the emergence of life that may be less deterministic than the features we have been discussing but more deterministic than downstream contingencies like the eukaryotic divergence ~2.5 billion years later between plants, animals and fungi.

7. First it was Hot, then it got Cool

The transition from hot to cool happens to cups of coffee. It happens to planets and it happens to life (Figs. 3 and 4). It is either the case that life emerges as soon as it can (in which case it will emerge when the environment is still at the hot end compatible with liquid water) or the emergence of life depends on a hot environment (in which case it will emerge *only* while the environment is at the hot end compatible with liquid water).

The connections between temperature and life are so fundamental that we can use them as guides in our estimates of what life forms we can reasonably expect to exist beyond the Earth. Hyperthermophiles may dominate not only the terrestrial tree of life but the trees of life of all planets in the universe.

Extant life is no longer a passive inhabitant of a given ambient temperature. The relation between life and temperature has become a quasi-deterministic one in the sense that although temperature has played a dominant role in constraining life, life has also been able to modify the temperature within limits set by the deterministic processes of planet formation and the evolution of stellar luminosity (Schwartzman 1999). The nature of this temperature regulation is a hotly debated central conjecture of the Gaia hypothesis (Schneider and Boston, 1991).

It may be the case that we are oblivious to the thermal straight-jacket that our biochemistry has to wear. The apparent quirkiness of biological evolution may only reflect our obsession with the changing fashions of phenotype rather than our biochemical foundations. Maybe the evolution of biochemistry today is just as thermally constrained and deterministic as it was 4 billion years ago. Maybe that is why our metabolic paths are so ancient.

8. References

Allègre, C.J., Manhès, G. and Göpel, C. (1995) The age of the Earth. Geochim. Cosmochim. Acta 59: 1445-1456.

Aragno, M. (1992) Aerobic, chemolithoautotrophic, thermophilic bacteria. In: J.K. Kristjansson (ed.) *Thermophilic Bacteria* CRC Press, Boca Baton, pp. 77-103.

Brock, T.D. (1978) *Thermophilic Microorganisms and Life at High Termperature,* Springer-Verlag, New York.

Brock, T.D., Madigan, M.T., Martinko, J. M. and Parker, J. (1994) *Biology of Microorganisms*, 7th Edition, Prentice Hall, Upper Saddle River, New Jersey

Canup, R.M. and Asphaug, E. (2001) Origin of the Moon in a giant impact near the end of the Earth's formation. Natur 412: 708-712.

deDuve, C. (1995) *Vital Dust,* Basic Books/HarperCollins, New York.

DiGiulio, M. (2000) The universal ancestor lived in a thermophilic or hyperthermophilic environment. J. Theor Biol. 203: 203-213.

Gogarten-Boekels, M., Hilario, E. and Gogarten, J.P. (1995) The effects of heavy meteorite bombardment on the early evolution - the emergence of the three domains of life. Orig. Life Evol. Biosph. 25: 251-264.

Grimaldo, S. and Cammarano, P. (1998) The root of the universal tree inferred from anciently duplicated genes encoding components of the protein-targeting machinery. J. Mol.Evol. 47: 508-516.

Halliday, A.N. (2000) Terrestrial accretion rates and the origin of the Moon. Earth Planet. Sci. Lett. 176: 17-30.

Halliday, A.N. (2001) Earth Science: In the beginning... Nature 409: 144-145.

Hartmann, W.K. and Davis, D.R. (1975) Satellite-sized planetesimals and lunar origin. Icarus 24: 504-515.

Hartmann, W.K., Ryder, G., Dones, L., and Grinspoon, D. (2000) The time-dependent intense bombardment of the primordial Earth/Moon system. In: R.M.Canup and K. Righter (eds.) *Origin of The Earth and Moon* University of Arizona Press, Tucson, pp. 493-512.

Hedges, B. (2002) The origin and evolution of model organisms. Nature Reviews Genetics 3: 838-849.

Iwabe, N., Kuma, K., Hasegawa, M., Osawa, S. and Miyata, T. (1989) Evolutionary relationship of archaebacteria, eubacteria, and eukaryotes inferred from phylogenetic trees of duplicated genes. Proc. Natl. Acad. Sciences, USA 86: 9355-9359.

Kimble, L.K., Mandelco, L., Woese, C.R. and Madigan, M.T. (1995) *Heliobacterium modesticaldum*, sp. nov., a thermophilic heliobacterium of hot springs and volcanic soils. Arch. Microbiol. 163: 259-267.

Knauth, L.P. (1992) Origin and diagenesis of cherts: an isotopic perspective. In: N. Clauer and S. Chaudhuri (eds.) *Isotopic Signatures and Sedimentary Records, Lecture Notes in Earth Sciences* #43, Springer-Verlag, N.Y. pp. 123-152.

Knauth, L. P. and Lowe, D.R. (2003) High Archean climatic temperature inferred from oxygen isotope geochemistry of cherts in the 3.5 Ga Swaziland Supergroup, South Africa. Bull. Geol. Soc. Am. (in press).

Kolb, E.W. and Turner, M.S. (1990) *The Early Universe* Academic Press Redwood City, CA.

Kristjansson, J.K. and Stetter, K.O. (1992) Thermophilic bacteria. In: J.K. Kristjansson (ed.) *Thermophilic Bacteria* CRC Press, Boca Baton pp. 1-18.

Kristjansson, J.K. (ed.) (1992) *Thermophilic Bacteria* CRC Press, Boca Baton

Lewis, J.S. (1997) *Physics and Chemistry of the Solar System,* Academic Press New York, p. 116.

Lineweaver, C. H. (1999) A younger age for the universe. Science 284: 1503-1507.

Lineweaver, C. H. (2001) An estimate of the age distribution of terrestrial planets in the universe: quantifying metallicity as a selection effect. Icarus 151: 307-313.

Madigan, M.T., Martinko, J. M. and Parker, J. (1997) *Brock Biology of Microorganisms*, 8th Edition, Prentice Hall, Upper Saddle River, New Jersey, p. 166.

Martin, R.D. (1993) Primate origins: plugging the gaps. Nature 363: 223-234.

Martin, W. and Russell, M.J. (2002) On the origins of cells: a hypothesis for the evolutionary transitions from abiotic geochemistry to chemoautotrophic prokaryotes, and from prokaryotes to nucleated cells. Phil. Trans. R. Soc. Lond. B 358, 59-85.

Matte-Tailliez, O., Brochier, C., Forterre, P. and Philippe, H. (2002) Archaeal phylogeny based on ribosomal proteins. Mol. Biol. Evol. 19: 631-639.

Pace, N.R. (1997) A molecular view of microbial diversity and the biosphere. Science 276: 734-740.

Schmidt, J.M. and Starr, M.P. (1989) *Planctomyces*. In: J.G. Holt (ed.) Bergey's Manual of Systematic Bacteriology, vol. 3, The Williams and Wilkins Co. Baltimore, pp. 1873, 1946-1958.

Schneider, S.H. and Boston, P.J. (eds.) (1991) *Scientists on Gaia*, MIT Press, Cambridge, MA.

Schopf. J.W. (1994) The oldest known records of life: Early Archean stromatilites, microfossils, and organic mattter. In: S. Bengston (ed.) *Early Life on Earth,* Columbia University Press, New York pp. 193-206.

Schwarz, A. (ed.) (2003) Origin of life and evolution of the biosphere, special edition of abstracts from the Oaxaca, Mexico meeting of the International Society for the Study of the Origin of Life (in press).

Schwartzman, D. (1999) *Life, Temperature, and the Earth: The Self-Organizing Biosphere*, Columbia University Press, NewYork.

Schwartzman, D. and Lineweaver, C.H. (2003) Precambrian surface temperatures and molecular phylogeny. In: R. Norris, C. Oliver and F. Stootman (eds.) *Bioastronomy 2002: Life Among the Stars* ASP Conf Series (in press).

Shock, E. I., Amend, J.P., Zolotov, M.Y. (2000) The early earth vs. the origin of life. In: R.M. Canup and K. Righter (eds.) *Origin of the Earth and Moon* University of Arizona Press, Tucson pp. 527-543.

Sirevag, R. (1992) Thermophilic autotrophs. In: J.K. Kristjansson (ed.) *Thermophilic Bacteria,* CRC Press, Boca Baton, pp. 195-219.

Sleep, N.H., Zahnle, K.J., Kasting, J.F. and Morowitz, H.J. (1989) Annihilation of ecosystems by large asteroid impacts on the early Earth. Nature 342: 139-142.

Sleep, N.H., Zahnle, K. and Neuhoff, P.S. (2001) Initiation of clement surface conditions on the early Earth. Proc. Natl. Acad. Sci. USA 98: 3666-3672.

Stetter, K.O. (1996) Hyperthermophilic procaryotes. FEMS Microbiol. Rev. 18: 149-158.

Stetter, K.O. (1998) Hyperthermophiles: isolation, classification, and properties. In: K. Horikoshi and W.D. Grant (eds.) *Extremophiles: Microbial Life in Extreme Environments* Wiley-Liss, New York pp. 1-24.

Stetter, K.O. (1999) Extremophiles and their adaptation to hot environments. FEBS Letters 452: 22-25.

Valley, J. W., Peck, W.H., King, E.M. and S.A. Wilde (2002.) A cool early Earth. Geology 30: 351-354.

Wiegel, J. (1992) The obligately anaerobic thermophile bacteria. In: J.K. Kristjansson (ed.) *Thermophilic Bacteria,* CRC Press, Boca Baton, pp. 105-184.

Woese, C., Kandler, O. and Wheelis, M.L. (1990) Towards a natural system of organisms: Proposal for the domains Archaea, Bacteria, and Eucarya. Proc. Natl. Acad. Science, USA 87: 4576-4579.

Biodata of **Yuval Ne'eman** author of ***"Paradigm Completion for Generalized Evolutionary Theory with Applications to Epistemology"***

Dr. Yuval Ne'eman was born in Tel Aviv (1925) and studied mechanical and electrical engineering at the Technion (Haifa) where he obtained his engineering degree in 1945. He joined the Hagana and stayed on in the Israel Defense Forces for 12 years, fighting as a field commander and later as a Vice Chief of Operations at the High Command. As colonel, Head of Defense Planning he laid the foundation of several contributions to the Defense forces. Head of Defense Planning he laid the foundations of the strategic doctrine applied by the IDF in 1952-1968. In 1958 Ne'eman returned to science, studying physics at Imperial College in London, while serving as Israel's Defense Attache in the UK, obtaining his PhD in 1962. Around 1960 the situation in Elementary Particle Physics was chaotic, with many new particles being discovered (about 100 by 1961) and no understanding of the pattern. Early in 1961, Ne'eman published his Unitary Symmetry classification (now known as "flavor SU(3)" or "The Eightfold Way") of the *hadrons*, namely the particles experiencing the "Strong' nuclear force, including both *baryons* (protons, neutrons, or hyperons such as Λ^0, Σ^+, Σ^0, Σ^- as etc – *material* particles, obeying Fermi-Dirac quantum statistics, i.e. the Pauli Principle) and *mesons* (radiation quanta, obeying Bose-Einstein quantum statistics). Ne'eman's scheme thus did for the particles what Mendeleyev's Periodic Chart of the Elements did for Chemistry or Linneus' for the Animal and Plant Kingdoms. The same scheme was independently discovered, though somewhat later and unpublished), by M. Gell-Mann. Early in 1962, and with H. Goldberg, Ne'eman conceived a mathematical method as Rutherford's atom explained the Periodic Chart) reproducing the classification and thus providing a structural model explaining it (just The structure involved fractionally charged constituents, the *quarks* (a term introduced by Gell-Mann, when, one year later, he and G. Zweig further developed the physical model). Dr. Ne'eman contributed to important further advances in particle physics, cosmology, astrophysics, and epistemology.

Since 1977, he has developed an interest and a parallel line of work, generalizing the theory of Evolution and formulation a universal paradigm, then also applying it to various areas, especially Social Anthropology and Evolutionary Epistemology.

Professor Ne'eman was the founder and Head of the Department of Physics and Astronomy at Tel-Aviv University (1965-1972), President of Tel-Aviv University (1971-1975), and served as Director of the Mortimer and Raymond Sackler Institute of Advanced Studies in that Institute (1979-1997). In 1969 he established the School of Engineering as founding dean (in 1997 he was elected President of the Israel Association of Engineers). He also founded the (1968) Center for Particle Theory at the University of Texas (Austin). He founded the Israel Space Agency (1983) and has chaired it since, and served on Israel's Atomic Energy Commission (1965-1984) and held a position of Scientific Director in the IAEC Soreq Establishment (1961-1963). Dr. Ne'eman was Israel's Chief Defense Scientist in 1974-1976. He served as President of Israel's Bureau of Standards in 1972-1976 and chaired the Steering Committee to the Med-Dead Conduit Project in 1977-1983. He was Israel's first Minister of Science and Development 1982-1984, then again in 1990-1992, when he also served as minister of Energy.

J. Seckbach (ed.), Origins, 249–260.

Professor Ne'eman has published over three hundred and fifty scientific papers and twenty books. He is a Member of Israel's National Academy of Sciences, a Foreign Associate of the National Academy of Sciences of the USA, A foreign Honorary Member of the American Academy of Arts and Sciences and a Member of the International Academy of Astronautics and several other Academies and Learned Societies. He has been awarded the Einstein Medal (Washington, 1969), the Wigner Medal (1982) the Israel prize (1969), and Birla Science Award (Hyderabad, 1998).

E-mail: **matildae@tauex.tau.ac.il**

PARADIGM COMPLETION FOR GENERALIZED EVOLUTIONARY THEORY WITH APPLICATION TO EPISTEMOLOGY

YUVAL NE'EMAN
School of Physics and Astronomy
Tel-Aviv University, Tel-Aviv, Israel 69978

1. Evolution Fully Generalized

Evolution was discovered and named in the XIXth Century within the (Darwinian) *biological* setting, but in the XXth, it spilled over in all directions. A solid* evolutionary theory of *nucleosynthesis* [Bethe (1967); Burbidge et al. (1987); Pagel (1997)], coupled together with *astrophysical* evolution, now explains the making of the *material content* of the observed universe, both as to the "cooking" of the chemical elements of which it is composed and as to the formation of the astronomical features (galaxies, stars, planets etc) in which this matter is organised. Further on in the same *past* direction, we encounter the Big Bang and now even have several (more or less speculative) theories of the evolutionary *making of universes*. [Carter (1993), Barrow and Tipler (1986), Guth (1983), Linde (1990), Harrison (1995), Smolin (1992)]. In the opposite (more *complex*) direction, *biological* evolution, after reaching man, has been continued by the evolution of human *societies* [Ne'eman (1980)] or *socio-anthropological* evolution; in this sector, the evolutionary levels are characterized by *technologies* (from the *Paleolithic* to the Age of *Information Technology*) as befits "*man the toolmaker*". This lead M. Bradie [Bradie (1986)] and Ruse [Ruse (1986)] to extend the application of the method to *Epistemology*, i.e. to the *making of science*, the latter being taken as the extension of man's *tools* (*literal Evolutionary E.)* .

D.T. Campbell and K. Popper [Popper (1972), Campbell (1974)] then realized that what was emerging was the *universality* of the evolutionary mechanism, whether the evolving population be universes, stars, living beings or even *ideas*! (*analogical E.E.*) – it all occurs via *blind variation* and *selective retension*, the essence of evolution. Note that some doubts about the "blindness" of the variations should have been assuaged after our reallocation of that role to *serendipity* [Kantorovich and Ne'eman (1989); Ne'eman (1993); Ne'eman (1999); Kantorovich (1993), and Kantorowitz (2000)].

----

* I use the adjective "solid" to imply that it has been checked in the laboratory.

2. Entropy: Gravity as Model

Earlier, Schroedinger [Schroedinger (1967)] had considered the overlap with thermodynamics and pointed out that evolution creates *order*, or *negative entropy* (which might have appeared paradoxical). However, as it involves *dissipative* systems, the *environment's rise in entropy more than compensates for this local loss,* thus maintaining the Second Law. We now have proof that a similar "compensation" mechanism occurs at the *fundamental* level, with the generation of order in (fermionic) matter being compensated by disorder in the (bosonic) field which supplies and replaces the dissipating energy. Growing a crystal by deposition in a saturated solution is a classical example, with gravity (and some atomic contributions to surface tension) supplying the ordering energy with a related generation of compensating (*positive*) entropy, "paying the price" of order.

Classical thermodynamics covers a physical region in which the short-ranged (nuclear) interactions (Strong and Weak) have been integrated out, together with the atomic and molecular electromagnetic bindings, leaving us with just *kinetic energy*, plus – perhaps -- some weak chemical potentials. All other parts of the fundamental Hamiltonian are included *indirectly* through the masses, angular momenta and various structural parameters at the molecular level. It is under such an assumption of *no-interaction* conditions that we present the elementary illustration demonstrating the relationship between entropy and the arrow of time: two pictures of a group of molecules, one [S] showing them *spread out* over a large volume, the other [D] showing them all in one relatively *dense* bunch. [D] is the relatively *more ordered* set up, [S] is the *less ordered* one.

With *chance* as the only intervening factor, and as the probability of many molecules accidentally converging towards and arriving at the same point is negligible, we conclude that [D] is the *earlier* take, [S] the *later* one, reached naturally as a result of the molecules' random motion.

It is instructive, at this point, to review the findings with respect to the only other identified contribution to *irreversibility,* the *Bekenstein* component [Bekenstein (1973)], corresponding to the action of *gravity* in its strongest *attractive* phase, namely in the formation of *black holes*, and thus in a region of phase space which is far from thermodynamics.

At first sight, it would seem that here too, order is generated, with all masses converging as in [D] onto the singularity at the center of the black hole (if it is a spherical one). Thus, as for [D] in our previous discussion, a black hole would represent the generation of order, i.e. negative entropy, a la Schroedinger. This approach, however, is wrong in that it follows the history of *matter* solely (as represented by the *energy--momentum tensor current density)* coupled to the gravitational field in the Hamiltonian and does not consider radiation and the gravitational field itself, with the tensions it exerts in its strong binding. I have shown elsewhere [Ne'eman (2000)] the special-relativistic approximation to Black-Hole physics, in which we assume that a black-hole is a *stellar object whose gravitational binding's (negative) self-energy has managed to "eat up" the entire invariant mass energy* (*k a numerical factor of order 1*) $Mc^2 - (k\, G_N\, M^2)/r = 0$, yielding the GR result for the Schwarzschild radius when $k \approx ½$. Considering the matter content solely would thus be a grave error. Let us consider this question using what has been learned about Black Hole Physics.

First, there is J. Bekenstein's intuitive formula [Bekenstein (1973)], namely *positive* entropy, proportional to the area of the black hole's envelope. This result was tested and confirmed by S. Hawking, using thermodynamical considerations and quantum barrier-penetration [Hawking (1975)]. What is this entropy? Full insight indeed came after Bekenstein's formula was *rederived* by C. Vafa and A. Strominger *within Quantum Gravity*, as described by String Theory [Strominger and Vafa (1996)]. This revealed where the *disorder* was: in the organisation of the *quanta of the gravitational field* ! The states counted by Strominger and Vafa are solitons and *topological realizations of the gravitational field* in its binding action! Thus, there is indeed a *negative* contribution to the total entropy coming from the "improvement" in the orderliness of the nuclei and electrons now imprisoned within the black hole, or even better, now stuck on its envelope (in view of the *holographic* interpretation of the conservation of quantum information [Susskind (1995)]. As against this negative increment, there is a (larger) positive contribution originating in the tensions created within the black hole's gravitational field quanta.

Note that in Cosmology, with yet another *arrow-of-time* (linked with the cosmological expansion), it was assumed that the dense state [D] being the most ordered, with the lowest entropy, a *contracting universe* would produce negative entropy, in violation of the Second Law. Prior to Bekenstein's identification of the black hole's contribution [Bekenstein (1973)], the conclusion used to be that *a collapsing universe would have an inverted time-arrow* and would then *become an expanding* one [Ne'eman (1970); Aharony and Ne'eman (1970)]. This picture ignored the contribution of the main actor in cosmology, namely *gravity*.

3. Evolution and Entropy: Measures of Complexity

Refocusing on Evolution, we are clearly again in regions in which the Hamiltonian contributes through *binding* components. These range from the role of Quantum Chromodynamics in *nucleosynthesis* to the biophysical contributions (mostly electromagnetic) making up nature's own *genetic engineering*. As with the area of a black hole's surface in the Bekenstein formula, we have to *identify* a *structural complexity* function, a *time-arrowed quantity* characterizing the action of the evolutionary *drive.*

Moreover, as against Schroedinger's negative entropy which relates to matter, the complexity function will represent the positive entropy produced by the tension within the binding fields.

Two approaches have been used to date, the more abstract [Li and Vitanyi (2000); Chaitin (1975); Bennett (1988); Fogelman (1991); Goertzel (1992); Szwast et al. (2002); Becker (2002)], inspired by Kolmogorov's treatment of information, and given by the length of the shortest program describing the system - and a pragmatic one, used in the biological domain, inspired by genetic studies. Here, effective measures of *complexity* have been abstracted from *experimental* requirements, e.g. in cases involving two species deriving from the same ancestry, estimating the time elapsed since that branching. This is done by *counting* the *number of mutations* which are not common to the two species, a linear procedure. In *nucleosynthesis*, it appears obvious that parametrizing the growth of *complexity* will involve the *advance* in the *Atomic Mass* and *Charge* Numbers reached, perhaps the path in the Segre diagram. As to the sign, here again it probably indicates positive entropy originating in the binding fields.

Once a useful and informative *level of complexity* has been properly defined, it should be appended to the *Second Law*, in an extension of thermodynamics, or better to a related formulation in Shannon's *Information Theory.*

4. Extinctions and a Balanced Evolutionary Paradigm

The importance and dimensions of the massive *extinctions* [Becker (2002)] occurring on the border between two geological eras has become clear in recent years. Survival of the *fittest* is replaced by survival of the *luckiest*. More specifically, it would appear that man's presence on earth is no less due to such *opportunities* created for the ("*scalawag*"**) small mammals by the catastrophic *extinction* of the *unlucky* dinosaurs -- than to the normal evolutionary *survival of the fittest.*

Darwinian evolutionists have essentially been biologists and have thus relegated the catastrophic changes in the environment to the boundary conditions. Both mutations and environmental catastrophic extinctions, however, are tychic*** interventions, except that mutations occur in a cybernetic program in very small steps (through errors entering in the routine procedure of copying the DNA molecules.) whereas catastrophes are single one-time rare events. Other examples of generalized evolution also point to the importance of the extinctions, as we shall see. We reformulate the evolutionary paradigm accordingly.

The components are (1) a population of N individual systems S -- with (2) each S controlled by its cybernetic program P{S} , with S existing in (3) an environment V, i.e. P{S}@V. P undergoes (4) a routine R^P which exposes it to random errors, so does (5) the environment V, which undergoes R^V. Chance T thus enters through two gates, namely M: T# (R^P{S}) → P'{S'} (a mutation in the system; "type M" for short) and E: T#(R^V) → V' (a "passive" mutation, a change in the environment, possibly a potential extinction; a "type E" mutation, for clarity). The new state of affairs is S'@V or S@V' which may or may not be as good and stable as the original S@V, depending on (6) the selection criteria C, acting like a sieve. Thus C [S'@V] → 0 describes a bad type M mutation, while C [S'@V] → S' is a good or indifferent one, C [S@V'] → 0 is a bad type E mutation, an extinction.

The characteristics of evolutionary processes are (a) the creation of order, (b) increasing complexity, (c) dissipation, (d) teleonomy (or an *apparent* teleology), (e) tinkering. We have discussed the inter-relationship between the first two. *Dissipation* implies being far from equilibrium, thus requiring a steady replacement procedure for the matter and energy exiting the system. A relatively simple example would be represented by a *tornado*, which is also a good example of the (rotational) order generated and of the external over-compensation in entropy (as experienced by the inhabitants of Japan or of the Caribbeans).

** scalawag: undersized and "worthless" animal profitting from a catastrophe which hit the dominating predators. In the American civil war, used of white republican southerners during Reconstruction.

*** tychic: "blind", random, derived from Tyche, Greek Goddess of Chance

Evolution always gives the impression of working towards an aim, of an intention behind any sequence. Surveying the previous steps S(n), S(n-1), S(n-2), .. ,S(n-m) while advancing from S(n) to the next, i.e. S(n+1) , one comes to think of the sequence of m+1 steps, as a machine fed with S(n--m) and now producing S(n+1). Kantorovich introduced the term "tinkering" to emphasize the "unengineered" character of the production.

Flying animals (birds) were not designed and produced – as airplanes were. Instead, the existence of feathers, which evolved as a warming device on some reptiles led to wings, etc. In a sense, tinkering also explains the difference between teleology (where there is an a-priori intention to produce the final state) and teleonomy (the real picture, in which the final state is selected by the small steps along the chain).

5. The Evolution of Human Society - the Age of Information as example.

Here the evolutionary levels are characterized by technologies – from the Paleolithic, etc., the Ages of Bronze and of Iron etc.. to the present Age of Information Technology – and I have shown elsewhere that the tychic elements enter via scientific discovery [Ne'eman (1980); Kantorovich and Ne'eman (1989); Ne'eman (1993); Ne'eman (1999)]. Let us analyze one such example.

In 1905, mathematical logicians were exposed to ridicule everywhere. Impressed by advances in logic (such as Boolean algebra) and by Cantor's Set Theory with its provision of a method of handling infinities, Russell and Whitehead had initiated an ambitious program [Crossley et al. (1972)] of axiomatization for the entirety of Mathematics – but the whole edifice suddenly seemed to collapse when they hit the "Russell-Whitehead Paradox", namely ("a includes b" means "b is a member of a") *does the set of all sets which do not include themselves include itself?* #

The unhappy logicians were forced back to *square one*, checking all their steps; they ended up in 1921 with the *Zermelo-Fraenkel* set of axioms and some very precise open questions. K. Goedel's *incompleteness* theorems threw additional light on the issues; further improvements were introduced by two doctoral students, J. v. Neumann (Budapest), who improved on the axioms, and A. Turing (Cambridge). The latter was asked, in this context, to check the concept of "computable functions" used by Goedel in his proof. Turing solved the problem by conceiving the Turing Machine, a programmable computer, which he later developed into a *universal Turing machine* – basically characterizing all present computers.

At the same time, Turing thereby also "solved" a famous problem in the Foundations of Mathematics, namely Hilbert's *Entscheidungsproblem* – by demonstrating it to be unsolvable (a conclusion independently reached by A. Church). The Second World War had meanwhile started and Turing was mobilized and assigned with other mathematicians and physicists to the Blechley Park center, where the Allies were trying to break German codes. On the other hand, von Neumann, being Jewish, had fled the Nazis and was now in the USA, involved in helping the Armed forces – first in calculating counter-battery fire and later in the context of the Manhattan Project.

An easier analog would be represented by a city contracting with a barber to have him shave all male citizen *who do not shave themselves*. Should the barber shave himself? The reader should try to check the barber's options as limited by his contract.

The need for much calculation, whether in cryptography or in artillery-control, was met by the construction of primitive computers – the "Bombe" in the UK and ENIAC in the USA. Within a few years computers were everywhere, and especially when they became transistorized, after some twenty years of vacuum tubes. The Age of Information had dawned, and this was due to a chain of events triggered by the Russell-Whitehead paradox [Ne'eman (1995), Davis (1994), Aspray (1994)].

Note that in the early XIXth Century, Charles Babbadge, helped by Lady Lovelace had launched (with the financial support of he British gonernment) the building of a calculating machine and later conceived the idea of a programmable machine (but never got to build this "Analytical Engine" project) and yet it fizzled out – one reason being that the status of technology was inappropriate, heavy mechanical gears instead of the electronics of the XXth century. Similar reasons explain the dead ends reached by such smart developers as Descartes, Pascal or Leibniz (the latter developed the binary number system for this purpose). A hundred years after Babbadge, his project was revived at Harvard by H. Aiken *and completed around 1940, but only as a calculator, not as a "logic machine".*

6. High-Energy Physics and the World Wide Web

The European Centre for Nuclear Research was founded after World War II as a scientific venture supported and exploited by 14 European countries. Scientifically, it managed throughout the Cold War to keep multinational nonfederated Europe advancing neck to neck with the two main world-powers, the USA (working mainly from Brookhaven, Fermi Lab., Stanford and Berkeley) and the USSR (with Dubna and Serpukhov-Protvino), with Japan and China joining the race from time to time. One great difficulty was due to the multinational composition itself, whether linguistically or at the technical level – each country with its preferred computerware or devices. Having to work together in large mixed experimental collaborations starting from the Seventies, a unified and elegant information system evolved and by the early Eighties various technical communications were exchanged between labs and accelerator facilities, finally crystallizing in a system now universal and known as "e-mail". Other features were later borrowed by a USA government Agency, ARPA and its ARPANET. With some interest displayed by the USA presidency (a formal initiative by Vice President Al. Gore), this has evolved into the World Wide Web, perhaps the most characteristical feature of an Information-shaped modern world. Again, a research program which was totally unrelated to a world-information issue ended up resolving it [White (1998); Berger (1996); Kouzes et al (1996)].

7. Twentieth Century Epistemology has Strong (de facto) Evolutionary Elements

Reviewing epistemology as it was understood and formulated in the XXth Century shows that several of the features characterizing an evolutionary theory were identified. Four key ideas were launched:

[a] Thomas Kuhn (1922-1996) stressed the role of *normal science* ($[R^P(S)]$ in our characterization) and of the *paradigm* (which we identify with the theoretical environment [V]) [Kuhn (1962)], to be replaced sometime ($T\#R^V \rightarrow V'$) by revolutionary science R^V and its resulting in a change of paradigm [V'].

[b] Sir Karl R. Popper (1902-1994) emphasized *falsification**, which consists in testing the theory in a region of its parameter space in which it has not yet been tested [Popper (1935)]. This is a change of environment (in parameter space) possibly leading to extinction.

[c] Imre Lakatos (1922-1974) organized the history of science [Lakatos (1991)] according to research "programs". This is very much teleonomy, the viewing in retrospcctive. Example: the title of Gordon Frazer's book "The Particle Century" [Fraser (1998)] is fitting, as the electron was identified in 1897 and the top quark in 1996, with nothing earlier and nothing yet since – but it represents retrospective viewing and is very much a matter of *teleonomy*.

[d] Paul Feyerabend (b. 1924) has emphasized *chance* and summarized with the phrase "*anything goes*" [Feyerabend (1977)]. The *tychic* interventions are indeed the key to evolutionary processes.

8. The discoveries towards the beginning of the XXth Century

Any textbook on Modern Physics will cover these experiments, and we list some examples in our bibliography [Thornton and Rex (1993); Serway et al. (1997); Segre (1980)]. Several of these experiments involved cathode rays and we start with these.

[a] Johann Hittorf (1824-1914) discovers (1875) that *cathode rays consist of negatively charged particles*. This was a repeat of the experiments of Julius Pluecker (1801-1868) with better vacuum technology $[R^P(S)]$. However, an object which was left inside by mistake $[T_M]$ threw a "shadow" on the anode, showing that the rays originate in the cathode, i.e. they are electrically negatively charged [S']. [electrons]

[b] Wilhelm C. Roentgen's (1845-1923) discovery of *X-rays* (1895). The routine $[R^P(S)]$ involved further study of cathode rays, using an evacuated Hittorf tube, placed inside a black cardboard box. Roentgen had also prepared a set of screens made of paper with a layer of barium-platinum cyanide, a $[R^P(S)]$ phosphorescent material which he intended to use later. Suddenly, however, he saw one such screen, laying by mistake near the box $[T_M]$, starting to phosphoresce. He reached with his hand and saw his bones.

[c] Radioactivity (1896) Henri Becquerel hears Poincaré reporting Roentgen's discovery to the Academie des Sciences and conjectures that X-rays are linked to cathode rays in the same manner that luminescent radiation is excited by exposing the relevant mineral to sunlight [roughly correct!]. He starts (routine R^P) research on luminescence, selecting out of his father's collection a sample of pechblende (a mineral containing uranium $[T_M]$) as luminescent mineral, exposing it daily to the sun, with a well-protected photographic film underneath $[R^P]$. The film was guaranteed to be fully protected from the Sun for an entire day – and yet $[T_M]$ one could observe the mineral's silhouette on the film. After several days, the weather changed and Becquerel postponed further exposures, leaving the mineral and film in a drawer together $[T_M]$.

* invalidation

When the weather improved, Becquerel took mineral and film out of the drawer – and observed to his surprise that the film had been exposed, indicating the presence of a new type of (unstimulated) hard radiation [S']. Note that the conjecture (linking the production of X-rays by cathode rays to the mechanism behind luminescence) was correct, though irrelevant to the new discovery.

(d) The Michelson — Morley experiment (1887)

The aether was introduced as the carrier of electromagnetic waves, whose existence had just been experimentally verified by H. Hertz and were now believed to include light as one sector. A.A. Michelson's idea was to measure (R^V P(S)@V) the earth's velocity in the aether, based on Galilean-Newtonian Kinematics (the conceptual environment V). However, this stretched the use of such kinematics (e.g. addition of velocities) very much beyond any previous check (R^V) and resulted in an extinction, (T_E #V → 0) namely the extinction of "unlucky" Newtonian Mechanics as a universally valid theory and its replacement in 1905 by a "scalawag" – namely Special Relativity.

(e) Max Planck's analysis of the spectrum of Black-Body radiation (1900)

Here too, there was a classical body of theory, a conceptual environment V with predictions ("the ultra-violet catastrophe") in the untested region [R^V] of ever shorter wavelengths and ever higher frequencies. The result was T_E #V → 0, the extinction of classical thermodynamics and its replacement by ("scalawag") Quantum Mechanics.

9. Summary and Conclusions

We have discussed the *scope* of evolution, from the making of universes to the growth of ideas, and functions such as *complexity* and *entropy*, characterizing evolutionary processes. We have then modified the basic paradigm of evolutionary theory so as to include the massive extinctions in transition layers between geological eras in the evolutionary history of life on earth. We then applied the improved paradigm to case studies in Evolutionary Socio-anthropology and Evolutionary Epistemology. The improvement in the paradigm is especially important in the latter.

In the example of the discoveries at the turn of the Century, the first three cases, namely the discoveries of cathode rays, X rays and of radioactivity are type M mutations in which an accident in the performance of a set program reveals new unknown phenomena. The last two, namely the eather-drift experiment and the black body radiation spectrum, are type E (extinctions), i.e. a body of theory (the conceptual environment of the moment) is suddenly demoted and becomes limited in its applicability to a restricted region in parameter space.

10. References

Aharony, Y. and Ne'eman, Y. (1970) Time Reversal Symmetry and the Oscillating Universe, International Journal of Theoretical Physics **3**, 437-441.

Aspray, W. (1994) The Mathematical Reception of the Modern Computer: John von Neumann and the IAS Computer, In: R. Herken (ed.) *The Universal Turing Machine, a Half-Century Survey II*, Springer Verlag, New York, p. 166.

Barrow, J.D. and Tipler, F.J. (1986) *The Anthhropic Cosmological Principle*, Clarendon Press, Oxford and New York.

Bekenstein, J.D. (1973) Black Holes and Entropy, Physical Review **D5**, 2333-2346.

Bennett, C. (1988) Logical Depth and Physical Complexity, In: R. Herken, (ed.) *The Universal Turing Machine: a Half-century Survey*. Oxford University Press, London, United Kingdom, pp. 227-258.

Berger, E. (1996) Birthplace of the Web, Fermi News, **19**, #16, 7.

Bethe, H. (1972) Nobel Lecture, in: *Nobel Lectures in Physics 1963-1970*, Elsevier Publishing, Amsterdam, The Netherlands, 209-236.

Bradie, M. (1986) Assessing Evolutionary Epistemology, Biology and Philosophy **1**, 401-459.

Becker, L. (2002) Repeated Blows, Scientific American **286**, #3, 62-69.

Burbidge E.M., Burbidge, G.R. Fowler, W.A. and Hoyle F. (1987) Synthesis of the Elements in Stars, Reviews of Modern Physics **29**, 547.

Campbell, D.T. (1974) Evolutionary Epistemology, In: P.A. Schilpp (ed.) *The Philosophy of Karl Popper*, 1 La Salle: Open Court Publishers, 413-463.

Carter, B. (1993) The Anthropic Selection Principle and the Ultra-Darwinian Synthesis, In: F. Bertola and U. Curi (eds.) *The Anthropic Principle*, Cambridge University Press, United Kingdom, 33-66.

Chaitin, G.J. (1975) A Theory of Program Size Formally Identical to Informaiton Theory, Journal of the Association for Computing Machinery **22**, 329-340.

Crossley, J.N. et al (1972) *What is mathematical logic*? Oxford Uuniversity Press, United Kingdom.

Davis, M. (1994) Mathematical Logic and the Origin of Modern Computers, In: R. Herken (ed.) *The* ***Universal Turing Machine, a Half-Century Survey II***, Springer Verlag, New York, p. **137.**

Feyeraband, P. (1977) *Against Method*, Schocken Books, New York.

Fraser G. (ed.) (1998) *The Particle Century*, Institute of Physics Publishers, United Kingdom.

Fogelman, F. (1991) *Les Theories de la Complexite* (*l'oeuvre d'Henri Atlan*) Ed. Seuil, Paris.

Goertzel, B. (1992) Self-Organizating Evolution, Journal of Social and Evolutionary Systems **15**, 7-53.

Guth, A.H. (1983) Inflationary universe: A possible solution to the horizon and flatness problems, Physical Review **D23**, 347; (1987) in *The Inflationary Universe*, Addison-Wesley Publishing, Reading, Massachusetts.

Harrison, E. (1995) The natural selection of universes containing intellegent life, Quarterly Journal of the Royal Astronomy Society **36**, 193-203.

Hawking, S. (1975) Particle Creation by Black Holes, Communications in Mathematical Physics **43**, 199-220.

Kantorovich, A. and Ne'eman, Y. (1989) Serendipity as a Source of Evolutionary Progress in Science, Studies in the History and Philosophy of Science **20**, 505-529.

Kantorovich, A. (1993) *Scientific Discovery – Logic and Tinkering*, State University New York Press, Albany, New York.

Kantorowitz, A. (2000) *From the Amoeba to Einstein,* University of Haifa and Zamora-Beitan Publishing, Haifa, Israel.

Kouzes, R.T, Myers, J.D., Wulf, W.A. (1996) Collaboratories: Doing Science on the Internet, Computer **29** (8), 40

Kuhn, T. (1962) *The Structure of Scientific Revolutions*, University of Chicago Press, Chicago, Illinois.

Lakatos, I. (1991) Falsification and the Methodology of Scientific Research, In: I. Lakatos and A. Musgrave (eds.) *Criticism and the Growth of Knowledge*, Cambridge University Press, United Kingdom, 91-196.

Li, M. and Vitanyi, P. (2002) *An Introduction to Kolmogorov Complexity and its Applications*. Springer Verlag [translation from Chinese].

Linde, A.D. (1990) *Inflation and Quantum Cosmology*, Academic Press, Boston, Massachusetts.

Ne'eman, Y. (1980) Science as Evolution and Transcendence, Acta Cien. Venez. **31**, 1-3.

Ne'eman, Y. (1993) Serendipity, science and society - an evolutionary view, in: Proc. Kon. Ned. Akad. v. Wetensch. **96**, 433-448.

Ne'eman, Y. (1999) *Order out of Randomness – science and human society in a generalized theory of evolution* (in Hebrew), Van Leer Jerusalem Institute and Hakibbutz Hame'uchad Publishers.

Ne'eman, Y. (2000) Heuristic Methodology for Horizons in GR and Cosmology, Gravitation and Cosmology **6**, Supplement, 30-33.

Ne'eman, Y. (1970) CP and CPT Symmetry Violations, Entropy and the Expanding Universe, International Journal of Theoretical Physics **3**, 1-6.

Ne'eman, Y. (1995) The sophism which ushered in the Age of Information Technology, In: R. Stavy and D. Tirosh (eds.) *Theory and Practice in Mathematics, Science and Technology Education* (in Hebrew), Tel-Aviv University, School of Education Publishing, Tel-Aviv, Israel, pp. 15-20.

Pagel B.E.J. (1997) *Nucleosynthesis and Chemical Evolution of Galaxies*, Cambridge University Press.

Popper, K. (1972) *Objective Knowledge*, Oxford Clarendon Press.

Popper, K. (1935) *Logik der Forschung.*
Ruse, M. (1986) *Taking Darwin Seriously*, Oxford-Blackwell Publishers.
Segre, E. (1980) *From X-rays to Quarks,* W.H. Freeman and Co. Publishers, New York.
Schroedinger, E. (1967) *What is life,* Cambridge University Press, United Kingdom.
Smolin, L. (1992) Did the Universe Evolve, Classical Quantum Gravity **9** (1), 173-191; (1997) *The Life of the Cosmos*, Weidenfeld and Nicholson Publishers, London.
Strominger A. and Vafa C. (1996) Microscopic origin of the Bekenstein-Hawking Entropy, Physics Letters **B379**, 99-104.
Susskind L. (1995) The world as a hologram, Journal of Mathematical Physics **36**, 6377- 6396.
Szwast Z., Sieniutycz S. and Shiner J.S. (2002) Complexity principle of extremality in evolution of living organisms by information-theoretic entropy, Chaos, Solitons and Fractals **13**, 1871-1888.
Thornton, S.T. and Rex A. (1993) *Modern Physics for Scientists and Engineers,* Saunders College and Harcourt, Brace Jovanovich Publishers; also Serway, R.A., Moses C. and Moyer C.A. (1997) *Modern Physics*, same publishers.
White, B. (1998) The World Wide Web and High Energy Physics, Physics Today, November, 30-36.

V. The First Cells

Biodata of **Radhey Gupta** author of "***The Outlines of Bacterial Evolution***"

Dr. Radhey S. Gupta is a Professor of Biochemistry in the Department of Biochemistry at McMaster University in Hamilton, Ontario (Canada). He received his Ph. D. in Molecular Biology in 1973 from the University of Bombay (Tata Institute of Fundamental Research). Dr. Gupta's research interests have covered a broad range of subjects including the study of drug resistant mutants in mammalian cells, heat shock proteins, structure-function studies on enzymes and novel aspects of mitochondrial proteins. He has published over 220 articles in peer-reviewed journals in different areas. For the past 10 years his primary research interest lies in using the molecular sequence data to understand the early evolutionary history of life. His work challenges the view that *Archaea* constitute the third domain of life and instead suggests that they have evolved from Gram-positive bacteria in response to antibiotic selection pressure. He has also proposed a detailed model for the origin of eukaryotic cell by fusion between an archaebacterium and a bacterium. His recent work has led to a logical deduction of the branching order of different main phyla within *Bacteria*, which has been a major unresolved issue, that is critical for understanding the early evolutionary history of life.

E-mail: **gupta@mcmaster.ca**

J. Seckbach (ed.), Origins, 263–279.

THE OUTLINES OF BACTERIAL EVOLUTION

RADHEY S. GUPTA
Department of Biochemistry, McMaster University
Hamilton, ON Canada L8N 3Z5

1. Introduction

Prokaryotic organisms are widely accepted to be the sole inhabitants of our planet for the first 2 billion years of life. Hence, an understanding of the evolutionary relationships among them is crucial for understanding the nature and origin of the first cell and of various metabolic and information transfer processes that characterizes the living organisms. Earlier efforts to understand the evolutionary relationships among them based on morphological, biochemical and physiological characteristics led to the identification of a number of different groups (e.g., Gliding bacteria, Sheathed bacteria, Spirochetes, Gram-negative aerobic rods and cocci, Gram-negative anaerobic bacteria, Methane producing bacteria, Gram-positive cocci, Endospore-forming rods and cocci, Actinomycetes) (Orla-Jensen 1909; Buchanan 1925; Kluyver and van Niel 1936; Stanier 1941; Stanier and van Niel 1962; Murray 1986a). However, it was unclear whether the species in these groups were evolutionarily related. There were no means available at the time to understand how these groups were related to each other or in what order they branched off from a common ancestor.

The use of molecular sequences for deducing evolutionary relationships gave new hope for understanding the genealogical relationships among prokaryotes. Detailed studies by Carl Woese and coworkers using 16S rRNA sequences led to the surprising finding that the prokaryotic organisms are of two different kinds, archaebacteria (or *Archaea*) and eubacteria (or *Bacteria*). These two groups differed markedly from each other in certain phylogenetic trees and also with regard to a number of biochemical characteristics. This has led to the currently widely accepted view that these groups have originated independently from a universal ancestor (Fox *et al.* 1980; Woese 1987; Woese et al. 1990). Although *Archaea* are distinct from *Bacteria* with regard to many characteristics, their origin and relationship to *Bacteria* is a controversial subject. This controversy has been discussed in detailed elsewhere (Gupta 1998a,b,c; Gupta 2000a; Lyons, 2002). In this chapter, I will only be considering the evolutionary relationships among *Bacteria*, which comprise the vast majority of the prokaryotes.

2. Bacterial Phylogeny: Current Views and Unresolved Issues

Our current understanding of the evolutionary relationships among *Bacteria* is largely based on 16S rRNA sequences (Woese 1987; Balows *et al.* 1992; Olsen et al. 1994; Ludwig and Klenk 2001). Based on their branching in the rRNA trees, a number of major groups or phyla within *Bacteria* have been identified. The second edition of Bergey's Manual of Systematic Bacteriology describes 23 phyla of culturable *Bacteria* which include: *Aquificae, Thermotogae, Thermodesulfobacteria, Deinococcus-Thermus, Chrysiogenetes, Chlorflexi, Thermomicrobia, Nitrospirae, Deferribacteres, Cyanobacteria, Chlorobi, Firmicutes* (low G+C gram-positive), *Actinobacteria* (high G+C gram-positive), *Bacteroidetes, Planctomycetes, Verrucomicrobia, Chlamydiae, Spirochaetes, Fibrobacteres, Acidobacteriae, Fusobacteria, Dictyoglomi* and *Proteobacteria* (Ludwig and Klenk 2001). Of these phyla, some consist of only one or a few species (e.g., *Thermodesulfobacteria, Thermomicrobia, Chrysiogenetes, Fibrobacteres Deferribacters)*, whereas others (viz., *Proteobacteria, Cyanobacteria, Spirochaetes,* and the low and high G+C gram-positive groups) are made up of several hundred to thousands of species accounting for more than 90-95% of the known bacteria. The main groups in the 16S rRNA trees are also readily apparent in phylogenetic trees based on other molecular sequences, including 23S rRNA, Hsp60, EF-Tu, RecA, ATPase, Hsp70, RNA polymerase, EF-G, etc. (Viale *et al.* 1994; Eisen 1995; Gupta 1995; Gupta *et al.* 1997; Brown and Doolittle 1997; Ludwig and Schleifer 1999; Ludwig and Klenk 2001), or trees based on combined data sets from different proteins (Brown et al. 2001; Wolf et al. 2001; Daubin et al. 2002).

All of the main groups within *Bacteria* were originally described when the sequence data was quite limited and these groups could be clearly distinguished from each other by long naked branches in the 16S rRNA trees. However, in recent years, rapid expansion in the number of sequence database entries has filled most of the naked branches that separated these groups (Ludwig and Schleifer 1999; Ludwig and Klenk 2001). As a result, it has now become increasingly imprecise to assign species to various groups based on their branching in the rRNA trees. An important limitation of the 16S rRNA or other phylogenetic trees has been that they are unable to resolve the relative branching order of different groups from a common ancestor (Woese 1991; Olsen and Woese 1993; Brendel *et al.* 1997; Ludwig and Schleifer 1999; Ludwig and Klenk 2001; Brown et al. 2001; Wolf et al. 2001; Daubin et al. 2002). This has led to a growing consensus that all (or most) of the main groups within *Bacteria* have branched off from a common ancestor at about a similar time (Doolittle 1999; Ludwig and Schleifer 1999; Ludwig and Klenk 2001). However, the branching order of different groups and how they are related to each other constitute critical issues that need be resolved in order to understand bacterial phylogeny (Gupta and Griffiths 2002).

One of the main reasons why phylogenetic trees have been unable to resolve the above critical aspects relates to their dependence upon a large number of variables and assumptions. Some of the variables and assumptions involved in the tree construction process are: reliability of the sequence data and alignment, regions of the sequences that are retained or excluded in phylogenetic analysis, number and range of species examined, order of addition of species in generating sequence alignment, differences in the evolutionary rates and base compositions of the species, phylogenetic methods employed, etc. (Woese 1987; Gupta 1998a; Ludwig and Klenk 2001). There are no standard criteria for generating optimal sequence alignment or for constructing phylogenetic trees. Often a change in only one of the above parameters can have a marked effect on the branching patterns of species in phylogenetic trees. Although, the effects of these variables on the relationships among closely related species is generally small, they can greatly alter the relative branching orders of the major lineages (Ludwig and Klenk 2001). The difficulty in controlling and assessing the effects of these variables on branching patterns in phylogenetic trees is widely recognized as exemplified by this statement from Woese (1991), "When it comes to defining the branching order among major groups, or to defining many of the lesser taxa within them, one is at the mercy of powerful and sophisticated tree construction algorithms and can consequently fall prey to their vagaries". Hence, to resolve these important issues in bacterial phylogeny, additional sequence-based methods that are minimally affected by the above variables are needed. One new approach that is showing great promise in these regards involves the use of shared conserved inserts or deletions (i.e., indel, also referred to as signature sequences) found in various proteins as phylogenetic markers (Gupta 1998a, 2000b, 2001). The rationale of this approach and the understanding of bacterial phylogeny that is emerging using this approach follows.

3. The Usefulness of Conserved Indels as Phylogenetic Markers to Define the Main Bacterial Taxa and Their Branching Orders

When a conserved indel (i.e., insertion or deletion) of defined length and sequence, is found at the same position in homologs from different species, then the simplest and most parsimonious explanation for this observation is that the indel was introduced only once during the course of evolution and then passed on or inherited by all descendants. This is a minimal assumption implicit in most evolutionary analyses. *Thus based on the presence or absence of a signature sequence, the species containing or lacking the signature can be divided into two distinct groups, which bear a specific evolutionary relationship to each other.* Further, if one could identify well-defined indels that have been introduced at critical branch-points during the course of evolution, then they can serve as useful milestones for evolutionary events. In these cases, it is expected that all species emerging from an ancestral cell in which the indel was first introduced will contain the indel, whereas all species that either existed prior to this event or which did not evolve from this ancestor will lack the indel.

To interpret whether a given indel is the result of an insertion or deletion event, and to understand its evolutionary significance concerning the ancestral or derived nature of the groups, a reference point is essential (Gupta 1998a, b). The root of the prokaryotic tree provides a useful reference point for such purposes. Based on duplicated gene sequences for elongation factor-1 and -2, and aminoacyl tRNA synthetases, the root of the prokaryotic tree has been placed between *Archaea* and *Bacteria* (Iwabe *et al.* 1989; Brown and Doolittle 1995; Gupta 1998a). Our analysis of these duplicated sequences has shown that this rooting lies more specifically between *Archaea* and Gram-positive bacteria (Gupta 1998a). The placement of rooting within these two groups of prokaryotes is also supported by other lines of evidence (Gupta 1998a). Both *Archaea* and Gram-positive bacteria lack a large indel in the Hsp70 protein, which is present in all Gram-negative bacteria (Fig. 1). This indel is also lacking in the MreB protein, which is homologous to the N-terminal half of Hsp70, where this indel is found. The homologs of MreB are present in different groups of prokaryotes (*Archaea*, Gram-positive and Gram-negative) and they all lack this indel. Since MreB is ancestral to Hsp70 and this indel is not present in this protein, this provides evidence that the Hsp70 homologs lacking this indel are ancestral to those containing it (Gupta and Singh 1992: Gupta 1998a). In other words, this indel represents an insertion in the Gram-negative bacteria rather than a deletion in the Gram-positive and *Archaea.* Based on this rooting, it is now possible to use the signature sequences found in various proteins to logically deduce the branching order of different groups of bacteria.

We have identified a large number of conserved indels in different proteins which provide valuable markers for understanding bacterial phylogeny (Gupta 1998a, 2000b; Griffiths and Gupta 2002). The identified signatures are of two kinds. One type of signatures, which we refer to as group-specific signatures, are unique for particular group of bacteria but generally not found in other bacteria. We have identified many such signatures for the Chlamydiae, Spirochetes, Cyanobacteria, *Deinococcus-Thermus*, low G+C Gram-positive and high G+C Gram-positive groups of bacteria (Gupta 1998a; Griffiths and Gupta 2002; unpublished results). An example of this kind of signature is provided in Fig. 2, where a 16 aa indel is present in MurA protein in all chlamydiae species, but not in other bacteria. A 1aa insert (a cysteine residue) in this position in various spirochetes provides a signature for the latter group as well. Such signatures are of much value for identifying different bacterial groups in clear molecular terms and for assignment of species to these groups (Griffiths and Gupta 2002). The second kinds of signatures are those where an indel is commonly present in several major groups of bacteria but absent in other groups. These signatures, which we refer to as the main line signatures, have been introduced at critical branch points during the course of bacterial evolution and they are of great value in understanding the branching order and inter-relationships among different groups (Gupta and Griffiths 2002; Gupta 2002). The signature in Hsp70 provides an example of this kind of signature (Fig. 1). The large indel in this case, is commonly present in all Gram-negative bacteria but not in any Gram-positive bacteria, indicating that it was likely introduced in a common ancestor of Gram-negative bacteria at the time when this group evolved or branched off from a Gram-positive ancestor. We have identified numerous main line signatures and our analyses indicate that they have been introduced at specific stages of bacterial evolution as shown in Fig. 3. Based upon the presence or absence of these signatures, most of the major groups within *Bacteria,* identified on the basis of 16S rRNA trees, can be clearly distinguished. Further, by simply following the presence or absence of these signatures in different groups of bacteria, it is possible to logically deduce their branching order from a common ancestor, which is as shown in Fig. 3 (Gupta 2001, 2002; Gupta and Griffiths 2002).

Group	Species	65					102
Archaea	Th. acidophilum	NPEGTIFAAKRKMG				TDYKFKVFDKE	FTPQQISAFILQKI
	Hal. marismortui	--DE--QSI--H--				QD--SVELDGE-	Y--E-V--M-----
	Hal. cutirubrum	--DQ--ASINAH--				EE--TVALGGD-	Y--EE---R-----
	Meth. mazei	--DN-VYSI--H--				EAM--VTLNG-D	Y---E---M----L
	Met. thermoauto.	---N--T-I--S--				--R-V--HG--	Y---E---------
Gram -positives	Strep. pyogenes	-- E-VISI-S---				-SE-VSANG--	Y---E---M---YL
	Strep. pneumoniae	-- D-VISI-S---				-SE-VSANG--	Y---E---M---YL
	Myc. leprae	-VDR--RSV--H--				S-WSIEIDG-K	Y-A-E---RV-M-L
	Myc. tuberculosis	-VDR-VRSV--H--				S-WSIEIDG-K	Y-APE---R--M-L
	Str. coelicolor	-VDR--RSV--H--				--W-VNLDG-D	-N-------V---L
	Bac. subtilis	-- N--MSI--H--				----VEIEG-D	Y---EV--I---HL
	Meg. elsdenii	---N--SSI--H--				ES-TVDIQG-K	Y---E---M----L
	M. pneumoniae	-- N--VSI--L--				-SN-VT-KNPD	GSTKEL--Q--SYL
	M. capricolum	-- NVVQSI-S---				-TS-VNLEG-D	YS-E----E--RYM
	M. genitalium	-- N--VSI--L--				-SN-V--TT--	LS-E-V--Q--SYL
	Clo. acetobutyl.	--DK--ISI-----				-AE-VAID--N	Y---E---M----L
	Clo. perfringens	--DK--MSI--H--				----VNIDG-D	LS--E---M----L
	L. lactis	-- E--ISI-S---				-SE-VSANG--	Y---E---M---NL
	Sta. aureus	-- N-VQSI--H--				----VDIEG-S	Y---E---M---NL
	T. maritima	---R--KSI-----				----VRID---	Y---E------K-L
Proteobacteria	E. coli	--QN-L--I--LI-	RRF	QD	EEVQRDVSIMPFKIIAAD	NG-AWVE-KGQK	MA-P----EV-L-M
	Hae. influenzae	--KN-L--I--LI-	---	ES	------IK----E-TR--	NG-AWVN-KGDK	LA-P----EV-K-M
	Sal. typhimurium	--QN-L--I--LI-	---	--	-----------Y---G--	NG-AWLD-KGQK	MA-P----EV-K-M
	Vib. cholera	--QN-L--I--LI-	---	E-	------IK---Y--VK--	NG-AWVEAKGQK	MAAP-V--EV-K-M
	Buch. Species	--KN-L--I--LI-	-K-	K-	D-----IK---YN-VNS-	NG-AWID-KKQK	MA-P----EV-K-M
	Nei. meningitidis	-AKN--Y----LI-	HK-	E-	K-----IES---E--K-N	NG-AWV-AQG--	LS-P----EV-R-M
	Pse. cepacia	--KN-L--V--LI-	---	EE	K---K-IGL--YS--K--	NG-AWVEGHGEK	MA-P-V--EAVR-M
	Rho. capsulatus	--TN-V--V--LI-	--T	T-	A--EK-KKLV-YN-VDGG	NG-AWVE-RGEK	-S-A-V--V----M
	Brad. japonicum	---R-F--V--LI-	--Y	D-	PM-EK-KKLV-Y--VK-S	NG-AWVEADGQT	YS-S-V-------M
	R. meliloti	---N-L--I--LI-	-T-	E-	PTT-K-KGMV-Y--VK--	NG-AWVEAHGTS	YS-S----M----M
	A. tumefaciens	--TN-L--V--LI-	--Y	E-	PT-EK-KALV--E-VKG-	NG-AWV-AQ--N	YS-S----M----M
	C. crescentus	--TN-L--I--LI-	-TA	S-	PV-EK-KGMV-YRSSR-R	AG-AWV-AHG-D	YS--EV-------M
	B. ovis	-----L--V--LL-	--Y	D-	PM-TK-KDLV-Y--VKG-	NG-AWVE-HG-K	YS-S----M----M
	Myxo. xanthus	---N-V-----LI-	-K-	DS	P-GKKAIGVS---VASSP	NG-AWVEIRG-G	YS-PEV--IV-M-M
	Hel. pylori	---K--YSI--I--	LM-	NE	DKAKEAEKRL-Y--VDRN	GA CAIEISG-I	Y---E---K--M-L
Other Gram -negatives	Chl. trachomatis	---K-LAST--FI-	-K-		S--ESEIKTV-Y-VAPNS	KG-AV-D-EQ-L	Y--EE-G-Q--M-M
	Chl. pneumoniae	---K-LGST--FI-	-KY		S--ASEIQTV-YTVTSGS	KG-AV-E-DG-Q	Y--EE-G-Q--M-M
	Cyto. aquatilis	--TK--ASI--F--	HT-		A-TTNESKRVSY-VVKGV	QQYSTVDIDGRL	Y-A-EL--MT---M
	P. gingivalis	--TK--YSI--F--	ETY		DQ-S-E-ERV---VVRG-	-NTPRVDIDGRL	Y---E---M----L
	Cb.tepidum	--KN---SI--F--	-KY		D--PNEKKLASYDVVN E	GGYA-V-IG--T	YS--E---M----M
	Aqu. aeolicus	D--N-VYES--FI-	-K-		NE- KEAKRVSY-VVPDE	KG-AA-DIPG-L	VR-EEVG-HV-R-L
	Tre. pallidum	---H--YSI--FI-	S--		N-LTGEAKKV-Y--VPQG	D-VRVE-EG-L	YST-E--------M
	Bor. burgdorferi	---N--YSI--F--	---		---ASEIKMV-Y--EKGL	NG-ARVNISNIKK	QMS-PE---AT-T-M
	Syn. sp. PCC7942	-A-N-VYSI--FI-	--W		-DTEAER-RVTYTCVPGK	DDTVNVTIR-RF	C---E---MV---L
	Sy. sp. PCC6803	-A-N-VYSI--FI-	--W		DDTVEER-RV-YNCVKGR	DDTVSVSIRGQS	Y---E---M----L
	Aph. halophytica	-A-N--YSI--FI-	W		DDTEQERNRVSYHCVPGR	DKTVDV-CWG-Q	Y---EL--M---TL
	Anab. variabilis	--QN-F--V--YI-	--Y		NELSPESKRV-YT-RKD-	VGNI-VARLN--	-AAEE---MV-K-L
	Cf.aurantiacus	---N-LYSV--FI-	-S-		D--TEERDV ---VVKGP	RN-VRIY-PQTNKE	YA--E---MV---L
	D. proteolyticus	--QA-L-EV--FI-	--W		D--KDEAARS--TVKEGP	GGSVRIE-DG-D	YA-E-V--EV-R-L
	D. radiodurans	--AA-L-EV--FI-	--W		D--KEEAARS--TVKEGP	SGSVRIE-NG-D	LA-E-V--EV-R-L
	Ther. aquaticus	-------EI--FI-	---		----EEAKRV-Y-VVPGP	DGGVRVE-KG-L	Y--EE---M--R-L

Figure 1. Excerpts from Hsp70 sequence alignments showing two different main line sequence signatures. The large box describes a 21-23 aa indel commonly present in all Gram-negative bacteria but not found in any *Archaea* or Gram-positive bacteria. The smaller box within the large box describes a 2 aa indel distinctive of various proteobacteria but not found in any other bacteria. The dashes in the alignment (-) indicate identity with the amino acids in the top line.

Group	Species	327		374
Proteobacteria	H. influenzae Rd	ETIFENRFMHIPELIRMGGKAEI		EGNTAVCHGVEQLSGTEVIATDLRA
	E. coli	--V-------V---S---AH---		-S--VI-----K---AQ-M------
	Sal. typhimurium	--V-------V---S---AR---		-S--VI-----T---AQ-M------
	V. cholerae	----------V---K---A----		----VI-GD-DR---AQ-M------
	X. fastidiosa	----------VD--L-L-A-IQ-		--H--IVQ---R---AP-M------
	Buch. sp. APS	--------IYTS------A-IK-		KN--II-Y-IPK-ISSN-FSS----
	Bru. melitensis	----------VQ--A-L-A-ISL		S-Q--TVE---R-K-AQ-M------
	Pas. multocida	-----------------------		-----I----DH---A--M------
	Pse. aeruginosa	--V-------VY-MN---AQILV		-----IVT--PK-K-AP-M------
	Nei. meningitidis	----------V---N---ANITT		-----FVQ---R---AV-K------
	Ral. solanacearum	----------VQ--N-L-ANIIT		-----AVT-------AT-M------
	Meso. loti	----------V---M-L-ANIKL		Q-TM-LVR-G-K-H-AQ-M------
	Ri. prowazekii	-N--------V---C---ADIVV		R--K--VR---M-K-A--M-S----
	A. tumefaciens	----------VQ--A-L-A-ISL		S-QM-RIE--TR-K-AP-M------
	Ca. crescentus	----------A---M-L-ADISV		S-GE-RVR--D--E-AQ-M------
	Sin. meliloti	----------VQ--A-L-A-ISL		S-Q--KVE--SR-K-AP-M------
	Myx. xanthus	-NI-------V---H-L-ADIT-		Q-P---VK--KG---AP-M------
	Camp. jejuni	-RL-------VS--L---ADIKL		N-HI-TIV-GKE-NAAD-M------
	Hel. pylori 26695	--L-------AS--Q-L-ANISL		KT-V-TIS-STE-T-SD-M------
Chlamydiae	Chl. trachomatis	--VH---LGYLKG-VK--AHCDL	FHECLSAKSCRYSTGN	HPHS--I--PTP-QA-DLVIP----
	Chl. muridarum	--VH---LGYLSG-AK--AHCDL	FHECLSAKSCRYSTGN	HPHS--I--PTP-QA-HLVIP----
	Chlam. caviae	--VH---LGYLRG-QQ--ASC-L	FYQCLSSKACRYATGN	FPHS--I---TP-RSSHLVIP----
	Chlam. pneumoniae	--VH---LGYLHG-QH--AECQL	FHQCLSTKACRYAIGN	FPHS--I--ATP-WASHLVIP----
	Chlam. abortus	--VH---LGYLRG-QK--ANC-L	FYQCLSSKACRYATGN	FPHS-II---TP-KASHLVIP----
	Chlam. psittaci	--VH---LGYLRG-QQ--ANC-L	FYQCLSSKACRYATGN	FPHS--I--ATP-KAS-LVIP----
	Chlam. felis	--VH---LGYLRG-QK--ANC-L	FYQCLSSKACRYATGN	FPHS-II---TP-KASQLVIP----
	Wad. chondrophila	--VY----GYTDT-KE--AEITL	FRQCLGGKECRFSSQA	FSHSLIVK--SP-T-R-INIP----
Other Gram negatives	Aqu. aeolicus	-N---H--H-AQ--N-L-ANITV		R----YVE---R-Y-S--YS-----
	Cb. tepidum	DR-YLE--N-----N-L-AHI--		RD-W-LV--PQE-T--K-MS-----
	Nostoc sp. PCC 7120	-SV----LRHAS--N-L-ADIRV		K----FVR--PL---AP--G-----
	Sy. sp. PCC6803	--V----LQ-VA--Q---ANIKL		K--A-FIQ--PF---AP-MS-----
	D. radiodurans	DPVYPD-LT-VA--H---ATITV		S-Y-Q-IQ- GT-HAAP-K-A----
Spirochetes	Tre. pallidum	-KM--S-MFFVDK--T--ARIIL	C	DPHR-LVS-PSA-H-SDLVSP-V--
	Bor. burgdorferi	-KM--S-MFFVDK--K--ARIVL	C	DPHRV-VT-KSS-K-NVLSSP-V--
	Trep. denticola	-KM--S-MFFVDK--G--ARITL	C	DPHR--IS-PSS-H-S-LVSP-V--
	T. maritima	-NV-KT--L-VD--K---ADI-V		S--V-IVK---K---AP-EG-----
Gram-positives	Bac. subtilis	--V-------AE-FR--N-DIK-		--RSVIIN-PV--Q-A--A------
	Bac. halodurans	D--YNA--K--D--R---ADIKV		--RS-IIN-KTK-Q-AK-R-S----
	Clo. acetobutylicum	----------V---K-F-ANIK-		D-RS--IV--KE-T-CSAR------
	Sta. aureus N315	--V-------VA-FK--NANINV		--RS-KLE-KS--Q-AQ-K------
	L. lactis	D--YQK-VN-V---A---ANISV		LDDRIIYDAPNE-T-SC-Q------
	Lis. innocua	D--YPS--K--A-IE-----FKL		--RS--VS-PVK-Q-SK-T------
	Str. coelicolor	--VY-S-LGFTSA-NQ--AHIQL	YRECLGGSDCRFGQRN	FLHS--VS-PTK-E-ADLVIP----
	Strep. pneumoniae	--V-----Q-LE-M --LHS--		IRD--RIV-GQP-Q-A--LS-----
	Strep. pyogenes	--V-----Q-LE-MR---LQS--		LRE--MI--GR--Q-AP-MS-----
	Cor. diphtheriae	-N---A--RFVD--V-L-AD-TV		D-HHV-MR---K--S-P-WSS-I--
	Myc. smegmatis	-NV--A--RFVE-M--L-AD-RT		D-HH--VR-IP---SAP-WSS-I--
	Myc. leprae	-NV--A--RFVE-M--L-AD-RT		D-HH--VR-LP---SAP-WCS-I--
	Myc. tuberculosis	-NV--A--RFVE-M--L-AD-RT		D-HH--VR-LP---SAP-WCS-I--

Figure 2. Partial alignment of MurA protein sequences showing two group specific signatures, one distinctive of chlamydiae (large box) and the other (smaller 1 aa indel) for the Spirochetes. In addition to the chlamydiae, the large indel is also present in *Streptomyces coelicolor*, which has likely acquired it by lateral gene transfer. Dashes (-) in the alignment show identity with the amino acid on the top line.

4. Testing the Reliability of the Indel Model Using Genome Sequence Data

The success of the indel approach in resolving the branching order of bacterial groups raises the obvious question as to how reliable are these results? There are two potential problems that could give rise to misleading results using the indel approach. First, it is the possible that the observed indels rather than being derived from a common ancestor, were introduced on multiple occasions in different phyla (or species) due to similar functional constraints on a protein. Second, the shared presence of an indel in different species can also occur if the indel was originally introduced in one species and then acquired by others by means of lateral gene transfer (LGT). The analysis of genome sequences, in recent years, has led to the notion that LGT among prokaryotic organisms is very common and it constitutes a major hurdle in understanding bacterial phylogeny (Doolittle 1999; Gogarten et al. 2002). Thus, it is very important to establish that the observed results have not been influenced by these factors.

The sequence data for completed bacterial genomes provides an objective means to test the reliability of the branching order deduced by means of indel. Presently, the genome sequences have been determined for 72 bacterial species (http://www.ncbi.nlm.nih.gov/PMGifs/Genomes/micr.html). These species belong to different groups on the basis of the 16S rRNA (Table 1). According to the indel model these groups have branched off from a common ancestor as shown in Fig. 3. This branching order makes very specific predictions as to which of the main line indels should be present or absent in different groups of bacteria. The model predicts that once a main line indel has been introduced, all species belonging to the latter branching groups should contain the indel, whereas all species from groups that branched off prior to the introduction of the indel, should be lacking the indel. If the deduced branching order is reliable, then the observed distribution of these indels should follow the prediction of the model. On the other hand, if these indels have been introduced either independently, or their distribution was affected by lateral gene transfers, then their presence or absence in different species will not be as predicted. In such a case, different groups of species or even individual species from different groups will either contain or lack these indels. Thus, by determining how closely the distribution of indels in bacterial genomes follows the predictions of the model, and how many exceptions or contradictions are observed, the reliability of the branching order could be objectively determined.

We have examined the presence of all 22 main line signatures shown in Fig. 3 in completed bacterial genomes by means of BLAST analysis and sequence alignments. The sequence alignments for most of the signatures have been published previously (Gupta 1998a, 2001). Results of these analyses indicating the presence or absence of various indels in different species are presented in Table 2. The Table lists the number of genomes where these proteins have been found and also the total number of species that are expected to contain or lack each of these indels based on the branching order shown in Figure 3. For example, for the signature sequence in Ala-tRNA synthetase, a protein found in all sequenced bacterial genomes, the model predicts that 38 of the 72 species should contain the indel whereas the remainder 34 should not have it. Similarly, the large 21-23 insert in Hsp70 protein, which according to the model has been

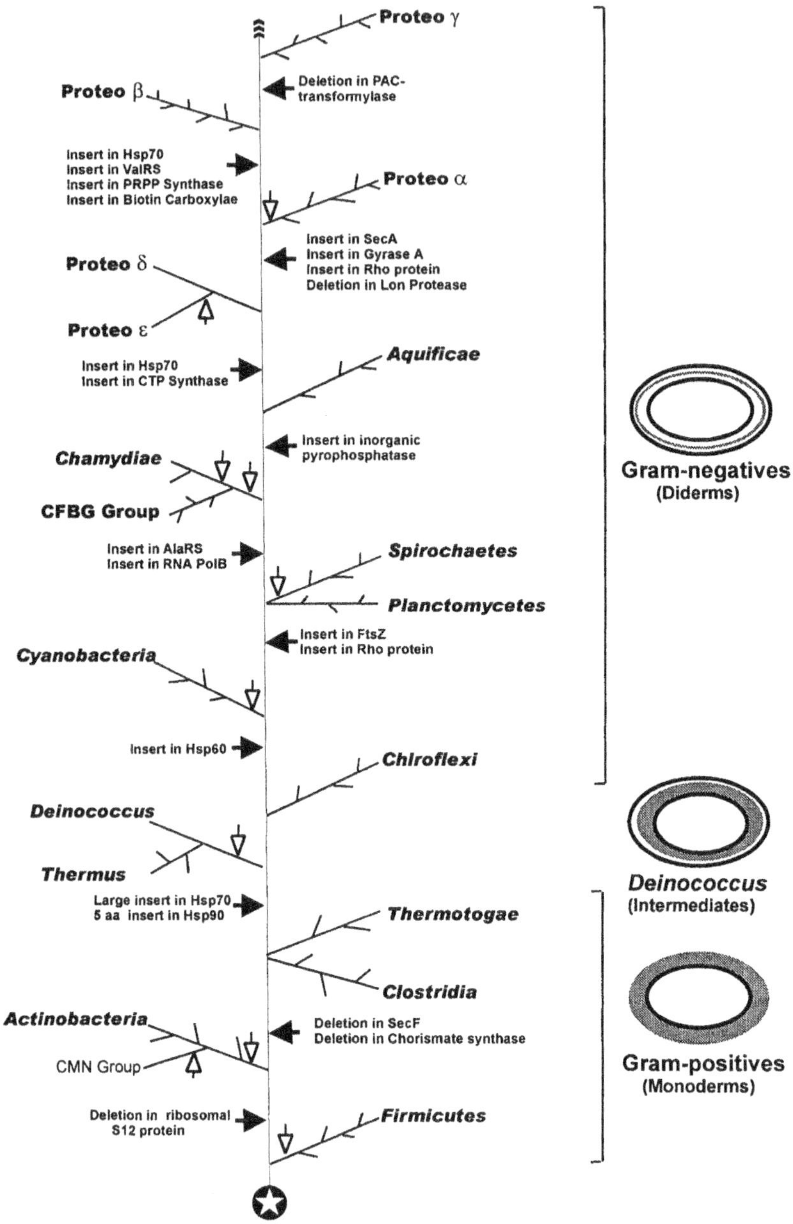

Figure 3. Branching order of *Bacterial* groups from a common ancestor based on conserved indels found in different proteins. The filled arrows depict the stages where the indicated main line signatures (Table 2) have been introduced. The open arrows denote the positions of many group-specific signatures. The observed branching order is strongly supported by sequence data for bacterial genomes and also the major differences in prokaryotic cell structures shown on the right.

TABLE 1. Different Groups of Bacteria Whose Genomes have been Sequenced

Firmicutes
Bacillus subtilis
Bacillus halodurans
Oceanabacillus iheyensis
Staphylococcus aureus N315
Staphylococcus aureus Mu50
Staphylococcus aureus MW2
Streptococcus pyogenes
Streptococcus pyogenes MGAS315
Streptococcus pyogenes MGAS8232
Streptococcus pneumoniae R6
Streptococcus pneumoniae TIGR4
Streptococcus agalactiae 2603
Mycoplasma genitalilum
Mycoplasma pneumoniae
Mycoplasma pulmonis
Ureaplasma urealyticus
Lactococcus lactis
Listeria innocua
Listeria monocytogenes

Actinobacteria
Mycobacterium tuberculosis H37Rv
Mycobacterium tuberculosis CDC1551
Mycobacterium leprae
Corynebacterium glutamicum
Streptomyces coelicolor

Clostridia-Thermotoga
Thermotoga martima
Clostridium acetobutylicum
Clostridium perfringens
Fusobacterium nucleatum
Thermoanaerobacter tengcongensis

Deinococcus-Thermus
Deinococcus radiodurans

Cyanobacteria
Synechocystis PCC6803
Nostoc sp. PCC7120
Thermosynechococcus elongatus

Spirochetes
Borrelia burgdorferi
Treponima pallidum

Chlamydia-CFBG
Chlamydia trachomatis
Chlamydia muridarum
Chlamidophyla pneumoniae CWL029
Chlmamidophyla pneumoniae J138
Chlmamidophyla pneumoniae AR39
Chlorobium tepidum TLS

Aquifex
Aquifex aeolicus

Proteobacteria-1 (δ,ε-subdivision)
Helicobacter pylori 26695
Helicobacter pylori J99
Campylobacter jejuni

Proteobacteria-2 (α-subdivision)
Rickettsia prowazekii
Caulobacter crescentus
Mesorhizobium loti
Agarobacterium tumefaciens-Dupont
Agarobacterium tumefaciens-Cereon
Rickettsia conorii
Sinorhizobium loti
Brucella melitensis

Proteobacteria-3 (β-subdivision)
Neisseria meningitidis MC58
Neisseria meningitidis Z2491
Ralstonia solanacearum

Proteobacteria-4 (γ-subdivision)
Echerichia coli K12
Echerichia coli O157:H7
Echerichia coli O157:H7 EDL933
Buchnera sp. APS
Buchnera aphidicola Sg
Pasteurella multocida
Pseudomonas aeruginosa
Vibrio cholerae
Xylella fastidiosa
Haemophilus influenzae
Yersinia pestsis C092
Yersina pestsis KIM
Salmonella typhimurium LT2
Salmonella typhi
Xanthomonas citri
Xanthomonas campestris
Xyellela fastidiosa

TABLE 2. Predicted versus Observed Distribution of Indels in Proteins from Bacterial Genomes

Protein	Signature description	Genomes with protein	No. of Genomes with indels Expected/Found	No. of Genomes lacking the indel Expected/Found	Exceptions Observed
Ribosome S12 protein	13 aa Low G+C indel	72	18/18	54/54	0
Hsp70/DnaK	21-23 aa G+/G- insert	72	44/44	28/28	0
Hsp90	5 aa G+/G- insert	40	10/10	30/30	0
Chorismate Synthase	15-17 aa deletion after *Actinobacteria*	64	19/22	45/42	3[a]
SecF protein	3-4 aa deletion after *Actinobacteria*	60	13/13	47/47	0
Hsp60/GroEL	1 aa insert after Deino.	70	43/43	27/27	0
FtsZ protein	1 aa insert after Cyano	64	35/35	29/29	0
Rho protein	2 aa insert after Cyano	58	40/41	18/17	1 (tm)[b]
Ala-tRNA Synthetase	4 aa inset after spirochetes	72	38/38	34/34	0
RNA Polymerase β- subunit	100-120 aa insert after spirochetes	72	38/38	34/34	0
Inorganic pyro-phosphatase	2 aa insert common to *Aquifex* and Proteobac.	50	32/32	18/18	0
Hsp70/DnaK	2 aa Proteo insert	72	31/31	41/41	0
CTP Synthetase	10 aa Proteo insert	65	31/31	34/34	0
Lon protease	1 aa αβγ-deletion	53	25/25	28/28	0
Rho Protein	3 aa αβγ-Proteo indel	58	28/28	30/30	0
DNA Gyrase A	26-34 aa αβγ-insert	72	28/28	44/44	0
SecA protein	7 aa αβγ-insert	72	28/28	44/44	0
HSP70/DnaK	4 aa βγ-Proteo insert	72	20/20	52/52	0
Val-tRNA Synthetase	37 aa βγ-Proteo insert	72	20/20	52/52	0
PRPP synthetase	1 aa βγ-Proteo insert	66	20/20	46/46	0
Biotin carboxylase	1 aa βγ-Proteo insert	63	18/18	45/45	0
PAC-transformylase	2 aa γ-proteo deletion	57	40/40	17/17	0

Protein abbreviations: PAC, 5-aminoimidazole-4 carboxamide formytransferase; PRPP, phoshoribosyl pyrophosphate.
[a] Smaller inserts, which could be of independent origin, are present in *Deinococcus, Aquifex aeolicus and Chlorobium tepidum.*
[b] *T. maritima* contains the indel, which is not expected.

introduced prior to the branching of *Deinococcus-Thermus* group (Fig. 3), should be present in 44 of the 72 species and absent in the remainder 28 species. The last few columns of Table 2 summarize the actual results obtained for different indels and the number of exceptions or contradictions that are observed. Some of these genes/proteins are not found in all species, hence the total number is not the same in every case. The results of these studies are remarkably clear (Table 2), as the presence or absence of these signatures in different genomes followed almost exactly the pattern as predicted by the model. In a total of 1416 observations involving the placement of indels in different species, only 4 exceptions or ambiguities are observed. The excellent correlation seen between the predicted and observed results in these analyses, and the ability of the model to predict with *remarkable accuracy* the presence or absence of various indels in different bacteria, provides strong evidence that the branching order of different groups as shown in Fig. 3 is reliable.

The evolutionary relationship among *Bacteria* that emerges based on signature sequences, independently, is also strongly supported by the major structural distinctions that exist within *Bacteria.* Bacteria have long been divided into two primary groups, Gram-positive and Gram-negative, based on their Gram-staining response (Gram 1884). All Gram-positives are bounded by a single unit lipid membrane and they generally contain a thick layer (20-80 nm) of peptidoglycan responsible for retaining the Gram-stain. A number of other bacteria bounded by *a single membrane*, but which stain Gram-negative due to either lack of the peptidoglycan layer (viz., mycoplasmas) or chemical composition of the cell wall (viz. *Spormusa, Megasphaera*), are also closely related to the gram-positive bacteria (Olsen et al. 1994). We have designated all bacteria bounded by a *single cell membrane* as 'Mononderm Bacteria' (Gupta 1998a). In contrast, all "true" Gram-negative bacteria, are bounded by both a cytoplasmic membrane as well as an outer cell membrane and they contain only a thin layer of peptidoglycan (2-3 nm) in between these two membranes. The presence of both inner and outer cell membranes in Gram-negative bacteria, whom we refer to as 'Diderm Bacteria' defines a new compartment in these cells, the periplasmic compartment. The Gram-positive and Gram-negative bacteria also differ with regard to many other characteristics including the presence or absence of lipopolysaccharides and teichoic acid in their cell envelopes. However, the distinction whether they are bounded by one or two membranes remains the most important structural distinction that exists within them (Stanier et al. 1976; Murray 1986a; Woese 1992; Gupta 1998a,b). The evolutionary relationship that emerges based on signature sequences shows that the bacterial groups bounded by a single membrane (i.e., gram-positive or monoderm bacteria) are phylogenetically distinct from all true gram-negative or diderm bacteria (Fig. 3). Of these two groups, the monoderm bacteria are indicated to be ancestral. The branching order of different group also places *Deinococcus-Thermus* group in an intermediate position between these two groups (Fig. 3). This placement is consistent with the unique characteristics of *Deinococcus*, which contains a thick peptidoglycan layer and shows positive gram-staining, but it is surrounded by both inner and outer cell membranes, similar to various gram-negative bacteria (Murray 1986b). Thus, in the evolutionary picture that is emerging based on signature sequences, a good concordance is observed between the cellular ultrastructural characteristics and molecular sequence data (Gupta 1998a, 2002; Gupta and Griffiths 2002), which is a major problem in most phylogenetic analyses (Woese, 1991; Olsen et al. 1994; Brown et al. 2001; Wolf et al. 2001; Daubin et al. 2002).

5. Indel Model Provides a Mean to Identify Lateral Gene Transfers

The analyses of genomic sequences in recent years have indicated that lateral gene transfers among different bacterial groups is widespread (Woese 1998; Jain *et al.* 1999; Doolittle 1999; Koonin *et al.* 2001; Gogarten *et al.* 2002). This has led to the notion either only very limited or no reliable phylogenetic inferences can be drawn concerning branching orders or interrelationships among bacterial groups (Woese 1998; Doolittle 1999; Ludwig and Klenk 2001; Wolf et al. 2001; Daubin et al. 2002). In this context, the high degree of concordance seen in our analyses between the predicted and observed results strongly indicate that the genes containing these indels, most of which are for highly conserved house-keeping functions, have not been affected or corrupted by factors such as LGTs. In addition, a large number of conserved indels that are specific for particular groups of bacteria, viz., cyanobacteria, *Deinococcus-Thermus,* chlamydiae, Spirochetes, *Firmicutes, Actinobacteria*, different groups of proteobacteria, have also been identified. These genes or proteins containing these conserved indels could be regarded as a stable core, analogous to the well-preserved fossils, which has been minimally altered by factors such as LGT during the course of evolution.

Based on the evolutionary model developed above, it is now possible to identify and logically interpret numerous instances of LGTs, a few examples of which are provided. We have previously described signature sequences consisting of a 55 aa deletion in the DnaJ protein, a 2 aa deletion in DNA polymerase I and a 10 aa insert in elongation factor Tu, that are commonly shared by various Cyanobacteria and the *Deinococcus-Thermus* group of species (Gupta and Johari 1998; Gupta 1998a). These two groups of bacteria are distinct from each other based on large number of other signatures (Fig. 3 and unpublished results) and their distinct branching in different phylogenetic trees (Viale *et al.* 1994; Eisen 1995; Gupta 1995, 1998a; Gupta *et al.* 1997; Brown and Doolittle 1997; Ludwig and Klenk 2001; Brown et al. 2001; Wolf et al. 2001; Daubin et al. 2002. Thus, the shared presence of these indels in these species is due to some ancient LGT events between these two groups of bacteria (Gupta and Johari 1998; Gupta 1998a). The signature sequence shown in Fig. 1 provides another clear example of LGT. In this case, the 16 aa insert in MurA protein, which is distinctive of the chlamydiae, is also present in *Streptomyces* species. In phylogenetic trees based on this protein, *Streptomyces* branch with high affinity with the Chlamydiae group rather than with other gram-positive bacteria (Griffiths and Gupta 2002). These results strongly indicate that the gene for this protein has been laterally transferred between these two groups. In another instance, a 17 aa conserved indel in the protein UDP-N-acetylglucosamine pyrophosphorylase (GlmU) is commonly shared by various archaebacteria and chlamydiae species (Griffiths and Gupta 2002). Although the GlmU is found in most other bacteria, they do not contain this insert. The presence of this uniquely shared indel between chlamydiae and archaebacteria suggests a specific relationship between these groups, exclusive of all other bacterial phyla. However, such a relationship is inconsistent with all other signatures (Fig. 3) and phylogenies, as well as various characteristics distinguishing *Archaea* and *Bacteria* (Woese et al. 1990; Brown and Doolittle, 1997; Gupta, 1998a). The most likely explanation of this shared signature is that the *glmU* gene was laterally transferred from an archaebacterium to a common ancestor of the chlamydiae group (Griffiths and Gupta 2002). This inference is supported by the fact that in phylogenetic trees based on GlmU, chlamydiae species are the closest relatives of archaebacteria (results not shown). Many other examples of LGTs, can be clearly identified using the present approach (Gupta 2000b and unpublished results). Thus, although LGT is of common occurrence in *Bacteria*, we do not find it be so pervasive or indiscriminate as to seriously jeopardise reliable determination of the phylogeny of *Bacteria*.

6. Summary

The genome sequences are enabling us to reliably deduce the evolutionary relationships among *Bacteria*, a problem central to understanding the origin of life. Global alignment of protein sequences have identified a large number of conserved indels (inserts or deletions) that are distinctive for either specific groups, or commonly shared by different groups of bacteria. These indels provide means to define all main groups within *Bacteria*, determined on the basis of 16S rRNA trees, in clear molecular terms. Based on the distribution of these indels it is now possible to reliably deduce that different bacterial phyla have branched off from a common ancestor in the following order: Low G+C gram-positive *(Firmicutes)* → High G+C gram-positive (*Actinobacteria*) → *Clostridia-Fusobacteria* → *Thermotoga* → *Deinococcus-Thermus-Green nonsulfur bacteria* → Cyanobacteria → *Spirochetes* → *Chlamydia-Cytophaga - Bacteriodes*- Green sulfur bacteria → *Aquifex* → *Proteobacteria-1* (ϵ and δ) → *Proteobacteria-2* (α) → *Proteobacteria-3* (β) and → *Proteobacteria* -4 (γ). Based on this branching order, the presence or absence of 22 indels in all 72 completed bacterial genomes was correctly predicted, with less than 5 exceptions seen in more than 1400 observations. The deduced relationship is also in accordance with the major morphological differences within *Bacteria*, with bacterial groups bounded by a single membrane (most Gram-positive bacteria) phylogenetically distinct from those bounded by both inner and outer cell membranes (all true Gram-negative bacteria). Contrary to the widespread notion, lateral gene transfer does not appear to be a serious analytical problem in deducing bacterial phylogeny.

7. Acknowledgements

The work from the author's laboratory was supported by research grants from the Canadian Institute of Health Research and the Natural Sciences and Engineering Research Council of Canada.

8. Reference List

Balows, A., Trüper, H.G., Dworkin, M., Harder, W., and Schleifer, K.H. (1992) *The Prokaryotes* New York: Springer-Verlag.

Brendel, V., Brocchieri, L., Sandler, S.J., Clark, A.J., and Karlin, S. (1997) Evolutionary comparisons of RecA-like proteins across all major kingdoms of living organisms *J. Mol. Evol.* **44**: 528-541.

Brown, J.R., Doolittle, W.F. (1995) Root of the universal tree of life based on ancient aminoacyl- tRNA synthetase gene duplications *Proc.Natl.Acad.Sci.USA* **92**: 2441-2445.

Brown, J.R., Doolittle, W.F. (1997) *Archaea* and the prokaryote-to-eukaryote transition *Microbiol. Mol. Biol. Rev.* **61**: 456-502.

Brown, J.R., Douady, C.J., Italia, M.J., Marshall, W.E., Stanhope, M.J. (2001) Universal trees based on large combined protein sequence data sets *Nat. Genet.* **28**: 281-285.

Buchanan, R.E. (1925) *General Systematic Bacteriology* Baltimore: Williams and Wilkins.

Daubin, V., Gouy, M., and Perriere, G. (2002) A phylogenomic approach to bacterial phylogeny: evidence of a core of genes sharing a common history *Genome Res.* **12**: 1080-1090.

Doolittle, W.F. (1999) Phylogenetic classification and the universal tree *Science* **284**: 2124-2128.

Eisen, J.A. (1995) The RecA protein as a model molecule for molecular systematic studies of bacteria: comparison of trees of RecAs and 16S rRNAs from the same species *J.Mol.Evol.* **41**: 1105-1123.

Fox, G.E., Stackebrandt, E., Hespell, R.B., Gibson, J., Maniloff, J., Dyer, T.A., Wolfe, R.S., Balch, W.E., Tanner, R.S., Magrum, L.J., Zablen, L.B., Blakemore, R., Gupta, R., Bonen, L., Lewis, B.J., Stahl, D.A., Luehrsen, K.R., Chen, K.N., and Woese, C.R. (1980) The phylogeny of prokaryotes *Science* **209**: 457-463.

Gogarten, J.P., Doolittle, W.F., and Lawrence, J.G. (2002) Prokaryotic evolution in light of gene transfer *Mol Biol Evol.* **19:** 2226-2238.

Gram, C. (1884) Ueber die isolierte farbung der Schizomyceten in Schnitt und Trockenpraparaten *Fortschr.Med.* **2**: 185-189.

Griffiths, E., Gupta, R.S. (2002) Protein signatures distinctive of chlamydial species: Horizontal transfer of cell wall biosynthesis genes *glmU* from Archaebacteria to Chlamydiae, and *murA* between Chlamydiae and *Streptomyces Microbiology* **148**: 2541-2549.

Gupta, R.S. (1995) Evolution of the chaperonin families (Hsp60, Hsp10 and Tcp-1) of proteins and the origin of eukaryotic cells *Mol. Microbiol.* **15:** 1-11.

Gupta, R.S. (1998a) Protein Phylogenies and Signature Sequences: A Reappraisal of Evolutionary Relationships Among Archaebacteria, Eubacteria, and Eukaryotes *Microbiol.Mol.Biol.Rev.* **62**: 1435-1491.

Gupta, R.S. (1998b) What are archaebacteria: Life's third domain or monoderm prokaryotes related to Gram-positive bacteria? A new proposal for the classification of prokaryotic organisms *Mol.Microbiol.* **29**: 695-708.

Gupta, R.S. (1998c) Life's third domain (*Archaea*): An established fact or an endangered paradigm? A new proposal for classification of organisms based on protein sequences and cell structure *Theor. Popul. Biol.* **54**: 91-104.

Gupta, R.S. (2000a) The *Natural* Evolutionary Relationships Among Prokaryotes *Crit.Rev.Microbiol.* **26**: 111-131.

Gupta, R.S. (2000b) The phylogeny of Proteobacteria: relationships to other eubacterial phyla and eukaryotes *FEMS Microbiol.Rev.* **24**: 367-402.

Gupta, R.S. (2001) The branching order and phylogenetic placement of species from completed bacterial genomes, based on conserved indels found in various proteins *International Microbiol.* **4**: 187-202.

Gupta, R.S. (2002) Phylogeny of Bacteria: Are we now close to understanding it? *ASM News* **68**: 284-291.

Gupta, R.S., Bustard, K., Falah, M., and Singh, D. (1997) Sequencing of heat shock protein 70 (DnaK) homologs from *Deinococcus proteolyticus* and *Thermomicrobium roseum* and their integration in a protein-based phylogeny of prokaryotes *J.Bacteriol.* **179**: 345-357.

Gupta, R.S., Griffiths, E. (2002) Critical Issues in Bacterial Phylogenies *Theor.Popul.Biol.* **61**: 423-434.

Gupta, R.S., Johari, V. (1998) Signature sequences in diverse proteins provide evidence of a close evolutionary relationship between the *Deinococcus-Thermus* group and Cyanobacteria *J.Mol.Evol.* **46**: 716-720.

Gupta, R.S., Singh, B. (1992) Cloning of the HSP70 gene from *Halobacterium marismortui*: relatedness of archaebacterial HSP70 to its eubacterial homologs and a model for the evolution of the HSP70 gene *J.Bacteriol.* **174**: 4594-4605.

Iwabe, N., Kuma, K., Hasegawa, M., Osawa, S., and Miyata, T. (1989) Evolutionary relationship of archaebacteria, eubacteria, and eukaryotes inferred from phylogenetic trees of duplicated genes *Proc.Natl.Acad.Sci.USA* **86**: 9355-9359.

Jain, R., Rivera, M., and Lake, J.A. (1999) The complexity hypothesis *Proc.Natl,Acad.Sci.USA* **96:** 3801-3806.

Kluyver, A.J., van Niel, C.B. (1936) Prospects for a natural system of classification of bacteria *Zentralblatt fur Bakteriologie, Parasitenkunde und Infektionskrankheiten II* **94**: 369-403.

Koonin, E.V., Makarova, K.S., and Aravind, L. (2001) Horizontal gene transfer in prokaryotes: quantification and classification *Annu.Rev.Microbiol.* **55** 709-742.

Ludwig, W., Klenk, H.-P. (2001) Overview: A phylogenetic backbone and taxonomic framework for prokaryotic systamatics. In: D.R. Boone and R.W. Castenholz (eds.) *Bergey's Manual of Systematic Bacteriology,* Berlin: Springer-Verlag, pp. 49-65.

Ludwig, W., Schleifer, K.H. (1999) Phylogeny of *Bacteria* beyond the 16S rRNA Standard *ASM News* **65**: 752-757.

Lyons, S.L. (2002) Thomas Kuhn is alive and well: the evolutionary relationships of simple life form-a paradigm under siege ? *Perspect.Biol.Med.* **45**: 359-376.

Murray, R.G.E. (1986b) Family II. *Deinococcaceae* Brooks and Murray 1981, 356[VP]. In: P.H.A. Sneath, N.S. Mair, M.E. Sharpe, and J.G. Holt (eds.) *Bergey's Manual of Systematic Bacteriology*, Baltimore: The Williams and Wilkins, pp. 1035-1043.

Murray, R.G.E. (1986a) The Higher Taxa, or, A Place for Everything? In: P.H.A. Sneath, N.S. Mair, M.E. Sharpe, and J.G. Holt (eds.) *Bergey's Manual of Systematic Bacteriology*, Baltimore: The Williams and Wilkins, pp. 31-34.

Olsen, G.J., Woese, C.R. (1993) Ribosomal RNA: a key to phylogeny. *FASEB Journal.* **7**: 113-123.

Olsen, G.J., Woese, C.R., and Overbeek, R. (1994) The winds of (evolutionary) change: breathing new life into microbiology. *J.Bacteriol.* **176**: 1-6.

Orla-Jensen, S. (1909) Die Hauptlinien de natürlichen Baketeriensystems nebst einer Uebersicht der Gärungsphenomene. *Zentr.Bakt.Parasitenk.II* **22**: 305-346.

Stanier, R.Y. (1941) The main outlines of bacterial classification *J.Bacteriol.* **42**: 437-466.

Stanier, R.Y., Adelberg, E.A., and Ingraham, J.L. (1976) *The Microbial World* New Jersey: Prentice-Hall Inc., pp. 1-871.

Stanier, R.Y., van Niel, C.B. (1962) The concept of a bacterium *Archiv fur Mikrobiologie* **42**: 17-35.

Viale, A.M., Arakaki, A.K., Soncini, F.C., and Ferreyra, R.G. (1994) Evolutionary relationships among eubacterial groups as inferred from GroEL (chaperonin) sequence comparisons *Int.J.Syst.Bacteriol.* **44**: 527-533.

Woese, C.R. (1987) Bacterial evolution. *Microbiol.Rev.* **51**: 221-271.

Woese, C.R. (1991) The use of ribosomal RNA in reconstructing evolutionary relationships among bacteria. In: R.K. Selander, A.G. Clark and T.S. Whittmay (eds.) *Evolution at Molecular Level*, Sunderland, MA: Sinauer Associates Inc., pp. 1-24.

Woese, C.R. (1992) Prokaryotc Systematics. The Evolution of a Science. In: A. Balows, H.G. Trüper, M. Dworkin, W. Harder and K.H. Schleifer (eds.) *The Prokaryotes*, New York: Springer-Verlag, pp. 3-18.

Woese, C.R. (1998) The universal ancestor *Proc.Natl.Acad.Sci.USA* **95**: 6854-6859.

Woese, C.R., Kandler, O., and Wheelis, M.L. (1990) Towards a natural system of organisms: proposal for the domains Archaea, Bacteria, and Eucarya *Proc.Natl.Acad.Sci.USA* **87**: 4576-4579.

Wolf, Y.I., Rogozin, I.B., Grishin, N.V., Tatusov, R.L., and Koonin, E.V. (2001) Genome trees constructed using five different approaches suggest new major bacterial clades *BMC Evol Biol.* **1**: 8

Biodata of **Helmut König** author (with co-authors H. Claus and E. Akca) of ***"Cell Wall Structures of Mesophilic, Thermophilic and Hyperthermophilic Archaea"***

Professor Helmut König is the Head of the Institute of Microbiology and Wine Research at the Johannes Gutenberg University, Mainz, Germany. He received his Ph.D. degree in 1979 in the Botanical Institute, University of Munich, Germany. After his Habilitation (at the University of Regensburg) and a position at the Institute of Applied Microbiology at the University of Ulm, Germany, he joined the University of Mainz in 1996.

E-mail: **hkoenig@uni-mainz.de**

Dr. Harald Claus is a scientific employee at the Institute of Microbiology and Wine Research at the Johannes Gutenberg-University in Mainz, Germany. He obtained his Ph.D. in 1983 at the Institute of Microbiology, Technical University of Darmstadt, Germany.

Mr. Erol Akca is a graduate student with Professor H. König at the Institute of Microbiology and Wine Research, Johannes Gutenberg University, Mainz, Germany. He intends to obtain his PhD at the end of 2003.

H. König

H. Claus

E. Akca

J. Seckbach (ed.), Origins, 281–298.

CELL WALL STRUCTURES OF MESOPHILIC, THERMOPHILIC AND HYPERTHERMOPHILIC ARCHAEA

HELMUT KÖNIG, HARALD CLAUS AND EROL AKCA
Institute of Microbiology and Wine Research,
Johannes Gutenberg-University, 55099 Mainz, Germany,

1. Introduction

At the end of the nineteen seventies a group of prokaryotes was recognized as a separate third domain of life with the Bacteria and Eukarya (Woese, 1987). This domain was named Archaea (Woese et al., 1990), because many species live in habitats reminding of the conditions of the early earth. The Archaea thrive in anaerobic niches, salt lakes, and marine hydrothermal systems and continental solfataras. The archaeal domain consists of three main phylogenetic lineages: the Crenarchaeota, the Euryarchaeota and the Korarchaeota. The methanogens, extreme halophiles, thermoplasmas, sulfate or sulfur metabolizing thermophiles and hyperthermophiles belong to the Archaea. It became obvious that the diversity of the Archaea is also reflected in a remarkable diversity of cell envelope types (Fig. 1; Kandler and König, 1985, 1993). A feature common to all Archaea is the lack of muramic acid and of a lipopolysaccharide-containing outer membrane. A universal cell wall polymer is missing. The cell walls of the Archaea are composed of different polymers such as a glutaminylglycan, heterosaccharide, methanochondroitin, pseudomurein, glycoprotein or a glycocalyx. The most common archaeal cell envelope is composed of a single protein or glycoprotein surface layer (S-layer), which is directly associated with the outside of the cytoplasmic membrane. The thermoplasmas lack a cell wall, but they possess a glycocalyx instead. Some cell envelope types are restricted to the Archaea, but similar building blocks are also found in natural polymeric compounds (e.g. connective tissue) of the other domains.
Already in the 1950s, significant differences from typical bacterial cell walls were established during cell envelope investigations of *Halobacterium* (Houwink, 1956). Later work with cells of *Sulfolobus* (Weiss, 1974), *Halococcus* (Brown and Cho, 1970) and *Methanosarcina* (Kandler and Hippe, 1977) also showed unusual structures. The S-layer glycoprotein of *Halobacterium salinarum* was the first glycoprotein discovered in prokaryotes (Mescher and Strominger, 1976). Initially, these novel cell wall structures were taken as curiosities. Their taxonomic significance was not realized prior to the publication of Archaea concept (Woese, 1987). The results of cell wall studies supported the new phylogenetic view.

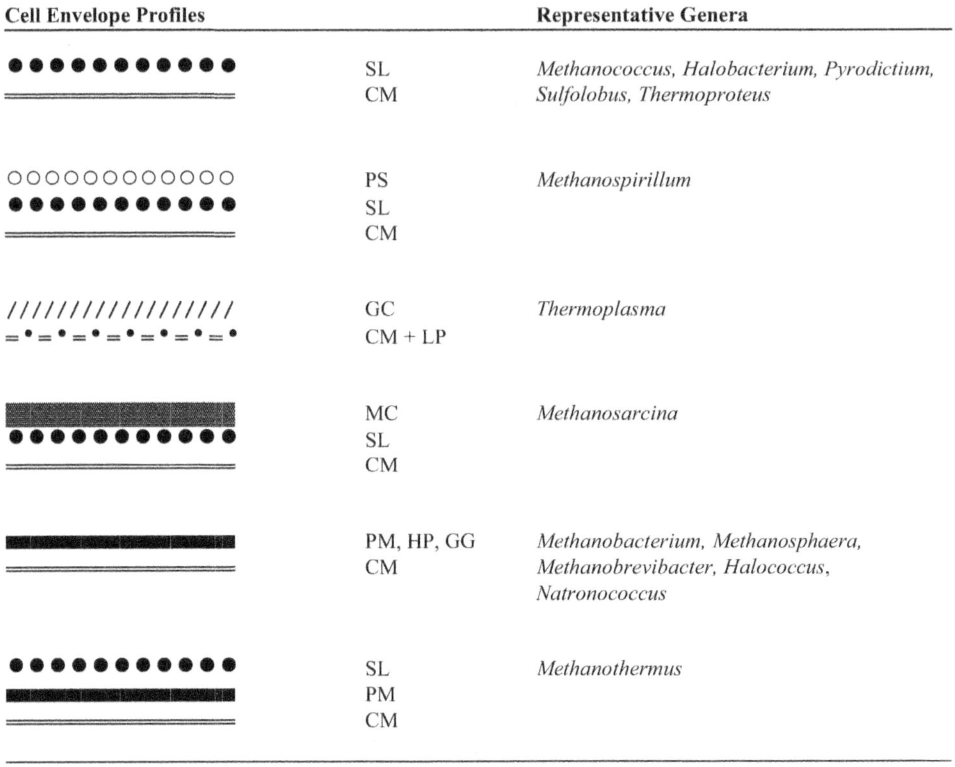

Figure. 1. Cell wall profiles of Archaea.
CM, cytoplasmic membrane; GC, glycocalyx; GG, glutaminylglycan; HP, heteropolysaccharide; LP, lipoglycan; MC, methanochondroitin; PM, pseudomurein; PS, protein sheath; SL, S-layer.

2. Diversity of Cell Wall Structures

2.1. GLUTAMINYLGLYCAN

The majority of the prokaryotic exopolymers are polysaccharides, but exopolymers composed of one type of amino acid, namely L- or D- glutamate, are formed, too. In general it seems that in Bacteria the α-linked polymers composed of glutamyl or glutaminyl residues possess the L-configuration, while in the case of the γ-linked exopeptides the glutamic acid residues have the D-configuration. The poly-γ-D-glutamyl polymers occur in the phylogenetically related genera *Bacillus*, *Sporosarcina*, and *Planococcus*. Similar polymers have been found in extremely halophilic cocci of the genus *Natronococcus* (Niemetz et al., 1997), which survive optimally in extremely alkaline and saline biotopes. The cell wall polymer of *Natronococcus occultus* is formed by a polyglutamine, which - in contrast to the eubacterial polymer - is glycosylated. The isolated cell wall polymer is composed of one amidated amino acid, namely L-glutamine, the two amino sugars glucosamine and galactosamine, the two uronic acids galacturonic acid and glucuronic acid, and the hexose glucose. In the intact polymer, the glutamine residues are linked via their γ-carboxylic group forming a chain of about 60 monomers. Two types of oligosaccharides are linked to the polyglutamine backbone (Fig. 2). One oligosaccharide consists of an *N*-acetylglucosamine pentasaccharide at the reducing end and a galacturonic acid oligosaccharide at the non-reducing end. The second oligosaccharide has an *N*-acetylgalactosamine disaccharide at the reducing end and a maltose unit at the non-reducing end.

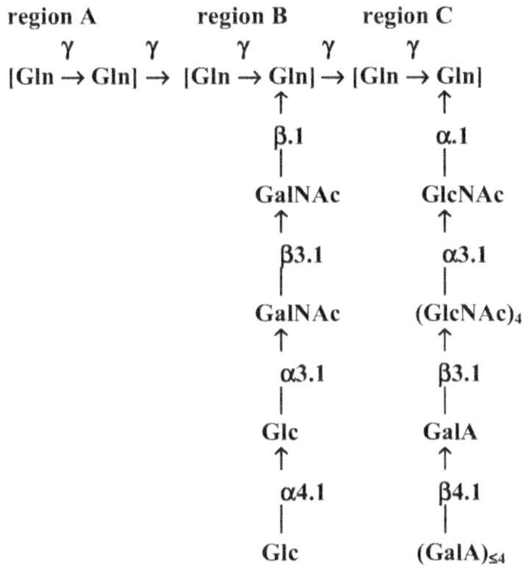

Figure 2. Proposed structure of the repeating units (region A-C) of the cell wall polymer of *Natronococcus occultus* (From Niemetz et al., 1997, modified).

Usually, N-linked oligosaccharides of bacterial and archaeal cell wall glycoconjugates are linked via galactosamine, glucose or rhamnose to the β-amide group of asparagine, while in the case of *Natronococcus occultus* the oligosaccharides are linked to the α-carboxylic groups of the polyglutamine backbone via an N-amide linkage. Therefore, the exopolymer of *Natronococcus occultus* represents a novel type of naturally occurring glycoconjugates.

In the case of *Bacillus anthracis* the polyglutamate polymer plays an important role in pathogenesis. Whether the structurally related glutaminylglucan of the natronococci is a pathogenic factor has yet to be investigated. They are not known as pathogens yet.

2.2. HETEROPOLYSACCHARIDE

The Gram-positive cells of *Halococcus morrhuae* are surrounded by an electron-dense cell wall with a thickness of about 50 nm. The cell wall polymer is composed of a complex heterosaccharide consisting of a mixture of neutral and amino sugars, uronic acids, glycine and an aminuronic acid (Reistadt, 1972; Schleifer et al., 1982). The carbohydrate

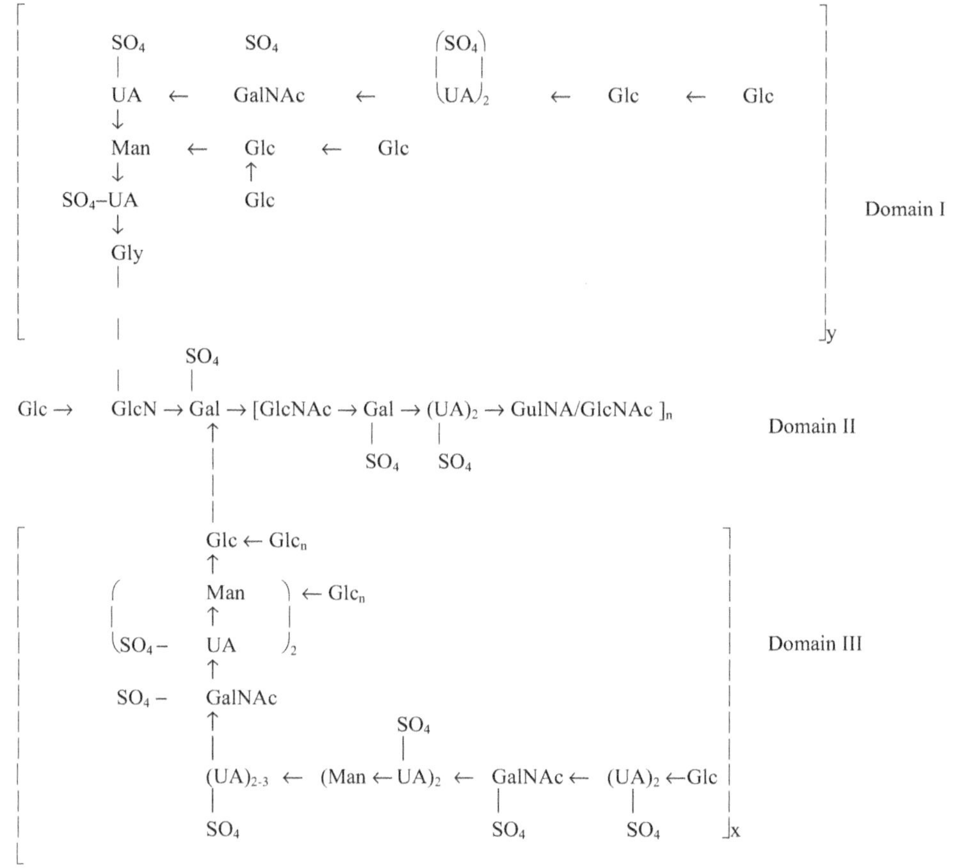

Figure 3. Proposed domain structure of the cell wall polymer of *Halococcus morrhuae* CCM 859 (From Schleifer et al., 1982; modified).

monomers are supposed to be arranged in three domains, which may be partly linked by *N*-glycyl-glucosaminyl bridges (Fig. 3).

2.3. METHANOCHONDROITIN

Cells of *Methanosarcina* form often large cuboid aggregates. The shape-maintaining component is a fibrillar nonsulfated polymer composed of an uronic acid and two *N*-acetylgalactosamine residues (Fig. 4). A trimer was identified as being the building block of the glycan (Kreisl and Kandler, 1986). This polymer forms a compact or sometimes loose matrix. Because of its reminiscence to eukaryotic tissue, especially to chondroitin sulfate it was named "methanochondroitin".

a) [→)-β-D-GlcA-(1→3)-β-D-GalNAc-(1→4)-β-D-GalNAc-(1→]$_n$.

b) [D-GlcA-(1→3)-GalNAc-(1→4)-]$_{23\text{-}25}$

c) [β-D-GlcA-(1→3)-β-D-GalNAc-(1→4)-β-D-GlcA-(1→]$_n$
(4 or 6 sulfate)

Figure 4: Proposed structure of the repeating units of the methanochondroitin (a), teichuronic acid of *Bacillus licheniformis* (b), chondroitin sulfate of animal connective tissue (c) (From Kandler and König, 1993)

Modifications of the methanochondroitin are due to the additional occurrence of galacturonic acid in the two species *Methanosarcina mazei* and *Methanosarcina* sp. G1. Electron microscopical investigations indicate that in some species an S-layer may be located between the cytoplasmic membrane and the methanochondroitin matrix. *Methanosarcina mazei* occurs in three morphological forms: single cells, laminae and packets (Mayerhofer et al., 1992). Since the methanochondroitin matrix is degraded during the life cycle leading to single disaggregated cells the methanochondroitin matrix is responsible for the aggregate formation (Boone and Mah, 1987).

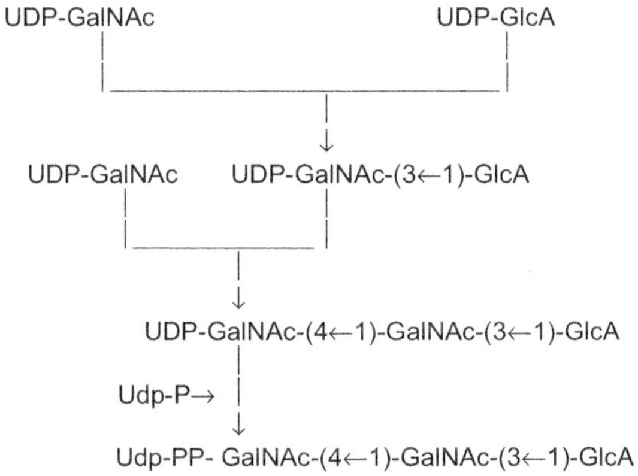

Figure 5. Proposed scheme of the biosynthesis of the methanochondroitin (From König et al., 1994): Udp = undecaprenyl.

The biosynthesis of methanochondroitin starts with the formation of UDP-*N*-acetylchondrosine and UDP-glucuronic acid, followed by the transfer of N-acetylchondrosine to UDP-N-acetylgalactosamine thus yielding a UDP-activated trisaccharide, which possess the same structure as the repeating unit of methanochondroitin (Fig. 5; König et al., 1994). The trisaccharide is subsequently linked to the lipid carrier undecaprenol via pyrophosphate, transferred through the cytoplasmic membrane and polymerized.

In substrate-depleted cultures of *Methanosarcina barkeri,* the cell wall matrix is lost yielding spontaneous protoplasts (Archer and King, 1984). During the life cycle of *Methanosarcina mazei*, the matrix material is degraded leading to a disaggregation of the cells. For complete disaggregation elevated concentrations of substrate (100 mM acetate or methanol) and divalent cations (Ca^{++}) were required (Boone and Mah, 1987).

2.4. PSEUDOMUREIN

The orders Methanobacteriales and Methanopyrales contain different genera with anaerobic rod- or lancet-shaped bacteria or cocci. They live under conditions, which reach from mesophilic to hyperthermophilic temperatures. The cells of this orders exhibit cell wall sacculi with a thickness of about 15 - 20 nm and a density of 1.39-1.46 g/cm^3 and a unit cell dimension of 4.5 x 10 x 21-22.5 **Å**. Conformational energy calculations suggested a similar secondary structure as for the eubacterial murein. Due to its chemical and three-dimensional structure this cell wall polymer has to be classified as a peptidoglycan, which means that glycan and peptides each constitute about 50% of this polymer. The archaeal peptidoglycan was named pseudomurein. The pseudomurein possesses structural similarities, but also differs significantly from the eubacterial murein (Fig. 6; Kandler and König, 1985, 1993). The glycan consists of alternating β(1 → 3)-linked N-acetyl-D-glucosamine and N-acetyl-L-talosaminuronic acid residues. The glycan strands are cross-linked by peptide subunits, which are often composed of a set of three L-amino acids: glutamic acid, alanine, and lysine.

```
(D-GalNAc)
-D-GlcNAc-(3←1)-β-L-NAcTalNA-
                             ↓
                             Glu (Asp)
                             ↓γ
                             Ala (Thr, Ser)
                             ↓ε
                             Lys←γ-Glu (— α,δ→ Orn)
                                       ↑
                                       Lys←γ- Glu
                                       ↑ε
                                       Ala (Thr, Ser)
                                       ↑γ
                                       Glu (Asp)
                                       ↑
                              -L-NAcTalNA-β-(1→3)-D-GlcNAc-
                                                     (D-GalNAc)
```

Figure 6. Proposed chemical structure of the pseudomurein (modifications in parentheses; from Kandler and König, 1993)

The main differences between pseudomurein and murein are the occurrence of talosaminuronic acid instead of murein, the presence of a β(1→3)-linkage instead of a β(1→4)-linkage of the glycan components, the partial replacement of glucosamine by galactosamine, the lack of D-amino acids, and the accumulation of ε- and γ-peptide bonds. Chemical modifications are found in the glycan as well as in the peptide moiety (Fig. 6).

The pathways of the biosynthesis of archaeal cell wall polymers exhibit similarities, but also striking differences to the well known biosynthetic pathways of structurally related bacterial and eukaryotic glycoconjugates. A pathway for the biosynthesis of the pseudomurein has been proposed (Fig. 7; König et al., 1994). The biosynthesis of the glycan intermediates starts with UDP-*N*-acetylglucosamine and UDP-*N*-acetyltalosaminuronic acid, which form the disaccharide UDP-GlcNAc(3←1)β-NAcTalNA. The biosynthesis of UDP-*N*-acetyltalosaminuronic acid is achieved by epimerization and oxidation of UDP-*N*-acetylgalactosamine, where UDP-*N*-acetylaltrosaminuronic acid forms an intermediate. In parallel a UDP-activated pentapeptide is formed. The biosynthesis of the pentapeptide starts with N^{α}-P-glutamic acid, wich is transformed to N^{α}-UDP-glutamic acid, to which the amino acid residues of alanine, lysine and glutamate are linked subsequently. The formation of a nucleotide-activated oligosaccharide and a nucleotide-activated pentapeptide are novel features of oligosaccharide and peptide biosynthesis showing that a common origin of the archaeal pseudomurein and the eubacterial murein is unlikely.

Prokaryotic cell wall components such as murein or lipopolysaccharides exhibit a series of biological activities. In the case of Archaea our knowledge is poor and it is restricted to the pseudomurein. The pseudomurein exhibits several biological activities. Like murein the pseudomurein is antigenic in animals. The immunological characterization with monoclonal antibodies has been revealed four distinctive determinants, namely the three glycan components GlcNAc, GalNAc, NAcTalNA and the C-terminal γ-glutamyl-alanine bond (Conway de Macario et al., 1983). Pseudomurein also causes somnogenic and pyrogenic effects. Intravenous injections of rabbits with suspensions of pseudomurein alter sleep and brain temperature (Johannsen et al., 1990). Acute inflammation was caused in a rat arthritis model (Stimpson et al., 1986).

As a consequence of its composition, which differs from the murein, the pseudomurein is resistant to cell-wall antibiotics such as β-lactams, which interact with the cytoplasmic intermediates. Pseudomurein is also insensitive against lysozymes and common proteases.

Methanobacterium bryantii undergoes spontaneous lysis when the NH_4^+ or Ni^{2+} ions are exhausted in the culture medium (Jarrel et al., 1982). In the presence of D-sorbitol, *Methanothermobacter thermoautotrophicus* forms protoplasts after addition of lysozyme to growing cells (Sauer et al., 1984). The target of the added lysozyme was not determined. An extracellular enzyme from a streptomycete was found to lyse *Methanobacterium formicicum* (Bush, 1985).

Lysis of the pseudomurein sacculi also leads to a disruption of the cells. An autolytic enzyme, which hydrolyzes the pseudomurein sacculi after depletion of the energy source could be induced in *Methanobacterium wolfei.*The lytic enzyme was found to be a peptidase hydrolysing the ε-bond between an alanine and a lysine residue of the peptide subunit (König et al., 1985; Kiener et al., 1987).

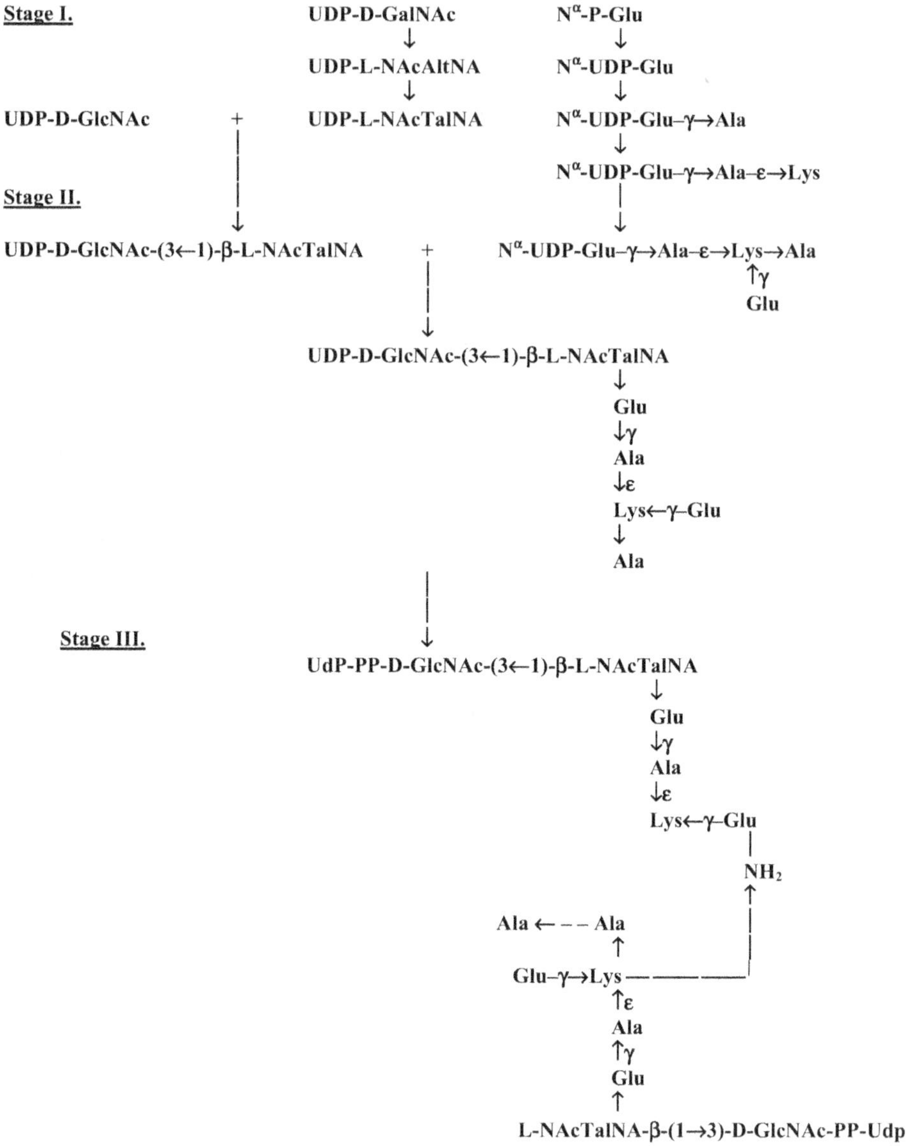

Figure 7. Proposed biosynthetic pathway of the pseudomurein (From König et al., 1994) Udp = undecaprenyl.

2.5. S-LAYER

Many Archaea possess proteinaceous surface layers (S-layers), which form two-dimensional regular arrays (Baumeister and Lembcke, 1992; Beveridge and Graham, 1991;

Kandler and König, 1985, 1993; Messner and Sleytr, 1992; Sumper and Wieland, 1995). The chemical structure of archaeal S-layer glycoproteins has been determined from a few archaeal species, e.g *Methanothermus fervidus* (Kärcher et al., 1993; cf. Kandler and König, 1993), *Halobacterium salinarum* and *Haloferax volcanii* (Lechner and Sumper, 1987; Sumper et al., 1990; cf. Sumper and Wieland, 1995).

The hyperthermophilic methanogen *Methanothermus fervidus* has a double-layered cell envelope. The pseudomurein is covered by an S-layer in hexagonal arrangement. The S-layer glycoprotein could be extracted with a solution of trichloroacetic acid and reverse-phase chromatography with aqueous formic acid as eluant. The mature protein consists of 593 amino acids. Compared to mesophilic S-layer proteins a significant higher amount of e.g., isoleucine, asparagine and cysteine and also 14% more β-sheet structure is present. The glycoprotein possesses also a typical leader peptide and 20 sequon structures as potential N-glycosylation sites. One type of oligosaccharide is present consisting of D-3-O-MetMan, D-Man and D-GalNAc (Kärcher et al., 1993). It is linked via *N*-acetylgalactosamine to asparagine residues of the peptide moiety (Fig. 8).

α-D-3-O-MetMan*p*-(1→6)-α-D-3-O-MetMan*p*-[(1→2)- α-D-Man*p*]$_3$-(1→4)-D-GalNAc

Figure 8. Proposed structure of the oligosaccharide of the S-Layer glycoprotein of *Methanothermus fervidus* (From Kärcher et al., 1993).

The surface glycoprotein of *Halobacterium salinarum* (formerly *Hb. halobium*; cf. Sumper and Wieland, 1995) was the first glycosylated protein detected in prokaryotes (Mescher and Strominger, 1976). The polypeptide possesses a stretch of 12 hydrophobic amino acids at the C-terminus, which functions as membrane anchor. Three different saccharide chains are linked to the peptide (Fig. 9). A sulfated pentasaccharide repeating unit forms a glycosaminoglycan chain, which is linked to the asparagine residue at the second N-terminal position of the peptide chain via an *N*-acetylgalactosamine. Asparaginylglucose is the linkage unit of ten sulfated glucose, glucuronic acid and iduronic acid containing oligosaccharides. In addition to the two types of *N*-linked glycan chains about 15 O-glycosidically linked glucosylgalactose disaccharides occur. The disaccharides form a cluster close to the transmembrane domain. The glycoprotein of *Halobacterium salinarum* is acidic because of the occurrence of more than 20% aspartic and glutamic acid residues and up to 50 mol of uronic acids and 50 mol sulfate residues per mol of peptide.

The mature polypeptide of the S-layer glycoprotein of *Haloferax volcanii* is composed of 794 amino acids. Like in the surface glycoprotein of *Halobacterium salinarum*, a hydrophobic stretch of about 20 amino acids at the C-terminus probably functions as transmembrane domain. Glucosyl-(1→2)-galactose disaccharides are assumed to be linked to threonine residues clustering close to the membrane anchor. Sulfated and amino sugar containing oligosaccharides are absent in the surface glycoprotein of *Haloferax volcanii*.

To maintain the rod shape, halobacteria require a high concentration of NaCl (>12%). Mg^{++} ions may form salt bridges between sulfate and uronic acid residues of the oligosaccharides of the S-layer glycoprotein (Wieland, 1988) required for the integrity of the S-layer subunits.

The biosynthesis of the glycan of the glycoprotein from *Halobacterium salinarum* has been studied in more detail (Sumper and Wieland, 1995; Wieland 1988), while in the case of *Methanothermus fervidus* only activated oligosaccharides have been isolated from cell extracts (König et al., 1994). In both species dolichol derivatives serve as lipid carrier, while in the case of the pseudomurein and methanochondroitin undecaprenyl pyrophosphate functions as lipid carrier. It is obvious that dolichy-P-(P) is the universal lipid carrier in glycoprotein synthesis in prokaryotes and eukaryotes (König et al., 1994).

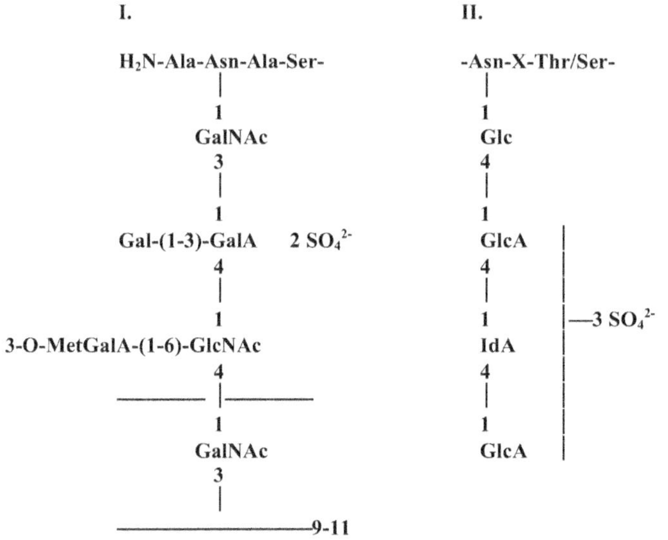

Figure 9. Proposed structure of the three glycan moieties of the S-layer glycoprotein of *Halobacterium salinarum* (From Wieland, 1988; modified). The building block composed of five different sugars (glycan I) is linearly repeated 10 - 12 times.

Depending on the glycan chain C_{60}-dolichyl-monophosphate and dolichylpyrophosphate serve as the lipid carriers for the glycosaminoglycan and the second sulfated oligosaccharide, respectively. The complete glycosaminoglycan including sulfation is synthesized at the lipid carrier and then transferred to the nascent protein. The linkage to the protein takes place at the cell surface. The acceptor peptide is Asn-Ala-Ser. Replacement of the serine residue of the consensus sequence by valine, leucine, and asparagine did not prevent N-glycosylation. N-glycosylation did not occur at Asn-479, when Ser-481 was removed (Zeitler et al., 1998), which indicates the presence of two different N-glycosyltransferases. In the case of the second sulfated oligosaccharide again completely sulfated lipid linked precursors are formed before transfer to the protein. Prior to the transfer of this saccharide chain to the cell surface some glucose residues are transiently methylated at carbon 3 (Sumper and Wieland, 1995; Wieland 1988).

Representing the outermost and only cell evelope layer, archaeal S-layers are directly exposed to the often extreme conditions of their surroundings. As a consequence, intrinsic resistance against archaeal environmental stresses like high salt, acidity and temperature are attributed to S-layers. The molecular mechanisms for this high stability are only poorly understood. As members of the genus *Methanococcus* (*Mc.*), *Methanothermococcus* (*Mtc.*) and *Methanocaldococcus* (*Mcc.*) include species living in mesophilic, thermophilic and hyperthermophilic environments they represent an ideal model system for studying their S-layers in focus of thermal adaptation (Akca et al., 2002). In order to get further hints about the molecular mechanisms of protein stabilization the primary and secondary structures of archaeal S-layer (glyco-)proteins were compared (Akca et al., 2002; cf. Claus et al., 2001, 2002). Akca et al., (2002) found an increase of charged amino acid residues and a reduction of polar residues in the S-layer proteins of the thermophilic and hyperthermophilic species as compared to their mesophilic counterparts (Table 1). As the overall hydrophobicity is even higher in the mesophilic strains this feature seems not to play a major role for adaptation to higher temperatures in the case of *Mtc. thermolithotrophicus* and *Mcc. jannaschii*. Thus, the increase in charged amino acids, especially of lysine, as found in the S-layer proteins of *Mtc. thermolithotrophicus* and *Mcc. jannaschii* could contribute to their increased thermal stability. The finding of a unique Ca^{++} binding site in the S-layer protein of *Mcc. jannaschii* points in the same direction (Akca et al., 2002). Similarly, an increase in charged residues can be observed in the S-layer proteins of *Methanosarcina mazei* (mesophilic) *Methanothermobacter thermoautotrophicus* (thermophilic) (derived from the corresponding gene) *Methanothermus fervidus* (hyperthermophilic) and *Archaeoglobus fulgidus* (hyperthermophilic). Interestingly, the S-layer glycoprotein of *Mt. fervidus* contains high amounts of Asn (Bröckl et al., 1991) and a basic isoelectric point. A significant feature of the S-layer proteins from *Mc. jannaschii* and other archaeal hyperthermophiles is the occurence of cysteine, which has been detected in only few S-layer proteins (Sára and Sleytr, 2000). Intramolecular disulfide-bridges may be another factor involved in the thermal stability of this surface protein. A prediction of the deduced secondary structure indicated a higher content of ordered structures, e.g., helical conformations in the S-layer proteins of the mesophilic *Mcc. voltae* and *Mc. vannielii* strains than in *Mtc. thermolithotrophicus* and *Mcc. jannaschii*, which in turn exhibit more loops (Akca, 2002; Claus et al., 2001, 2002). Similarity exists between the S-layer proteins of *Archaeoglobus fulgidus, Methanothermus fervidus, Methanothermobacter thermoautotrophicus* (gene *MTH719*) and *Methanosarcina. mazei,* which shows the highest portion of ß-sheets. The obtained data suggest that ionic interactions, e.g., salt bridges seem to be play a major role in protein stabilization at high temperatures. Despite the differences in the growth optima and the predominance of some amino acids the primary structures of S-layers revealed also a significant degree of identity between phylogenetically related Archaea. These observations indicate that protein sequences of S-layers have been conserved during the evolution from hyperthermophilic to mesophilic life.

Signal sequences for transcription and translation of the S-layer genes from members of the three genera *Methanococcus*, *Methanothermococcus* and *Methanocaldococcus* were described (Akca et al., 2002). In the case of *Mcc. jannaschii* a TATA Box was located 19 nucleotides and the "factor B recognition element" (BRE) 29 nucleotides upstream from the transcription start. The translation start and the Shine-Dalgarno sequence are localized downstream beginning at nucleotides 45 and 33, respectively. The TATA-Box corresponded completely to the consensus sequence AA/TT TTATA of methanogenic archaea (Thomm, 1996). In contrast to *Mcc. jannaschii,* several tandem

Table 1. Chemical characteristics of selected archaeal S-layer proteins.

Species[a]	Characteristics of amino acid composition						
	Nonpolar[a]	Polar[b]	Acidic[b]	Basic[b]	Aliphatic Index[c]	Hydropathicity[c]	pI[c]
Methanococcus voltae (M59200)	48.5	24.2	18.6	8.6	94.16	-0.0091	4.15
Methanococcus vannielii (AJ308553)	52.5	24.3	14.8	8.0	95.51	-0.061	4.25
Methanthermoococcus thermolithotrophicus (AJ308554)	48.5	23.5	17.9	10.1	100.14	-0.079	4.30
Methanocaldococcus jannaschii (AJ311636)	45.0	23.3	20.5	11.2	92.42	-0.296	4.27
Methanosarcina mazei (X77929)	40.2	46.8	5.2	7.7	68.22	-0.315	8.9
Methanothermobacter thermoautotrophicus (AAB85224)	46.5	39.4	6.1	7.9	93.03	-0.086	8.75
Methanothermus fervidus (X58297)	43.7	36.9	9.1	10.2	91.10	-0.236	8.47
Archaeoglobus fulgidus (AF1413)	43.9	34.0	12.9	9.2	72.28	-0.331	4.68
Staphylothermus marinus (U57967)	47.6	37.5	9.1	5.8	108.39	0.169	4.45
Halobacterium salinarum (J02767)	40.6	33.4	21.6	4.4	72.86	-0.543	3.60
Haloferax volcanii (M62816)	43.0	35.2	19.0	2.7	80.17	-0.275	3.44
Haloarcula japonica (D87290)	43.0	30.2	23.5	3.2	73.94	-0.462	3.40
Pyrococcus abyssi (NT01PA0829)	48.3	26.0	15.0	10.7	93.74	-0.119	4.68
Pyrococcus horikoshii (NT01PH1418)	45.3	24.1	18.2	12.3	92.92	-0.308	4.62

[a]accession no.;[b]mol (%) calculated after Karlson (1994); [c]calculated with ProtPARAM Tool.

promoters have been described for the S-layer gene of *Mc. voltae* (Kansy et al., 1994). The BRE Box, which binds the transcription factor TFB only differs in one nucleotide from the consensus sequence GGAAA. The ribosome binding site 5′-AGGAGAU-3′, usually localized 3-9 nucleotides in front of the translation start point (Dalgaard and Garrett, 1993), was found to be complementary to a region at the 3′ terminus of the 16 S rRNA of *Mcc. jannaschii*. Translation is terminated with a series of stop codons, which is a common feature of methanogenic Archaea (Dalgaard and Garrett, 1993). They are followed by a poly A / poly T sequence, which probably leads to formation of a hair-pin structure and terminates transcription. Similar promotor regions were suggested for the mesophilic *Methanococcus vannielii* and the thermophilic *Methanothermococcus thermolithotrophicus*, however it seems that they are located at different positions within the gene sequence (Akca et al., 2002).

The first successful crystallization of a complete hyperthermophilic surface (glyco-)protein under microgravity conditions was described (Evrard et al., 1999). Peptide domains from the S-layer protein of *Methanocarcina mazei* were also successfully crystallized (Jing et al., 2002). The limited number of crystal structures from thermophiles and hyperthermophiles has hampered detailed structural comparisons with mesophiles. Up to now it was not possible to get crystals for X-ray studies from S-layer (glyco-) proteins. We selected the S-layer glycoprotein of *Methanothermus fervidus* for our crystallization experiments. This species has been isolated from an Icelandic hot solfatara field. It grows optimally at 85 °C and its S-layer may serve as a model to study the molecular strategies for survival at high temperatures. The gene (*slg*A) encoding the S-layer glycoprotein has been sequenced (Bröckl et al., 1991) and the chemical structure of the heterosaccharide has been elucidated (Kärcher et al., 1993). The mature peptide is predicted to consist of 593 amino acids resulting in a molecular mass of 65 kDa (Bröckl et al., 1991). With mass spectroscopy (MALDI) a molecular mass of 83 kDa was determined for the the mature glycoprotein indicating that the glycan moiety accounts for 22 % of the mass. The derived amino acid composition contains high amounts of Asn. The heterosaccharide is composed of 3-O-methylmannose, mannose and N-acetylgalactosamine. For X-ray analysis it is essential to get crystals of high quality. Crystallization experiments were conducted under microgravity conditions (Evrard et al., 1999) using the Advanced Protein Crystallization Facilities developed by Dornier. The crystals were succesfully grown during the flight STS-05 of the space shuttle Discovery. Hanging drop reactors (HD-80) were used. One of the crystals with dimensions of 30 x 20 x 5 μm was selected for X-ray analysis. The diffraction experiments were performed at the EMBL DESY synchrotron facility in Hamburg. The crystal system is monoclinic and has the space group C2. Meanwhile, we obtained also crystals from *Bacillus sphaericus* during the flight STS-101.

2.6. LIPOGLYCAN

Members of the thermoacidophilic genus *Thermoplasma* like the mycoplasmas lack a cell envelope. The stabilization of the cells is most likely maintained by a mannose-rich lipoglycan (Langworthy et al., 1982) and a glycoprotein (Yang and Haug, 1979). These two constituents are located in the cytoplasmic membrane and their glycan chains are directed to the outside of the cell forming a glycocalyx.

3. Perspectives

The cell envelopes of the Archaea are often directly exposed to extreme environmental conditions and they can not be stabilized by cellular factors. Adaptation to extreme environments has not led to similar cell surface structures. *Halobacterium* sp. and *Halococcus* sp. thrive in saturated salt brines, but they possess quite different cell wall polymers such as glycoprotein, heterosaccharide or glutaminylglycan. The same is true for hyperthermophilic Archaea. They may serve as models to elucidate survival strategies of natural compounds and may give clues about molecular mechanisms of resistance against high temperature, low and high pH and high salt concentrations (Evrard et al., 1999).

These investigations may lead to applications of new biomaterials. S-layers represent the most common cell surface layer of Archaea. They are the simplest biological membranes found in nature. A wide spectrum of applications for S-layers has already emerged. Isolated S-layer subunits assemble into monomolecular crystalline arrays in suspension, on surfaces and interfaces. These lattices have functional groups on the surface in an identical position and orientation in a nanometer range.The features have led to different applications such as ultrafiltration membranes, immobilization matrices for functional molecules, affinity microcarriers and biosensors, conjugate vaccines, carriers for Langmuir-Blodgett films and reconstituted biological membranes and regular arragement of technical elements in molecular nanotechnology (Sleytr et al., 1994). In the past the applied investigations have nearly exclusively been performed with eubacterial S-layers. Since archaeal S-layer (glyco-)proteins are often resistant under extreme conditions, a new spectrum of future developments with archaeal S-layer (glyco-)proteins should be found.

The genes of some archaeal S-layer proteins have been characterized (Bröckl et al., 1991; Lechner and Sumper, 1987; Sumper et al., 1990; Yao et al., 1994; cf. Akca et al., 2002; cf. Claus et al., 2001, 2002). From the species mentioned above in some cases (cf. http://www.cbs.dtu.dk/GenomeAtlas/Archaea/index.html) e.g. of *Methanothermobacter thermoautotrophicus* strain ΔH (Smith et al., 1997) the complete genome sequence has been published. So far no genes of enzymes envolved in the cell wall biosynthesis have been unambigously assigned. The knowledge of the complete genome may be helpful in the identification of special enzymes involved in the biosynthesis and degradation of cell wall polymers.

4. Acknowledgement

We would like to thank the Deutsche Forschungsgemeinschaft, the "Ministerium für Bildung, Wissenschaft, Forschung und Technologie" (Bonn) as well as the "Deutsches Zentrum für Luft- und Raumfahrt" (Bonn), Germany, for supporting this work, and the European Space Agency for providing the flight opportunity.

5. References

Akca, E., Claus, H., Schultz, N., Karbach, G., Schlott, B., Debaerdemaeker, T., and König, H. (2002) Genes and derived amino acid sequences of S-layer proteins from mesophilic, thermophilic and extremely thermophilic methanococci. Extremophiles 6, 351-358.

Archer, D. B., and King, N. R. (1984) Isolation of gas vesicles from *Methanosarcina barkeri*. J. Gen. Microbiol. 130, 167-172.

Baumeister, W., and Lembcke, G. (1992) Structural features of archaebacterial cell envelopes. J. Bioenergetics and Biomembranes 24, 567-575.

Beveridge, T. J., and Graham, L. L. (1991) Surface layers of bacteria. Microbiol. Rev. 55, 684 - 705.

Boone, D. R., and Mah, R. A. (1987) Effects of calcium, magnesium, pH, and extent of growth on the morphology of *Methanosarcina barkeri* S-6. Appl. Environ. Microbiol. 53, 1699-1700.

Bröckl, G., Behr, M., Fabry, S., Hensel, R., Kaudewitz, H., Biendl, E., and König, H. (1991) Analysis and nucleotide sequence ot the genes encoding the surface-layer glycoproteins of the hyperthermophilic methanogens *Methanothermus fervidus* and *Methanothermus sociabilis*. Eur. J. Biochem. 199, 147-152.

Brown, A. D., and Cho, K. J. (1970) The walls of extremely halophilic cocci. Gram-positive bacteria lacking muramic acid. J. Gen. Microbiol. 62, 267-270.

Bush, J. W. (1985) Enzymatic lysis of the pseudomurein containing methanogen *Methanobacterium formicicum*. J. Bacteriol. 163, 27-36.

Claus, H., Akca, E., Schultz, N., Karbach, G., Schlott. B., Debaerdemaeker, T., Declercq, J.-P., and König, H., (2001) Surface (glyco-)proteins: primary structure and crystallization under microgravity conditions. Proceedings of the First European Workshop on Exo/Astrobiology, pp. 313 - 320. ESA Publication Division Noordwijk, The Netherlands.

Claus, H., Akca, E., Debaerdemaeker, T., Evrard, C., Declercq, J-P., and König, H. (2002) Primary structure of selected archaeal mesophilic and extremely thermophilic outer surface layer proteins. System. Appl. Microbiol. 25, 3-12.

Conway de Macario, E., Macario, A. J. L., Magarinos, M. C., König, H., and Kandler, O. (1983) Dissecting the antigenic mosaic of the archaebacterium *Methanobacterium thermoautotrophicum* by monoclonal antibodies of defined molecular specificity. Proc. Natl. Acad. Sci. U.S.A. 80, 6346-6350.

Dalgaard, J. Z., and Garrett, R. A. (1993) Archaeal hyperthermophile genes. In: M. Kates, D. J. Kushner, and A. T. Matheson (eds.). The Biochemistry of Archaea (Archaebacteria), Elsevier, Amsterdam. pp. 535-563.

Evrard, Ch., Declercq, J-P., Debaerdemaeker, T., and König, H. (1999) The first successful crystallization of a prokaryotic extremely thermophilic outer surface layer glycoprotein. Z. Kristallogr. 214, 427-429.

Houwink, A. L. (1956) Flagella, gas vacuoles and cell-wall structure in *Halobacterium halobium*; an electron microscope study. J. Gen. Microbiol. 15, 146-150.

Jarrel, K. F., Colvin, J. R., and Sprott, G. D. (1982) Spontaneous protoplast formation in *Methanobacterium bryantii*. J. Bacteriol. 149, 346-353.

Jing, H., Takagi, J., Liu, J., Lindgren, S., Zang, R., Joachimiak, A., Wang, J., and Springer, T. A. (2002) Archaeal surface layer proteins contain β propeller, PKD, and β helix domains and are related to metazoan cell surface proteins. Structure 10, 1453-1464.

Johannsen, L., Toth, L. A., Rosenthal, R. S., Opp, M. R., Obál Jr., F., Crady, A. B., and Krueger, J. M. (1990) Somnogenic, pyrogenic, and hematologic effects of bacterial peptidoglycan. Am. J. Physiol. 258, R182-R186.

Kärcher, U., Schröder, H., Haslinger, E., Allmaier, G., Schreiner, R., Wieland, F., Haselbeck, A., and König, H. (1993) Primary structure of the heterosaccharide of the surface glycoprotein of *Methanothermus fervidus*., J. Biol. Chem. 268, 26821-26826.

Kandler, O., and Hippe, H. (1977) Lack of peptidoglycan in the cell walls of *Methanosarcina barkeri*. Arch. Microbiol. 113, 57-60.

Kandler, O., and König, H. (1985) Cell envelopes of archaebacteria. In: C. R. Woese and R. S. Wolfe. (eds.). The Bacteria. A Treatise on Structure and Function. Archaebacteria, Academic Press, New York. Vol. VIII, pp 413-457.

Kandler, O., and König, H. (1993) Cell envelopes of Archaea: structure and chemistry. In: M. Kates, D. J. Kushner, A. T. Matheson (eds.). The Biochemistry of Archaea (Archaebacteria), Elsevier Science Publishers, Amsterdam. pp. 223-259.

Kansy, W. J., Carinato, M. E., Monteggia, L. M., and Konisky, J. (1994) In vivo transcripts of the S-layer encoding structural gene of the archaeon *Methanococcus voltae*. Gene 148, 131-135.

Karlson, P., Doenecke, D., and Koolman, J. (1994) Kurzes Lehrbuch der Biochemie für Mediziner und Naturwissenschaftler. G. Thieme Verlag, Stuttgart.

Kiener, A., König, H., Winter, J., and Leisinger, T. (1987) Purification and use of *Methanobacterium wolfei* pseudomurein endopeptidase for lysis of *Methanobacterium thermoautotrophicum*.. J. Bacteriol. 169, 1010-1016.

König, H., Semmler, R., Lerp, C., and Winter, J. (1985) Evidence for the occurrence of autolytic enzymes in *Methanobacterium wolfei*. Arch. Microbiol. 141, 177-180.

König, H., Hartmann, E., and Kärcher, U. (1994) Pathways and principles of the biosynthesis of methanobacterial cell wall polymers. System. Appl. Microbiol. 16, 510-517.

Kreisl, P., and Kandler, O, (1986) Chemical structure of the cell wall polymer of *Methanosarcina*. System. Appl. Microbiol. 7, 293-299.

Langworthy, T. A., Tornabene, T. G., and Holzer, G. (1982) Lipids of archaebacteria. Zentralbl. Bakteriol., Mikrobiol. Hyg., Abt. 1, Orig. C3, 228-244.

Lechner, J., and Sumper, M. (1987) The primary structure of a procaryotic glycoprotein. Cloning and sequencing of the cell surface glycoprotein gene of halobacteria. J. Biol. Chem. 262, 9724-9729.

Mayerhofer, L. E., Macario, A. L. J., Conway de Macario, E. (1992) Lamina, a novel multicellular form of *Methanocarcina mazei* S-6. J. Bacteriol. 174, 309-314.

Mescher, M. F., Strominger, J. L. (1976) Purification and characterisation of a prokaryotic glycoprotein from the cell envelope of *Halobacterium salinarium*. J. Biol. Chem. 251, 2005-2014.

Messner, P., and Sleytr, U. B. (1992) Crystalline bacterial cell-surface layers. Adv. Microb. Physiol. 33, 213-274.

Messner, P., Allmaier, G., Schäffer, C., Wugeditsch, T., Lortal, S., König, H., Niemetz, R., and Dorner, M. (1997) Biochemistry of S-layers. FEMS Microbiol. Rev. 20, 25-46.

Niemetz, R., Kärcher, U., Kandler, O., Tindall, B. J., and König, H. (1997) The cell wall polymer of the extremely halophilic archaeon *Natronococcus occultus*. Eur. J. Biochem. 249, 905-911.

Reistadt, R. (1972) Cell wall of an extremely halophilic coccus. Investigation of ninhydrin-positive compounds. Arch. Microbiol. 82, 24-30.

Sára, M., and Sleytr, U. B. (2000) S-layer proteins. J. Bact. 182, 859-868.

Sauer, F. D., Mahadevan, S., and Erfle, J. D. (1984) Methane synthesis by membrane vesicles and a cytoplasmic cofactor isolated from *Methanobacterium thermoautotrophicum*. Biochem. J. 221, 61-69.

Schleifer, K. H., Steber, J., and Mayer, H. (1982) Chemical composition and structure of the cell wall of *Halococcus morrhuae*. Zentralbl. Bakt., Mikrobiol. Hyg., Serie C3, 171-178.

Sleytr, U. B., Sára, M., Messner, P., and Pum, D. (1994) Two-dimensional protein crystals (S-layers): fundamentals and applications. J. Cell. Biochem. 56, 171-176.

Smith, D. R. and 36 other authors (1997) Complete genome sequence of *Methanobacterium thermoautotrophicum* ΔH: functional and comparative genomics. J. Bacteriol. 179, 7135 -7155.

Stimpson, S. A., Brown, R. R., Anderle, S. K., Klapper, D. G., Clark, R. L., Cromartie, W. J., and Schwab, J. H. (1986) Arthropathic properties of cell wall polymers from normal flora bacteria. Infect. Immun. 51, 240-249.

Sumper, M., Berg, E., Mengele, R., and Strobel, I. (1990) Primary structure and glycosylation of the S-layer protein of *Haloferax volcanii*. J. Bacteriol. 172, 7111-7118.

Sumper, M., and Wieland, F. T. (1995) Bacterial glycoproteins. In: J. Montreuil, J. F. G. Vliegenthart, and H. Schachter (eds.). Glycoproteins, Elsevier, Amsterdam, pp. 455-473.

Thomm, M. (1996) Archaeal transcription factors and their role in transcription inititation. FEMS Microbiol. Rev. 18,159-171.

Wieland, F. T. (1988) The cell surfaces of halobacteria. In: F. Rodriguez-Valera (ed.). Halophilic Bacteria, CRC Press Inc., Boca Raton. Vol. II., pp. 55-65.

Weiss, L. R. (1974) Subunit cell wall of *Sulfolobus acidocaldarius*. J. Bacteriol. 118, 275-284.

Woese, C. R. (1987) Bacterial evolution, Microbiol. Rev. 51, 221-271.

Woese, C. R., Kandler, O., and Wheelis, M. L. (1990) Towards a natural system of organisms: Proposal for the domains Archaea, Bacteria, Eucarya. Proc. Natl. Acad. Sci. USA 87, 4576-4579.

Yang, L. L., and Haug, A. (1979) Purification and partial characterization of a procaryotic glycoprotein from the plasma membran of *Thermoplasma acidophilum*. Biochim. Biophys. Acta 556, 277-365.

Yao, R. Macario, A. J. L., and Conway de Macario, E. (1994) An archaeal S-layer gene homolog with repetitive subunits. Biochim. Biophys. Acta 1219, 697-700.

Zeitler, R., Hochmuth, E., Deutzmann, R., and Sumper, M. (1998) Exchange of Ser-4 for Val, Leu and Asn in the sequon Asn-Ala-Ser does not prevent N-glycosylation of the cell surface glycoprotein from *Halobacterium halobium*. Glycobiology 8, 1157-1164.

Biodata of **Derek R. Lovley** author of ***"Potential Role of Dissimilatory Iron Reduction in The Early Evolution of Microbial Respiration"***

Dr. Derek R. Lovley is a Professor at the Department of Microbiology, University of Massachusetts-Amherst, Amherst MA. He was honored as a Distinguished University Professor at his Department. Dr. Lovley obtained his Ph.D. (1982) at the Michigan State University in Microbiology. Among his awards are: the Division K Lecturer for the American Society for Microbiology, Fellow, American Academy of Microbiology, Grant Winner, Popular Science, U.S. Geological Survey Mendenhall Lecturer (highest scientific honor of USGS).

E-mail: **dlovley@mcrobio.umass.edu**

J. Seckbach (ed.), Origins, 299–313.

POTENTIAL ROLE OF DISSIMILATORY IRON REDUCTION IN THE EARLY EVOLUTION OF MICROBIAL RESPIRATION

DEREK R. LOVLEY
Department of Microbiology, University of Massachusetts-Amherst, Amherst MA 01003 USA

1. Introduction

In the absence of time travel, it may be impossible to ever definitively determine the mechanisms by which early life evolved on Earth. Proposals for early forms of microbial respiration should be consistent with mechanisms of energy conservation that would be possible under geochemical conditions likely to have been present at the time that life emerged. Furthermore, it might be expected that this early form of respiration would be highly conserved in those microorganisms most closely related to the last common ancestor(s) of modern life. As detailed below, of all the known forms of microbial respiration, electron transfer to Fe(III) best fits these geological and microbiological criteria.

Until recently, speculation about the earliest forms of microbial respiration has focused on virtually every known form of microbial respiration other than Fe(III) reduction. One reason for Fe(III) being ignored in the past is that Fe(III) reduction has only recently been recognized as a form of energy conservation (Lovley 1991). Although it has been known for many years that Fe(III) could be reduced during anaerobic growth of some microorganisms, these Fe(III)-reducing microorganisms had a primarily fermentative metabolism and only reduced Fe(III) as a side reaction in their metabolism (Lovley 1987). There was no evidence that this Fe(III) reduction yielded energy to support cell growth and it was often considered that Fe(III) was reduced by non-enzymatic reactions in the cultures. The ability of microorganisms to conserve energy to support growth with Fe(III) serving as the electron acceptor was first reported in the 1980s (Balashova and Zavarzin 1980; Lovley et al. 1987). Unlike other more well-studied forms of anaerobic respiration, the biochemical mechanisms for dissimilatory Fe(III) reduction are still largely unknown.

Microbial Fe(III) reduction is generally considered to be an environmentally significant process on modern Earth. Fe(III) is typically the most abundant potential electron acceptor for organic matter oxidation in most soils and sediments (Lovley 1991, 2000). Therefore, with the development of anoxic conditions, microbial Fe(III) reduction can become important in organic matter oxidation in environments such as flooded soils, aquatic sediments, or subsurface environments (Lovley 1991, 2000). In addition to contributing to the degradation of naturally occurring organic matter, microbial Fe(III) reduction can play an important role in the removal of organic contaminants from subsurface environments (Lovley 1995; Anderson and Lovley

1997). This is apparent, for example, in petroleum-contaminated aquifers in which Fe(III)-reducing microorganisms effectively remove aromatic hydrocarbons from groundwater (Lovley et al. 1989; Lovley and Lonergan 1990; Lovley et al. 1994; Anderson et al. 1998; Rooney-Varga et al. 1999; Snoeyenbos-West et al. 2000). Some Fe(III)-reducing microorganisms, such as *Geobacter* species, can substitute toxic metals, such as U(VI), for Fe(III) in their respiration and in the process aid in the bioremediation of metal-contaminated groundwaters (Lovley et al. 1991; Finneran et al. 2002; Holmes et al. 2002). Another practical use of Fe(III)-reducing microorganisms is their ability to transfer electrons onto electrodes, making it possible to harvest electricity from anoxic aquatic sediments and other sources of waste organic matter (Bond et al. 2002; Bond and Lovley 2003).

Speculation about the potential role of Fe(III)-reducing microorganisms on early Earth began with the discovery that microbial reduction of Fe(III) oxides could produce copious quantities of the magnetic mineral magnetite (Lovley et al. 1987; Lovley 1990). This provided a potential explanation for the magnetite accumulations in Precambrian Banded Iron Formations, which geological evidence suggested was formed as the result of the oxidation of organic matter to carbon dioxide, coupled to the reduction of Fe(III) oxide to magnetite (Baur et al. 1985; Walker 1987). The discovery of magnetite, presumably the result of microbial activity, deep in the Earth's surface led to the proposal of a deep, hot biosphere below the surface of Earth, and possibly other planets (Gold 1992). Magnetite with a morphology similar to that produced by Fe(III)-reducing microorganisms in a Martian meteorite was suggested to provide evidence for previous life on Mars (McKay et al. 1996). These discoveries led to the further investigation of whether Fe(III) reduction might be an important process in extreme environments, such as those found on early Earth.

2. Hyperthermophilic Fe(III)-Reducing Microorganisms

2.1. Fe(III) REDUCTION IS A HIGHLY CONSERVED CHARACTERISTIC OF HYPERTHERMOPHILES

It is often speculated that life emerged on a hot, early Earth (Baross and Hoffman 1985; Pace 1991; Holm 1992; Bock and Goode 1996). A common justification for this is that, based on 16S rDNA sequences, all of the extant microorganisms that are most closely related to the last common ancestor(s) of modern life are hyperthermophilic *Archaea* and *Bacteria*. An alternative explanation is that hyperthermophiles were the only life forms that survived some catastrophic, high temperature event that took place well after life had evolved. However, if it is accepted that the physiological characteristics of the most deeply branching extant organisms represent those found in the last common ancestor(s), then current microbiological evidence suggests that the earliest microorganisms were hyperthermophiles.

Using this same logic, speculations can also be made about the forms of respiration found in the last common ancestor(s). All of the commonly considered types of microbial respiration including oxygen reduction, nitrate reduction, sulfate reduction, and methanogenesis are found in representative deeply branching hyperthermophilic microorganisms (Stetter 1996) and thus each has been regarded as a potential early form of respiration by one or more investigators. Until 1998 (Vargas et al. 1998), it was not recognized that hyperthermophiles could use Fe(III) as an electron acceptor as previous investigations into the potential for hyperthermophiles to reduce Fe(III) had yielded negative results (Stetter 1996).

However, when a diversity of hypethermophiles that had been isolated on electron acceptors other than Fe(III) were tested for their ability to reduce Fe(III) with hydrogen as an electron donor, all of the organisms were capable of Fe(III) reduction (Vargas et al. 1998). Many of these organisms could also reduce extracellular quinones, such as humic acids (Lovley et al. 2000), which is a common feature of mesophilic Fe(III) reducers (Lovley et al. 1996; Lovley et al. 1998). Furthermore, those hyperthermophiles that were investigated in more detail for their ability to grow with Fe(III) serving as the sole electron acceptor, conserved energy to support growth from Fe(III) reduction. Most remarkable in this regard was the hyperthermophilic bacterium *Thermotoga maritima*. This organism was previously regarded to have a fermentative metabolism and even though it could transfer electrons to S°, S° reduction did not appear to be an energy-conserving form of respiration in *T. maritima*. In contrast, with Fe(III) as the electron acceptor, *T. maritima* grew with hydrogen as the sole electron donor and Fe(III) as the electron acceptor (Vargas et al. 1998). Hyperthermophiles such as *Archaeoglobus fulgidus* that do not reduce S° do reduce Fe(III) (Vargas et al. 1998), and as detailed below, providing Fe(III) as an electron acceptor also expands the metabolic capabilities of some other hyperthermophilic *Archaea*. These results suggest that Fe(III) reduction is more closely associated with central metabolism in some hyperthermophiles than reduction of S° or other electron acceptors. Every hyperthermophile that has been tested to date can reduce Fe(III), representing a wide phylogenetic diversity within the hyperthermophiles (Fig. 1). Thus, other than the ability to grow at high temperature, the capacity for Fe(III) reduction may be amongst the most highly conserved metabolic characteristics of hyperthermophiles.

2.2. NOVEL FORMS OF HYPERTHERMOPHILIC METABOLISM LINKED TO Fe(III) REDUCTION

Another example in which provision of Fe(III) as an electron acceptor was found to increase the metabolic capability of a hyperthermophile is the expanded range of carbon metabolism in *Ferroglobus placidus* in the presence of Fe(III). In the initial characterization of *F. placidus* it was concluded that it did not use organic electron donors to support growth with its known electron acceptors, nitrate and thiosulfate (Hafenbradl et al. 1996). However, when Fe(III) oxide was provided as an electron acceptor *F. placidus* could grow with acetate as the electron donor and Fe(III) as the electron acceptor (Tor et al. 2001). Acetate was oxidized to carbon dioxide with the following stoichiometry:

$$\text{acetate}^- + 8\ \text{Fe(III)} + 4\ H_2O \longrightarrow 2\ HCO_3^- + 8\ \text{Fe(II)} + 9\ H^+ \quad (1)$$

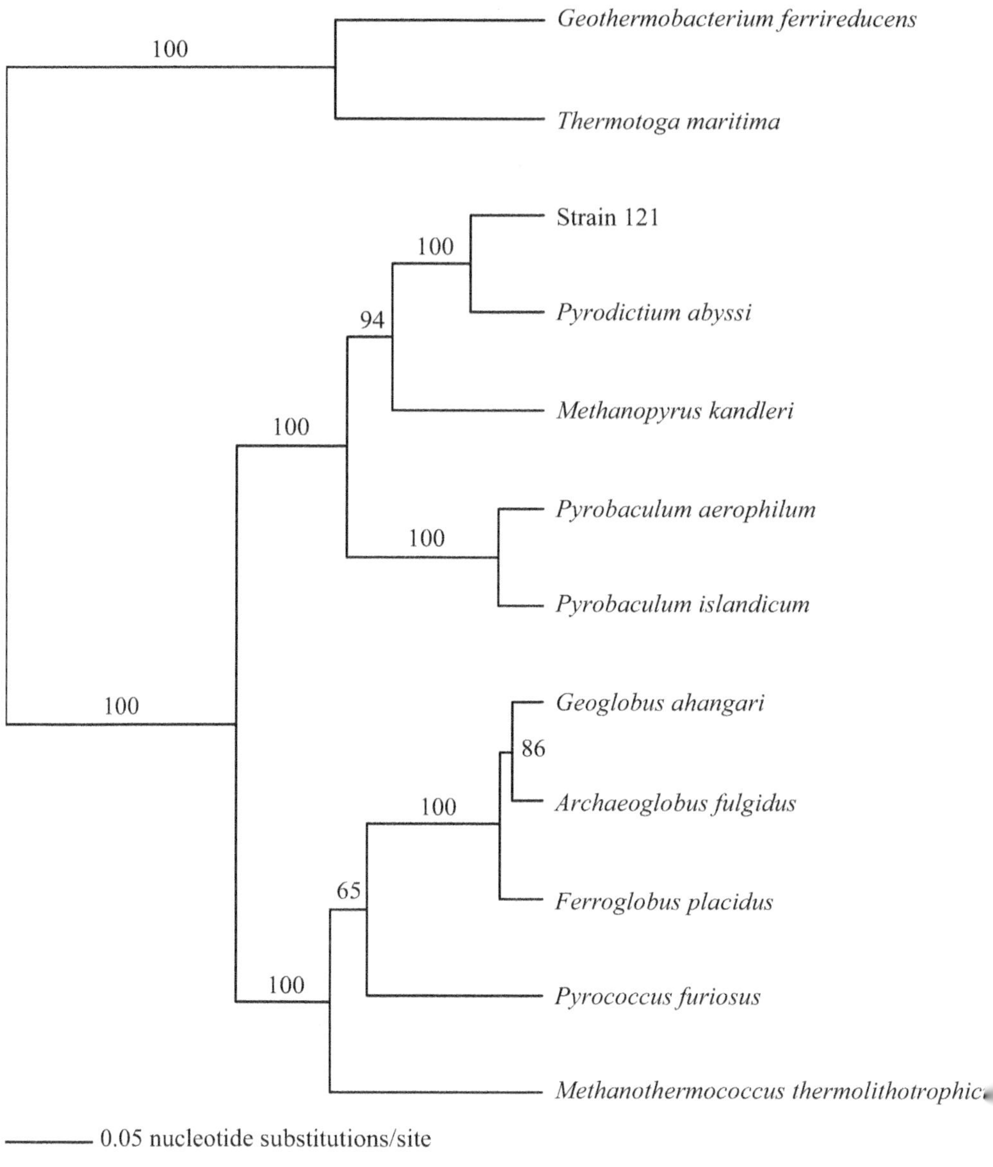

Figure 1. Phylogenetic tree constructed by maximum likelihood analysis showing the phylogenetic relationship of known Fe(III)-reducing hyperthermophiles. Bootstrap analysis was performed with 100 replicates. *Thermotoga maritima* served as an outgroup.

A novel isolate from the Guaymas Basin in the Gulf of California, *Geoglobus ahangari*, oxidized acetate in a similar manner (Tor et al. 2001; Kashefi et al. 2002b). These represent the only documented cases of anaerobic acetate oxidation in hyperthermophilic microorganisms. Although there have been instances in which it has been suggested that hyperthermophiles might anaerobically oxidize acetate with sulfate or sulfite as the electron acceptor (Volkl et al. 1993; Huber et al. 1997), quantitative data on acetate oxidation were not presented and subsequent attempts to grow these other organisms on acetate have been unsuccessful (Afshar et al. 1998; Tor et al. 2001). In fact, prior to the report of hyperthermophiles that could oxidize acetate with the reduction of Fe(III), it was considered that acetate produced in hot (i.e. > 80°C) microbial ecosystems would need to diffuse into cooler environments before microorganisms could metabolize it (Slobodkin et al. 1999b). Acetate is expected to be a key intermediate in the anaerobic degradation of organic matter in hot, microbial ecosystems (Tor et al. 2003), just as it is in cooler environments (Lovley and Chapelle 1995). The finding that there are hyperthermophiles that can oxidize acetate with Fe(III) suggests that acetate produced within hot microbial ecosystems can serve as an important energy source within these ecosystems, rather than being exported as an energy source for surrounding cooler environments. It seems likely that as the diversity of hyperthermophiles is further explored, hyperthermophiles capable of anaerobically oxidizing acetate with electron acceptors other than Fe(III) will also be found.

Further investigation of the metabolism of *F. placidus* and *G. ahangari* demonstrated that the capacity of these hyperthermophiles to oxidize organic compounds not previously known to be microbially degraded in hot microbial ecosystems extended beyond acetate. For example, *F. placidus* grew with a variety of aromatic compounds as the electron donor and Fe(III) oxide as the sole electron acceptor (Tor and Lovley 2001). To date, *F. placidus* is still the only hyperthermophile known to have this capability. The stoichiometry of benzoate and phenol uptake and Fe(III) reduction demonstrated that *F. placidus* completely oxidized these aromatic compounds to carbon dioxide with Fe(III) serving as the sole electron acceptor. Other aromatic compounds supporting growth included 4-hydroxybenzoate, benzaldehyde, *p*-hydroxybenzaldehyde, and *t*-cinnamic acid (3-phenyl-2-propenoic acid). *F. placidus* only oxidized aromatic compounds with Fe(III). Nitrate, which serves as an electron acceptor for growth on hydrogen or Fe(II), did not support growth on aromatic compounds.

Long-chain fatty acids are a significant component of organic matter in many environments. *G. ahangari* oxidized palmitate and stearate with Fe(III) as the electron acceptor (Kashefi et al. 2002b). The stoichiometry of Fe(III) reduction suggested that the long-chain fatty acids were completely oxidized to carbon dioxide. *G. ahangari* is the only hyperthermophile known to oxidize long-chain fatty acids.

The ability of Fe(III) reducers such as *F. placidus* and *G. ahangari* to oxidize organic acids and aromatic compounds, suggests for the first time that the complete oxidation of complex organic matter back to carbon dioxide may be possible in hot microbial ecosystems. Consortia of fermentative microorganisms and Fe(III) reducers should be able to oxidize the major components of organic matter. Since Fe(III) is likely to be available as an electron acceptor in many hot microbial ecosystems, including the deep hot subsurface (Gold 1992), petroleum reservoirs (Greene et al.

1997; Slobodkin et al. 1999a), terrestrial hot springs (Brock et al. 1976), and hydrothermal marine sediments (Jannasch 1995; Karl 1995), this may be a fairly common phenomenon. As discussed below, early Earth represents another environment in which such metabolism may have been common.

2.3. CULTURING "UNCULTURABLE" HYPERTHERMOPHILES WITH FE(III)

Many of the microorganisms that live in hot microbial ecosystems and have been detected via analysis of 16S rDNA sequences have yet to be cultured. Given the finding that the capacity for Fe(III) reduction is so highly conserved among hyperthermophiles, a likely strategy for recovering many of these "as-yet-uncultured" organisms is to use Fe(III) as the electron acceptor for isolation. This was evident in a study of microorganisms living in Calcite Springs in Yellowstone National Park. A novel bacterium, *Geothermobacterium ferrireducens*, was isolated with a novel procedure in which Fe(III) oxide was incorporated into solidified medium (Kashefi et al. 2002a). The 16S rDNA sequence of *G. ferrireducens* was closely related to 16S rDNA sequences that were abundant in Calcite Spring (Hugenholtz et al. 1998). The likely physiology of these previously uncultured microorganisms was in doubt because of a lack of close relatives in culture. *G. ferrireducens* grows exclusively with hydrogen as the electron donor and Fe(III) oxide as the electron acceptor (Kashefi et al. 2002a). Thus, one likely reason organisms in this phylogenetic clade had not been previously cultured is that few studies have employed Fe(III) as an electron acceptor. In a similar manner, *G. ahangari*, described above, only uses Fe(III) as an electron acceptor (Kashefi et al. 2002b). Another Fe(III) reducer, strain 121, which was isolated from a marine hydrothermal vent and has the highest upper temperature limit for growth of any known organism (Kashefi and Lovley 2003) also exclusively uses Fe(III). These results suggest that much of the vast diversity of hyperthermophilic microorganisms that have not yet been cultured may be microorganisms that use Fe(III), but not other electron acceptors such as sulfur forms, nitrate, oxygen, or carbon dioxide that have been used in the vast majority of previous isolation attempts.

2.4. GEOLOGICAL SIGNATURES FROM MICROBIAL METAL REDUCTION

Hyperthermophiles can reduce poorly crystalline Fe(III) oxide to ultra-fine grained crystals of the magnetic mineral magnetite (Kashefi and Lovley 2000). This is similar to the production of magnetite during microbial Fe(III) reduction at cooler temperatures (Lovley et al. 1987; Lovley 1990). Although accumulations of ultra-fine grained magnetite have been considered evidence for the activity of Fe(III)-reducing microorganisms in extreme environments (Gold 1992; McKay et al. 1996), at the present time there is no reliable way to distinguish the ultra-fine grained magnetite produced as the result of microbial Fe(III) reduction from magnetite produced by abiotic oxidation of Fe(II). The possibility that Fe(III) reducers might isotopically fractionate iron during Fe(III) reduction (Beard et al. 1999) might ultimately provide a method for distinguishing between these different mechanisms of magnetite formation. If so, then magnetite in ancient rocks might serve as a geological signature of the activity of hyperthermophiles on early Earth. However, strategies for accounting for abiological mechanisms for fractionation (Anbar et al. 2000) must first be developed.

In addition to Fe(III), hyperthermophiles can reduce a variety of other metals. The ability of a diversity of hyperthermophiles to reductively precipitate gold via reduction of soluble Au(III) to Au(0) might be responsible for the formation of some gold deposits (Kashefi et al. 2001). The reduction of alternative metals in hypethermophiles has been studied most intensively in *Pyrobaculum islandicum* which, in addition to Fe(III), can reduce U(VI), Tc(VII), Cr(VI), Co(III), Mn(IV), and Au(III) with hydrogen as the electron donor. The reduced products of some of these metals, most notably uranium, might provide a geological signature for the activity of hyperthermophilic Fe(III) reducers. For example, the precipitation of uranium as the result of the reduction of soluble U(VI) to insoluble U(IV) at temperatures of ca. 100°C is considered to have led to the formation of typical sandstone-type uranium deposits (Hostetler and Garrels 1962). Given that the lignite organic matter often associated with U(IV) deposits does not abiotically reduce U(VI) at temperatures below 120°C (Nakashima et al. 1984), the accumulation of the U(IV) minerals in environments that are considered to have been at 80-120° C at the time of U(VI) reduction could reasonably be assumed to be a geological signature of hyperthermophilic microorganisms. For example, it is tempting to speculate that the large U(IV) accumulations that formed the naturally generated Oklo nuclear reactor in the Precambrian period (Brookins 1990) was the result of the activity of hyperthermophiles reducing U(VI).

3. Hydrogen Oxidation Coupled to Fe(III) Reduction as the First Form as Microbial Respiration

Prebiotic Earth has been described as a "giant photoelectrochemical cell" (Russell and Hall 2002). Numerous studies (see Cairns-Smith et al. 1992; Russell and Hall 2002 for reviews) have suggested that high levels of ultraviolet radiation impinging upon the Archaean sea, which contained high levels of dissolved Fe(II), resulted in the formation of Fe(III) and hydrogen gas as follows:

$$2\ \mathrm{Fe(II)} + 2\ \mathrm{H^+} \xrightarrow{hv} 2\ \mathrm{Fe(III)} + \mathrm{H_2} \quad (2)$$

The H_2 formed in this manner was probably primarily lost to the atmosphere, whereas the Fe(III) would have precipitated as insoluble Fe(III) oxides, forming a "positive electrode" (Russell and Hall 2002). Geologically produced hydrogen emanating from the subsurface represented a "negative electrode". Other potential electron acceptors for hydrogen oxidation, such as oxygen, nitrate, or sulfate were probably not abundant (Cameron 1982; Walker and Brimblecombe 1985; Eastoe et al. 1990). If this geological scenario is correct, then conditions were highly favorable for the development of a biological entity that could take advantage of the energy available from hydrogen oxidation coupled to Fe(III) reduction.

In fact, it has been proposed that the development of the first membrane system capable of electron transport and energy conservation through a chemiosmotic mechanism started by catalyzing the oxidation of hydrogen with the reduction of Fe(III) (Russell et al. 1998; Russell and Hall 2002). Initially, such membranes were probably inorganic, possibly comprised of iron sulfide minerals, but eventually evolved into lipid membranes as life became more similar to known forms of extant life (Russell et al. 1998; Russell and Hall 2002). If so, then it seems likely that this earliest of organic life forms would also have been a hydrogen-oxidizing Fe(III)-reducing entity. This concept, originated by geologists, and based upon the best current information on the geochemical conditions on early Earth, matches well with the extrapolation from the physiology of hyperthermophilic *Bacteria* and *Archaea* that the last common ancestor(s) were hydrogen-oxidizing Fe(III) reducers. Although the last common ancestor(s) was likely to have been a metabolically sophisticated respiratory organism (Pace 1991; de Duve 1995; Woese 1998), it is probable that the more primitive microorganisms that preceded the last common ancestor would also have had the capacity to transfer electrons to extracellular Fe(III).

The next major advance in the development of microbial communities based on Fe(III) reduction is likely to have coincided with the evolution of photosynthesis. The earliest photosynthesis may have relied on Fe(II) as an electron donor (Hartman 1984; Widdel et al. 1993; Ehrenreich and Widdel 1994) or could have been oxygenic photosynthesis. The emergence of photosynthesis would have greatly increased the availability of organic matter while also generating large quantities of Fe(III), either from the direct oxidation of Fe(II) in photosynthesis or from oxygen abiotically oxidizing Fe(II). Thus, it is expected that as photosynthesis became important hydrogen-based Fe(III)-reducing communities were increasingly supplanted by Fe(III)-reducing microorganisms fueled primarily by organic matter oxidation. Such heterotrophic Fe(III)-reducing communities appear to have been prevalent by the early Precambrian period as evidenced by the accumulation of massive amounts of magnetite in the Precambrian Banded Iron Formations that can be attributed to the activity of Fe(III)-reducing microorganisms (Lovley 1991, 2000). The geological record indicates that this oxidation of organic matter coupled to Fe(III) reduction was globally significant before other mechanisms for anaerobic or aerobic oxidation of organic matter to carbon dioxide (Walker 1987).

3.1. WHY Fe(III) REDUCTION RATHER THAN CARBON DIOXIDE REDUCTION AS THE FIRST FORM OF RESPIRATION

Although many commonly considered electron acceptors such as oxygen, nitrate, and sulfate are not likely to have been prevalent when life evolved, carbon dioxide was. Thus, it is important to consider whether electron transfer to carbon dioxide, rather than Fe(III) reduction, could have been the first form of microbial respiration. Modern forms of energy conservation with hydrogen as the electron donor and carbon dioxide as the electron acceptor are methane production and acetogenesis. Both of these processes are biochemically complex in part due to the fact that there is little energy available from these reactions as well as kinetic constraints on carbon dioxide reduction. In contrast, Fe(III) is readily reduced by a wide variety of redox-active molecules. For example, it is much more likely that the iron-sulfide membranes proposed to have led to the evolution of life could have transferred electrons to Fe(III) than onto carbon dioxide. Even highly evolved, extant methanogens may preferentially transfer electrons to Fe(III) rather than produce methane (Bond and Lovley 2002). When Fe(III) is available, methanogens can maintain hydrogen concentrations at levels so low that methane production from hydrogen is not thermodynamically favorable (Bond and Lovley 2002). Thus, it seems unlikely methanogenesis could have prevailed over Fe(III) reduction in less highly regulated primitive life forms.

A less compelling, but notable argument against methanogenesis or acetogenesis being one of the earliest forms of respiration is that none of the deeply branching hyperthermophiles are acetogens and few have the capacity for methane production. There are no methanogens among the *Bacteria*. In contrast, as discussed above, the ability to oxidize hydrogen with the reduction of Fe(III) is highly conserved among deeply branching hyperthermophilic *Bacteria* as well as *Archaea*.

3.2. EVIDENCE THAT HYDROGEN-BASED MICROBIAL ECOSYSTEMS CAN EXIST

A microbial community fueled primarily by hydrogen and using Fe(III) as the sole electron acceptor, as proposed for the earliest life, is unlikely to be found on modern Earth because 1) most environments even in the subsurface contain organic matter capable of supporting microbial growth and 2) environments devoid of organic matter and containing hydrogen are now restricted to deep subsurface environments in which there is no mechanism for generating significant quantities of Fe(III). However, the concept that hydrogen could have served as the primary energy source for hydrogen-based microbial ecosystems on Early Earth has been supported by the discovery of a hydrogen-based community in deeply buried igneous subsurface rock in Idaho (Chapelle et al. 2002). Molecular analyses indicated that over 90% of the microorganisms in this environment were hydrogen-oxidizing, methane-producing microorganisms. Geochemical data demonstrated that hydrogen was present at levels capable of supporting methanogenesis, but that significant quantities of organic matter were not available. Although there have been other claims for hydrogen-based microbial communities, subsequent geochemical and microbiological analyses have suggested that organic matter was the primary energy source in those other environments (Anderson et al. 1998). However, there is now at least one site in which a hydrogen-based microbial exists.

4. Future Directions and Conclusions

There is so little geological and biological data available from early Earth that is relevant to the evolution of the first forms of microbial respiration that the topic of which form of respiration evolved first is certain to be debated for some time. However, a strong case can be made for hydrogen oxidation coupled to Fe(III) reduction being the first form of microbial respiration because: 1) it is likely that Fe(III) was abundant at the time life evolved, whereas other commonly considered electron acceptors such as oxygen, nitrate, or sulfate were not; 2) it is also likely that hydrogen was present in sufficient quantities to serve as an electron donor for Fe(III) reduction; 3)

Fe(III) reduction can be catalyzed by simpler biochemical mechanisms than those required for reduction of carbon dioxide, the only other abundant electron acceptor available; and 4) the capacity for hydrogen oxidation coupled to Fe(III) reduction is a much more highly conserved metabolic capability than the reduction of other known electron acceptors among the hyperthermophilic *Bacteria* and *Archaea* that are most closely related to the last common ancestor(s).

The finding that deeply branching hyperthermophiles have the capacity for Fe(III) reduction leads to the question of whether the same mechanisms for Fe(III) reduction are conserved in Fe(III)-reducing mesophilic microorganisms in the *Bacteria*, which are considered to have evolved later. At present, there have only been preliminary investigations into the mechanisms of Fe(III) reduction in hyperthermophiles (Childers and Lovley 2001). Although there has been considerably more study of Fe(III) reduction in mesophilic *Bacteria*, the mechanisms for electron transport to Fe(III) have yet to be elucidated (Lovley 2000).

The finding that the capacity for Fe(III) reduction is so highly conserved in hyperthermophiles suggests that Fe(III) reduction is still an important process in modern, hot microbial ecosystems. Many hot environments, which are cool enough to support life, are at anoxic/oxic boundaries in which hot, anoxic, Fe(II)-rich hydrothermal fluids contact cooler, oxic environments. The oxidation of Fe(II) at the anoxic/oxic boundary can produce significant quantities of Fe(III) which precipitates onto the surrounding hot sediments where it can potentially serve as an electron acceptor (Jannasch 1995; Karl 1995). Thus, although, Fe(III) reduction in hot, microbial ecosystems has received little attention, a better understanding of this process is likely to provide significant insight into microbial activity in hot environments such as those that may have led to the evolution of early life.

5. Acknowledgements

I thank Dawn Holmes for preparing the phylogenetic tree and Kazem Kashefi and Jason Tor for sharing unpublished data. This research was supported by the National Science Foundation LExEn Program grant MCB-0085365.

6. References

Afshar, S., Kim, C., Monbouquette, H. G. and Schroder, I. (1998). Effect of tungstate on nitrate reduction by the hyperthermophilic archaeon *Pyrobaculum aerophilum*. Appl. Environ. Microbiol. 64, 3004-3008.

Anbar, A., Roe, J., Barling, J. and Nealson, K. H. (2000). Nonbiological fractionation of iron isotopes. Science. 288, 126-128.

Anderson, R. T., Chapelle, F. H. and Lovley, D. R. (1998). Evidence against hydrogen-based microbial ecosystems in basalt aquifers. Science. 281, 976-977.

Anderson, R. T. and Lovley, D. R. (1997). Ecology and biogeochemistry of in situ groundwater bioremediation. Adv. Microbial Ecol. 15, 289-350.

Anderson, R. T., Rooney-Varga, J., Gaw, C. V. and Lovley, D. R. (1998). Anaerobic benzene oxidation in the Fe(III)-reduction zone of petroleum-contaminated aquifers. Environ. Sci. Technol. 32, 1222-1229.

Balashova, V. V. and Zavarzin, G. A. (1980). Anaerobic reduction of ferric iron by hydrogen bacteria. Microbiology. 48, 635-639.

Baross, J. A. and Hoffman, S. E. (1985). Submarine hydrothermal vents and associated gradient environments as sites for the origin and evolution of life. Origins of Life. 15, 327-345.

Baur, M. E., Hayes, J. M., Studley, S. A. and Walter, M. R. (1985). Millimeter-scale variations of stable isotope abundances in carbonates from banded iron-formations in the Hamersley Group of Western Australia. Econ. Geol. 80, 270-282.

Beard, B. L., Johnson, C. M., Cox, L., Sun, H., Nealson, K. H. and Aguilar, C. (1999). Iron isotope biosignatures. Science 285, 1889-1892.

Bock, G. R. and Goode, J. A. (1996). *Evolution of hydrothermal ecosystems on Earth (and Mars?).* West Sussex, England, John Wiley & Sons Ltd.

Bond, D. R., Holmes, D. E., Tender, L. M. and Lovley, D. R. (2002). Electrode-reducing microorganisms harvesting energy from marine sediments. Science 295, 483-485.

Bond, D. R. and Lovley, D. R. (2002). Reduction of Fe(III) oxide by methanogens in the presence and absence of extracellular quinones. Environ. Microbiol. 4, 115-124.

Bond, D. R. and Lovley, D. R. (2003). Electricity production by *Geobacter sulfurreducens* attached to electrodes. Appl. Environ. Microbiol. 69, 1548-1555.

Brock, T. D., S., C., Petersen, S. and Mosser, J. L. (1976). Biogeochemistry and bacteriology of ferrous iron oxidation in geothermal habitats. Geochim. Cosmochim. Acta 40, 493-500.

Brookins, D. G. (1990). Radionuclide behavior at the Oklo nuclear reactor, Gabon. Waste Manage. 10, 285-296.

Cairns-Smith, A. G., Hall, A. J. and Russell, M. J. (1992). Mineral theories of the origin of life and an iron sulfide example. Orig. Life Evol. Biosphere.22, 161-180.

Cameron, E. M. (1982). Sulphate and sulphate reduction in early Precambrian oceans. Nature 296, 145-148.

Chapelle, F. H., O'Neill, K., Bradley, P. M., Methé, B. A., Ciufo, S. A., Knobel, L. L. and Lovley, D. R. (2002). A hydrogen-based subsurface community dominated by methanogens. Nature 415, 312-316.

Childers, S. and Lovley, D. R. (2001). Differences in Fe(III) reduction in the hyperthermophilic archaeon, *Pyrobaculum islandicum*, versus mesophilic Fe(III)-reducing bacteria. FEMS Microbiol. Lett. 195, 253-258.

de Duve, C. (1995). *Vital Dust.* New York, Basic Books.

Ehrenreich, A. and Widdel, F. (1994). Anaerobic oxidation of ferrous iron by purple bacteria, new type of phototrophic metabolism. Appl. Environ. Microbiol. 60, 4517-4526.

Eastoe, C. J., Gustin, M. S., Hurlbut, D. F. and Orr, R. L. (1990). Sulfur isotopes in early Proterozoic volcanogenic massive sulfide deposits: new data from Arizona and implications for ocean chemistry. Precambrian Res. 46, 353-364.

Finneran, K., Anderson, R. T., Nevin, K. P. and Lovley, D. R. (2002). Bioremediation of uranium-contaminated aquifers with microbial U(VI) reduction. Soil and Sediment Contamination. 11, 339-357.

Gold, T. (1992). The deep, hot biosphere. Proc. Natl. Acad. Sci. USA. 89, 6045-6049.

Greene, A. C., Patel, B. K. C. and Sheehy, A. J. (1997). *Deferribacter thermophilus* gen. nov., sp. nov., a novel thermophilic manganese- and iron-reducing bacterium isolated from a petroleum reservoir. Int.. J. Syst. Bacteriol. 47, 505-509.

Hafenbradl, D., Keller, M., Dirmeier, R., Rachel, R., Robnagel, P., Burggraf, S., Huber, H. and Stetter, K. O. (1996). *Ferroglobus placidus* gen nov., sp. nov., a novel hyperthermophilic archaeum that oxidizes Fe^{+2} at neutral pH under anoxic conditions. Arch. Microbiol. 166, 308-314.

Hartman, H. (1984). The evolution of photosynthesis and microbial mats: a speculation on the banded iron formations. In: Y. Cohen, R.W. Castenholz, and H.O. Halverson (eds.) *Microbial Mats: Stromatolites*. New York, Alan R. Liss, Inc.: 449-453.

Holm, N. G. (1992). Why are hydrothermal systems proposed as plausible environments for the origin of life? Origins Life Evol. Biosphere.22, 5-14.

Holmes, D. E., Finneran, K. T. and Lovley, D. R. (2002). Enrichment of *Geobacteraceae* associated with stimulation of dissimilatory metal reduction in uranium-contaminated aquifer sediments. Appl. Environ. Microbiol. 68, 2300-2306.

Hostetler, P. B. and Garrels, R. M. (1962). Transportation and precipitation of uranium and vanadium at low temperatures with special reference to sandstone-type uranium. Econ. Geol. 57, 137-167.

Huber, H., Jannasch, H., Rachel, R., Fuchs, T. and Stetter, K. O. (1997). *Archaeoglobus veneficus* sp. nov., a novel facultative chemolithotrophic hyperthermophilic sulfite reducer, isolated from abyssal black smokers. System. Appl. Microbiol. 20, 374-380.

Hugenholtz, P., Pitulle, C., Hershberger, K. L. and Pace, N. R. (1998). Novel division level bacterial diversity in a Yellowstone hot spring. J. Bacteriol. 180, 366-376.

Jannasch, H. W. (1995). Microbial interactions with hydrothermal fluids. *Seafloor Hyrothermal Systems: Physical, Chemical, Biological, and Geological Interactions*, Geophysical Monograph 91. 273-296.

Karl, D. M. (1995). Ecology of free-living hydrothermal vent microbial communities. *The Microbiology of Deep-Sea Hydrothermal Vents*. D. M. Karl. New York, CRC Press: 35-124.

Kashefi, K., Holmes, D. E., Reysenbach, A.-L. and Lovley, D. R. (2002a). Use of Fe(III) as an electron acceptor to recover previously uncultured hyperthermophiles: isolation and characterization of *Geothermobacterium ferrireducens*, gen., nov., sp. nov. Appl. Environ. Microbiol. 68, 1735-1742.

Kashefi, K. and Lovley, D. R. (2000). Reduction of Fe(III), Mn(IV), and toxic metals at 100 °C by *Pyrobaculum islandicum*. Appl. Environ. Microbiol. 66, 1050-1056.

Kashefi, K. and Lovley, D. R. (2003). Extending the upper temperature limit for life. Science (submitted).

Kashefi, K., Tor, J., Nevin, K. P. and Lovley, D. R. (2001). Reductive precipitation of gold by dissimilatory Fe(III)-reducing *Bacteria* and *Archaea*. Appl. Environ. Microbiol. 67, 3275-3279.

Kashefi, K., Tor, J. M., Holmes, D. E., VanPraagh, C. V. G., Reysenbach, A.-L. and Lovley, D. R. (2002b). *Geoglobus ahangari*, gen. nov., sp. nov., a novel hyperthermophilic Archaeum capable of oxidizing organic acids and growing autotrophically on hydrogen with Fe(III) serving as the sole electron acceptor. Int. J. Syst. Evol. Microbiol. 52, 719-728.

Lovley, D. R. (1987). Organic matter mineralization with the reduction of ferric iron: A review. Geomicrobiol. J. 5, 375-399.

Lovley, D. R. (1990). Magnetite formation during microbial dissimilatory iron reduction. *Iron Biominerals*. R. B. Frankel and R. P. Blakemore. New York, Plenum Press: 151-166.

Lovley, D. R. (1991). Dissimilatory Fe(III) and Mn(IV) reduction. Microbiol. Rev. 55, 259-287.

Lovley, D. R. (1995). Bioremediation of organic and metal contaminants with dissimilatory metal reduction. J. Industr. Microbiol. 14, 85-93.

Lovley, D. R. (2000). Fe(III) and Mn(IV) Reduction. *Environmental Microbe-Metal Interactions*. D. R. Lovley. Washington, D.C., ASM Press: 3-30.

Lovley, D. R., Baedecker, M. J., Lonergan, D. J., Cozzarelli, I. M., Phillips, E. J. P. and Siegel, D. I. (1989). Oxidation of aromatic contaminants coupled to microbial iron reduction. Nature 339, 297-299.

Lovley, D. R. and Chapelle, F. H. (1995). Deep subsurface microbial processes. Rev. Geophsy. 33, 365-381.

Lovley, D. R., Coates, J. D., Blunt-Harris, E. L., Phillips, E. J. P. and Woodward, J. C. (1996). Humic substances as electron acceptors for microbial respiration. Nature. 382, 445-448.

Lovley, D. R., Fraga, J. L., Blunt-Harris, E. L., Hayes, L. A., Phillips, E. J. P. and Coates, J. D. (1998). Humic substances as a mediator for microbially catalyzed metal reduction. Acta Hydrochim. Hydrobiol. 26, 152-157.

Lovley, D. R., Kashefi, K., Vargas, M., Tor, J. M. and Blunt-Harris, E. L. (2000). Reduction of humic substances and Fe(III) by hyperthermophilic microorganisms. Chem. Geol. 169, 289-298.

Lovley, D. R. and Lonergan, D. J. (1990). Anaerobic oxidation of toluene, phenol, and p-cresol by the dissimilatory iron-reducing organism, GS-15. Appl. Environ. Microbiol. 56, 1858-1864.

Lovley, D. R., Phillips, E. J. P., Gorby, Y. A. and Landa, E. R. (1991). Microbial reduction of uranium. Nature. 350, 413-416.

Lovley, D. R., Stolz, J. F., Nord, G. L. and Phillips, E. J. P. (1987). Anaerobic production of magnetite by a dissimilatory iron-reducing microorganism. Nature. 330, 252-254.

Lovley, D. R., Woodward, J. C. and Chapelle, F. H. (1994). Stimulated anoxic biodegradation of aromatic hydrocarbons using Fe(III) ligands. Nature. 370, 128-131.

McKay, D. S., Gibson Jr., E. K., Thomas-Deprta, K. L., Vali, H., Romanek, C. S., Clement, S. J., Chillier, X. D. F., Maechling, C. R. and Zare, R. N. (1996). Search for past life on Mars: possible relic biogenic activity in Martian meteorite ALH84001. Science. 273, 924-930.

Nakashima, S., Disnar, J. R., Perruchot, A. and Trichet, J. (1984). Experimental study of mechanisms of fixation and reduction of uranium by sedimentary organic matter under diagenetic or hydrothermal conditions. Geochim. Cosmochim. Acta. 48, 2321-2329.

Pace, N. R. (1991). Origin of life-facing up to the physical setting. Cell. 65, 531-533.

Rooney-Varga, J. N., Anderson, R. T., Fraga, J. L., Ringelberg, D. and Lovley, D. R. (1999). Microbial communities associated with anaerobic benzene mineralization in a petroleum-contaminated aquifer. Appl. Environ. Microbiol. 65, 3056-3063.

Russell, M. J., Daia, D. E. and Hall, A. J. (1998). The emergence of life from FeS bubbles at alkaline hot springs in an acid ocean. *Thermophiles: The Keys to Molecular Evolution and the Origin of Life?* J. Wiegel and M. W. W. Adams. Philadelphia, PA, Taylor & Francis Ltd.: 77-126.

Russell, M. J. and Hall, A. J. (2002). Chemiosmotic coupling and transition element clusters in the onset of life and photosynthesis. Geochem. News 113, 6-12.

Slobodkin, A., Jeanthon, C., L'Haridon, S., Nazina, T. and Miroshnichenko, M. (1999a). Dissimilatory reduction of Fe(III) by thermophilic bacteria and archaea in deep subsurface petroleum reservoirs in Western Siberia. Curr. Microbiol. 39, 99-102.

Slobodkin, A. I., Zavarzina, D. G., Sokolova, T. G. and Bonch-Osmolovskaya, E. A. (1999b). Dissimilatory reduction of inorganic electron acceptors by thermophilic anaerobic prokaryotes. Microbiology. 68, 600-622.

Snoeyenbos-West, O. L., Nevin, K. P. and Lovley, D. R. (2000). Stimulation of dissimilatory Fe(III) reduction results in a predominance of *Geobacter* species in a variety of sandy aquifers. Microbial Ecol. 39, 153-167.

Stetter, K. O. (1996). Hyperthermophilic procaryotes. FEMS Microbiol. Rev. 18, 149-158.

Tor, J. M., Amend, J. P. and Lovley, D. R. (2003). Metabolism of organic compounds in anaerobic, hydrothermal sulfate-reducing sediments. Environ. Microbiol. (in press).

Tor, J. M., Kashefi, K. and Lovley, D. R. (2001). Acetate oxidation coupled to Fe(III) reduction in hyperthermophilic microorganisms. Appl. Environ. Microbiol. 67, 1363-1365.

Tor, J. M. and Lovley, D. R. (2001). Anaerobic degradation of aromatic compounds coupled to Fe(III) reduction by *Ferroglobus placidus*. Environ. Microbiol. 3, 281-287.

Vargas, M., Kashefi, K., Blunt-Harris, E. L. and Lovley, D. R. (1998). Microbiological evidence for Fe(III) reduction on early Earth. Nature. 395, 65-67.

Volkl, P., Huber, R., Drobner, E., Rachel, R., Burggraf, S., Trincone, A. and Stetter, K. O. (1993). *Pyrobaculum aerophilum* sp. nov, a novel nitrate-reducing hyperthermophilic Archaeum. Appl. Environ. Microbiol. 59, 2918-2926

Walker, J. C. G. (1987). Was the Archaean biosphere upside down? Nature. 329, 710-712.

Walker, J. C. G. and Brimblecombe, P. (1985). Iron and sulfur in the pre-biologic ocean. Precambrian Research 28, 205-222.

Widdel, F., Schnell, S., Heising, S., Ehrenreich, A., Assmus, B. and Schink, B. (1993). Ferrous iron oxidation by anoxygenic phototrophic bacteria. Nature. 362, 834-835.

Woese, C. (1998). The universal ancestor. Proc. Natl. Acad. Sci. USA. 95, 6854-6859.

Biodata of **Yuko H. Itoh** author (with co-author T. Itoh) of ***"The Evolution of Lipids in Prokaryotes"***

Dr. Yuko H. Itoh is a Professor in the Department of Bioinformatics, Faculty of Engineering, Soka University, Tangi-cho, Hachioji, Tokyo, Japan. She received her PhD (1975) in Medical Science from Keio University, Tokyo, on 'Specific inhibitors for RNA dependent RNA polyhmerase.

E-mail: **Yhitoh@t.soka.ac.jp**

Dr. Toshihiro Itoh is a Professor at the Division of Chemistry, Center for Natural Science, Kitasato University, Sagamihara, Kanagawa, Japan. He obtained his PhD (1975) in Agricultural Science from Nagoya University, Aichi, Japan. His thesis was on "Isolation and characterization of a novel phospholipid, pyrophosphatidic acid, from yeast."

E-mail: **itoht@class.kitasato-u.ac.jp**

T. Itoh and Y.H. Itoh

J. Seckbach (ed.), Origins, 315–328.

THE EVOLUTION OF LIPIDS IN PROKARYOTES

Y. H. ITOH[1], and T. ITOH[2]
[1]*Department of Bioengineering, Faculty of Engineering, Soka University, 1-236 Tangi-cho, Hachioji, Tokyo 192-8577, Japan.*
[2]*Division of Chemistry, Center for Natural Science, Kitasato University, 1-15-1 Kitasato, Sagamihara, Kanagawa 228-8555, Japan.*

1. What is a Prokaryote?

The basic structure of all living organisms on the earth is the cell. There are two categories of organisms, prokaryotes and eukaryotes, based on cellular organization such as chromosomal structures and the intracellular conformations. The most obvious difference of prokaryotes from eukaryotes is the absence of a nuclear membrane, which means that the nucleoplasm (genophore) is not separated from the cytoplasm by a unit–membrane system. The prokaryotic cells also lack membrane-bound subcellular organelles except for the thylakoids of cyanobacteria (oxygenic photosynthetic bacteria), anoxygenic phototrophs, nitrofyers, and methanotrophs, and for the nuclear envelope of some *Planctomycetes* species. Prokaryotes usually exist as single cells or simple associations of similar cells. The size of the smallest dimension of prokaryotic cells is usually 0.2-10.0 μm while that of eukaryotic cells is usually 2.0 μm or much larger. Before the development of DNA technology, it was difficult to distinguish prokaryotic organisms from microscopic eukaryotes without electron microscopy of thin sections. Now it is possible to determine and compare sequences of the small subunit rRNA (ss rRNA) genes from a significant number of prokaryotes and eukaryotes; such phylogenetic classification of microorganisms is still in progress.

Prokaryotes are divided into two fundamentally different kinds, Archaea (also called Archaebacteria or Archaebacteria) and Bacteria (also called eubacteria) by the molecular evolutional studies of the small subunit rRNA started by Woese and Fox in the 1970s. The majority of trees based on phylogenetic analyses using conserved amino acid sequences of some proteins support a closer relationship between the Archaea and the Eukarya than between the Archaea and the Bacteria. Especially, many archaeal information processing functions are similar to the eukaryotic mechanisms and not to the bacterial ones. It is now generally accepted that the Archaea, the Bacteria and the Eukarya (also called the eukaryotes) evolved by separation of their major evolutionary pathways from a common ancestral cell (Figure 1) . However, the exact location where the deepest branching occurs is not clear (Woese et al., 1990; Woese, 2000). Because there are many thermophilic microorganisms near the root of the universal phylogenetic tree, the common ancestral cell may well have been a thermophilic microorganism.

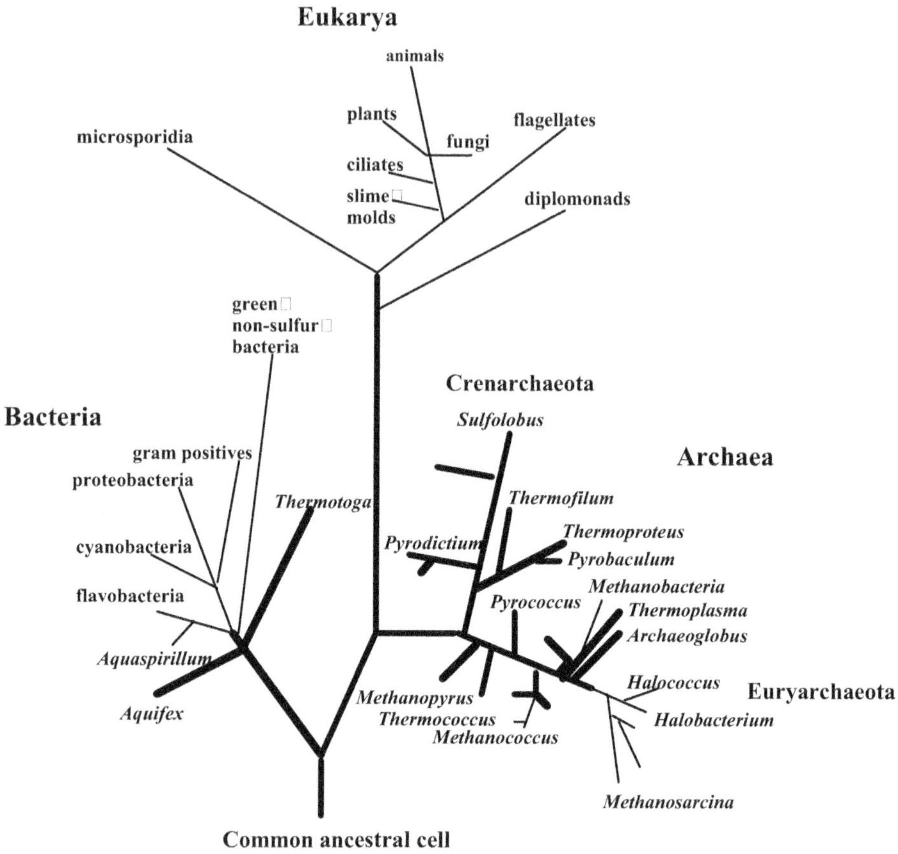

Figure 1. Universal phylogenetic tree based on small subunit rRNA gene sequences (Modified from Woese et al., 1990). Bold lines lead to hyperthermophiles (Modified from Baross et al., 1996).

The archaeal membrane lipids differ significantly from the eukaryal and the bacterial ones (Figure 2). All archaeal strains have unique membrane phospholipids, which consist of exclusively di-, and tetraether glycerol and long chain isoprenoid alcohols instead of fatty acids linked by ester linkages in the bacterial and eukaryal glycerophospholipids. One of the more remarkable differences between the archaeal lipids and the bacterial and eukaryal lipids is the stereochemical opposite structure of the glycerophosphate backbone of the phospholipids. The stereostructure of the glycerophosphate backbone of archaeal phospholipids is *sn*-glycerol-1-phosphate (G-1-P); on the other hand, the bacterial and eukaryal counterpart is the enantiomeric *sn*-glycerol-3-phosphate (G-3-P) (Langworthy and Pond, 1986; Koga et al., 1998)

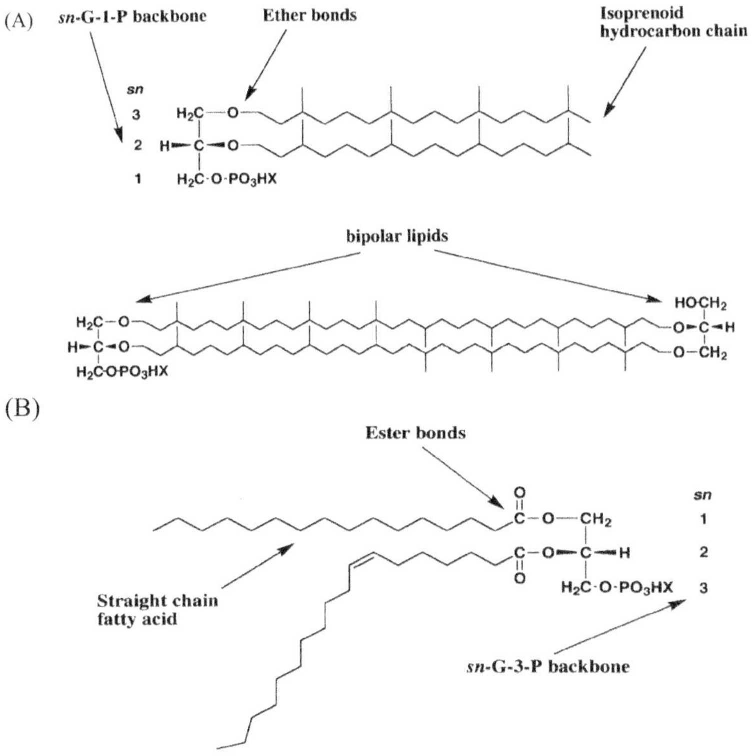

x = choline, ethanolamine, serine, glycerol, etc.

Figure 2. Structures of archaeal polar lipids (A) and bacterial and eukaryal polar lipids (B).

In this chapter, the lipids as the essential structural component of a cell will be described from the point of view of the evolution of their structures and functions.

2. Structures and Functions of Lipids in Prokaryotes

Lipid molecules on the earth display considerable structural and functional diversities at the present time. The molecules of lipids contain long chain fatty acids or a similar structure of hydrocarbon chains, which exist *in vivo* or derived from living organisms. Any group of organic compounds that includes triacylglycerides, sterols, terpens, waxes, and hydrocarbons consists of the simple lipids. The typical complex lipids are the diacylglycerides that contain additional elements such as phosphorus, nitrogen, or sulfur, or small hydrophilic carbon compounds such as sugars, ethanolamine, serine, choline, or *myo*-inositol. The lipids containing phosphate groups are called the phospholipids. The major functions of lipids are as follows: 1) components of cell membranes (cytoplasmic membrane and membranes of subcellular organelles), 2) components of cell walls such as lipopolysaccharides (LPS) in Gram negative bacteria, 3) carriers of energy or reservoir of structural building blocks such as glycogen and poly-β-hydroxyalkanoates (PHA) in many Archaea and Bacteria, 4) cofactors for several enzymes, and 5) mediators of signal transduction.

The existence of a cell is necessary for the transfer of genetic information from parents to offspring for conservation of the species. The cell membrane is the essential structural component of a cell. The amphiphilic property of glycerolipids is vital for the cell membrane lipids. The most fundamental function of the membranes of a cell is to separate an inner water layer from an outer water layer to form selective-permeability barriers.

Many amphiphilic lipids can form monolayers at the air-water or oil-water interface. In the case of the amphiphilic molecules with one hydrocarbon tail, they can also form micelles in which hydrophobic tails collect in the sphere of a hydrocarbon and polar heads project into the surrounding water. On the other hand, amphiphilic molecules with two hydrocarbon tails can form lipid bilayers. Glycerol and sphingosine are the most suitable molecules to form membrane lipids, which have two parallel or antiparallel hydrophobic tails and one polar head. The major structures of the cell membrane are usually lipid bilayers, which are constructed with complex lipids such as glycero- or sphingo-phospholipids (Figure 3). The polar heads of these lipids are placed on the outside and inside of the membranes.

Glycerolipids are widely distributed in Archaea, Bacteria, and Eukarya; however, sphingolipids are limited to Eukarya and a small number of bacterial genera. The lipid composition of a cytoplasmic membrane in Bacteria is simpler than that of Eukarya but the number of molecular species of the bacterial membrane

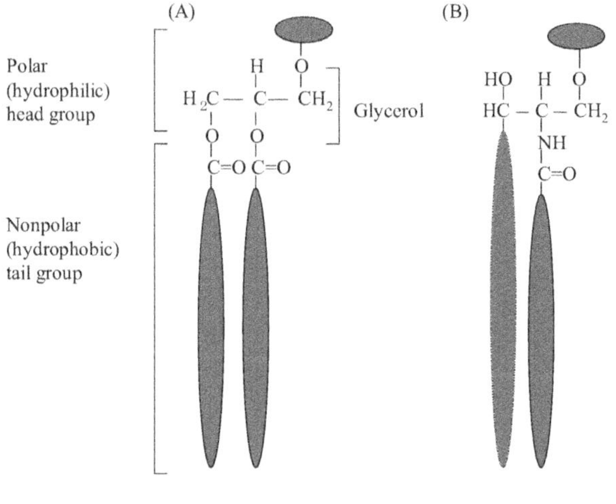

Figure 3. Basic structures of glycerolipid (A), and sphingolipid (B).

phospholipids nearly equal one hundred. Such variety depends on the diversity of the polar head groups and variations in the size of the hydrocarbon chains and the number of double bonds in fatty acids. Unique structures of phospholipids in the Archaea are described in the next section. Functions of the cytoplasmic membrane in prokaryotes are complicated because they lack any subcellular organelles and major metabolic functions occur in and on the cytoplasmic membrane. The membrane is composed of 40 % lipids and 60 % proteins; some of the latter are enzymes many of which are involved in one way or another in the transport of substance into and out of the cell. In contrast to the appearances of diagrammatic drawings the cytoplasmic membrane is actually quite fluid with the combination of phospholipids and proteins, which gives significant freedom to move about the membrane surface. It was described as a fluid mosaic model by Singer and Nicolson (1972), in which the bilayer unit membrane composed of phospholipids with the hydrophobic groups are directed inward and the hydrophilic groups toward the outside. The inner surface of the membrane faces the cytoplasm and the outer surface faces the environment where the cell associates with water. Some proteins may also associate quite firmly with the surface of the membrane and act as membrane-bound proteins. The membranes of Eukarya have sterols to stabilize its structure and make it less flexible by increasing the rigidity and amphiphilic character. Some bacterial cells contain hopanoids instead of sterols with the same roles. The structure of the C_{30} hopanoid (diploptene) closely resembles cholesterol in rings *a* through *c* (Figure 4). As far as is known, hopanoids are not found in any species of Archaea. However, the precursors of hopanoids in Bacteria are also phylogenetic precursors of sterols and might contribute to archaeal membrane lipid biosynthesis with fewer steps than that of the *n*-acylphospholipids biosynthesis (Rohmer et al. 1979; Ourisson and Nakatani 1994).

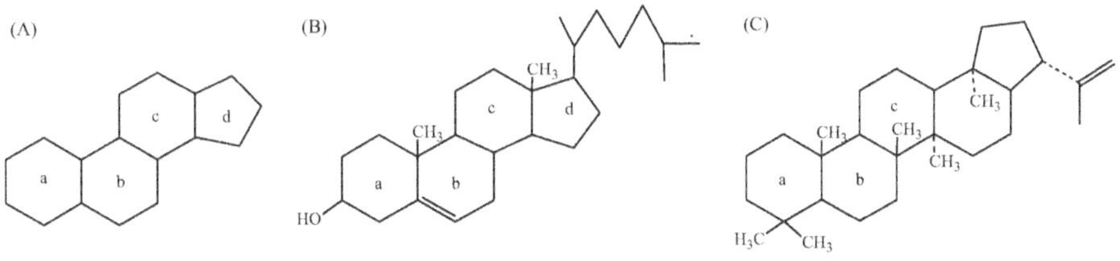

Figure 4. Sterols and hopanoids: (A) General structure of a sterol, (B) cholesterol, and (C) diploptene.

Not only the cytoplasmic membranes, but also cell walls contain compound lipids such as the LPS in Gram-negative bacteria. Lipid A is the lipid part of the LPS and the endotoxic properties of lipid A mainly affect the mesodermal cells of animals. During the biosynthesis of the peptidoglycan, C_{55} undecaprenol diphosphate acts as a lipid intermediate to be combined with the *N*-acetylmuramic acid pentapeptide and *N*-acetylglucosamine for producing the cross-bridges of the peptidoglycan.

A lipid like compound poly-β-hydroxybutyric acid (PHB) is one of the PHA and the most common inclusion body in prokaryotic organisms. The monomers of β-hydroxybutyric acid connected by ester linkages; forming the long polymer, then aggregate into granules. A thin non-unit membrane consisting of lipid separating them from the cytoplasm bound most granules or other inclusions. The functions of inclusions are almost always a storage depot for carbon and energy or a reservoir of structural building blocks.

From 1975, the general idea of functional lipids has been generally accepted in addition to that of the structural lipid (Michell, 1975). Especially, phospholipids and phospholipases are the leading parts of the modulators or transducers of signal-transduction, especially in Eukarya.

3. The Differences between Archaeal Lipids and Bacterial Lipids

The almost 5 million or more species on earth probably originated from a common ancestral cell. Before biological evolution occurred, biological materials had been synthesized through chemical evolution. Existing organisms carry the records of chemical evolution and the evolution of life from the birth of the universe. It is interesting to think about the structure of lipids, especially membrane lipids, from a common ancestral cell. To determine the lipids of the common ancestral cell, it should be useful to investigate the structure of archaeal lipids compared with the bacterial lipids. Figure 2 shows the differences between the archaeal glycerolipids and bacterial glycerolipids. Glycerophospholipids are the major components of the bacterial membranes. The non-polar tails of the lipids of Bacteria are long-chain fatty acid molecules attached to a polar head group such as *sn*-glycerol-3-phosphate by ester bonds. The hydrocarbon chains of a fatty acyl group contain 12 to 24 carbon atoms. About half of the fatty acyl groups are saturated and the other halves are unsaturated or polyunsaturated. On the basis of the fatty acids biosynthetic relationships, membrane fatty acids can be divided into two families, the straight-chain fatty acid family and the branched-chain fatty acid family. The latter includes iso-, anti-iso-, and ω-alicyclic fatty acids with or without a substitution such as unsaturation and hydroxylation. These fatty acids are not as common as the straight-chain fatty acids in Bacteria, but are still significant and diagnostic for certain species or strains. The fatty acids composition has been used for taxonomic classification and identification of Bacteria. Table 1 shows an example of the fatty acids composition in the genus *Thermus* (Donato et al., 1990), which are thermophilic bacteria growing optimally about 66 – 75 °C. The degree of saturation of the hydrocarbon chains and the length of the chain itself influence the melting point of the fatty acid. A variety of glycerophospholipids are derived from many ionic and polar substituents such as choline, ethanolamine, serine, glycerol, phosphatidylglycerol or *myo*-inositol attached to the phosphate group of the glycerophospholipids (Figure 2[B]).

On the other hand, archaeal lipids are composed of isoprenoid alcohols attached to a *sn*-glycerol-1-phosphate by ether bonds (Figure 2[A]). Isoprenoid chains in archaeal lipids vary in size from C_{15} to C_{40}. In the case of C_{15} to C_{25} or cyclic C_{40} used to form a membrane, a lipid bilayer is present (Kates 1972). These isoprenoid chains are distributed in a specific manner in the three major classes of archaeal strains (Figure 5). The hypothetical models of membranes of three groups of Archaea are shown in Figure 5(B).

TABLE 1. Fatty acid composition (%) of the polar lipids of *Thermus* strains at the optimum growth temperature.

(constructed by the authors based on the paper of Donato et al., 1990)

Strains	**Growth temp. (°C)**	**iC14**	**nC14**	**iC15**	**aC15**	**nC15**	**C15:1**	**iC16**	**nC16**	**C16:1**	**iC17**	**aC17**	**nC17**	**A**	**B**
T. aquaticus	73	-	-	26.2	3.8	-	-	7.1	7.4	-	46.6	7.4	1.0	-	-
T. thermophilis	73	<1.0	-	30.1	7.6	<1.0	-	4.4	1.0	-	44.9	10.5	-	-	-
T. filiformis	73	<1.0	-	5.6	22.5	-	-	8.9	1.7	-	9.2	48.6	1.0	<1.0	-
Thermus sp. .SPS-8	73	-	-	55.6	7.4	2.1	-	1.5	<1.0	-	29.2	2.9	-	-	-
Thermus sp. VI-5b2	70	<1.0	-	60.5	8.2	3.5	-	2.4	-	-	22.1	2.1	-	-	-
Thermus sp. FQ-3	70	1.3	-	34.2	7.3	<1.0	-	15.7	1.0	<1.0	32.3	5.9	<1.0	<1.0	-
Thermus sp. LFC-1	73	-	-	25.1	10.2	<1.0	-	4.9	<1.0	-	41.9	14.7	-	-	-
Thermus sp. RQ-1	73	1.1	-	13.6	11.0	-	-	15.9	2.6	<1.0	30.2	22.7	<1.0	-	<1.0
Thermus sp. RQ-2	73	1.0	2.4	38.8	11.8	3.2	1.3	3.7	3.0	5.4	14.8	3.3	2.9	1.3	<1.0
T. ruber	60	<1.0	<1.0	69.6	6.6	1.2	-	1.3	1.7	1.3	12.3	2.3	1.8	-	-
T. rubens	60	-	-	59.6	4.0	1.6	-	1.5	1.9	1.3	21.4	3.9	2.4	-	-

Abbreviations for fatty acid methyl esters: i, iso-branched; a, anteiso-branched; n, strait chain; C15:1, pentadecanoic acid; C16:1, hexadecenoic acid.

A, B, unidentified fatty acids, with chain lengths between C17 and C19.

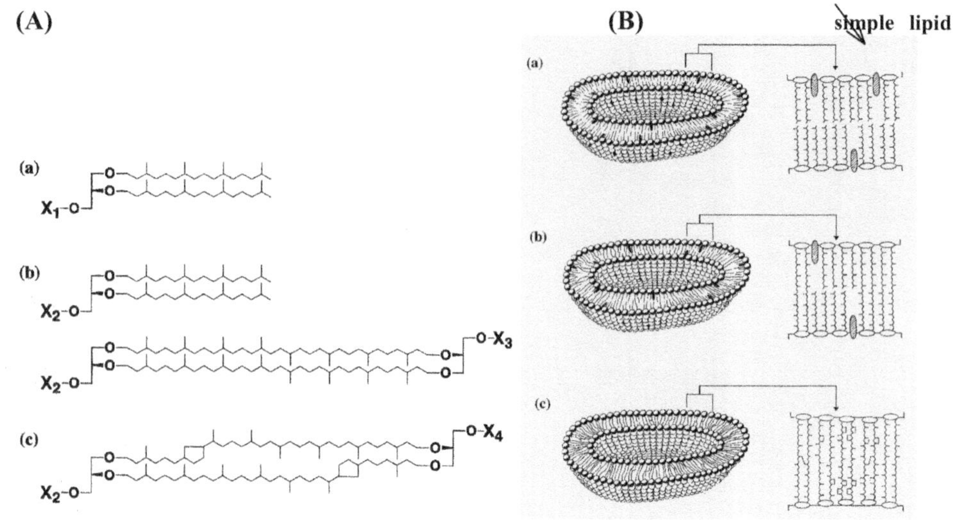

Figure 5. Typical major lipid structures (A) and schematic diagrams (B) of the membranes of three groups of Archaea; (a) halophilic Archaea, (b) methanogenic Archaea, and (c) thermophilic Archaea. X_1: phosphate containing polar heads, or sugar residues. X_2: phosphate and phosphoamino containing polar heads. X_3 : sugar residues. X_4 : polyol, and / or sugar residues, and sulfate.

The Archaea are divided into two phyla, Crenarchaeota and Euryarchaeota by phylogenetic classification. They predominantly occur in extreme environments similar to those that may have existed in early earth, such as anaerobic, hypersaline, or hydrothermally and geothermally heated areas. Above all, thermophilic Archaea are distributed in both phyla, Crenarchaeota and Euryarchaeota, and are present near the root of the phylogenetic tree (Figure 1, bold lines). Lipids of halophiles are mainly composed of C_{20} and some C_{25} isoprenoid alcohols to form the diether bilayers as shown in Figure 5(A) (a) (Kates, 1992). Lipids of methanogens contain C_{20} and C_{40} isoprenoid alcohols also shown in Figure 5(A) (b) (Koga et al., 1993). The tetraether type glycerophospholipids with C_{40} isoprenoid chains are major membrane lipids widely distributed in thermophilic archaeal cells (Figure 5(A) (c)). In this type of the core lipids, isoprenoid alcohol side chains from each glycerol molecule are covalently bonded together; a bipolar tetraether monolayer instead of a lipid bilayer can be formed (Langworthy and Pond 1986). Lipid monolayers of thermophilic Archaea are quite resistant to peeling apart. Variations in the typical major lipid structures of *Sulfolobus acidocaldarius* are shown in Figure 6. There are two types of the tetraether core lipids, caldarchaeol (glycerol 2, 3 - dialkyl - *sn* - glycerol tetraether) and calditoglycerocaldarchaeol (glycerol 2, 3 - dialkyl- *sn* - calditol tetraether) in most strains of the order *Sulfolobales*. Calditol [2 - hydroxymethyl - 1 - (2, 3 - dihydroxypropoxy) 2, 3, 4, 5 - cyclopentanetetraol] is a polyol with an ether linkage in the molecule (Sugai et al., 1995). Calditoglycero- caldarchaeol was found to be a major core component that constituted 70 – 80 % of all core lipids of them except one *Metallosphaera* strain in our study (Itoh et al., 2001a). The polar head groups which consist of small sugars, inositolphosphate, and sulfate are combined with the core lipids to produce big varieties of their membrane lipids.

A C_{40} isoprenoid chain can contain from zero to a maximum of four cyclopentane rings (De Rosa et al., 1980). The cyclization number of C_{40} isoprenoid chains in thermophilic Archaea influences the fluidity of lipids, whereas the number of carbons and the degree of unsaturation in fatty acids do so in Bacteria and Eukarya. The cyclization number of the C_{40} isoprenoid chains significantly changed depending on the growth temperature of strains of *Thermoplasma acidophilum* (Uda et al., 2001) and *S. acidocaldarius* (De Rosa et al., 1980). The optimum growth temperature of *Pyrococcus horikoshii* is 10 °C higher than the maximum growth temperature of *S. acidocaldarius;* however, the cyclization number of the core lipids in this strain was smaller than expected (Sugai et al., 2000; Itoh et al., 2001b). It is known that the number of carbons and the degree of unsaturation in fatty acids influence the fluidity of the lipids in Bacteria and Eukarya. The thermophilic archaeal lipids contain cyclopentane rings instead of double bonds for changing the fluidity to adapt to the growth conditions. The rigid structure of the cyclopentane rings in the isoprenoid chains makes the membrane less fluid.

In addition to the cyclization of the tetraether lipids, covalent bonding of the two C_{40} isoprenoid chains was found in hyperthermophiles. This is called the C_{40} H-form (Morii et al., 1998; Sugai et al., 2000). These characteristic structures of the lipids seem to contribute to their fundamental physiological roles in hyperthermophiles. These results show that the hyperthermophilic archaeal tetraether lipids are the simplest and most stable membrane lipids in existing living organisms.

Lipids	Structure of lipids
GL-8	Glc (1-1) - O -
GL-7	Gal (β1-1) -O -
GL-6b	Qui -Gal (1-1) - O -
GL-6a	Xyl -Gal (1-1) - O -
GL-5	Unidentified
GL-3	Glc (1-3) -Glc (1-1) - O -
GL-2	Glc (β1-3) -Gal (β1-1) -O -
GL-1a	Glc (β1-4') -O -
PL-1	-O- PO_3 - Ino
GL-4b	O_3S -O -Glc (β1-3) -Gal (β1-1) -O -
GL-4a	O_3S -O -Glc (β1-4') -O -
GPL-2	Glc (β1-3) -Gal (β1-1) -O - -O- PO_3 - Ino
GPL-1	Glc (β1-4') -O - -O- PO_3 - Ino
GPSL-2	O_3S -O -Glc (β1-3) -Gal (β1-1) -O - -O- PO_3 - Ino
GPSL-1	O_3S -O -Glc (β1-4') -O - -O- PO_3 - Ino

Figure 6 continued

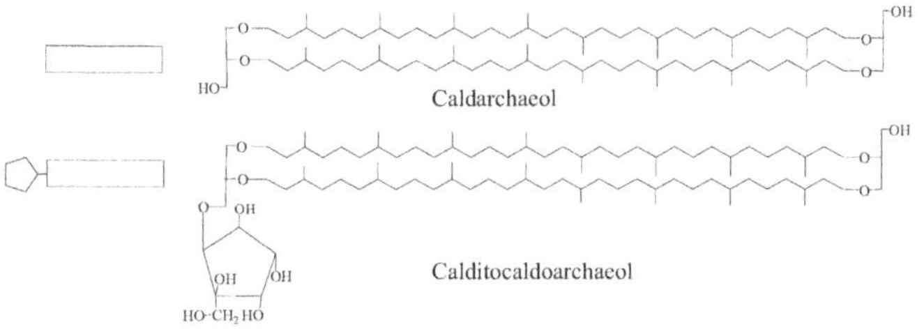

Figure 6. Diagram of the typical major lipid structures of *Sulfolobus acidocaldarius* and their tetraether core lipids. Abbreviations: Glc, glucose; Gal, galactose; Qui, quinobose; Xyl, xylose; SO_3, sulfate; PO_3, phosphate; Ino, inositol.

Thus, the stereostructure of the glycerophosphate backbones of phospholipids of Archaea and of Bacteria are entirely opposite (Kates, 1972; Langworthy and Pond, 1986; Koga et al.1998). The stereochemical differences between the G-1-P archaeal lipids and G-3-P bacterial and eukaryal lipids might have occurred by the function of some proteins long after the first cell was developed by the reactions of small organic molecules. We propose that the structure of lipids of the common ancestral cell may have been similar to those of hyperthermophilic Archaea.

As described in section 2, sterols contribute to increasing the rigidity and amphiphilic character of the eukaryal membranes, whereas they are normally absent in the bacterial and archaeal membranes except in some methanotrophic strains. Ourisson and Nakatani (1994) reported that derivatives of one triterpene family (the hopanes) are widely distributed in Bacteria, acting as re-inforcers in their membranes, the same role as sterols in Eukarya. These researchers also suggest that these molecules seem to be the evolutional precursors of archaeal lipids. In the case of thermophilic Archaea, C_{40} tetraethers provide stability in an excellent way (Rohmer et al. 1979). The major lipid properties of Archaea, Bacteria, and Eukarya are listed in Table 2.

TABLE 2. Major lipid properties of Archaea, Bacteria, and Eukarya

	Archaea	Bacteria	Eukarya
Glycerolipids	+	+	+
Hydrocarbon chain	Isoprenoid	Fatty acid	Fatty acid
Size of hydrocarbon	$C_{15\text{-}25}/C_{40}$	$C_{12\text{-}24}$	$C_{12\text{-}24}$
Hydrocarbon bonding type	Ether	Ester	Ester
Position in glycerol	*sn*-2, 3	*sn*-1, 2	*sn*-1, 2
Phospholipids	+	+	+
Glycolipids	+	+	+
Phosphoglycolipids	+	+	+
Sulfoglycolipids	+	-	+
Phosphosulfoglycolipids	+	-	-
Sulfolipids	-	-/(+)	+
Sphingolipids	-	-/(+)	+
Triterpene Family	+	+	+
Steroids	-	-	+

4. The Evolution of Lipids

Before the evolution of life occurred, biological materials had been synthesized through chemical evolution. In the case of lipids for cell membranes, an amphiphilic property is necessary for the molecule. Glycerol could be selected as the most suitable molecule, which can hold two hydrophobic long chains and one polar head. The isoprenoid chains are synthesized by fewer steps than fatty acid chains including almost the same number of carbons. The tetraether monolayer lipids with a covalent bond in the center of a molecule seem to be the simplest and most stable. A common ancestral cell membrane might have been constructed with molecules such as the H-form caldarchaeol.

It is interesting that not only the G-1-P dehydrogenase gene was found and the enzyme was purified from *Methanothermobacter thermautotrophicus* (formerly classified as *Methanobacter thermautotrophicum*) (Koga et al., 1998) but also the G-3-P dehydrogenase gene ortholog was found and a slight activity was detected in the same strain (Nishihara et al., 1999).

The draft sequence of the human genome was opened to the public on June 26, 2000. The genomes of 98 species of prokaryotes including 16 Archaea and 82 Bacteria also have been completed and presented in public databases as of November 2002. Next, interest will focus on understanding the function of their gene products. The number of sequences in the archaeal strains is still insufficient to be used to compare the sequences in the eukaryal and bacterial domains. Based on archaeal whole genome analyses, less than one third of the open reading frames identified with the genes of known functions up to now. However, the phylogenetic relationships of the genes among living organisms and the evolution of those genes will be uncovered in the near future. We will probably be able to discuss the evolution of lipids from an ancestral cell more definitively after functional analysis of archaeal genomes has progressed.

5. References

Baross, J.A., and Holden, J. F. (1996) Overview of hyperthermophiles and their heat-shock proteins. In: M.W.W. Adams (ed.) *From Enzymes and proteins from hyperthermophilic microorganisms*. Academic Press, New York, pp. 1-34.

De Rosa, M. Gambacorta, A. Nicolaus, B. Sodano, S. and Bu'Lock, J.D. (1980) Structural regulations in tetraether lipids of Caldariella and their biosynthetic and phyletic implications. Phytochem. **19**, 833-836

Donato, M. M., Seleiro, E. A., and Da Costa, M. S. (1990) Polar lipid and fatty acid composition of strains of the genus *Thermus*. System. Appl. Microbiol. **13**, 234-239.

Itoh, Y.H. Kurosawa, N. Uda, I.Sugai, A. Tanoue, S. Itoh, T. Horiuchi, T. and Itoh, T. (2001a) *Metallosphaera sedula* TA-2, a calditoglycerocaldoarchaeol deletion strain of a thermophilic archaeon. Extremophiles **5**, 241-245.

Itoh, Y.H. Sugai, A., Uda, I., and Itoh, T. (2001b) The evolution of lipids. Adv. Space Res. **28**, 719-724.

Kates, M. (1972) Ether linked lipids in extremely halophilic bacteria. In: F. Synder (ed.) *From Ether Lipids, Chemistry and Biology*. Academic Press, New York, pp. 351-398.

Kates, M. (1992) Archaebacterial lipids: structure, biosynthesis and function. In: M.J. Danson, D.W. Hough, and G. G. Lunt (eds.) *From The Archaebacteria: Biochemistry and Biotechnology*, Biochem. Soc. Symp. **58**, Portland Press, London and Chapel Hill, pp. 51-72.

Koga, Y. Nishihara, M Morii, H. and Akagawa-Matsushita, M. (1993) .The ether polar lipids of methanogenic bacteria. Structure, comparative aspects, and biosynthesis. Microbiol. Rev. **57**, 164 -182.

Koga, Y. Kyuragi, T. Nishihara, M. Sone, N. (1998) Did archaeal and bacterial cells arise independently from noncellular precursors? A hypothesis stating that the advent of membrane phospholipid with enantiomeric glycerophosphate backbones caused the separation of the two lines of descent. J. Mol. Evol. **46**,54-63.

Langworthy, T.A. and Pond, J.L. (1986) Archaebacterial ether lipids and chemotaxonomy. Sys. Appl. Microbiol. **7**, 253-257.

Michell, R. H. (1975) Inositol phospholipids and cell surface receptor function. Biochim. Biophys. Acta **415**, 81-147.

Morii, H. Eguchi, T. Nishihara, M. Kakinuma, K. Koig, H. and Koga, Y. (1998) A novel ether core lipid with H-shaped C_{80}-isoprenoid hydrocarbon chain from the hyperthermophilic methanogen *Methanothermus fervidus*. Biochem. Biophys. Acta **1390**, 339-345,.

Nishihara, M. Yamazaki, T. Oshima, T. and Koga, Y. (1999) *sn*-glycerol-1-phosphate-forming activities in Archaea: separation of archaeal phospholipid biosynthesis and glycerol catabolism by glycerophosphate enantiomers. J. Bacter. **181**, 1330 -1333.

Ourisson, G. and Nakatani Y. (1994) The terpenoid theory of the origin of cellular life: the evolution of terpenoids to cholesterol. Chem. and Biol. **1**, 11-23.

Rohmer, M. Bouvier P. and Ourisson, G. (1979) Molecular evolution of biomembranes: Structural equivalents and phylogenetic precursors of sterols. Proc. Natl. Acad. Sci. U.S.A. **76**, 847-851.

Singer, S.J. and Nicolson, G.L. (1972) The fluid mosaic model of the structure of cell membranes. Science **175**, 720-731,.

Sugai, A. Sakuma, R. Fukuda, I. Kurosawa, N. Itoh Y. H. Kon, K. Ando, S. and Itoh, T. (1995) The structure of the core polyol of the ether lipids from *Sulfolobus acidocaldarius*. Lipids **30**, 339-344.

Sugai, A. Masuchi, Y. Uda, I. Itoh, T. and Itoh Y.H. (2000) Core lipids of hyperthermophilic archaeon, *Pyrococcus horikoshii* OT3. J. Jpn. Oil Chem. Soc. **49**, 695-700.

Uda, I. Sugai, A. Itoh Y. H. and Itoh, T. (2001) Variation in molecular species of polar lipids from *Thermoplasma acidophilum* depends on growth temperature. Lipids **36**, 103-105.

Woese, C. R. Kandler, O. and Wheelis, M.L. (1990) Towards a natural system of organisms: Proposal for the domains Archaea, Bacteria, and Eukarya. Proc. Natl. Acad. Sci. U.S.A. **87**, 4576-4579.

Woese, C. R. (2000) Interpreting the universal phylogenetic tree. Proc. Natl. Acad. Sci. U.S.A. **97**, 8392-8396.

Biodata of **Józef Kaźmierczak** the author (with co-author Stephan Kempe) of ***"Calcium Build-up in the Precambrian Sea: A Major Promoter in the Evolution of Eukaryotic Life"***

Dr. Józef Kaźmierczak is head of Biogeology Division at the Institute of Paleobiology Polish Academy of Sciences in Warsaw. He obtained his Ph.D. in 1971 from the Faculty of Geology of Warsaw University and D.Sc. in 1983 from the Faculty of Biology and Earth Sciences of the Jagiellonian University (Cracow). His research interests cover biosedimentology, geomicrobiology and evolution of biomineralization processes in space and time. He is currently carrying comparative studies on modern and fossil (particularly Archean) carbonate and siliceous deposits produced by cyanobacterial mats and biofilms.

E-mail: **jkaz@twarda.pan.pl**

Dr. Stephan Kempe is Professor for Physical Geology and Global Cycles at the Institute of Applied Geosciences, University of Technology Darmstadt. He obtained his Dr. rer. nat. in 1976 and his Dr. Sc. in 1984 from the Institute of Geology and Paleontology of the Hamburg University. He participated in research projects on biogeochemical cycles, marine geochemistry, and environmental geology. He served as coordinator of the SCOPE/UNEP project: "Transport of Carbon and Minerals in Major World Rivers", and was leader of oceanographic and geological expeditions to the Pacific, North Sea, Black Sea, Lake Van (Turkey) and various crater lakes (Indonesia, Tonga, Hawaii).

E-mail: **kempe@geo.tu-darmstadt.de**

J. Seckbach (ed.), Origins, 329–345.

CALCIUM BUILD-UP IN THE PRECAMBRIAN SEA

A Major Promoter In The Evolution Of Eukaryotic Life

J. KAŹMIERCZAK[1] **and S. KEMPE**[2]
[1]*Institute of Paleobiology, Polish Academy of Sciences, Twarda 51/55, PL-00818, Warszawa, Poland and* [2]*Institute for Applied Geosciences, TU-Darmstadt, Schnittspahnstr. 9, D-64287 Darmstadt, Germany*

1. Introduction

Across a time envelope of 3 billion years, the Ca^{2+} concentration in the ocean is assumed to have increased 1,000 to 100,000 times (Kempe and Degens, 1985). It has been proposed that the evolution of eukaryotic life is related to this progressive Ca^{2+} build-up in the Precambrian sea (Kaźmierczak and Degens, 1986; Kempe *et al.* 1989; Kempe and Kaźmierczak, 1994). Five Ca^{2+} mediated major evolutionary events can be recognized: (i) eukaryogenesis, (ii) cell aggregation, (iii) cell-to-cell adhesion, (iv) multicellularity, and (v) biocalcification. Geochemical, physiological and paleontological data are presented in support of this concept.

To suggest that a single factor, that is, Ca^{2+} content in the Precambrian sea, is the prime agent behind the origin and evolution of eukaryotes is provocative - to say the least. To most of us the biochemistry of the living cell, and the morphological diversity of marine biota through geologic time are too manifold to readily accept such a simple regulating device. A wealth of published data in the field of biology and geology, in particular the research on Precambrian microfossils, Ca^{2+} physiology, structural role of metal ions in molecular biology, biocalcification, and global geochemistry, actually support this claim. As Johann Wolfgang von Goethe once remarked '*That world history has to be rewritten from time to time, there is in our days no doubt left, indeed. Such a necessity does not come about because much of what has taken place has been discovered anew, but because novel opinions are presented, for reason that the comrade of a progressive time is brought to standpoints, from where the past can be overlooked in his own way.*' We share the same feeling and in that light we will re-examine the origin and development of the eukaryotic life.

2. Earliest Eukaryotes: Paleontological view

When it comes to earliest eukaryotes, the question of their nature and time of origin is still unresolved (Cloud, 1976; Cloud and Glaessner, 1982; Runnegar, 1982; Brocks *et al.*, 1999; Schopf, 1999). Some authors proposed a very long, entirely prokaryotic period in biospheric evolution encompassing the Archean and most of the Proterozoic (Knoll and Barghoorn, 1975; Schopf, 1978; Schopf and Oehler, 1976; Awramik, 1982).

Others extended the eukaryotic record back into the early Proterozoic and even Archean (Kaźmierczak, 1976, 1981; Durham, 1978; Hofmann and Chen, 1981; Kauffman and Steidtmann, 1981; Robbins *et al.*, 1985). According to Pflug (1978), organic structures of possibly eukaryotic origin (yeast?) may even occur in the oldest terrestrial rocks from the 3.8 billion years old Isua Formation, Greenland. The reason why no consent can be reached has principally to do with the different kinds of yardsticks employed in measuring and classifying early microfossils (Cloud, 1976; Awramik, 1982; Kaźmierczak, 1981; Pflug, 1982). Most Precambrian microfossils are characterized by spherical and filamentous shapes of very modest diagnostic value. For this reason, cell size is often the only tool for separating microbial eukaryotes from prokaryotes (Knoll and Barghoorn, 1975; Fairchild, 1985). However, the ranges in cell size do overlap (prokaryotes: 0.2-55 μm, average 2-4 μm, largest 680 μm; eukaryotes 1-350 μm, average 20 μm). Thus, a clear assignment is difficult to make. Yet, it is predominantly the criterion of cell size that led most authors to assume a rather late appearance of eukaryotic organisms, i.e., at about 1.3 to 1.4 billion years before present and with certainty only close to the end of the Precambrian, ca. 680 million years ago (the Ediacaran metazoans) (Knoll and Barghoorn, 1975; Francis *et al.*, 1978; Cloud and Glaessner, 1982). The situation changed dramatically after discoveries of acritarch-like microfossils in strata ca. 2.1 b.y old (Han and Runnegar, 1992) and recent findings of biomarkers indicative for the presence eukaryotic organisms in deposits ca. 2.7 b.y old (Brocks *et al.* 1999). Overviews by Knoll (1992, 1999) and Altermann (2002) well illustrate the progress in search for the traces of oldest eukaryotes achieved during the last thirty years of investigations.

3. Chemistry of the Precambrian Sea: the Soda Ocean Model

The mineral spectrum of marine evaporites has in the opinion of some authors not markedly changed since the late Precambrian (Braitsch, 1962; Walther, 1972; Siever, 1974; Ala, 1976). Even massive extraction of rock salt during the Permian could only slightly change the ocean's salinity by about 3 to 4 permil, which in any event was severe enough to cause mass extinction (Stevens, 1977). All this seems to suggest that the composition of seawater had remained relatively constant throughout the Phanerozoic (Holland *et al.,* 1986) or only with small deviations from what is known today (Hardie, 1996). Not so for the Precambrian. One viewpoint is that early ocean was acidic, high in calcium and bicarbonate, and quite low in sodium and chloride (Maisonneuve, 1982; Walker, 1983; MacLeod *et al.*, 1994). In the opinion of Morse and Mackenzie (1998), the Hadean ocean could had been a NaCl-dominated aqueous solution with higher than today's dissolved inorganic carbon (DIC) and alkalinity concentrations and possibly lower calcium concentration. This assumption rests on the notion that the atmosphere at that time must have been loaded with the greenhouse gas CO_2 to compensate for a 'faint' early Sun (Hart, 1978; Kuhn and Kasting, 1983). An entirely different scenario, that is alkaline conditions, high pH values (~10), very low calcium concentrations, and high sodium and dissolved carbonate contents has been proposed (Kempe and Degens, 1985; Kempe and Kaźmierczak, 1994, 2002) - comp. Figs. 1 and 2. It is based on the existence of modern soda lakes which form in areas where fresh volcanic rocks are available for weathering (Kempe and Kaźmierczak,

2003). These modern equivalents of the ancient ocean are by no means marginal phenomena as Lake Van, the largest soda lake existing today, illustrate. It is by volume (573 km^3) the fourth largest closed water body on Earth with an alkalinity of 151.2 to 155.8 meq/l and a pH of ca. 9.7 (Kempe *et al.*, 1991).

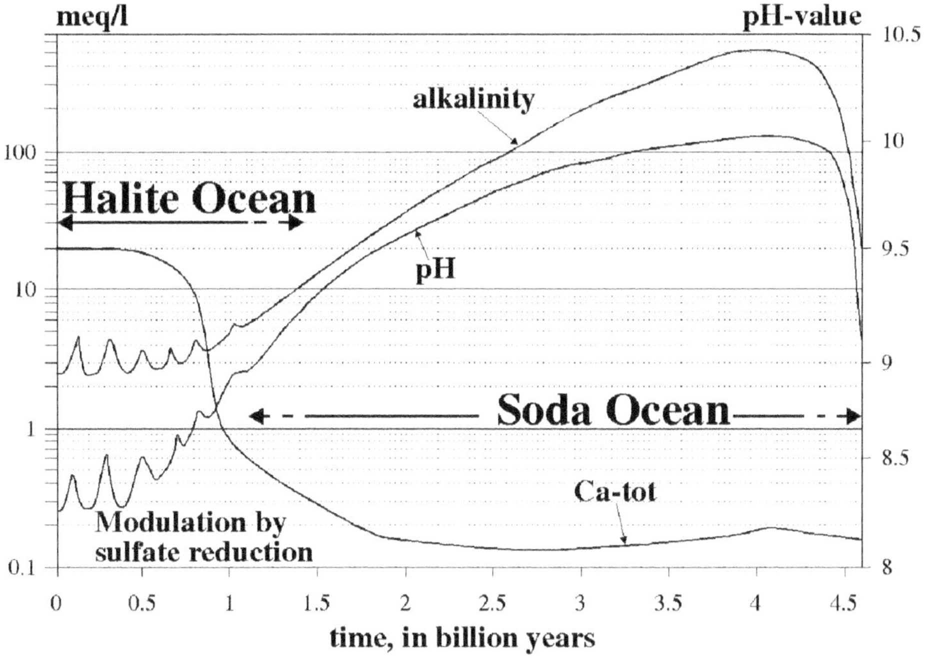

Figure 1. Conceptual model of ocean development through Earth history (after Kempe and Kaźmierczak, 1994). Three phases can be recognized: (1) The early and fast consumption of atmospheric CO_2 by silicate weathering. Weathering is greatly aided by availability of fresh volcanic ashes and fine grained impact debris blankets. The ocean turns - from a mildly acidic start - alkaline within a few million years after the first stable ocean came into existence. (2) For about 3 Ga the ocean was of a high alkalinity and pH and Earth had a low atmospheric PCO_2. The possibility existed that the ocean froze and thawed several times(for example after impacts). Freezing may have provided the mechanism by which protocell membranes could have formed when fluid inclusions in the ice were compacted into vesicles of decreasing size (Kempe and Kaźmierczak, 2002). The slow removal of sodium and carbonates became with the onset of plate tectonics through subduction and growing continents onto which marine carbonates and organic carbon were accreted. These processes determined the long duration of the soda ocean phase of Earth history. (3) Finally the alkalinity dropped to a level where total Ca^{2+} rose dramatically, thereby exerting a high stress on the biochemical machinery of living systems. The ocean is now dominated by halite (NaCl) (which has been present earlier as well). Due to the increase in O_2 in the atmosphere, the ocean can also gain in sulphate. Therefore, in stagnating basins, alkalinity can be produced short-term in the Phanerozoic ocean by sulphate reduction (schematic wiggles in the alkalinity and pH curves) thus providing a mechanism to cause mass extinctions (by spontaneous impact induced overturn of anaerobic oceans). For details see Kempe and Kaźmierczak (1994).

The 'soda ocean model', as it is termed, suggests that CO_2 from early degassing is readily sequestered by weathering of silicates and concentrated in the sea in amounts more than a thousand times that found in modern ocean because sodium carbonate species are highly soluble. Consequently, PCO_2 in the air was low, being a kinetic equilibrium of volcanic net input and silicate weathering consumption. This statement is in line with the suggestion of Henderson-Sellers and Cogley (1982), who postulated a 'run-away greenhouse' effect should CO_2 exceed. The subsequent evolution from an early soda- to the present halite-dominated sea was brought about by: (1) slow subduction of 'soda' water trapped in the pores of sediments, (ii) formation of organic matter, and (iii) long-term storage of marine carbonates on evolving continents. In addition, percolation of seawater through the oceanic crust modified carbonate mineral equilibria by fixing Mg^{2+} in crustal material (serpentinization)

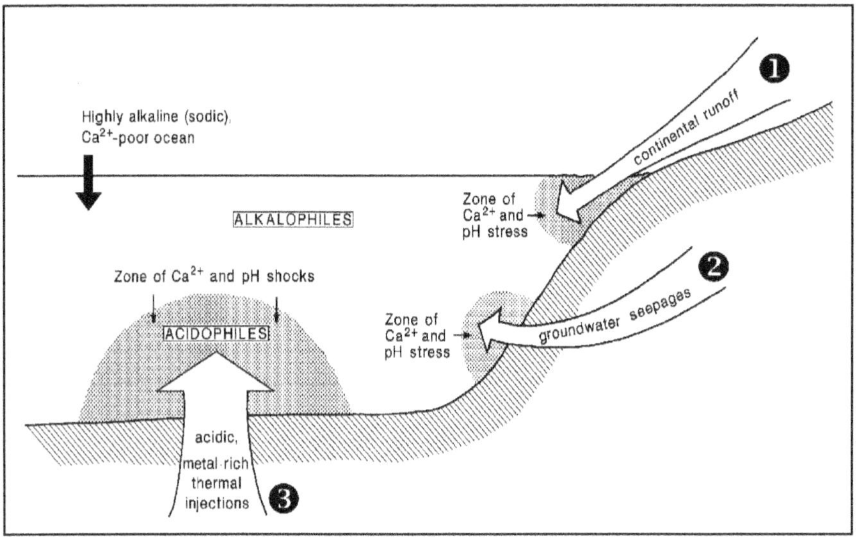

Figure 2. Main areas of calcium and pH stress in the early Precambrian highly alkaline (sodic) ocean): ❶ and ❷, Ca^{2+}- and pH-stressed areas located predominantly within photic zone; ❸, Ca^{2+}- and pH-stressed areas located in deeper waters, predominantly within disphotic and aphotic zones. The contact of acidophilic life with highly basic waters after establishment of a soda ocean is considered to be crucial for the subsequent evolution of living systems.

and releasing Ca^{2+} from the mineral to the aqueous phase. The gradual depletion of dissolved carbonates caused a drop in pH in the Precambrian sea. Roughly 1 billion years before present, a kind of titration point was probably reached, where Ca^{2+} and Mg^{2+} substituted Na^+ as cations balancing the dissolved carbonates. In turn, pH values decreased rapidly and Ca^{2+} could build up in the late Proterozoic ocean. The living cell was suddenly faced with an enormous Ca^{2+} potential 'knocking' at the cell wall.

4. Cellular Ca^{2+}

Calcium is the most ubiquitous metal ion in cellular systems (Williams, 1976). Its universal role as messenger and regulator is accepted (Kretsinger, 1977a; Borle, 1981; Campbell, 1983; Carafoli and Penniston, 1985; Cheung, 1987 – comp. Fig. 3). The wide scope of contributions relating to Ca^{2+} in molecular biology, physiology and biochemistry can be found, for instance, in practically every issue of biological periodicals of the past two decades, not to mention all the books and symposia that were recently concerned with this remarkable cation.

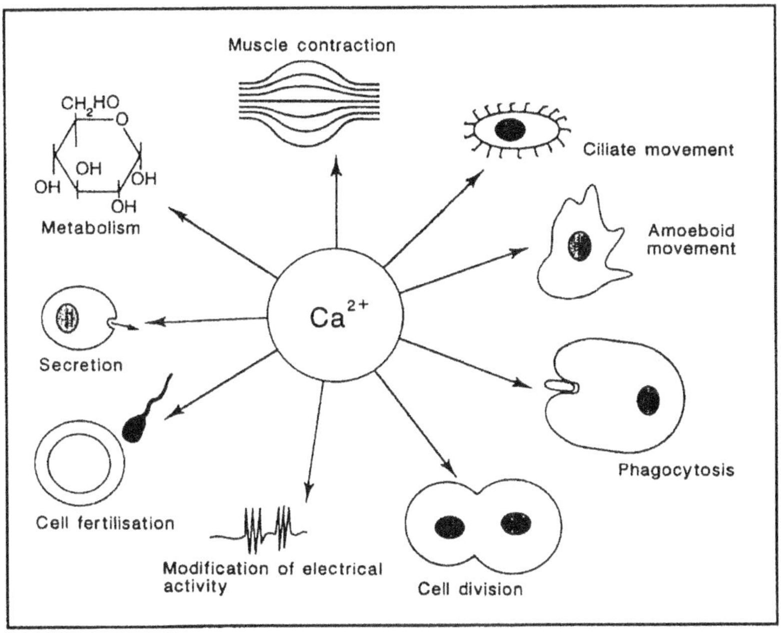

Figure 3. Examples of processes activated by an increase in intracellular free Ca^{2+}. Metabolic and secretory pathways involved in Ca^{2+} extrusion processes are instrumental in biologically controlled (enzymatic) biocalcification (after Campbell, 1983, slightly modified).

The main element of the Ca^{2+} control system is the upkeep of an extremely low free cytosolic calcium content, about 10^{-6} to 10^{-7} M. The maintenance of this level appears to be an ancestral and universal condition of life (e.g., Muller, 1996), one which initially evolved to avoid formation of insoluble $Ca_2(PO_4)_3$ in the cytosol, so that cells could utilize inorganic phosphate as a source of energy (Kretsinger, 1977a, b; 1983; Lowenstam and Margulis, 1980). Ca^{2+} sustenance is a formidable task, considering that there is a large intracellular and extracellular Ca^{2+} pool which exceeds the free calcium content in the cytosol by orders of magnitude. The crux of the problem has to do with the ways and means of releasing Ca^{2+} from the various pools down a steep concentration gradient into the cytosol. At the plasma membrane the existence of calcium channels seems to be an effective device for regulating the flow of Ca^{2+}. However, the kind of mechanism involved in gating, that is the opening and closing of a voltage-sensitive ion channel, is still unresolved. Whether other ions, carbohydrates, proteins, phospholipids and related membrane constituents do participate, is still debated. Biological membranes exhibit selective ion exchange characteristics which are principally controlled by phospholipids (Matheja and Degens, 1971) whereby divalent ions become more strongly adsorbed than monovalent ions: for instance, an addition of Ca^{2+} will release both H^+ and K^+. It is of note that uptake of metal ions proceeds at very low concentrations and even against a concentration gradient. In principle, the outstanding ion exchange properties of phosphates are related to the structural order of P-O bonds both in inorganic and organic polyphosphates. It is concluded that the coordinative bonding of metal ions with the PO_4 groups of phospholipids in cellular membranes is identical to that observed in phosphate minerals. Chains composed of phosphate tetrahedra and metal ion coordination polyhedra are cross-linked into a planar pattern. A stable surface geometry arises with well-defined holes and channels of the type illustrated in Engel *et al.* (1985). The interchangeable nature of metal ions causes membranes to act as dynamic molecular sieve. Pore size and sieving qualities of such fabrics is thus a function of type and concentration of metal ions. Inasmuch as gating of voltage-sensitive ion channels is in the last instance controlled by enzymes or high Ca^{2+}-affine proteins that remove or deposit ions along the membrane surface, conformational changes and a periodic pulsation of the lattice is the consequence. Oscillations of ionic constituents have been observed (Chance and Yoshioka, 1966), and electron density distribution perpendicular through a membrane can only be reconciled by assuming the presence of metal ions between the lipid bi-leaflets. Intracellular level of Ca^{2+} determines the opening and closing of gap junctions (e.g., Erulkar, 1981; Unwin and Henderson, 1984).

Calcium shapes the geometry of channels and regulates transfer of molecules. At normal cytosolic Ca^{2+} levels, channels are highly permeable to a wide range of molecular sizes with an upper size limit for peptide molecules of about 1,200 to 1,900 daltons; a rise in Ca^{2+} to 5 x 10^{-5} M in the junctional locale will drastically lower permeability (Rose *et al.*, 1977). Basically, 'metal-phosphate gates' make the plasma membrane relatively impermeable to extracellular Ca^{2+} which in seawater is 10^{-2} M, that is 100,000 times cytosolic level. Binding of Ca^{2+} to specific sites at the internal face of membranes alters conformation of the membrane constituents, thus allowing transfer or passage of hydrophilic substances across the lipid barrier. High-affinity Ca^{2+} proteins and glycoproteins confer specificity for binding, removal and transportation of Ca^{2+} (Elbrink and Bihler, 1975). Increase in calcium conductance and damage to the plasma membrane could disrupt this permeability barrier with a consequent influx of Ca^{2+}. For example, a variety of membrane-active toxins are activated only in the presence of high extracellular Ca^{2+}, but not at a low concentration (Schanne *et al.*, 1979). A Ca^{2+} gradient across the plasma membrane is thus needed for the toxicity of widely different agents. Alternatively, it is conceivable that a lethal influx of Ca^{2+} down the steep electrochemical gradient between outside and inside of the cell is the primary factor (e.g. Gupta and Pushkala, 1999). In addition to the cytotoxic effects, the reason for maintaining a very low Ca^{2+} level in all cells is the low solubility of calcium phosphates in an aqueous medium. To prevent intracellular precipitation of the highly insoluble salts, most cells are equipped with pumps "invented" for the speedy removal of deleterious calcium.

High-affinity calcium binding proteins serve hereby as carriers for shipping the Ca^{2+} surplus to the external environment. The shipping system is located within the endoplasmic reticulum and the Golgi apparatus (e.g., McFadden *et al.,* 1986). Calmodulin, an intracellular receptor protein, may serve as a major regulator of cellular calcium concentrations by activating an assortment of pumps. Calmodulin has been identified in almost all eukaryotic cells. Prokaryotic cells do not contain calmodulin but some, e.g., *Escherichia coli,* utilize precursor protein resembling that of the EF-hand structure typical for calmodulin (Varnaman *et al.,*1968).

Irvine and Moor (1986), Berridge (1993), and Berridge *et al.* (1998) have thrown light on receptor-regulated Ca^{2+} mobilisation into the cytosol of sea urchin eggs involving inositol trisphosphate ($Ins(1,4,5)P_3$) and inositol tetrakisphosphate ($Ins(1,3,4,5)P_4$). It is conceivable that this pathway may also operate in other cells. Another mechanism proposed is that Ca^{2+} release from endoplasmic reticulum is mediated by a guanine nucleotide (Gill *et al.*, 1986). Against this geochemical and physiological background, we like to project inferred stages in the development of the eukaryotic life in the course of the Precambrian (Fig. 4).

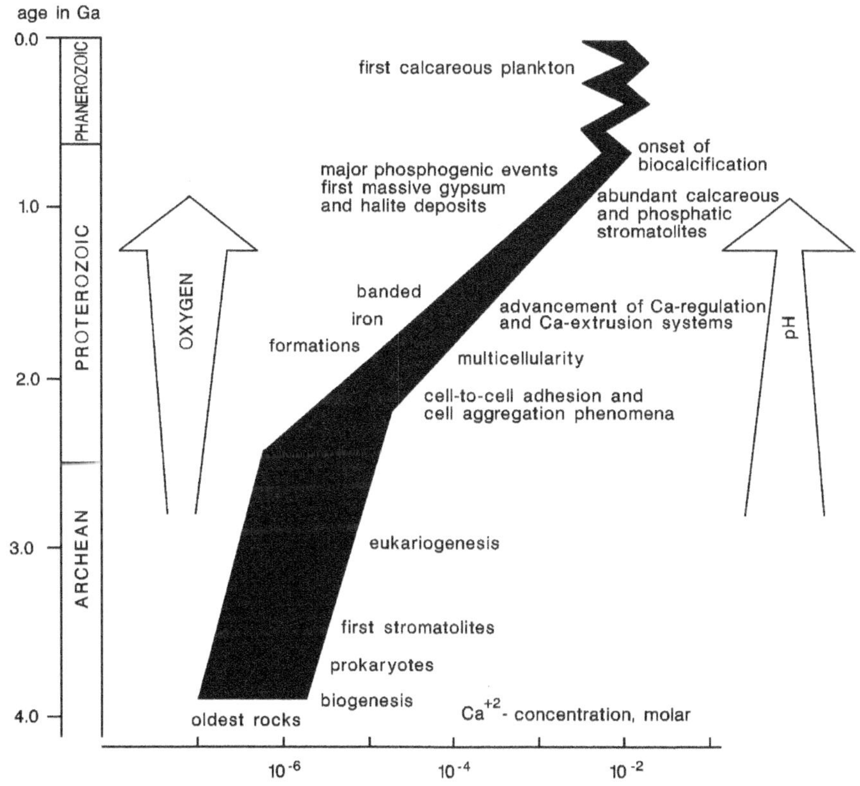

Figure 4. Diagrammatic presentation of major biological and geochemical events in the Earth's history shown in relation to temporal changes of Ca^{2+} and oxygen concentrations and pH values as implicated by the soda ocean model of Kempe and Degens (1985) (after Kaźmierczak and Degens, 1986, modified).

5. Eukaryogenesis

Of the two modern theories of eukaryotic cell origin, i.e., the autogenous (Taylor, 1976, for review) and the serial endosymbiotic theory (Margulis, 1981; McFadden, 2001), the latter, particularly when connected with theories of lateral gene transfer (Katz, 2002) and rare genomic events (Gribaldo and Philippe, 2002), is gaining recently more acceptance. Accordingly, chloroplast and mitochondrion symbionts were acquired by means of endocytosis, a process which is calcium dependant with a maximum uptake at 10^{-4} M external Ca^{2+} level (Prusch and Minck, 1985; Young, 1985). A number of nuclear activities are Ca^{2+} mediated: DNA synthesis (Boynton *et al.*, 1980), regulation of cell proliferation (Sasaki and Hidaka, 1982), assembly and disassembly of microtubules in chromosome movement during mitosis (Marcum *et al.,* 1978) or activation of histone kinase (Weisman *et al.*, 1978). Chromosome breakage at times of external calcium deficiency underscores the need of Ca^{2+} to maintain chromosome configuration (Bajaj *et al.*, 1971). High external Ca^{2+} raises free cytoplasmic Ca but not the nuclear one, thus indicating that nuclear Ca^{2+} is regulated independently of cytoplasmic Ca by gating mechanism in the nuclear envelope (Williams *et al.*, 1985), most likely via calmodulin. These observations may mean that the shift of the genetic material to the cell interior in proto-eukaryotes, and its membranization was probably an adaptive response to protect the genetic apparatus against high dosis of Ca^{2+} building up in the ambience. This could have happened already during the Archean because critical mass limit for a dividing eukaryotic cell to initiate DNA synthesis is in the order of 26 μm^3 (Calvert and Dawes, 1984) corresponding - in case of a spherical cell- to a cell diameter of 3.5 μm. Thus, even the most ancient carbonaceous microspheres described from the 3.4 billion years old Swaziland System of South Africa (Knoll and Barghoorn, 1977) having a size range 5-28 μm, could well be remnants of eukaryotic unicells.

6. Cell Aggregation

It has been known for same time that in highly Ca^{2+}-depleted media cells in tissues and gastrulating embryos fall to adhere; in turn, cell aggregates disintegrated. More recent experiments reveal the same pattern. For instance, a Ca^{2+} level not lower than 10^{-5} M is needed to keep slime mold cells tightly aggregated. A similar Ca^{2+} concentration is required to let yeast flocculate (Esser and Kües, 1983; Rose, 1984). It is assumed that Ca^{2+} is coordinating element between anionic groups of adjacent cells or a co-factor in activating the binding capacity of specific glycoproteins to carbohydrates. Inasmuch as flocculated yeast is richer in proline, lysine and aspartate compared to non-flocculated yeast the glycoprotein concept is favored. In the freshwater volvocacean *Gonium pectorale* the Ca^{2+} level for optimal growth of 4-cell colony is 100 times lower than that of 16-cell colony. Should Ca^{2+} fall below a critical value only individual *Gonium* cells are present (Groves and Kostir, 1961). For the coenobial green algae *Coelastrum* unicells exist below 10^{-4} M and colonies above that level. However, when Ca^{2+} was raised to above 10^{-2} M *Coelastrum* growth became inhibited (toxic effect) (Chan, 1976). Transposing the findings to the Precambrian we conclude from the presence of the potential oldest coenobial algae, the volvocacean-like *Eosphaera* (Kaźmierczak, 1979), in the ca. 1.9 billion years old Gunflint Iron Formation and the *Botryococcus*-like aggregates found in the similarly old Michigamme Formation (LaBerge *et al.*, 1984), that the critical Ca^{2+} levels of 10^{-5} to 10^{-4} M needed for inducing cell aggregation in algal protists were established by that time (Fig. 2). The Gunflint *Eosphera* and *Eosphaera*-like structures, when compared to well-defined Devonian volvocaceans *Eovolvox*, gave a count of 64 and 128 cells for a coenobium versus 256 and 512 for the Devonian counterpart (Kaźmierczak, 1981). Both fossils are separated by a time span of at least 1.5 billion years, during which Ca^{2+} content in the sea increased probably by about two orders of magnitude (Figs. 1 and 4). High abundance and variety of coenobial and colonial algae during late Proterozoic, especially the Vendian (e.g., Tynni and Uutela, 1984; Xue Yao-song *et al.*, 1995; Burzin, 1999; Xiao Shuhai, 2002) runs parallel to the trend in the evolution of unicellular algae to increase cell size and/or cell sculpture evoked by an excess of Ca^{2+} and other metals (Kaźmierczak *et al.*, 1985; Kaźmierczak and Degens, 1986; Kempe and Kaźmierczak, 1994). The intensive aggregation of cells during late Proterozoic time was seemingly enhanced not only by high Ca^{2+}, but by phosphate in the ocean too, as documented by large phosphorite deposits (Cook and Shergold, 1984). Phosphate is known as a potential enhancer of Ca^{2+} migration into cells (Cotmore *et al.*, 1971).

7. Cell-to-Cell Adhesion and Cell Fusion Phenomena

Ca^{2+} and other divalent cations can induce cell-to-cell adhesion and cell fusion phenomena (e.g., Lucy, 1978; Burger and Misevic, 1985; McConachie and O'Day, 1986). For example, Ca^{2+} triggered agglutination of cells, membrane lysis and cell fusion leads to the formation of multicellular syncytia and polykaryons. Cell fusion can also be catalyzed by pre-incubating the cells in a high-pH medium at elevated temperature (37°C) and by Ca^{2+} addition (Toister and Loyter, 1971). Such data suggest that local influx of Ca^{2+} to a highly basic Precambrian ocean via hydrothermal vents or river run-off (comp. Fig. 2) could promote various cell fusion phenomena which for heterotrophic cells might have led to origin of polykaryon, that is a multinuclear organisation of the type observed in same ciliates, or in the case of algal cell aggregates to coenocytic organisation or morphogenetic phenomena as found in siphonalean algae (Goodwin and Brière, 1994). Several nonseptate or sparingly septate branched filamentous microfossils of the type found in ca. 2 billion years old Gunflint sediments (e.g., *Archaeorestis magna*, *Xenotrix incorrecta* - Awramik and Barghoorn, 1977) and in the 1.3 billion years old Beck Spring Formation (Cloud, 1976), may actually represent coenocytic algae. The same may be true for many of the megascopic carbonaceous compressions including a great variety of oblong and ribbon-like films known as: vendotaenids, beltinids, moranids, etc. They range in age from the early Proterozoic (ca. 2.0 b.y. ago) into the Phanerozoic and, similarly as many Precambrian unicellular and colonial algae, display generally a tendency to increase in size with time (Hofmann, 1985).

8. Multicellularity

As in algal cell aggregates the integrity of cells in metazoan tissues is strictly controlled by extracellular Ca^{2+} concentration. Experiments carried out on embryos and tissues of various animals demonstrate that in most of them cells failed to adhere when the extracellular Ca^{2+} level is below 10^{-4} M (e.g., Franchi and Camatini, 1985; Volberg *et al.*, 1986). Others cease growth when external Ca^{2+} is lower than 10^{-4} M. The requirement of sponges for extracellular Ca^{2+} in order to keep cells together, are lower. In *Microciona* and *Haliclona*, cells dissociate when Ca^{2+} in the medium falls below 10^{-5} M and they re-aggregate when Ca^{2+} is raised above that level (Kretsinger, 1977b). Generally, external Ca levels between 10^{-5} and 10^{-4} M appear to be the chemical prerequisite for the origin of multicellular life and its subsequent evolution. Such Ca^{2+} levels could have been reached in the marine realm quite early, probably during the transition of Archean and Proterozoic. It is related to alkalinity decrease, consumption of dissolved Ca-complexing polymers and intensified riverine input of Ca^{2+} as a result of rapid cratonization of the lithosphere at that time (e.g., Veizer, 1985).

These shifts in Ca^{2+} concentrations promoted not only cell contact phenomena but fostered Ca-regulation and Ca-extrusion systems in eukaryotes, resulting in evolution of a whole spectrum of calcium-binding and calcium-modulating proteins exercising sophisticated functioning in metazoan cell systems. In addition to aforementioned coenobial algal eukaryotes, burrows of metazoans (Kauffman and Steidtmann, 1981) and fecal pellets-like structures (Robbins *et al.*, 1985) have been described from rock formations at least 2 billion years old. This implies that the earliest metazoans and the first aggregates of eukaryotic algal cells are of about the same age. Metazoan evolution is tightly linked to the presence of molecular oxygen in air and surface water, and the biosynthesis of a structural protein, i.e. collagen, the most dominant extracellular fibrous protein in all metazoans (Towe, 1970). Biosynthesis of collagen requires molecular oxygen. In turn, without O_2 in the ambience, no metazoans. For a long time early Earth was considered to have an anoxic and possibly reducing atmosphere. Only at about the end of the Precambrian oxygen levels > 1% present atmospheric level (PAL) were assumed. A careful evaluation of the geological, paleontological and geochemical evidence now favours an early oxidized atmosphere, instead (Carver, 1981; Clemmey and Badham, 1982; Towe, 1983; Ohmoto, 1996). Thus, there appears to be no longer a restriction in placing the origin of eukaryotes close to the Archean/Proterozoic boundary, or even earlier. Collagen is the only protein having both hydroxyproline and hydroxylysine which are hydroxylation products of proline and lysine, respectively. Its precursor protein is believed to exist in chrysophytan and chlorophytan phytoflagellates which could possibly be the ancestral groups for the first metazoans (Lamport, 1977; 2001). The question comes up whether collagenous proteins were solely invented to strengthen the metazoan body, or whether they served primarily in metal ion detoxication and only later were adopted as body building material. The idea is advanced that Ca^{2+} build-up in the sea led to two main lineages among protists: one characterized by cell walls rich in polysaccharides, and the other, by cell walls containing collagen in addition. The first divergence generated in time the higher plants, whereas the second one evolved to the metazoans. We tentatively propose that the rapid rise in Ca^{2+} and dissolved phosphates in the sea gave those metazoans a selective advantage which were capable of counteracting this stress by increased collagen production, inasmuch as hydroxyamino acids are capable to scavenge Ca^{2+} (Addadi and Weiner, 1985) and phosphates are known to accelerate Ca^{2+} migration into cells (Cotmore *et al.*, 1971). The excretion of high Ca^{2+}-affine organic substances invented for fast removal and neutralization (detoxication) of the dangerous Ca^{2+} excess from cells and tissues started a process commonly referred to as biocalcification (Kaźmierczak *et al.*, 1985; Kaźmierczak and Degens, 1986; Kempe and Kaźmierczak, 1994).

In our opinion the most determining events in the evolution of multicellular life were: (i) synthesis of collagen, (ii) biocalcification, and (iii) large scale exodus of algal plant precursors and animals from sea to land. The saying that collagen is the 'tape and glue of the metazoan world' (Towe, 1970) pointedly emphasizes the structural and physiological significance of the Ca^{2+} stress phenomenon, - biocalcification, principally involving the deposition of calcium carbonates (aragonite, calcite), and calcium phosphates (carbonate apatite: i.e., dahllite), is responsible for shell, bone and teeth formation and is thus - even macroscopically - a witness for Ca^{2+} regulation in cellular systems (reviewed in: Degens, 1976), and exodus of higher forms of life from a Ca^{2+}-charged sea to rivers, lakes, soils and atmospheric exposure, especially during the Ordovician and Silurian, could possibly be judged as an escape of organisms from a Ca^{2+} stressed environment. Fishes, for instance, are first reported from freshwater environments and they only reluctantly invaded the sea considering their frequent return to the freshwater habitat.

The principal biochemical blue-print of life was designed during the Precambrian. We propose that the development of the eukaryotic cell - as manifested in this blue-print - has been promoted by the gradual build-up of Ca^{2+} in the sea across three billion years of Earth history. Biocalcification across many taxa during Vendian time may be looked upon as a kind of global Ca^{2+} titration scheme. Subsequent changes in skeletal morphology and thickness of marine invertebrates or calcareous plankton reflect, in our opinion, just fine tuning in evolution, but again promoted by Ca^{2+} fluctuation in the world ocean (Kaźmierczak *et al.*, 1985; Degens *et al.*, 1985; Kempe and Kaźmierczak, 1994).

9. Acknowledgments

Supported by the Foundation for Polish Science (J.K.) and the Deutsche Forschungsgemeinschaft (S. K).

10. References

Addadi, L. and Weiner, S. (1985) Interaction between acidic proteins and crystals: stereochemical equirements in biomineralization. Proceedings of the National Academy of Sciences USA **82**, 4110- 4114.

Ala, M. A. (1976) Salt diapirism in South Iran. American Association of Petroleum Geologists Bulletin **58**, 1758-1770.

Altermann, W. (2002) The evolution of life and its impact on sedimentation, In: W. Altermann and P. L. Corcoran (eds.) Special Publication of the International Association of Sedimentologists 33, *Precambrian Sedimentary Environments: A Modern Approach to Ancient Depositional Systems.* Oxford: Blackwell Science Ltd., pp. 15-32.

Awramik, S. M. (1982) The Pre-Phanerozoic fossil record, In: H. D. Holland and M. Schidlowski (eds.) *Mineral Deposits and the Evolution of the Biosphere.* Berlin: Springer-Verlag, Berlin, pp. 67-82.

Awramik, S. M. and Barghoorn, E. S. (1977) The Gunflint microbiota. Precambrian Research **5**, 121-142. Bajaj, Y. P. S., Rasmussen, H. P. and Adams, M. W. (1971) Electronmicroprobe analysis of isolated plant cells. Journal of Experimental Botany **22**, 749-752.

Berridge, M. J. (1993) Inositol trisphosphate and calcium signalling. Nature **361**, 315-325.

Berridge, M.J, Bootman, M. D. and Lipp, P. (1998) Calcium – a life and death signal. Nature **395**, 645-648.

Borle, A. B. (1981) Control, modulation and regulation on cell calcium. Reviews of Physiology Biochemistry and_Pharmacology **90**, 14-153.

Boynton, A. L., Whitfield, J. F. and MacManus, J. P. (1980) Calmodulin stimulates DNA synthesis by rat liver cells. Biochemical and Biophysical Research Communications **95**, 745-749.

Braitsch, O. (1962) Entstehung und Stoffbestand der Salzlagerstätten, In: W. v. Engelhardt and J. Zeman (eds.), *Mineralogie und Petrographie in Einzeldarstellungen,* vol. 3, Springer-Verlag, Berlin, pp. 212-243. Brocks, J. J., Logan, G. A., Buick, R. and Summons R. E. (1999) Archean molecular fossils and the early riseof eukaryotes. Science **285**, 1033-1036.

Burger, M. M. and Misevic, G. (1985) Cell encounter: Molecular and biological aspects of initial contacts, In: H. J. Marthy (ed.), Cellular and Molecular Control of Direct Interactions, Plenum Publishing Corporation, New York, pp. 3-26.

Burzin, M. B. (1999) Coloniality of Precambrian planktonic unicellate microorganisms. Zhurnal Obshchei Biologii **60**, 41-48.

Calvert, G. R. and Dawes, I. W. (1984) Cell size control of development in *Saccharomyces cerevisiae.* Nature **312**, 489- 491.

Campbell, A. K. (1983) *Intracellular Calcium. Its Universal Role as Regulator.* J. Wiley and Sons, Chichester.

Carafoli, E. and Penniston, J. T. (1985) The calcium signal. Scientific American **253/5**, 50-58.

Carver, J. H. (1981) Prebiotic atmospheric oxygen level. Nature **292**, 136-138.

Chan, K. Y. (1976) Control of colony formation in *Coelastrum microporum*_(Chlorococcales, Chlorophyta). Phycologia **15**, 149-154.

Chance, B. and Yoshioka, T. (1966) Sustained oscillations of ionic constituents of mitochondria. Archives of Biochemistry and Biophysics **117**, 451-464.

Cheung, W. Y. (ed.) (1987) *Calcium and Cell Function.* Molecular Biology v. 7, Academic Press, Orlando, Florida.

Clemmey, H. and Badham, N. (1982) Oxygen in the Precambrian atmosphere: an evaluation of the geological evidence. Geology **10**, 141-146. Cloud, P. (1976) Beginnings of biospheric evolution and their geochemical consequences. Paleobiology **2**, 351- 387.

Cloud, P. and Glaessner, J.H. (1982) The Ediacarian period and system: Metazoa inherit the Earth. Science **17**, 783-792.

Cook, P. and Shergold, J. H. (1984) Phosphorus, phosphorites and skeletal evolution at the Precambrian/Cambrian boundary. Nature **308**, 231-236.

Cotmore, J. M., Nichols, J. Jr. and Wuthier, R. E. (1971) Phospholipid calcium phosphate complex: enhanced calcium migration in the presence of phosphate. Nature **172**, 1339-1341.

Degens, E. T. (1976) Molecular mechanisms on carbonate, phosphate, and silica deposition in the living cell.Topics in Current Chemistry **64**, 1-112.

Degens, E. T., Kaźmierczak, J., Ittekott, V. (1985) Cellular response to Ca^{2+}-stress and its geological implications. Acta Palaeontologica Polonica **30**, 115-135.

Durham, J. W. (1978) The probably metazoan biota of the Precambrian as indicated by the subsequent record. Annual Review of Earth and Planetary Sciences **6**, 21-42.

Elbrink, J. and Bihler, I. (1975) Membrane transport: Its relation to cellular metabolic rates. Science **188**, 1177-1184

Engel, A., Massalski, A., Schindler, H., Dorset, D. L. and Rosenbusch, J. P. (1985) Pores in channel triplets merge into single outlets in *Escherichia*_*coli* outer membrane. Nature **317**, 643-645.

Erulkar, S. D. (1981) The versatile role of calcium in biological systems. Interdisciplinary Science Reviews **6**, 322-332.

Esser, K. and Kües, U. (1983) Flocculation and its implication for biotechnology. Process Biochemistry **18**, 21- 23.

Fairchild, T. R. (1985) Size as criterion for distinguishing probable eukaryotic unicells in silicified Precambrian microfloras, in: VIII Congress of Brasilian Paleontology: 1983, sér. Geol., no. 27,

Paleont./Estratigr. **2**, 315- 320.
Franchi, E. and Camatini, M. (1985) Evidence that a Ca^{2+} chelator and a calmodulin blocker interfere with the structure of inter-Sertoli junctions. Tissue and Cell **17**, 13-25.
Francis, S., Margulis, L. and Barghoorn, E. S. (1978) On the experimental silicification of microorganisms. II. On the time of appearance of eukaryotic organisms in the fossil record. Precambrian Research **6**, 65-100.
Gill, D., Ueda, T., Chueh, S.-H. and Noel, M. W. (1986) Ca^{2+} release from endoplasmic reticulum is mediated by a guanine nucleotide regulatory mechanism. Nature **320**, 461-464.
Goodwin, B. C. and Brière, C. (1994) Mechanics of the cytoskeleton and morphogenesis of *Acetabularia*. International Review of Cytology 150, 225-242.
Gribaldo, S. and Philippe, H. 2002. Ancient phylogenetic relationships. Theoretical Population Biology **61**, 391- 408.
Groves, L. B.,Jr. and Kostir, W. J. (1961) Some factors affecting the formation of colonies in *Gonium pectorale*. Ohio Journal of Science **61**, 321-323.
Gupta P. D. and Pushkala, K. (1999) Importance of the role of calcium in programmed cell death: a review. Cytobios **99**, 83-95.
Han, T.-M. and Runnegar, B. (1992) Megascopic eukaryotic algae from 2.1-billion-year-old Negaunee Iron Formation, Michigan. Science **257**, 232-235.
Hardie, L. A. 1996. Secular variation in seawater chemistry: An explanation for the coupled secular variation in the mineralogies of marine limestones and potash evaporites over the past 600 m.y. Geology **24**, 279-283.
Hart, M. H. (1978) The evolution of the atmosphere of the Earth. Icarus **33**,23-39.
Henderson-Sellers, A. and Cogley, J. G. (1982) The Earth's early hydrosphere. Nature **298**, 832-835.
Hofmann, H. J. (1985) Precambrian carbonaceous megafossils, In: D. F. Toomey and M. H. Nitecki (eds.) *Paleoalgology: Contemporary Research and Applications*, Springer-Verlag, Berlin, pp. 20-33.
Hofmann, H. J. and Chen, J. (1981) Carbonaceous megafossils from the Precambrian (1800 Ma) near Jixian, northern China. Canadian Journal of Earth Sciences **18**, 443-447.
Holland, H. D., Lazar, B. and McCaffrey, M. (1986) Evolution of the atmosphere and oceans. Nature **320**, 27-33.
Irvine, R. F. and Moor, R. M. (1986) Micro-injection of inositol 1,3,4,5-tetrakisphosphate activates sea urchin eggs by a mechanism dependent on external Ca^{2+}. Biochemical Journal **240**, 917-920.
Katz, L. A. (2002) Lateral gene transfers and the evolution of eukaryotes: theories and data. International Journal of Systematics and Evolutionary Microbiology **52**, 1893-1900.
Kauffman, E. G. and Steidtmann, J. R. (1981) Are these the oldest metazoan trace fossils? Journal of Paleontology **55**, 923-947.
Kaźmierczak, J. (1976). Devonian and modern relatives of the Precambrian *Eosphaera*: possible significance for the early eukaryotes. Lethaia **9**, 39-50.
Kaźmierczak, J. (1979) The eukaryotic nature of *Eosphaera*-like ferriferous structures from the Precambrian Gunflint Iron Formation, Canada: A comparative study. Precambrian Research **9**, 1-22.
Kaźmierczak, J. (1981) The biology and evolutionary significance of Devonian volvocaceans and their Precambrian relatives. Acta Palaeontologica Polonica **26**, 299-337.
Kaźmierczak, J. and Degens, E. T. (1986) Calcium and the early eukaryotes. Mitteilungen aus dem Geologisch- Paläontologischen Institut der Universität Hamburg **61**, 1-20.
Kaźmierczak, J., Ittekkot, V. and Degens, E. T. (1985) Biocalcification through time: environmental challenge and cellular response. Paläontologische Zeitschrift **59**, 15-33.
Kempe, S. and Degens, E. T. (1985) An early soda ocean? Chemical Geology **53**, 95-108.
Kempe, S. and Kaźmierczak, J. (1994) The role of alkalinity in the evolution of ocean chemistry, organization of living systems and biocalcification processes, In: F. Doumenge (ed.), *Past and Present Biomineralization Processes.Considerations about the Carbonate Cycle*, Bulletin de l'Institute Oceanographique, Monaco, no. spec. **13**, 61-117.
Kempe, S. and Kaźmierczak, J. (2002) Biogenesis and early life on Earth and Europa: Favoured by an alkaline ocean? Astrobiology **2**, 123-130.
Kempe, S. and Kaźmierczak, J. (2003) Modern soda lakes: Model environments for an early alkaline ocean, In: T. Müller and H. Müller (eds.) *Modelling in Natural Sciences – Design, Validation and Case Studies*. Springer-Verlag, Berlin, pp. 309-322.
Kempe, S., Kaźmierczak, J., and Degens, E.T. (1989) The soda ocean concept and its bearing on biotic and crustal evolution, In: R. E. Crick (ed.), Origin*, Evolution and Modern Aspects of Biomineralization in Plants and Animals*. Proceedings of the 5th International Symposium on Biomineralization, Arlington, Texas, May, 1986, Plenum Press, New York, pp. 29-43.

Kempe, S., Kaźmierczak, J., Landmann, G., Konuk, T., Reimer, A. and Lipp, A. (1991) Largest known microbialites discovered in Lake Van, Turkey. Nature, **349**, 605-608.

Knoll, A. H. (1992) The early evolution of eukaryotes. Science **256**, 622-627.

Knoll, A. H. (1999) A new molecular window of early life. Science **285**, 1025-1026.

Knoll, A. H. and Barghoorn, E. S. (1975) Precambrian eucaryotic organisms: A reassessment of the evidence. Science **190**, 52-54.

Knoll, A. H. and Barghoorn, E. S. (1977) Archean microfossils showing cell division from the Swaziland System of South Africa. Science **198**, 396-398.

Kretsinger. R. H. (1977a) Evolution of the informational role of calcium in eukaryotes, In: R. H. Wasserman *et al.* (eds.) *Calcium Binding Proteins and Calcium Function,* North Holland Publ., New York, pp. 63-72.

Kretsinger, R. H. (1977b) Why does calcium play an informational role unique in biological systems? In: B. Pullman and N. Goldbleum (eds.), *Metal-Ligand Interactions in OrganicChemistry and Biochemistry (9th Jerusalem Symposium).* Part 2, E. Lidel Publishing Company, Dordrecht, pp. 257-263.

Kretsinger, R. R. (1983) A comparison of the roles of calcium in biomineralization and in cytosolic signaling, In: P. Westbroek and E. W. De Jong (eds.), *Biomineralization and Biological Metal Accumulation*, Reidel Publishing Company, Dordrecht, pp. 123-131.

Kuhn, W. R. and Kasting, J. F. (1983) Effects of increased CO_2 concentrations on surface temperature of the early Earth. Nature **301**, 53-55.

LaBerge G. L., Robbins, E. I. and Schmidt, R. G. (1984) New microfossil evidence of eukaryotes in early Proterozoic rocks (1.9 B. y.) from the Lake Superior Region, North America, in: 27th International Geological Congress, Moscow, August 4-14, 1984, Abstracts vol. 1, section 01 to 03t p. 241. Moscow: Nauka, Moscow, p. 241.

Lamport, D. T. A. (1977) Structure, biosynthesis and significance of cell wall glycoproteins. Recent Advances in Phytochemistry **11**, 79-112.

Lamport, D. T. A. (2001) Life behind cell walls: paradigm lost, paradigm regained. Cellular and Molecular Life Sciences **58**, 1363-1385.

Lowenstam, H. and Margulis, L. (1980) Evolutionary prerequisites for early Phanerozoic calcareous skeletons. BioSystems **12**, 27-41.

Lucy, J. A. (1978) Mechanism of chemically induced cell fusion. Cell Surface Reviews **5**, 267-304.

Maisonneuve, J. (1982) The composition of the Precambrian ocean waters. Sedimentary Geology **31**, 1-11.

Marcum, J. M., Dedman, J. R., Brinsky, D. R. and Means, A. R. (1978) Control of microtubule assembly-disassembly by calcium-dependent regulator protein. Proceedings of the National Academy of Sciences USA **75**, 3771-3775.

Margulis, L. (1981) *Symbiosis and Cell Evolution.* W. H. Freeman and Company, San Francisco.

Matheja, J. and Degens, E. T. (1971) Structural Molecular Biology of Phosphates. Fortschritte der Evolutionsforschung **5**, 1-180.

McConachie, D. R. and O'Day, D. H. (1986) The immediate induction of extensive cell fusion by Ca^{2+} addition in *Dictyostelium discoideum.* Biochemistry and Cell Biology **64**, 1281-1287.

McFadden, G. J. (2001) Primary and secondary endosymbiosis and the origin of plastids. Journal of Phycology **37**, 951-959.

McFadden, G. J., Preisig, H. R. and Melkonian, M. (1986) Golgi apparatus activity and membrane flow during scale biogenesis in the green flagellate *Scherffelia dubia* (Prasinophyceae). II. Cell wall secretion and assembly. Protoplasma **131**, 293-302.

McLeod, G., McKeown, C., Hall, A. J. and Russell M. J. (1994) Hydrothermal and oceanic pH conditions of possible relevance to the origin of life. Origins of Life and Evolution of the Biosphere **24**, 19-41.

Morse, J. W. and Mackenzie, F. T. (1998) Hadean ocean carbonate geochemistry. Aquatic Geochemistry 4, 301-319.

Muller, A. W. J. (1996) Hypothesis: the termosynthesis model for the origin of life and the emergence of regulation by Ca^{2+}, In: D. K. Apps (ed.), *Essays in Biochemistry,* v. 31, Portland Press, London, pp.103-119.

Ohmoto, H. (1996) Evidence in pre 2.2 Ga paleosols for the early evolution of atmospheric oxygenated terrestrial biota. Geology **24**, 1135-1138.

Pflug, H. D. (1978) Yeast-like microfossil detected in oldest sediments of the Earth. Naturwissenschaften **65**, 611-615.

Pflug, H. D. (1982) Early diversification of life in the Archean. Zentralblatt für Bakteriologie und Hygiene I. Abt. Orig. C **3**, 53-64.

Prusch, R. D. and Minck, D. R. (1985) Chemical stimulation of phagocytosis in *Amoeba proteus* and the influence of externall ca cium. Cell and Tissue Research **242**, 557-564.

Robbins, E. T., Porter, K. G. and Haberyan, K. A. (1985) Pellet microfossils: possible evidence for metazoan life in early Proterozoic time. Proceedings of the National Academy of Sciences USA **82**, 5809-5813.

Rose, A. H. (1984) Physiology of cell aggregation: flocculation by *Saccharomyces cerevisiae* as a model system, In: K. C. Marshall (ed.), *Microbial Adhesion and Aggregation*, Springer-Verlag, Berlin, pp. 323-335.

Rose, B., Simpson, I. and Loewenstein W. R. (1977) Calcium ion produces graded changes in permeability of membrane channels in cell junction. Nature **267**, 625-627.

Runnegar, B. (1982) The Cambrian explosion: animals or fossils? Journal of the Geological Society of Australia **29**, 395-411.

Sasaki, Y. and Hidaka, H. (1982) Calmodulin and cell proliferation. Biochemical and Biophysical Research Communications **104**, 451-456.

Schanne, F. A. X., Kanne, A. B., Young, E. E. and Farber, J. L. (1979) Calcium dependence of toxic cell death. Science **206**, 700-702.

Schopf, J. W. (1978) The evolution of the earliest cells. Scientific American **239/3**, 85-102.

Schopf, J. W. (1999) *Cradle of Life. The Discovery of Earth's Earliest Fossils*. Princeton University Press, Princeton, New Jersey.

Schopf, J. W. and Oehler, D. Z. (1976) How old are the eukaryotes? Science **193**, 47-49.

Siever, R. (1974) The steady state of the Earth's crust, atmosphere and oceans. Scientific American **230/6,** 72-79.

Stevens, C. H. (1977) Was development of brackish oceans a factor in Permian extinctions? Geological Society of America Bulletin **88**, 133-138.

Taylor, F. J. R. (1976) Autogenous theories for the origin of eukaryotes. Taxon **25**, 377-390.

Toister, Z. and Loyter, A. (1971) Ca^{2+}-induced fusion of avian erythrocytes. Biochimica et Biophysica Acta **242**, 719-724.

Towe, K. M. (1970) Oxygen-collagen priority and the early metazoan fossil record. Proceedings of the National Academy of Sciences USA **65**, 781-788.

Towe, K. M. (1983) Precambrian atmospheric oxygen and banded iron formations: A delayed ocean model. Precambrian Research **20**, 161-170.

Tynni, R. and Uutela, A. (1984) Microfossils from the Precambrian Muhos Formation in Western Finland. Geological Survey of Finland Bulletin **330**, 5-38.

Unwin, N. and Henderson, R. (1984) The structure of proteins in biological membranes. Scientific American **250/2**, 56-66.

Varnaman, T. C., Wakil, S. J. and Hill, L. J. (1968) The complete amino acid sequence of the acyl carrier protein of *Escherichia coli*. Journal of Biological Chemistry **243**, 6420- 6431.

Veizer, J. (1985) Carbonates and ancient oceans: isotopic and chemical record on time scales of 10^7 - 10^9 years, In: E. T. Sunquist and W. S. Broecker (eds.), *The Carbon Cycle and Atmospheric CO_2: Natural Variations Archean to Present*, The American Geophysical Union Geophysical Monographs **32**, 595-601.

Volberg, T., Geiger, B., Kartenbeck, J. and Franke, W. W. (1986) Changes ,in membrane-microfilament interaction in intercellular adherens junctions upon removal of extracellular Ca^{2+} ions. The Journal of Cell Biology **102**, 1832-1842.

Walker, J. C. G. (1983) Possible limits on the composition of the Archean ocean. Nature **302**, 518-520.

Walther, H. W. (1972) Über Salzdiapire in SE-Iran. Geologisches Jahrbuch **90**, 359-388.

Weisman, D. M., Singh, T. J. and Wang, J. H. (1978) The modulator-dependent protein kinase. Journal of Biological Chemistry **253**, 3387-3390.

Williams, D. A., Fogarty, K. E., Tsien, R. Y. and Fay, F. S. (1985) Calcium gradients in single smooth muscle cells revealed by the digital imaging microscope using Fura-2. Nature **318**, 558-561.

Williams, R. J. P. (1976) Calcium chemistry and its relation to biological function, In: D. J. Duncan (ed.), *Calcium in Biological Systems*, Symposia of the Society of Experimental Biology **30**. Cambridge: Cambridge University Press, Cambridge, pp. 1-17.

Xiao Shuai (2002) Mitotic topologies and mechanics of Neoproterozoic algae and animal embryo. Paleobiology **28**, 244-250

Xue Yao-song, Tang Tian-fu, Yu Cong-liu and Zhou Chuan-ming (1995) Large spheroidal Chlorophyta fossils from Doushantuo Formation phosphoric sequence (late Sinian), central Guizhou, south China. Acta Palaeontologica Sinica **34**, 688-706.

Young, J. D. E. (1985) Role of ionic events in the triggering of phagocytosis. Journal of Theoretical Biology **116**, 475-478.

Biodata of **Philip Bell** author of ***"The Viral Eukaryogenesis Theory"***

Dr. P. J. L. Bell is a postdoctoral research fellow at Macquarie University, New South Wales, Australia and a founding Director of Microbiogen Pty Ltd. (a Biotechnology company in Sydney, Australia.) He obtained his Ph.D. in 1999 from the University of New South Wales, Australia. His research interests are: Evolution, Genetics and Molecular Biology and he is specializing in genetics of industrial yeast strains and the development of improved yeast strains for pharmaceutical, baking, brewing and fuel alcohol industries. Dr. Bell is also the discoverer of epicoccone, a novel fluorescent dye produced by the fungus *Epicoccum nigrum* that has applications in proteomics, cell tracking and flowcytometry.

E-mail: **pbell@rna.bio.mq.edu.au**

J. Seckbach (ed.), Origins, 347–367.

THE VIRAL EUKARYOGENESIS THEORY

P. J. L. BELL
Macquarie University, New South Wales, 2109, Australia

1. Introduction

The five kingdom classification scheme of life embraced the Darwinian concept of common descent and placed all life forms on a single phylogenetic tree. In this scheme, cells without nuclei were placed at the base of the tree in kingdom Monera. Kingdom Monera gave rise to kingdom Protista (unicellular organism containing nuclei), which in turn gave rise to the plant, animal, and fungal kingdoms.

Analysis of ribosomal RNA gene sequences overturned the five kingdom scheme and led to the 'Woesian revolution' in which all cellular life is divided into three evolutionarily distant primary lineages (Woese et al., 1990). The three primary lineages were established as domains, termed the Bacteria, the Archaea, and the Eukarya. Whole genome analysis of Bacteria and Archaea has subsequently shown that lateral gene transfer is a major force in evolution, and thus phylogenetic analysis of single gene families may generate trees that do not conform to the three domain structure. Whole genome analysis of eukaryal genomes has also revealed that they are a complex chimera, or mosaic, of bacterial and archaeal genes (Doolittle 1998; Lake et al., 1999; Nowitzki et al., 1998; Ribeiro and Golding 1998; Rivera et al., 1998; Rosenthal et al., 1997). Despite problems with lateral gene transfer, the three domain system is robust, since many phenotypic characteristics such as lipid composition, the presence of peptidoglycan, and cytoskeletal structures support the uniqueness of the three domains (Doolittle, 1999; Forterre and Philippe, 1999).

The three domain classification scheme continues to embrace the Darwinian concept that all life shares a common origin, which is supported by the recognizable homology between the ribosomal apparatus of all cellular organisms. It is, however, currently uncertain how each of the three domains evolved from the last ancestor of cellular life.

In one school of thought, domains Archaea and Bacteria were first to diverge from the common ancestor, and the Eukarya subsequently arose from domain Archaea. This school of thought underpins the viral eukaryogenesis theory, where it is proposed that domain Eukarya evolved from an archaeal ancestor in which a complex DNA virus established a persistent presence. Through a process of gene transfer from the archaeal genome to the viral genome, it is proposed that the virus was able to take over genetic control of the cell and evolve into the nucleus. The derivation of the nucleus from a complex DNA virus is proposed as an explanation for the origin of many of the unique aspects of cellular design found in the Eukarya.

2. Unique Aspects of Cellular Design in Domain Eukarya

The doubling times and metabolic capacity of an eukaryote such as *Saccharomyces cerevisiae*, an archaeon such as *Archaeoglobus fulgidus*, and a bacterium such as *Escherichia coli* are comparable, and the compact size of their genomes reflects the fact that these single celled microbes are subjected to similar selection pressures for rapid replication and efficient nutrient utilization. Furthermore, yeast, like both *E. coli* and *A. fulgidus*, is a heterotroph that absorbs simple nutrients from the environment. Despite these ecological similarities, and similarities in genome size (*S. cerevisiae*: 5885 genes (Goffeau et al., 1996), *E. coli*: 4288 genes (Blattner et al., 1997), *A. fulgidus*, 2436 genes (Klenk et al., 1997)), the genetic and cellular design of the yeast differs from *E. coli* and *A. fulgidus* in several fundamental aspects.

For example, the genome of *S. cerevisiae* is organized into 16 linear chromosomes whereas the genomes of *E. coli* and *A. fulgidus* are organized into single circular chromosomes. One central problem for linear replicons such as yeast chromosomes, but not circular replicons, arises from the fact that all known DNA polymerase enzymes require a 3' OH group on which to add the next nucleotide (Kipling, 1995). As a result, synthesis of linear chromosomes cannot start *de novo,* and without a mechanism to overcome this, every round of DNA synthesis will result in the inexorable loss of chromosomal DNA (Kipling, 1995). Replication of eukaryal chromosomes has therefore required the evolution of specialized telomeres not generally found in the prokaryotes. It should be noted that although some bacterial genera such as *Borrelia* and *Streptomyces* have linear chromosomes (Casjens, 1999; Lin et al., 1993; Hinnebusch and Tilly, 1993), linearity does not seem to be ancestral in the Bacteria because the phylogenetic position of these species suggests that they are derived from progenitors with circular replicons (Casjens, 1999). It has also been proposed that the linear DNA structures in *Borrelia* have been derived through horizontal transfer from a virus (Hinnebusch and Barbour, 1991).

Yeast chromosomes, also unlike prokaryotic chromosomes, are separated from the cytoplasm by the nuclear membrane. As a consequence, the mRNA of the Eukarya is synthesized within the nucleus and transported into the cytoplasm for translation. Transport across the nuclear membrane is mediated via the nuclear pore complexes (NPC) which are composed of multiple copies of ~ 100 distinct polypeptides (Gant and Wilson, 1997). Some translation products from the cytoplasm must also be transported via the NPCs into the nucleus to perform essential functions such as transcription and DNA replication. Although bacteria from the genus *Gemmata* possess a membrane bound region encompassing the genome (Wang et al., 2002), these do not resemble the eukaryal nuclear membrane in structure or function since the bacterial nuclear bodies contain both the genetic material and the translation apparatus (Lindsay et al., 2001).

The structure and processing of mRNA in yeast and other members of the Eukarya domain is also fundamentally different to that observed in the Bacteria or the Archaea. The mRNA of Eukarya is generally capped, polyadenylated and processed after transcription and prior to extrusion into the cytoplasm. Once in the cytoplasm, eukaryal ribosomes recognize the mRNA cap and translation is initiated at the first ATG encountered by the ribosome (Kozak, 1999). By contrast, archaeal and bacterial mRNA is not capped, contains only small poly A tails, and is not subject to extensive post-transcriptional processing (Keeling et al., 1994; Sarkar, 1997). In both Archaea and

Bacteria, translation of mRNA does not require a cap and is initiated at ribosome binding sites encoded within the mRNA.

Eukaryal genes encoding proteins are also generally transcribed independently of each other (e.g. Hinnebusch, 1988; Struhl, 1995), whereas archaeal and bacterial genes involved in similar pathways are often organized into co-transcribed operons (Keeling et al., 1994). Although a limited number of polycistronic operons exist in Eukarya such as nematodes (Blumenthal, 1995) and trypanosomes (Imboden et al., 1987), in both these cases, the individual coding regions are spliced to generate individual monocistronic mRNA transcripts that are transported into the cytoplasm. This contrasts with the archaeal and bacterial operons where the polycistronic transcripts are not spliced, but rather each of the proteins of the operon is translated from the single transcript. Given the phylogenetic distribution of eukaryal polycistronic transcripts, it would appear that polycistronic operons are a limited phenomenon in the Eukarya which has evolved independently in the different lineages (Blumenthal, 1995).

During cellular replication, the yeast nucleus undergoes a complex mitotic process where chromosomes are independently replicated, condensed and segregated (Gant and Wilson, 1997). In higher Eukarya the nuclear membrane disintegrates during mitosis, and is re-assembled at telophase when the chromosomes de-condense. Most eukaryal species including yeast can also enter into meiosis where the diploid cell goes through two mitosis-like cell divisions to generate four haploid cells with two different mating types. By contrast mitosis in Archaea and Bacteria is much simpler and involves DNA replication followed by nucleoid partition, which is in turn followed by cell division (Bernander, 1998). No prokaryotic processes have been described that resemble either eukaryal mitosis or meiosis.

Yeast has several other features of cellular design which are not found in either the bacterial or archaeal domains. These include the cytoskeleton, which is essential for phagocytosis and chromosome segregation, membrane fusion processes which are essential for endocytosis and mating, and several organelles such as the mitochondria, endoplasmic reticulum, Golgi apparatus, vacuole, and lysosomes which are essential for a number of metabolic functions.

If *S. cerevisiae*, *E. coli*, and *A. fulgidus* ultimately descend from a common ancestor, the following crucial question arises. How did the Eukarya evolve to have such a different cellular design to the other two domains? To address this question it is important to address the evolutionary origin of the three domains.

3. Evolutionary Origin of the Three Domains

In the progenote hypothesis, (Woese, 1987; Woese, 1998; Woese, 2000), the last universal ancestor of the three domains of life was not a single organism, but rather a communal consortium of progenotes unlike modern cells. In particular their component parts had different ancestries, and the progenotes were subject to high levels of lateral gene transfer. In this hypothesis the universal ancestor was a communal, loosely knit, diverse conglomerate of cells that evolved as a unit (Woese, 1998). After the conglomerate split into several distinct communities the three primary lineages became established, and each evolved into the modern cellular lineages.

In an alternative hypothesis, the last universal ancestor is placed in the eukaryal lineage, and thus the last common ancestor was of eukaryal-like design. This hypothesis was proposed originally by Reanney, (1974) and has recently revived by several authors and supported using phylogenetic analysis (Brinkmann and Philippe, 1999; Forterre and Philippe, 1999; Penny and Poole, 1999). By placing the ancestor of all living organisms in the eukaryal branch of the phylogenetic tree, the 'prokaryote' genetic design is a later innovation which was selected for from the more complex eukaryal design by processes such as thermoreduction and r-selection (Forterre, 1995; Forterre and Philippe, 1999; Penny and Poole, 1999; Poole et al., 1998).

In a third model, several lines of evidence are used to support the idea that the last common ancestor was of archaeal/bacterial design. For example, studies on duplicated genes that exist in all three domains provide evidence that the first bifurcation in the tree of life separated the bacterial lineage from the archaeal/eukaryal lineage (Gogarten et al., 1989; Gribaldo and Cammarano, 1998; Iwabe et al., 1989). Elucidation of complete genomic sequences has also suggested that the last common ancestor was an organism with metabolic networks and genetic machinery similar to those of extant prokaryotic organisms (Kyrpides et al., 1999). In addition, the observation that the bacterial and archaeal ribosomal genes are structured into operons of similar structure (Keeling et al., 1994) but possess distantly related sequences, can be used to argue that the Bacteria and the Archaea shared an ancient common ancestor in which these genes were already assembled into an operon. Also consistent with the first bifurcation occurring between the Bacteria and the Archaea is the observation that many eukaryal information processing genes are more closely related to archaeal genes than they are to the bacterial equivalents (Brown and Doolittle, 1997; Bult et al., 1996; Kletzin, 1992; Koonin et al., 1997; Snel et al., 1999). Since genes involved in the complex transcriptional and translational machinery appear to be less prone to lateral gene transfer (Jain et al., 1999; Rivera et al., 1998), results such as these can be used to argue that the Archaea and the Eukarya are sister groups, and have thus diverged from each other after their common ancestor diverged from the Bacteria.

Finally, the fossil record also supports the early origin of cells of prokaryotic design compared to cells of eukaryal design. The first fossils of prokaryotes indicate that cells of this design were present in the order of 3500 million years ago (Schopf, 1993). Molecular fossils of biological lipids preserved in ancient shales indicate that Eukarya may have evolved by 2700 million years ago which is 500 million to 1 billion years earlier than the extant fossil record (Brocks et al., 1999), but still 700 million years after the first prokaryotic fossils.

By placing the ancestor of all living organisms in between the bacterial and archaeal branch of the phylogenetic tree, the eukaryal design is a later innovation which evolved from the more simple prokaryotic design. The tree shown in Figure 1 is proposed as a model to represent the evolution of the three domains in the viral eukaryogenesis theory. In this scheme the last common ancestor of all life was a non-compartmentalized simple unicellular organism. Accordingly, the similarities in genetic design observed between the Bacteria (e.g. *E. coli*) and the Archaea (e.g. *A. fulgidus*) are primarily due to their descent from a common ancestor which possessed these characteristics. In this model eukaryogenesis is the process by which a lineage of archaeal genetic design evolved into a lineage with eukaryal design (Figure 1B).

The fundamental differences between archaeal and eukaryal design make the process of eukaryogenesis a challenging evolutionary problem, especially since no contemporary organisms are thought to possess cellular architectures that are intermediate between the Archaea and the Eukarya.

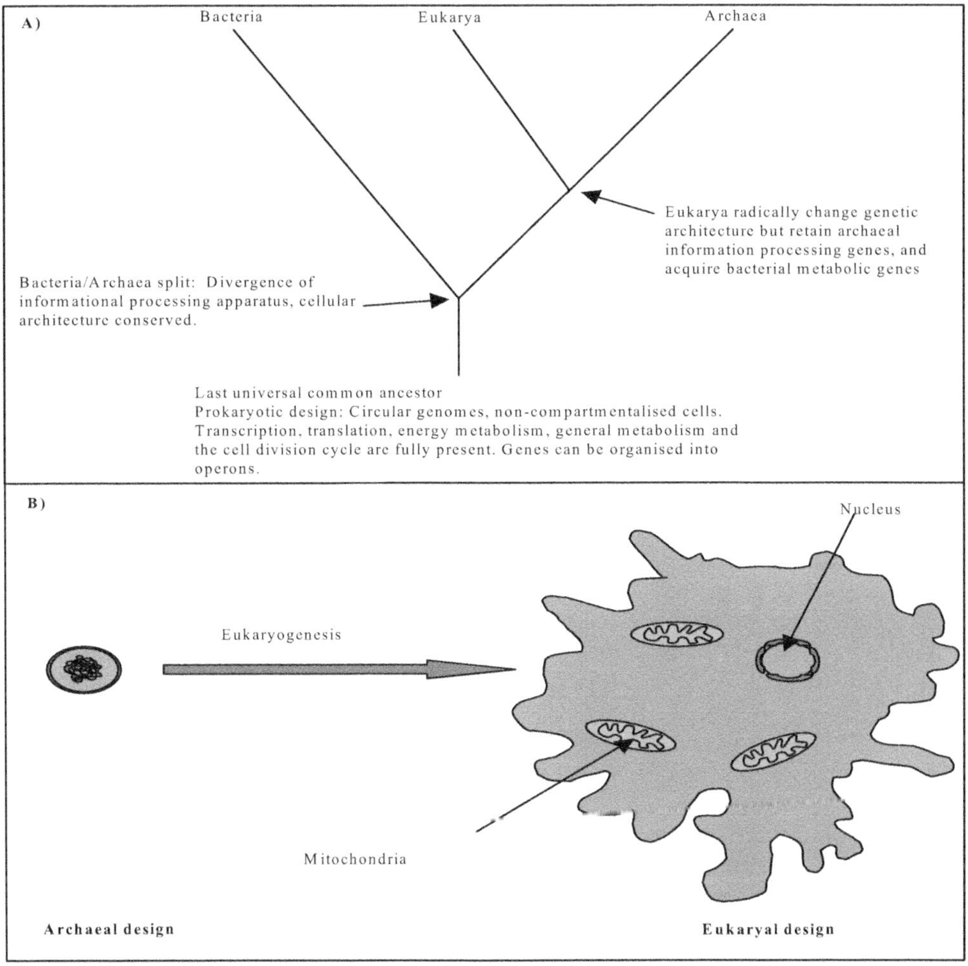

Figure 1. Proposed evolution of the three cellular domains of life. A) If the root of the universal tree lies between the Bacteria and the Archaea, the Eukarya and the Archaea are sister groups that share a common ancestor. The shared genetic architecture between the archaeal and bacterial domains is a consequence of their descent from a prokaryotic common ancestor. B) Eukaryogenesis requires a radical change in architecture including the evolution of the nuclear membrane, mRNA capping, mRNA splicing, linear chromosomes, telomeres, meiosis, mitosis, sex, mitochondria, cytoskeleton, and the endoplasmic reticulum.

4. Endosymbiotic Theories and Eukaryogenesis

Several models have been proposed to explain the origin of the Eukarya including models in which the Eukarya arose *via* symbiosis, primary fusion events, or phagocytosis between the Archaea and the Bacteria (Cavalier-Smith, 1987; Gupta, 1998; Gupta and Golding, 1996; Hartman and Federov, 2002; Lake and Rivera, 1994; Lopez-Garcia and Moreira, 1999; Margulis, 1996; Martin and Müller, 1998; Moreira and Lopez-Garcia, 1998; Ribeiro and Golding, 1998; Sogin, 1991; Vellai et al., 1998; Zillig, 1991).

Endosymbiotic based models such as those outlined by Margulis (1996), propose that the mitochondria and other organelles of the Eukarya arose from free living cells that were incorporated into the ancestral host cell to the mutual advantage of both organisms. These endosymbiotic theories for the origin of organelles are convincing when biochemical and genetic similarities exist between the organelle and the free-living organism (See Martin, 1999 for discussion). Thus the correspondences between the mitochondria and some of the free-living alpha-proteobacteria provide a strong argument for the evolution of the mitochondria from a free-living bacteria (e.g. Gray et al., 1999). Similarly the extensive biochemical, genetic and phylogenetic correspondences between the chloroplasts and free living prokaryotic algae provide a very strong argument that the chloroplasts are derived from an endosymbiosis event between an ancestral eukaryote and a prokaryotic alga. Using the same rationale, the origin of the nucleus is difficult to explain since the nucleus is biochemically and genetically unlike any prokaryote (See Martin, 1999 for discussion).

In the viral eukaryogenesis theory, it is proposed that the nucleus is biochemically unlike any free living Archaea or Bacteria because it is derived from a complex DNA virus rather than a cellular organism. Thus the nucleus does not share many biochemical similarities with free-living cells, but rather shares biochemical and genetic similarities with a class of complex DNA viruses.

5. The Viruses: an Introduction

Although viruses are very diverse in size, structure, and genome organization, they share a number of common characteristics. These include an obligate intracellular life cycle, protein encapsulation of nucleic acid, an evolutionary history independent of its host, and a lack of translational machinery (Strauss et al., 1996). The lack of translational machinery prevents the placement of viruses into any universal tree of life based on ribosomal RNA sequences, but does not imply that they do not have significant effects on the evolution of the three cellular domains. In particular, viruses appear to have a role in horizontal gene transfer and maintenance of biodiversity (e.g. Fuhrman, 1999). It has also recently been speculated that they may also have played a fundamental role in the original transition from RNA based cells into DNA based cells (Forterre, 2001).

The first appearance of viruses in the evolution of life is uncertain and direct evidence for viruses at any early stage of the earth's history is unlikely to exist due to the limitations of the fossil record. However, today viruses are very abundant. For example, in marine environments, they typically number ten billion per litre, and play a vital role in bacterial and algal biodiversity by limiting the abundance of any one species

(Fuhrman, 1999). Indeed, it has been shown that viruses are consistently the most abundant biological entities (about 5-25 times the bacterial abundance) in a wide variety of marine environments (Fuhrman, 1999). Furthermore, the types of viruses present are diverse and viral abundance is dynamic, reflecting the fact that viruses are active members of the ecosystem (Fuhrman, 1999).

Three major theories have been proposed to explain the origin of the viruses, and all are necessarily speculative (reviewed in Strauss et al., 1996). The first theory, the regressive theory of virus origins, proposes that viruses are highly degenerate intracellular parasites that evolved from microorganisms that progressively lost biosynthetic abilities. Present day examples of intracellular parasitism (e.g. *Chlamydia*) could be viewed as modern day intermediates in such an evolutionary pathway. In the second theory, it is proposed that viruses originated from host cell RNA or DNA. This theory postulates that viruses arose from normal cellular components that gained the ability to replicate autonomously, and thereby evolve independently of the host. In the third theory, it is proposed that viruses evolved along with the most primitive molecules containing self-replicative abilities, and thus descend from pre-cellular organisms of a putative RNA world.

None of these three theories restrict the origin and evolution of the viruses to modern times. Both the regressive theory for the origin of viruses, and the host cell RNA/DNA theories are compatible with the appearance of viruses relatively soon after the evolution of cellular life. Since the first prokaryotic fossils that exist are 3.5 billion years old, viruses could have evolved at about this time and thus have a very ancient origin that predated the oldest of the eukaryal fossils. The third 'RNA world' model predicts that virus-like organisms would have evolved hand in hand with the earliest cellular life, and would thus have a very ancient origin predating that of the cellular domains.

6. Large Eukaryal DNA Viruses share Common Features with the Eukaryal Nucleus

A survey of viral species indicates that some of the large eukaryal DNA viruses share biochemical and genetic similarities with the nucleus that are not found in any prokaryotic cells. In particular, the Pox viruses and the African Swine Fever Virus (ASF virus) share several characteristics with the nucleus (Table 1).

The Pox viruses and the ASF virus replicate directly in their host's cytoplasm and are capable of producing capped and polyadenylated mRNA. Their mRNA is translated by host ribosomes located in the cytoplasm, and like all viruses, they cannot sustain themselves outside of a cell because they are not capable of metabolic functions such as the production of ATP, lipids, nucleotides, and amino acids.

Purified virons of both the poxvirus and the ASF virus contain a large number of polypeptide species with enzymatic activity (Blasco, 1995). This assemblage includes an early transcription system that is independent of the nucleus and can carry out synthesis of mRNAs that are polyadenylated, capped and methylated (Moss, 1996). This early transcription system is packaged within the core of the infectious poxvirus particle, providing a mechanism for the synthesis of viral mRNA that is extruded from the core into the cytoplasm for translation (Moss, 1996).

TABLE 1. Comparison between nuclei, the pox viruses and free-living prokaryotes

	Nucleus	Pox virus	Prokaryote
Membrane bound structures	√	√	√
ATP production	X	X	√
Linear chromosomes	√	√	X
Cytoplasmic replication	√	√	X
Disassemble membrane during replication	√	√	X
Membrane assembled at ER	√	√	X
Transcription not translation	√	√	X
mRNA export across membrane	√	√	X
Capping of mRNA	√	√	X
Import of proteins	√	√	X

Analysis of the poxviruses, the ASF virus, and several of the large eukaryal viruses, has identified a common set of shared conserved genes. This observation has been used to argue that at least four viral families (poxviridae, asfaviriridae, iridoviridae and phycodnaviridae) are descended from a common ancestral virus (Iyer et al., 2001). The proposed ancestor of these viruses (the NCLDV viruses) is tentatively reconstructed as a nucleocytoplasmic DNA virus with an icosohedral capsid, which encoded complex systems for DNA replication and transcription (Iyer et al., 2001). At least 31 genes are proposed to be present in their common ancestor, including, a family B DNA polymerase, an ATP dependent ligase, a type II topoisomerase, an mRNA capping enzyme, three RNA polymerase subunits, and several nucleotide metabolism genes such as dUTPase, ribonucleotide reductase subunits. Together these suggest that the NCLDV ancestor had an elaborate system for genome replication and expression of capped and polyadenylated mRNA (Iyer et al., 2001). Since the hosts of these viruses range from single celled algae to mammalian cells, it appears likely that the NCLDV ancestor appeared early in the evolution of the Eukarya.

It is proposed in the viral eukaryogenesis theory that the eukaryotic nucleus is descended from an ancient NCLDV-like virus that infected an archaeal host.

7. Phylogenetic Analysis of Core Viral Genes

Although the NCLDV viruses and the nucleus share features of genetic design such as the ability to cap mRNA, this does not demonstrate that the nucleus is descended from an NCLDV virus as proposed in the viral eukaryogenesis theory. There are at least two alternative explanations other than the proposed descent of the nucleus from a virus. Firstly large NCLDV viruses may share similarities in genetic design with the nucleus because the viruses have evolved to replicate in the eukaryal cytoplasm and have adapted to the eukaryal cytoplasm by borrowing genetic machinery from their host. In this case the similarities are secondary to the adaptation of the viruses to the eukaryal host. Alternatively it is possible that the nucleus and the virus do share a common ancestor, but it is the complex DNA viruses that are descended from a simplified nucleus that has escaped from the cytoplasm to become a parasitic virus. This would be a novel hypothesis for the origin of the viruses, and one that has not been proposed so far.

One way to address these questions is to analyze the phylogeny of the viral and nuclear genes that were present in the proposed NCLDV viral ancestor. By examining their phylogenetic origin, it may be possible to determine whether the NCLDV viruses obtained their genetic machinery from their eukaryal hosts, or whether the core gene assemblage of the viruses existed prior to the evolution and divergence of the Eukarya.

The mRNA capping enzyme is a key member of the core viral gene assemblage found in the proposed NCLDV ancestor. mRNA capping entails GMP transfer from GTP to a 5' phosphate RNA end to form the structure G(5')ppp(5')N, and is similar to reactions occurring in strand joining by polynucleotide ligases (Shuman and Schwer, 1995). Sequence conservation is found between the capping enzymes and ATP dependent ligases, and it has been proposed that cellular and DNA virus capping enzymes, together with ATP dependent ligases constitute a covalent nucleotidyltransferase superfamily which evolved from a common ancestor (Shuman and Schwer, 1995). Significantly, guanylytransferases (mRNA capping enzymes) have not been described in either the bacterial or archaeal domains and thus presumably arose in either the eukaryal lineage or in a non-cellular lineage such as the complex cytoplasmic DNA viruses.

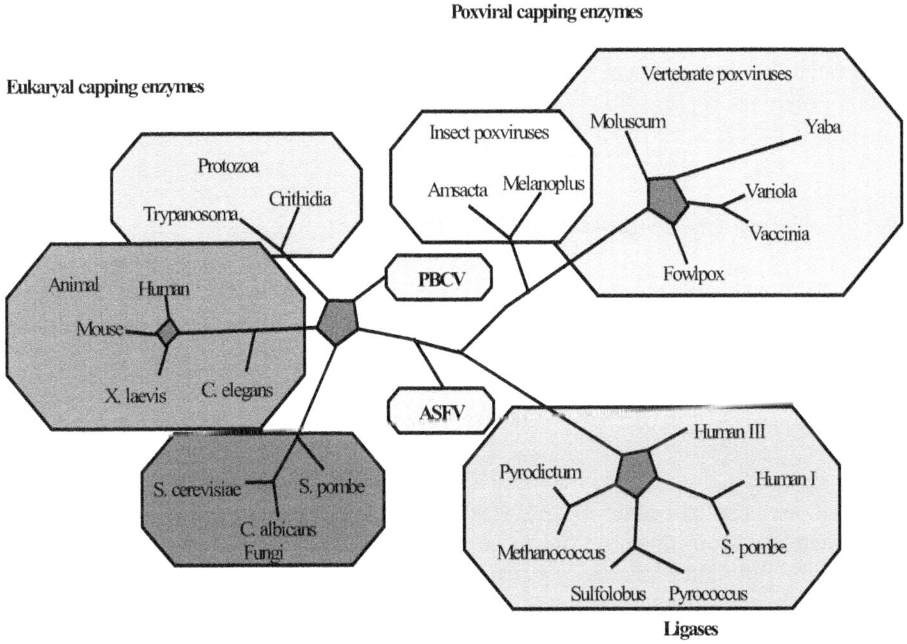

Figure 2. Unrooted phylogenetic tree of the guanylytransferase domain of mRNA capping enzymes and ATP dependent ligases based on conserved regions of the nucleotidyl transferase superfamily (Bell, 2001). Trees were constructed using protein distance, proteins parsimony and maximum likelihood methods. Branching patterns differed slightly depending upon the method used, but all methods generated trees with the similar overall structures. Differences in tree topology between the three different methods are represented by shaded polygons.

A phylogenetic tree based on the alignment of the conserved co-linear motifs of the nucleotidyltransferase superfamily has been previously constructed (Bell, 2001). The analysis generated the tree shown in Figure 2. When the root of the tree is positioned between the ligases and the guanylytranferases, the poxviral guanylytransferases diverge from the main trunk of the guanylytransferase lineage before the divergence of the Eukarya, suggesting the poxviral capping genes are not derived from their eukaryal hosts, but rather derive from a common ancestral gene that existed prior to the emergence of the Eukarya.

Phylogenetic analysis of other core replicative proteins of the NCLDV viruses show similar phylogenetic patterns to the mRNA capping enzyme. Thus a subunit of the poxviral DNA-dependent RNA polymerase apparently originated prior to the separation of the three types of eukaryal polymerases (Sonntag and Darai, 1996) and therefore presumably predates the divergence of the eukaryotes, since all Eukarya possess the three types of polymerase. Similarly, the DNA topoisomerase gene of the ASF virus was present very early in the viral genome, apparently before protozoa, yeast and metazoa diverged (Garcia-Beato et al., 1992). Finally DNA polymerases from complex DNA viruses appear to have diverged from the base of the tree that leads to the Eukarya, leading to the suggestion that the DNA virus replication proteins gave rise to those of the Eukarya (Villarreal and DeFillippis, 2000) which supports the viral eukaryogenesis theory. In a similar phylogenetic study on DNA polymerases, it was suggested that the eukaryal DNA polymerase appears to be most closely related to viral DNA polymerases, and this was used to independently propose the hypothesis that the nucleus is descended from a virus (Takemura, 2001).

8. The Viral Eukaryogenesis Theory

In the viral eukaryogenesis theory, it is proposed that the early Eukarya evolved from a consortium of three different organisms; an archaeal host, a bacterial endosymbiont, and a complex DNA virus that infected the archaeal host.

8.1 ARCHAEAL, BACTERIAL, AND VIRAL ANCESTORS OF THE EUKARYA

In the syntrophic hypothesis (Moreira and Lopez-Garcia, 1998) it was proposed, that the original archaeal ancestor of the Eukarya was a methanogenic Euryarchaeota. As discussed by Moreira and Lopez-Garcia (1998), a methanogenic Euryarchaeota was chosen for the model primarily because of several Eukarya-Euryarchaeota connections including histones, nucleosomes, DNA topoisomerases and lipids. For similar reasons, it is proposed in the viral eukaryogenesis theory that the archaeal host of the cytoplasmic virus was a member of the methanogenic Euryarchaeota. In particular, the existence of proteins with a histone fold within the methanogenic euryarchaeota and the Eukarya supports a methanogenic ancestor for the Eukarya. It is further proposed that the cellular architecture of the ancestral methanogen was essentially modern and similar in several ways to the '*Methanoplasma elizabethii*' described by Rose and Pirt (1981). Of particular importance to the viral eukaryogenesis theory, an ancestral host like '*M.*

elizabethii' lacks a cell wall and thus could be readily infected by viruses that utilize membrane fusion to enter into their hosts.

The single study reported on the '*M. elizabethii'* was conducted to examine the break down of fatty acids into methane. In that study it was found that the methanogen grows as part of a syntrophic consortium of both Bacteria and Archaea. It is proposed in the viral eukaryogenesis theory, that a bacterial syntroph, like those observed in the methanogenic consortium, was the ancestor of the mitochondria and present at the origin of the Eukarya. Due to the many biochemical, genetic and phylogenetic linkages between the mitochondria and the alpha-proteobacteria, it is proposed that the original syntrophic bacterium that evolved into the mitochondria was an alpha-proteobacteria. Its role in the consortium was to breakdown complex organic compounds into hydrogen and carbon dioxide that could then be transferred to the methanogen. The close proximity of the two allowed the waste products of the bacterial fermentation to be removed, limiting feedback inhibition of the bacterial metabolism, and providing the methanogen with a source of essential nutrients.

It is proposed that the viral ancestor of the nucleus was an NCLDV-like virus of a cell wall less methanogen. The phylogenetic analysis of the NCLDV DNA polymerase, DNA ligase, and DNA topoisomerase show they are more closely related to archaeal rather than bacterial genes, supporting a possible archaeal host for the ancestor of these viruses. In addition, the recent complete sequencing of the SIRV1 virus of the thermophilic archaeon *Sulfolobus* supports the existence of complex NCLDV-like viruses in the Archaea. The complete sequence of SIRV1 reveals that this archaeal virus shares structural features in common with the poxviruses, asfaviruses and the phycodnaviruses (Blum et al., 2001).

The ancestral virus would have possessed a large linear chromosome with telomeres similar to those observed in the poxviruses. Like the poxvirus, the ancestral virus could enter the host cell by membrane fusion and would have encoded its own replicative apparatus. This apparatus included a DNA polymerase, an RNA polymerase, and an mRNA capping enzyme. The virus could replicate independently of the host genome and once it entered a cell, it would synthesize capped and polyadenylated mRNA which it extruded from the viral core into the host cytoplasm. Once into the host cytoplasm, the capped and polyadenylated mRNA was preferentially translated by the host ribosomal machinery into viral proteins. Presumably, the virus also encoded a factor that could interact with the host ribosomes, modifying them in such a way as to ensure that they selectively recognized capped viral mRNA, rather than uncapped host mRNA, ensuring selective translation of the viral transcripts.

8.2. LYTIC INFECTION CYCLE

It is proposed that the ancestral NCLDV virus, like the poxvirus, existed in two forms; an extracellular enveloped form and an intracellular mature form (Figure 3). The intracellular mature form possessed a DNA core that was surrounded by two membranes (Moss, 1996) derived from the host cell membrane system. The two membranes (outer and inner) were composed of lipoprotein bilayers, and surrounded the core containing the DNA chromosome (Moss, 1996). The infectious extracellular enveloped virus contained an additional lipoprotein bilayer surrounding the outer membrane and this membrane contained several unique proteins that promoted fusion of the virus to the plasma membrane of the host cell.

It is proposed that the complex DNA virus replicated in the methanogen via a replication cycle similar to that of the poxvirus (Figure 3). One major difference between the proposed replication cycle of the ancestral archaeal NCLDV virus and the poxvirus replication cycle is the lack of a nucleus in the host methanogen. It should be noted that DNA bacteriophages with complexities similar to the pox virus (e.g. T4 bacteriophage ~200 kb genome, ~300 genes), are capable of infecting prokaryotes such as *E. coli* which lack a nucleus. Furthermore, the presence of several key information processing genes in the core assemblage of the proposed NCLDV ancestor supports the idea that the ancestral virus was capable of autonomous replication within the host cytoplasm. Finally, it has been shown that poxviruses can replicate in enucleated mammalian cells (Villarreal et al., 1984) it therfore seems likely that the replication cycle of the ancient virus did not require a nucleus.

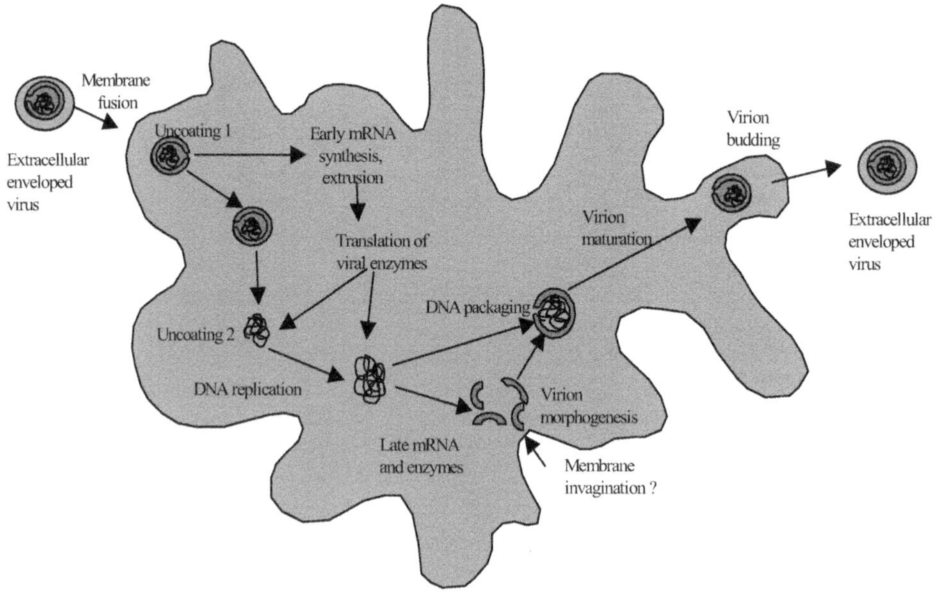

Figure 3. Replication cycle of the viral ancestor of the nucleus. The extracellular enveloped virus fuses with the plasma membrane. Initial uncoating is associated with loss of virion proteins and lipid but the genome remains sequestered in the core. The virion transcriptional machinery synthesizes capped and polyadenylated mRNA which is extruded across viral membrane into the host cytoplasm. The mRNA includes transcripts for the virion uncoating. After uncoating, the poxviral DNA replicates and late viral genes are expressed. The viral cresents for packaging of the progeny viruses are assembled in the cytoplasm, possibly from invaginations of the outer archaeal membrane. DNA is packaged into the maturing virions, and essential enzymes are transported into the maturing virions. Virions are pushed through the host membrane, regenerating the extracellular enveloped virus.

Due to the presence of the outer envelope surrounding the viral core, the fusion of the viral envelope to the archaeal cell membrane would release intact membrane bound virion core into the host cytoplasm (Moss, 1996). After entry into the host cell, the virus would extrude capped and polyadenylated mRNA into the host cytoplasm, and this mRNA would be translated by the host ribosomes into viral proteins including those required to disassemble the viral particle. The viral genome and selected enzymes would be packaged into a membrane bound core that would bud through the host membrane, regenerating the infectious virus.

8.3. ESTABLISHMENT OF A PERSISTENT VIRAL INFECTION

A crucial step in the evolution from complex DNA virus to nucleus was the establishment of a persistent viral presence in the host cytoplasm (Figure 4). This persistent infection, like the persistent infections of Friend erythroleukemia cells with vaccinia virus (Pogo and Friend, 1982), would have allowed the virus replicate to for many generations without killing the host. By displaying viral membrane fusion proteins on the host membrane, the methanogen membrane could have fused with its own or other membranes. If the filamentous, branched methanoge had grown around one of the bacterial members of the consortia and the membranes fused, the bacterium could become a captive of the infected host cell (Figure 4).

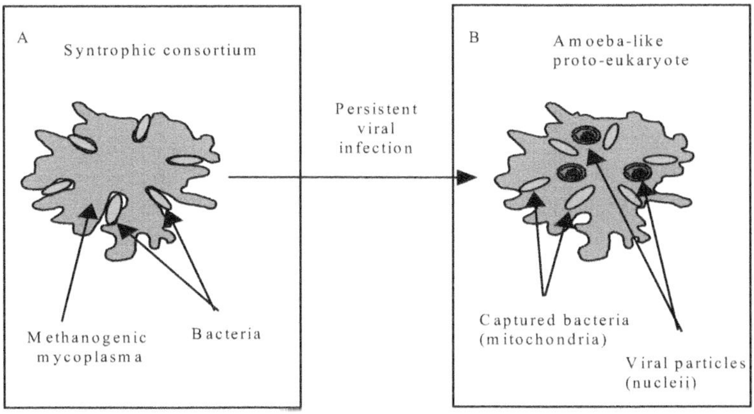

Figure 4. Establishment of a persistent infection by virus generates an amoeba-like ancestor of the Eukarya. A) Methanogen uses H_2 and CO_2 produced by syntrophic bacterium to generate methane. Filamentous growth habit of methanogen allows it to maintain close contact with bacteria ensuring efficient transfer of H_2 and CO_2 from bacteria to methanogen. B) Virus establishes persistent infection in host Archaea and produces capped and polyadenylated mRNA which is preferentially translated by host ribosomes. Viral fusion proteins in host membrane allows capture of bacteria in a phagocytosis-like process. Captured bacterial members provide hydrogen, ATP, lipids, amino acids and nucleotides to the methanogen cytoplasm, thus allowing replication of the virus and further growth of the host.

Once the NCLDV virus had established a persistent presence in the cytoplasm of the host, host genes could have been laterally transferred to the viral genome, allowing the virus to gradually usurp some of the functions of the host chromosome. Ultimately it is proposed that the viral genome acquired a complete set of genes for translation from the host, allowing the virus to direct its own protein synthesis. Under these circumstances it is likely the archaeal chromosome and its genes would have become less important to the consortium because the bacterial endosymbionts could provide adequate energy and raw materials to support growth and replication of the consortium. It is proposed that one consequence of this dependence on bacterial metabolism was a change from archaeal ether linked lipid membranes to eukaryal/bacterial ester linked lipid membranes.

9. Evolution of the Eukaryal Genome

The viral eukaryogenesis theory predicts that the eukaryal nucleus should possess genes from three sources. One set of genes will be those derived from the archaeal host during the transition from virus to autonomous nucleus. In general, these will include DNA processing genes (e.g. DNA polymerases, RNA polymerases, DNA ligases etc), RNA processing genes (eg rRNA, tRNA and ribosomal proteins) and genes required to maintain the structural integrity of the host DNA (e.g. histones).

A second set of genes will be those derived from the bacterial endosymbionts. Laterally transferred bacterial genes that confered a selective advantage on the host are not likely to have been information processing genes because the bacterial information processing system was incompatible with the archaeal/viral system. Instead beneficial bacterial genes would be those genes essential for the processes that the bacterial endosymbionts performed. Thus in the viral eukaryogenesis theory, eukaryal biosynthetic genes would be derived primarily from bacterial endosymbionts (the mitochondria) or bacterial prey. This is consistent with the phylogenetic evidence where it is found that many metabolic genes, such as those of the tricarboxcylic acid cycle, are derived from the mitochondria or other bacteria (e.g. Schnarrenberger and Martin, 2002). In plants, this gene acquisition process has been extended to the chloroplasts, where the transfer of genes from the chloroplast to the nucleus appears to be an ongoing process (e.g. Henze et al., 1995).

The third source of eukaryal genes will be those derived from the ancient NCLDV virus. These viral genes would predominantly be those that were essential to the replication of the virus, and are expected to be unique to eukaryal and viral genomes. A recent analysis of the proteins found in eukaryal cells has confirmed that there are over 300 eukaryal genes that have no significant homology to proteins in the Archaea or the Bacteria (Hartman and Fedorov, 2002). Prominent amongst these genes are nuclear specific genes involved in the synthesis, processing and transport of RNAs out of the nucleus into the cytoplasm. These genes include transcription factors, poly(A) polymerases, an mRNA capping enzyme, nuclear pore proteins, spliceosomal proteins, and proteins associated with the RNA polymerases (Hartman and Federov, 2002). Other genes include those associated with the plasma membrane, endoplasmic reticulum, translational apparatus, and the cell cycle.

The existence of a significant number of unique eukaryal genes that are fundamental to the operation of the Eukarya suggests that the eukaryal cell is not simply a chimera of a bacterium and an archaeaon, but rather may have involved a third organism (Hartman and Fedorov, 2002). In the model of Hartman and Federov, this third organism was a 'chronocyte' cell not of either the archaeal or bacterial domains but rather from an RNA based lineage (Hartman and Federov, 2002). In the viral eukaryogenesis theory, it is proposed that these unique genes were derived from the viral ancestor of the nucleus.

10. Conclusions and Perspectives

The viral eukaryogenesis theory describes an evolutionary process that leads to the appearance of an amoeba-like eukaryote (Figure 4). The poorly developed general metabolic capacity of the organism, combined with its ability to engulf bacterial cells would have provided the selective principle for the capture and maintenance of mitochondrial and chloroplast ancestors. The dependence upon the capture of bacteria for raw materials would also have ensured that all the earliest Eukarya would have possessed ancestors of the mitochondria, as well as complex membrane fusion based processes. The novel predatory life style also ensured that the Eukarya evolved along a different evolutionary trajectory to the Bacteria and the Archaea which trend towards biochemical complexity and rapid replication (eg Forterre, 1995). In particular, the evolutionary pressures on the evolution of the Eukarya would have been towards organismal complexity to facilitate the predatory life style.

Acceptance or rejection of the viral eukaryogenesis theory will rely on the accumulation of biochemical, genetic and physiological evidence that either supports or refutes the model. Since all the organisms involved in the viral eukaryogenesis theory have extant representatives, it should be possible to examine the biochemistry and genetics of these organisms, and determine whether they are consistent with the viral eukaryogenesis theory. In particular, if the theory is valid, comparative analysis of nuclear and viral replicative processes should reveal many similarities that are not observed in either of the prokaryotic domains.

The viral eukaryogenesis theory provides straightforward explanations for many of the complex changes that occurred in the transition from archaeal genomic design to eukaryal genomic design. Specifically, it provides a rationale for the evolution of a chimeric genome, mRNA capping, mRNA transport across membranes, the separation of transcription from translation, the evolution of linear chromosomes, and the evolution of membrane fusion processes.

The viral eukaryogenesis theory does not, in its present form, provide an adequate explanation for the origin of the endoplasmic reticulum, nor for the evolution of the cytoskeleton. However, these inadequacies may arise because of the limited knowledge of the biodiversity of life on earth. For example, only one reference to a methanogen lacking a cell wall appears to exist, and this has been studied only in its role as a member of a consortium for the breakdown of fatty acids into methane. Very little is known about its basic biology, and thus such questions as to how it maintains its filamentous form without a cell wall (a rudimentary cytoskeleton?) is completely unknown. In addition, with regard to the origin of the endoplasmic reticulum, it should be noted that the endoplasmic reticulum is essential in the replication of the nucleus and the NCLDV viruses such as the pox and ASF viruses. Significantly, both viral and individual eukaryal chromosomes are wrapped by membranes derived from the endoplasmic reticulum (Andres et al., 1998; Gant and Wilson, 1997; Tolonen et al., 2001). If the viral eukaryogenesis theory is valid and the NCLDV viruses and the nucleus share a common ancestry, then the processes by which their genetic material is wrapped at the endoplasmic reticulum should be related, and this may provide a clue to the origin of the endoplasmic reticulum.

Finally, no organisms thought to be intermediate between the Archaea and the Eukarya have yet been found. If the viral eukaryogenesis theory is valid, it is likely that intermediates do exist, but they have not been recognized. These intermediates would not be single organisms but rather would be Archaea that were persistently infected with complex DNA viruses. To the author's knowledge, no systematic studies have been performed on viruses of methanogens without cell walls. However, in studies on thermophilic Archaea such as *Sulfolobus*, it is found that many DNA viruses are capable of infecting Archaea (Rice et al., 2001) and some of these, such as the SNDV virus, establish persistent infections in their host (Arnold et al., 2000). It seems likely therefore that when the methanogens without cell walls are studied, they will be found to be infected with complex DNA viruses. The discovery of a complex DNA virus of similar genetic architecture to the poxvirus and capable of capping its mRNA would provide an opportunity to thoroughly test the viral eukaryogenesis theory.

11. References

Andres, G., Garcia-Escudero, R., Simon-Mateo, C. and Vinuela, E. (1998) African Swine Fever virus is enveloped by a two membraned collapsed cisterna derived from the endoplasmic reticulum. J. Virol. **72**: 8988-9001.

Arnold, H.P., Zeise, U. and Zillig, W. (2000) SNDV, a novel virus of the extremely thermophilic and acidophilic archaeon *Sulfolobus*. Virology **272**: 409-416.

Bell, P.J.L. (2001) Viral eukaryogenesis: was the ancestor of the nucleus a complex DNA virus? J. Mol. Evol. **53**: 251-265.

Bernander, R. (1998) Archaea and the cell cycle. Mol. Microbiol. **29**:955-961.

Blasco, R. (1995) Evolution of poxviruses and African swine fever virus. In: A. Gibbs, C.H. Calisher, and F. Garcia-Arenal (eds.) Molecular basis of virus evolution. Cambridge: Cambridge University Press. pp. 255-269.

Blattner, F.R., Plunkett, G. 3rd, Bloch, C.A., Perna, N.T., Burland, V., Riley, M., Collado-Vides, J., Glasner, J.D., Rode, C.K., Mayhew, G.F., Gregor, J., Davis, N.W., Kirkpatrick, H.A., Goeden, M.A., Rose, D.J., Mau, B. and Shao, Y. (1997) The complete sequence of *Escherichia coli* K-12. Science **277**: 1453-1474.

Blum, H., Zillig, W., Mallok, S., Domdey, H. and Prangishvilli, D. (2001) The genome of the archaeal virus SIRV has features in common with genomes of eukaryal viruses. Virology **281**: 6-9.

Blumenthal, T. (1995) Trans-splicing and polycistronic transcription in *Caenorhabditis elegans* Trends Genet. **11**: 132-136.

Brinkmann, H. and Philippe, H. (1999) Archaea sister group to Bacteria? Indications from tree construction artifacts in ancient phylogenies. Mol. Biol. Evol. **16**: 817-825.

Brocks, J.J., Logan, G.A., Buik, R. and Summons, R.E. (1999) Archean molecular fossils and the early rise of the eukaryotes. Science **285**: 1033-1036.

Brown, J.R. and Doolittle, W.F. (1997) Archaea and the prokaryote-to-eukaryote transition. Microbiol. Mol. Biol. Rev. **61**: 456-502.

Bult, C.J, White, O., Olsen, G.J., Zhou, L., Fleischmann, R.D., Sutton, G.G., Blake, J.A., Fitzgerald, L.M., Clayton, R.A., Gocayne, J.D., Kerlavage, A.R., Dougherty, B.A., Tomb, J.F., Adams, M.D., Reich, C.I., Overbeek, R., Kirkness, E.F., Weinstock, K.G., Merrick, J.M., Glodek, A., Scott, J.L., Geohagen, N.S.M. and Venter, J.C. (1996) Complete genome sequence of the methanogenic archaeon, *Methanococcus jannaschii*. Science **273**: 1058-1073.

Casjens, S. (1999) Evolution of the linear DNA replicons of the *Borrelia* spirochetes. Curr. Opin. Microbiol. **2**: 529-534.

Cavalier-Smith, T. (1987) The origin of eukaryote and archaebacterial cells. Ann. N. Y. Acad. Sci. **503**: 17-54.

Doolittle, W.F. (1998) You are what you eat: a gene transfer ratchet could account for bacterial genes in the eukaryote genome. Trends Genet. **14**: 307-311.

Doolittle, W.F. (1999) Phylogenetic classification and the universal tree. Science **284**: 2124-2128.

Forterre, P. (1995) Thermoreduction, a hypothesis for the origin of prokaryotes. C. R. Acad. Sci. III **318**: 415-422.

Forterre, P.F. and Philippe, H. (1999) Where is the root of the universal tree of life. Bioessays **21**: 871-879
Forterre, P. (2001) Genomics and early cellular evolution. The origin of the DNA world. Life Sciences **324**: 1067-1026.
Fuhrman, J.A. (1999) Marine viruses and their biogeochemical and ecological effects. Nature **399**: 541-548
Gant, T.M. and Wilson, K.L. (1997) Nuclear assembly. Ann. Rev. Cell. Dev. Biol. **13**: 669-695.
Garcia-Beato, R., Freue, J.M.P, Lopez-Otin, C., Blasco, R., Vinuela, E. and Salas, M.L. (1992) A gene homologous to topoisomerase II in African Swine Fever Virus. Virology **188**: 938-947.
Goffeau, A., Barrell, B.G., Bussey, H., Davis, R.W., Dujon, B., Feldmann, H., Galibert, F., Hoheisel, J.D., Jacq, C., Johnston, M., Louis, E.J., Mewes, H.W., Murakami, Y., Philippsen, P., Tettelin, H. and Oliver, S.G. (1996) Life with 6000 genes. Science **546**: 563-567.
Gogarten, J.P., Kibak, H., Dittrich, P., Taiz, P., Bowman, E.J., Manolson, M.F., Poole, P.J., Data, T., Oshima, T., Konishi, J., Denca, K. and Yoshida, M. (1989) Evolutionary relationship of archaebacteria, eubacteria and eukaryotes inferred from phylogenetic trees of duplicated genes. Proc. Natl. Acad. Sci. USA **86**: 9355-9359.
Gray, MW., Burger, G. and Lang, B.F. (1999) Mitochondrial evolution. Science **283**: 1476-1481.
Gribaldo, S. and Cammarano, P. (1998) The root of the universal tree of life inferred from anciently duplicated genes encoding components of the protein-targeting machinery. J. Mol. Evol. **47**: 508-516.
Gupta, R.S. (1998) Protein phylogenies and signature sequences: a reappraisal of evolutionary relationships among archaeabacteria, eubacteria, and eukaryotes. Microbiol. Mol. Biol. Rev. **62**: 1435-1491.
Gupta, R.S. and Golding, G.B. (1996) The origin of the eukaryotic cell. Trends Biochem. Sci. **21**: 166-171.
Hartman, H. and Federov, A. (2002) The origin of the eukaryotic cell: a genomic investigation. Proc. Natl. Acad. Sci. USA **99**: 1420-1425.
Henze, K., Badr, A., Wettern, M., Cerff, R. and Martin, W. (1995) A nuclear gene of eubacterial origin in *Euglena gracilis* reflects cryptic endosymbioses during protist evolution. Proc. Natl. Acad. Sci. USA **92**: 9122-9126.
Hinnebusch, A.G. (1988) Mechanisms of gene regulation in the general control of amino acid biosynthesis in *Saccharomyces cerevisiae*. Microbiol. Rev. **52**: 248-273.
Hinnebusch, J. and Barbour, A.G. (1991) Linear plasmids of *Borrelia burgdorferi* have telomeric structure and sequence similar to those of a eukaryotic virus. J. Bacteriol. **173**: 7233-7239.
Hinnebusch, J. and Tilly, K. (1993) Linear plasmids and chromosomes in bacteria. Mol. Microbiol. **10**: 917-922.
Imboden, M.A, Laird, P.W., Affolter, M. and Seebeck, T. (1987) Transcription of the intergenic regions of the tubulin gene cluster of *Trypanosoma brucei*: evidence for a polycistronic transcription unit in a eukaryote. Nuc. Acid. Res. **15**: 7357-7368.
Iwabe, N., Kuma, K., Hasegawa, M., Osawa, S. and Miyata, T. (1989) Evolutionary relationship of archaeabacteria, eubacteria and eukaryotes inferred from phylogenetic trees of duplicated genes. Proc. Natl. Acad. Sci. USA, **86**: 9355-9359.
Iyer, L.M., Aravind, L. and Koonin, E.V. (2001) Common origin of four diverse families of large eukaryotic DNA viruses. J. Virol. **75**: 11720-11734.
Jain, R., Rivera, M.C. and Lake, J.A. (1999) Horizontal transfer amongst genome: the complexity hypothesis. Proc. Natl. Acad. Sci. USA **96**: 3801-3806.
Keeling, P.J., Charlebois, R.L. and Doolittle, W.F. (1994) Archaebacterial genomes: eubacterial form and eukaryotic content. Curr. Opin. Genet. Dev. **4**: 816-822.
Kipling, D. (1995) The Telomere. Oxford University Press Inc. New York
Klenk, H.P., Clayton, R.A., Tomb, J.F., White, O., Nelson, K.E, et al. (1997) The complete genome of the hyperthermophilic sulphate-reducing archaeon *Archaeoglobus fulgidus*. Nature **390**: 364-370.
Kletzin, A. (1992) Molecular characterisation of a DNA ligase gene of the extremely thermophilic archaeon *Desulfurolobus ambivalens* shows close phylogenetic relationship to eukaryotic ligases. Nuc. Acid. Res. **20**: 5389-5396.
Koonin, E.V., Mushegian, A.R., Galperin, M.Y. and Walker, D.R. (1997) Comparison of archaeal and bacterial genomes: computer analysis of protein sequences predicts novel functions and suggests a chimeric origin for the archaea. Mol. Microbiol. **25**: 619-637.
Kozak, M. (1999) Initiation of translation in prokaryotes and eukaryotes. Gene **234**: 187-208.
Kyrpides, N., Overbeek, R. and Ouzounis, C. (1999) Universal protein families and the functional content of the last universal ancestor. J. Mol. Evol. **49**: 413-423.
Lake, J.A. and Rivera, M.C. (1994) Was the nucleus the first endosymbiont? Proc. Natl. Acad. Sci. USA **91**: 2880-2881.
Lake, J.A., Jain, R. and Rivera, M.C. (1999) Mix and match in the tree of life. Science **283**: 2027-2028.

Lin, Y.S., Kieser, H.M., Hopwood, D.A. and Chen, C.W. (1993) The chromosomal DNA of *Streptomyces lividans* 66 is linear. Mol. Microbiol. **10**: 923-933.

Lindsay, M.R., Webb, R.I., Strous, M., Jetten, M.S., Butler, M.K., Forde, R.J. and Fuerst, J.A. (2001) Cell compartmentalisation in planctomycetes: novel types of structural organisation for the bacterial cell. Arch. Microbiol. **175**: 413-429.

Lopez-Garcia. P. and Moreira, D. (1999) Metabolic symbiosis at the origin of eukaryotes. Trends Biochem. Sci. **24**: 88-93.

Margulis, L. (1996) Archaeal-eubacterial mergers in the origin of eukarya - phylogenetic classification of life. Proc Natl Acad Sci USA **93**: 1071-1076.

Martin, W. (1999) A briefly argued case that mitochondria and plastids are descendants of endosymbionts, but that the nuclear compartment is not. Proc. R. Soc. Lond. B. **266**: 1387-1395.

Martin, W. and Müller, M. (1998) The hydrogen hypothesis for the first eukaryote. Nature **392**: 37-41.

Moreira, D. and Lopez-Garcia, P. (1998) Symbiosis between methanogenic archaea and delta-proteobacteria as the origin of eukaryotes: the syntrophic hypothesis. J. Mol. Evol. **47**: 517-530.

Moss, B. (1996) Poxviridae: The viruses and their replication. In: B.N. Fields, D.M., Knipe and P.M. Howley (eds.) Fundamental Virology (Third ed.). Philadelphia: Lippencott-Raven Publishing. pp. 1164-1185

Nowitzki, U., Flechner, A., Kellerman, J., Hasehawa, M., Schnarrenberger, C. and Martin, W. (1998) Eubacterial origin of nuclear genes for chloroplast and cytosolic glucose-6-phosphate isomerase from spinach: sampling eubacterial gene diversity in eukaryotic chromosomes through symbiosis. Gene **214**: 205-213

Penny, D. and Poole, A. (1999) The nature of the last universal common ancestor. Curr. Opin.Gen. Dev. **9**: 672-677.

Pogo, B.G.-T. and Friend, C. (1982) Persistent infection of Friend erythroleukemia cells with vaccinia virus. Proc. Natl. Aca. Sci. USA **79**: 4805-4809.

Poole, A.M., Jeffares, D.C. and Penny, D. (1998) The path from the RNA world. J. Mol. Evol. **46**: 1-17.

Reanney, D.C. (1974) On the origin of the prokaryotes. J. Theor. Biol. **48**: 243-251.

Ribeiro, S. and Golding, G.B. (1998) The mosaic nature of the eukaryotic nucleus. Mol. Biol. Evol. **15**: 779-788.

Rice, G., Stedman, K., Snider, J., Wiedenheft, B., Willits, D., Brumfield, S., McDermott, T. and Young, M.J. (2001) Viruses from extreme thermal environments. Proc. Natl. Acad. Sci. USA. **98**: 13341-13345.

Rivera, M.C., Jain, R., Moore, J.E. and Lake, J.A. (1998) Genomic evidence for two functionally distinct gene classes. Proc. Natl. Acad. Sci. USA **95**: 6239-6244.

Rose, C.S. and Pirt, S.J. (1981) Conversion to fatty acids and methane: Role of two mycoplasmal agents. J. Bacteriol. **147**: 248-252.

Rosenthal, B., Mai, Z., Caplivski, D., Ghosh, S., de la Vega, H., Graf, T. and Samuelson, J. (1997) Evidence for the bacterial origin of genes encoding fermentation enzymes of the amitochondriate protozoan parasite *Entamoeba histolytica*. J. Bacteriol. **179**: 3736-3745.

Sarkar, N. (1997) Polyadenylation of mRNA in prokaryotes. Ann. Rev. Biochem. **66**: 173-197.

Schnarrenberger, C. and Martin, W. (2002) Evolution of the citric acid cycle and the glyoxylate cycle of higher plants- A case study of endosymbiotic gene transfer. Eur. J. Biochem. **269**: 868-883.

Schopf, W.J. (1993) Microfossils of the early archean apex chert: New evidence of the antiquity of life. Science **260**: 640-646.

Shuman, S. and Schwer, B. (1995) RNA capping enzyme and DNA ligase: a superfamily of covalent nucleotidyltransferases. Mol. Microbiol. **17**: 405-410.

Snel, B., Bork, P. and Huyen, M.A. (1999) Genome phylogeny based on gene content. Nat. Genet. **21**: 108-110.

Sogin, M.L. (1991) Early evolution and the origin of the eukaryotes. Curr. Opin. Genet. Dev. **1**: 457-463.

Sonntag, K.-C. and Darai, G. (1996) Evolution of viral DNA-dependent RNA polymerases. Virus Genes **11**: 271-284.

Strauss, E.G., Strauss, J.H. and Levine, A.J. (1996). Virus evolution. In: B.N. Fields, D.M., Knipe and P.M. Howley (eds.) Fundamental Virology (Third ed.). Philadelphia: Lippencott-Raven Publishing. pp. 153-173.

Struhl, K. (1995) Yeast transcriptional regulatory mechanisms. Ann. Rev. Gen. **29**: 651-674.

Takemura, M. (2001) Poxviruses and the origin of the eukaryotic nucleus. J. Mol. Evol. **52**: 419-425.

Tolonen, N., Doglio, L., Schleich, S. and Locker, J.K. (2001) Vaccinia virus DNA replication occurs in endoplasmic reticulum-enclosed cytoplasmic mini-nuclei. Mol. Biol. Cell. **12**: 2031-2043.

Vellai, T., Takacs, K. and Vida, G. (1998) A new aspect to the origin and evolution of eukaryotes. J. Mol. Evol. **46**: 499-507.

Villarreal, E.C., Roseman, N.A. and Hruby, D.E. (1984) Isolation of vaccinia virus mutants capable of replicating independently of the host cell nucleus. J. Virol. **51**: 359-366.
Villarreal, L.P. and DeFillippis, V.R. (2000) A hypothesis for DNA viruses as the origin of eukaryotic replication proteins. J. Virol. **74**: 7079-7084.
Wang, J., Jenkins, C., Webb, R.I. and Fuerst, J.A. (2002) Isolation of *Gemmata*-like and *Isophaera*-like planctomycte bacteria from soil and freshwater. Appl. Environ. Micro. **68**: 417-422.
Woese, C.R. (1987) Bacterial evolution. Microbiol. Rev. **51**: 221-271.
Woese, C.R. (1998) The universal ancestor. Proc. Natl. Acad. Sci. USA **95**: 6854-6859.
Woese, C.R. (2000) Interpreting the universal phylogenetic tree. Proc. Natl. Acad. Sci. USA **97**: 8392-8396.
Woese, C.R., Kandler, O. and Wheelis, M.L. (1990) Towards a natural system of organisms: proposal for the domains Archaea, Bacteria, and Eucarya. Proc. Natl. Acad. Sci. U S A **87**: 4576-4579.
Zillig, W. (1991) Comparative biochemistry of Archaea and Bacteria. Curr. Biol. **1**: 544-551.

VI. Extremophiles & Biodiversity

Life in the Extreme Environments

[TIBTECH. 11, p. 358 (1993)]

We thank Elsevier Publisher for permission to use this picture.

Biodata of **Joseph Seckbach**, editor all volumes of the **"C**ellular **O**rigins, **L**ife in **E**xtreme Habitats and Astrobiology (COLE)**"** book series, published by Kluwer. He is the author of ***"Introduction to the Extremophiles"*** (with co-author A. Oren) chapter and of the ***"Introduction"*** to this book.

Dr. Joseph Seckbach is the initiator and chief editor of ***Cellular Origins, Life in Extreme Habitats and Astrobiology*** **(COLE)** book series, see: **http://www.wkap.nl/prod/s/COLE** and author of several chapters in this series. He is the co-author (with R. Ikan) of the ***Chemistry Lexicon*** (1991, 1999). Dr. Seckbach earned his Ph.D. from the University of Chicago, Chicago, IL (1965) and spent his postdoctoral years in the Division of Biology at Caltech (Pasadena, CA). Then he headed at the University of California at Los Angeles (UCLA) a team for searching of extraterrestrial life. Seckbach has been appointed to the faculty of the Hebrew University (Jerusalem, Israel) and performed algal research and taught Biological courses. He spent his sabbatical periods in Tübingen (Germany), UCLA and Harvard University. He served at Louisiana State University (LSU), (1997/1998) as the first selected occupant of the John P. Laborde endowed Chair for the Louisiana Sea Grant and Technology transfer, and as a visiting Professor in the Department of Life Sciences at LSU (Baton Rouge, LA).

Among his publications are books, scientific articles concerning plant ferritin (phytoferritin), cellular evolution, acidothermophilic algae, and life in extreme environments. He also edited and translated several popular books. His recent interest is in the field of enigmatic microorganisms and life in extreme environments.

E-mail: **seckbach@huji.ac.il**

Professor Aharon Oren is a member of the Division of Microbial and Molecular Ecology, the Institute of Life Sciences, The Hebrew University of Jerusalem. He earned his Ph.D. at the Hebrew University in1978, and spent a post-doctoral period at the University of Illinois at Urbana-Champaign. His research interests are in ecology, physiology, biochemistry and taxonomy of halophilic microorganisms and the microbiology of hypersaline environments. He has edited a book on "***Microbiology and Biogeochemistry of Hypersaline Environments***" (CRC Press, Boca Raton, 1999), and recently wrote "*Halophilic Microorganisms and their Environments*", the fifth volume in the series "***Cellular Origin and Life in Extreme Habitats***", edited by Joseph Seckbach and published by Kluwer Academic Publishers, Dordrecht, the Netherlands, 2002.

E-mail: **orena@cc.huji.ac.il**

J. Seckbach

A. Oren

J. Seckbach (ed.), Origins, 373–393.

INTRODUCTION TO THE EXTREMOPHILES

JOSEPH SECKBACH[1] and AHARON OREN[2]
[1] P.O. Box 1132, Efrat 90435, Israel. [2] Division of Microbial and Molecular Ecology, The Institute of Life Sciences, and The Moshe Shilo Minerva Center for Marine Biogeochemistry, The Hebrew University of Jerusalem, Jerusalem 91904 Israel

1. Extremophiles – Life at the Limits

When we think about life on Earth we generally imagine a world inhabited by plants, animals and microorganisms that grow and maintain complex interrelationships under the cozy, convenient conditions for life found in most ecosystems, terrestrial as well as marine. The temperatures are typically between 10-40°C, the pH is close to neutrality, pressure is close to one atmosphere, water is available plentifully, and levels of harmful radiation are low. These are the kind of "normal" environments that harbor the greatest biological diversity, including of higher plants and animals that we commonly associate with life on our planet.

However, not all habitable environments on Earth are equally suitable for the support of such complex ecosystems. A great variation in abiotic conditions is found with respect to factors such as temperature, pH, water availability, pressure, etc. Compared to the kind of environments described above, a hot spring with a temperature above 80°C and a pH below 2 may surely be considered an environment hostile to life. The same is true for a soda lake saturated with salt, in which the availability of water is strongly reduced, and in addition the pH of 10-11 or higher severely stresses any organism that attempts to colonize it. There are many different kinds of such "extreme environments". Some are geographically restricted, such as the black smoker deep-sea hydrothermal vents (presenting a combination of high temperature and high hydrostatic pressure), or acidic hot springs. Others are far more abundant, such as, the deep-sea environment with its low temperature (1-4°C) and its several hundreds of atmospheres of hydrostatic pressure. Another case is the hot subsurface environment whose exploration has started only a few years ago, and which now appears to be inhabited by a variety of heat- and pressure-tolerant microorganisms.

All those environments mentioned are colonized by life forms, in most cases by prokaryotic microorganisms only, but sometimes eukaryotic microorganisms and even macroorganisms may be found as well (e.g. the brine shrimp *Artemia salina* in hypersaline lakes). When we refer to such microorganisms as extremophiles we use the "normal" environments that support higher organisms, including man, as our reference point. When adopting this anthropocentric view we should not forget that these seemingly "hostile" environments are often inhabited by surprisingly diverse microbial communities, and also that the microbes found in such niches are often perfectly adapted to life under the "unusual" harsh conditions in which they live. More than that, their adaptation is often so extensive that they are unable to live in the more "moderate" environments that we normally associate with life on Earth. An organism such as *Natrialba magadii* that inhabits the salt-saturated and extremely alkaline Lake Magadi (Kenya) will not grow when the pH is lowered to values below 8.5, and will lyse when suspended in solutions that contain less than 90 g/l salt. For such an organism an encounter with a high water activity, neutral pH environment will signify immediate death. Similarly, hyperthermophilic Archaea that grow optimally at temperatures above 100°C will experience any temperature below 70-80°C as "freezing" conditions, under which they may be able to survive for some time, but not to actively metabolize and multiply. Many extremophiles experience "relatively normal" environments as uninhabitable and are not able to exist in milder habitats.

It is becoming more and more clear that almost any habitat on Earth that is compatible with the stability of biomolecules is colonized at least by some type of microorganism adapted to the specific conditions of that environment, conditions that are often extremely hostile or even deadly to all other forms of life.

We can divide these extremophiles according to the nature of the stressful (at least, as experienced as stressful by most living organisms) factor(s) in their environment: high temperature (thermophiles and hyperthermophiles), low temperature (psychrophiles), low pH (acidophiles), high pH (alkaliphiles), high salt concentrations (halophiles), dry conditions (xerophiles), and high pressure (barophiles or piezophiles). Table 1 summarizes some of the most spectacular cases of the "record holders" with respect to life under extremes of temperature, pH, salinity, and harmful radiation. More information on these organisms and on the nature of the adaptations enabling them to thrive under such unusual and seemingly hostile conditions will be given in the sections below.

Many of these extremophiles are highly specialized organisms, each being adapted to a very specific set of conditions. The price many of these microorganisms have to pay is their low level of adaptability to changes in their environmental conditions. An extreme halophile such as *Halobacterium salinarum*, an organism perfectly adapted to life at 200-250 g/l salt, will immediately lyse the moment the salt concentration is lowered to values below 150 g/l. Some extreme acidophiles, e.g. *Picrophilus*, will not survive even a short exposure to pH values above 4. Such record holders for life under physical or chemical extremes are generally specialists adapted to life in a constant environment, and have few possibilities to adjust to changes in the conditions that may occur in their environment. However, other extremophiles are far more flexible. Thus, some halophilic representatives of the domain Bacteria can rapidly adjust to drastic changes in the salinity of their medium, and some (e.g. *Halomonas* and *Chromohalobacter* species) can live in a wide range of salt concentrations, from near-zero to close to saturation (Oren, 2002; Ventosa et al., 1998). Maybe we should consider these, the truly versatile organisms as the real champions among the extremophiles.

Research on extremophiles has many important and intriguing aspects. First of all, the extremophiles teach us the limits of life "as we know it", and extend our understanding of the biodiversity on Earth. Second, the elucidation of the mechanisms behind the ability of extremophilic microorganisms to withstand otherwise hostile conditions provides not only a more profound insight in the functioning of living cells, but may also lead to interesting biotechnological applications and economic exploitation of the extremophiles. The term "Superbugs" for such all-powerful extremophiles, coined a number of years ago by Horikoshi and Grant (1991) signifies both the abilities of these organisms to live in unusual environments and the possibility of their biotechnological exploitation. Third, the understanding of the unusual properties of the extremophiles leads to questions about their origin. Have the existing extremophiles adapted recently to the unusual environments in which they live, or may they be vestiges of ancient types of organisms that evolved on early Earth when conditions were much more extreme than they are today? The study of the extremophiles has therefore important implications for our views on the origin of life. Fourth, the understanding of the limits of life on Earth, as displayed by the extremophiles, may provide us with clues on the possibility of the existence, now or in the past, of similar life elsewhere in the Universe. The science of astrobiology or exobiology therefore derives many of its tools from the investigation of life in extreme environments on Earth.

TABLE 1. "Record-holding" extremophiles and their environmental limits. The table is based in part on data presented by Madigan (2000).

Environmental factor	**Organism**	**Habitat**	**Phylogenetic affiliation**	**Tolerance to stress**
High temperature	*Pyrolobus fumarii*	Hot undersea hydrothermal vents	Archaea - Crenarchaeota	Maximum 113°C, Optimum 106°C, Minimum 90°C
Low temperature	*Polaromonas vacuolata*	Sea-ice	Bacteria	Minimum 0°C Optimum 4°C, Maximum 12°C
Hydrostatic pressure	Strain MT41	Mariana Trench	Bacteria	Maximum >100 MPa, Optimum 70 MPa, Minimum 50 MPa
Low pH	*Picrophilus oshimae*	Acidic hot springs	Archaea - Euryarchaeota	Minimum pH -0.06, Optimum pH 0.7, Maximum pH 4 (is also thermophilic)
High pH	*Natronobacterium gregoryi*	Soda lakes	Archaea - Euryarchaeota	Maximum pH 12, Optimum pH 10, Minimum pH 8.5 (is also halophilic)
High salt concentration	*Halobacterium salinarum*	Salt lakes, salted hides, salted fish	Archaea - Euryarchaeota	Maximum NaCl saturation, Optimum 250 g/l salt, Minimum 150 g/l salt
Ultraviolet and ionizing radiation	*Deinococcus radiodurans*	Isolated from ground meat; true habitat unknown	Bacteria	Resistant to 1.5 kGy gamma radiation and to 1500 J/m^2 of ultraviolet radiation

(1 MPa = about 10 atmospheres)

Many aspects of the biology of the extremophiles, the origin of life on Earth, and the possibility of the existence of life on other planets are discussed in the different chapters in this book. The present chapter intends to give a broad general introduction to the subject of extremophilic microorganisms. The subject has been reviewed many times in the past in a much more complete way that what was had been possible here. The book on "Microbial life in extreme environments" edited by Don Kushner in 1978 still can be considered as an exemplary work, even though much new information has accumulated since, and the discovery of even more extremophilic microorganisms that were unknown at the time has shifted the recognized upper temperature limit and the lower pH limit for life to values undreamed of at the time. Other valuable reference volumes and review articles are the 1986 book on "Microbes in extreme environments" (Herbert and Codd, 1986), two books edited by Horikoshi and Grant (1991, 1998), including the already mentioned book on "Superbugs", two books edited by Seckbach (1999, 2000), the book by Wharton (2002), and review articles such as those by Kirstjansson and Hreggvidsson (1995), Madigan and Marrs (1997), Madigan (2000), Seckbach and Oren (2000), and Rothschild and Mancinelli (2001). In addition, there are many specialized reviews on specific groups of extremophiles, some of which are cited elsewhere in this chapter. Further information on the world of extremophiles can be found in the book series on "Origins, Life in Extreme Habitats and Astrobiology (COLE)," at: http://www.wkap.nl/prod/s/COLE.

2. Thermophiles and Hyperthermophiles

Temperature is one of the most important environmental factors governing survival and growth of microorganisms. Every organism has its characteristic minimum, optimum, and maximum temperatures for growth.

There are many hot environments in nature that are colonized by high-temperature adapted microorganisms (Brock, 1978). Hot springs are often found in areas with volcanic activity, including the deep-sea hydrothermal vents and active seamounts where volcanic lava is emitted directly onto the seafloor. The upper temperature of terrestrial hot springs is determined by the boiling temperature of water. Environments with liquid water at temperatures exceeding 100°C can be encountered in the marine environment, where the boiling temperature of water is increased according to the in situ hydrostatic pressure. Many deep sea hydrothermal vents emit water at a temperature of around 350°C. Such "black smokers" form "chimneys" built of metal sulfides that precipitate when the metal-rich hydrothermal fluid mixes with the cold seawater. A temperature gradient is established between the chimney and the ambient seawater (around 2°C), and in such temperature gradients a range of thermophilic and hyperthermophilic microorganisms may find their niche. Other hot environments are the result of solar heating, geothermal activity, combustion, and human activities. Among the man-made high-temperature biotopes we may mention compost piles, smoldering coal refuse piles, and geothermal power plants (Stetter, 1996). An environment that has not been extensively explored thus far for the presence of thermophilic microorganisms is the deep, hot subterranean biosphere, which may well be inhabited by a variety of thermophiles and hyperthermophiles that use inorganic and organic chemicals as energy sources (Onstott et al., 1999). If indeed the upper limit for life is somewhere between 113°C (the highest temperature known thus far to support microbial growth) and 150°C (a hypothetical value based on our understanding of the thermal stability of biological molecules – see Section 12), thermophilic life may be possible in most areas of the earth's crust down to depths between 5 and 10 km (Gold, 1992).

Microorganisms can be classified into psychrophiles (cold-loving organisms), mesophiles (organisms thriving at "normal" temperatures from 20°C up to about 40-45°C), thermophiles, and hyperthermophiles according to the temperature range enabling growth. As temperature increases, we find progressively fewer groups of organisms that can cope with the stress of high temperatures. As nature presents us with a continuum of temperature ranges and organisms adapted to growing in them, a strict definition of "thermophiles" and "hyperthermophiles" is not possible; organisms that grow at temperatures exceeding 80°C are generally termed hyperthermophilic (Rothschild and Mancinelli, 2001). The upper limit for eukaryotic life (fungi, algae, protozoa) is about 60°C. The recently described amoeba *Echinamoeba thermarum*, isolated from hot springs and growing up to 57°C, is the most thermotolerant member of the "Protozoa" known (Baumgartner et al., 2003). The maximum temperature at which photosynthesis occurs (by unicellular cyanobacteria, i.e. members of the domain Bacteria) is 70-74°C., while *Cyanidium caldarium*, a thermoacidophile rhodophytan, photosynthesizes at an upper temperature of 57°C. More information on photosynthesis at different environmental extremes can be found in the review article by Oren and Seckbach (2001). Certain heterotrophic and chemoautotrophic members of the domain Bacteria tolerate higher temperatures, up to about 95°C.

The most thermophilic organisms known all belong to the domain Archaea. In fact, many representatives of the archaeal domain grow at temperatures exceeding 70-80°C (Stetter, 1996). Such hyperthermophiles may survive for long times at "mesophilic" temperatures, but will not grow unless the temperature is increased to levels more agreeable to these organisms.

At the moment the most thermophilic organisms known is the crenarchaeote *Pyrolobus fumarii*. It was isolated from the wall of a deep-sea black smoker, and grows upper limit temperature above 100°C are e.g. *Pyrococcus woesei* (maximum 104.8°C), *Pyrodictium occultum* and *Pyrodictium brockii* (optimum 105°C), and *Hyperthermus* optimally at 106°C with a maximum growth temperature of 113°C. No growth is possible below 90°C. The cells can survive autoclaving for an hour at 121°C (Blöchl et al., 1997; Madigan and Oren, 1999; Stetter, 1998). Other Archaea that grow with an *butylicus* (maximum 108°C), all anaerobic chemoautotrophic or heterotrophic species isolated from environments such as marine solfataric mud and hydrothermal vents (Stetter, 1996; Stetter et al., 1983). There is also an interesting methanogen, *Methanopyrus kandleri*, isolated from hydrothermal vents, that can grow at temperatures up to 110°C.

To what extent the upper temperature limit for life can be extended above 113°C is not clear, and the true upper limit is still unknown. However, evidence was recently presented suggesting that certain iron-reducing bacteria can be grown at somewhat higher temperatures (Kashefi and Lovley, Nature, submitted for publication, as cited in the chapter by Lovley in the present volume) (Lovley, 2003).

3. Psychrophiles

The oceans, which cover about three quarters of the surface of the Earth, have an average temperature of 5°C, and water in the depths of the oceans averages 1-2°C. A substantial part of the biosphere is thus constantly found at low temperatures. In addition, part of the continents are characterized by low temperatures. Thus, in Antarctica and in the permafrost of Siberia we can find cold-loving (psychrophilic) microorganisms that are adapted to life in severe cold.

Liquid water is absolutely required for life: when the intracellular water solidifies, growth is suspended. The freezing point of the intracellular water therefore determines the lower temperature for growth. This freezing point can be lowered to some extent to values below 0°C by the presence of high concentrations of solutes, which act as "antifreeze" solutions.

A great diversity of psychrophilic or psychrotolerant microorganisms inhabits cold environments worldwide. Thus, snow and ice can become colonized by eukaryotic snow algae such as *Chlamydomonas nivalis* that live at temperatures around 1°C. Such pigmented snow algae may color the snow green, orange, or red. Sea ice still may contain large numbers of living bacteria, often present in pockets of liquid water within the ice (Staley and Gosink, 1999). Bacterial numbers in permanently cold areas may be surprisingly high. It was recently ascertained that Antarctic soils at in situ temperatures varying from -0.5 to +3.8°C may contain between 3×10^6 and 2×10^9 living bacterial cells per gram, as based on bioluminescent ATP determination, values that are more than four orders of magnitude higher than previously suggested (Cowan et al., 2002). High bacterial numbers may be encountered in permafrost soils as well (Vorobyova et al., 1997). Such microbial communities in permafrost may be surprisingly active: measurements of incorporation of ^{14}C-labeled acetate into the lipid fraction enabled the estimation of minimum doubling times of the community of 1 day at 5°C, 20 days at -10°C, and about 160 days at -20°C (Rivkina et al., 2000).

The most psychrophilic microorganism extant in culture today is probably *Polaromonas vacuolata*, a member of the γ-Proteobacteria branch of the Bacteria isolated from Antarctic marine waters. It grows between 0-12°C with an optimum at 4°C (Irgens et al., 1996),

4. Acidophiles

Microbial life is possible over a wide range of pH values, from below zero to above twelve, i.e., within a range of more than twelve orders of magnitude in proton concentration. Organisms living near the lower end of the pH scale are termed acidophiles, organisms that require a high pH environment for growth are named alkaliphiles.

Acidic environments may originate abiotically, e.g. the solfatara fields in volcanic areas, or may be biologically generated. Thus, oxidation of sulfide, elemental sulfur, pyrite and other similar compounds by chemoautotrophic bacteria such as *Thiobacillus thiooxidans* or *Thiobacillus ferrooxidans* leads to the accumulation of sulfuric acid with a concomitant decrease in pH. Acidity may also result from the production of organic acids in fermentation or from the oxidation of ferrous iron: *Thiobacillus ferrooxidans* generates acid by oxidation of Fe^{2+} to Fe^{3+}, which precipitates as $Fe(OH)_3$ with the release of protons.

We find acidophilic microorganisms in each of the three domains of life: Archaea, Bacteria, and Eukarya. Among the eukaryotes there are both phototrophic and heterotrophic extremely acidophilic and/or acid-tolerant representatives. The unicellular green alga *Dunaliella acidophila* requires strongly acidic conditions of pH 0-3, with an optimum at pH 1.0 (Pick, 1999). *Cyanidium caldarium*, a unicellular thermoacidophilic alga (Rhodophyta), has its growth optimum at pH 2-3 and tolerates pH values as low as 0.2 while growing at temperatures up to 57°C (Seckbach, 1994, 1999). This acidophilic alga tolerates even 1 N sulfuric acid. There are also a few fungi that grow at pH values near zero, notably species of the genera *Acontium, Cephalosporium*, and *Trichosporon*.

As already mentioned, *Thiobacillus* species and related organisms are good examples of acidophilic representatives of the bacterial domain. Thus, *Thiobacillus ferrooxidans* and *Leptospirillum ferrooxidans* grow optimally at pH 2-4. *Thiobacillus thiooxidans* grows at pH values as low as 0.5, and its growth ceases above pH 4-6.

The prize for the most acidophilic and most acid-tolerant microorganisms again goes to a representative of the archaeal domain: the coccoid hyperacidophilic species *Picrophilus oshimae* and *Picrophilus torridus* grow down to pH -0.06, which is the lowest pH known to date to support growth, and they have their optimum pH at about 0.5. Above pH 4 the cells lyse. In addition to their requirement for strongly acidic conditions these Archaea also need high temperatures (optimum 60°C) for growth. *Picrophilus* spp. have been isolated from Japanese solfataras. They are aerobic heterotrophs that grow well on yeast extract as carbon and energy source (Schleper et al., 1995a, 1995b, 1996). Many other Archaea are (thermo)acidophilic (Stetter, 1999), a well-known example being *Thermoplasma acidophilum*, which grows optimally at pH 1.8-2 with a minimum at about pH 0.4.

5. Alkaliphiles

Alkaliphilic bacteria that grow at pH values above 9-10 and sometimes higher can easily be isolated from soil and other environments. One such organism, *Bacillus alcalophilus*, which grows at pH values beyond 10, has become a popular object for the study of the molecular and bioenergetic aspects of adaptation to high pH values. There are also more extreme alkaliphiles, such are often found in soda lakes in which the pH can reach values as high as 12. A great variety of alkaliphiles exists representing many different types of metabolism, so that even extremely alkaline lakes can support complete cycling of elements such as carbon, nitrogen and sulfur (Zavarzin and Zhilina, 2000; Zhilina and Zavarzin, 1994).

Obligate alkaliphiles also occur in the archaeal domain. Species of the haloalkaliphilic genera *Natronobacterium* and *Natronococcus* are characteristic inhabitants of hypersaline soda lakes (e.g. Lake Magadi in Kenya, the Wadi Natrun lakes in Egypt, and certain lakes in China and Tibet) (Oren, 2002). These organisms can live up to about pH 12, and are thereby probably the life forms that can grow at the highest pH of all living organisms.

6. Halophiles

Requirement for salt is common in marine bacteria, which spend their lives in an environment that contains about 30-35 g/l salts. Extremely halophilic microorganisms can withstand much higher salt concentrations, up to NaCl saturation, i.e. over 300 g/l salt. Even the Dead Sea, on the border between Israel and Jordan (total dissolved salt concentration about 340 g/l, including more than 1.9 M Mg^{2+} and 0.4 M Ca^{2+}) harbors a variety of halophilic microorganisms adapted to the special conditions prevailing in that unique habitat. The presence of haloalkaliphilic extreme halophiles such as *Natronobacterium* and *Natronococcus* in hypersaline soda lakes has already been mentioned in the previous section. Salt lakes such as the Dead Sea, Great Salt Lake (Utah), hypersaline alkaline lakes, and NaCl-saturated crystallizer ponds for the production of solar salterns are often colored red. This reddish appearance is a result of the carotenoids and rhodopsins present in very dense communities of halophilic Archaea of the family Halobacteriaceae in the brines (Oren, 2000a, 2000b, 2002). Some species of unicellular green algae of the genus *Dunaliella* (Chlorophyta) may also contribute to the coloration of the brine. These algae may be colored orange-red as a result of their massive accumulation of β-carotene.

Also the domain Bacteria contains many species of halophilic or extremely halotolerant species, including aerobic heterotrophs, fermentative species, sulfate reducers, as well as oxygenic phototrophs (cyanobacteria) and anoxygenic photosynthetic sulfur bacteria (*Halochromatium*, *Halorhodospira*). An in-depth overview of the halophilic microorganisms and their environments was recently given by Oren (2002).

7. Barophiles

As already mentioned in Section 3, the ocean is the largest ecosystem on Earth, and, with a mean depth of almost 4 km and a maximum depth of exceeding 11 km much of it is found under high pressure. For every 10 m depth in the water column the hydrostatic pressure increases by about 1 atmosphere (0.1 MPa). The pressure at the deepest point in the sea – the bottom of the Mariana Trench near the Philippines, is about 110 MPa. Compared to our understanding of the thermophilic, acidophilic, alkaliphilic and halophilic microorganisms we know relatively little about the pressure-loving (barophilic [weight] or piezophilic [pressure]) bacteria that inhabit the deep sea. The main reason is that specialized expensive equipment is required to grow and handle microorganisms under hundreds of atmospheres of pressure. Only in recent years have the first truly barophilic bacteria been isolated and characterized. Barophilic bacteria are generally polyextremophiles, being psychrophilic in addition, as expected for organisms that live not only under high hydrostatic pressure but at low temperatures as well, such as found in the deep sea.

Truly barophilic (rather than barotolerant) bacteria are generally defined as such bacteria that require a minimum a pressure of at least 0.1 MPa and that grow optimally between 10-50 MPa or higher. Thus, strains related to the genera *Shewanella* and *Moritella* (γ-branch of the Proteobacteria), isolated from a depth of nearly 11 km in the Mariana Trench, grew optimally at pressures of 70 and 80 MPa, respectively, and did not show growth below 50 MPa (Kato et al., 1998). More barophilic strains belonging to the *Shewanella* group that grow optimally at 50-70 MPa and 10°C have been recovered from the deep sea of the Japan Trench and the Philippine Trench (Kato and Bartlett, 1997). Barophiles are also found within the archaeal domain, as exemplified by *Themococcus barophilus*, a hyperthermophile isolated from a Mid-Atlantic Ridge hydrothermal vent, that grows at temperatures up to 100°C and requires high pressure when grown at the highest temperautres (Marteinsson et al., 1999) (see also Section 10).

The most barophilic strain described thus far is probably the (not yet named) strain MT41, isolated by Yayanos et al. (1981) from the Mariana Trench. The upper pressure limit at which microbial growth has been observed is about 130 MPa (Yayanos, 1986). For an additional discussion on deep-sea Bacteria, see Yayanos (2000).

8. Radiation Resistance

Electromagnetic radiation in the wavelength range of visible light and near-infrared (400-1,100 nm) at moderate intensities is generally beneficial or non-harmful to life. Photosynthetic processes depend on such radiation energy (in the visible for oxygenic phototrophs and also near-infrared for anoxygenic phototrophs). Phototrophic organisms are responsible for nearly all primary production on Earth. However, electromagnetic radiation has also harmful effects, especially when the radiation is of short wavelength (ultraviolet radiation, gamma radiation) and of high intensity.

It is therefore not surprising that organisms exposed to high doses of radiation have developed protective mechanisms, and some microorganisms are markedly resistant to high doses of electromagnetic radiation. Levels of ultraviolet radiation that reached the earth's surface were undoubtedly much larger during the early evolution of life than they are today. The UV-shielding stratospheric ozone layer originated relatively late, after photosynthetically-released oxygen had accumulated to sufficiently high levels in the atmosphere.

Short-wavelength UV radiation and other forms of ionizing radiation such as gamma radiation cause the formation of breaks in DNA molecules and bring about additional kinds of damage. It is therefore not surprising that all living organisms possess at least some kind of repair mechanism to counteract the effect of harmful radiation. Another strategy used by certain microorganisms is to accumulate UV-absorbing substances that convert the harmful energy into heat before it can reach the sensitive targets within the cell. Many cyanobacteria produce such UV-absorbing substances (scytonemin, mycosporine-like amino acids). This strategy of increasing radiation resistance by means of UV-absorbing pigments is reviewed by Castenholz in this volume (Castenholz, 2003).

The most radiation-resistant of all known organisms is the terrestrial bacterium *Deinococcus radiodurans*. Its cells retain their viability at doses of UV radiation more than 20 times as high as such doses that are lethal to *Escherichia coli*, and it is 200 times more resistant to gamma radiation. *Deinococcus* resists up to 30,000 Gy (3 MRad) of gamma radiation. For comparison, humans are killed by less than 5 Gy. It is not quite clear what the natural niche of *D. radiodurans* may be, and why it has developed such an extreme radiation tolerance. It has been postulated that the phenomenon may have some connection with the drought tolerance of the species. Radiation and desiccation induce similar DNA damages, and the radiation resistance of *D. radiodurans* is accompanied by a high desiccation resistance. For further discussion on microbes at high levels of radiation see Rothschild (1999).

9. Tolerance to Matric Water Stress: Xerophilic Microorganisms

Low water availability can be caused by the presence of high concentrations of salts or other solutes (osmotic water stress, see Section 6) or by drought (matric water stress). The availability of water can be expressed as water activity (a_w), defined as the ratio of the water vapor pressure over a sample to that over pure water. It equals the relative humidity of the air in equilibrium with the sample in a closed system. The lowest water activity that still supports growth of microorganisms is about 0.61, a value at which the drought-loving (xerophilic) fungus *Xeromyces bisporus* is still able to grow. For comparison, the water activity of saturated NaCl solutions, the environment in which *Halobacterium* and other halophilic Archaea of the family Halobacteriaceae live, is about 0.75.

The ability of microorganisms to survive and even grow in low water environments is surprising. Even the driest environments on Earth such as hot deserts or the dry valleys of Antarctica support a variety of microorganisms (Cowan et al., 2002).

10. Microorganisms Adapted Simultaneously to More Than One Type of Stress

The sections above cover most types of stressful conditions under which microorganisms can live. The list can easily be extended by including e.g. life at extremely low nutrient concentrations, high oxygen tensions, presence of high concentrations of toxic heavy metal ions, etc.

It is important to note that there are many kinds of microorganism, considered extremophilic according to the criteria used above, that are exposed simultaneously to more than one environmental extreme, and many of these are optimally adapted to life under multiple stress factors. The term "polyextremophiles" has been coined by Rothschild and Mancinelli (2001) to describe such organisms. Table 2 gives a representative list of microorganisms adapted to life at more than one environmental extreme.

The combination "thermoacidophile" is quite commonly found, as many Archaea thrive in hot acidic springs and volcanic sites. Thus, the aerobic chemolithotroph *Sulfolobus* grows optimally at 75°C at pH 2-3 while oxidizing sulfide or elemental sulfur using oxygen as electron acceptor. *Acidianus* grows at pH 2 at temperatures as high as 95°C, either aerobically using elemental sulfur as energy source and oxygen as electron acceptor or anaerobically using hydrogen as electron donor and elemental sulfur as electron acceptor. *Thermoplasma*, *Metallosphaera*, *Sulfurisphaera*, and *Stychiolobus* are other examples of such thermoacidophiles (Fuchs et al., 1996; Huber et al., 1989; Kurosawa et al., 1998; Stetter, 1998). The oxygenic phototrophic eukaryote *Cyanidium* and its cohorts are also adapted to life at low pH (1-4) and high temperatures (45-57°C) (Seckbach, 1994, 2000).

A cell wall-less archaeon related to *Ferroplasma acidiphilum* and provisionally designated with the grammatically incorrect name "Ferroplasma acidarmanus" has been described to grow at pH 0-0.5 at 40°C in acid mine drainage in Iron Mountain, California, where it thrives in a brew of sulfuric acid with iron concentrations over 100 g/l and tens of grams per liter of copper, arsenic, cadmium and zinc (Edwards et al., 2000).

TABLE 2. Examples of extremophilic microorganisms ("polyextremophiles") growing at a combination of more than one environmental extreme.

Environmental factors	**Organism**	**Habitat**	**Phylogenetic affiliation**	**Tolerance to stress**
High temperature, Low pH	*Thermoplasma acidophilum*	Solfatara fields, Coal refuse piles	Archaea - Euryarchaeota	Maximum temperature 63°C; Minimum pH 0.5
	Picrophilus oshimae	Acidic hot springs	Archaea - Euryarchaeota	Minimum pH -0.06; Maximum temperature 65°C
	Metallosphaera spp.	Solfataras, Uranium mines	Archaea - Crenarchaeota	Maximum temperature 80°C; Minimum pH 1.0
	Sulfurisphaera	Acid hot spring	Archaea - Crenarchaeota	Maximum temperature 92°C; Minimum pH 1
	Cyanidium caldarium	Acidic hot springs	Eukarya	Maximum temperature 57°C; Minimum pH 0.2
High pH, High salt concentration	*Natronobacterium gregoryi*	Soda lakes	Archaea - Euryarchaeota	Maximum pH 12; NaCl saturation
High temperature, High pH	*Thermococcus alcaliphilus*	Shallow marine hydrothermal springs	Archaea - Euryarchaeota	Maximum temperature 90°C; Maximum pH 10.5
High temperature, High pressure	*Thermococcus barophilus*	Thermal vent, Mid-Atlantic Ridge	Archaea - Euryarchaeota	Maximum temperature 100°C; Requires pressure of 15-17.5 MPa at the highest temperatures
Low temperature, High pressure	Deep-sea bacteria	Deep sea environments	Bacteria	Live at 2-4°C and pressures of 50-110 MPa

Other "polyextremophiles" live in a combination of high temperature and high pH ("thermoalkaliphiles"). An example is *Thermococcus alcaliphilus*, which has a temperature range of 56-90°C and grows optimally at pH 9.0, with growth being possible up to pH 10.5 (Keller et al., 1995). Barophilic thermophiles exit in deep-sea hot vents, such as *Thermococcus barophilus* which grows at atmospheric pressure at temperatures up to 95°C. At higher temperatures (95-100°C) growth is only possible at elevated pressures of 15-17.5 MPa (Marteinsson et al., 1999). Many hypersaline lakes are characterized by high pH values. Such soda lakes may be saturated with salt while reaching pH values as high as 11-12. Haloalkaliphiles (Archaea as well as Bacteria) thrive in such environments. The deep sea as a biotope is not only cold (0-4°C), but organisms living in it are also exposed to high hydrostatic pressures and are adapted to those, often to the extent that they are unable to grow at atmospheric pressure.

11. Extremophiles and the Phylogenetic Tree of Life

The above paragraphs show that extremophilic microorganisms are found in all three domains of life: Archaea, Bacteria, and Eukarya. We have encountered xerophilic fungi and halophilic, acidophilic and thermoacidophilic unicellular algae. However, most extremophiles are prokaryotes. Alkaliphiles are widespread in different groups of the domain Bacteria, and acidophiles and thermophiles also abound there. The true diversity of thermophiles within the domain Bacteria is surely much larger than those thermophilic species extant in culture. Analyses of 16S rDNA sequences recovered from Obsidian Pool, Yellowstone National Park, USA (74-93°C) have shown that this extremely hot environment harbors representatives of many groups of Bacteria, including species that belong to novel candidate divisions (phyla) within the Bacteria domain that have as yet no cultured representatives (Hugenholtz et al., 1998). The isolation of those bacteria that harbor these sequences is now a major challenge to the microbiologists.

In spite of the impressive diversity of extremophiles among the Bacteria, it is clear that many of the most extremophilic microorganisms belong to the domain Archaea and that the archaeal domain contains the largest percentage of hyperextremophiles. Among the Archaea we find the greatest variety of thermophiles, including those that grow at the very highest temperatures known to support life. Both the Crenarchaeota and the Euryarchaeota phyla contain large numbers of thermophiles. Sequencing of 16S rDNA clones obtained from Obsidian Pool in Yellowstone has not only provided us with information on the presence of novel phyla of Bacteria (see above), but species belonging to a new phylum of Archaea are present as well: the "Korarchaeota", none of which have been brought into culture as yet (Barns et al., 1996). A fourth proposed phylum of thermophilic Archaea, the "Nanoarchaeota", is thus far represented by a single species, designated "Nanoarchaeum equitans", a very small (about 0.4 μm diameter) symbiont of another extreme thermophile, *Ignicoccus* (Crenarchaeota) that grows at 90°C (Huber et al., 2002). As explained in the next section, the Archaea are especially suited to withstand extremes of temperature and other environmental extremes as well, thanks to the high stability of the lipids of which their cell membranes are composed.

It may sometimes seem that the archaeal domain may consist almost solely of extremophiles. This is especially true when we consider the methanogens to be extremophilic on account of their high sensitivity toward molecular oxygen. Moreover, at least some methanogens are thermophilic as well. However, it is now becoming more and more clear that (yet uncultured) Archaea abound in "conventional" environments of mesophilic temperature, near-neutral pH, etc. A substantial part of the bacterioplankton in the oceans consists of Archaea, belonging to the Euryarchaeota as well as to the Crenarchaeota - a phylum earlier considered as thermophilic par excellence (Karner et al., 2001). We still know little about the properties of these abundant non-extremophilic marine Archaea. It was recently ascertained that the elusive marine Crenarchaeota most probably lead a (chemo)autotrophic way of life, as shown by the pattern of incorporation of $^{13}CO_2$ into their biphytanyl membrane lipids (Wuchter et al., 2003).

12. How to Be an Extremophile? Mechanisms Enabling Life in Extreme Environments

After having found such a staggering variety of extremophiles that populate even the most inhospitable environments found on Earth, many questions can be asked about the mechanisms enabling these organisms not only to survive in extremes of temperature, pH, and other stressful factors, but also to grow and multiply under these conditions. How do they do it, and what can we learn from their properties about the essence of the processes of life itself?

We can divide the mechanisms behind extremophilic behavior (to the extent those mechanisms are known) into two groups. One strategy that is possible only in some cases is "to keep the environment out", to establish firm barriers between the hostile outside world and the cell's cytoplasm, so that the intracellular mechanisms can function without needing to be especially adapted. Acidophiles and alkaliphiles use this kind of strategy: they maintain their intracellular pH at values close to neutral in all cases. The special adaptations enabling them to grow at pH values around zero or above ten must then be sought in the properties of the cytoplasmic membrane that establishes and maintains such large pH gradients and is responsible for intracellular pH homeostasis. A very low permeability of the membrane to protons and absence of uncontrolled proton movement are first prerequisites for an acidophile or an alkaliphile. In fact these properties are not unusual, as a low proton permeability is required for any cell whose bioenergetic processes are based on the generation and exploitation of proton gradients over the membrane. The acidophiles need to have efficient outward proton pumps that enable protons to be removed from the cell into the medium against a concentration gradient of four, five or more orders of magnitude. The alkaliphiles have a different problem: when bioenergetic processes such as respiratory or photosynthetic electron transport based phosphorylation depend on primary outward proton pumps in the cell membrane, the pH gradient over the membrane (acidic inside) is of the "wrong" orientation. Alkaliphiles appear to solve this problem by balancing the "inverted" pH gradient by strong membrane potentials (negative inside) and often also by using sodium ions rather than protons to couple bioenergetic transformations. All these adaptations leading to the establishment of intracellular homeostasis are of no use in the case of extracellular enzymes excreted by these extremophiles: these have to be stable and active at the low or the high pH of the environment in which the cell lives. Only few structural data are available on such enzymes that may shed light on the mechanism(s) of their low or high pH adaptation.

In other cases such homeostasis of the cell's cytoplasm at "normal", non-extremophilic conditions is not possible. A unicellular organism cannot regulate its temperature as warm-blooded animals can, so all components of psychrophilic and (hyper)thermophilic microorganisms must be stable and functional at the temperature of their environment. The same is true for barophiles. In the case of resistance to high levels of short-wave radiation, protection by special compounds in the outer cell layers is sometimes possible, as shown by the case of scytonemin in the extracellular sheaths of certain cyanobacteria (Castenholz, 2003). On the other hand, there are no mechanisms for *Deinococcus radiodurans* to prevent high levels of shortwave radiation reaching its DNA, and the unusually high radiation tolerance is therefore based on the existence of highly efficient DNA repair mechanisms.

Most proteins of mesophilic cells would denature at temperatures of 100°C and higher, at which the hyperthermophiles thrive. All their enzymes should therefore be stable and functional at high temperatures. There is still no simple unifying model that explains what makes an enzyme molecule heat-stable. Many thermophilic proteins have been purified and characterized. Ionic interactions are often of increased importance in such proteins, as appears from the number and extent of ion-pair networks. However, the correlation between the extent of ionic interactions and thermophilic behavior is not universal, and there are substantial differences in this respect between different proteins (Hough and Danson, 1999). Some thermophilic proteins have found interesting biotechnological applications. Famous examples are the DNA polymerases of *Thermus aquaticus* (Bacteria) (*Taq* polymerase) or *Pyrococcus furiosus* (Archaea) (*Pfu* polymerase). An excellent review on these and other "extremozymes" was given by Hough and Danson (1999).

The nucleic acids DNA and RNA form another heat-sensitive component of cells. Although a high content of guanine and cytosine stabilizes DNA against thermal denaturation, the genomic G+C content is not correlated with optimum growth temperature, and hyperthermophiles often have surprisingly low G+C contents in their DNA. In spite of this fact the DNA of hyperthermophiles appears to be very stable in vivo. Salts can significantly stabilize DNA against thermodegradation (Marguet and Forterre, 1998). On the other hand, striking correlations were found between the G+C content of ribosomal and transfer RNA stems and the optimum growth temperature (Galtier and Lobry, 1997).

The earlier claim that some hydrothermal vent bacteria can be grown at temperatures of at least 250°C (Baross and Deming, 1983) was probably based on an artefact (Trent et al., 1984). Because of the limited stability of amino acids, peptide bonds, phosphodiester bonds, and other chemical bonds, the upper temperature of life may be expected not to exceed 150°C (White, 1984).

The lowest temperature at which psychrophiles can grow is probably determined by the availability of liquid water. Formation of ice crystals inside the cell will halt its growth and may also structurally damage membranes and other cell components. Our understanding of the mechanisms enabling enzymes to function at low temperatures is increasing now the first cold-active enzymes have been crystallized, and the number of genes of cold-stable enzymes that have been cloned and sequenced is rapidly growing. The data available did not yet lead to a consistent picture of the molecular basis of psychrophily. Changes observed in frequency of particular molecular bonds and amino acid side chains in cold-active proteins from psychrophiles include: 1, increased numbers of polar and less hydrophobic residues; 2, additional glycine residues and low arginine/lysine ratios; 3, fewer hydrogen bonds, aromatic interactions, and ion pairs; 4, lack of salt bridges; 5, presence of additional surface loops with increased polar residues and/or decreased proline content; 6, modified α-helix dipole interactions; 7, reduced hydrophobic interactions between subunits. No single cold-active enzyme displays all these features, and each has a suite of changes (relative to its mesophilic/thermophilic counterparts) that confer the necessary conformational flexibility at the active site, at the expense of the enzyme's activity at higher temperatures (Russell, 2000). Psychrophilic enzymes must also resist cold denaturation, which is caused in part by diminished hydrophobic interactions at low temperatures. Freezing is also experienced by many microorganisms, and the ability to tolerate and survive this additional stress is not restricted to psychrophiles.

Life at high hydrostatic pressure also requires special adaptations of the enzymes, as increased pressure decreases the binding capacity of enzymes for their substrates. Protein synthesis and membrane transport are often pressure-sensitive in non-barophilic microorganisms. Relatively few proteins appear to be pressure-controlled in barophiles. In the barophile *Photobacterium* sp. strain SS9 it was found that the expression of *ompH* and *ompL*, genes encoding outer membrane porin proteins, depends on the pressure. The OmpH protein is maximally abundant at the pressure optimum of 28 MPa, while the OmpL protein is produced in greatest quantity at 0.1 MPa. Several additional genes were found whose expression was regulated by pressure or that appear to be critical to baroadaptation (Kato and Bartlett, 1997).

The great diversity of halophilic microorganisms presents us with an interesting case in which different groups have solved the problem of life at high salt concentrations in different ways. No halophilic microorganism can maintain a diluted cytoplasm of low osmotic pressure because of the high permeability of the cytoplasmic membrane to water. Some groups of halophiles accumulate salt (mainly KCl) to balance the high salt concentrations present in their medium. This strategy is used by the Halobacteriaceae (Archaea), by the anaerobic fermentative Halanaerobiales (Bacteria), and by *Salinibacter*, a recently discovered aerobic extremely halophilic representative of the Bacteria (Oren, 2000a, 2000b, 2002, 2003). This type of adaptation requires far-reaching changes of the intracellular enzymes to be stable and active in the presence of molar concentrations of KCl. Such halophilic proteins are characterized by a high excess of acidic amino acids, and low contents of basic and hydrophobic amino acids (Dennis and Shimmin, 1997; Lanyi, 1974; Oren, 2000a, 2002; Mevarech et al., 2000). Other halophiles (aerobic Bacteria, halophilic methanogenic Archaea, as well as halophilic unicellular eukaryotic algae and halophilic fungi) exclude salts from their cells as much as possible, and instead synthesize or accumulate organic solutes to provide the necessary osmotic balance. This strategy requires far less adaptation of the intracellular proteins as these organic "compatible" solutes do not greatly interfere with the functioning of "normal" enzymes (Oren, 2000a; Ventosa et al., 1998). It should be noted that the extracellular enzymes of such "low-salt-in" halophiles still have to be functional at high salt.

The properties of the cytoplasmic membrane determine in many cases the limits of life of extremophiles. The permeability of the membrane to protons and other ions has to be carefully controlled under all growth conditions, and the proper fluidity should be maintained under all circumstances. Psychrophilic members of the domain Bacteria have membranes with a very high content of unsaturated fatty acids. Thus, *Polaromonas vacuolata* lipids contain 74-79% of 16:1 ω7c, the highest level reported in any species, with in addition 7-9% of 18:1, and no more than 14-17% of saturated fatty acids (16:0) (Irgens et al., 1996). Some psychrophilic Bacteria (and barophilic Bacteria as well) also contain polyunsaturated fatty acids. For example, the Antarctic sea ice species of the genus *Psychroflexus* possess fatty acids with four and five double bonds (Bowman et al., 1998).

It has been postulated that the proton permeability of the cytoplasmic membrane may be the major factor that determines the maximum growth temperature of any species. The proton permeability of membranes appears to be kept constant at low values, independent of the optimum growth temperature of the organism investigated and of the type of lipids (archaeal or bacterial), from psychrophiles to extreme thermophiles. This explains why the most thermophilic microorganisms are all Archaea: the archaeal ether lipid membranes, and especially the monolayer types, are extremely impermeable to protons even at high temperatures, and therefore cells that possess this type of membrane are more suitable to life and survival in extremely hot environments (van de Vosseberg et al., 1998).

13. Extremophiles and the Origin of Life

Our knowledge of the extremophiles and the physico-chemical parameters that determine the limits of life on Earth today is of great importance for our understanding of how life may have originated on our planet and possibly elsewhere in the Universe as well.

Life has existed on Earth at least for 3.8 billion years. It may well be assumed that it originated when climatic conditions on our planet were very different from those today. It has often been postulated that the first living cells were anaerobic thermophiles or hyperthermophiles that evolved in environments resembling today's hydrothermal vents. If this is true, we may agree with the claim by Rothschild and Mancinelli (2001) that "If life arose in a high-temperature, anoxic hydrothermal vent, any environment that deviates from that is "extreme". Others have proposed that life could have started below the terrestrial surface since the upper land was inhospitable during the early history. Under the upper surface or in depths underneath terrestrial or marine habitats the microbes may have evolved, protected from the celestial bodies bombarding young Earth and sealed from the lethal levels of ultraviolet radiation at the surface.

The small-subunit rRNA-based phylogenetic tree of life provides considerable support for a "hot" origin of life. If indeed the root of the tree should be sought somewhere between the Archaea and the Bacteria, all known organisms that are located in the tree close to the root are thermophiles. This is true for many representatives of the Crenarchaeota, notably for *Methanopyrus*, an extremely thermophilic methanogen, phylogenetically unrelated to the euryarchaeotal methanogens, which is located in the tree close to the presumed root. Another group of thermophiles close to the root is the already mentioned Korarchaeota, whose 16S rDNA sequences were recovered from Obsidian Pool in Yellowstone National Park (Barns et al., 1996) but are yet to be brought into culture. The fact that also those members of the domain Bacteria that are located closest to the root (*Aquifex*, *Hydrogenobacter*, *Thermus*, and relatives) are markedly thermophilic as well, albeit less so than the most thermophilic Archaea, is also an argument in favor of a "hot" origin of life. However, there is no general consensus that life indeed originated in hot environments and that the first microbes were thermophiles. Theories that favor environments of more moderate temperatures for the origin of life may be equally convincing (see e.g. Bada and Lazcano, 2002). Whatever the relation between today's extremophiles and early life may be, the extremophiles provide us with intriguing research objects, enable us to obtain information on the basic processes of life, show us the almost unlimited potential of life to adapt to environmental conditions, provide us with economically valuable products, and even may give us some insight into the possibilities for life elsewhere in the Universe.

14. Acknowledgements

We thank Professors Richard W. Castenholz and David A. Wharton for their constructive and critical reviewing of our manuscript.

15. References

Bada, J.F., and Lazcano, A. 2002. Some like it hot, but not the first biomolecules. Science 296: 1982-1983.

Barns, S.M., Delwiche, C.F., Palmer, J.D., and Pace, N.R. (1996) Perspectives on archaeal diversity, thermophily and monophyly from environmental rRNA sequences. Proc. Natl. Acad. Sci. USA 93: 9188-9193.

Baross, J.A., and Deming, J.W. 1983. Growth of "black smoker" bacteria at temperatures of at least 250°C. Nature 303: 423-426.

Baumgartner, M., Yapi, A., Gröbner-Ferreira, R. and Stetter, K.O. 2003. Cultivation and properties of *Echinamoeba thermarum* n. sp., an extremely thermophilic amoeba thriving in hot springs. Extremophiles, in press.

Blöchl, E., Rachel, R., Burggraf, S., Hafenbradl, D., Jannasch, H.W., and Stetter, K.O. 1997. *Pyrolobus fumarii*, gen. and sp. nov., represents a novel group of archaea, extending the upper temperature limit for life to 113 °C, Extremophiles 1: 14-21.

Bowman, J.P., McCammon, S.A., Tewis, T., Skerrat, J.H., Brown, J.L., Nichols, D.S., and McMeekin, T.A. 1998. *Psychroflexus torquis* gen. nov., sp. nov., a psychrophilic species from Antarctic sea ice, and reclassification of *Flavobacterium gondwanense* (Dobson et al. 1993) as *Psychroflexus gondwanense* gen. nov., comb. Nov. Microbiology UK 144: 1601-1609.

Brock, T.D. (1978) Thermophilic microorganisms and life at high temperatures. Springer-Verlag, New York.

Castenholz, R.W. 2003. Phototrophic bacteria under UV stress. This volume.

Cowan, D.A., Russell, N.J., Mamais, A., and Sheppard, D.M. 2002. Antarctic Dry Valley mineral soils contain unexpectedly high levels of microbial biomass. Extremophiles 6: 431-436.

Dennis, P.P., and Shimmin, L.C. 1997. Evolutionary divergence and salinity-mediated selection in halophilic archaea. Microbiol. Mol. Biol. Rev. 61: 90-104.

Edwards, K.J., Bond, P.L., Gihring, T.M., and Banfield, J.F. 2000. An archaeal iron oxidizing acidophile important in acid mine drainage. Science 287: 1796-1799.

Fuchs, T., Huber, H., Teiner, T, Burggraf, S., and Stetter, K.O. 1996. *Metallosphaera prunae*, sp. nov., a novel metal-mobilizing, thermoacidophilic archaeum, isolated from an uranium mine in Germany. Syst. Appl. Microbiol. 18: 560-566.

Galtier, N., and Lobry, J.R. 1997. Relationships between genomic G+C content, RNA secondary structures, and optimal growth temperature in prokaryotes. J. Mol. Evol. 44: 632-636.

Gold, T. 1992. The deep, hot biosphere. Proc. Natl. Acad. Sci. USA 89: 6045-6049.

Herbert, R.H., and Codd, G.A. (eds.) 1986. Microbes in extreme environments. Academic Press, London.

Horikoshi, K., and Grant, W.D. (eds.) 1991. Superbugs. Microorganisms in extreme environments. Japan Scientific Societies Press, Tokyo; Springer-Verlag, Berlin.

Horikoshi, K. and Grant, W.D. (eds.) 1998. Extremophiles, Microbial life in extreme environments, Wiley-Liss Publishers, New York.

Hough, D.W., and Danson, M.J. 1999. Extremozymes. Curr. Opin. Chem. Biol. 3: 39-46.

Huber, G., Spinnler, C., Gambacorta, A., and Stetter, K.O. 1989. *Metallosphaera sedula* gen. and sp. nov. represents a new genus of aerobic, metal-mobilizing, thermoacidophilic archaebacteria. Syst. Appl. Microbiol. 12: 38-47.

Huber, H., Hohn, M.J., Rachel, R., Fuchs, T., Wimmer, V.C., and Stetter, K.O. 2002. A new phylum of

Archaea represented by a nanosized hyperthermophilic symbiont. Nature 417: 63-67.
Hugenholtz, P., Pitulle, C., Hershberger, K.L., and Pace, N.R. 1998. Novel division level bacterial diversity in a Yellowstone hot spring. J. Bacteriol. 180: 366-376.
Irgens, R.L., Gosink, J.J., and Staley, J.T. 1996. *Polaromonas vacuolata* gen. nov., sp. nov., a psychrophilic marine, has-vacuolate bacterium from Antarctica. Int. J. Syst. Bacteriol. 46: 822-826.
Karner, M.B., DeLong, E.F., and Karl, D.M. 2001. Archaeal dominance in the mesopelagic zone of the Pacific Ocean. Nature 409: 507-510.
Kato, C., and Bartlett, D.H. 1997. The molecular biology of barophilic bacteria. Extremophiles 1: 111-116.
Kato, D., Li, L., Nogi, Y., Nakamura, Y., Tamaoka, J., and Horikoshi, K. 1998. Extremely barophilic bacteria isolated from the Mariana Trench, Challenger Deep, at a depth of 11,000 meters. Appl. Environ. Microbiol. 64: 1510-1513.
Keller, M., Braun, F.-J., Dirmeier, R., Hafenbradl, D., Burggraf, S., and Stetter, K.O. 1995. *Thermococcus alcaliphilus* sp. nov., a new hyperthermophilic archaeum growing on polysulfide at alkaline pH. Arch. Microbiol. 164: 390-395.
Kristjansson, J. K., and Hreggvidsson, G.O. 1995. Ecology and habitats of extremophiles. World J. Microbiol. Biotechnol. 11: 17-25.
Kurosawa, N., Itoh, Y.H., Iwae, T., Sugai, A., Uda, I., Kimura, N., Horiuchi, T., and Itoh, T. 1998. *Sulfurisphaera ohwakuensis* gen. nov., sp. nov., a novel extremely thermophilic extremophilic acidophile of the order Sulfolobales. Int. J. Syst. Bacteriol. 48: 451-456.
Kushner, D.J. (ed.) 1978. Microbial life in extreme environments. Academic Press, London.
Lanyi, J.K. 1974. Salt-dependent properties of proteins from extremely halophilic bacteria. Bacteriol. Rev. 38: 272-290.
Lovley, D.R. 2003. Potential role of dissimilatory iron reduction in the early evolution of microbial respiration. This volume.
Madigan, M.T. 2000. Bacterial habitats in extreme environments, In: Seckbach, J. (ed.), Journey to diverse microbial worlds. Adaptation to exotic environments. Kluwer Academic Publishers, Dordrecht. pp. 63-72
Madigan, M.T., and Marrs, B.L. 1997. Extremophiles. Sci. Am. 276 (4): 66-71.
Madigan, M.T. and Oren, A. 1999. Thermophilic and halophilic extremophiles. Curr. Opinion Microbiol. 2: 265-269.
Marguet, E., and Forterre, P. 1998. Protection of DNA by salts against thermodegradation at temperatures typical for hyperthermophiles. Extremophiles 2: 115-122.
Marteinsson, V.F., Birrien, J.-L., Reysenbach, A.-L., Vernet, M., Marie, D., Gambacorta, A., Messner, P., Sleytr, U.B., and Prieur, D. 1999. *Thermococcus barophilus* sp. nov., a new barophilic and hyperthermophilic archaeon isolated under high hydrostatic pressure from a deep-sea hydrothermal vent. Int. J. Syst. Bacteriol. 49: 351-359.
Mevarech, M., Frolow, F., and Gloss, L.M. 2000. Halophilic enzymes: proteins with a grain of salt. Biophys. Chem. 86: 155-164.
Onstott, T.C., Phelps, T.J., Kieft, T., Colwell, F.S., Balkwill, D.L., Fredrickson, J.K., and Brockman, F.J. 1999. A global perspective on the microbial abundance and activity in the deep subsurface, pp. 489-500 In: Seckbach, J. (ed.), Enigmatic microorganisms and life in extreme environments. Kluwer Academic Publishers, Dordrecht.
Oren, A. 2000a. Life at high salt concentrations. In: Dworkin, M., Falkow, S., Rosenberg, E., Schleifer, K.-H., and Stackebrandt, E. (eds.), The Prokaryotes. A handbook on the biology of bacteria: ecophysiology, isolation, identification, applications. 3rd. ed. Springer-Verlag, New York (electronic publication) (release 3.1).
Oren, A. 2000b. The order Halobacteriales. In: Dworkin, M., Falkow, S., Rosenberg, E., Schleifer, K.-H., and Stackebrandt, E. (eds.), The Prokaryotes. A handbook on the biology of bacteria: ecophysiology, isolation, identification, applications. 3rd. ed. Springer-Verlag, New York (electronic publication) (release 3.2).
Oren, A, 2002. Halophilic microorganisms and their environments. Kluwer Academic Publishers, Dordrecht.
Oren, A., and Seckbach, J. 2001. Oxygenic photosynthetic microorganisms in extreme environments. In: Elster, J., Seckbach, J., Vincent, W., and Lhotsky, O. (eds.), Algae and extreme environments - ecology and physiology. Nova Hedwigia Beiheft 123: 13-31.
Pick, U. 1999. *Dunaliella acidophila* - a most extreme acidophilic alga, In: Seckbach, J. (ed.), Enigmatic microorganisms and life in extreme environments. Kluwer Academic Publishers, Dordrecht. pp. 467-478
Rivkina, E. M., Friedmann, E.I., McKay, C.P., and Gilichinsky, D.A. 2000. Metabolic activity of permafrost bacteria below the freezing point. Appl. Environ. Microbiol. 66: 3230-3233.
Rothschild, L. J. (1999) Microbes and radiation, In: J. Seckbach (ed.) Enigmatic microorganisms and life in extreme environments. Kluwer Academic Publishers, Dordrecht. pp. 549-562.

Rothschild, L. J., and Mancinelli, R. L. 2001. Life in extreme environments. Nature 409: 1092-1101.
Russell, N.J. 2000. Toward a molecular understanding of cold activity of enzymes. Extremophiles 4: 83-90.
Schleper, C., Pühler, G, Kühlmorgen, B. and Zilllig, W. 1995a. Life at extremely low pH. Nature 375: 741-742.
Schleper, C., Pühler, G., Holz, I., Gambacorta, A., Janovic, D., Santarius, U., Klenk, H.P., and Zillig, W. 1995b. *Picrophilus* gen. nov., fam. nov.: a novel aerobic, heterotrophic, thermoacidophilic genus and family comprising archaea capable of growth around pH 0. J. Bacteriol. 177: 7050-7059.
Schleper, C., Pühler, G., Klenk, H.-P., and Zillig, W. 1996. *Picrophilus oshimae* and *Picrophilus torridus* fam. nov., gen. nov., sp. nov., two species of hyperacidophilic, thermophilic, heterotrophic, aerobic archaea. Int. J. Syst. Bacteriol. 46: 814-816.
Seckbach, J. (ed.) 1994. Evolutionary pathways and enigmatic algae: *Cyanidium caldarium* (Rhodophyta) and related cells. Kluwer Academic Publishers, Dordrecht.
Seckbach, J. (ed.) 1999. Enigmatic microorganisms and life in extreme environments. Kluwer Academic Publishers, Dordrecht.
Seckbach, J. (ed.) 2000. Journey to diverse microbial worlds. Adaptation to exotic environments. Kluwer Academic Publishers, Dordrecht.
Seckbach, J., and Oren, A. 2000. A vista into the diverse microbial world: an introduction to microbes at the edge of life In: Seckbach, J. (ed.), Journey to diverse microbial worlds. Adaptation to exotic environments. Kluwer Academic Publishers, Dordrecht., pp. 5-13.
Staley, J.T., and Gosink, J.J. 1999. Poles apart: biodiversity and biogeography of sea ice bacteria. Ann. Rev. Microbiol. 53: 189-215.
Stetter, K.O. 1996. Hyperthermophilic procaryotes. FEMS Microbiol. Rev. 18: 149-158.
Stetter, K.O. 1998. Hyperthermophiles: isolation, classification and properties, In: Horikoshi, K., and Grant, W.D. (eds.), Extremophiles: Microbial life in extreme environments. Wiley-Liss, Inc., New York. pp. 1-24
Stetter, K.O., Konig, H., and Stackebrandt, E. 1983. *Pyrodictium* gen. nov., a new genus of submarine disc-shaped sulphur reducing archaebacteria growing optimally at 105°C. Syst. Appl. Microbiol. 4: 535-551.
Trent, J.D., Chastain, R.A., and Yayanos, A.A. 1984. Possible artefactual basis for apparent bacterial growth at 250°C. Nature 307: 737-740.
van de Vosseberg, J.L.C.M., Driessen, A.J.M., and Konings, W.N. 1998. The essence of being extremophilic: the role of the unique archaeal membrane lipids. Extremophiles 2: 163-170.
Ventosa, A., Nieto, J.J., and Oren, A. 1998. Biology of aerobic moderately halophilic bacteria. Microbiol. Mol. Biol. Rev. 62: 504-544.
Vorobyova, E., Soina, V., Gorlenko, M., Minkovskaya, N., Zalinova, N., Mamukelashvili, A., Gilichinsky, D., Rivkina, E., and Vishnivetskaya, T. 1997. The deep cold biosphere: facts and hypothesis. FEMS Microbiol. Rev. 20: 277-290.
Wharton, D.A. 2002. Life at the limits: Organisms in extreme environments. Cambridge University Press, Cambridge.
White, R.H. 1984. Hydrolytic stability of biomolecules at high temperatures and its implication for life at 250°C. Nature 310: 430-432.
Wuchter, C., Schouten, S., Boschker, H.T.S., and Sinninghe Damsté, J.S. 2003. Bicarbonate uptake by marine Crenarchaeota. FEMS Microbiol. Lett. 219: 203-207.
Yayanos, A.A. 1986. Evolutional and ecological implications of the properties of deep-sea barophilic bacteria. Proc. Natl. Acad. Sci. USA 83: 9542-9546.
Yayanos, A. A. (2000) Deep sea bacteria, In: J. Seckbach (ed.) Journey to diverse microbial worlds. Adaptation to exotic environments. Kluwer Academic Publishers, Dordrecht. pp. 161-174.
Yayanos, A.A., Dietz, A.S., and van Boxtel, R. 1981. Obligately barophilic bacterium from the Mariana Trench. Proc. Natl. Acad. Sci. USA 78, 5212-5215.
Zavarzin, G.A., and Zhilina, T.N. 2000. Anaerobic chemotrophic alkaliphiles, In: Seckbach, J. (ed.) Journey to diverse microbial worlds. Adaptation to exotic environments. Kluwer Academic Publishers, Dordrecht. pp. 193-208
Zhilina, T.A., and Zavarzin, G.A. 1994. Alkaliphilic anaerobic community at pH 10. Curr. Microbiol. 29: 109-112.

Biodata of **V.V. Kevbrin** author (with co-authors Ch. S. Romanek and J. Wiegel) of ***"Alkalithermophiles: a double challenge from extreme environments"***

Dr. Vadim .V. Kevbrin is a senior scientist at the Institute of Microbiology RAS. in Moscow, Russia. He obtained his Cand. Sci. in Microbiology in 1994. The thesis was devoted to anaerobic halophilic bacteria developing at the cyanobacterial communities of marine origin. Since then his interest shifted toward alkaliphilic microorganisms. He isolated and described few novel taxons of alkaliphiles different systematic and trophic groups. During 1.5 year staying as a visiting scientist at the Prof. Wiegel's lab at the University of Georgia he worked with alkalithermophiles, a new and perspective group of microorganisms.

E-mail: **kevbrin@inmi.host.ru**

Dr. Christopher S. Romanek is an Associate Professor at the University of Georgia and shares a joint appointment in the Department of Geology and the Savannah River Ecology Laboratory. His area of specialty is low-temperature geochemistry, particularly the application of stable isotopes to problems in the earth and environmental sciences. Dr. Romanek received his Ph.D. in geology from Texas A&M University in 1991, where he focused on the stable isotope fractionation of carbon in carbonate minerals and the fluids from which they precipitate. Prior to arriving at the University of Georgia in 1995, Dr. Romanek held a National Research Council Fellowship at NASA's Johnson Space Center in Houston, TX, where he developed laser microprobe instrumentation and studied the low-temperature geochemistry of extraterrestrial materials. Most recently, Dr. Romanek has conducted research on carbon isotope fractionation in microbial systems and the expressions of microbiological activity that are recorded in modern and ancient sedimentary environments.

Dr. Jürgen Wiegel is a Professor of microbiology at the University of Georgia USA. He obtained his Ph.D. and Dr. Sci. in 1973 and 1982 respectively, at the University of Göttingen, Germany. His two main research interests are the anaerobic degradation of chlorinated aromatic compounds, and anaerobic thermophiles (ecology, phylogeny/taxonomy, physiology and biotechnological applications, industrial enzymes). Dr. Wiegel has described several new thermophilic anaerobes, some of which have been patented, e.g., for ethanol production, and includes the novel group alkalithermophilic anaerobes growing optimally above pH 8.5 and 55^0C with doubling times as short as 10 minutes. His studies take place among other sites also in Yellowstone National Park (Wyoming, USA) and at Kamchatka, (Russia). websites www.uga.edu/mib/people/wiegel.htm;www.gly.uga.edu/kamchatka (Kamchatkaproject)

E-mail: **jwiegel@uga.edu**

J. Seckbach (ed.), Origins, 395–412.

ALKALITHERMOPHILES: A DOUBLE CHALLENGE FROM EXTREME ENVIRONMENTS

V.V. KEVBRIN[1], C. S. ROMANEK[2] AND J. WIEGEL[3]
[1]Institute of Microbiology RAS, Prospect 60-letija Octiabria, 7/2, 117312 Moscow, Russia; [2]Savannah River Ecology Lab, Aiken, NC and Department of Geology, University of Georgia, Athens, GA 30602; [3]Department of Microbiology, University of Georgia, Athens, GA 30602-2605, USA

1. Introduction

The study of extremophilic microorganisms, in short extremophiles, has increased drastically over the last few years. An illustration for this increased interest is the establishment of the new International Society for Extremophiles and the recently introduced journal Extremophiles. Microorganisms are named extremophiles, when they are well adapted to and grow optimally at environmental and physicochemical parameters unsuitable for the typical and widely studied, mesophilic microorganisms such as *Escherichia coli, Bacillus subtilis* and *Neurospora crassa,* to name a few.

Despite the acceleration of descriptions of novel species, most of the described extremophiles are characterized only by one distinctive 'extreme'. In this chapter, we discuss a subgroup of 'multi-extremophiles' coined the alkalithermophiles (also referred to as thermoalkaliphiles). They are of interest to the scope of this book for two reasons: 1) ancestral alkalithermophiles could have been one of the earliest forms of life as some geochemical models and geological evidence suggest that the ocean of Early Earth has been alkaline in nature and capable of supporting primitive alkaliphilic microorganisms, and 2) alkalithermophiles can be regarded as one type of model organism for the study of possible extraterrestrial life. We believe alkalithermophilic microorganisms are one of the possible types of organisms that could have evolved on Mars, if life ever arose there (see below). Based on reasoning as discussed elsewhere (Wiegel and Adams, 1998), the authors believe that life probably originated not in hyperthermobiotic environments but on mineral surfaces in moderate thermobiotic (e.g., 60-85°C range), relatively shallow pools at the edges of the early Earth's oceans. The drastic changes of physico-chemical parameters over space and time in such environment would have provided the necessary dynamic conditions for frequent association and dissociation of prebiotic and biotic structures and thus changing selection pressures to lead to superior surviving combinations (Shock et al., 1998; Baross, 1998; Miller and Lazcano, 1998). These assumed selection conditions proposedly lead to a 'bush-like origin' of life as suggested by Kandler (1998) and thus is different from the frequently assumed quasi monophylogenetic progenote. Thus, some form of alkalithermophiles can be proposed as logical descendants of hypothetical early life forms.

Alkalithermophiles are alkaliphilic thermophiles (thermophilic alkaliphiles). A simplified definition for thermophiles and alkaliphiles is given in Table 1. Different definitions have been proposed for both extremes (Wiegel 1986, 1998a, 2002 and literature cited therein).

TABLE 1. Marginal data for simplified definitions of thermophiles and alkaliphiles or when combined for alkalithermophiles.

Thermotolerant	T_{min} --	T_{opt} <50°C	T_{max} <60°C
Thermophiles	T_{min} --	T_{opt} >50°C	T_{max} >60°C
Extreme thermophiles	T_{min} usually >60°C	T_{opt} >75°C	T_{max} >85°C
Alkalitolerant*	pH_{min} <7.0	pH_{opt} <8.5	pH_{max} >9.5
Alkaliphiles*		pH_{opt} ≥8.5	pH_{max} >9.5
Facultative	pH_{min} <7.0		
Obligate	pH_{min} >7.0		

* For thermophilic / psychrophilic microorganisms these values depend on the $pH^{temperature}$, i.e., at which temperature the pH values were measured and the pH meter calibrated. The above values are for pH^{25C}. For detailed discussion of pH determinations at elevated temperatures see Wiegel, 1998a, 2002 and literature cited therein.

In this chapter we present the diversity of the presently known alkalithermophiles in two figures depicting the relationship of the optimal growth temperature versus the optimal growth pH (Fig. 1 and 2). In agreement with the definitions in table 1, we included microorganisms exhibiting both temperature optima of 50°C and above and pH optima of 8.5 and above, but also some interesting species with a slightly lower pH optimum (For a more detailed description of their properties see the reviews Wiegel, 1998a; 2002 or original descriptions).

2. Habitats and Isolation

Although alkalithermophiles require alkaline conditions and elevated temperatures for optimal growth, this does not necessarily restrict the distribution of alkalithermophiles to very distinct niches where both conditions are provided, niches such as alkaline hot springs, alkaline (soda) lakes with geothermal or solar heat sources. The more general question behind this statement is an old issue, and was already addressed by Beijerinck (cited by Baas-Becking, 1934) in the form "Everything is everywhere, but the milieu selects". In other words, the question is, can alkalithermophiles be isolated from everywhere and therefore are regarded as ubiquitous microorganisms? Because of lack of data, this question cannot be answered for the alkalithermophiles with certainty. The mere presence of an amplified 16S rRNA sequence when using molecular techniques or visualization of a few (introduced) cells among 10^{10} CFU's does not constrain the viable habitat boundaries for a microorganism. The definition of a habitat (as indicated by the Latin root of the word) requires growth of an organism in that environment. On the other hand, spore forming thermophiles including spore forming alkalithermophiles should be "found" everywhere. One also can speculate that most aerobes are more or less ubiquitous. Many microorganisms while adsorbed to dust particles, fine sand grains, and soil particles can withstand various states of dryness. As such they can be distributed worldwide by wind and rain (e.g., Bacteria have hitchhiked on fine grains of sand from the Sakhara and on ash particles from Mt. Helena over the ocean; and in fine water droplets that are carried into the atmosphere by storm). Based on such observations, "all" Bacteria should be more or less distributed around the globe. However, the match between physico-chemical and geochemical conditions of the receiving environments and the physiological capabilities of the introduced microorganism will determine the survival and growth of the introduced microorganism. With respect to a potential evolution through a 'forced' adaptations, it then becomes a question of how different the environmental conditions are from those required to allow the introduced microorganism maintenance metabolism and basic metabolic functions or may be even growth – regardless of how slow -- so it is able to accumulate and "select" over time beneficial mutations for a better fit in the new environment.

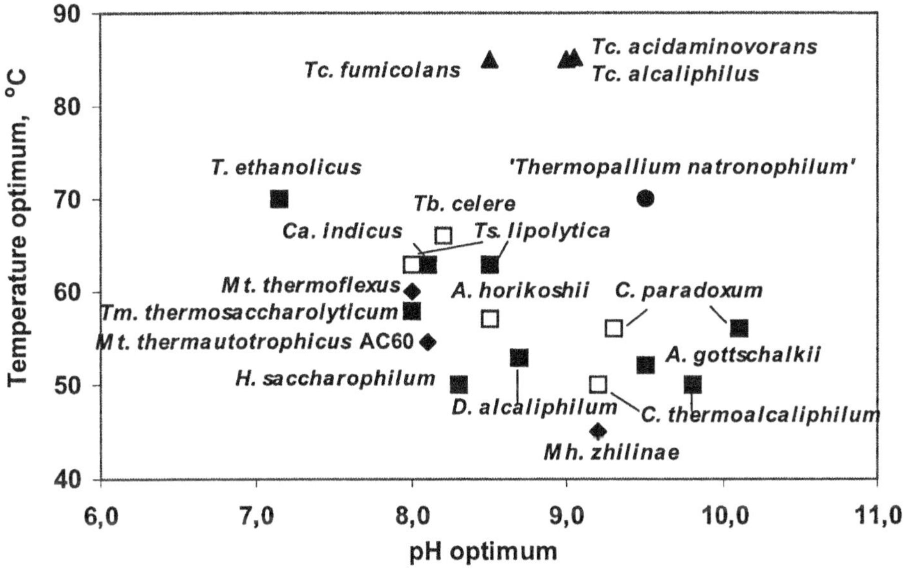

Figure 1. Anaerobic alkalithermophiles graphed according to their temperature and pH optima. Triangles, diamonds, squares and circles represent Archaea, methanogenic Archaea, the Gram-type positive (Firmicutes) and the Gram-type negative alkalithermophiles, respectively. Solid symbols represent pHs determined at room temperature or unknown temperature. Open symbols are for the "true" pHs measured at appropriate temperatures: for *Clostridium thermoalcaliphilum*, *C. paradoxum*, *A. horikoshii*, *Ts. lipolytica* and *Tb. celere* at 50, 55, 60, 60, and 66 °C, respectively. For readability genera have been abbreviated: A = Anaerobranca; C = Clostridium; Ca = Caloramator; D = Desulfotomaculum; H = Halonatronum; Mt = Methanothermobacter; Mh = Methanohalophilus: T. = Thermoanaerobacter, Tb = Thermobrachium, Tc. = Thermococcus; Tm = Thermoanaerobacterium: Ts = Thermosyntropha. Note that *H. saccharophilum* has an extreme wide temperature optimum between 36 and 55°C, whereas *T. ethanolicus* has an extreme wide pH optimum between 5.8 and 8.5 but that only a medium value could be used in the graph.

Although only a few (compared to mesophiles; Grant et al., 1990) alkalithermophiles have been validly described, a variety of distribution patterns have been observed. The following cases can be distinguished based on spatial relationships: the microorganisms are found **1)** only in one very restricted location (= narrow biogeography) but various environmental niches (relaxed biogeochemistry), e.g., the non spore forming *Anaerobranca horikoshii* is found in both slightly alkaline and acidic springs but only at a specific location in Yellowstone National Park containing both type of springs next to each other, **2)** only in one type of environment but at different geographical locations (= narrow and restricted biogeochemistry but relaxed biogeography), e.g. the sporulating *C. paradoxum* is found only in sewage sludge but in all tested sewage samples from various continents, and 3) ubiquitously distributed (= relaxed biogeography and biogeochemistry), e.g., non-sporulating *Thermobrachium celere* in thermobiotic, mesobiotic, alkaline and slightly acidic sediments from various continents. However, for nearly half of the known anaerobic alkalithermophiles only one or two strains from a single location have been described, and since negative isolation results are usually not reported, it is not known whether or not attempts to isolate them from other locations or other environments were made and failed.

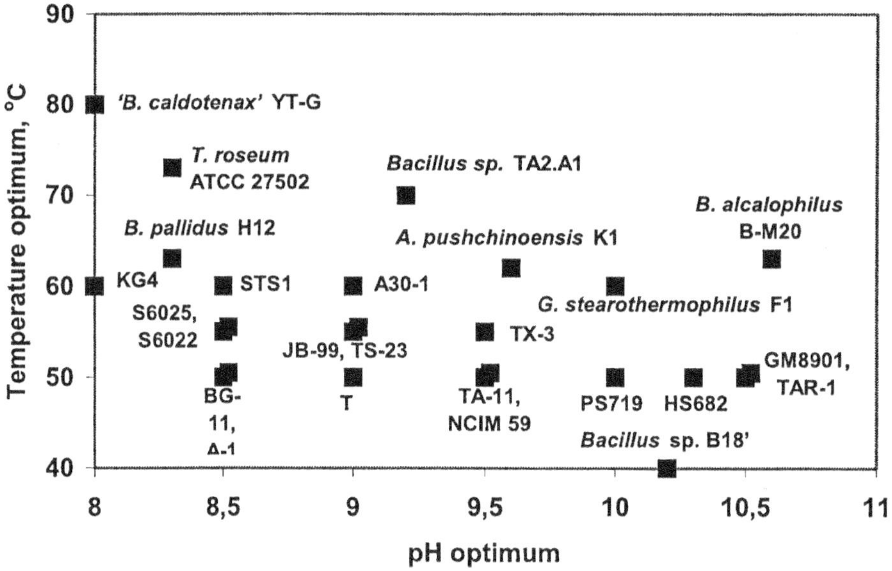

Figure 2. Aerobic and facultatively aerobic alkalithermophiles graphed according to their temperature and pH optima. *A* = *Anoxybacillus;* B = *Bacillus; G* = *Geobacillus; T* = *Thermomicrobium;*. The following strains belong to: KG4 = *Anoxybacillus* sp.; STS1 = *'B. thermoalcaliphilus'*; HS682 = *Thermoactinomyces* sp.; A-1 = *Thermoactinomyces sacchari*; S6025 = *B. thermocloaceae*; S6022 = *Spharobacter thermophilus*; T = *Thermoactinomyces* sp.; PS719, NCIM 59, TX-3, JB-99, A30-1, BG-11 = *Bacillus* sp.

Hot springs, fumaroles, and steam vents exhibit various pH, salt concentrations and temperature gradients, and they are spread all over the world but occur at higher frequency in the volcanic regions (Waring, 1965). Alkalinity in hot springs is caused by bleaching carbonate or silicate-bearing rocks by hot water. Alkaline hot springs with pH values above 8.5 are much less common than slightly to strong acidic (pH 6.0 to below pH 1) springs and mud pools and thus less frequently studied. This could also be the reason that alkalithermophiles have only recently been described. Examples of habitats ideal for isolation of alkalithermophilic microorganisms are the large, warm soda lakes of the East African Rift (Kenya, Tanzania, and Ethiopia), which have high concentrations of sodium carbonates (up to saturation) and pH values of 10-11. They are heated from above by tropical sun and from below by hot alkaline springs. They represent fairly stable water bodies on a geological time scale (Varnam, 2000). A variety of alkaliphiles have been isolated and characterized from these locations (Jones et al., 1998; Zavarzin et al., 1999; Martins et al., 2001) including anaerobic alkalithermophiles from Lake Bogoria: *Thermosyntropha lipolytica* (Svetlitshnyi et al., 1996), *Anaerobranca gottschalkii* (Prowe and Antranikian, 2001) and some unidentified strains. In respect to alkalithermophiles, the less studied soda lakes of Asia have more profound continental, seasonal climatology (Zavarzin et al., 1999). Northern Egypt has a set of desert alkaline soda lakes in the Wadi Natrun area, which due to their lower elevation are fed by underground water from the river Nile. They have an intensive microbial flora (Imhoff et al., 1979) and are known as a source for the isolation of various mesophilic alkaliphiles. To date, no alkalithermophiles from these lakes have been validly published although alkalithermophilic, *Bacillus*-like strains have been isolated (unpublished).

Surprisingly, the most alkalithermophilic anaerobe reported with a T_{opt} above 50°C *C. paradoxum* (Li et al., 1993) comes not from an alkaline hot spring but from mesobiotic, neutrophilic sewage sludges. So far, it has not been detected in sources that have not received gray water or sewage. Apparently these Bacteria are well adapted to this niche as *C. paradoxum* and alike have been found in all tested sewage sludges from three continents. The sewage facilities were found to contain per ml of anaerobic sludge up to 1000 colony forming units of anaerobic alkalithermophiles similar to *C. paradoxum* and *C. thermoalcaliphilum* (MPNs for aerobes alkalithermophiles have not yet been reported). During microbial decomposition of the proteinaceous components, alkaline pH's in the microgranula of the sludge can easily occur due to ammonia formation (ammonification of proteinaceous components). In that sense, animal manure containing material, especially under prolonged storage, leads frequently to ammonia release with concomitant local rise of the pH, and several alkalithermophiles have been isolated from manure samples. Similar environments are present by decomposed plant materials (e.g., compost piles). Bacteria with a restricted biogeography but relaxed biogeochemistry are represented by *Anaerobranca horikoshii* (Engle et al., 1995). This anaerobic alkalithermophile has been isolated from slightly alkaline as well as slightly acidic hot springs of different geochemistry and temperatures, however, so far only from a very restricted area in Yellowstone National Park (Wyoming, USA). It could not be isolated from samples of other geothermal areas in Japan, New Zealand and Iceland. Strains of *Thermobrachium celere* are an example of a ubiquitous alkalithermophile; strains have been isolated from mesobiotic and thermobiotic, slightly acidic and alkaline, and from somewhat 'pristine' and polluted environments located at several continents (Engle et al., 1996). An example of an aerobic alkalithermophile isolated from non-alkaline samples is *Bacillus* sp. Strain JB-99 isolated from slightly acidic (pH 5.8) sugar cane molasses (Johnvesly and Naik, 2001).

From the viewpoint of evolution and adaptation arises the question "is there a difference between alkalithermophiles isolated from alkaline thermobiotic environments and those from mesobiotic environments?" Collected data indicates an interesting tendency observed for the anaerobic alkalithermophiles. Heterotrophic strains from mesobiotic environments exhibit short doubling times of less than 30 minutes and include the extremely short doubling times of 16 and 10 min for *C. paradoxum* and *T. celere,* respectively. However, all the strains isolated from thermobiotic environments exhibit longer doubling times of 30 min or more including the various tested strains of *T. celere* from thermobiotic sources. The chemolithotrophic bacterial and archaeal strains do not fall within these doubling times. The isolation and characterization of additional chemolithotrophic and archaeal species are required to be able to make a statement whether the alkalithermophilic strains from mesobiotic environments grow faster than those from thermobiotic habitats. One adaptational, evolutionary explanation for the observed tendency of faster growth rates among the heterotrophic strains from mesobiotic environments, could be, that in a macroscopically mesophilic and neutrophilic environment (i.e., measuring bulk conditions), suitable growth conditions exist only in micro-niches and for short periods of time and thus require the alkaliphilic thermophilic Bacteria to have a fast growth response. In contrast, in the thermobiotic and alkaline environment, usually requirements for a fast growth response do not exist. However, the question whether the environmental conditions select for these fast growing, alkalithermophilic microorganisms or whether these Bacteria responded through evolution to live in theses niches is presently difficult to answer.

3. Alkalithermophiles: Potential Models of Primitive Microorganism in the Late Hadean Sea and on Mars

The scarcity of described alkalithermophiles is somewhat surprising if one considers the possibility that the Earth's earliest ocean may have been thermoalkaline in nature. The sedimentary rock record of the Precambrian Era has been studied intensively to gain insights into the mode and pace of microbial evolution (see seminal work of Schopf, 1992 and literature cited therein). Conflicting hypotheses have emerged regarding ocean chemistry during this early period of earth's history. Scientifically defendable arguments exist for the early stabilization of an ocean chemistry similar to that observed today (Holland, 1984; Walker et al., 1983). These models are founded on empirical and theoretical arguments as well as relict sedimentary materials and structures that are diagnostic of specific environmental conditions. Alternative models suggest that the early ocean was chemically distinct and evolved over geologic time to attain a modern ocean chemistry. Various authors (e.g., Kempe and Degens, 1985); Kempe et al., 1989; Kempe and Kazmierczak, 1997) have hypothesized that the chemistry of the early Precambrian ocean was similar to that of soda lakes (Na-CO_3 rich), which commonly occur in modern rift valley systems (e.g. East Africa). With the weathering of an evolving continental crust, the Precambrian ocean acquired the modern chemistry of NaCl brine by 2.2 Ga. Their arguments are based on the nature of distinct core materials available for weathering in the Precambrian landscape (calc-alkaline magmatites and komatiites) and the greater P_{CO2} of the Precambrian atmosphere. Morse and Marion

(1999) came to a similar conclusion when modeling fluid chemistry during the evolution of hypothetical solutions on the surface of Mars.

Under conditions where alkalinity was greater than the sum of dissolved Ca and Mg, alkali carbonate brines developed with a pH in excess of 9 (Eugster and Hardie, 1978; Drever, 1988). Alternatively, Morse and Mackenzie (1998) hypothesized that the earliest Hadean ocean (4.3 to 3.8 Ga) was hot (70 to 100°C) and slightly acidic (pH 5.8 $\pm$ 0.2), and evolved to the present day chemistry of oceans by about 3.8 Ga. Although other chemistries are possible depending on the initial Ca concentration of the fluid, they concluded the early Hadean ocean was probably more alkaline than oceans are today. Other models suggest that alkaline springs may have contributed dissolved solids to the Hadean oceans, facilitating the evolution of life (e.g., Morse and Mackenzie, 1998; Russel, 1996; Russel et al., 1998; Schwarzman, 1998). Regardless of whether the early ocean was a sodium carbonate or NaCl dominated fluid, but if this period was a time of high divergent evolution, one can assume that alkalithermophiles or alkalitolerant Bacteria were an important component of the terrestrial biosphere. Speculation extends to the role of alkalithermophiles in anaerobic dissimilatory iron reduction in the origin of the Precambrian Banded Iron Formation. We have recently found (Slobodkin and Wiegel, 1997; Wiegel et al., 2003) that resting cells of chemolithotrophic Fe(III) reducing thermophiles formed magnetite at pH 11 at geologically and ecologically relevant rates of around 1 mol x ml^{-1} and day^{-1}.

A significant body of research suggests that liquid water may have existed on the surface of Mars early in the history of the planet in the form of seeps, springs, floods or oceans (Zuber et al., 2000). Observations indicate that a Martian analog to the Earth's saltpans and saline lakes may have existed in basins intermittently in the past (Moore and Bullock, 1999). Thus, salt-tolerant alkalithermophiles could thrive in such extraterrestrial environments and so we believe that some of the alkalithermophiles with their special physiological properties such as short doubling times --as short as ten minutes for *Thermobrachium celere*-- make them good models for extraterrestrial life as well as models to simulate origin of life processes.

4. Adaptive Mechanisms

Life at simultaneous high pH and temperature requires special adaptive mechanisms, which during the course of evolution would be both adaptive and essential for life supporting processes. Few publications focus on how these Bacteria cope with extremes of pH and temperature. It is assumed that alkalithermophiles combine adaptive mechanisms from both alkaliphiles and thermophiles. To date, no alkalithermophiles have been described that grow at combinations of pH value greater than 12 and temperatures around or above 100°C. Such conditions probably would require different protective mechanisms than are presently known for alkaliphiles, e.g., novel mechanisms to prevent hydrolysis of proteins at a rate too fast for permitting continued growth. However, the existence of such organisms can not totally be ruled out, considering that at the opposite side of the pH scale the aerobic archaeal acidothermophiles *Picrophilus* species exist which grow optimally at pH values around 0.5 and temperatures above 60 °C (Schleper et al., 1996) or *Acidianus infernus* growing optimally at 90°C between pH 1.5 and 2 (Segerer et al. 1986). If such "hyperalkalithermophiles" were to exist, the ΔpH between internal and media pH would have to be greater than the usually observed 1 to 2 pH units or the bacterium would have to tolerate a higher internal pH (above 9.6) than ever has been reported (Krulwich, 2000; Cook et al., 1966).

The alkalitolerant/alkalithermophilic anaerobic Bacteria also contain examples of Bacteria with wide pH optima, such as the ethanologenic, glycolytic *Thermoanaerobacter ethanolicus* (pH^{25C}_{min} 4.4 and pH^{25C}_{max} 9.8). While growing at its temperature optimum of 69°C, its doubling time does not exhibit any pH-dependence between pH^{25C} 5.8 and 8.5 (Wiegel and Ljungdahl, 1981). An evolutionary question is then whether alkaliphilic and alkalitolerant thermophiles growing over an unusual wide pH range, contain different sets of gene products for the limiting anabolic and catabolic steps and which are specifically expressed when growing at the higher or lower pH ranges. This question is mainly relevant for the anaerobes, which do not maintain an internal pH stasis as do alkaliphilic aerobes do. An analogous scenario has been proposed as an evolutionary and adaptive mechanism for thermophiles growing over an extended temperature range of more than 35°C (Wiegel, 1990, 1998b). Although not unequivocally proven, there are indications that these Bacteria contain two sets of genes for some critical gene products for which either the stability and activity or their expression limit the respective growth either at the lower or higher temperature range. In general these thermophiles exhibit biphasic responses when graphing doubling times versus growth temperature. For alkaliphiles growing over a wide pH range, graphs of growth rates versus media pH show also biphasic or triphasic curves with more or less pronounced intermediary plateaus. Further studies are required to substantiate this hypothesis. Out of this discussion arises an evolutionary question "can alkaliphiles or extremely alkalitolerant microorganisms arise today through horizontal gene transfer or modification of duplicated genes (Olendzenski and Gogarten, 1998), that is by gaining genes specifically expressing alkaline stable cell components and enzymes active at high pH which will enable the microorganisms to grow at an extended pH range and eventually become an alkaliphile? Such genes would constitute a second set for critical gene products able to work at the extremes where the original gene products do not function properly anymore. To the knowledge of the authors, no experimental approaches to evolve alkali(thermo)philes from neutrophiles have been reported.

Basic knowledge about molecular mechanisms of alkaliphily comes mainly from studies of three strictly aerobic and mesophilic bacilli, *Bacillus pseudofirmus* OF4 (Krulwich, 1995, 2000; Krulwich et al., 2001), *Bacillus halodurans* C-125 (Horikoshi, 1999), *Bacillus cohnii* YN-2000 (Yumoto, 2002), *Bacillus alcalophilus* (Guffanti et al., 1981a,b; Hoffman and Dimroth, 1991). Similar studies of alkalithermophiles include only a few publications on the unidentified aerobic *Bacillus* sp. TA2.A1 (Peddie et al., 1999, 2000; Olsson et al., 2003) and two obligatorily anaerobic Bacteria with fermentative metabolism, *Clostridium paradoxum* (Cook et al., 1996) and *Anaerobranca gottschalkii* (Prowe et al., 1996). The genome of the latter species has recently been sequenced, thus, more information will be available in the near future, so comparisons can be made with the genomes of mesophilic aerobic alkaliphiles (Takami and Horikoshi, 2000).

There seems to be a major difference between the aerobic and anaerobic alkaliphiles. Whereas the aerobic alkaliphiles exhibit pH homeostasis, and are even able to raise or lower the external media pH to obtain optimal growth conditions (Horikoshi 1991), the anaerobic alkaliphiles change their intracellular pH with the extracellular pH and thus, do not maintain a pH homeostasis. It appears that the shortest doubling time for growth is observed when a maximal ΔpH occurs, and when the ΔpH diminished growth ceases as well (Cook et al., 1996). The absence of pH homeostasis was also found for anaerobic rumen Bacteria (Russell, 1991). Based on this observation, the argument can be made that the anaerobic system to deal with alkaliphilic growth conditions is simpler and thus the mechanisms in anaerobes evolved prior to those in aerobes. This is in agreement with the widely accepted assumption that the anaerobic ancestors evolved before aerobic ones due to the lack of oxygen in the atmosphere at the time life originated on early Earth (Russel et al., 1998; Baross, 1998).

All mesophilic as well as thermophilic alkaliphiles must cope with a reversed pH gradient (more acidic inside the cell than outside) and the consequences of this for their energy metabolism (Krulwhich et al., 2001 and literature cited therein). In addition, a well-established correlation between Na^+-dependence and alkaliphily for mesophilic alkaliphiles is also applicable to alkalithermophiles. All alkalithermophiles require at least low levels of Na^+ for optimal growth and metabolism, although the required Na^+ and balancing K^+-ion concentrations vary strongly among the species. A sodium dependence can be expected because sodium ions play a role in both types of microorganisms, for aerobic alkaliphiles in pH homeostasis, solute uptake and motility (Krulwich et al., 2001) and for some thermophiles in the sodium motive force (Δs). The latter one has an energetic advantage over a proton motive force (Δp) at elevated temperatures because the cytoplasmic membrane is much less permeable for sodium ions than for protons as temperature increases (Driessen et al., 1996; van de Vossenberg et al., 1995). It seems that the evolution of sodium ion dependent energy processes in alkaliphiles was not accidental since natural alkalinity is frequently caused by sodium carbonates and sodium is one of the most abundant and widely distributed elements of the earth's crust.

Like mesophilic alkaliphiles, the alkalithermophilic *C. paradoxum* exhibits a low Δp (near 35 mV) at simultaneous high intracellular ATP concentration (near 1 mM). Although not well characterized, it is assumed that sodium motive force (Δs) plays a significant role in substrate uptake and ATP-generation (Cook et al., 1996). For the alkalithermophilic *Anaerobranca gottschalkii* strain LBS3, (Prowe and Antranikian, 2001), it has been shown that the uptake of most amino acids was caused by an artificially imposed sodium gradient (Prowe et al., 1996). Leucine uptake was strictly coupled to sodium symport and sodium could not be replaced for lithium, that is usually interchangeable with sodium. Also, the membrane-bound ATPase was found to be Na^+ - translocating. While Δp was not determined, it appears that the bioenergetics of *Anaerobranca gottschalkii* relies on sodium ions. Oppositely, the aerobic alkaliphile *Bacillus* sp. TA2.A1, despite a low Δp, has a conventional proton-coupled F_1F_0-ATPase (Cook et al. 2003). As the external pH is increased from 7.5 to 10.0, the H^+/ATP stoichiometry increased from 2.0 to 5.7, which is consistent with earlier observations for other aerobic alkaliphiles (Krulwich, 1995). Yet, this *Bacillus* uses a Δs to transport glutamate and sucrose into the cell (Peddie et al., 1999, 2000). One can speculate for anaerobes that lack respiration, sodium-based energy transduction would be preferential because substrate-level phosphorylation produces less protons compared to oxidative phosphorylation. Thus, it appears that anaerobic alkaliphiles have been restricted in their capability to vary proton stoichiometry to compensate for a low Δp as it has been suggested for aerobic alkaliphiles (Krulwich, 1995).

Another question is, whether in comparison to mesophilic alkaliphiles and neutrophilic thermophiles, specific fatty acid profiles have evolved in alkalithermophiles as protective mechanisms against damage by alkaline and elevated temperature conditions. A slight tendency for a predominance of i-C-15 fatty acids in the tested anaerobic alkalithermophiles has been observed (Li et al., 1993; Prowe and Antranikian, 2001; Wiegel, unpublished). However, since the investigated species all belong phylogenetically to the Firmicutes branch, the tendency might be just due to the taxomic correlation of the isolates. Furthermore, the fatty acid profile from *C. paradoxum* changed notably with the growth pH without revealing any tendencies in comparison to the other profiles. Thus, so far there is no indication that special traits have evolved in these extremophiles in respect to cell lipids.

In summary, based on a rather restricted set of data, no special adaptations have been observed for the alkalithermophiles compared to the mesophilic alkaliphiles or neutral thermophiles, respectively. More detailed studies on additional and phylogenetically different microorganisms are required to elucidate whether alkalithermophiles have special adaptive features or whether it is generally true that they just combine known mechanisms from both mesophilic alkaliphiles and neutrophilic thermophiles to adapt to the multiple stressors. The major observed differences among the present isolates can be attributed to the aerobic versus anaerobic lifestyle of the microorganisms.

5. Diversity of Alkalithermophiles

5.1. AEROBES

5.1.1. Bacteria

In 1972, Heinen and Heinen described three strains of thermophilic, spore forming Bacteria (Heinen and Heinen, 1972). All three originated from slightly alkaline hot springs of Yellowstone National Park in the USA. One of them (‘*Bacillus caldotenax*’) is a slight alkalithermophile. Characteristically, this first report of an aerobic alkalithermophile was devoted to the isolation and description of extracellular enzymes (amylase, phosphatase and proteases and it appears that the isolation of aerobic alkalithermophilic microorganisms was and is driven by the desire to obtain useful industrial enzymes which are alkaline and temperature stable, exhibiting a long shelf life, and have high specific activities at alkaline pH and elevated temperatures (Niehaus et al., 1999; Bertoldo and Antranikian, 2002 and literature cited therein). This could explain why many of the alkaliphilic strains isolated in the past have not been identified at the species level, i.e., they are described simply as *Bacillus* sp. without further phylogenetic analysis (Fig. 2) and that almost all of the isolates produce some useful enzymes (Horikoshi, 1999; Gupta et al., 2002).

The majority of the isolates belong to the Gram-type positive *Bacillaceae* (Firmicutes) and *Actinomycetes* which includes the related genus *Sphaerobacter* (Fig. 1). The exception is the Gram-type negative, pink pigmented *Thermomicrobium roseum* isolated from Yellowstone National Park. It is phylogenetically a deep branching bacterium with an unusual cell wall (Merkel et al., 1980). The nearest relatives are the non-green sulfur Bacteria. It exhibits a narrow pH range from pH 7.0 to8.7.

5.1.2. *Archaea*
No aerobic alkalithermophilic Archaea have been found so far.

5.2. ANAEROBES

5.2.1. Bacteria
Anaerobic alkalithermophiles, were only described very recently with the first true alkalithermophilic obligate anaerobes, *C. paradoxum* (Li et al., 1993) and *C. thermoalcaliphilum* (Li et al., 1994). The search for these alkalithermophiles was again initiated by the desire to find industrially useful enzymes among the unexplored biodiversity of extremophiles. The list of validly published anaerobic alkalithermophiles has been broadened since then (Fig. 1) (Wiegel, 1998a, 2002). Most of the described anaerobic alkalithermophiles are Gram-type positive proteolytic or glycolytic Bacteria (for the use of the systematic term "Gram type" see Wiegel, 1981). *Thermosyntropha lipolytica* represents a syntrophic (with a hydrogen-utilizer) lipolytic, non-glycolytic example among the anaerobic alkalithermophiles. To date, only one anaerobic Gram type negative true alkalithermophile has been obtained, the so far not validly published '*Thermopallium natronophilum*' (pH_{opt} around 9.5; T_{opt} around 70°C, Meijer et al., 1996). The most alkaliphilic isolate among the thermophiles is *C. paradoxum* with a $pH^{25C}{}_{opt}$ of 10.1 (=pH^{55C} 9.3) and a T_{opt} around 56°C.

5.2.2. Archaea
No true alkalithermophilic methanogens have been described. However, there are three archaeal alkalithermophiles within the genus *Thermococcus,* two of them were isolated from shallow marine vents: *T. alcaliphilus* (Keller et al., 1995) and *T. acidaminovorans* (Dirmeier et al., 1998) with a pH_{opt} of around 9.0 and a T_{opt} of 85°C can be regarded as one of the most extreme alkalithermophiles. The three species represent within the group of the alkalithermophiles the species with the highest T_{opt} and $T_{max.}$

5.3. HALOALKALITHERMOPHILES

Bacteria: An adaptation to grow in environments representing three extreme environmental conditions, alkaline pH, elevated temperatures and high salinity is an evolutionary interesting combination. However, based on the scarcity of such isolates, it seems to be difficult for microorganisms to evolve mechanisms to overcome the obstacles of such triple extreme conditions. To date, although most alkalithermophiles require Na-ion for growth and energy metabolism, only two halophilic alkalithermophiles with optimal NaCl concentration above 3% have been described.

One is the strict anaerobic, spore forming *Halonatronum saccharophilum* isolated from Lake Magadi (Kenya), with optima for NaCl 7-12% w/v (halophilic), pH^{RT} 8.0-8.5 (slight alkaliphilic) and temperature 36-55 °C (thermophilic) (Zhilina et al., 2001). The other one is the facultatively aerobic, moderate halophilic, and a moderate alkalithermophile and, so far, not validly published *'Bacillus thermoalcaliphilus'*. The optima (growth ranges in brackets) 6-7 % (wt/vol) (0 – 11%), NaCl, pH^{RT} 8.0-9.0 (6 – 12) and 60 °C (T_{max} 69°C), respectively (Sarkar et al., 1988; Sarkar, 1991).

5.3.1. Archaea

Surprisingly to date there are no alkalithermophilic archaeal halophiles despite of their numerous aerobic alkalimesophilic representatives. The optimal sodium ion concentration for the above mentioned halophilic alkalithermophiles does not exceed 6%, wt/vol. This is five-fold lower than what can be observed with mesophilic haloalkaliphilic Archaea. Again, it is unclear at this time whether this reflects conditions prohibiting microbial growth or just a limitation in the performed exploration of this group of Bacteria and Archaea.

5.4. NARROW PHYLOGENY AND PHYSIOLOGY

The authors hypothesize that the relatively narrow phylogenetic and physiological diversity of alkalithermophiles is not due to a restricted evolution, but rather to a restricted isolation scheme. The presence of Archaea, Gram-type positive, and Gram-type negative Bacteria, suggests that alkalithermophiles do not represent a specific evolutionary branch. This is in agreement with the hypothesis that present day alkalithermophiles could be both descendants of early life forms, but also that some could be more recent adaptations to otherwise "empty" niches.

6. Are Protein Stability the Limiting Features for Extending the Growth Range for Alkalithermophiles?

Whereas some mesophiles can grow above pH 12, there are no thermophiles able to grow at pH^{25C} above 11.5 and 70°C. The alkaline protease from *Bacillus* sp. B18' exhibiting maximal activity at pH 12-13 and 70°C (Fujiwara et al., 1991, 1993) and the exoamylase (Amyl I) from *Bacillus* sp. GM8901 exhibiting a pH optimum of 11-12 at the temperature optimum of 60 °C and withstanding up to pH 13 (Kim et al., 1995) are two examples which demonstrate that microbial proteins /enzymes can be stable under such harsh. Thus a general protein instability may not be a factor in the inability to isolate alkalithermophiles able to grow optimally at pH values above 11 and temperatures above 80°C. This does not exclude the possibility of limitations due to that some obligatory required enzymes have not (yet?) evolved into enzymes with such thermo- and alkali stability. Future research should focus on 1) possible unrecognized physiological restrictions, which could exclude the existence of these microorganisms, and 2) novel conditions for isolation and cultivation.

7. Summary

Alkalithermophiles represent an exciting group of extremophilic microorganisms, which contain representatives of both, Bacteria and Archaea. Their adaptations towards high pH and elevated temperature draws attention not only as a source of industrially valuable enzymes but also for studying adaptive mechanisms to extreme environmental parameters.

The authors suggest that some of the alkalithermophiles can function as model organisms for primordial life forms when Earth's environment may have been extreme in pH and temperature. Some of the alkalithermophiles, e.g. chemolithotrophic, CO-oxidizing iron reducers, could have been one of the earliest microbial life forms on this planet. In short, many opportunities exist for studies to answer many unanswered questions regarding this interesting subgroup of extremophiles. For example, no obligate chemolithotrophic or acetogenic alkalithermophiles are known. Last not least alkalithermophiles have a high potential for biotechnological applications, especially as source for industrial enzymes

Furthermore, the authors are convinced that the boundaries for life have not yet been found in respect to combined elevated temperatures and alkaline pH (i.e. the highest growth temperatures at pH 12 or the most alkaline pH for growth at 100°C). For example, is it possible that microorganisms exist which grow optimally around 100°C and at a pH^{100C} above 11 or even 12? It can be expected that in the near future further aerobic and anaerobic alkalithermophiles with exciting properties will be isolated and described, when further extreme environments will be analyzed using traditional and novel microbial culture techniques and molecular survey methods such as metagenomics. These studies will with high probability extend the boundaries for conditions under which life can thrive. Subsequently, this will lead to new theories of how life could have evolved on early Earth and whether it could presently (or in the past) exists in extraterrestrial habitats.

8. Acknowledgements

We thank James T. Staley for providing the original citation of Baas Becking and Anna Kallistova for providing some references. Part of the reported research was funded by industrial grants to JW and by the National Science Foundation grants NSF 0238407 and NSF Int 0211100.

9. References

Baas-Becking, L.M.G. (1934) *Geobiologie of inleiding tot de miliekunde,* W.P. van Stockum and N.V. Zoon, The Hague, The Netherlands.

Baross, J.A. (1998) Do the geological and geochemical records of the Early Earth support the prediction from global phylogenetic models of a thermophilic, In: J. Wiegel and M.W.W. Adams (eds.) *Thermophiles: The Keys to Molecular Evolution and the Origin of Life?* Taylor and Francis, London, pp. 3-18.

Bertoldo, C. and Antranikian, G. (2002) Starch-hydrolyzing enzymes from thermophilic Archaea and Bacteria. Curr. Opin. Chem. Biol. **6,** 151-160.

Cook, G.M., Russell, J.B., Reichert, A. and Wiegel, J. (1996) The intracellular pH of *Clostridium paradoxum*, an anaerobic, alkaliphilic, and thermophilic bacterium. Appl. Environ. Microbiol. **62**, 4576-4579.

Dirmeier, R., Keller, M., Hafenbradl, D., Braun, F.J., Rachel, R., Burggraf, S. and Stetter, K.O. (1998) *Thermococcus acidaminovorans* sp. nov., a new hyperthermophilic alkalophilic archaeon growing on amino acids. Extremophiles **2**, 109-114.

Drever, J. I. (1988) *The Geochemistry of Natural Waters*, 2nd ed., Prentice Hall, Englewood Cliffs, NJ.

Driessen, A.J.M., van de Vossenberg, J.L.C.M. and Konings, W.N. (1996) Membrane composition and ion-permeability in extremophiles. FEMS Microbiol. Rev. **18**, 139-148.

Engle, M., Li, Y., Rainey, F., DeBlois, S., Mai, V., Reichert, A., Mayer, F., Messmer, P. and Wiegel, J. (1996) *Thermobrachium celere*, gen. nov., sp. nov., a fast growing thermophilic, alkalitolerant, and proteolytic obligate anaerobe. Int. J. Syst. Bacteriol. **46**, 1025-1033.

Engle, M., Li, Y., Woese, C. and Wiegel, J. (1995) Isolation and characterization of a novel alkalitolerant thermophile, *Anaerobranca horikoshi* gen. nov. sp. nov. Int. J. Syst. Bacteriol. **45**, 454-461.

Eugster, H.P. and Hardie, L.A. (1978) Saline Lakes, In: A. Lerman (ed.) *Lakes - Chemistry, Geology, Physics*, Springer-Verlag, N.Y., pp. 237-293.

Fujiwara, N., Masui, A. and Imanaka, T. (1993) Purification and properties of the highly thermostable alkaline protease from an alkaliphilic and thermophilic *Bacillus* sp. J. Biotechnol. **30**, 245-256.

Fujiwara, N., Yamamoto, K. and Masui, A. (1991) Utilization of a thermostable alkaline protease from an alkalophilic thermophile for the recovery of silver from used X-ray film. J. Ferment. Bioeng. **72**, 306-308.

Godfroy, A., Meunier, J.R., Guezennec, J., Lesongeur, F., Raguenes, G., Rembault, A. and Barbier, G. (1996) *Thermococcus funicolans* sp. nov., a new hyperthermophilic archeaon isolated from a deep-sea hydrothermal vent in the north Fiji Basin. Int. J. Syst. Bacteriol. **46**, 1113-1119.

Grant, W.D., Mwatha, W.E., and Jones, B.E. (1990) Alakliphiles:ecology, diversity and applications. FEMS Microbiol. Rev. **75**, 255-270.

Guffanti, A.A., Bornstein, R.F. and Krulwich T.A. (1981) Oxidative phosphorylation by membrane vesicles from *Bacillus alcalophilus*. Biochim. Biophys. Acta **635,** 619-630.

Guffanti, A.A., Cohn D.E., Kaback H.R. and Krulwich T.A. (1981) Relationship between the Na+/H+ antiporter and Na+/substrate symport in *Bacillus alcalophilus*. Proc. Natl. Acad. Sci. USA **78,** 1481-1484.

Gupta, R., Beg, Q.K. and Lorenz, P. (2002) Bacterial alkaline proteases: molecular approaches and industrial applications. Appl. Microbiol. Biotechnol. **59**, 15-32

Heinen, U.J. and Heinen, W. (1972) Characteristics and properties of a caldoactive bacterium producing extracellular enzymes and two related strains. Arch. Mikrobiol. **82**, 1-23.

Hoffmann, A., and Dimroth P. (1991) The electrochemical proton potential of *Bacillus alcalophilus*. Eur. J. Biochem. **201,** 467-473.

Holland, H.D. (1984) *The Chemical Evolution of the Atmosphere and Oceans*, Princeton University Press, Princeton, N.J.

Horikoshi, K. (1991) *Microorganisms in alkaline environments*, Kodansha-VCH, Tokyo, Japan.

Horikoshi, K. (1999) Alkaliphiles: Some applications of their products for biotechnology. Microbiol. Mol. Biol. Rev. **63**, 735-750.

Imhoff, J.F., Sahl, H.G., Soliman, G.S.H. and Trüper, H.G. (1979) The Wadi Natrun: Chemical composition and microbial mass developments in alkaline brines of eutrophic desert lakes. Geomicrobiol. J. **1**, 219-234.

Jones, B.E., Grant, W.D., Duckworth, A.W. and Owenson, G.G. (1998) Microbial diversity of soda lakes. Extremophiles **2**, 191-200.

Kandler, O. (1998) The early diversification of life and the origin of the three domains: A proposal, In: J. Wiegel and M.W.W. Adams (eds.) *Thermophiles: The Keys to Molecular Evolution and the Origin of Life?* Taylor and Francis, London, pp. 19-31.

Keller, M., Braun, F.J., Dirmeier, R., Hafenbradl, D., Burggraf, S., Rachel, R. and Stetter, K. (1995) *Thermococcus alcaliphilus* sp. nov., a new hyperthermophilic archaeum growing on polysulfide at alkaline pH. Arch. Microbiol. **164**, 390-395.

Kempe, S. and Degens, E.T. (1985) An Early soda ocean? Chem. Geol. **53**, 95-108.

Kempe, S. and Kazmierczak, J. (1997) A terrestrial model for an alkaline martian hydrosphere. Planet. Space Sci. **45**, 1493-1499.

Kempe, S., Kazmierczak, J. and Degens, E.T. (1989) The soda ocean concept and its bearing on biotic evolution, **In: R.E. Crick (ed.)** *Origin, Evolution and Modern Aspects of Biomineralization in Plants and Animals*, Plenum Press, N.Y., pp. 29-39.

Kim, T.U., Gu, B.G., Jeong, J.Y., Byun, S.M. and Shin, Y.C. (1995) Purification and characterization of a maltotetraose-forming alkaline α-amylase from an alkalophilic *Bacillus* strain, GM8901. Appl. Environ. Microbiol. **61**, 3105 – 3112.

Krulwich, T. A. (1995) Alkaliphiles: Basic molecular problems of pH tolerance and bioenergetics. Mol. Microbiol. **15**, 403-410.

Krulwich, T.A. (2000) Alkaliphilic prokaryotes, In: M. Dworkin, S. Falkow, E. Rosenberg, K-H Schleifer and E. Stackebrandt (eds.) *The Prokaryotes: An Evolving Electronic Database for the Microbiological Community*, 3rd edn., Springer-Verlag, Berlin, version 3.1 (www.prokaryotes.com), ISBN 0-387-14254.

Krulwich, T.A., Ito, M. and Guffanti, A.A. (2001) The Na^+ - dependence of alkaliphily in *Bacillus*. Biochim. Biophys. Acta **1505**, 158-168.

Li, Y., Engle, M., Mandelco, L. and Wiegel, J. (1994) *Clostridium thermoalcaliphilum* sp. nov., an anaerobic and thermotolerant facultative alkaliphile. Int. J. Syst. Bacteriol. **44,** 111-118.

Li, Y., Mandelco, L. and Wiegel, J. (1993) Isolation and characterization of a moderately thermophilic anaerobic alkaliphile, *Clostridium paradoxum,* sp. nov. Int. J. Syst. Bacteriol. **43**, 450-460.

Martins, R.F., Davids, W., Abu Al-Soud, W., Levander, F., Radstrom, P. and Hatti-Kaul, R. (2001) Starch-hydrolyzing Bacteria from Ethiopian soda lakes. Extremophiles **5**, 135 - 144.

Meijer, D., van Steenbergen, R., and Jones, B. (1996) Isolation and characterization of a novel alkaliphilic thermophilic bacterium from the soda lake environment belonging to the order *Thermotogales,* In: Abstracts, Thermophiles '96, International conference on "*The Biology, Ecology, and Biotechnology of Thermophilic Microorganisms*," September 1996, Athens, GA, University of Georgia, Athens, pp. 1A4.

Merkel, G.J., Durham, D.R., and Perry, J.J. 1980 The atypical cell wall composition of *Thermomicrobium roseum*. Can. J. Microbiol. **26**, 556-559

Miller, S.L. and Lazcano, A. (1998) Facing up to chemical realities: Life did not begin at the growth temperature of hyperthermophiles, In: J. Wiegel and M.W.W. Adams, (eds.) *Thermophiles: The Keys to Molecular Evolution and the Origin of Life?* Taylor and Francis, London, pp. 127-133.

Moore, J.M. and Bullock, M.A. (1999) Experimental studies of Mars analog brines. J. Geophys. Res. Planets **104**, 21925-21934.

Morse, J.M. and Mackenzie, F.T. (1998) Hadean ocean carbonate geochemistry. Aquatic Geochem. **4**, 301-319.

Morse, J.W. and Marion, G.M. (1999) The role of carbonates in the evolution of early martian oceans. Amer. J. Sci. **299**, 738-761.

Niehaus, F., Bertoldo, C., Kähler M. and Antranikian, G. (1999) Extremophiles as a source of novel enzymes for industrial application. Appl. Microbiol. Biotechnol. **51**, 711-729.

Olendzenski, L. and Gogarten, J.P. (1998) Deciphering the molecular record for the early evolution of Life: Gene duplication and horizontal gene transfer, In: J. Wiegel and M.W.W. Adams (eds.) *Thermophiles: The Keys to Molecular Evolution and the Origin of Life?* Taylor and Francis, London, pp. 165-176.

Olsson, K., Keis, S., Morgan, H.W., Dimroth, P. and Cook, G.M. (2003) Bioenergetic properties of the thermoalkaliphilic *Bacillus* sp. strain TA2.A1. J. Bacteriol. **185**, 461-465.

Peddie, C.J., Cook, G.M. and Morgan, H.W. (1999) Sodium-dependent glutamate uptake by an alkaliphilic, thermophilic strain, *Bacillus* strain, TA2.A1. J. Bacteriol. **181**, 3172-3177.

Peddie, C.J., Cook, G.M. and Morgan, H.W. (2000) Sucrose transport by the alkaliphilic, thermophilic *Bacillus* sp strain TA2.A1 is dependent on a sodium gradient. Extremophiles **4**, 291-296.

Prowe, S.G. and Antranikian, G. (2001) *Anaerobranca gottschalkii* sp. nov., a novel thermoalkaliphilic bacterium that grows anaerobically at high pH and temperature. Int. J. Syst. Evol. Microbiol. **51**, 457-465.

Prowe, S.G., van de Vossenberg, J.L.C.M., Driessen, A.J.M., Antranikian, G. and Konings, W.N. (1996) Sodium-coupled energy transduction in the newly isolated thermoalkaliphilic strain LBS3. J. Bacteriol. **178**, 4099-4104.

Russel, J.B. (1991) Intracellular pH of acid tolerant ruminal bacteria. Appl. Environ. Microbiol. **57**, 3383-3384.

Russell M.J. (1996) The generation at hot springs of sedimentary ore deposits, microbialites and life. Ore Geology Reviews **10**, 199-214.

Russell, M.J., Adaia, D.E., and Hall, A.J. (1998) The emergence of life from FeS bubbles at alkaline hot springs in an acid ocean, In: J. Wiegel and M.W.W. Adams (eds.) *Thermophiles: The Keys to Molecular Evolution and Origin of Life?* Taylor and Francis, London, pp. 77-126.

Sarkar, A. (1991) Isolation and characterization of thermophilic, alkaliphilic, cellulose-degrading *Bacillus thermoalcaliphilus* sp. nov from termite (*Odontotermes obesus*) mound soil of a semiarid area. Geomicrobiol. J. **9**, 225-232.

Sarkar, A., Varma, A. and Sarkar, A. (1988) Influence of cellulolytic organisms associated with a termite (*Odontotermes obesus*), on carbon mobility in a semiarid ecosystem. Arid Soil Res. Rehab. **2**, 75-84.

Schleper, C., Pühler, G., Klenk, H.P. and Zillig, W. (1996) *Picrophilus oshimae* and *Picrophilus torridus* fam. nov., gen. nov., sp. nov., two species of hyperacidophilic, thermophilic, heterotrophic, aerobic Archaea. Int. J. Syst. Bacteriol. **46**, 814-816.

Schopf, J.W. (1992) Paleobiology of the Archean, In: J.W. Schopf and C. Klein (eds.) *The Proterozoic Biosphere*, Cambridge University Press, NY., pp. 25-39.

Schwarzman, D.W. (1998) Life was thermophilic for the first two-thirds of earth history, In: J. Wiegel and M.W.W. Adams (eds.) *Thermophiles: The Keys to Molecular Evolution and Origin of Life?* Taylor and Francis, London, pp. 33-43.

Segerer, A., Neuner, A.M., Kristjansson, J.K. and Stetter, K.O. (1986) *Acidianus infernus* gen. nov., sp. nov., and *Acidianus brierleyi* comb. nov.: facultatively aerobic, extremely acidophilic thermophilic sulfur-metabolizing archaebacteria. Int. J. Syst. Bacteriol. **36**, 559-564.

Shock, E.L., McCollom, T. and Schulte, M.D. (1998) The emergence of metabolism form within hydrothermal systems, In: J. Wiegel and M.W.W. Adams (eds.) *Thermophiles: The Keys to Molecular Evolution and the Origin of Life?* Taylor and Francis, London, pp. 59-76.

Slobodkin, A. and Wiegel, J. (1997) Fe(III) as an electron acceptor for H_2 oxidation in thermophilic anaerobic enrichment cultures from geothermal areas. Extremophiles **2**, 106-109.

Svetlitshnyi, V., Rainey, F. and Wiegel, J. (1996) *Thermosyntropha lipolytica* gen. nov., sp. nov., a lipolytic, anaerobic, organoheterotrophic, alkalitolerant thermophile utilizing short- and long chain fatty acids in syntrophic co-culture with a methanogen. Int. J. Syst. Bacteriol. **46**, 1131-1137.

Takami, H. and Horikoshi, K. (2000) Analysis of the genome of an alkaliphilic *Bacillus* strain from an industrial point of view. Extremophiles **4**, 99-108.

van de Vossenberg, J.L.M.C., Ubbink-Kok, T., Elferink, M.G.L.,Driesen, A.J.M. and Konings, N. (1995) Ion permeability of the cytoplasmic membrane limits the maximum growth temperature of bacteria and Archaea. Mol. Microbiol. **18**, 925-932.

Varnam, A.H. (2000) *Environmental Microbiology,* ASM Press, Washington D.C.

Walker, J.G.G. (1977) *Evolution of the Atmosphere*, Macmillan, N.Y.

Walker, K.R., Shanmugan, G., and Ruppel, S.C. (1983) A model for carbonate to terrigenous clastic sequences. Bull. Geol. Soc. Am. **94**, 700–712.

Waring, G.A. (1965) *Thermal springs of the United States and other countries of the world. A summary,* Geological Survey Professional Paper 492, United States Government printing Office, Washington D.C.

Wiegel., J. (1981) Distinction between the Gram reaction and the Gram type of bacteria. Int. J. Syst. Bacteriol. **31**, 88.

Wiegel, J. (1986) Methods for isolation and study of thermophiles, In: T. D. Brock (ed.) *Thermophiles: General, Molecular and Applied Microbiology*. John Wiley and Sons, New York, pp. 17-37.

Wiegel, J. (1990) Temperature spans for growth: A hypothesis and discussion. FEMS Microbiol. Rev. **75**, 155-170.

Wiegel, J. (1998a) Anaerobic alkalithermophiles, a novel group of extremophiles. Extremophiles **2**, 257-267.

Wiegel, J. (1998b) Lateral gene exchange, an evolutionary mechanism for extending the upper or lower temperature limits for growth of an microorganism? A hypothesis, In: J. Wiegel and M.W.W. Adams (eds.) *Thermophiles: The keys to molecular evolution and the origin of life?* Taylor and Francis, London, pp. 175-185.

Wiegel, J. (2002) Thermophiles: Anaerobic alkalithermophiles. **In: G. Bitton (ed.)** *Encyclopedia of Environmental Microbiology,* John Wiley & Sons Inc, New York, pp. 3127-3140.

Wiegel, J. and Adams, W.W.W. (eds.) (1998) *Thermophiles: The keys to molecular evolution and the origin of life?* Taylor and Francis, London.

Wiegel, J., Ayres, K., and Hanel, J. (2003) Chemolithoautotrophic thermophilic iron(III)-reducer, In: L.G. Ljungdahl, M.W.W. Adams, M. Johnson, G. Ferry, and L. Barton (eds.) *Biology and Physiology of Anaerobic Bacteria,* Springer Verlag, N.Y., pp. 235-251.

Wiegel, J. and Ljungdahl, L.G. (1981) *Thermoanaerobacter ethanolicus* gen. nov., spec. nov., a new, extreme thermophilic, anaerobic bacterium. Arch. Microbiol. **128,** 343-348.

Yumoto, Y. (2002) Bioenergetics of alkaliphilic *Bacillus* spp. J. Biosci. Bioeng. **93**, 342-353.

Zavarzin, G.A., Zhilina, T.N. and Kevbrin, V.V. (1999) The alkaliphilic microbial community and its functional diversity. Microbiology (RU) **68**, 503-521.

Zhilina, T.N., Garnova, E.S., Tourova, T.P., Kostrikina, N.A. and Zavarzin, G.A. (2001) *Halonatronum saccharophilum* gen. nov sp nov.: A new haloalkaliphilic bacterium of the order Haloanaerobiales from Lake Magadi. Microbiology (RU) **70**, 64-72.

Zuber, M.T., Slomon, S.C., Phillips, R.J., et al. (2000) Internal structure and early thermal evolution of Mars form Mars Global Surveyor topography and gravity. Science **287**, 1788-1793.

Biodata of **Aharon Oren** author of ***"Halophilic Micro-organisms: Physiology and Phylogeny"***

Professor Aharon Oren is a member of the Division of Microbial and Molecular Ecology, the Institute of Life Sciences, The Hebrew University of Jerusalem. He earned his Ph.D. at the Hebrew University in1978, and spent a post-doctoral period at the University of Illinois at Urbana-Champaign. His research interests are in ecology, physiology, biochemistry and taxonomy of halophilic microorganisms and the microbiology of hypersaline environments. He has edited a book on "***Microbiology and Biogeochemistry of Hypersaline Environments***" (CRC Press, Boca Raton, 1999), and recently wrote "*Halophilic Microorganisms and their Environments*", the fifth volume in the series "***Cellular Origin and Life in Extreme Habitats***", edited by Joseph Seckbach and published by Kluwer Academic Publishers, Dordrecht, the Netherlands, 2002.

E-mail: **orena@cc.huji.ac.il**

J. Seckbach (ed.), Origins, 413–426.

HALOPHILIC MICROORGANISMS: PHYSIOLOGY AND PHYLOGENY

AHARON OREN
Division of Microbial and Molecular Ecology,
The Institute of Life Sciences,
and the Moshe Shilo Minerva Center for Marine Biogeochemistry,
The Hebrew University of Jerusalem, 91904 Jerusalem, Israel

1. Introduction

In an essay published in 1991, Hans Trüper and his coworkers explored the possible correlation between the phylogenetic affiliation of halophilic prokaryotes and the ways these organisms cope with the presence of high salt concentrations. Realizing that "on a first view there is apparently little correlation between taxonomy and halophily", the authors provided a survey that mainly centered on the *Proteobacteria* and the *Firmicutes*. The conclusions obtained were:

1. "All eubacteria [at least those included in the survey] that gain energy from photosynthesis or respiration and are capable of haloadaptation are able to accumulate and/or synthesize organic compatible solutes".
2. "Extreme halophily in eubacteria is always accompanied by glycine betaine synthesis".
3. "Throughout the phylogenetic system of the prokaryotes in the modern sense there exist halophilic and halotolerant genera or species in close (phylogenetic) relationship with non-halophilic and non-halotolerant ones. Therefore halophily and halotolerance as such may be of little value for the study of phylogenetic relations".
4. "Archaeobacteria [*Archaea*] and anaerobic fermenting eubacteria are incapable of synthesizing organic compatible solutes".

The authors wisely added that "The regularities observed may, however, not stand for long, as there are still those unidentified solutes which may change the whole picture. Also, we are quite convinced that further search will present more unknown compatible solutes."

Now, more than ten years later, our understanding of the modes of osmotic adaptation of halophilic microorganisms has greatly increased. In addition, new groups of halophiles have been identified, both within the bacterial and within the archaeal domain. The connection between the phylogenetic position of the different halophilic microorganisms and their mode of osmotic adaptation therefore deserves a critical reexamination.

The present chapter intends to provide a survey of the nature of haloadaptation in microorganisms, followed by a discussion of the phylogenetic and evolutionary implications of adaptation to high salt concentrations. The overall picture that emerges is far from clear. There are quite a few cases in which different strategies of osmotic adaptation occur in groups of organisms that are phylogenetically closely related. The implications of these observations for our views on the origin and evolution of haloadaptation will be explored in the paragraphs below.

2. Halophiles in the Tree of Life

Halophilic microorganisms can be found in all three domains of life: *Archaea*, *Bacteria*, and *Eukarya* (Oren, 1999a, 2002). In each of these domains we encounter representatives that can grow up to the highest salt concentrations (NaCl saturation).

No correlation is obvious between the place of an organism within the universal tree of life and its ability to grow at high salt concentrations. Halophilic microorganisms are most often interspersed between non-halophilic relatives. Thus, within the domain *Archaea* we find methanogenic halophiles in the families *Methanospirillaceae* and *Methanosarcinaceae*. Some of these can grow at salt concentrations up to 300 g l^{-1}, and are phylogenetically closely related to non-halophilic methanogens. Within the domain *Bacteria* there are halophiles of widely varying physiological properties: we find aerobic and anaerobic chemoheterotrophs, photoautotrophic and photoheterotrophic species, as well as chemolithotrophs in many different branches within the domain (Ollivier et al., 1994; Oren, 1999a).

There are three notable cases of phylogenetically coherent groups of microorganisms that are composed entirely or almost entirely of halophiles:

1. The order *Halobacteriales* with a single family, the *Halobacteriaceae* (domain *Archaea*). This order mostly consists of red, extremely halophilic aerobic microorganisms. The order forms a branch within the *Euryarchaeota*, branching off close to the *Methanomicrobiales/Methanosarcinales*.
2. The family *Halomonadaceae* (domain *Bacteria*, *γ-Proteobacteria*). Most species are moderately halophilic, metabolically versatile aerobes or facultative anaerobes.
3. The order *Halanaerobiales*, families *Halanaerobiaceae* and *Halobacteroidaceae* (domain *Bacteria*, low G+C branch of the *Firmicutes*). All representatives are obligate anaerobes. Most have a fermentative metabolism, but some are homoacetogens, and a genus that lives by anaerobic respiration (*Selenihalanaerobacter*) has recently been identified as well (Switzer-Blum et al., 2001).

Within the domain *Eukarya* we find relatively few true halophiles or extremely halotolerant microorganisms. However, green algae of the genus *Dunaliella* are ubiquitously present in waters ranging in salinity from slightly brackish to saturated sodium chloride brines.

3. Osmotic Adaptation of Microorganisms – General Considerations

Halophilic microorganisms live in high-salt, low-water activity environments. As biological membranes are permeable to water, the intracellular solute concentrations should be maintained at a level at least osmotically equivalent with the salt concentration in the medium. Moreover, at least a minimal turgor pressure has to be present within the cells to allow expansion and growth.

The microbial world has devised two fundamentally different strategies to achieve the required osmotic adjustment. One strategy (the "high-salt-in" mode of life) is based on the accumulation of inorganic ions - mainly K^+ and Cl^- - at high concentrations within the cells' cytoplasm. The aerobic halophilic *Archaea* of the order *Halobacteriales* are the best-known group that uses this way of coping with life at high salt. As discussed below, there also are a few types of *Bacteria* that lead a similar mode of life: the anaerobic *Halanaerobiales* (low G+C branch of the *Firmicutes*) and the aerobic genus *Salinibacter* (*Cytophaga – Flavobacterium – Bacteroides* branch). In such organisms all intracellular enzymes should be active in the presence of molar concentrations of salt, and this requires special adaptations of the entire intracellular enzymatic machinery. A limited adaptability to changing conditions is the price such cells have to pay for being able to grow at the highest salt concentrations, as most of their enzymes will not be functional in a low-salt environment.

An alternative strategy, and one that allows a much larger flexibility to changing salt concentrations, is the use of low-molecular-weight organic osmotic solutes (also known as "compatible solutes"), while keeping the intracellular ionic concentrations at a level as low as possible. These solutes are often accumulated intracellularly to molar concentrations. By virtue of being very poor enzyme inhibitors, they protect enzymes against the inhibition which would otherwise occur in solutions of low water availability. A wide range of such solutes has been identified. They include simple sugars, polyols, and amino acid derivatives (see Section 6). Use of compatible solutes is widespread within the domain *Bacteria* as well as in halophilic *Eukarya* (see Section 6), but is also found in certain *Archaea* (see Section 5). The concentrations of such osmotic solutes are regulated in accordance with the external salt concentration, and they are adjusted as required following changes in the outside salinity.

4. Osmotic Adaptation in the *Halobacteriales*

The aerobic *Archaea* of the order *Halobacteriales* (family *Halobacteriaceae*, including such genera as *Halobacterium*, *Haloferax*, *Haloarcula*, and *Halorubrum*) contain molar concentrations of K^+ and Cl^- in their cytoplasm, while maintaining their intracellular Na^+ concentration at levels considerably lower than that of their medium. K^+ concentrations as high as 4.6 M have been measured in *Halobacterium* cells grown in 4 M Na^+ (Christian and Waltho, 1962). A somewhat lower value (2.1 M) was obtained by Shporer and Civan (1977) based on NMR measurements in *Halobacterium* cells grown in medium containing 4.3 M Na^+. Much lower K^+ concentrations of around 1.1 M were estimated to occur in the cytoplasm of *Natronococcus occultus* grown in 3.4 M Na^+

(Desmarais et al., 1997). This organism is known to also accumulate an organic compatible solute, 2-sulfotrehalose (Desmarais et al., 1997; see also Martin et al., 1999).

The cells actively extrude Na^+ by means of a Na^+/H^+ antiporter located in the cell membrane. K^+ enters in part passively via a uniport system, driven by the negative-inside membrane potential. In addition, multiple K^+ transport systems have been detected in the genome of *Halobacterium* strain NRC-1 (Ng et al., 2000), including the ATP-driven KdpABC system and the low-affinity K^+ transporter TrkAH. Chloride is taken up from the medium, both by the light-driven chloride pump halorhodopsin and by light-independent symport with Na^+ ions (Duschl and Wagner, 1986).

The presence of molar KCl concentrations within the cells requires far-going adaptations of the entire intracellular machinery, as the presence of molar concentrations of salts is generally devastating to proteins and other macromolecules. Hydrophobic interactions are enhanced, causing aggregation of proteins and/or collapse of their tertiary structure. High ion concentrations interfere with essential electrostatic interactions within or between macromolecules by charge shielding.

Already in the 1970s it was reported that the bulk protein of *Halobacterium* and related organisms has an unusually high excess of acidic amino acids (aspartate, glutamate) over basic amino acids (lysine, arginine) (Reistad, 1970). Most proteins of *Halobacterium* NRC-1 have predicted isoelectric points within the range of 3.5-5.0 (Kennedy et al., 2001; Ng et al., 2000). Analysis of the *Halobacterium* NRC-1 genome showed aspartate to be much more abundant than glutamate, and lysine to be much more underrepresented than arginine (Ng et al., 2000). A low content of hydrophobic amino acid residues is another prominent feature of the proteins of the *Halobacteriales*. The content of the borderline hydrophobic amino acids serine and threonine is often increased (Lanyi, 1974).

Most proteins of the *Halobacteriales* denature when suspended in low salt solutions (Eisenberg et al., 1992; Mevarech et al., 2000). "Salting-out" salts stabilize, while "salting-in" salts inactivate halophilic enzymes (Lanyi, 1974). A model, called the "solvation-stabilization model" has been formulated to explain the halophilic behavior of such proteins. According to this model, the folded protein binds relatively large amounts of salt and water in KCl and in NaCl, and the excess of acidic amino acids in the protein composition could provide favored sites for specific water and ion binding to the tertiary or quaternary structure (Ebel et al., 1999). In the absence of high salt concentrations the proteins' tertiary and quaternary structure is altered, often irreversibly so. As a result, *Halobacterium* and its relatives have a very limited potential to adapt to lowered external salt concentrations.

Dennis and Shimmin (1997) have provided an in-depth discussion of the evolutionary processes that may have been involved in the development of salt-tolerant and salt-requiring enzymes. They compared the genes for ribosomal proteins L1 and L11 of six species of *Halobacteriaceae* belonging to three genera with the corresponding genes of enteric bacteria (*Escherichia*, *Serratia*, *Proteus*). A similar comparison was performed between the Fe/Mn superoxide dismutase of these halophiles with the corresponding enzyme of six *Mycobacterium* species. Over the regions analyzed, the halophilic ribosomal protein genes are 71 to 81% identical at the nucleotide level, and at the amino acid level the corresponding proteins are 65 to 81%

identical. The comparisons indicate that halophilic genes accumulate two to three times as many non-synonymous nucleotide substitutions as the homologous non-halophilic genes. The most striking feature is the high proportion of nonsynonymous nucleotide substitutions that result in replacement of serine positions, using both the TCN and the AGY codon families, and of replacements involving acidic amino acids. The evolutionary modifications required to reengineer a protein so that it becomes halophilic appear to involve the introduction of additional acidic residues onto the surface of the protein. Acidic residues are more hydrated than other amino acids and can coordinate the organization of a hydrated salt ion network at the surface of the protein. Dennis and Shimmin suggest that many of these substitutions, resulting in amino acid replacements, represent "evolutionary tinkering" at sites which influence the hydrophobic and surface hydration properties of the proteins. Environmental salinity may have provided the driving force for fixation of these nonsynonymous substitutions in natural halobacterial populations. In an altered environment a nonsynonymous substitution that was previously deleterious may suddenly become advantageous and therefore have an increased probability of being maintained in the population. If environmental fluctuation occurs extremely rapidly, the fittest proteins are likely to be those that are functional over the widest range of conditions (although not necessarily the most efficient under any particular condition). When fluctuations occur at a rate similar to the rate of substitution, nonsynonymous substitutions leading to increased fitness under specific conditions may be fixed before the environment alters, thus leading to a high nonsynonymous-to-synonymous-substitution ratio.

The far-reaching differences in amino acid composition of halophilic proteins as compared to non-halophilic proteins suggest that the change of a non-halophilic protein into a halophilic one and vice versa is not simple to achieve. However, this may not always be the case. In a recent paper, Bergqvist et al. (2002) documented that the TATA binding protein (a general transcription factor found both in *Archaea* and in *Eukarya*) of the marine hyperthermophilic archaeon *Pyrococcus woesei* (an organism not otherwise known for its halophilic properties) optimally binds to DNA at high salt (and high temperature). No more than three mutations of glutamate to alanine residues are required to turn the protein into a non-halophilic one. If such a phenomenon may turn out to be more general, it implies that a halophilic phenotype could be rapidly acquired in evolutionary time.

Analysis of the genome of *Halobacterium* NRC-1 showed that many predicted proteins were similar to those of the Gram-positive bacterium *Bacillus subtilis*, and in addition a large number of unique homologs were found with the radiation-resistant bacterium *Deinococcus radiodurans* (Ng et al., 2000; Kennedy et al., 2001). This suggested that NRC-1 may have acquired a substantial number of genes from certain *Bacteria*. Genes with bacterial character include many of those of the aerobic respiratory chain: ten *nuo* genes, encoding subunits of NADH dehydrogenase, part of the *cox* genes which encode subunits of cytochrome oxidase, and *men* genes for menaquinone biosynthesis. These genes are clustered in several regions of the genome. The majority of the *nuo* genes are most similar to the corresponding genes from the cyanobacterium *Synechocystis* PCC 6803. One possibility is that anaerobic halophiles may have adapted to an oxidizing atmosphere by acquiring the electron transport chain through lateral transfer events from bacterial organisms capable of aerobic respiration. If this was indeed the case, it shows again that adaptation of proteins to function in the presence of extremely high ionic concentrations may be less "difficult" than previously assumed. A second possibility is that a common ancestor of the halophiles and methanogens may have harbored an ancient respiration system similar to that of the bacteria that was lost by the present-day methanogens (Kennedy et al., 2001).

5. Organic Osmotic Solutes in the Domain Archaea

The fact that the halophilic *Archaea* of the order *Halobacteriales* use KCl for osmotic stabilization does not imply that organic solutes are unknown in the archaeal domain. Some alkaliphilic members of the *Halobacteriales* (*Natronomonas pharaonis*, *Natrialba magadii*, *Natronobacterium gregoryi*, *Natronococcus occultus*) contain an organic solute, 2-sulfotrehalose. Intracellular sulfotrehalose concentrations are regulated according to the salt concentration of the medium. *Natronococcus occultus* cells grown in 3.4 M NaCl contain about 1 M sulfotrehalose (Desmarais et al., 1997). K^+ ions neutralize the negative charge of sulfotrehalose (Martin et al., 1999).

It was recently shown that *Halobacterium salinarum* has both a binding protein and a transducer protein that participate in chemotaxis toward and transport of compatible solutes of the betaine family. Moreover, ^{13}C-NMR analysis of extracts of cells grown in the presence of 10 mM betaine showed considerable amounts of betaine to be present intracellularly. L-glutamate was detected in cells grown witout betaine (Kokoeva et al., 2002).

Organic osmotic compounds are accumulated by methanogenic *Archaea* as well. Methanogens of the genus *Methanohalophilus* grow up to 125-175 g l^{-1} salt, and they synthesize or accumulate a range of compatible solutes, including glycine betaine, β-glutamine, β-glutamate, *N*ε-acetyl-β-lysine, and others (Lai and Gunsalus, 1992; Lai et al., 1991; Menaia et al., 1993; Robertson et al., 1992).

6. Osmotic Adaptation in the Major Groups of Halophilic *Bacteria* (*Proteobacteria*, *Firmicutes*)

There have been many attempts in the past to assess the intracellular ionic concentrations of different halophilic and halotolerant members of the domain *Bacteria*. The reports are often confusing, and this is first of all caused by methodological problems inherent to the accurate estimation of intracellular volumes, necessary for the calculation of the true cytoplasmic ion concentrations. Critical reviews of the available data were given by Ventosa et al. (1998) and by Oren (1999b). The results obtained may also strongly depend on the growth conditions and the physiological state of the cells.

A few general trends are clear: (1), Potassium ions are generally accumulated to a few tenths of a molar; K^+ contributes thus relatively little to the osmotic balance. (2), The intracellular Na^+ concentration is generally lower than that that of the medium. All halophiles examined possess mechanisms to extrude Na^+ ions from the cells. Na^+ extrusion is generally based on activity of Na^+/H^+ antiporters, but respiration-driven primary sodium pumps are also encountered in a number of species (Ventosa et al., 1998). (3), The sum of the intracellular $Na^+ + K^+$ is insufficient to balance the osmotic pressure of the medium. The possession of a large excess of acidic amino acids in the bulk protein, such as encountered in the aerobic *Archaea* of the family *Halobacteriaceae*, is not a common feature of most halophilic *Bacteria*. However, in some cases, such as in *Halomonas elongata* (*γ-Proteobacteria*), the excess of acidic over basic amino acids was documented to be somewhat higher than in non-halophiles, while still much lower than in *Halobacterium* and its relatives.

The main osmotically active compounds within the cytoplasm of most aerobic heterotrophic and both aerobic and anaerobic photrophic *Bacteria* are organic compatible solutes rather than inorganic ions. Organic osmotic solutes are also used by eukaryotic halophilic microorganisms. The accumulation of molar concentrations of glycerol within the cytoplasm of the halotolerant unicellular alga *Dunaliella* is a well-known example. Compatible solutes are polar, highly soluble molecules. Most are either uncharged or zwitterionic at the physiological pH. The long list of compounds identified as osmotic solutes in different microorganisms can be divided into a number of categories: (1), polyols such as glycerol, arabitol, mannitol, erythritol; (2), simple sugars (sucrose, trehalose) and heterosides such as glucosylglycerol; (3), trimethylammonium compounds (glycine betaine, glutamate betaine) and dimethylsulfonium compounds (dimethylsulfoniopropionate); (4), amino acids (proline, glutamate, glutamine, and derivatives); (5), *N*-acetylated diamino acids (*N*δ-acetylornithine, *N*ε-acetyllysine); (6), ectoine (1,4,5,6-tetrahydro-2-methyl-4-pyrimidine carboxylic acid) and its β-hydroxy derivative; (7), glutamine amide derivatives (*N*α-carbamoylglutamine amide; *N*α-acetyl-glutaminylglutamine amide).

In spite of the great structural diversity among these compatible solutes, some generalizations can be made relating to their use (Galinski, 1993, 1995): (1), Disaccharides such as sucrose and trehalose have a limited potential as compatible solutes, but they are often found in combination with other osmotic solutes. (2), *De novo* biosynthesis of polyols, while common in algae and in fungi, is seldom observed in halophilic prokaryotes. (3), With the exception of glycine betaine, all nitrogen-containing compatible solutes are derived from the glutamate or the aspartate branch of amino acid biosynthesis. (4), All solutes that are employed at concentrations above 0.5 M are polar, highly soluble molecules which carry no net charge.

Compatible solutes are strong water structure formers and as such they are to a large extent excluded from the hydration shell of proteins (Galinski, 1995). This "preferential exclusion" probably defines their function as effective stabilizers of the hydration shell of proteins (Arakawa and Timasheff, 1985; Timasheff, 1992). According to the calculations of Bolen and coworkers, reduced exposure of the peptide backbone towards the solvent in the presence of compatible solutes specifically stabilizes the protein structure. The osmotic property selected for is the unfavorable interaction between the osmolyte and the peptide backbone ("osmophobic effect") (Bolen and Baskakov, 2001; Liu and Bolen, 1995).

7. Osmotic Adaptation in Anaerobic Fermentative Halophiles (Order *Halanaerobiales*)

The notion that all halophilic *Bacteria* use organic compatible solutes to achieve osmotic balance was first challenged with the discovery of obligately anaerobic, fermentative halophiles. The order *Halanaerobiales*, families *Halanaerobiaceae* and *Halobacteroidaceae*, is phylogenetically related to the low G+C Gram-positive branch of the bacterial domain. However, its representatives display many physiological and biochemical properties previously known only from the aerobic *Archaea* of the order *Halobacteriales*. All attempts to detect significant amounts of organic osmotic solutes in these anaerobes have failed thus far. On the other hand, measurements of intracellular ionic concentrations within the cells of *Halanaerobium praevalens*, *Halanaerobium acetethylicum*, and *Halobacteroides halobius* showed concentrations of K^+, Na^+, and Cl^- to be sufficiently high to allow osmotic equilibrium of the cells' cytoplasm with the hypersaline medium (Oren, 1986; Rengpipat et al., 1988). X-ray microanalysis with the transmission electron microscope has been used to probe the intracellular ionic concentrations of individual cells of *Halanaerobium praevalens*. In exponentially growing cells, K^+ was the major cation. However, in stationary phase cells the ionic content and the K^+/Na^+ ratio of individual cells was highly variable (Oren et al., 1997). The use of inorganic ions rather than organic compatible solutes by the halophilic fermentative *Bacteria* has been explained on the basis of bioenergetic constraints: as little energy is generated in the course of the fermentative dissimilatory pathways, this group apparently has chosen for the energetically less costly accumulation of K^+ and Cl^- rather than for the expensive production of organic solutes (Oren, 1999b).

As may be expected from the intracellular ionic concentrations, the cellular enzymes are functional at high salt concentrations. The enzymes tested (including alcohol dehydrogenase, glyceraldehyde-3-phosphate dehydrogenase, pyruvate dehydrogenase, carbon monoxide dehydrogenase, hydrogenase, and the fatty acid synthetase complex) function well in molar concentrations of KCl or NaCl, and activity is generally poor or abolished altogether in the absence of salt (Oren and Gurevich, 1993; Rengpipat et al., 1988). To achieve such salt-tolerant and salt-dependent behavior, the proteins of at least some representatives of the group (*Halanaerobium praevalens*, *Halobacteroides halobius*) show an excess of acidic over basic amino acids almost as high as found in the aerobic *Archaea* of the order *Halobacteriales* (Oren, 1986). However, the analysis of a partial genome library of the thermophilic *Halothermothrix orenii* (family *Halanaerobiaceae*) did not suggest high levels of acidic amino acids to be present (Mijts and Patel, 2001).

8. *Salinibacter*, an Extremely Halophilic Aerobic Member of the *Bacteria* with an Unusual Physiology

Another case in which a member of the *Bacteria* uses KCl rather than organic osmotic solutes to achieve osmotic balance, while having adapted its intracellular enzymatic machinery to the presence of high salt concentrations, is the recently discovered *Salinibacter ruber*. *Salinibacter* is a red-pigmented rod-shaped aerobic organotroph, isolated from saltern crystallizer ponds in Spain. Its presence in NaCl-saturated saltern ponds was first inferred from 16S rDNA library studies. Bacterial sequences were recovered that belonged to the *Cytophaga/Flavobacterium/ Bacteroides* phylum. Their closest known relative was *Rhodothermus marinus*, a marine thermophilic bacterium from undersea hot springs. Fluorescence *in situ* hybridization showed this phylotype to be abundant: between 5 and 27% of the prokaryotes in the crystallizer ponds along the coast of Spain, on Mallorca and on the Canary Islands belonged to this type (Antón et al., 2000).

Isolates harboring 16S rDNA nearly identical to the above-described phylotype have now been obtained, and the species *Salinibacter ruber* has been formally described (Antón et al., 2002; see also Oren et al., 2002a). *Salinibacter* requires at least 100 g l^{-1} NaCl for growth, and optimal growth is achieved between 150 and 230 g l^{-1}. It is thus one of the most salt-requiring organisms known within the domain *Bacteria*.

A search for intracellular organic osmotic solutes within *Salinibacter* cells, using techniqes such as HPLC and ^{13}C-NMR, showed glutamate, glycine betaine, and *N*α-acetyllysine to be present at low concentrations only. Quantitatively these compounds contribute little to the overall osmotic balance. On the other hand, extremely high K^+ concentrations were detected intracellularly, with K^+/protein ratios being in the same range as in the *Halobacteriales*. X-ray microanalysis in the electron microscope confirmed the presence of high intracellular concentrations of K^+ as well as of Cl^- (Oren et al., 2002b). It is thus suggested that the mode of salt adaptation of *Salinibacter* resembles that of the aerobic halophilic *Archaea*, being based on accumulation of KCl rather than on organic osmotic solutes. In accordance with the high intracellular salt concentration, the bulk protein of *Salinibacter* shows a high excess of acidic over basic amino acids, accompanied by a low content of hydrophobic amino acids (Oren and Mana, 2002). Tests of salt dependence of selected enzymes showed different patterns. Some enzymes (NAD-dependent isocitrate dehydrogenase, the fatty acid synthetase complex) functioned optimally at 0.5-2 M KCl, and showed little activity in the absence of salt. The NADP-dependent isocitrate dehydrogenase activity was largely independent of the KCl or NaCl concentration over the range 0-3 M. NAD-dependent malate dehydrogenase was most active at low salt, but in 3-3.5 M KCl or NaCl about 25% of the optimal rates were measured. There are probably two NAD-dependent glutamate dehydrogenase activities, one that is inhibited by molar salt concentrations, and one that is inactive in the absence of salt and is activated by molar concentations of NaCl (Oren and Mana, 2002).

9. Halophiles and the Origin of Life

A final question to be addressed is whether adaptation to high salt concentrations may be an ancient feature of life. Did the first organisms that inhabited Earth have halophilic properties? There are several arguments in favor of a halophilic origin of life. Dundas (1974, 1998) has extensively speculated on this idea. The theory is based on the Oparin-Haldane hypothesis of the origin of life within a "primordial soup" of abiotically formed organic molecules. When such organic compounds accumulated in seawater, drying-up of the salty water concentrated the salts together with the organic building blocks of life. If indeed the Precambrian oceans were 1.5-2 times as salty as the present-day seas, as has recently been suggested (Knauth, 1998), salt saturation was achieved much more rapidly than at present. Chemical evolution may therefore have proceeded in a supersaturated salty brine under conditions of a very low water activity. Being a halophile may have been advantageous for a hypothetic primordial cell for several reasons. First of all, osmotic sensitivity is highly reduced in salt-saturated brines, as increases and decreases in osmotic pressure of the medium may be effectively buffered by crystallization of excess salt and by dissolution of salt crystals, respectively. Moreover, salt protects DNA from strand breaks (Friefelder and Trumbo, 1969), thus reducing the need for complex highly effective enzymatic repair mechanisms. The fact that halophilic behavior is found in a wide variety of physiological types of organisms may be used as an argument that adaptation to high salt may be an ancient property. We know aerobic halophiles as well as anaerobes, heterotrophs as well as phototrophs and chemoautotrophs.

There is, however, little phylogenetic evidence for a "hypersaline" origin of life. Most halophiles known today are located on distant, relatively "recent" branches of the universal phylogenic (small-subunit rRNA based) tree of life. There are no true halophiles close to the supposed root of the tree. Models based on protein phylogenies and signature sequences (Gupta, 1998) also do not place the halophilic Archaea close to the hypothetic universal ancestor. An even more compelling argument against the possible origin of life in high salt is the variety of strategies used by the present-day halophiles to cope with the presence of high salt concentrations in their medium. These strategies range from the exclusion of salts and production and/or uptake of organic osmotic solutes to the complete adaptation of the intracellular machinery to function at molar concentrations of potassium chloride. These two strategies appear in phylogenetically unrelated branches of both the archaeal and the bacterial domain; the "high-salt-in" option has never yet been identified in any representative of the *Eukarya*. If indeed adaptation to the presence of high salt would have been an ancient characteristic of life, one might expect a much greater uniformity in the ways modern organisms cope with salt in their medium. The great variety of modes used by the present-day halophiles to deal with the high salinity in their environment rather suggests that adaptation to life at high salt concentrations may have been reinvented many times during the more recent evolution of the three domains of life.

10. References

Antón, J., Rosselló-Mora, R., Rodríguez-Valera, F., and Amann, R. 2000. Extremely halophilic *Bacteria* in crystallizer ponds from solar salterns. Appl. Environ. Microbiol. 66: 3052-3057.
Antón, J., Oren, A., Benlloch, S., Rodríguez-Valera, F., Amann, R., and Rosselló-Mora, R. 2002. *Salinibacter ruber* gen. nov., sp. nov., a novel extreme halophilic bacterium from saltern crystallizer ponds. Int. J. Syst. Evol. Microbiol. 52: 485-491.
Arakawa, T., and Timasheff, S.N. 1985. The stabilization of proteins by osmolytes. Biophys. J. 47: 411-414.
Bergqvist, S., Williams, M.A., O'Brien, R., and Ladbury, J.E. 2002. Reversal of halophilicity in a protein-DNA interaction by limited mutation strategy. Structure 10: 629-637.
Bolen, D.W., and Baskakov, I.V. 2001. The osmophobic effect: natural selection of a thermodynamic force in protein folding. J. Mol. Biol. 310: 955-963.
Christian, J.H.B., and Waltho, J.A. 1962. Solute concentrations within cells of halophilic and non-halophilic bacteria. Biochim. Biophys. Acta 65: 506-508.
Dennis, P.P., and Shimmin, L.C. 1997. Evolutionary divergence and salinity-mediated selection in halophilic archaea. Microbiol. Mol. Biol. Rev. 61: 90-104.
Desmarais, D., Jablonski, P.E., Fedarko, N.S., and Roberts, M.F. 1997. 2-Sulfotrehalose, a novel osmolyte in haloalkaliphilic archaea. J. Bacteriol. 179: 3146-3153.
Dundas, I. 1974. Halobacteria. BioSystems 6: 66-67.
Dundas, I. 1998. Was the environment for primordial life hypersaline? Extremophiles 2: 375-377.
Duschl, A., and Wagner, G. 1986. Primary and secondary chloride transport in *Halobacterium halobium*. J. Bacteriol. 168: 548-552.
Ebel, C., Faou, P., Franzetti, B., Kernel, B., Madern, D., Pascu, M., Pfister, C., Richard, S., and Zaccai, G. 1999. Molecular interactions in extreme halophiles – the solvation-stabilization hypothesis for halophilic proteins, In: A. Oren. (ed.) *Microbiology and biogeochemistry of hypersaline environments*. CRC Press, Boca Raton. pp. 227-237
Eisenberg, H., Mevarech, M., and Zaccai, G. 1992. Biochemical, structural, and molecular genetic aspects of halophilism. Adv. Prot. Chem. 43: 1-62.
Friefelder, D., and Trumbo, B. 1969. Matching of simple-strand breaks to form double-strand breaks in DNA. Biopolymers 7: 681-693.
Galinski, E.A. 1993. Compatible solutes of halophilic eubacteria: molecular principles, water-solute interactions, stress protection. Experientia 49: 487-496.
Galinski, E.A. 1995. Osmoadaptation in bacteria. Adv. Microb. Physiol. 37: 273-328.
Gupta, R,S, 1998. Protein phylogenies and signature sequences: a reappraisal of evolutionary relationships among Archaebacteria, Eubacteria, and Eukaryotes. Microbiol. Mol. Biol. Rev. 62: 1435-1491.
Kennedy, S.P., Ng, W.V., Salzberg, S.L., Hood, L., and DasSarma, S. 2001. Understanding the adaptation of *Halobacterium* species NRC-1 to its extreme environment through computational analysis of its genome sequence. Genome Res. 11: 1641-1650.
Knauth, L.P. 1998. Salinity history of the Earth's early ocean. Nature 345: 554-555.
Kokoeva, M.V., Storch, K.-F., Klein, C., and Oesterhelt, D. 2002. A novel mode of sensory transduction in archaea: binding protein-mediated chemotaxis towards osmoprotectants and amino acids. EMBO J. 21: 2312-2322.
Lai, M.-C., and Gunsalus, R.P. 1992. Glycine betaine and potassium ions are the major compatible solutes in the extremely halophilic methanogen *Methanohalophilus* strain Z7302. J. Bacteriol. 174: 7474-7477.
Lai, M., Sowers, K.R., Robertson, D.E., Roberts, M.F., and Gunsalus, R.P. 1991. Distribution of compatible solutes in the halophilic methanogenic archaebacteria. J. Bacteriol. 173: 5352-5358.
Lanyi, J.K. 1974. Salt-dependent properties of proteins from extremely halophilic bacteria. Bacteriol. Rev. 38: 272-290.
Liu, Y., and Bolen, D.W. 1995. The peptide backbone plays a dominant role in protein stabilization by naturally occurring osmolytes. Biochemistry 34: 12884-12891.
Martin, D.D., Ciulla, R.A., and Roberts, M.F. 1999. Osmoadaptation in Archaea. Appl. Environ. Microbiol. 65: 1815-1825.
Mevarech, M., Frolow, F., and Gloss, L.M. 2000. Halophilic enzymes: proteins with a grain of salt. Biophys. Chem. 86: 155-164.
Menaia, J.A.G.F., Duarte, J.C., and Boone, D.R. 1993. Osmotic adaptation of moderately halophilic

methanogenic archaeobacteria, and detection of cytosolic *N,N*-dimethylglycine. Experientia 49: 1047-1054.

Mijts, B.N., and Patel, B.K.C. 2001. Random sequence analysis of genomic DNA of an anaerobic, thermophilic, halophilic bacterium, *Halothermothrix orenii*. Extremophiles 5: 61-69.

Ng, W.V., Kennedy, S.P., Mahairas, G.G., Berquist, B., Pan, M., Shukla, H.D., Lasky, S.R., Baliga, N.S., Thorsson, V., Sbrogna, J., Swartzell, S., Weir, D., Hall, J., Dahl, T.A., Welti, R., Goo, Y.A., Leithauser, B., Keller, K., Cruz, R., Danson, M.J., Hough, D.W., Maddocks, D.G., Jablonski, P.E., Krebs, M.P., Angevine, C.M., Dale, H., Isenberger, T.A., Peck, R.F., Pohlschroder, M., Spudich, J.L., Jong, K.-H., Alam, M., Freitas, T., Hou, S., Daniels, C.J., Dennis, P.P., Omer, A.D., Ebhardt, H., Lowe, T.M., Liang, P., Riley, M., Hood, L., and DasSarma, S. 2000. Genome sequence of *Halobacterium* species NRC-1. Proc. Natl. Acad. Sci. USA 97: 12176-12181.

Ollivier, B., Caumette, P., Garcia, J.-L., and Mah, R.A. 1994. Anaerobic bacteria from hypersaline environments. Microbiol. Rev. 58: 27-38.

Oren, A. 1986. Intracellular salt concentration of the anaerobic halophilic eubacteria *Haloanaerobium praevalens* and *Halobacteroides halobius*. Can. J. Microbiol. 32: 4-9.

Oren, A. 1999a. Life at high salt concentrations, In: M. Dworkin,, S. Falkow, E. Rosenberg,, K-H. Schleifer and E. Stackebrandt (eds.) *The Prokaryotes*. A handbook on the biology of bacteria: ecophysiology, isolation, identification, applications. 3rd. ed. Springer-Verlag, New York (electronic publication).

Oren, A. 1999b. Bioenergetic aspects of halophilism. Microbiol. Mol. Biol. Rev. 63: 334-348.

Oren, A. 2002. Halophilic microorganisms and their environments. Volume 5 of *Cellular Origin, Life in Extreme Habitats and Astrobiology* , J. Seckbach ed.). Kluwer Academic Publishers, 577 pp.

Oren, A., and Gurevich, P. 1993. The fatty acid synthetase of *Haloanaerobium praevalens* is not inhibited by salt. FEMS Microbiol. Lett. 108: 287-290.

Oren, A., and Mana, L. 2002. Salt dependence of selected enzymes of *Salinibacter ruber*, an extremely halophilic bacterium. Extremophiles 6: 217-223.

Oren, A., Heldal, M., and Norland, S. 1997. X-ray microanalysis of intracellular ions in the anaerobic halophilic eubacterium *Haloanaerobium praevalens*. Can. J. Microbiol. 43: 588-592.

Oren, A., Rodríguez-Valera, F., Antón, J., Benlloch, S., Rosselló-Mora, R., Amann, R., Coleman, J., and Russell, N.J. 2002a. Red, extremely halophilic, but not archaeal: the physiology and ecology of *Salinibacter ruber*, a Bacterium isolated from saltern crystallizer ponds, In: A. Ventosa (ed.) *Halophilic microorganisms*. Springer-Verlag, Berlin., in press.

Oren, A., Heldal, M., Norland, S., and Galinski, E.A. 2002b. Intracellular ion and organic solute concentrations of the extremely halophilic bacterium *Salinibacter ruber*. Extremophiles, in press.

Rengpipat, S., Lowe, S.E., and Zeikus, J.G. 1988. Effect of extreme salt concentrations on the physiology and biochemistry of *Halobacteroides acetoethylicus*. J. Bacteriol. 170: 3065-3071.

Reistad, R. 1970. On the composition and nature of the bulk protein of extremely halophilic bacteria. Arch. Mikrobiol. 71: 353-360.

Robertson, D.E., Lai, M.-C., Gunsalus, R.F., and Roberts, M.F. 1992. Composition, variation, and dynamics of major osmotic solutes in *Methanohalophilus* strain FDF1. Appl. Environ. Microbiol. 58: 2438-2443.

Shporer, M., and Civan, M.M. 1977. Pulsed nuclear magnetic resonance study of ^{39}K within halobacteria. J. Membr. Biol. 33: 385-400.

Switzer Blum, J., Stolz, J.F., Oren, A., and Oremland, R.S. 2001. *Selenihalanaerobacter shriftii* gen. Nov., sp. Nov., a halophilic anaerobe from Dead Sea sediments that respires selenate. Arch. Microbiol. 175: 208-219.

Timasheff, S.N. 1992. A physicochemical basis for the selection of osmolytes by nature, In: G. N. Somero, C. B. Osmond and C. L. Bolis (eds.) *Water and life*. Springer-Verlag, New York. pp. 70-86.

Trüper, H.G., Severin, J., Wohlfarth, A., Müller, E., and Galinski, E.A. 1991. Halophily, taxonomy, phylogeny and nomenclature, pp. 3-7 In: Rodriguez-Valera, F. ed.), General and applied aspects of halophilic microorganisms. Plenum Press, New York.

Ventosa, A., Nieto, J.J., and Oren, A. 1998. Biology of moderately halophilic aerobic bacteria. Microbiol. Mol. Biol. Rev. 62: 504-544.

Biodata of **Vera Soina** author (with co-author E.A. Vorobyova) of "***Adaptation of Bacteria to the Terrestrial Permafrost Environment: A Biomodel for Astrobiology***"

Ms. Vera Sergeevna Soina is a post-doctoral senior researcher at the Soil Biology Department of Soil Science Faculty in Moscow State Lomonosov University. She obtained her Master degree in Soil Science and Soil Microbiology (in 1972) at the Moscow State University. She posses the scientific degree of Ph.D. candidate of Biological Sciences. Among her projects are: Formation of resting cells by asporogenous microorganisms as a strategy of long-term survival in the environment; Cytological aspects of structural adaptation of microorganisms from soil and subsoil environments; Microbial ecosystems in Permafrost as a model system for potential preservation of Life on Mars. Ms. Soina teaches undergraduate students in the Soil Biology Department.

E-mail: **soina@yandex.ru**

J. Seckbach (ed.), Origins, 427–444.

ADAPTATION OF BACTERIA TO THE TERRESTRIAL PERMAFROST ENVIRONMENT:
A Biomodel for Astrobiology

V.S. SOINA AND E.A. VOROBYOVA
Soil Science Department, Moscow Lomonosov State University
Vorobyovy Gory, Moscow, 119899, Russia

1. Introduction

Current data indicate that Earth permafrost from polar region is a most static and balanced environment, where microbial communities have survived for at least some millions of years. The diversity of cells and microbial activities after thawing of permafrost sediments are different from those of soils, but similar to those in unfrozen subterranean deposits. The abundance of bacterial biomass is also comparable in frozen and unfrozen sediments (Vorobyova et al., 1997, 2001; Gilichinsky, 2001).

The high degree of preservation of viable microorganisms in subsoil sediments doesn't seem so paradoxical, because soil and deep soil deposits are regarded as heterogeneous natural substrates, which contain organic-mineral complexes, immobilized microbial cells and organic macromolecules. Such a system provides effective protection against various external extreme factors: temperature fluctuations, low nutrient availability, radiation, etc., which could be incompatible with viability of microorganisms (Zvyagintsev and Golimbet, 1983; Zvyagintsev et al., 1985).

In permafrost sediments, chemical and physical properties of organic-mineral complexes are considerably altered in comparison with unfrozen deposits, due to ice formation. The solid phase (ice) makes up 92-97% of the total water volume in permafrost. Such a distinct cryotexture provides a closed system without any water infiltration, creating favorable conditions for long-term cryopreservation of microorganisms. Simultaneously with the solid water phase (ice), 3 to 8 % of the water remains in unfrozen state in permafrost, usually as thin films enveloping ice grains, mineral and organic particles (Gilichinsky, 2001).

It can be anticipated that the high density of the cells in microbial communities, the constant low temperatures, the cryogenic structure of the sediments, the presence of unfrozen thin films of water, and the physico-chemical processes under freezing in sediments (concentration of salt solutions, transfer of liquid moisture and ions, vapor and gases) aid in the preservation of various groups of microorganisms. Thus, Earth permafrost is now considered not as an extreme environment but as a mostly balanced biotope, especially for bacteria. This allows us to consider terrestrial permafrost bacteria as a biomodel for extraterrestrial life on extinct space cryogenic bodies (Soina et al., 1995; Vorobyova et al., 1997; Gilichinsky, 2001). On the whole, Earth permafrost sediments, the age of which is defined as the time of stay at subzero temperatures from several thousands up to a few million years, represent a unique opportunity to investigate the duration of life preservation.

The strategy for bacterial survival under freezing in permafrost sediments must therefore include special mechanisms for adaptation to long-term action of subzero temperatures. Detection of considerable numbers of viable microorganisms in frozen sediments (10^2- 10^7 colony forming units (CFU/g dry weight) has raised the question of the possibility of metabolic activity in such a habitat. According to published data that provide both indirect (Vorobyova and Soina, 1994; Vorobyova et al., 1996; Rivkina et al., 1998; Gilichinsky, 2001) and direct (Rivkina et al., 2000; Gilichinsky, 2001) evidence of slow metabolism in permafrost, we cannot exclude bacterial activity at a low level. The answer in this case could be psychrophily, since constant low temperatures (-7^0C - 20^0C) is one of the cardinal characteristics of the permafrost. However, it seems that psychrophily would not explain the mechanisms of prolonged bacterial survival in permafrost, because true psychrophiles are known to be very sensitive to increased temperature, and they will not survive the thawing of frozen samples that is applied when isolating of viable bacteria from permafrost on nutrient media (Achwood-Smith, 1970; Heckly, 1978; Ray, 1984; Friedmann, 1994).

Studies of Arctic and Antarctic permafrost sediments of different lithology and age showed that the salient features of microbial communities in such cold habitats are the high rate of cell proliferation after thawing and the abundance of psychrotrophic bacteria (Gilichinsky et al., 1993; Soina and Vorobyova, 1994; Vorobyova et al., 1996; 1997; Gilichinsky, 2001). Strict psychrophiles, mesophilic and non-strict thermophilic bacteria were also isolated from frozen sediments, but they were not predominant forms. Psychrotrophs are known to be less sensitive to the elevated above-zero temperatures than strict psychrophiles, and rapidly become active after thawing of permafrost (Nelson and Parkinson, 1978; Friedmann, 1994).

It may be assumed that during freezing of deposits such cold- tolerant forms may transit to a reversible state of low metabolic activity and develop anti-stress mechanisms ensuring rapid restoration of their biological activity after thawing of the sediments. Further endogenous regulatory processes under low temperatures can stimulate the transition of the cells to a hypometabolic state or deep resting state- anabiosis. Such cell differentiation may be characterized by special structural and biochemical properties that can distinguish such forms from the metabolic active cells.

It is well known that a common survival strategy of bacteria in nature is the formation of dormant cells, examples of which are specialized resting forms such as spores and cysts, known only for a limited number of bacterial species. However, most bacteria isolated from permafrost sediments were found to be non-spore-forming (Soina and Vorobyova, 1994; 1996; Vorobyova et al., 1997; 2001; Gilichinsky, 2001), so it is very important to understand how they survive long-term action of low temperatures and ice.

Study of extraterrestrial life needs proper interpretation of possible signs of life that require the creation of comprehensive modeling databases on terrestrial objects. It seems very useful to include in such databases cytomorphological characteristics of intact bacterial cells both *in situ* in thawed samples of permafrost, or cells isolated from permafrost and exposed to various external stress factors, obtained by application of advanced electron microscopy techniques (ESEM, SEM, TEM). Microscopical study of cytomorphological features of permafrost bacteria contributes to our understanding of survival mechanisms under long-term natural cryoconservation, and of differentiation of the cells by variation in physiological status (active, hypometabolic, dormant, or irreversible loss of viability), which can be recognized *in situ*.

2. Application of electron microscopy methods for morphological and ultrastructural study of bacterial cells *in situ*.

Study of fine structure of bacterial cells by method of transmission electron microscopy (TEM) in native samples of frozen sediments immediately after thawing at room temperature showed that bacteria with different types of cell walls (Gram- positive and Gram- negative type) preserved their integrity after long-term exposure to subzero temperatures. These data correlated with the high number of viable cells isolated and counted on nutrient media (Soina and Vorobyova, 1994; Soina et al., 1995; Vorobyova et al., 1997; Vorobyova et al., 2001).

Bacterial cells released and concentrated from subsoil permafrost sediments of various types and age had similar features. The majority of the cells did not reveal any signs of lysis, which could be expected after thawing of the samples. Most of the intact cells were represented by small forms (0.2 - 0.5 μm in diameter), both as single cells or in conglomerates, and contained surface fibrillar capsular layers covered with organomineral particles which are likely to ensure protection against external mechanical destruction by ice (Fig. 1). As seen in Fig. 1, the bacterial cells *in situ* exhibited undamaged cell walls, nucleoids and cytoplasm. The ability to form conglomerates was lost after incubation on nutrient media at room temperature, but the capsular surface layers remained after numerous passages of isolated bacterial strains.

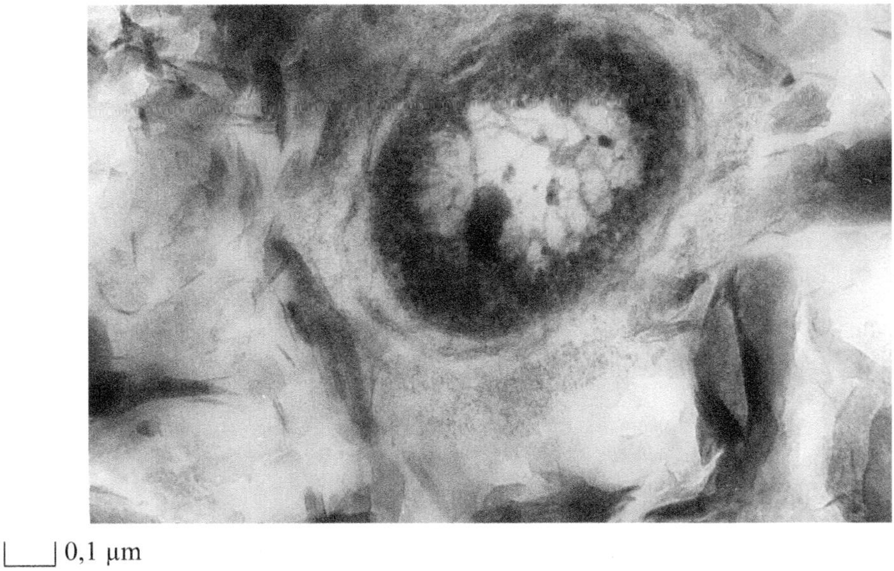

Figure 1. Ultrastructure of "dwarf" cell concentrated from the sample of permafrost (age 0.2 mln years). The cell is surrounded by an external fibrous layer and organomineral particles.

Other cells surrounded by organomineral particles were larger in diameter (up to 0.8-1.0 μm) and revealed capsular layers, thickened cell walls and an electron-dense cytoplasm, resembling dormant forms like cysts (Fig. 2). Some cells were L-shaped and lacked cell walls.

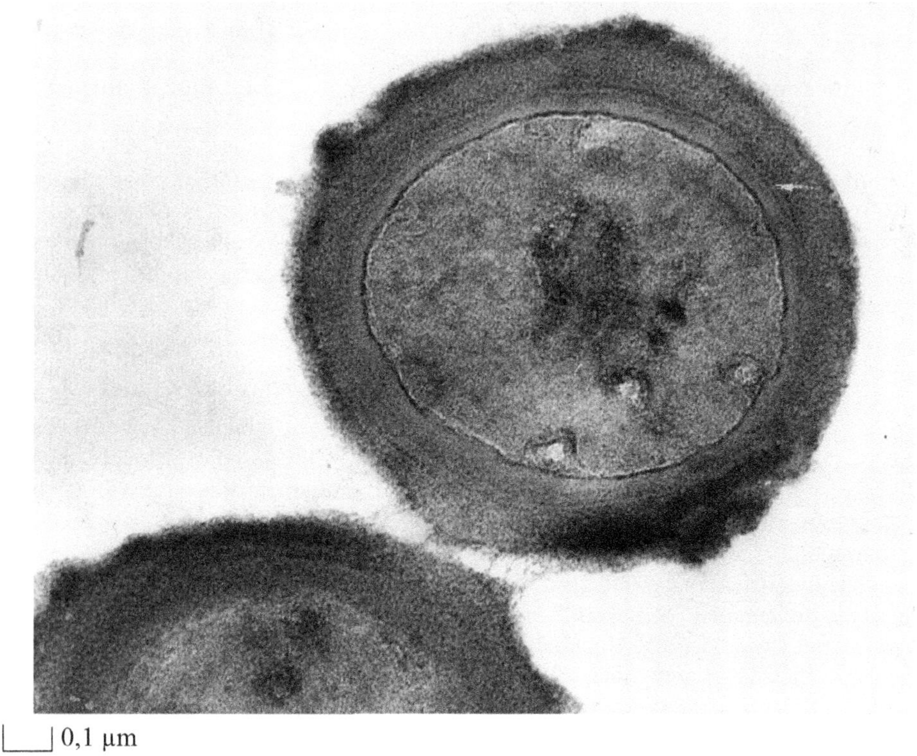

Figure 2. Aggregates of cyst- like cells concentrated from the sample of permafrost sediment (age of permafrost 1.8-2.0 mln years) and viewed in TEM.

TEM studies also revealed a limited number of typical spore-forming cells or mature spores. This fact can be explained by imperfect method for separating and concentrating microbial cells from soil and subsoil layers for direct imaging by TEM, when spores and large cells may be lost to the discarded pellet after exhaustive centrifugal washing procedures (Bae et al., 1972). On the other hand, the ability to form specialized resting forms (spores and cysts) is inherent to a limited number of bacterial genera, therefore the investigated permafrost sediments could initially have been populated mainly by non-spore forming bacteria.

Improved understanding of the environmental properties in permafrost and the structural state of bacterial cells *in situ* in frozen sediments could also be achieved with environmental scanning electron microscopy (ESEM). One advantage in using ESEM, in contrast to conventional SEM, is the ability to directly observe hydrated and dehydrated biological samples without the need of conventional drying and coating. ESEM provides a fast and clear view of microbial morphology and biofilms containing bacteria in relation to the substratum (Litle et al., 1991; Camerone and Donald, 1994). One of the ESEM accessories is the cooling stage that can cool (or heat) a specimen. The cooling stage allows reactions to be slowed for observation. As viewed in ESEM, bacteria frozen in native samples were lying in a distinct gel sheath (biofilm) and presumably were found on the surface of microaggregates of organomineral particles, but not on the ice inclusions (Fig. 3).

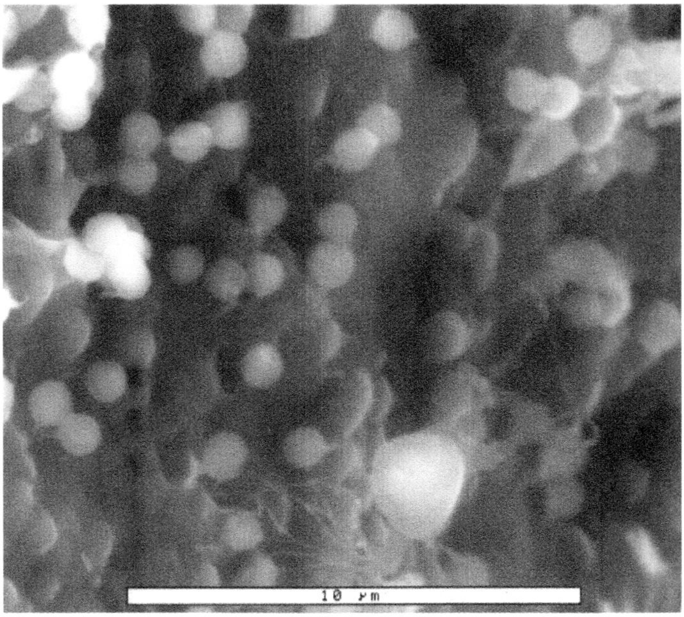

Figure 3. The arrangement of bacterial cells viewed directly in permafrost sediment by ESEM (age of permafrost 1.8-2.0 mln years).

The bacterial cells observed were presumably cocci of small size (0.4 –0.8 µm in diameter). Direct observation of bacterial morphology in frozen samples did not reveal any visible lesions of the cells or especially shrunken forms. Also, no mechanical disruptions, or shrinkage of the cells in the gel were evident in the images in ESEM in the experiments in which native samples of permafrost were frozen at the range of temperatures from -25°C to -50°C. Thus, these results are correlated with data on internal structural stability of bacterial cells after initial thawing of the samples as viewed in thin sections of cells in TEM (Soina and Vorobyova, 1994; Soina et al., 1995). The presence of viable microorganisms in subsoil permafrost give rise to one more view on the microstructure of the cryogenic subterranean environment and its possible role as one of the factor determining preservation of microbial communities.

Formation of cryogenic microstructures starts from freezing of the inter-aggregate pore solution, which increases the concentration of solutes in the pores. Water in intra-aggregate pores freezes at lower temperatures than in the inter-aggregate environment and migrates to centers of ice crystallization. This leads to a decrease in the thickness of the unfrozen water films inside the aggregates and to the strengthening of structural bounds between particles, which stimulates the aggregation process (Ershov et al., 1988). Microbial cells are able to migrate with unfrozen water, and as viewed in ESEM, they presumably are arranged on aggregated organomineral constituents of frozen sediments. Liquid unfrozen water with dissolved components and also water vapor and other gases take part in ion transfer in frozen samples. A significant role of ion transfer is played by a liquid mobile film of water on the surface of ice grains. The negative temperatures as well as the freezing and thawing processes induce gradual changes in the relative number of exchangeable ions, which depend on the duration of freezing. Prolonged freezing leads to an increase in exchangeable Ca^{2+} and Mg^{2+} in frozen sediments (Ostroumov, 1993). The unfrozen water films on the ice surface are too small (7- 10 nm) to be viewed directly in ESEM due to insufficient instrument resolution (~ 20 nm). At the same time the elucidation of the role of unfrozen water in permafrost samples in the preservation of microorganisms or their slow metabolism must undoubtedly be the subject of further investigations.

Investigation of permafrost bacterial communities *in situ* by ESEM and TEM methods suggests that bacterial cells may be protected against the stress of long-term freezing and subsequent thawing by the gel sheath in which they are embedded and by their thick cell walls.

An amorphous organic mass was usually observed in ultrathin sections of soils and unfrozen subsoil sediments (Robert and Chenu, 1992). It is frequently suggested that this mass consists of extracellular polysaccharides - a constituent of major importance in microbial functions in soils and subsoil sediments due to its close binding with organomineral matter and slow degradation in nature. The role of surface capsules around the cells in permafrost seems also important for bacterial survival. In contrast to isolates from unfrozen subsoil sediments and soil, the capsule remains as a structural component around the cells after numerous further passages on nutrient media, suggesting its importance to the survivability in the cold environment. It is plausible to assume that in native permafrost sediments extracellular polysaccharides may also provide preservation of bacterial cells. They serve as possible reservoir of unfrozen water; as a source of reserve nutritional substances necessary for restoration of bacterial activity after thawing; as a specific surface cell structure that promotes better adsorption of the cells on organic-mineral particles; as cryoprotector, preventing mechanical damage of cells by ice. Unfortunately, this concept lacks convincing experimental evidence and needs careful evaluation.

Application of ESEM to direct examination of bacteria in native samples in the frozen and melted state, or in pure cultures confirmed the structural stability of bacterial isolates from permafrost under those conditions. The permafrost temperatures of -7^0C to -20^0C in Arctic and Antarctic sediments fall within the range of temperatures where intracellular ice formation can be observed upon rapid freezing, but the low rates of freezing in such biotopes as frozen subsoil sediments and small size of bacterial cells apparently can prevent intracellular ice formation. According to published data in which differential scanning calorimetry was used for examination of thermal events under freezing and thawing it was suggested that the cytoplasm of bacteria is probably not frozen in the permafrost environment (McGrath et al., 1994). If biological activity can occur in ice-cemented sediments, the presence of intracellular water would undoubtedly be necessary. TEM and ESEM investigations of cell structures of permafrost bacteria subjected to freezing and thawing in model experiments can provide possibilities for study of the mechanism of cellular injury or stability under freezing.

On the whole, the absence of considerable cell structure lesions that could be expected after long-term freezing also suggests that if such lesions occur at all, they are not numerous or lethal, and they can be rapidly repaired during the short-term thawing of the samples prior to the cell fixation necessary for observation in TEM.

3. Survival and cytomorphological characteristics of permafrost bacteria under temperature fluctuations

It is plausible to assume that soil bacteria or bacterial isolates from unfrozen subsoil environments are also able to withstand damage during freezing, due to small size and the presence of polysaccharide capsules, similar to bacteria from permafrost. However, the main difference between these habitats is the possibility that permafrost, through the level of exposure and the exposure time (permafrost age), selects for microorganisms with a higher stability to long term freezing or repeated freezing - thawing cycles, and the readiness to quick restoration of their activity after thawing. As mentioned above the possibility of development of adaptive mechanisms can be provided for by the physical and chemical properties of cryogenic sediments.

It is well known that apart from the specificity of bacterial cells themselves several factors apparently affect the survival of bacteria under extreme cold conditions, such as the freezing and thawing rates, minimal temperatures, and the composition of the suspending medium (MacLeod and Calcott, 1976; McGrath et al., 1994). Additionally, the cells can be adapted to the cold environment by biosynthesis of cold shock and antifreeze proteins, membrane stabilizers, and so called "compatible" solutes (MacLeod and Calcott, 1976; Ray, 1984; Franks et al., 1990; McGrath et al., 1994). For permafrost bacteria it is reasonable to expect that under the conditions in which solutes are concentrated upon freezing in sediments the cells first may have to be adapted by accumulating intracellular "compatible" solutes – osmoprotectants, that serve both as osmo- and cryoprotectors. Bacterial cells can normally accumulate such osmoprotectant solutes. The osmoprotectant solutes can be either constitutive (i.e., normally accumulated by the cells) or inducible (i.e., their production can be induced under water potential stress). Gram-positive bacteria can produce both constitutive and inducible solutes while Gram-negative - presumably inducible solutes (Harris, 1981; Ray, 1984; Robert and Chenu, 1992). These compounds that act as "compatible" solutes in osmotically stressed cells can protect proteins in freeze- thawing processes (Measures, 1975; Carpenter, 1993), and also perform functions such as stabilization of membranes and intracellular structures, prevention of cell volume decrease to the critical state by water removal from the cell, and prevention of freezing of the cytoplasm (McGrath et al., 1994).

The development of a mechanism for accumulation of osmoprotectants under conditions of prolonged freezing in sediments was judged according to the resistance of isolated bacterial strains to increased NaCl concentrations in the nutrient medium. The bacterial strains isolated from frozen sediments and belonging to the genera

Micrococcus, *Arthrobacter*, *Bacillus*, *Exiguobacterium* and *Pseudomonas* displayed growth in liquid TSB medium with 8%, 12% and 20% NaCl. *Arthrobacter* sp. strain 24-13 revealed a higher capacity for growth even in growth medium with 20% NaCl (Fig. 4).

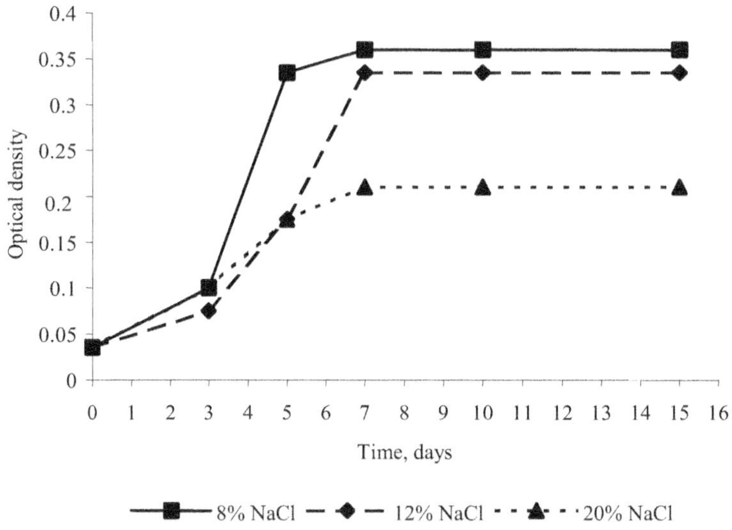

Figure 4. Growth of *Arthrobacter* sp. strain 24-13, isolated from a sample of syngenetically frozen sediments, (age 0.2 mln years) in liquid TSB medium with different content of NaCl.

All other investigated strains revealed weak growth, but preserved their viability at 20% NaCl in the growth medium for one month. On the other hand, it is well known that bacterial strains maintained frozen in culture collections become very sensitive in laboratory experiments to increased content of salts, in particular to NaCl, and also to other chemical agents (Ray, 1984; Sidyakina, 1985). It is possible that the resistance of permafrost bacterial strains to NaCl is explained by their adaptation to the long-term impact of the osmotic stress within frozen sediments.

To study the survival of bacteria in native permafrost sediments under freezing and thawing processes, the samples of two kinds of permafrost sediments and tundra soil were subjected to repeating cycles of freezing at -20^{0}C in a refrigerator and thawing at room temperature (+ 23^{0}C). After 10, 30 and 60 cycles the survivors were allowed to grow on nutrient media and colony-forming units (CFU) were counted to characterize the viability of bacteria following such stress. The results shown in Fig. 5 indicate little change in sensitivity (percentage of viable cells of initial count in unfrozen tundra soil or after a single thawing of frozen sediments) of bacteria subjected to repeated freezing and thawing action, both in permafrost and tundra soil samples.

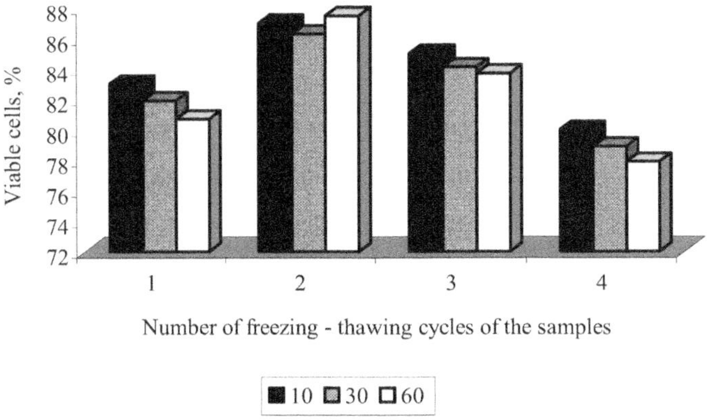

Figure 5. Effect of repeating cycles of freezing and thawing on viability of bacteria in different permafrost sediments and in the sample of tundra soil. (1- sample of epigenetically frozen sediments, age ten thousand years, 2 – sample of syngenetically frozen sediments, age 1.8 – 2.0 mln years, 3 - sample of syngenetically frozen sediments, age 0.2 mln years, 4 – sample of tundra soil)

Nevertheless, the sensitivity of bacterial communities in tundra soil was higher than in the samples of permafrost. Some differences were revealed for the samples of syngenetically and epigenetically frozen layers. In epigenetically frozen sediments where freezing occurred from the top after sedimentation, and bacteria were cryoconserved only once and at a low rate, bacteria were more sensitive to such stress than those from syngenetically frozen sediments (Gilichinsky et al., 1993). In syncryogenic sediments where sedimentation occurred concurrently with freezing from below, bacterial communities underwent freezing and thawing processes before the transition of the sediments to permafrost state (Gilichinsky, 2001) and thus revealed higher stability to repeated cycles of freezing and thawing in a model experiment (Fig. 5, samples # 2, 3). A high percentage of viable bacteria from the initial number preserved in the sample of ancient syncryogenic sediment indicates that long- duration sub zero temperature (1.8- 2.0 mln years) can facilitate the development of anti-stress mechanisms and increase the degree of bacterial stability to such stress. It is interesting to note that the increased number of CFU as seen for sample # 2 (Fig. 5) after 60 cycles of repeated freezing and thawing was attributed to a limited number of bacterial species, able to withstand temperature fluctuations.

TEM and ESEM observations of bacterial cells *in situ* in the sample of this sediment did not reveal any considerable lesions of the cells, and the ultrastructure of the cells was similar to that viewed *in situ* before repeated freezing and thawing. Among the isolates which preserved their viability after 60 cycles of freezing and thawing were spore forming and non-spore forming bacteria with both types of cell wall structure (Gram-positive and Gram-negative). The ultrastructure of five isolates, belonging to the genera *Bacillus*, *Arthrobacter*, *Micrococcus* (Gram-positive type of cell wall), *Pseudomonas* and *Aeromonas* (Gram- negative type of cell wall) was studied after 30 cycles of freezing at -20^{0}C in a refrigerator and following thawing at room temperature in water and soil suspensions without cryoprotectors. The time between each freezing and thawing was 3 days. The viability of bacteria was determined by number of culturable cells (CFU) on TSA medium in percentages of the initial number before exposing to freezing and thawing.

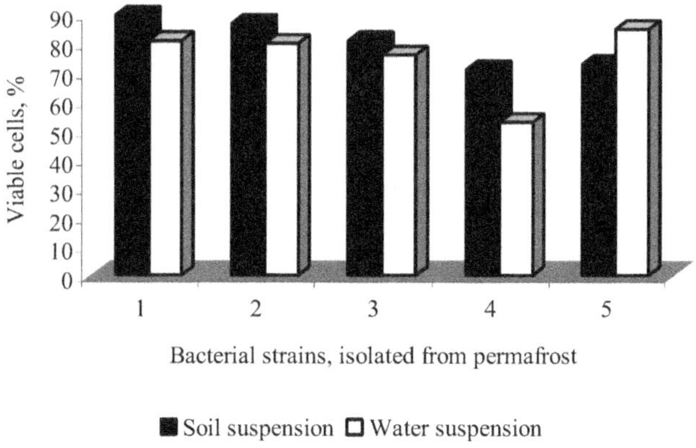

Figure 6. The viability of bacteria isolated from syncryogenic sediment (age of 1.8- 2.0 mln years) after repeating cycles of freezing and thawing (1- *Bacillus* sp. 39- 3, 2- *Arthrobacter* sp. 45-13, 3 - *Micrococcus* sp. 16-2, 4- *Aeromonas* sp. 38-3, 5 - *Pseudomonas* sp. 23-10.

Results as represented in Fig. 6 indicate that both in water and soil suspensions bacteria survived in the native sample of syncryogenic sediment after 60 cycles of freezing and thawing, and revealed a high survival capacity in pure cultures without any additional cryoprotectors: 78-90% in soil suspensions, and 60-87% in water suspensions. Generally the stability in soil suspension was higher than in water suspensions, with the exception of the Gram-negative bacterium *Pseudomonas* sp. 23-10.

The ultrastructure of bacterial cells after repeated cycles of freezing and thawing, as viewed in TEM, also didn't reveal any visible lesions in cell structure. Cells were characterized by a high density of the cytoplasm and stable cell wall structure. Similar studies, but under freezing in liquid nitrogen revealed that soil bacteria with both types of cell walls could successfully survive in water and soil suspensions. Gram- negative *Pseudomonas* spp. demonstrated higher survival (100%) in water than Gram- positive *Arthrobacter* spp. (71%). This experiment allowed only a single thawing after preservation in the frozen state (Sidyakina et al., 1992).

To investigate the ultrastructure of cells in frozen state, the biomass of the same five bacterial strains at the stationary phase of growth and the biomass of a yeast strain also isolated from permafrost for comparison, was exposed to freezing to -90^0C in sealed vials at a controlled rate of -5^0C/min. After 24 hours cells were subjected to freeze-substitution with 2% (wt/vol) osmium tetroxide in anhydrous acetone in sealed vials and held at - 80^0C in an ultra - cold freezer for 72 h (Graham and Beveridge, 1990). Then the vials were allowed to thaw to room temperature and the biomass was prepared by routine method for observation in TEM (Glauert, 1980).

Experiments in which cells were frozen at -90^{0}C with subsequent substitution of ice showed that no obvious internal ice formation occurred in any of the bacteria investigated, and no cavities in the cytoplasm or shrunken and squeezed cytoplasm by ice crystals were observed. The cytoplasm in all Gram-positive bacterial strains seems to be highly concentrated and densely packed with aggregates of ribosome. In some strains osmiophilic globules can be seen concentrated near the center of the cells in thin sections (Fig. 7).

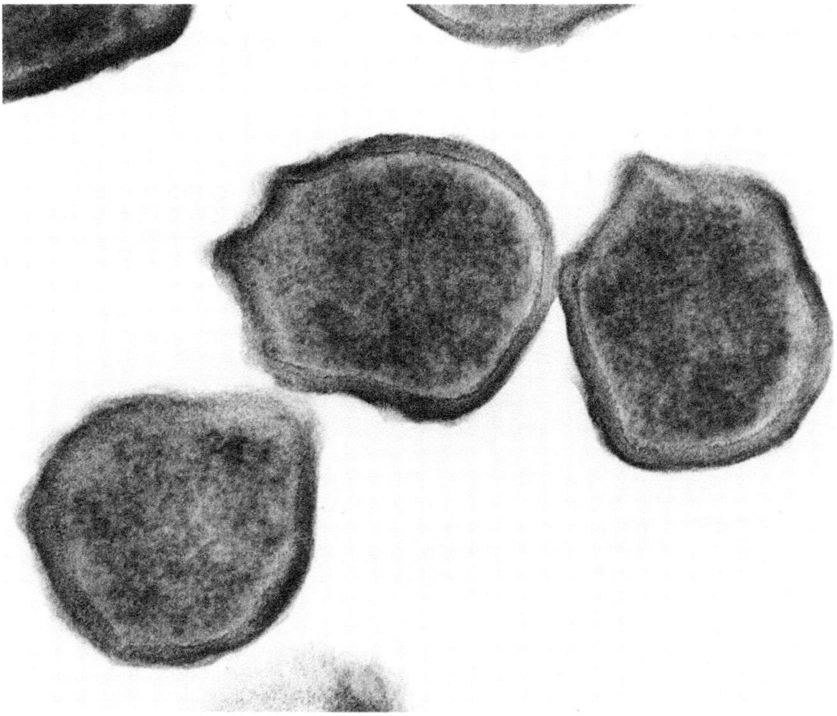

0,1 μm

Figure 7. Ultrastructure of the cells of *Arthrobacter* sp. 24-13 under freezing of the biomass at -90^{0}C in sealed vials and subsequent freeze-substitution with 2% (wt/vol) osmium tetroxide in anhydrous acetone.

In the Gram-negative strain of *Pseudomonas* sp. strain 23-10 the density of cytoplasm was as high as in the Gram-positive strains. Cell walls also seemed to be stable. The behavior of the cell walls in the Gram-negative strain *Aeromonas* sp. 38-3 differed from the others. The cell wall formed stretched structures together with the cytoplasm membrane at several sites. Between stretched sites of walls and cytoplasm, accumulation of thin granular material could be seen (Fig. 8, arrow). It is difficult now to explain such responses of the cell walls, but the possibly influence of osmotic stress can play a certain role. In contrast to the behavior of bacteria, the cells of the yeast strain were considerably shrunken when frozen to -90^{0}C (plasmolysis was not revealed and cell walls shrank together with the protoplast), and large cavities which could be the sites of ice crystals were seen inside the cells.

Investigations of the ultrastructure of permafrost strains in the frozen state using TEM and freeze-substitution methods confirmed previous assumptions: sites of ice formation were not revealed in the cells frozen at -90^{0}C, the cytoplasm seemed to be highly concentrated and was densely packed with osmiophilic globules that can testify to an increased content of amino compounds known to form chemical bonds with osmium tetroxide upon fixation of the cells. Also, amino compounds are regarded as osmoprotec-

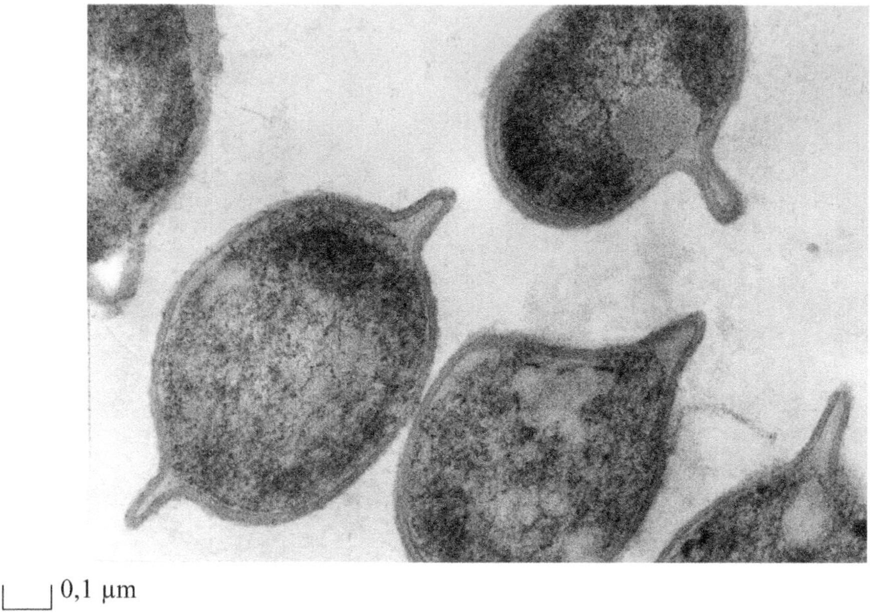

Figure 8. Ultrastructure of the cells of *Aeromonas* sp. 38-3 following freezing of the biomass at -90^{0}C in sealed vials and subsequent freeze-substitution with 2% (wt/vol) osmium tetroxide in anhydrous acetone.

tant compatible solutes (McGrath et al., 1994). Thus, previous suggestions referring to the possibility of accumulation in bacterial cells of "compatible" solutes that can prevent freezing of the cytoplasm, may find confirmation in our investigations of frozen cells in TEM.

As mentioned above, cells of the yeast strain isolated from permafrost were not very stable under freezing, and shrinkage and sites of ice formation were revealed that correspond to the well known behavior of typical yeast strains under freezing (Morris et al., 1988). These results can serve as one more example of the different response of eukaryotes and prokaryotes to freezing. It seems that lack of ice crystal formation in various bacterial cells, together with other biochemical and structural peculiarities which must be the subject of further investigations, give an advantage to prokaryotes in survival under prolonged freezing.

It is generally believed that the stability of cytoplasmic and other membranes is the important factor that permits bacterial survival under extreme cold conditions (Calcott and MacLeod, 1975; MacLeod and Calcott, 1976; Heckly, 1978). The data presented reveal a high viability of permafrost strains after repeating cycles of freezing and thawing, testifying to the highly developed possibilities of membrane repair. It was postulated that Gram- positive bacteria generally survive better after freezing than Gram-negative, which have an external membrane as a component of the cell wall (Sidyakina, 1985). In contrast to typical strains, permafrost bacteria with both types of cell walls displayed a similarly high stability after 30 cycles of freezing and thawing in soil and water suspensions.

The above results do not exclude the presence of bacteria sensitive to thawing processes in permafrost, which give poor growth on rich nutrient media after initial melting of the samples. Any procedures to increase their viability by further plating failed. Other bacteria, which revealed high viability appear to possess considerable resistance to temperature fluctuations. Undoubtedly, populations of soil bacteria, or bacteria from unfrozen subsoil sediments also contain a sub-population of cells stable to cold shock, but according to the results presented here the fraction of damaged cells in such a population under freezing increases as is expressed in the decreasing of number of CFU on nutrient media. On the contrary, the fraction of intact cells of bacterial isolates from permafrost sediments can exceed those of a similar quality of soil bacteria by 10-15%. It seems likely that high survival after repeating cycles of freezing and thawing of permafrost isolates may be regarded as a result of cold adaptation in this unique habitat, but more evidence should be obtained by detailed investigation of possible lesions under thawing in a wide range of permafrost isolates in comparison with the behavior of typical soil strains.

Study of cell differentiation of both Gram-positive and Gram-negative bacterial isolates from permafrost revealed that some of them growing on nutrient media can be very sensitive to increased temperatures (+37^{0}C), as manifested by damage of surface capsular layers, cell walls and even lysis of the cells (Soina and Vorobyova, 1996). However, under low cultivation temperatures (+4^{0}C) the cells as viewed in TEM showed a thickened cell wall (20-28 nm), an electron-dense cytoplasm, and a decrease in size of the cells. Comparison of the ultrastructure of bacterial cells grown under low temperature on nutrient media generally revealed similar cellular structure resembling cyst-like forms when viewed *in situ*, or after freezing and thawing (Demkina et al., 2000; Mulukin et al., 2001).

4. Conclusions

At the present time it is still unknown whether life is preserved on Mars or on other planets. It is suggested that life could be preserved on Mars in the deep subsurface layers at least on the microbial (prokaryote) level. Advances in investigation of extreme environments on Earth, where microbial survival takes place, open great possibilities for subsurface bacteria to be used as models in exobiological studies. The relatively high survival of bacteria in the most ancient permafrost sediments indicate a high structural and biochemical stability of those, so called "primitive" forms. It is now evident that survivors are adapted to prolonged subzero temperatures in another way than psychrophily, because among the isolates cold- tolerant forms were predominant. The cell types of such cold- tolerant forms revealed high structural stability following freezing and thawing, and were ready to restore cell proliferation on nutrient media.

The most surprising fact is the high survival in permafrost of non-spore forming bacteria. The occurrence of non-spore forming bacteria in permafrost can be explained by their entering into a resting state in which they survive starvation and long-term freezing for hundreds of thousands of years at least. The formation of resting (anabiotic) cystlike refractile cells in cultures of non-spore forming bacteria of the genera *Arthrobacter* and *Micrococcus* isolated from Arctic permafrost sediments was revealed under special growth conditions that have been controlled by a specific autoregulatory system including stimulators of autolysis (d_2 factor) and autoinducers of anabiosis (d_1 factor) (Demkina et al., 2000; Mulukin et al., 2001). An imbalance of these factors in favor of the d_1 factor ($d_1 > d_2$) facilitated the formation of cyst- like refractile cells. Cyst-like refractile cells possessed all the features of resting forms: viability, very low or absent metabolic activity, high resistance to unfavorable factors, and a specific ultrastructure, resembling the bacterial cells examined directly in permafrost sediments *in situ*, or in cultures of the isolates stored at +4 °C, in cell suspensions subjected to repeated freezing-thawing stress. Hence, comparative analysis of thin sections as "portraits" of resting cells in long-stored cultures and *in situ* led us to suggest that non-spore-forming bacteria in permafrost sediments can exist in the form of cyst- like cells.

It is interesting to note that bacteria of the genera *Arthrobacter* and *Micrococcus*, isolated from permafrost subsoil, produce greater amounts of the d_1 extracellular factor that favors the transition of cells to anabiosis than closely related typical strains (Mulukin et al., 2001). Therefore, due to the greater capability for synthesizing these factors, permafrost isolates of these non-spore-forming bacteria could rapidly enter the deeply dormant state and form cyst- like cells, which are resistant to unfavorable conditions. Also, the cells of *Micrococcus* sp. permafrost isolate, while growing exponentially with lower rates than typical strains, possessed an elevated resistance to cyanide, a well-known inhibitor of cytochrome *a* oxidase (Vorobyova and Soina, 1994; Mulukin et al., 2001). The constitutively low growth rate of the permafrost micrococci, which was independent of the medium composition, correlated with increased resistance to stresses. Probably it is one of the factors that promote the survival of these and other bacteria under extreme cold conditions. It seems that long term freezing in subsoil sediments provides better preservation of cell structures of bacteria than those from soil and unfrozen sediments and promotes the development of specific physiological peculiarities that increase survival of bacteria in so called "extreme" environment such as permafrost. Thus, permafrost bacteria may be regarded as useful model organisms in different exobiological experiments.

5. Acknowledgments

The authors express their gratitude to Dr. D.A. Gilichinsky (Institute of Physico-chemical and Biological Problems of Soil Science, Russian Academy of Sciences, Russia) and Dr. J. McGrath and Dr. J. Tiedje (Center for Microbial Ecology, Michigan State University, USA) for their cooperation and enabling the possibility to investigate unique permafrost sediments by methods of electron microscopy.

This research was supported by the Russian Foundation for Basic Researches (grant no. 02-04-4985) and in part by grants NSF BIR 912 0006 and NSF INT 9315089.

6. References

Achwood – Smith, M.J. (1970) Effects of low temperatures on microorganisms, plants and cold-blooded animals, In: A.J Smith (ed.) *Current Trends in Cryobiology*. Plenum Press, New York- London, pp. 5-43.

Bae, H.S., Cota-Robles, E., and Casida, E.J. (1972) Microflora of soil as viewed by transmission electron microscopy. Appl. Microbiol. 23, 637-648.

Calcott, P.H. and MacLeod, R.A. (1975) The survival of *Escherichia coli* from freeze-thaw damage: permeability barrier damage and viability. Can. J. Microbiol. 21, 1724-1732.

Camerone, R.E. and Donald, A.M. (1994) Minimizing sample evaporation in the environmental scanning electron microscope. J. Microscopy 173, 227-237.

Carpenter J.F. (1993) Stabilization of proteins during freezing and dehydration: application of lesions from nature. Cryobiology 30, 220-221.

Demkina, E., Soina, V.S., El-Registan, G.I., and Zvyagintsev, D.G. (2000) Reproductive resting forms of *Arthrobacter globiformis*. Mikrobiologiya (Engl. trsl.) 69, 309-313.

Ershov E.D., Lebedenko U.P., Chuvilin, E.M. and Yashin, O.M. (1988) Mechanisms of cryogenic processes in freezing deposits, In: E. Ershov (ed.) *Microstructure of Frozen Rocks*. Moscow State University, Moscow, Russia (in Russian), pp.6-38.

Franks, F., Mathias, S.F., and Hatley, R.V. (1990) Water temperature and life, In: M. Laws and F. Franks (eds.) *Life at Low Temperatures*. Phil. Trans. R. Soc. London, pp.517-534.

Friedmann, I. (1994) Permafrost as microbial habitat, In: D.A.Gilichinsky (ed.) *Viable Microorganisms in Permafrost*. Puschino, Russia, pp. 21-26.

Gilichinsky D.A., Soina V.S., and Petrova M.A. (1993) Cryoprotective properties of water in the Earth cryolithosphere and its role in exobiology. Origins of Life and Evolution of the Biosphere 23, 65-75.

Gilichinsky, D.A. (2001) Permafrost model of extraterrestrial habitat, In: G. Horneck (ed.) *Astrobiology IX*. Springer -Verlag, Heidelberg, pp. 271 -295.

Glauert, A.M. (1980) Fixation, dehydration and embedding of biological specimens, In: A.M.Glauert (ed.), *Practical Methods in Electron microscopy*. North Holland Publishing Co., Amsterdam, pp. 324-350.

Graham L.L., and Beveridge T.J. (1990) Evaluation of freeze- substitution and conventional embedding protocols for routine electron-microscopic processing of eubacteria. J. Bacteriol. 172, 2141-2149.

Harris, R.F. (1981) Effect of water potential on microbial growth and activity, In: J.F. Parr, W.R. Gardner, and L.F.Elliot (eds.), *Water Potential Relations in Soil Microbiology*. Soil Sci. Soc. of America, Madison, Wisconsin, pp. 31-45.

Heckly, R.J. (1978) Preservation of microorganisms. Adv. Appl. Microbiol. 24, 1-53.

Litle, B., Wagner, P., Ray, R., Pope, R., and Scheetz, R. (1991) Biofilms: an ESEM evaluation of artifacts introduced during SEM preparation. J. Industrial Microb. 8, 213-222.

MacLeod, R. and Calcott, P.H. (1976) Cold shock and freezing damage to microbes, In: T.R.Gray (ed.) *The Survival of Vegetative Microbes*. Cambridge Univ. Press, Cambridge, pp. 81-110.

McGrath, J., Wagner, S., and Gilichinsky, D. (1994) Cryobiological studies of ancient microorganisms isolated from the Siberian permafrost, In: D.A.Gilichinsky (ed.) *Viable Microorganisms in Permafrost*. Puschino, Russian Academy of Sciences, Russia, pp. 48-67.

Measures J.C. (1975) Role of amino acids in osmoregulation of non-halophilic bacteria. Nature 257, 398-400.

Morris, G.J., Conlson, G.F., and Clarke, K.J. (1988) Freezing injury on *Saccharomyces cerevisiae*: the effect of growth conditions. Cryobiology 25, 471-482.

Mulukin, A. L., Demkina, E. V., Kozlova, A. N., Soina, V. S., and El-Registan, G. I. (2001) Synthesis of anabiosis autoinducers by non-spore-forming bacteria as a mechanism regulating their activity in soil and subsoil sedimentary rocks. Mikrobiologiya (Engl. trsl.) 70, 535-541.

Nelson, L.M. and Parkinson, D. (1978) Effect of freezing and thawing on survival of three bacteria isolated from an Arctic soil. Can. J. Microbiol. 24, 1468-1474.

Ostroumov, E.V. (1993) Unfrozen water types in frozen soils. Proc. of Joint Russian -American Seminar on Cryopedology and Global Change, Puschino, Russian Academy of Sciences, Russia, pp. 248-256.

Ray, B. (1984) Reversible freeze-injury, In: A. Hurta and A. Nasin (eds.) *Repairable Lesions in Microorganisms*, Academic Press, London, pp. 237-270.

Rivkina, E., Gilichinsky, D., Wagner, S., Tiedje, J., and McGrath, J. (1998) Biogeochemical activity of anaerobic microorganisms from buried permafrost sediments. Geomicrobiology 15, 187-193.

Rivkina, E., Friedman, I., McKay, C., and Gilichinsky, D. (2000) Microbial activity of permafrost bacteria below the freezing point. Appl. Environ. Microbiol. 66, 3230-3234.

Robert, M. and Chenu, C. (1992) Interactions between soil minerals and microorganisms, In: G. Stotzky and J.M. Bolag (eds.) *Soil Biochemistry*. Marcel Dekker Inc., New York, pp. 307-404.

Sidyakina, T.M. (ed.) (1985) *Cryopreservation of Microorganisms*, Puschino, Russian Academy of Science, Russia (in Russian), pp.1-35.

Sidyakina, T.M., Lozitskaya, N.D., Dobrovolskaya, T.G., and Kalakoutskii, L.V. (1992) Cryopreservation of various types of soil bacteria and mixtures thereof. Cryobiology 29, 274-280.

Soina, V.S. and Vorobyova, E.A. (1994) Preservation of microbial cell structures in the permafrost, In: D.A.Gilichinsky (ed.) *Viable Microorganisms in Permafrost*. Pushchino, Russian Academy of Science, Russia, pp. 99-115.

Soina, V.S., Vorobyova, E.V., Zvyagintsev, D.G., and Gilichinsky, D.A. (1995) Preservation of cell structures in permafrost: a model for exobiology. Adv. Space Res. 15, 237-242.

Soina, V.S. and Vorobyova, E.A. (1996) Role of cell differentiation in high tolerance by prokaryotes of long-term preservation in permafrost. Adv. Space Res. 18, 97-101.

Vorobyova, E.A., and Soina, V.S. (1994) Evaluation of the biochemical activity of permafrost deposits during thawing, In: D.A.Giluchinsky (ed.) *Viable Microorganisms in Permafrost*. Pushchino, Russian Academy of Sciences, Russia, pp. 99-115.

Vorobyova, E.A., Soina, V.S., and Mulukin, A.L. (1996) Microorganisms and enzyme activity in permafrost after removing of long- term cold stress. Adv. Space Res. 18, 103-108.

Vorobyova, E., Soina, V., Gorlenko, M., Zalinuva, N., Mamukelashvili, A., Gilichinsky, D., Rivkina, E., and Vishnevetskaya, T. (1997) The deep cold biosphere: facts and hypothesis. FEMS Microbiol. Rev. 20, 277-290.

Vorobyova, E.A., Minkovsky, N., Mamukelashvili, A., Zvyagintsev, D.G., Soina V.S., Polanskaya, L., and Gilichinsky, D. A. (2001) Microorganisms and biomarkers in permafrost, In: R.Paepe and V.Melnikov (eds) *Permafrost Responce on Economic Development, Environmental Security and Natural Resourses*. Kluwer Academic Publishers, Netherlands, pp. 527-541.

Zvyagintsev, D.G., Gilichinsky, D.A., Blagodatsky, S.A., Vorobyova, E.A., Khlebnikova, G.M., Arkhangelov, A.A., and Kudryavtseva, N.N. (1985) Survival time of microorganisms in permanently frozen sedimentary rock and buried soils. Mikrobiologiya 54, 131-136 (in Russian).

Zvyagintsev, D.A., and Golimbet, V.E. (1983) Kinetics of cell number, biomass and productivity of microbial communities in soils. Uspechi mikrobiologii 18, 215-231 (in Russian).

Biodata of **Richard Castenholz** author of ***"Phototrophic Bacteria Under UV Stress"***

Dr. Richard W. Castenholz is a professor emeritus in the Center for Ecology and Evolutionary Biology at the University of Oregon. He received his B.S. in Botany at the University of Michigan (1952) and his Ph.D. in Botany at Washington State University (1957). He has been a faculty member at the University of Oregon ever since. His early research was on the ecology of both freshwater and marine epilithic diatoms, but in the late 1960s he switched almost entirely to the study of cyanobacteria and anoxygenic phototrophic bacteria of hot spring mats. During this period *Chloroflexus* and *Heliothrix* were first described by Pierson and Castenholz (1974) and Pierson et al. (1985), respectively. Later this developed into a study of phototrophic prokaryotes in microbial mats of other freshwater and marine habitats, including mats dominated by cyanobacteria in Antarctica. The research questions he has asked include: how do various microorganisms tolerate and adapt to environmental extremes? These extremes include high and low temperature, low pH, high salinity, desiccation, toxic levels of sulfide, and high solar irradiance (particularly UV radiation).

In these studies he and colleagues have characterized scytonemin, a UV-sunscreen pigment in sheaths surrounding many highly exposed cyanobacteria, and have shown its role in increasing fitness under UV stress. They have also shown a vital escape response of motile cyanobacteria to UV radiation in the context of soft microbial mats, both in temperate and polar environments.

Throughout the last 20 years, Dr. Castenholz has also been trying (with others) to unravel the extremely disordered taxonomy of cyanobacteria, and consequently was heavily involved as an author and co-editor in the recent edition of the Bergey's Manual of Systematic Bacteriology (2001). The Culture Collection of Microorganisms from Extreme Environments (CCMEE) is also an important asset of his lab here at Oregon (http://cultures.uoregon.edu).

E-mail: **rcasten@darkwing.uoregon.edu**

J. Seckbach (ed.), Origins, 445–461.

PHOTOTROPHIC BACTERIA UNDER UV STRESS

RICHARD W. CASTENHOLZ
Center for Ecology and Evolutionary Biology
University of Oregon, Eugene, OR 97403, USA

1. Introduction

Ultraviolet radiation (UVR) that is relevant to reception by Earth's inner atmosphere, past and present, is arbitrarily classified into three regions [i.e. 400-320 (or 315) nm = UVA; 320 (or 315)-280 nm = UVB; 280-~190 nm = UVC].

There have been numerous reviews in recent years on the effects of ultraviolet radiation (UVR) on microorganisms (e.g. Castenholz and Garcia-Pichel 2000, Vincent and Neale 2000, Whitehead et al. 2000, Roy 2000, Häder 2001, Cockell 2001, Day and Neale 2002, Shick and Dunlap 2002), so this addition may seem superfluous. It will be brief and concise, and limited mainly to photosynthetic prokaryotes. It is well known that even normal environmental exposures to UVR in habitats in which the affected organisms are growing may cause damage that can often be rectified by one or more repair mechanisms if the organisms are metabolically active. Even microorganisms that produce UV protective compounds, such as mycosporine-like amino acids (MAAs) and the shielding pigment, scytonemin, demonstrate temporary damage at least during mid-day levels of summer solar radiation when active, but perhaps less so when dormant (Brenowitz and Castenholz 1997, Miller et al. 1998, Dillon et al. 2003).

With respect to origins of early life it is very likely that anoxygenic phototrophic *Bacteria* with a single photosystem preceded the oxygenic *Bacteria* with two photosystems now exemplified by the cyanobacteria. However, in the Archean or early Proterozoic Eons when the prevailing phototrophic life forms presumably evolved, the exposure to solar irradiance, a necessity for phototrophic growth, also carried the consequences of exposure to a significantly higher level of UV radiation as a result of O_2 being absent or extremely low (see Pierson 1994, Garcia-Pichel 1998). Without some form of protection or escape mechanism, mutation rate would probably have accelerated, but it is unlikely that benefits would accrue instead of deleterious effects. Although somewhat controversial until recently, it now seems certain that there was a near absence of atmospheric O_2 in the late Archean (or earlier than 2.4 x 10^9 yr ago) (see Wiechert 2002). UVC (~190-280 nm) was most certainly relevant during the Archean and early Proterozoic Eons, although it cannot presently be detected at sea level or even at high terrestrial elevations, since it is absorbed by O_2/O_3 in the stratosphere. During the Archean Eon (3.8-2.5 x 10^9 yr ago) an atmospheric O_2 level of 10^{-5} PAL (Present Atmospheric Level), which is likely according to Kasting (1987), would not be sufficient to produce ozone or attenuate UVC or UVB. It is possible, however, that sulfur vapor composed of S_8 and other elemental sulfur molecules may have provided an effective UV screen in a warm primitive anoxic atmosphere (Kasting et al. 1989). If no such screen was present, UVC levels of 1-4 W m^{-2} over the UVC range may have reached the Earth's surface during the Archean and early Proterozoic (Kasting 1987, Cockell 2001). Although daylength was shorter at intermediate latitudes at that time (e.g. 14 h maximum) because of a more rapid Earth rotation, the estimated effects of UVC and UVB were probably still much greater than at present (Cockell 1998, 2001). Estimates of the relative damage of incoming UVA, UVB, and UVC to a selected cyanobacterium during the Archean, Proterozoic, and Phanerozoic are shown in Fig. 6 of Garcia-Pichel (1998). Cockell (1998) has also discussed the likely biological effects of UVR on early Earth. In today's world, although UV radiation reaching Earth's surface is less due to O_2 and ozone absorption in the stratosphere, in aerobic habitats it may result in serious cell damage through the production of reactive forms of oxygen.

This presentation, then, will merely attempt to summarize the various UV tolerance and avoidance strategies that are currently utilized by phototrophic bacteria. There is at least the possibility that some of these same strategies may have evolved during the early Precambrian when the UVC and UVB fluxes and diurnal doses reaching the Earth's surface were greater than present (see Cockell 2001). Passive "strategies", such as burial at shallow depths in sediments, or sacrificial death of surface mat layers that protect lower layers of microorganisms, or the possible concentrations of ferrous iron (Fe II) in anoxic waters that would reduce UVR exposure, or simply growing at depth in water, will not be discussed here. It is certain that phototrophic microorganisms evolved their own biochemical and behavioral methods of UVR "tolerance" and "escape", and it is these that will be considered.

2. The Negative Effects of UVR and Physiological and Biochemical Strategies of Counterbalance

For a concise listing of negative UV effects on various aspects of metabolism, DNA stability, behavior, and development in cyanobacteria see Table 1 in Castenholz and Garcia-Pichel (2000) and Häder (2001) and Jagger (1985) for other organisms. Although the photon fluence rate at sea level currently grades from high UVA at 400 nm to nil in the UVC range, the detrimental effects per photon mount exponentially with shortening wavelengths (Jagger 1985, Whitehead et al. 2000). Because of the relatively more lethal effects of highly energetic UVC radiation, and because of the reception of this spectral region on Earth only in the Archean and early Proterozoic, it will be discussed first.

2.1. UVC

UVC emitting lamps (germicidal) have been used in many past experiments with bacteria, viruses, DNA and proteins (see Jagger 1985). Most damage is through direct absorption by DNA (absorption maximum ~260 nm). Among photosynthetic microorganisms, one of earliest examinations of the effect of UV radiation was of a coccoid cyanobacterium by Van Baalen (1968). In this he showed that UVC (254 nm max) negatively affected short-term photosynthetic rate and survival in *Agmenellum quadruplicatum* (now *Synechococcus* sp. PCC 73109), but that the DNA damage could be cured by photoreactivation at wavelengths from ~395-450 nm (Van Baalen and O'Donnell 1972).

One of the most notable microorganisms with respect to tolerance of high UVC radiation is *Chloroflexus aurantiacus*, the photoheterotrophic, anoxygenic bacterium that was first isolated and described by Pierson and Castenholz (1974). Values of continuous UVC radiation within the possible Archean limits mentioned above (0.01 W m^{-2}) allowed growth rates similar to those of controls under anoxic conditions (Pierson et al. 1993). *Escherichia coli*, however, was strongly inhibited by this intensity. Growth was even measurable in the presence of UVC irradiance of over 0.6 W m^{-2}, although the yield was less than about 10% of that of the controls. Oxic conditions decreased the yield further but only in the presence of UVC. The growth of *Chloroflexus* continued for over 50 h at intensities of over 0.02 W m^{-2}. Thus, this strain has very high UVC tolerance, but it is not presumed that this could occur indefinitely without DNA damage and deleterious accumulation of mutations (Pierson et al. 1993). Therefore, external mechanisms must be invoked for phototrophic life, utilizing solar irradiance but excluding high levels of deleterious UVR. It is hypothesized that shallow waters and sediments during the Archean or early Proterozoic may have carried a load of ferrous or ferric iron which have major UVR absorption capacities, but still allow passage of wavelengths required for photosynthesis. Therefore, this external element (together with intrinsic repair capability or vertical migratory escape behavior in microbial mats) may have allowed the continual growth and livelihood of this type of organism during the eras of higher UVC and UVB. Pierson et al. (1993) hypothesized, therefore, that UVR was not likely a significant deterrent to the success or unhampered evolution of early phototrophic bacteria on Earth.

Cyanobacteria that synthesize sheaths and inhabit exposed, near-surface habitats usually also synthesize the UVR-absorbing pigment, scytonemin, which has major absorption maxima in the UVC and UVA spectral regions, but also substantial absorption in the UVB (Proteau et al. 1993) (Fig. 1 and 2). The major absorption in the UVC is perhaps an intrinsic property of scytonemin that was of benefit during the early Precambrian and was retained as an innate property of the unique structure of this low molecular weight compound (m.w. 544) (Proteau et al. 1993). O_2 is required for scytonemin synthesis, but the internal production of O_2 by cyanobacterial ancestors presumably occurred long before O_2 accumulation in the waters and atmosphere. Although at present scytonemin functions primarily as a screen and protection from UVA radiation, Dillon and Castenholz (1999) found that it afforded considerable protection against UVC damage to photosynthesis in scytonemin-containing *Calothrix* and *Chroococcidiopsis* cultures when 0.5-1.0 W m^{-2} UVC radiation was added artificially to natural outdoor solar irradiance. Although many cellular compounds absorb in the UVC region, the presence of scytonemin in the external shield provided protection that the glycan sheath without scytonemin did not.

2. 2. UVB

UVB causes damage that is similar to that caused by UVC, is less potent, but more extensively studied. Therefore, these effects are discussed here rather than under the UVC section above. UVB includes the wavelengths of most concern to humans, and the apparent recent increase on Earth is due to periodic and regional stratospheric losses of ozone. UVB (280-315/320 nm) reaches the Earth's surface in strengths that cause substantial damage to many cellular components, including DNA as a direct target forming various photoproducts (e.g. mainly cyclobutane - pyrimidine dimers, but

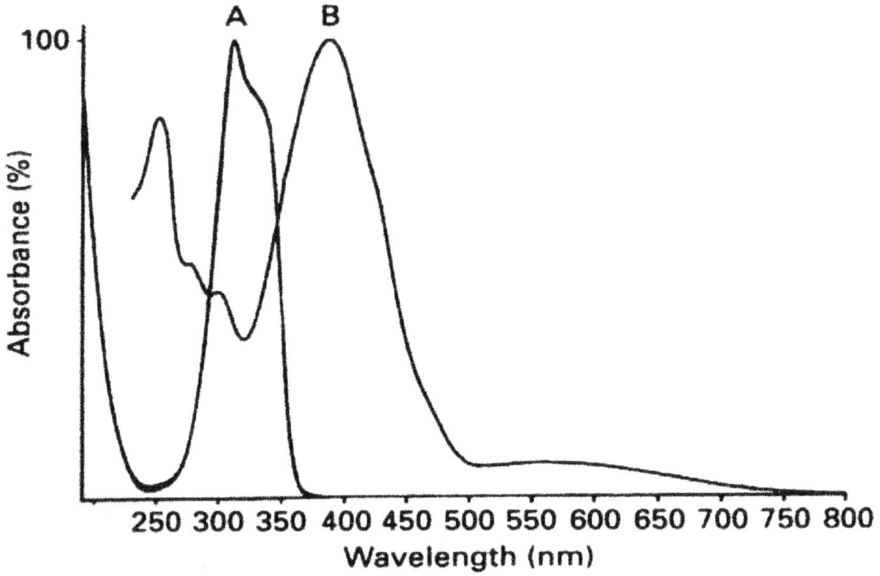

Figure 1. The absorption spectrum of oligo- saccharide-MAA in H_2O (A) and scytonemin, (B) in tetrahydro-furan (by permission of S. Scherer ; Ehling-Schulz and Scherer 1999).

also pyrimidine (6-4) pyrimidone products, and cytosine products) (Ravanat et al. 2001) Purine bases are also photoreactive. Other impairments include accelerated degradati on of photosystem II proteins (e.g. D1) and destruction of the light-harvesting phycobiliproteins, since these have an absorbance peak in the UVB in addition to the visible region (Lao and Glazer 1996, Wingard et al. 1997). Also, various activities such as synthesis of chlorophyll, energy transfer in light harvesting, nitrogen fixation, RUBISCO and ATP synthase activities, nitrate and ammonium uptake, nitrogen fixation, motility and photoorientation, and cell differentiation are negatively affected by UVB radiation (e.g. Häder 1984, Sass et al. 1997, Choi et al. 1999, Castenholz and Garcia-Pichel, 2000). Effects may be more severe when cells are under the stress of low nutrient availability. In photosynthetic microorganisms (especially cyanobacteria) UVB damage to light-harvesting complexes may exceed damage to DNA (Lao and Glazer 1996). D1/D2 proteins of PS II reaction center may be destroyed not only directly by UVR but also indirectly by reactive oxygen produced by high intensity violet/blue light or UV radiation. De novo synthesis of D1/D2 proteins is an essential part of the repair process (Sass et al. 1997). Proteins in general may be photooxidized by UVB radiation since tryptophan, tyrosine, phenylalanine and histidine absorb in the 290-315 nm range (MacDonald et al. 2003). In one case, photooxidative inhibition and damage by

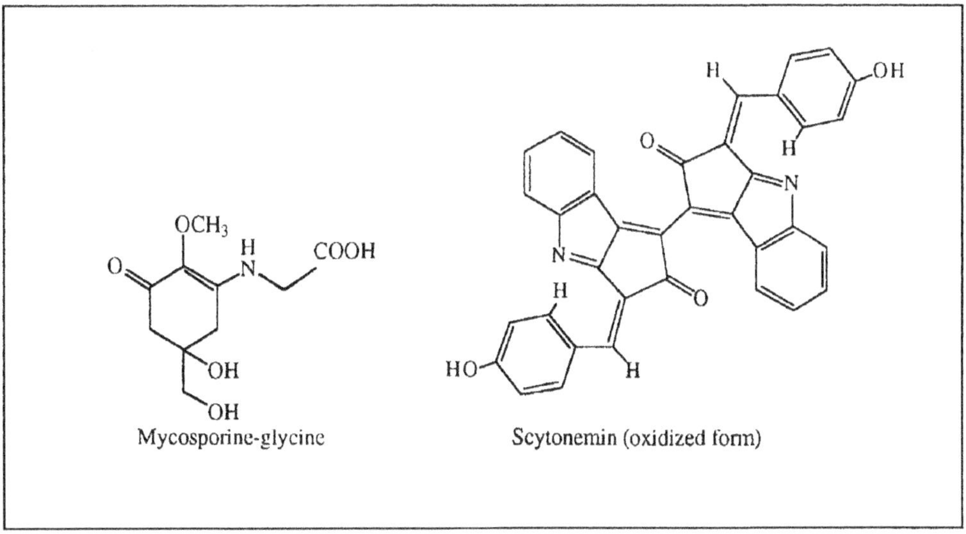

Figure 2. The chemical structure of one type of MAA and scytonemin in the oxidized form as in living cyanobacteria (from Castenholz and Garcia-Pichel 2000).

UVB was reversed after about two weeks of continuous exposure in a species of *Anabaena* (He et al. 2002). MacDonald et al. (2003) found that a naturally high ratio of visible light to UVB resulted in a greater tolerance of UVB without short-term inhibition of PS II or growth in *Synechococcus* PCC 7942. The exposure of the same unicellular cyanobacterium (*Anacystis nidulans* R-2 = *Synechococcus* PC 7942) to UVB and UVA resulted in the production of several "UV-shock" proteins, some of which may function as enzymes that scavenge reactive oxygen molecules (Shibata et al. 1991). Another study, again using *Synechococcus* PCC 7942, has demonstrated that an ATP-dependent Clp protease is essential for acclimation to UVB, and results in the rise of a modified D1 PS II protein (D1:2) that is more resistant to UVB stress (Porankiewicz et al. 1998). The *psbA* genes (*psbA2 and psbA3*) that code for a more UVB tolerant, PS II D1 protein may be activated in *Synechocystis* PCC 6803 under even low levels of UVB exposure (Máté et al. 1998, Campbell et al. 1998). With whole genome profiling of some phototrophic prokaryotes now possible, transcripts from genes involved in various light (or UVR) intensity transitions can be probed with full genome DNA microarrays (see Gill et al. 2002).

UVB affects shallow benthic phototrophs, e.g. intertidal flats, shallow lagoons and lakes, also photoplankton, especially in clear oceanic waters where harmful intensities may reach 20-30 m (much less in turbid or waters rich in chromophoric organic matter), as well as exposed terrestrial crusts and epi- and endo-lithic pellicles of cyanobacteria or microalgae. There is a particularly large literature on the effects of UVR (mainly UVB) on phyto- and cyano-plankton which includes diatoms, dinoflagellates, and nano- and pico-plankton composed of small eukaryotes and the well described *Synechococcus* and *Prochlorococcus* species of the cyanobacteria. Effects of UVB in lakes and the sea are particularly difficult to assess because of vertical and horizontal mixing and, thus, like other field studies requires extrapolation of short term daylight results to full days, weeks and seasons (see Day and Neale 2002). Nevertheless, work in the Southern Ocean by various investigators have estimated that photoplankton productivity has been reduced by ~0.2 to 2-4% as a result of increased UVB due to periodic ozone depletion. However, estimates of this type differ greatly (see Day and Neale 2002 for references). As with UVC, some of the negative effects on DNA (formation of cyclobutane pyrimidine dimers and 6-4 photoproducts which block DNA and RNA polymerases) may be repaired by photoreactivation, utilizing UVA and/or violet/blue wavelengths that are absorbed by chromophores of photolyase, a DNA-bound enzyme that restores the intact bases, and also by nucleotide excision repair (Jagger 1985, Sancar 1994, 1996). Increases in temperature have been shown to enhance the DNA photoactivation repair rates in algae (Pakker et al. 2000).

Although scytonemin has substantial absorbance in the UVB spectral region, the protective proficiency against UVB has not been tested. Mycosporine-like amino acid derivatives (MAAs) are known, not only in cyanobacteria, but also in numerous microalgae and marine organisms (Shick and Dunlap 2002, Castenholz and Garcia-Pichel 2000). In invertebrates, MAAs are acquired secondarily. MAAs are low molecular weight, water soluble compounds that are condensation derivatives of a cyclohexenone ring and amino acid (or imino alcohol residues) (Fig. 2). Although different species of MAAs absorb maximally at various wavelengths ranging from 310 to 360 nm (i.e. from near UVB into the UVA region; Fig. 1) there is no certainty that they are important protective compounds in most cyanobacteria. They are synthesized primarily when exposed to UVB wavelengths. However, in all but one of the cyanobacteria examined, these compounds are localized in the cytoplasm and at best could serve as alternative targets for perhaps only about 10-30% of UVB photons penetrating a rather large unicell (Garcia-Pichel and Castenholz 1993, Garcia-Pichel et al. 1993). In cells of small diameter (< 10 μm), the beneficial effect would be negligible (Garcia-Pichel 1994). Nevertheless, the tight association of MAA synthesis with UVB exposure in many microorganisms strongly suggests that MAAs are important in some way in UVB protection. Perhaps, though, MAAs are a product of UVB exposure with another, unknown function (see discussion in Shick and Dunlap 2002). In any case, it is apparent that intracellular MAAs cannot provide complete protection from UVB/A radiation. Species differences are quite apparent. For example, even though *Oscillatoria priestleyi* apparently contained some MAA, it was more negatively affected in culture by UVB than its compatriot species in Antarctic ponds, *Phormidium murrayi;* the latter (without MAAs) was more tolerant of UVB (Quesada and Vincent 1997). However, the *Oscillatoria* species was one that escapes from surface irradiance by vertical movements into microbial mats (see later under behavioral responses and Nadeau et al. 1999), whereas the less motile *Phormidium* does not. In terrestrial *Nostoc commune*, however, MAAs are bound to oligosaccharides in the inner external glycan sheath (glycocalyx), with scytonemin in the outer portion (Ehling-Schulz et al. 1997, Ehling-Schulz and Scherer 1999). In the case of the *Nostoc*, then, there is no doubt that MAAs, together with scytonemin, could provide a nearly perfect protective shield against UVB as well as UVA. Although absorbed UVR energy must also be dissipated, this poses no cellular problem when the compound is located in the extracellular sheath. However, within the cytoplasm absorption of UVR could result in transfer of energy to sensitive molecules. Many cyanobacteria synthesize one or more MAAs with or without an extracellular sheath that contains scytonemin. However, essentially all known sheathed scytonemin producers are capable of making MAAs either constitutively or by UVB induction (Garcia-Pichel and Castenholz 1993).

The induction of various MAAs is mainly through the absorption of UVB radiation (Ehling-Schulz et al. 1997, Sinha et al. 2001). However, Portwich and Garcia-Pichel (1999) have shown that either UVB or osmotic stress could induce MAA synthesis in a strain of *Chlorogloeopsis.* They also presented evidence that the photoreceptor involved in the induction of MAA synthesis is a pterin with a distinct absorption peak at 310 nm (Portwich and Garcia-Pichel 2000).

Shick and Dunlap (2002) suggest that gadusols, which are intermediates in the shikimate pathway of MAA synthesis, may have been early absorbers of UVC/B radiation and may also have served as strong antioxidants. Although high levels of MAAs accumulate in cyanobacterial cells of hypersaline environments (Oren 1997), the role of MAAs outside of these habitats is not osmoprotection.

MAAs and scytonemin are both absent in many cyanobacteria [(e.g. *Mastigocladus* (=*Fischerella*), *Synechococcus* spp. and many *Leptolyngbya* (*Phormidium*) types from hot springs, vertically migrating types in soft microbial mats, and many others in the Culture Collection of Microorganisms from Extreme Environments at the University of Oregon] (Castenholz and Garcia-Pichel 2000, and unpublished data). It is these that must have other means of tolerating UVR or avoiding it. In the case of exposure to full solar irradiance without escape possibilities, it is apparent that only efficient repair mechanisms can make up for damage done during daylight hours of clear skies with sun elevation near the zenith (Miller et al. 1998). Most repair may take place during darkness or during periods of low light as in morning or before dusk. Cyanobacteria are known to have a very efficient daytime photoreactivation system as compared to *Escherichia coli* (Castenholz and Garcia-Pichel 2000). There is, however, little known of UVB effects and tolerance capability outside of the laboratory, with the exception of many papers on oceanic photoplankton, both cyanobacterial and eukaryotic. Protection by accumulated MAAs has been shown clearly in marine dinoflagellates, but the exact nature of the protection is unclear, and it is certainly incomplete (see Shick and Dunlap 2002)

2. 3. UVA

The detrimental effects of UVA may be both direct and indirect. Nucleotides, such as NADH or NADPH (essential components of anoxygenic and oxygenic phototrophs), with absorption maxima at ~340 nm, would be examples of primary targets of UVA radiation. DNA and other targets of UVB/C are also affected to varying degrees by UVA, but not as much as by UVB. UVA radiation affects many phenomena negatively, but the exact mode is not known in most cases. UVA photooxidation mainly involves the guanine components of DNA as a result of singlet oxygen generation (Ravanat et al. 2001), in which case a photosensitizing compound must be involved. Photosynthesis, growth rate and chlorophyll synthesis are all affected negatively. However, the main effects may be through the production of reactive oxygen species that are also produced through high intensity violet/blue light. The main products are singlet oxygen ($^1O_2$), and peroxy radicals, and free radical reactions (e.g. the production of OH•). Unsaturated lipids in cell membranes (including phospholipids) are targets of oxidative damage induced by UVA and visible light (Girotti 2001). However, the importance of UVA relative to UVB remains unevaluated. It should be remembered, however, that UVA has an important positive value in photoreactivation of UV-damaged DNA.

UVA (with UVB excluded) may result in very dramatic decreases in photosynthetic rate in various cyanobacteria, such as thermophilic *Synechococcus* (Miller et al. 1998), marine *Oscillatoria* and *Spirulina* (Kruschel and Castenholz 1998), Antarctic *Oscillatoria* (Nadeau et al. 1999), and thermophilic *Fischerella* (unpublished data) in outdoor experiments in which the effects of visible irradiance alone, vis + UVA, and vis +UVA and B were compared. The specific nature or explanation of this UVA inhibition has not been investigated.

There are, however, some very effective ways that photosynthetic prokaryotes may prevent or alleviate the various detrimental effects of UVA (320-400 nm) and shorter wavelength visible radiation (400-~500 nm). A high carotenoid content may act as an effective antioxidant or "quencher" of reactive oxygen, inhibit lipid peroxidation and stabilize membranes (Britton 1995, Niyogi 1999). Photons in excess of those that can be used in photosynthesis are nevertheless absorbed by tetrapyrroles (e.g. chlorophylls and phycobilins) and will result at least in triplet chlorophyll (^{3}Chl) and associated $^1O_2$ production in phototrophs in oxic environments. Specific carotenoids (e.g. myxoxanthophyll, echininone, zeaxanthin) may be particularly effective quenchers through the thermal dissipation of excitation energy of chlorophyll and singlet oxygen. It is probable that a high ratio of effective carotenoids to photosensitizer (i.e. tetrapyrrole) may be extremely important, particularly at suboptimal temperatures when repair and synthetic processes are slow (see Castenholz and Garcia-Pichel 2000). The protective xanthophyll cycle apparently occurs in eukaryotes *in vivo*, but not in prokaryotic phototrophs (Josue and Frank 2002). However, some of the same carotenoids that are thought to be protective in eukaryotes are also part of the cyanobacterial complement (e.g. zeaxanthin). This xanthophyll alone may be sufficient for photoprotection (Niyogi 1999). One of the most conspicuous aspects of high light intensity regimes in natural communities of photosynthetic prokaryotes (e.g. hot spring mats) is the yellow or orange coloration in summer in regions of high light intensity with a very high ratio of carotenoids to chlorophyll and phycobilins (e.g. Norris et al. 2002, Vincent et al. 1993). However, the effect of carotenoids as a UVR screen is minimal, since almost all of carotenoid absorption is in the violet/blue/green region of the visible spectrum in prokaryotes, and in some cases may function mainly as major light harvesting complexes (Castenholz and Garcia-Pichel 2000).

Superoxide radical anions are also produced with high light intensity and increased levels of superoxide dismutase (SOD) may result, for example increased synthesis of Mn SOD above constitutive levels of Fe SOD in cyanobacteria (see Castenholz and Garcia-Pichel 2000). H_2O_2, a product of SOD activity, is considered to be relatively harmless, but in *Synechocystis* sp. PCC 6803, peroxide decomposition is catalyzed by catalase-peroxidase and a thiol-specific peroxidase (Tichy and Vermaas 1999). Tocopherols, ascorbates, and glutathione are also used as detoxifying compounds (Niyogi 1999).

An important preventative and passive method of protection from UVA is by the production of scytonemin in sheaths of cyanobacteria, but this does not occur in all cyanobacteria. Only those that possess sheaths (or a glycocalyx) and are commonly exposed to UVR (even low UVR) synthesize scytonemin. This includes numerous surface layers of microbial mats in flats infrequently covered by tidal waters (e.g. Cockell and Rothschild 1999), lower temperature (<~ 45°C) hot spring mats and crusts (e.g. Brenowitz and Castenholz 1997), terrestrial mats and crusts (e.g. Garcia-Pichel and Belnap 1996), shallow and clear oligotrophic fresh waters (Johnson and Castenholz 2000), cyanolichens (Büdel et al. 1997) and many other exposed habitats (Garcia-Pichel and Castenholz 1991, Castenholz and Garcia-Pichel 2000). Scytonemin, as mentioned earlier, has major absorption peaks at ~250 and 370 nm (in vivo) (Fig. 1). It is a unique dimeric indole alkaloid with a molecular weight of 544 (Proteau et al. 1993; Fig. 2). In mature surface filaments of cyanobacteria scytonemin-rich sheaths may screen out >95 % of the UVA photons. This passive method of preventing or slowing the inhibition or destruction caused by UVA radiation apparently results in great benefits in terms of fitness and survival, particularly in habitats where desiccation or suboptimal growth conditions prevail (Dillon et al. 2003, Dillon and Castenholz 2003, Brenowitz and Castenholz 1997, Garcia-Pichel et al. 1992).

The synthesis of scytonemin is primarily a result of exposure to UVA although violet/blue wavelengths have some effect (Garcia-Pichel and Castenholz 1991), but some rock-inhabiting cyanobacteria synthesize scytonemin constitutively (unpublished data). Stresses such as increased temperature and photooxidative conditions, in conjunction with UVA, caused a synergistic increase in scytonemin production in one species of *Chroococcidiopsis* (Dillon et al. 2002). In addition, a moderate osmotic stress, without UVR or blue light, resulted in some scytonemin synthesis.

3. Behavioral Counterbalances to UVR Exposure

Although much is known about vertical movements of gas-vacuolate cyanobacteria, purple bacteria, and green bacteria in lakes, and the role that light intensity often plays in buoyancy regulation (Walsby 1994), apparently no work that focuses on the UV portion of the spectrum has been published.

However, the vertical migration by gliding motility of cyanobacteria in soft microbial mats and sediments in response to light and UV intensity has been documented (for a summary see Castenholz and Garcia-Pichel 2000). The downward movement of filamentous cyanobacteria (0.4-> 1 mm) in dense microbial mats has been shown to be a response to high solar irradiance, particularly UVR, both UVB and/or UVA (Garcia-Pichel et al. 1994, Ramsing and Prufert-Bebout 1994, Bebout and Garcia-Pichel 1995, Kruschel and Castenholz 1998, Nadeau et al. 1999). A lessening of UVR intensities (or darkness) allows the return towards the mat surface or to the surface unless UVR intensities again stop the movement. These observations and data are from saline or hypersaline mats in middle latitudes and from an Antarctic salt pond. In most of these cases it has been shown that failure to migrate from the surface during periods of high solar UV irradiance would result in extreme inhibition of photosynthesis or death (Garcia-Pichel et al. 1994, Kruschel and Castenholz 1998, Nadeau et al. 1999). Filamentous and unicellular cyanobacteria have been shown to move vertically in soft hot spring mats also as a response to high light intensity, but at the time of the work UVR was not taken into consideration (Castenholz 1968, Richardson and Castenholz 1987, Ramsing et al. 1997). In almost all of the cases cited, the cyanobacteria in question did not possess MAAs nor did they have the ability to withstand large doses of UVR. Thus, escape was the only tactic for survival. Garcia-Pichel et al. (1994) have shown in one case by the use of micro light sensors, that the gradual descent (to <0.5 mm) and ascent of *Spirulina* and *Oscillatoria* in a soft hypersaline mat would result in maximum photosynthetic rates during most daylight hours. Thus, the daytime descent may not be simply an escape but a finely regulated optimization of irradiance available for photosynthesis. Migrating cyanobacteria also retained high cellular contents of chlorophyll *a* and phycobilins. This condition would be distinctly advantageous during overcast periods of low light intensity and during early morning and evening when maximal photon capture would be required for adequate photosynthetic rates. These same pigments would also act as photosensitizers if the cells were exposed to high solar irradiance. In more sedentary cyanobacteria these pigments are regulated downward thus increasing the ratio of carotenoids to photosensitizers.

4. Effects of UVR Reduction or Exclusion on Phototrophic Communities

The long-term (3-5 weeks) influence of UVR on a complex lotic community colonizing artificial substrate was studied by Bothwell et al. (1993, 1994). With enhanced UVR many changes occurred in these communities of several trophic levels. Greater benthic diatom biomass accumulated with enhanced UVB even though both UVA and UVB could reduce photosynthesis and growth of these diatoms. Chironomid larvae, which ordinarily exhibited a great grazing pressure on the diatom community, however, had a greater sensitivity to UVB. It is probable that the +UV treatment would be closer to natural conditions in the field.

When epiphytic cyanobacterial mats from tropical mangrove communities were subjected to long term +UV and -UV treatments (i.e. 27 days with laboratory UVB +UVA + PAR, or UVA + PAR, or PAR only) large differences in species arrangement and abundance occurred (Sheridan 2001). With UVA and UVB the top layer of scytonemin-containing, diazotrophic *Nostoc commune* and *Scytonema* sp. were maintained as in the natural mat. When UVR was excluded, the more UV-tolerant *Nostoc* was overrun or overgrown by a less UV-tolerant species of *Phormidium* which was not capable of diazotrophy. Therefore, overall N_2-fixation decreased greatly.

In Yellowstone National Park cyanobacterial mats in two alkaline geothermal streams (40-47°C) were covered for 1-3 months with filters that excluded or transmitted UVR (Norris et al. 2002). Over some adjacent areas, 25% transmission neutral density screens were also used. There were no apparent changes in general bacterial or cyanobacterial community composition during the summer with or without high or low UVR, as assessed by DGGE (denaturing gradient gel electrophoresis). Although the bacterial composition of these communities was apparently stable, surface layers of cyanobacteria protected by filters from UV radiation were not as competent photosynthetically as those that had been maintained under UVR. This decrease in competence was expressed as a temporary loss of the ability to perform at a maximum rate under full solar irradiance (including UVR). It is probable that changes at the level of gene expression were occurring in the dominant species rather than changes in species composition.

5. Summary and Conclusions

The influences of UV radiation on microbial populations have been studied with increasing intensity in recent years. One reason is the awareness of decreasing regional ozone levels in the stratosphere which result in increased UVB flux reaching the Earth's surface, and also because microbial populations and species may exhibit a more immediate and measurable sensitivity to small increases in UVR than larger macrophytes and metazoans. As it turns out, some cyanobacteria, probably representing the oldest oxygenic inhabitants of the planet, have already evolved many methods (or complex strategies) for coping with both present levels and possibly higher levels of UVR that occurred in the early Precambrian and might again occur in the future (Table 1).

Since cyanobacteria have invaded (or remain as relicts) in a large number of extreme environments, including shallow water and terrestrial surfaces, they currently must cope often with high solar irradiance in which UVR can be the most detrimental factor. They have done so by evolving sunscreen pigments that envelope the cell and work even when cells are at rest, by synthesizing internal compounds such as mycosporine-like amino acids (the true value of which is still not well evaluated), by possessing efficient systems for repair of damaged DNA and for replacement of UV damaged compounds, by gene regulation and also by using directed gliding motility in soft microbial mats or sediments for escaping the diurnally high intensities of solar irradiance (Table 1).

TABLE 1. Multiple methods of tolerating UV radiation by cyanobacteria A= sheathed types, B= motile types (filamentous and unicellular), C= no motility and no sheaths. A combination of many methods are used by all species. Motile hormogonia of type A may avoid UVR.

METHOD	A	B	C
Scytonemin	yes	no	no
MAAs	most-yes	many-yes	many-yes
Motility-avoidance	no	yes	no
Chlorophyll-phycobilin-carotenoid regulation	yes	yes	yes
Carotenoid quenching	yes	yes	yes
SOD, peroxidase, etc	yes	yes	yes
Repair - synthesis	yes	yes	yes

6. Acknowledgments

I would like to thank the US National Science Foundation and the NASA Astrobiology Institute for support over several years, as well as many students and colleagues for their collaboration and input. I also thank Jack Liu for the photograph on my biodata page.

7. References

Bebout, B.M. and Garcia-Pichel, F. (1995) UVB-induced vertical migrations of cyanobacteria in a microbial mat. Appl. Environ. Microbiol. **61,** 4215-4222.

Bothwell, M.L., Sherbot, D., Roberg, A.C. and Daley, R.J. (1993) Influence of natural ultraviolet radiation on lotic periphytic diatom community growth, biomass accrual, and species composition: short-term versus long-term effects. J. Phycol. **29**, 24-35.

Bothwell, M.L., Sherbot, D.M.J. and Pollock, C.M. (1994) Ecosystem response to solar ultraviolet-B radiation: influence of trophic level interactions. Science **265**, 97-100.

Brenowitz, S. and Castenholz, R.W. (1997) Long-term effects of UV and visible irradiance on natural populations of a scytonemin-containing cyanobacterium (*Calothrix* sp.). FEMS Microbiol. Ecol. **24**, 343-352.

Britton, G. (1995) Structure and properties of carotenoids in relation to function. FASEB J. **9**, 1551-1558.

Büdel, B., Karsten, U. and Garcia-Pichel, F. (1997) Ultraviolet-absorbing scytonemin and mycosporine-like amino acid derivatives in exposed, rock-inhabiting cyanobacterial lichens. Oecologia **112**, 165-172.

Campbell, D., Eriksson, M-J., Öquist, G., Gustafsson, P. and Clarke, A.K. (1998) The cyanobacterium *Synechococcus* resists UV-B by exchanging photosystem II reaction center D1 proteins. Proc. Natl. Acad. Sci. USA **95**, 364-369.

Castenholz, R.W. (1968) The behavior of *Oscillatoria terebriformis* in hot springs. J. Phycol. **4**, 132-139.

Castenholz, R.W. and Garcia-Pichel, F. (2000) Cyanobacterial responses to UV-radiation. In: B.A. Whitton and M. Potts (eds.) *The Ecology of Cyanobacteria.* Kluwer Academic. Publishers. Dordrecht, pp. 591-611.

Choi, J-S., Chung, Y-H., Moon, Y-J., Kim, C., Watanabe, M., Song, P-S., Joe, C-O., Bogorad, L. and Park, Y.M. (1999) Photomovement of the gliding cyanobacterium *Synechocystis* sp. PCC 6803. Photochem. Photobiol. **70**, 95-102.

Cockell, C.S. (1998) The biological effects of high ultraviolet radiation on early Earth: a theoretical evalulation. J. Theor. Biol. **193**, 717-729.

Cockell, C.S. (2001) A photobiological history of Earth. In: C.S. Cockell and A.R. Blaustein (eds.) *Ecosystems, Evolution, and Ultraviolet Radiation.* Springer-Verlag, New York, pp. 1-35.

Cockell, C.S. and Rothschild, L.J. (1999) The effects of UV radiation A and B on diurnal variation in photosynthesis in three taxonomically and ecologically diverse microbial mats. Photochem. Photobiol. **69**, 203-210.

Day, T.A. and Neale, P.J. (2002) Effects of UV-B radiation on terrestrial and aquatic primary producers. Ann. Rev. Ecol. Syst. **33**, 371-396.

Dillon, J.G. and Castenholz, R.W. (1999) Scytonemin, a cyanobacterial sheath pigment, protects against UVC radiation: implications for early photosynthetic life. J. Phycol. **35**, 673-681.

Dillon, J. G., Tatsumi, C.M., Tandingan, P.G. and Castenholz, R.W. (2002) Effect of environmental factors on the synthesis of scytonemin, a UV-screening pigment, in a cyanobacterium (*Chroococcidiopsis* sp.). Arch. Microbiol. **177**, 322-331.

Dillon, J.G. and Castenholz, R.W. (2003) The synthesis of the UV-screening pigment, scytonemin, and photosynthetic performance in isolates from closely related natural populations of cyanobacteria (*Calothrix* sp.). Environ. Microbiol. **5** (in press).

Dillon, J.G., Miller, S.R. and Castenholz, R.W. (2003) UV-acclimation responses in natural populations of cyanobacteria (*Calothrix* sp.). Environ. Microbiol. **5** (in press).

Ehling-Schulz, M., Bilger, W. and Scherer, S. (1997) UV-B-induced synthesis of photoprotective pigments and extracellular polysaccharides in the terrestrial cyanobacterium *Nostoc commune.* J. Bacteriol. **179**, 1940-1945.

Ehling-Schulz, M. and Scherer, S. (1999) UV protection in cyanobacteria. Eur. J. Phycol. **34**, 329-338.

Garcia-Pichel, F. (1994) A model for internal self-shading in planktonic microorganisms and its implications for the usefulness of sunscreens. Limnol. Oceanogr. **39**, 1704-1717.

Garcia-Pichel, F. (1998) Solar ultraviolet and the evolutionary history of cyanobacteria. Origins Life Evol. Biosphere **28**, 321-347.

Garcia-Pichel, F. and Castenholz, R.W. (1991) Charcaterization and biological implications of scytonemin, a cyanobacterial sheath pigment. J. Phycol. **27**, 395-409.

Garcia-Pichel, F., Sherry, N.D., and Castenholz, R.W. (1992) Evidence for an ultraviolet sunscreen role of the extracellular pigment scytonemin in the terrestrial cyanobacterium *Chlrogloeopsis* sp. Photochem. Photobiol. **56**, 17-23.

Garcia-Pichel, F. and Castenholz, R.W. (1993) Occurrence of UV-absorbing, mycosporine-like compounds among cyanobacterial isolates and an estimate of their screening capacity. Appl. Environ. Microbiol. **59**, 163-169.

Garcia-Pichel, F., Wingard, C.E. and Castenholz, R.W. (1993) Evidence regarding the UV-sunscreen role of a mycosporine-like compund in the cyanobacterium *Gloeocapsa* sp. Appl. Environ. Microbiol. **59**, 170-176.

Garcia-Pichel, F. and Belnap, J. (1996) Microenvironments and microscale productivity of cyanobacterial desert crusts. J. Phycol. **32**, 774-782.

Garcia-Pichel, F., Mechling, M. and Castenholz, R.W. (1994) Diel migrations of microorganisms within a benthic, hypersaline mat community. Appl. Environ. Microbiol. **60**, 1500-1511.

Gill, R.T., Katsoulakis, E., Schmitt, W., Taroncher-Oldenburg, G., Misra, J. and Stephanopoulos, G. (2002) Genome-wide dynamic transcriptional profiling of the light-to-dark transition in *Synechocystis* sp. strain PCC 6803. J. Bacteriol. **184**, 3671-3681.

Girotti, A.W. (2001) Photooxidized oxidation of membrane lipids: reaction pathways, cytotoxic effects, and cytoprotective mechanisms. J. Photochem. Photobiol. *B* **63**, 103-113.

Häder, D-P. (1984) Effects of UV-B on motility and photoorientation in the cyanobacterium, *Phormidium uncinatum*. Arch. Microbiol. **140**, 34-39.

Häder, D-P. (2001) Ultraviolet radiation and aquatic microbial ecosystems. In: C.S. Cockell and A.R. Blaustein (eds.) *Ecosystems, Evolution, and Ultraviolet Radiation*. Springer-Verlag, New York, pp. 150-169.

He, Y-Y, Klisch, M. and Häder, D-P. (2002) Adaptation of cyanobacteria to UV-B stress correlated with oxidative stress and oxidative damage. Photochem. Photobiol. **76**, 188-196.

Jagger, J. (1985) *Solar-UV Action on Living Cells*. Praeger, New York.

Johnson, A.C. and Castenholz, R.W. (2000) Preliminary observations of the benthic cyanobacteria of Waldo Lake and their potential contribution to lake productivity. Lake and Reservoir Management **16**, 85-90.

Josue, J.S. and Frank, H.A. (2002) Direct determination of the S_1 excited-state energies of xanthophylls by low-temperature fluorescence spectroscopy. J. Phys. Chem. A **106**, 2815-2824.

Kasting, J. F. (1987) Theoretical constraints on oxygen and carbon dioxide concentrations in the Precambrian atmosphere. Precambrian Res. **34**, 205-229.

Kasting, J.F., Zahnle, K.J., Pinto, J.P. and Young, A.T. (1989) Sulfur, ultraviolet radiation, and the early evolution of life. Origins Life Evol. Biosphere **19**, 95-108.

Kruschel, C. and Castenholz, R.W. (1998) The effect of solar UV and visible irradiance on the vertical movements of cyanobacteria in microbial mats of hypersaline waters. FEMS Microbiol. Ecol. **27**, 53-72.

Lao, K. and Glazer, A.N. (1996) Ultraviolet-B photodestruction of a light-harvesting complex. Proc. Natl. Acad. Sci. USA **93**, 5258-5263.

Máté, Z., Sass, L., Szekeres, M., Vass, I. and Nagy, F. (1998) UV-B-induced differential transcription of *psbA* genes encoding the D1 protein of photosystem II in the cyanobacterium *Synechocystis* 6803. J. Biol. Chem. **273**, 17439-17444.

MacDonald, T.M., Dubois, L., Smith, L.C. and Campbell, D.A. (2003) Sensitivity of cyanobacterial antenna, reaction center and CO_2 assimilation transcripts and proteins to moderate UVB: light acclimation potentiates resistance to UVB. Photochem. Photobiol. **77**, 405-412.

Miller, S.R., Wingard, C.E. and Castenholz, R.W. (1998) Effects of visible light and UV radiation on photosynthesis in a population of a hot spring cyanobacterium, a *Synechococcus* sp., subjected to high-temperature stress. Appl. Environ. Microbiol. **64**, 3893-3899.

Nadeau, T-L., Howard-Williams, H. and Castenholz, R.W. (1999) Effects of solar UV and visible irradiance on photosynthesis and vertcal migration of *Oscillatoria* sp (cyanobacteria) in an Antarctic microbial mat. Aquat. Microbial Ecol. **20**, 231-243.

Niyogi, K.K. (1999) Photoprotection revisited: genetic and molecular approaches. Ann. Rev. Plant Physiol. Plant Molec. Biol. **50**, 333-359.

Norris, T. B., McDermott, T.R. and Castenholz, R.W. (2002) The long-term effects of UV exclusion on the microbial composition and photosynthetic competence of bacteria in hot-spring microbial mats. FEMS Microbiol, Ecol. **39**, 193-209.

Oren, A. (1997) Mycosporine-like amino acids as osmotic solutes in a community of halophilic cyanobacteria. Geomicrobiol. J. **14**, 231-240.

Pakker, H., Martins, R.S.T., Boelen, P., Buma, A.G.J., Nikaido, O. and Breeman, A.M. (2000) Effects of temperature on the photoreactivation of ultraviolet-B-induced DNA damage in *Palmaria palmata* (Rhodophyta). J. Phycol. **36**, 334-341.

Pierson, B.K. (1994) The emergence, diversification, and role of photosynthetic eubacteria. In: S. Bengtson (ed.) *Early Life on Earth*. Columbia Univ. Press, N.Y., pp. 161-180.

Pierson, B.K. and Castenholz, R.W. (1974) A phototrophic gliding filamentous bacterium of hot springs, *Chloroflexus aurantiacus*, gen. and sp. nov. Arch. Microbiol. **100**, 5-24.

Pierson, B. K., Mitchell, H.K. and Ruff-Roberts, A. L. (1993) *Chloroflexus aurantiacus* and ultraviolet radiation: implications for Archean shallow-water stromatolites. Origins Life Evol. Biosphere **23**, 243-260.

Porankiewicz, J., Schelin, J. and Clarke, A.K. (1998) The ATP-dependent Clp protease is essential for acclimation to UV-B and low temperature in the cyanobacterium *Synechococcus*. Mol. Microbiol. **29**, 275-283.

Portwich, A. and Garcia-Pichel, F. (1999) Ultraviolet and osmotic stresses induce and regulate the synthesis of mycosporines in the cyanobacterium *Chlorogloeopsis* PCC 6912. Arch. Microbiol. **172**, 187-192.

Portwich, A. and Garcia-Pichel, F. (2000) A novel prokaryotic UVB photoreceptor in the cyanobacterium *Chlorogloeopsis* PCC 6912. Photochem. Photobiol. **71**, 493-498.

Proteau, P.J., Gerwick, W.H., Garcia-Pichel, F. and Castenholz, R.W. (1993) The structure of scytonemin, an ultraviolet sunscreen pigment from the sheaths of cyanobacteria. Experientia **49**, 825-829.

Quesada, A. and Vincent, W.F. (1997) Strategies of adaptation by Antarctic cyanobacteria to ultraviolet radiation. Eur. J. Phycol. **32**, 335-342.

Ramsing, N.B. and Prufert-Bebout, L. (1994) Motility of *Microcoleus chthonoplastes* subjected to different light intensities quantified by digital image analysis. In: L.J. Stal and P. Caumette (eds.) *Microbial Mats. Structure, Development and Environmental Significance.* NATO ASI Series Vol. 35, Springer-Verlag, Berlin, pp. 183-191.

Ramsing, N.B., Ferris, M.J. and Ward, D.M. (1997) Light-induced motility of thermophilic *Synechococcus* isolates from Octopus Spring, Yellowstone National Park. Appl. Environ. Microbiol. **63**, 2347-2354.

Ravanat, J-L., Douki, T. and Cadet, J. (2001) Direct and indirect effects of UV radiation on DNA and its components. J. Photochem. Photobiol. B **63**, 88-102.

Richardson, L.L. and Castenholz, R.W. (1987) Diel vertical movements of the cyanobacterium *Oscillatoria terebriformis* in a sulfide-rich hot spring microbial mat. Appl. Environ. Microbiol. **53**, 2142-2150.

Roy, S. (2000) Strategies for the minimisation of UV-induced damage. In: S. de Mora, S. Demers and M. Vernet (eds.) *The effects of UV Radiation in the Marine Environment.* Cambridge Univ. Press, Cambridge. pp. 176-205.

Sancar, A. (1994) Mechanisms of DNA excision repair. Science **266**, 1954-1956.

Sancar, A. (1996) No "end of history" for photolyases. Science **272**, 48-49.

Sass, L., Spetea, C., Máté, Z., Nagy, F. and Vass, I. (1997) Repair of UVB induced damage of photosystem II via *de novo* synthesis of the D1 and D2 reaction centre subunits in *Synechocystis* sp. PCC 6803. Photosynth. Res. **54**, 55-62.

Sheridan, R.P. (2001) The role of ultraviolet radiation in maintaining the three-dimensional structure of a cyanobacterial mat community and facilitating nitrogen fixation. J. Phycol. **37**, 731-737.

Shibata, H., Katsuya, B. and Ochiai, H. (1991) Near-UV irradiation induces shock proteins in *Anacystis nidulans* R-2; possible role of active oxygen. Plant Cell Physiol. **32**, 771-776.

Shick, J.M. and Dunlap, W.C. (2002) Mycosporine-like amino acids and related gadusols: biosynthesis, accumulation, and UV-protective functions in aquatic organisms. Ann. Rev. Physiol. **64**, 223-262.

Sinha, R.P., Klisch, M., Helbling, E.W. and Häder, D-P. (2001) Induction of mycosporine-like amino acids (MAAs) in cyanobacteria by solar ultraviolet-B radiation. J. Photochem. Photobiol. B **60**, 129-135.

Tichy, M. and Vermaas, W. (1999) In vivo role of catalase-peroxidase in *Synechocystis* sp. strain PCC 6803. J. Bacteriol. **181**, 1875-1882.

Van Baalen, C. (1968) The effects of ultrviolet irradiation on a coccoid blue-green alga: survival, photosynthesis, and photoreactivation. Plant Physiol. **43**, 1689-1695.

Van Baalen, C. and O'Donnell, K. (1972) Action spectra for ultraviolet killing and photoreactivation in the blue-green alga *Agmenellum quadruplicatum*. Photochem. Photobiol. **15**, 269-274.

Vincent, W.F. and Neale, P.J. (2000) Mechanisms of UV damage to aquatic organisms. In: S. de Mora, S. Demers and M. Vernet (eds.) *The Effects of UV Radiation in the Marine Environment.* Cambridge Univ. Press, Cambridge. pp. 149-176.

Vincent, W.F., Downes, M.T., Castenholz, R.W. and Howard-Williams, C. (1993) Community structure and pigment organisation of cyanobacteria-dominated microbial mats in Antarctica. Eur. J. Phycol. **28**, 213-221.

Walsby, A.E. (1994) Gas vesicles. Microbiol. Rev. **58**, 94-144.

Whitehead, R.F., de Mora, S.J. and Demers, S. (2000) Enhanced UV radiation – a new problem for the marine environment. In: S. de Mora, S. Demers and M. Vernet (eds.) *The effects of UV Radiation in the Marine Environment.* Cambridge University Press, Cambridge. pp. 1-34.

Wiechert, U.H. (2002) Earth's early atmosphere. Science **298**, 2341-2342.

Wingard, C.E., Schiller, J.R. and Castenholz, R.W. (1997) Evidence regarding the possible role of c-phycoerythrin in ultraviolet-B tolerance in a thermophilic cyanobacterium. Photochem. Photobiol. **65**, 833-842.

Biodata of **Ricardo Amils** author (with co-authors E. González-Toril, F. Gómez, D. Fernández-Remolar, N. Rodríguez, M. Malki, A. Aguilera and L.A. Amaral-Zettler) of ***"Importance Of Chemolithotrophy For Early Life On Earth: The Tinto River (Iberian Pyritic Belt) Case"***

Dr. Ricardo Amils is a Professor of Microbiology, Molecular Biology at the Center of Molecular Biology, University Autonoma de Madrid and is a Senior Scientist Associated to the Center of Astrobiology (INTA-CSIC). He obtained his Ph.D. in 1973 from the University Autonoma de Barcelona. Spain. His main research activities are: Functional Evolution of Ribosomes, Ribosomal Reconstitution, Biomining, Bioremediation, Specific Metal Sequenstering, Microbial Ecology of Extremophilic Habitats, and the Molecular Biology of Extremophiles. He published more than 130 papers (referred articles in journals and chapters of books).

E-mail: **ricardo.amils@uam.es**

J. Seckbach (ed.), Origins, 463–480.

IMPORTANCE OF CHEMOLITHOTROPHY FOR EARLY LIFE ON EARTH:
The Tinto River (Iberian Pyritic Belt) Case

R. AMILS[1,2], E. GONZÁLEZ-TORIL[2], F. GÓMEZ[1], D. FERNÁNDEZ-REMOLAR[1], N. RODRÍGUEZ[1], M. MALKI[2], J. ZULUAGA[3], A. AGUILERA[1], L.A. AMARAL-ZETTLER[4].

[1]Centro de Astrobiología (CSIC-INTA), Torrejón de Ardoz, 28855 Madrid, Spain, [2]Centro de Biología Molecular (CSIC-UAM), U. Autónoma de Madrid, Cantoblanco, 28049 Madrid, Spain, [3] Departamento de Química Física Aplicada, U. Autónoma de Madrid, Cantoblanco, 28049, Spain, and [4]Marine Biological Laboratory, Woods Hole, Massachusetts 02543, USA.

1. Introduction

The discovery of subsurface chemolithotrophic microorganisms participating in a radiation-independent carbon cycle has opened an interesting perspective in astrobiology (Godd, 1992; Bachofen et al., 1998; Pedersen, 2000; Chapelle et al., 2002; Gómez and Amils, 2002). Although hydrogen is the preferred chemolithotrophic substrate due to its high energy potential, there is a growing list of alternative sources of lithotrophic energy (S^{2-}, S^0, Fe^{2+}), which amplifies the metabolic possibilities (versatility) of this energy conservation system. Pyrite, with its wide distribution in our planet, is an important chemolithotrophic substrate because both of its components, sulfide and ferrous iron, can be used by sulfur- and iron-oxidizing microbes as a source of energy. Furthermore, sulfur- and iron-oxidizers together with sulfur- and iron-reducing microorganisms have a fundamental role in the operation of two important biogeochemical cycles, the sulfur and the iron cycles. Until recently, much attention has been paid to sulfur microbiology, probably because from a bioenergetic point of view much more energy can be obtained from sulfur reduced compounds than from ferrous iron. Not to mention that many experts did not believe that iron oxidation could produce sufficient energy to sustain microbial growth. The first acidophilic strict chemolithotroph, *Acidithiobacillus ferrooxidans* (formerly *Thiobacillus ferrooxidans*), was isolated from an acidic pond in a coal mine more than fifty years ago (Colmer et al., 1950). Although *At. ferrooxidans* can obtain energy oxidizing both sulfide and ferrous iron, much more attention was paid to the sulfur oxidation reaction, for the bioenergetic reasons stated above and because of its correlation with the generation of acid mine drainage, a problem of environmental concern.

The discovery that some strict chemolithotrophs like *Leptospirillum ferrooxidans* or *Ferroplasma* spp. could grow using ferrous iron as their only source of energy, and that these microorganisms are mainly responsible for metal bioleaching processes and acid mine drainage, has completely changed this perspective (Golyshina et al., 2000; Edwards et al., 2000). Furthermore, it is now well established that iron can be oxidized anaerobically in the absence of oxygen, coupled to anoxygenic photosynthesis or to the respiration using nitrate as an electron acceptor (Widdel et al., 1993; Benz et al., 1998).

2. Extreme Acidic Environments

Extremophiles, organisms able to grow in extreme conditions, have recently attracted considerable attention, not only because they prove that life is robust and adaptable, thus increasing the probability of finding life elsewhere in the universe (Morrison, 2001), but due to their biotechnological potential (Rothschild and Mancinelli, 2001). Most of the best-characterized extreme environments on Earth correspond to geophysical constraints (temperature, pressure, ionic strength, radiation, etc.) in which opportunistic microorganisms have developed diverse adaptive strategies. However, some extreme acidic habitats are unique in that they are actually produced by the metabolism activity of chemolithotrophic microorganisms and not a geophysical constraint of the habitat.

Acidic, metal-rich environments have two major origins. The first one is associated with volcanic activities. The acidity in these locations may derive from the microbial oxidation of elemental sulfur:

$$S^o + 11/2\,O_2 + H_2O \rightarrow SO_4^{2-} + 2H^+ \qquad \text{(reaction 1)}$$

which is produced as a result of the condensation reaction between sulfur containing volcanic gases:

$$2\,H_2S + SO_2 \rightarrow 3\,S^o + H_2O \qquad \text{(reaction 2)}$$

Mineral sulfides are also susceptible to microbial oxidation producing sulfuric acid, contributing to the total acidity of these environments (see below). In these acidic, sulfur-rich locations, commonly known as solfataras, temperature gradients are easily formed. These sites may therefore be colonized by a variety of acidophilic microorganisms with different optimal temperatures (Johnson, 1998).

Acidic, metal-rich environments can be also found associated with mining activities. Mining of metals and coal exposes sulfidic minerals to the combined action of water and oxygen, which facilitate bacterial attack. The most abundant sulfidic mineral, pyrite, is of particular interest in this context. The biooxidation of pyrite has been studied in detail. The process occurs in several steps, with the overall reaction being:

$$4FeS_2 + 14H_2O + 15O_2 \rightarrow 4Fe(OH)_3 + 8SO_4^{2-} + 16H^+ \qquad \text{(reaction 3)}$$

These environments vary greatly in their physico-chemical characteristics and also in their microbial ecology. High temperatures may occur as a consequence of biological activity, facilitating the colonization by thermophilic acidophiles. Acidic ecosystems associated with mining activities are, at the geological and evolutionary scale, very young. However some metal mines have a rather long history in some regions of the world. Sites such as Tinto River are known to have been exploited by chalcolithic miners 5000 years ago (Avery, 1974; Leblanc et al., 2000). The recent microbial characterization of this habitat uncovered a natural complex ecosystem which was in operation long before any mining activities in the area (López-Archilla et al., 2001; González-Toril et al., 2002, Amils et al., 2003; Fernández-Remolar et al., 2003).

3. Acidic Chemolithotrophy

The mechanisms by which microbes can obtain energy oxidizing sulfidic minerals, a process of biotechnological interest known as bioleaching, have been rather controversial for many years (Ehrlich, 2001). Two basic mechanisms, concerning the relative relationship between the mineral substrate and the microbial catalyst, the direct and the indirect attack, have been considered the explanation of most reported bioleaching observations (Silverman and Ehrlich, 1964). In the direct attack, microorganisms solubilize mineral sulfides by attaching to their surface, facilitating an enzymatic oxidation by channeling electrons from the mineral to an appropriate electron acceptor (respiration). In the indirect attack, microorganisms are mainly involved in the regeneration of ferric iron, a strong oxidant, in the solution. This soluble ferric iron is responsible for the oxidation of exposed sulfidic minerals and other reduced chemolithotrophic substrates.

Many experiments have been performed to clarify this issue with important technological and environmental implications. The demonstration that ferric iron present in the cell envelope and/or in the exopolymers of leaching microorganisms was responsible for the electron transfer from the sulfidic minerals to the electron transport chain (Gehrke et al., 1995) shed light on this controversial issue. Based on this observation Sand and coworkers have recently proposed that because ferric iron is ultimately responsible for the oxidation of metal sulfides in both the direct and the indirect attack, there is no basic difference between these mechanisms (Sand et al., 1995).

The difference seems to exist at the level of the chemical attack mechanism, which is dependent on the structure of the sulfidic substrates. Three metal sulfides, pyrite, molybdenite and tungstenite, which can be only oxidized by ferric iron, undergo oxidation through the so-called thiosulfate mechanism:

$$FeS_2 + 6Fe^{3+} + 3H_2O \rightarrow S_2O_3^{2-} + 7Fe^{2+} + 6H^+ \qquad \text{(reaction 4)}$$
$$S_2O_3^{2-} + 8Fe^{3+} + 5H_2O \rightarrow 2SO_4^{2-} + 8Fe^{2+} + 10H^+ \qquad \text{(reaction 5)}$$

with sulfate as the main product (Sand et al., 2001). Most other sulfides (e.g., chalcopyrite, sphalerite, galena, etc.) are susceptible to a proton acid attack as well as to ferric iron oxidation. They are oxidized through the so-called polysulfide mechanism (Sand et al., 2001):

$$8MS + 8Fe^{3+} + 8H^+ \rightarrow 8M^{2+} + 4H_2S_n + 8Fe^{2+} \quad (n \geq 2) \qquad \text{(reaction 6)}$$
$$4H_2S_n + 8Fe^{3+} \rightarrow S_8^0 + 8Fe^{2+} + 8H^+ \qquad \text{(reaction 7)}$$

In this case elemental sulfur is produced which can be further microbial oxidized (e.g., by *Acidithiobacillus* spp.) to sulfuric acid, following equation 1. Most authors currently agree with these basic mechanisms for metal sulfide bioleaching (Ehrlich, 2001; Hansford and Vargas, 2001), in which iron and not sulfur plays a central and critical role.

4. The Tinto River, Physico-Chemical Parameters

The Tinto River (Iberian Pyritic Belt, Southwestern Spain) is an unusual ecosystem due to its acidity (mean pH 2.3), size (100 km long), high concentration of heavy metals (Fe, Cu, As, Zn, Cr) and an unexpected level of microbial diversity. The Tinto River springs up in the center of the Iberian Pyritic Belt, at Peña de Hierro, and flows into the Atlantic Ocean at Huelva. The river gives its name to an important mining district, which has been in operation for more than five thousand years (Avery, 1974).

The Iberian Pyritic Belt is a 250 km long and 25-70 km wide geological entity that is embedded in the South-Portuguese geotectonic zone of the Iberian Peninsula. The Iberian Pyritic Belt is one of the largest sulfide deposits of the world. Massive bodies of iron and copper sulfides, as well as minor quantities of lead and zinc, constitute the main mineral ores (Boulter, 1996; Leistel et al., 1998). Its formation took place during the Hercynian orogenesis by hydrothermalism (Boulter, 1996; Lescuyer et al., 1998). The basin of the river covers an area of around 1700 km^2. According to the geological and chemical characteristics of the Tinto River Basin, three main zones can be defined: the northern (from Peña de Hierro to Niebla), the transitional (from Niebla to San Juan del Puerto) and the estuary (from San Juan del Puerto to the Atlantic Ocean). According to climatic parameters the northern area corresponds to a subhumid lower mesomediterranean to upper thermomediterranean climate, whereas the transitional and the estuarine areas can be included in a subhumid to dry thermomediterranean climate with semiarid conditions (Rivas-Martínez, 1987; Asensi and Diaez, 1987). The climograms of the basin show a characteristic bimodality, with humid and temperate seasons alternating with dry and warm seasons.

TABLE 1 Comparison of physicochemical parameters of the Tinto River with rivers of the same geographical area. Concentrations in mg/L, nd: not determined (adapted from Amils et al., 2003)

	pH	Fe	Zn	Cu	SO_4^{2-}
Guadiamar	7.8	12.1	0.97	0.07	312.5
Partido	7.8	24.8	0.11	0.14	89.8
Agrio	4.3	50.0	33.90	1.97	nd
Odiel (mean)	3.6	170.0	136.0	42.3	3876.0
Tinto (mean)	2.3	2261.0	225.0	109.0	10110.0

Table 1 shows representative physico-chemical parameters measured in the Tinto River compared to those of local rivers. It should be noted that the low pH and correspondent high concentration of ferric iron in the Tinto River is at least one order of magnitude higher than the rest of the rivers listed, two of which, the Odiel and the Agrio Rivers are also associated with mining activities (López-Archilla and Amils, 1999; López-Archilla et al., 2001). The measured redox potentials in the Tinto River range from +280 mV to +610 mV, and the oxygen content varies from saturation to complete anoxic conditions.

An important characteristic of the Tinto system is its constant acidic pH all along the river. This is a direct consequence of the strong buffer capacity of ferric iron. When the river is diluted by rain or by neutral tributaries, hydrolysis of ferric iron generates free protons, which maintain the pH of the water:

$$Fe^{3+} + 3\,H_2O \Leftrightarrow Fe(OH)_3 + 3H^+ \qquad \text{(reaction 8)}$$

The reaction is reversible. When intense evaporation occurs during the hot summers, protons are consumed by the dissolution of ferric hydroxide precipitates. Due to this buffering capacity of ferric iron, a pH of around 2.3 remains constant all along the river, requiring only sufficient ferric iron in solution. The constant pH of the Tinto ecosystem may be extremely important for the maintenance of the high level of microbial diversity of the habitat.

5. The Tinto River, Microbial Diversity

To elucidate the qualitative and quantitative degree of biodiversity in the Tinto ecosystem conventional microbiological methods (isolation from enrichment cultures and phenotypic characterization) and molecular ecology techniques (DGGE and FISH) have been extensively used (López-Archilla et al., 2001; González-Toril, 2002; Amaral-Zettler et al., 2002; González-Toril et al, in press). The different microbial populations found in the Tinto River can be grouped according to their ecological role as primary producers, decomposers and consumers.

5.1. PRIMARY PRODUCERS

Chemolithotrophic prokaryotes and photosynthetic protists are the main primary producers in the Tinto ecosystem (Table 2). Concerning chemolithotrophs, strict iron-oxidizing prokaryotes, mainly *At. ferrooxidans*, *Leptospirillum ferrooxidans* and *Ferroplasma* spp., have been isolated and identified in different locations along the Tinto River. Quantification using specific fluorescent probes (FISH) has shown that *L. ferrooxidans* and *At. ferrooxidans* are the main chemolithotrophs present in most of the sampling stations along the river (Gonzalez-Toril, 2002; González-Toril et al., submitted). Iron-oxidizing Archaea (*Ferroplasma* spp.) were found in low numbers, and only in the upper part of the river.

Of the sulfur-oxidizing prokaryotes only *At. ferrooxidans* was found in high numbers (this bacteria can oxidize both sulfur and ferrous iron). Other sulfur-oxidizing members of the *Acidithiobacillus* genus, such as *At. thiooxidans*, were detected only as minor components of the system (López-Archilla et al., 2001; González-Toril, 2002). Hybridization experiments using specific fluorescent probes have shown that large numbers of metabolically active *At. ferrooxidans* are found in the anoxic parts of the river. In addition, different *At. ferrooxidans* isolates from the Tinto are able to reduce ferric iron through anaerobic respiration of reduced sulfur compounds (Ehrlich, 2002). This is probably the main role of *At. ferrooxidans* in the Tinto ecosystem. Anaerobic respiration of ferrous iron by *L. ferrooxidans* has also been observed (Mejias, personal communication), which strongly suggests fully operative aerobic and anaerobic iron cycles in the Tinto ecosystem.

Photosynthetic protists (algae) accounted for the greatest proportion of biomass (over 65%) of the Tinto ecosystem. Members of the Bacillariophyceae (Diatoms), Euglenozoa (*Euglena*), Chlorophyta (*Chlamydomonas, Chlorella*, *Klebsormidium* and *Zignema*), and Rhodophyta (*Galdieria*) have been identified and some of them isolated and characterized (López-Archilla et al., 1993; Moreira et al., 1994; López-Archilla et al., 2001, Amaral -Zettler et al., 2002). The use of molecular ecology techniques, mainly sequence analysis of amplified rRNA genes, identified algae closely related to those characterized phenotypically (*Chlamydomonas*, *Euglena, Klebsormidium*, *Zignema* and *Chlorella*), emphasizing the high degree of eukaryotic diversity existing in the extreme conditions of the Tinto River (Amaral-Zettler et al., 2002; Aguilera et al., submitted).

5.2. DECOMPOSERS (HETEROTROPHS)

A large number of heterotrophic bacteria have been isolated from enrichment cultures. Some of the isolates were identified as members of the *Acidiphilium* genus. Members of this genus have frequently been found associated with chemolithotrophic bacteria, especially iron-oxidizers. All known species of the genus *Acidiphilium* are facultative anaerobic respiring, able to couple the oxidation of organic substrates with the reduction of ferric iron. Furthermore, some *Acidophilium* spp. can reduce iron in the presence of oxygen (Bridge and Johnson, 2000). Quantification with specific probes had shown that this type of bacteria appears in significant numbers in the Tinto ecosystem, meaning that they are important elements of the Tinto iron cycle (González-Toril, 2002). Molecular ecology techniques have also detected members of the *Ferrimicrobium* and *Acidimicrobium* genera. These Gram-positive bacteria were originally described as iron oxidizers and more recently as ferric iron reducers (anaerobic respiring) using reduced carbon compounds as electron donors (Bridge and Johnson, 2000). Their relative concentration in the water column is rather low.

Hybridization experiments strongly suggest the presence of sulfate reducing bacteria in the Tinto River (González-Toril, 2002). To date, these microorganisms have not been isolated from the most acidic part of the river, although they have been isolated from the estuary zone, the least acidic section of the Tinto ecosystem (pH between 4 and 6, depending on the tide). Johnson has recently described the isolation of acidophilic sulfate reducing bacteria in acid mine drainage ecosystems (Johnson, 1998; Johnson, 1999), making it reasonable to assume that these important elements of the sulfur cycle may exist in the acidic zone of the Tinto River.

TABLE 2 Prokaryotic diversity in the Tinto ecosystem. Type of lithotrophic metabolism (S: sulfur, Fe: iron); isolation from enrichment cultures, detection by DGGE and quantification by FISH using specific probes.

Microorganism	S	Fe	Isolation	DGGE	FISH
L. ferrooxidans	-	+	dif. strains	+	+++
At. ferrooxidans	+	+	dif. strains	+	+++
Acidiphilium spp.	-	+	dif. strains	+	++
Ferroplasma sp.	-	+	-	+	+/-
Ferrimicrobium sp.	+	+	-	+	+/-
Acidimicrobium sp.	+	+	-	+	+/-
Sulfate reducers	+	-	-	+	+/-
At. thiooxidans	+	-	+	-	-

Within the decomposers, fungi were very abundant and exhibited great diversity, including yeast and filamentous forms. A high percentage of the hyphomycetes isolates were able to grow in the Tinto River conditions. Some of the isolated yeast species can also be found in other less extreme aquatic environments, but the isolated Dematiaceae seem to be specific to the extreme conditions of the habitat, since they are rarely present in neutral fresh waters (López-Archilla et al., 2001). Many of the fungi that have been detected using molecular techniques have unique sequences that probably correspond to novel genera (López-Archilla et al., 1995; López-Archilla et al., 2001; Amaral-Zettler et al., 2002).

5.3. CONSUMERS

Among the eukaryotes, heterotrophic protists constitute the major consumer group in the Tinto ecosystem. Different flagellates, ciliates (phylum Ciliophora), amoebae of the class Lobosea (phylum Rhizopoda) and some representatives of the class Heliozoa (phylum Actinopoda) have been observed, mainly associated with biofilms (López-Archilla et al., 2001). Molecular analysis detected also heterotrophic protists lineages, such as cercomonads, vahlkampfiid amoebae and stramenopiles that eluded phenotypic detection until now. The phylogenetic placement of several other sequences is consistent with the idea that they represent new eukaryotic lineages (Amaral-Zettler et al., 2002).

6. Bioinduced Iron Formations in the Tinto Ecosystem

Most of the biomass is located on the riverbed and the surface of the rocks forming dense biofilms, composed mainly of filamentous algae and fungi in which prokaryotes are trapped. Heterotrophic protists have been also found associated with these biofilms. Significant iron mineral precipitation is observed on the surface of these biofilms, generating iron stromatolites, which grow seasonally following the exposure of the biofilms to the water (Fig. 1). These iron formations strongly support that the Tinto

River corresponds to a natural and not to an industrially contaminated habitat (Amils et al., 2000; Amils et al., 2003; Fernández-Remolar et al., 2003).

As mentioned, mining activities during the last 5000 years have altered the system (Avery, 1974), but evidence of its great antiquity has been found in massive laminated iron beds in three iron formations occupying different elevations above the present river. The most extensive formation (Alto de la Mesa), more than 35 m above the river, is organized into two different lithostratigraphic units, one with paraconglomerate and massive facies and the other with laminar-stromatolitic, massive microbiolithic, in which some fossil plants can be easily identified. It was first reported by Phillips who suggested its sedimentary origin (Phillips, 1881). The morphology of the outcrops and the strata thickness and continuity suggest that an extensive fluvial environment produced this iron rich formation.

An intermediate unit (Nerva), at 20-30 m above the river, consists of terrace sediments of goethitic composition that occupy the valley of the modern river. Their characteristic facies are thick dark greenish and reddish massive ironstones with laminar beds. The presence of laminar and dome-shaped structures might be of biogenic origin. These characteristics suggest that an acidic meandering river could give rise to this formation. The youngest unit, 1-2 m above the river, corresponds to abandoned river bars that are being entrenched by the modern fluvial system. Their facies are mainly represented by strata of conglomerates and paraconglomerates with a massive matrix of iron oxide and badly sorted boulders. Secondary massive laminar iron facies can also be observed locally. These sediments seem to have been formed in an acidic high-energy flow that experienced a sequence of sedimentation events.

Figure 1. Iron-bioformation from Berrocal.

The age of these formations is not known. No reliable geochronological analysis record is currently available to provide consistent results for dating these formations. However, if the climatic parameters that generate the actual sediments are considered and the paleoclimatical record is taken into account, a preliminary estimate of their stratigraphical position can be established. Considering the chronostratigraphical position of a well characterized local alteration profile (Bonares "laterite") (Rodríguez-

Vidal et al., 1985), an upper Pliocene to lower Pleistocene age may be inferred for the first ironstone fluvial unit and the second fluvial terrace formation. A late Pleistocene to early Holocene origin may be inferred for the third formation (Clemente et al., 1985). These massive laminated iron formations corresponding to old terraces of the river predate the oldest mining activity reported in the area, and are similar to the laminar structures which are being formed currently in the river (Amils et al., 2000; Fernández-Remolar et al., 2003).

Until recently, it had been generally accepted that the extreme conditions found in the Tinto River were the direct consequence of the mining activities performed in the area during the last 5000 years (Geen et al., 1997; Leblanc et al., 2000; Davis Jr. et al., 2000; Elbaz-Poulichet et al., 2001). Although it is obvious that the mining activity had to have had an important impact in the Tinto system, the existence of ancient iron-rich deposits formed prior to any known mining activity, under hydrochemical conditions similar to the modern deposits, can be considered a strong argument in favor of a natural origin of the river.

7. The Tinto River Model System

Although the presence of sulfur-oxidizing bacteria (*At. ferrooxidans* and *At. thiooxidans*) and sulfate reducing bacteria strongly suggests the operation of a sulfur cycle in the Tinto ecosystem, the identification in the Tinto River of iron-oxidizing (*L. ferrooxidans*, *At. ferrooxidans*, *Ferroplasma* spp., *Ferromicrobium* sp. and *Acidimicrobium* spp.) and iron-reducing prokaryotes (*Acidiphilium* spp., *At. ferrooxidans*, *Ferromicrobium* sp. and *Acidimicrobium* spp.), which can operate not only in aerobic but also in anaerobic conditions, highlight the existence of a fully operative iron cycle in the different conditions found in the river (Figure 2).

Considering the geomicrobiological characteristics of the Tinto ecosystem we postulate that the river is predominantly under the control of iron (González-Toril et al., 2002). Iron is the main product of bioleaching of pyrite and other iron bearing sulfidic minerals, like chalcopyrite, both of which are present in high concentrations in the Iberian Pyritic Belt. The activity of iron oxidizing prokaryotes is responsible for both the solubilization of sulfidic minerals and the correspondent high concentrations of ferric iron, sulfate and protons in the water column (equations 4, 5, 6 and 7). Ferric iron is a strong oxidant able to oxidize exposed sulfidic minerals. Sulfur oxidizing microorganisms are responsible for the generation of sulfuric acid from the oxidation of the elemental sulfur produced by the polysulfide mechanism attack of acid-soluble sulfides (equation 1).

Iron has different properties of ecological interest, which give the Tinto system an interesting perspective. Iron is not only a source of energy for iron oxidizing prokaryotes, but can also be used as an electron acceptor for anaerobic respiration. In addition, ferric iron is responsible for the constant pH of the ecosystem. The dilution of the river by rain and/or neutral tributaries promotes the precipitation of ferric hydroxide and the generation of protons, which maintain the acidity of the river. Although this reaction is reversible, the dilution factor is stronger than evaporation, so an important part of the iron remains precipitated along the river, giving rise to the iron bioformation. Accordingly, the concentration of soluble iron decreases gradually from its origin to the mouth of the river. Furthermore, acidic ferric iron solutions readily absorb harmful UV radiation, thus protecting the organisms growing in its waters (Gómez and Amils, 2002).

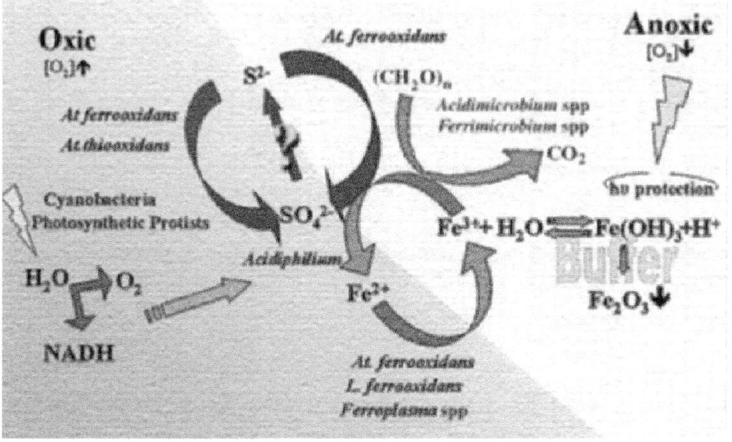

Figure 2. Geomicrobiological model of the Tinto River. The role of the different microorganisms identified and isolated from the river is shown associated to the iron and sulfur cycles operating in the Tinto ecosystem.

This iron-controlled scenario seems reasonable for the chemolithotrophic prokaryotes found in the Tinto ecosystem. However, given that eukaryotic diversity in the Tinto River is much greater than prokaryotic diversity (López-Archilla et al., 2001; Amaral-Zettler et al., 2002), and that most of the primary production of the system derives from the activity of photosynthetic protists, what is the advantage for eukaryotes to develop in an extreme acidic environment? A possible answer to this question may be linked to the limited availability of iron in a neutral pH world.

Although iron is an extremely important element for life (Erhlich, 2002), it is a limiting factor for growth at neutral pH (Price, 1968; Archibald, 1983; Martin, 1990; Boyd et al., 2000). Organisms have developed very specific and elaborate mechanisms to trap iron anywhere they find it (Reid et al., 1993; Braun and Killmann, 1999). Why is this so, if iron is one of the most abundant elements on Earth? In an oxidized atmosphere at neutral pH, soluble iron is rapidly oxidized into insoluble compounds, which are incorporated into anaerobic sediments where sulfate-reducing bacteria may further transform them into pyrite, an even less reactive iron mineral. The geological recycling of these sediments and the microbiology associated with the iron cycle are the only ways to reintroduce this critical element into the biosphere. A possible advantage for the eukaryotes thriving in the extreme conditions of the Tinto is an unlimited iron supply provided by the chemolithotrophs growing on the rich iron sulfides of the Iberian Pyritic Belt (Gómez et al., 2003).

The availability of iron and other heavy metals is considered so important that a model has recently been proposed for an anoxic Proterozoic ocean deprived of iron and other heavy metals as a consequence of sulfate reducing activity. The limitation of these critical elements promoted a global nitrogen cycle crisis, impairing the dissemination of photosynthetic protists during the Proterozoic and the overall evolution of eukaryotes (Anbar and Knoll, 2002).

Has this iron-limiting scenario been a constant in the history of life on Earth? According to different authors, the Archean oceans held important concentrations of soluble iron (Holland, 1973; Anbar and Knoll, 2002). Concerning the compatibility of soluble iron and oxygen, there are two different schools defending rather antagonistic models. According to Ohmoto's Archean oxygenic atmospheric model, most of the iron would have been precipitated due to the fast kinetics of iron oxidation at neutral pH (Ohmoto, 1997). A different situation would be found in a hypothetical acidic ocean (see below), in which iron oxidation by oxygen would be very slow. If we use Holland's conventional model (Holland, 1978; Holland 1984; Kasting, 1993), in which most of the Archean was dominated by an atmosphere with an extremely low concentration of oxygen, the possibility of developing the concept of an Archean iron world becomes feasible.

8. Archaean Iron World

Assuming Holland's Archean atmospheric model and that the absence or low concentration of oxygen precluded the appearance of sulfate reducing activity, the iron concentrations in the Archean may have been as high as 50 μM (Holland, 1973), three orders of magnitude higher than in the actual oxygenated seawater (Anbar and Knoll, 2002). Under these conditions it is reasonable to postulate the appearance of microorganisms able to obtain energy oxidizing iron under anaerobic conditions. This activity, in turn, would have promoted the appearance of iron-reducing metabolism, thus maintaining an operative Archean iron cycle. Furthermore, different authors consider anaerobic respiration using ferric iron as an electron acceptor to be a very primitive mechanism for energy conservation (Lovley, 2000). In our case we stress the advantage of an operative full biological iron cycle instead of linking the oxidation of iron to non-predictive external abiotic reactions. A possible advantage of a microbial iron cycle over a photosynthetic one is its independence from radiation (Gómez and Amils, 2002). It has been suggested that the sulfur cycle was not fully operative at that time. Although microbial sulfur oxidation was probably active (Anbar and Knoll, 2002), different authors consider that sulfate reducing activity was not operative until the Proterozoic, probably as a consequence of the low concentration of sulfate in the Archean (Canfield, 1998).

Is there any evidence for an Archean iron world? Different reports suggest that the presence of high concentrations of iron in the Archean oceans was the result of intense submarine magmatic and hydrothermal activity (Holland, 1973; de Ronde et al., 1984; Barley et al., 1997; Vargas et al., 1998). A high concentration of ferrous iron in solution would have necessarily required a lower concentration of hydrogen sulfide to avoid the formation of insoluble pyrite. The specific geological markers for the Archean, the controversial Banded Iron Formations (BIFs), are fully compatible with this iron world scenario. The massive BIFs deposited from about 3.8 to 2 Gy are attributed to episodic oxidation of oceanic ferrous iron, resulting in precipitates of hematite and mixed oxide magnetite (Cloud, 1973; Holland, 1973), although alternative mechanisms not linked to the photochemical generation of oxygen have been described (Trendall and Morris,

1983; de Duve, 1987; Widdel et al., 1993; Fenchel et al., 1998; Benz et al., 1998). Of the different interpretations offered for the BIFs generation, those related to microbial iron metabolism are not only valid alternatives, but the direct consequence of an operative Archean iron world. "Red Beds", extensive iron oxide deposits beginning at about 2 Gy, provide evidence of continental weathering and therefore of oxygen accumulation in the atmosphere to levels of perhaps 0.01 present atmospheric level. The roughly coincidental disappearance of BIFs and appearance of Red Beds, marked a transition period towards the end of the Archean and the beginning of the Proterozoic and suggests the nearly complete disappearance of reduced metal species from seawater, resulting in a slightly oxidizing atmosphere (Anbar and Knoll, 2002). Interestingly enough, an Archean iron world, based on an operative metabolic iron cycle, does not require a high concentration of soluble sources of reduced iron, because chemolithotrophic microorganisms can efficiently oxidize insoluble iron mineral, similarly to iron reducing microorganisms (Lovley, 2000; Blake and Johnson, 2000).

Concerning the existence of an effective iron cycle that could perform in the anoxic conditions prevailing in the bottom of the Archean oceans and lakes, micropaleontology cannot offer much more information than the products of the microbial lithotrophic metabolism. Unfortunately, isotopic iron fractionation does not seem to be a valuable tool to demonstrate iron metabolism in old sedimentary rocks. However, we have an extant ecosystem, the Tinto River, in which most of the basic questions concerning the geomicrobiology of a putative Archean iron world can be addressed.

9. The Tinto River as a Model for the Archean Iron World

Using specific hybridization probes we can attribute more than 80% of the prokaryotic activity of the Tinto ecosystem to well characterized microorganisms which are basically involved in the iron cycle (González-Toril, 2002; González-Toril et al., submitted). In addition, these activities produce iron bioformation, which over time become fossilized (iron stromatolites). These bioformation show a mineral composition and a banding pattern that reflect both the geomicrobiology of the system and the climate change record, similarly to the Archean BIFs.

Concerning the oxygen, which was most probably absent or present in very low concentration during the Archean period, we know that the Tinto´s iron cycle can operate not only in the aerobic conditions that dominate the surface of the river, but also in the anoxic conditions that prevail at many sites along the river. Furthermore, the Tinto River is an excellent model system to gain insight into how an iron world operated not only in the absence (Archean), but also at increasing concentrations of oxygen (Proterozoic and Phanerozoic). Probably the most interesting areas of the Tinto ecosystem are the ones at the fluctuating interface between oxic and anoxic conditions.

As to the acidic pH, one of the most outstanding characteristics of the Tinto ecosystem is a direct consequence of the physico-chemical properties of ferric iron, the main metabolic product of pyrite oxidation. If ferrous iron, in soluble and/or insoluble mineral form, as well as iron-oxidizing bacteria is present in the same system, ferric iron will be produced in spite of the oxygen concentration. Ferric iron will precipitate as ferric hydroxide releasing protons according to equation 8, maintaining a constant acidic pH in the water column, as long as enough ferric iron remains in solution. The Tinto's geomicrobiology is able to maintain enough soluble iron in the system to keep its acidic pH constant despite the dilutions introduced by neutral tributaries and seasonal flooding consequence of its Mediterranean climatic regime (González-Toril et al., 2002).

Although the history of iron in the biosphere is still an open question, we would like to suggest that the Tinto ecosystem, as well as other iron-rich acidic environments, are relics of an Archean iron world. Obviously the actual conditions in which the Tinto ecosystem operates are different from the ones prevailing in the Archean, but the properties of the microorganisms identified and characterized so far allow us to extrapolate their performance in the Archean. It should be underlined that we are dealing with an extant ecosystem. Conveniently addressed questions would facilitate a more detailed characterization of the system, which should help to clarify its origin and the role of the different components of the habitat in different evolutionary scenarios.

10. Conclusions

The geomicrobiological characterization of the Tinto River (Iberian Pyritic Belt, Southwestern Spain) strongly suggests that the extreme conditions of the system (acidity and high concentration of heavy metals) are the result of the metabolic activity of chemolithotrophic prokaryotes associated with the iron and sulfur cycles. The Tinto ecosystem is under the control of iron, which is used not only for energy conservation but also to maintain a constant acidic pH and to protect microorganisms from radiation. The analysis of iron-rich stromatolitic formations generated by precipitation of iron minerals on the surface of biofilms, showing structures similar to ancient iron-rich deposits formed prior to any known mining activity in the area, imply a natural origin for the river. An Archaean iron world model based on the properties of the Tinto ecosystem is presented.

11. Acknowledgements

This work was supported by grants BIO99-0184 (CICYT), BXX2000-1385 (DGICYT) 07M/0038/2001 (CAM) and institutional grants to the Centro de Astrobiología and the Centro de Biología Molecular (F. Areces).

12. References

Amaral-Zettler, L.A., Gómez, F., Zettler, E., Keenan, B.G., Amils, R., Sogin, M.L. (2002) Eukaryotic diversity in Spain's River of Fire. *Nature* **417**, 137.

Amils, R., González-Toril, E., Gómez, F., Fernández-Remolar, D., Rodríguez, N. (2000) Geomicrobiology of an extreme acidic environment: the Tinto River case. *2000 Spring Meeting, American Geophysical Society*, Abstract B22B-05.

Amils, R., Gónzalez-Toril, E., Fernández-Remolar, D., Gómez, F., Rodríguez, N., Durán, C. (2003) Interaction of the sulfur and iron cycles, the Tinto River case. *Rev/Views in Environmental Science and Biotechnology*, in press.

Anbar, A.D., Knoll, A.H. (2002) Proterozoic ocean chemistry and evolution: a bioinorganic bridge? *Science* **297**, 1137- 1142.

Archibald, F. (1983) *Lactobacillus plantarum*, an organism not requiring iron. *FEMS Microbiology Letters* **19**, 29-????.

Asensi, A., Diaez, B. (1987) Andalucía Occidental, In: M. Peinado and S. Rivas-Martínez (eds.) *La vegetación de España.* U. de Alcalá de Henares, Colección Aula Abierta **3**, 199-230.

Avery, D. (1974) *Not on Queen Victoria's Birthday*, Collins, London.

Bachofen, R., Ferloni, P., Flynn, L. (1998) Review: microorganisms in the subsurface. *Microbiological Research* **153**, 1-22.

Barley, M.E., Pickard, A.L., Sylvester, P.J. (1997) Emplacement of a large igneous province as a possible cause of banded iron formation 2.45 billion years ago. *Nature* **385**, 55-58.

Benz, M., Brune, A., Schink, B. (1998) Anaerobic and aerobic oxidation of ferrous iron at neutral pH by chemoheterotrophic nitrate-reducing bacteria. *Archievs of Microbiology* **169**, 159-165.

Blake II, R., Johnson, D.B. (2000) Phylogenetic and biochemical diversity among acidophilic bacteria that respire iron, In: D.R. Lovely (ed.) *Environmental microbe-metal interactions*, ASM Press, Washington, pp. 53- 78.

Boulter, C.A. (1996) Did both extensional tectonics and magmas act as major drivers of convection cells during the formation of the Iberian Pyrite Belt massive sulfide deposits? *Journal of the Geological Society of London* **153**, 181-184.

Boyd, W. P. et al. 2000. A mesoscale phytoplankton bloom in the polar Southern Ocean stimulated by iron fertilization. *Nature* **407**, 695-702.

Braun, V., Killmann, H. (1999) Bacterial solutions to the iron-supply problems. *Trends in Biochemical Sciences* **24**, 104-109.

Bridge, T.A.M., Johnson, D.B. (2000) Reductive dissolution of ferric iron minerals by *Acidiphilium* SHJ. *Geomicrobiology Journal* **17**, 193-206.

Canfield, D.E. (1998) A new model for Proterozoic ocean chemistry. *Nature* **396**, 450.

Chapelle, F.H., O'Nelly; K., Bradley, P.M., Methé, B.A., Ciufo, S.A., Knobel, L.L., Lovley, D.R. (2002) A hydrogen-based subsurface microbial community dominated by methanogens. *Nature* **415**, 312-314.

Clemente, L., Menanteau, L., Rodríguez-Vidal, J. (1985) Los depósitos holocenos en el estuario de los ríos Tinto y Odiel (Huelva, España). *Actas I Reunión del Cuaternario Ibérico* **I**, 339-353.

Cloud, P. (1973) Paleoecological significance of the Banded Iron Formations. *Economic Geology* **68**, 1135-1143.

Colmer, A.R., Temple, K.L., Hinkle, H.E. (1950) An iron-oxidizing bacterium from the acid drainage of some bituminous coal mines. *Journal of Bacteriology* **59**, 317-328.

Davis Jr., R.A., Welty, A.T., Borrego, J., Morales, J.A., Pendon, J.G., Ryan, J.G. (2000) Rio Tinto estuary (Spain): 5000 years of pollution. *Environmental Geology* **39**, 1107-1116.

de Duve, C. (1987) Selection by differential molecular survival: a possible mechanism of early chemical evolution. *Proceedings of the National Academy of Science USA* **84**, 8253-8256.

de Ronde, C.E., De Wit, M.J., Spooner, E.T.C. (1984) Early Archaean (<3.2 Ga) Fe-oxide-rich, hydrothermal discharge vents in the Barbeton greenstone belt, South Africa. *Geolog. Soc. Am. .Bull.* **106**, 86-104.

Edwards, K.J., Bond, P.I., Gihrin, T.M., Banfield, J.F. (2000) An archaeal iron oxidizing extreme acidophile important in acidic mine drainage. *Science* **287**, 1796-1798.

Ehrlich, H.L. (2001) Past, present and future of biohydrometallurgy. *Hydrometallurgy* **59**, 127-134.

Ehrlich, H.L. (2002) *Geomicrobiology*, fourth edition, Marcel Dekker, Inc., New York

Elbaz-Poulichet, F., Braungardt, C., Achterberg, E., Morley, N., Cossa, D., Beckers, J.M., Nomérange, P., Cruzado, A., Leblanc, M. (2001) Metal biogeochemistry in the Tinto-Odiel rivers (Southern Spain) and in the Gulf of Cadiz: a synthesis of the results of TOROS project. *Continental Shelf Research* **21**, 1961-1973.

Fenchel, T., King, G.M., Blackburn, T.H. (1998) *Bacterial Biogeochemistry:The Ecophysiology of Mineral Cycling*, 2nd edition. Academic Press, San Diego.

Fernández-Remolar, D., Rodríguez, N., Gómez, F., Amils, R. (2003) The geological record of an acidic environment driven by the iron hydrochemistry: the Tinto River system. *Journal of Geophysical Research, Planets*, in press.

Gehrke, T., Hallmann, R., Sand, W. (1995) Importance of exopolymers from *Thiobacillus ferrooxidans* and *Leptospirillum ferrooxidans* for bioleaching, In: T. Vargas, C.A. Jerez., K.V. Wiertz, and H. Toledo (eds.) *Biohydrometallurgical Processing*, vol 1, Universidad de Chile, Santiago, pp. 1-11.

Geen, A. van, Adkins, J.F., Boyle, E.A., Nelson, C.H., Palanques, A. (1997). A 120-year record of widespread contamination from mining of the Iberian pyrite belt. *Geology* **25**, 291- 294

Godd, T. (1992) The deep hot biosphere. *Proceedings of the National Academy of Science USA* **89**, 6045-6049.

Golyshina, O.V., Pivovarova, T.A., Karavaiko, G.I., Kondratèva, T.F., Moore, E.R.B., Abraham, W.R., Lündsford, H., Timmis, K.N., Yakimov, M.M., Golyshin, P.N. (2000) *Ferroplasma acidiphilium* gen. nov., sp. nov., an acidophilic, autotrophic, ferrous-iron-oxidizing, cell-wall-lacking, mesophilic member of the *Ferroplasmaceae* fa. nov., comprising a distinct lineage of the Archaea. *Int. J. Syst. Evol. Microbiol.* **50**, 997-1006.

Gómez, F., Amils, R. (2002) Life as an environmental transformer, In: B. Montesinos, A. Jiménez and E.F. Guinan (eds.) *The Eevolving Sun and its Influence on Planetary Environments*, The Astronomical Society of the Pacific Conference Series, San Francisco, vol 269, pp. 339-352.

Gómez F, Fernández-Remolar, D González-Toril E., Amils R (2003) The Tinto River, an extreme Gaian enviroment, In: L. Margulis, J. Miller, P. Boston, S. Schneider and E. Crist (eds.) *Gaia 2002*. MIT Press, Boston, in press.

González-Toril, E. (2002) Ecología molecular de la comunidad microbiana de un ambiente extremo: el Río Tinto. Ph.D. Thesis, U. Autónoma de Madrid, Spain.

González-Toril, E., Gómez, F., Rodríguez, N., Fernández-Remolar, D., Zuluaga, J., Marín, I., Amils, R. (2002) Geomicrobiology of the Tinto River, a model of interest for biohydrometallurgy. *Hydrometallurgy* in press.

Hansford, G.S., Vargas, T. (2001) Chemical and electrochemical basis of bioleaching processes. *Hydrometallurgy* **59**, 135-145.

Holland, H.D. (1973) The oceans: a possible source of iron in iron-formations. *Economic Geology* **68**, 1169-1172.

Holland, H.D. (1978) *The Cchemistry of the Atmosphere and Oceans*. Wiley, New York.

Holland, H.D. (1984) *TheChemical Evolution of the Atmosphere and Oceans*. Priceton University Press, New York.

Johnson, D.B. (1998) Biodiversity and ecology of acidophilic microorganisms. *FEMS Microbiological Ecology* **27**, 307-317.

Johnson, D.B. (1999) Importance of microbial ecology in the development of new mineral technologies, In: R. Amils and A. Ballester (eds.) *Biohydrometallurgy and the Environment toward the Mining of the 21st century*, vol A.. Elsevier, Amsterdam, pp. 645-656.

Kasting, J.F. (1993) Earth´s early atmosphere. *Science* **259**, 920-926.

Leblanc, M., Morales, J.A., Borrego, J., Elbaz-Poulichet, F. (2000) A 4,500-year-old mining pollution in Southwestern Spain: long-term implications for modern mining pollution. *Economy Geology* **95**, 655-662.

Leistel, J.M., Marcoux, E., Thiéblemont, D., Quesada, C., Sánchez, A., Almodóvar, G.R., Pascual, E., Sáez, R. (1998) The volcanic-hosted massive sulphide deposits of the Iberian Pyrite Belt. *Mineralium Deposita* **33**, 2-30.

Lescuyer, J.L., Leistel, J.M., Marcoux, E., Milési, J.P., Thiéblemont, D. (1998) Late Devonian-Early Carboniferous peak sulphide mineralization in the Western Hercynides. *Mineralium Deposita* **33**, 208-220.

López-Archilla, A.I., Marín, I., Amils, R. (1993) Bioleaching and interrelated acidophilic microorganisms from Río Tinto, Spain. *Geomicrobiology Journal* **11**, 223-233.

López-Archilla, A.I., Marín, I., González, A., Amils, R. (1995) Identification of fungi from an acidic river, In: L. Rossen, V. Rubio, M.T. Dawson, and J. Frisvard (eds.) *Fungal identification techniques*. European Commission, Luxemburg, pp. 202-211.

López- Archilla, A.I., Amils, R. (1999) Microbial community composition and ecology of an acidic aquatic environment. *Microbial Ecology* **41**, 20-35.

López-Archilla, A.I., Marín, I., Amils, R. (2001) Microbial community composition and ecology of an acidic aquatic environment: the Tinto River, Spain. *Microbial Ecology* **41**, 20-35.

Lovley, D.R. (2000) Fe(III) and Mn(IV) reduction, In: D.R. Lovley (ed.) *Environmental microbe-metal interactions*, ASM Press, Washington, pp. 3- 30.

Martin, J.H. (1990) Glacial-interglacial CO_2 chamge: the iron hypothesis. *Paleoceanography* **5**, 1-13.

Moreira, D., López-Archilla, A.I., Amils, R., Marín, I. (1994) Characterization of two new thermoacidophilic microalgae: genome organization and comparison with *Galdieria sulphurica*. *FEMS Microbiology Letters* **122**, 109-114.

Morrison, D. (2001) The NASA Astrobiology Program. *Astrobiology* **1**, 3-14.

Ohmoto, H. (1997) When did the earth´s atmosphere become oxic? *Geochemical News* **93**, 13.

Pedersen, K. (2000) Exploration of deep intraterrestrial microbial life: current perspectives. *FEMS Microbiol. Letters* **185**, 9-16.

Phillips, J.A. (1881) Note on the ocurrence of remains of recent plants in brown iron-ore. *Journal of the Geological Society of London* **27**, 1-5.

Price, C.A. (1968) Iron compounds and plant nutrition. *Annual Review of Plant Physiology* **19**, 239-248.

Reid, R.T., Live, D.H., Faulkner, D.J., Buttler, A. (1993) A siderophore from a marine bacterium with an exceptional ferric ion affinity constant. *Nature* **366**, 455-458.

Rivas-Martínez, S. (1987) Nociones sobre fitosociología, biogeografía y bioclimatología, In: M. Peinado and S. Rivas-Martínez (eds.) *La vegetación de España*. U. de Alcalá de Henares, Colección Aula Abierta **3**, 19-45.

Rodríguez-Vidal, J., Mayoral, E., Pedón, J.G. (1985) Aportaciones paleoambientales al tránsito Plio-Pleistoceno en el litoral de Huelva. *Actas I Reunión del Cuaternario Ibérico* **I**, 447-459.

Rothschild, L.J., Mancinelli, R.L. (2001) Life in extreme environments *Nature* **409**, 1092-1101.

Sand, W., Gehrke, T., Hallmann, R., Schippers, A. (1995) Sulfur chemistry, biofilm and the (in)direct attack mechanism- a critical evaluation of bacterial leaching. *Applied Microbiology and Biotechnology* **43**, 961-966.

Sand, W., Gehrke, T., Jozsa, P.G., Schippers, A. (2001) (Bio)chemistry of bacterial leaching-direct vs. indirect bioleaching. *Hydrometallurgy* **59**, 159-175.

Silverman, M.P., Ehrlich, H.L. (1964) Microbial formation and degradation of minerals. *Advances in Applied Microbiology* **6**, 153-206.

Trendall, A.F., Morris, R.C. (1993) *Iron-Formation: Facts and Problems,* Elsevier, Amsterdam.

Vargas, M., Kashefi, K., Blunt-Harris, E.L., Lovely, D.R. (1998) Microbiological evidence for Fe(III) reduction on early Earth. *Nature* **395**, 65- 67.

Widdel, F., Schnell, S., Heising, S., Ehrenreich, A., Assmus, B., Schink, B. (1993) Ferrous iron oxidation by anoxygenic phototrophic bacteria. *Nature* **362**, 834-836.

Biodata of **Warwick F. Vincent** author (with co-authors D. Meuller, P. Van Hove and C. Howard-Williams) of ***"Glacial Periods on Early Earth and Implications for the Evolution of Life"***

Dr. Warwick F. Vincent is Professor of Limnology and Canada Research Chair in Aquatic Ecosystem Studies at Laval University, Québec City, Canada. He obtained his Ph.D. in Ecology from the University of California, Davis, in 1977. He is a Fellow of the Royal Society of Canada and an Honorary Fellow of the Royal Society of New Zealand. His current research concerns underwater light, the ecology of Cyanobacteria and the effects of climate on high latitude lakes, wetlands, rivers and coastal seas.

E-mail: **warwick.vincent@bio.ulaval.ca**

Mr. Derek Mueller obtained his Master's degree at McGill University working on cryoconite communities on glaciers in the McMurdo Dry Valleys, Antarctica, and on Axel Heiberg Island in the Canadian High Arctic. He is presently a doctoral student at Laval University, working on the microbial ecology and ice dynamics of the Ward Hunt Ice Shelf, Canada.

Mr. Patrick Van Hove obtained his Bachelor's degree in Biochemistry from Laval University. He is conducting his doctoral research on the molecular ecology of High Arctic lakes with Dr. Warwick F. Vincent (Université Laval, Québec, Canada) and Dr. Annick Wilmotte (Université de Liège, Belgium). He has a special interest in extreme environments and Astrobiology.

Dr. Clive Howard-Williams is General Manager (Freshwater) of New Zealand's National Institute of Water and Atmospheric Research. His Ph.D. is in Botany from the University of London. He is a Fellow of the Royal Society of New Zealand and has conducted limnological research in many parts of the world, from the tropics to the Arctic and Antarctica. He is a Vice-president of the Scientific Committee on Antarctic Research.

E-mail: **c.howard-williams@niwa.co.nz**

W.F. Vincent

D. Mueller

P. Van Hove

C. Howard-Williams

J. Seckbach (ed.), Origins, 481–501.

GLACIAL PERIODS ON EARLY EARTH AND IMPLICATIONS FOR THE EVOLUTION OF LIFE

WARWICK F. VINCENT[1], DEREK MUELLER[1], PATRICK VAN HOVE[1] AND CLIVE HOWARD-WILLIAMS[2]
[1]Centre d'études nordiques, Université Laval, Québec City, QC G1K 7P4, Canada. [2]National Institute of Water and Atmospheric Research, Christchurch, New Zealand.

1. Introduction

Modern-day thermophiles and their habitats such as hydrothermal vents, surface hot springs and other high temperature environments are often viewed as analogs for Precambrian microbial ecosystems. The prebiotic Earth is thought to have been extremely hot, with the origin of life delayed by several hundred million years of cooling to temperatures that allowed the stable formation of biomolecules and ultimately cells (Schwartzman 1999, and references therein). Similarly, from analysis of 16S rDNA phylogenies, the tree of life is often considered to be rooted in high temperature (hyperthermophilic) microbes (Woese 1987). There is new geological evidence, however, that the early Earth may have cooled much more rapidly from its original Hadean state than was previously surmised (Valley et al. 2002). Molecular arguments concerning the hyperthermal roots of life now seem less certain (Doolittle 1999, Brochier and Philippe 2002). Furthermore, there is mounting support for the view that the Precambrian biosphere experienced extreme low temperature conditions at several intervals during the Paleo- and Neoproterozoic, and perhaps even during the earliest steps in the emergence and evolution of life (Nisbet and Fowler 1999).

The timing, duration and extent of cooling during the Precambrian are subjects of considerable discussion and debate, and views range from localized glacial activity to complete freeze-up of the ocean surface. Extreme cold has largely been considered as a negative factor that would cause severe inhibition of Precambrian biological processes, eventually leading to a widespread loss of species and perhaps even the total extinction of surface life (Williams et al. 1998). On the other hand, little attention has been given to the success and biodiversity of microbial communities that thrive in a variety of extreme cold environments in the modern-day polar regions and which indicate the potential for growth, survival and evolution despite prolonged subzero temperatures.

In the present article, we first briefly summarize the geological, geochemical and modeling evidence for freeze-up episodes during the Precambrian and the various scenarios that have been surmised from this evidence. We then examine the diverse range of microbial cryo-ecosystems that occur today in ice-bound and other extreme cold environments and that could provide analogs of low temperature systems on early

Earth. We place special emphasis on ice shelf ecosystems in the Arctic and Antarctic, a focus of our current research, and conclude by examining some of the evolutionary implications of extreme cold in the microbial habitats of the early Earth.

2. Earth's Earliest Freeze-up

Traditionally, the Precambrian has been thought of as a period characterized by prolonged extreme high temperatures at the surface of the planet and in the overlying atmosphere. During Earth's accretion phase, 4.56-4.45 Ga (10^9 years) before the present, there was sufficient heat from meteoritic impacts, collision with a Mars-size body, the formation of the metallic core, and high radioactivity, to melt the entire planet (Valley et al. 2002). Depending on the magnitude of ongoing bombardment by meteorites and the nature of the atmosphere, these hyperthermal, non-aqueous conditions may have persisted up until 3.6-3.8 Ga before the present, the date of the earliest known water-laid sediments. However, ancient zircon crystals (zirconium silicate, a component of granitic rocks that forms in the presence of water) have recently been discovered, and their analysis suggests that the early Earth may have cooled much more rapidly (Wilde et al. 2001). These crystals have been dated at 4.404 Ga and their isotopic signatures ($\delta^{18}O$) imply surface temperatures below 200 ^{o}C that would allow the existence of oceans at that time, some 500 Ma earlier than in previous models (Valley et al. 2002).

Could this early ocean have eventually cooled to freezing at or before the emergence of life? Some calculations suggest that freezing could have occurred even during the accretion phase of early Earth. Mukhin and Pimenov (2002) conclude that there would have been very high rates of convective cooling of surface impact craters, and that the early growing Earth could never have supported a steam atmosphere. Low partial pressures of CO_2 and H_2O in the overlying atmosphere would result in minimal greenhouse warming, and 90% of the Earth's surface would be below the freezing point of water. In this scenario, prebiotic Earth would be characterized by impact crater hot spots amidst a vast expanse of cold traps, conditions that could have favored the preservation and accumulation of organic molecule precursors formed during impact events (Mukhin et al. 1989).

Although the first scenarios of the origin of life invoked high temperature photochemical reactions, it is now known that biomolecules are highly unstable when subjected to prolonged heating (Levy and Miller 1998, Bada and Lazcano 2002) and it is unlikely that an RNA world could have persisted in hot water environments (Moulton et al. 2000). On the other hand, nucleic acids are preserved for long periods of time in the cold; for example, intact DNA has been extracted from ancient glacial ice (reviewed in Priscu and Christner 2003) while decomposition rates are rapid under warmer conditions. Even the ice-formation process itself could have favored early steps in the emergence of life. Freeze-up causes a concentration of solutes that can result in polymerization and the formation of biomolecule-like entities including oligonucleotides (Kanavarioti et al. 2001). This is analogous to the solute-concentrating processes that occur in clays, aerosols and evaporating water that can similarly lead to the synthesis of organic polymers.

The radiative output from the sun has gradually increased over Earth's history and was some 25% lower in the early Precambrian relative to today. Sagan and Mullen (1972) concluded that such a low incident radiative flux would result in a complete freeze-up of the Earth's oceans, yet noted the evidence of liquid water conditions in the Precambrian. This 'early faint sun paradox' is resolved by assuming that the concentration of greenhouse gases, especially CO_2 and CH_4, were massively higher relative to today. There is debate, however, as to whether the initial high CO_2 concentrations could be maintained in the atmosphere of early Earth, particularly in the presence of liquid water that would dissolve the gas to form carbonates (Mukhin and Pimenov 2002). An atmosphere instead dominated by carbon monoxide, perhaps over a frozen Earth, appears to be more compatible with current notions of prebiotic synthesis and the origin of life (Miyakawa et al. 2002).

Schwartzman (1999) argues that the overall constraint on the emergence and evolution of life has been the rate of weathering reactions, specifically the reaction of CO_2 with silicates in the Earth's crust, for example for plagioclase, a common mineral in basalt:

$$CaAl_2Si_2O_8 + 2CO_2 + 3H_2O \rightarrow Ca(HCO_3)_2 + Al_2Si_2O_5(OH)_4 \text{ (kaolinite)} \qquad (1)$$

He suggests that the consumption of atmospheric CO_2 by these reactions would have accelerated once biological communities emerged. In particular, the soils produced in developing terrestrial ecosystems would cause accelerated weathering by trapping pore water within a matrix that has a large specific surface area for chemical reactions. The resultant drop in atmospheric CO_2 and associated cooling could in turn have gradually set the stage for the evolution of less thermotolerant life-forms. However, the early emergence and development of ecosystems based on heat-loving prokaryotes now seems less certain.

Several biological lines of evidence have been used to argue the hyperthermal origins of life. The earliest apparent fossils resemble modern day cyanobacteria, such as oscillatorians that occur in hot springs (Schopf and Klein 1992). However, oscillatorians are also widely distributed in extreme cold environments, and in many parts of the north and south polar regions they are the ecosystem dominants (Vincent 2000). Deeply rooted Bacteria and Archaea were thought to be largely thermophiles and consistent with hot surface conditions on the Archaean Earth (Woese 1987). However, Archaea are now known to contain many extreme cold water representatives, for example in Antarctic surface waters (Massana et al. 1998) as well as throughout the frigid deep ocean. Recent studies on the latter suggest that these cold water Archaea (Crenarchaeota) may have a total population size of 10^{28} cells, making them one of the ocean's most abundant cell types (Karner et al. 2001).

A hyperthermophilic root for the tree of life based on molecular clock analyses of 16S rDNA is often cited by researchers working on high-temperature environments such as geothermal soils (Huber et al. 2000) and deep subsurface rocks (Pedersen 1997), however these calculations now seem questionable. This type of analysis assumes linear evolution and hierarchical branching, and the method is flawed if there are deeply branching relationships (Gribaldo and Philippe 2002). Lateral gene transfer is increasingly viewed as a major feature of prokaryotic evolution, with strong evidence of transfer between distantly related organisms, even between the bacterial and archaeal domains (Nelson et al. 1999). This common pattern of genetic exchange throws doubt on past interpretations of the ribosomal DNA tree (Doolittle 1999, Wolf et al. 2002). Phylogenies based on other genes do not support a 'hot start' to life and there is evidence that thermophily arose relatively late in evolutionary time. Studies of the GC content of rRNA in different organisms suggest a mesophilic common ancestor (Galtier et al. 1999). Analyses of the thermal stability and structure of biomolecules also seem more consistent with a 'cold start' hypothesis (Forterre 2002). Furthermore, new records and a re-analysis of the rDNA database suggests that a non-thermophilic group, the Planctomycetes, were the earliest bacteria to emerge (Brochier and Philippe 2002).

If the early Earth did indeed experience widespread glaciation there would need to be some mechanism to allow eventual thawing and open water conditions. One suggestion is that periodic bombardments could provide such energy (Bada et al. 1994). Small impacts could have pierced any surface crusts of ice over the oceans leading to the venting of greenhouse gases into the atmosphere and accelerated warming. Larger impacts could result in sufficient energy to cause complete melting. Bada et al. (1994) calculate that impacts of bolides greater than 100 km diameter would release enough thermal energy to melt 300 m of ice cover on a frozen ocean.

3. Glaciations during the Paleoproterozoic

The earliest geological evidence of widespread glaciation on Precambrian Earth is from rocks of the Paleoproterozoic era that are dated at about 2.4 Ga before the present (Fig. 1). Glacially derived sediments of this age occur in the Transvaal Supergroup of southern Africa, and paleomagnetic studies indicate that they formed within the tropics at a paleolatitude of 11 ± 6 degrees. These beds are interspersed with volcanic flows and are overlaid by a 100 m thick layer of banded iron formations. Further above is a 45 m stratum of manganese-rich carbonate (the Kalahari Manganese Field, the world's largest land-based reserve of Mn), overlaid by dolomite.

Kirschvink et al. (2000) have interpreted the Transvaal sequence as indicating that the global climate experienced a rapid and massive set of changes during the Paleoproterozoic. Specifically, they argue that global cooling caused run-away albedo conditions in which extreme cooling led to a complete freeze-up of the surface oceans, a so-called 'snowball Earth' event ('une Terre boule de neige', Kirschvink 2002) that persisted for tens of millions of years. By this scenario, the prolonged ice cover and cessation of thermohaline circulation prevented the re-oxygenation of the bottom waters of the ocean, leading to anoxia and accumulation of reduced iron (Fe II) and manganese (Mn II). Eventual warming caused by the gradual atmospheric accumulation of CO_2 from volcanoes allowed the melting of ice, and triggered a cyanobacterial bloom and photosynthetic oxygen production that in turn caused the massive precipitation of oxidized iron (Fe III), then manganese (Mn IV). Kirschvink et al. (2000) suggest that the resultant drop in trace metal availability may have resulted in competition for these elements and a divergence in the use of Fe or Mn by the enzyme superoxide dismutase. Their phylogenetic analysis of the amino acid sequences of this enzyme showed that such divergence may have taken place well after the initial separation of the Bacteria and Archaea lineages, perhaps consistent with a snowball Earth scenario.

4. Glaciations during the Neoproterozoic

Several lines of evidence indicate that extreme cooling also occurred towards the end of the Precambrian resulting in a series of snowball Earth events between 0.6 and 0.8 Ga before the present. Glaciogenic rocks from the Neoproterozoic are found today on most continents, indicative of widespread glaciation at that time. Paleomagnetic analysis of some of these rocks show that they have low paleolatitudes, suggesting that glacial conditions extended all the way to the tropics where the continental landmasses were centered (Hoffman and Schrag 2002). The break up of the supercontinent Rodinia at that time has been suggested as a potential mechanism favoring global cooling. Such a break up could have accelerated rock weathering, the consumption of atmospheric CO_2 by reactions such as Eq. 1., and may also have increased the availability of habitat for CO_2-fixing stromatolites in shallow seas (Hoffman et al. 1998). The localization of landmasses in the tropics would have also raised planetary albedo, further enhancing the tendency towards global glaciation (Kirschvink 1992).

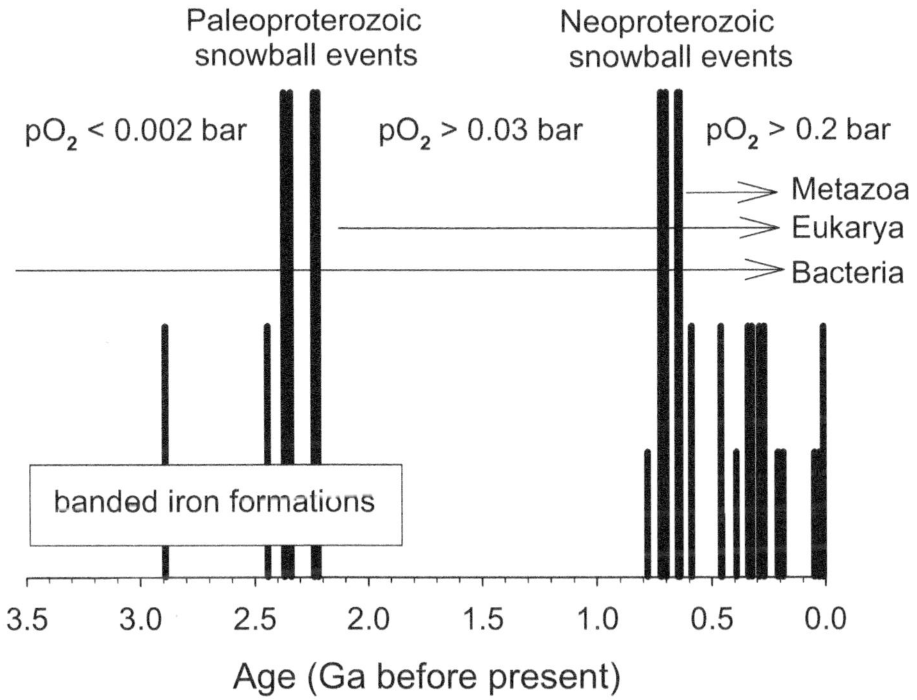

Figure 1. Glacial events during the last 3.5 Ga (vertical bars). Banded iron formations occurred continuously up to 1.86 Ma before the present, and also occur in Neoproterozoic glaciogenic rock formations. Modified from Hoffman and Schrag (2002).

A detailed analysis of the C-isotopic geochemistry of Neoproterozoic sequences seems consistent with snowball Earth events. The sequences show an unprecedented decrease in $\delta^{13}C$, followed by a slow return to normal levels (Hoffman et al. 1998). As the Earth got colder, the $\delta^{13}C$ of the oceans, and subsequently carbonate rocks, would drop due to a reduction in primary productivity. During the ensuing glaciation, the atmosphere and ocean would take on an isotopic signal similar to CO_2 from volcanic outgassing (Hoffman et al. 1998). Since carbonate production was curtailed by the global glaciations, the stratigraphic record includes lengthy gaps during these epochs. Also during this time, the ice cover would reduce the exposure of rock to silicate-weathering and would interrupt the hydrological cycle, allowing little precipitation. These reduced losses would cause volcano-derived CO_2 to gradually build up in the atmosphere, ultimately culminating in a strong greenhouse effect and rapid melting and break-up of the sea ice. After ocean-atmosphere communication was restored, the carbonates would take on a slightly more positive signal as atmospheric carbon was hydrated and precipitated, followed by a gradual, yet slight decrease due to Rayleigh distillation effects (Hoffman et al. 1998). It is also speculated that the drop in the carbonate $\delta^{13}C$ signature immediately prior to global freeze-up could have been the result of slow leakage of methane produced in anoxic oceanic basins (Schrag et al. 2002).

The Neoproterozoic glaciogenic sequences are overlain by thick caps of carbonate, specifically meters to tens of meters of dolostone and limestone. These are thought to represent the rain-out and subsequent carbonate precipitation of inorganic carbon that had built up in the atmosphere and ocean (Hoffman et al. 1998). These cap rocks are usually found above sequences showing glacial disturbance and are geographically widespread implying a global-scale process. The rocks indicate rapid depositional processes and severe perturbation of the global carbon cycle (Hoffman et al. 1998).

The analysis of banded iron formations, deposits of Fe(III) and chert, provides further evidence of large-scale freeze-up. These deposits are common in the stratigraphic record of early Earth, disappear after 1.86 Ga before the present and then reappear only within the Neoproterozoic glacial deposits (Fig. 1). This reappearance in the record after a 1.5 Ga absence is consistent with stagnation of the ocean and anoxia beneath the thick sea ice of a Neoproterozoic snowball Earth. The accumulation of reduced iron in the absence of reduced sulfur, and the subsequent precipitation of Fe(III) by oxygenic photosynthesis would have occurred during the ice-melt phase (Kirschvink 1992).

A variety of modeling analyses lend some support to the Neoproterozoic snowball Earth hypothesis. Application of a simple energy budget model shows the precarious balance between the two endpoints of complete global freeze-up and an ice-free Earth, and also illustrates the hysteresis in CO_2 conditions that control the planetary shift between these two disparate conditions (Fig. 2). The results of a GCM simulation using a 50 m slab ocean shows that under late Precambrian conditions of a 6% reduced solar constant, sea-ice and subfreezing temperatures occur from the poles to the Equator once atmospheric CO_2 concentrations fall below a threshold of 1700 ppm, (Jenkins and Smith 1999). Further calculations incorporating the latent heat of freezing suggest that the resultant ice could be relatively thin in the tropics and would allow some penetration of sunlight for photosynthesis into the seawaters beneath (McKay 2000). More recent modeling analyses based on equatorial temperatures of -30 °C and a spectral model for sunlight absorption suggest equilibrium ice thicknesses of the order of 1 km; however, under conditions of slightly warmer surface temperatures and low albedo (for example associated with the transport of sediment to ice shelf surfaces in shallow seas; see Section 6.1, below) the simulated ice thickness is less than 1 m (Warren et al. 2002).

An ice sheet model has also been run using estimates of the Neoproterozoic climate regime and the results imply a band of open water in the tropics during snowball Earth events (Hyde et al. 2000). This 'slushball Earth' model is consistent with some observations but has been criticized as having an unrealistic distribution of continental landmasses; furthermore, the incomplete freeze-up that it predicts would not allow the ocean-wide anoxia that has been invoked to explain to the mineral sequences (Hoffman and Schrag 2002).

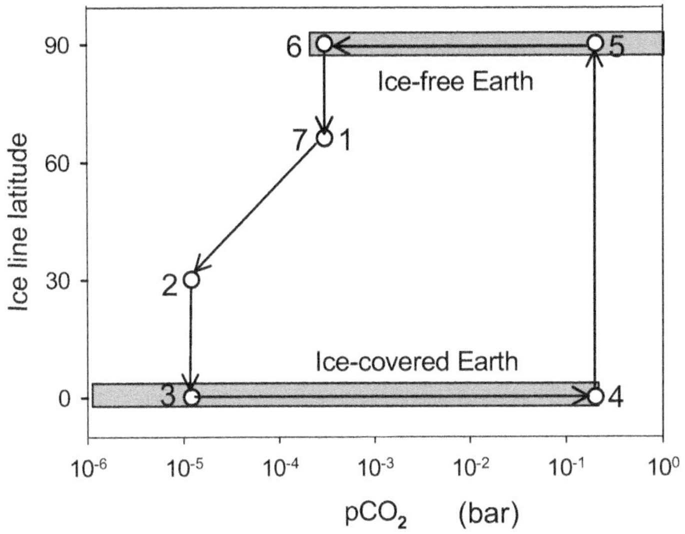

Figure 2. The postulated snowball Earth cycle. A decrease in pCO_2 from current values (1) results in ice expansion to lower latitudes (2) and a run-away albedo effect, ultimately causing a complete freeze-up to the Equator (3). A massive increase in pCO_2, for example via prolonged volcanic activity, would be required to melt the thick ice cover (4). Temperatures then rise to extreme values that preclude ice even at the highest latitudes (5); only after prolonged CO_2 drawdown (6) would cooling allow polar ice to the extent observed today (7). Based on Fig. 6 in Hoffman and Schrag (2002).

An alternative explanation of glaciogenic deposits at tropical latitudes is that the obliquity of the Earth (the tilt of the Earth's axis of rotation relative to its orbital plane) shifted to a large angle (Williams et al. 1998). With an obliquity greater than 54°, mean annual temperatures would be lower at the equator than at the poles and glaciation would be more likely at lower latitudes. Williams (1993) argued that an abrupt decrease in obliquity after 600 Ma resulted in an amelioration of climate that set the stage for the rise of metazoans. The obliquity model, however, has little direct support and seems inconsistent with several aspects of the geological record (Hoffman and Schrag 2002).

The snowball Earth hypothesis has stimulated intense debate in the Earth sciences community and there is currently no consensus about the extent of Neoproterozoic glaciation. Although complete freeze-up appears to reconcile several disparate types of data including the carbonate stratigraphy, the $\delta^{13}C$ isotopic record, glaciogenic rock sequences and the banded iron formations, there are an increasing number of results that do not seem concordant. These include sedimentological data from northern Norway (Arnaud and Eyles 2002), stratigraphic analysis of Neoproterozoic glacial deposits in Oman (Leather et al. 2002), new analyses of the carbon and strontium isotopic records (Kennedy et al. 2001), evidence of glacial sediment deposition from free-floating icebergs (Condon et al. 2002), and modeling simulations of the Neoproterozoic climate and ocean with different land mass configurations (Poulsen et al. 2002).

One of the persistent challenges to the snowball Earth hypothesis is the argument that global freeze-up would result in the extinction of all living organisms, while the fossil record implies a continuity of life (Hoffman and Schrag 2002). Open water environments such as polynyas, flaws and tidecracks provide rich environments for diverse life-forms in today's polar regions. For example, more than 250 protist taxa have been identified in the North Water polynya between Ellesmere Island and Greenland (Lovejoy et al. 2002). Openings of this sort would clearly allow marine life to continue despite extensive glaciation, and would also allow anoxia elsewhere. Relatively thin ice cover in the tropics would similarly allow photosynthetic activity and complex food webs in seawaters beneath (McKay 2000), although other modeling scenarios imply that the ice could be too thick to allow such photosynthesis (Warren et al. 2002)

Earth scientists appear to be accepting the point of view that prokaryotes thrive in many extreme cold environments today on Earth, but still seem unaware that such environments can also harbor much more sophisticated organisms including eukaryotic algae and multicellular animals (metazoans). These modern-day cryo-ecosystems occur throughout alpine and polar regions, and can also be found in the cold temperate zones.

5. Modern-day Cryo-ecosystems

Micro-organisms are known to survive and even flourish in a wide range of present-day ice environments. Most bacteria and many eukaryotes are able to maintain prolonged dormancy under conditions of complete freeze-up, and in fact many microbes are routinely stored in liquid nitrogen (-196 °C). Conditions during the freezing process are critically important. Severe physical damage can be caused by ice crystal formation and there are osmotic and other chemical stresses caused by the freeze-concentration of solutes. Additionally, the mechanical stresses imposed during rethawing can result in cellular damage and mortality (Vincent 1988). However, many organisms have cellular and biochemical strategies that allow them to mitigate and withstand these effects. Such organisms are capable of rapid recovery after thawing and may also maintain substantial enzymatic and other biochemical activities at near-freezing temperatures (Deming and Huston 2000, Rothschild and Mancinelli 2001).

Snow provides a habitat for many types of microbial communities in the polar regions (Vincent 1988) as well as at temperate latitudes (Jones et al. 2001). These communities are often dominated by eukaryotic algae that are rich in UV-protecting carotenoids (Hoham and Ling 2000). Their primary production can in turn support metazoan grazers such as tardigrades, collembolans and ice worms. The latter, unusual annelids that are phylogenetically related to leeches, are adapted to freeze-thaw transitions and complete their life cycles despite temperatures that rarely rise above 0 °C (Shain et al. 2001). An unrelated group of 'ice worms' (polychaetes) has been discovered growing on methane hydrates in frigid deep waters of the ocean (Fisher et al. 2000), suggesting another kind of chemoautotrophic ecosystem, in addition to hydrothermal vents, that could potentially survive during major freeze-up events.

Observations by the explorer Nordenskiöld on the Greenland ice cap in the 19th century revealed that 'a brown polycellular alga' (mostly cyanobacteria but also eukaryotes) bound together the sediments in water-filled depressions in the ice; these communities in turn accelerated local melting and the further development of microbial mats (Vincent 2000). These so-called cryoconite communities occur widely on glaciers in the Arctic and Antarctica as well as in alpine environments elsewhere, and support micro-invertebrates such as nematodes, rotifers and tardigrades (Mueller et al. 2001).

Sea ice covers more than 25 million km^2 of the ocean surface each year and provides a dynamic, highly structured environment for microbial communities as well as an associated food web (Vincent 1988, and references therein). The biomass dominants in these microbial ecosystems are eukaryotic algae (mostly pennate diatoms) that live in the extensive brine channels between the elongate ice crystals. These environments are also the habitat for a variety of other microbial extremophiles including many new taxa of bacteria that have been isolated and characterized over the last decade (Staley and Gosink 1999, Thomas and Dieckmann 2002, Bowman 2003).

Freshwater ice lacks the brine channeling of sea ice and contains a much less developed microbial flora. Nonetheless, microbial consortia of cyanobacteria, micro-algae and heterotrophic bacteria have been observed in the ice on a variety of polar and alpine lakes (Priscu and Fritsen 1998, Priscu and Christner 2003). An unusual type of cryo-environment has been recently described in the McMurdo Dry Valleys, a benthic layer of brine-ice slush (< -10 °C) overlain by 19 m of lake ice (Doran et al. 2002).

Several other extreme low temperature environments have been described in recent years including streams beneath glaciers (Skidmore et al. 2000), the junction between ice crystals in the Antarctic ice cap (Price 2000), ancient permafrost soils (Rivkina et al. 2000) and supercooled water in ice clouds (Sattler et al. 2001). These underscore the broad success of microbial communities in surviving freeze-up and near-freezing growth conditions. One environment in particular, the thick sea ice that occurs today as ice shelves in the polar regions, is of particular relevance to discussions about Precambrian glaciation. These Arctic and Antarctic ice shelves provide the habitat for several types of microbial mat community. The mats are dominated by prokaryotes, but also provide local refugia for the growth and development of more complex organisms including eukaryotic microalgae and microinvertebrates.

6. Ice Shelf Cryo-ecosystems

Thick perennial sea ice is likely to have been especially widespread during glacial periods on Earth, including global freeze-up events of the Precambrian. The microbial communities living in this type of environment today are thus of great interest as analogs for cryo-ecosystems of the past. Microbial studies have now been conducted on two landfast, thick-ice (10-80 m) systems: the McMurdo Ice Shelf in the Ross Sea sector of

Antarctica, and the Ward Hunt Ice Shelf in high Arctic Canada (Fig. 3). These environments are completely frozen throughout most of the year, but both contain rich biological communities dominated by cyanobacteria that resume photosynthesis and growth during the periods of meltwater generation each summer.

6.1. MCMURDO ICE SHELF

The McMurdo Ice Shelf is composed of 1200 km^2 of marine-derived ice at the western edge of the Ross Ice Shelf. The ablation zone consists of two types of environment (Howard-Williams et al. 1990). About 30% of its area is characterized by an undulating topography, covered by a layer of moraine and marine sediments 10 to 20 cm thick. This material is thought to be transported up from the seabed by anchor ice formation or from zones of ice grounding on the sea floor (Debenham 1920) and results in low surface albedo values that influence the equilibrium thickness of ice (see Section 4 above) as well as the local topography (Fig. 3). The ice shelf undulations range up to 20 m, with the hollows occupied by melt ponds ranging from 1- 30 000 m^2 in area. The remainder of the ice shelf surface is characterized by a flatter overall relief, but with turrets of ice up to 0.5 m tall. Sediment cover is patchy in this region and the ponds tend to be small and shallow. In late summer, flowing waters are common in this 'pinnacle ice' region, including fast-flowing rivers up to several km long. In some areas there are elongate, north-south oriented parallel ponds, with no visible outflows.

The meltwater ponds of the McMurdo Ice Shelf encompass a broad range of salinities that reflect the variable influence of the underlying sea during ice formation, and redissolution of salts during the production of melt waters. In a set of extensive transect surveys (Howard-Williams et al. 1990), pinnacle ice melt waters ranged from 57 to 4000 $\mu S\ cm^{-1}$, while the undulating ponds spanned a much wider range from fresh (130 $\mu S\ cm^{-1}$) to hypersaline (56 000 $\mu S\ cm^{-1}$). On parts of the ice shelf there are large deposits of mirabilite ($Na_2SO_4.10H_2O$) which form salts beds up to 1 m thick. These are thought to be derived from trapped pockets of seawater brines during the basal freezing of the ice, with the brines or precipitated salts then transported to the surface via cracks in the ice, or via the ongoing process of basal freezing and surface ablation. The ice shelf thus supports microhabitats for halophilic as well as low salinity microbes.

Many of the meltwater ponds on the McMurdo Ice Shelf contain a benthic layer of microbial 'ice-mats' (Fig. 3, Vincent 1988, Hawes et al. 1993). In the pinnacle ice region, the mats form a loosely bound community of cyanobacterial filaments and diatoms amongst the thin layer of gravel and sediments on the ice. In the undulating ice ponds, however, the communities are composed of cohesive, mucilaginous biofilms up to 50 mm thick. These microbial mats are often brightly pigmented with a surface orange, pink or brown layer rich in photoprotective carotenoids overlying a blue-green layer enriched in light-harvesting pigments, especially phycocyanin and chlorophyll *a*. In some of the more saline habitats, these aerobic upper layers are underlain by thin film of purple sulfur bacteria and then a deeper, black anaerobic zone. The upper layers are dominated by cyanobacteria, typically oscillatorians, although in some ponds there are also high populations of *Nostoc commune* and *N. microscopicum*. Profiles of spectral irradiance through the mat have shown that the surface carotenoid-rich layer screens out UV and blue wavelengths and that the oscillatorians in the photosynthetically most

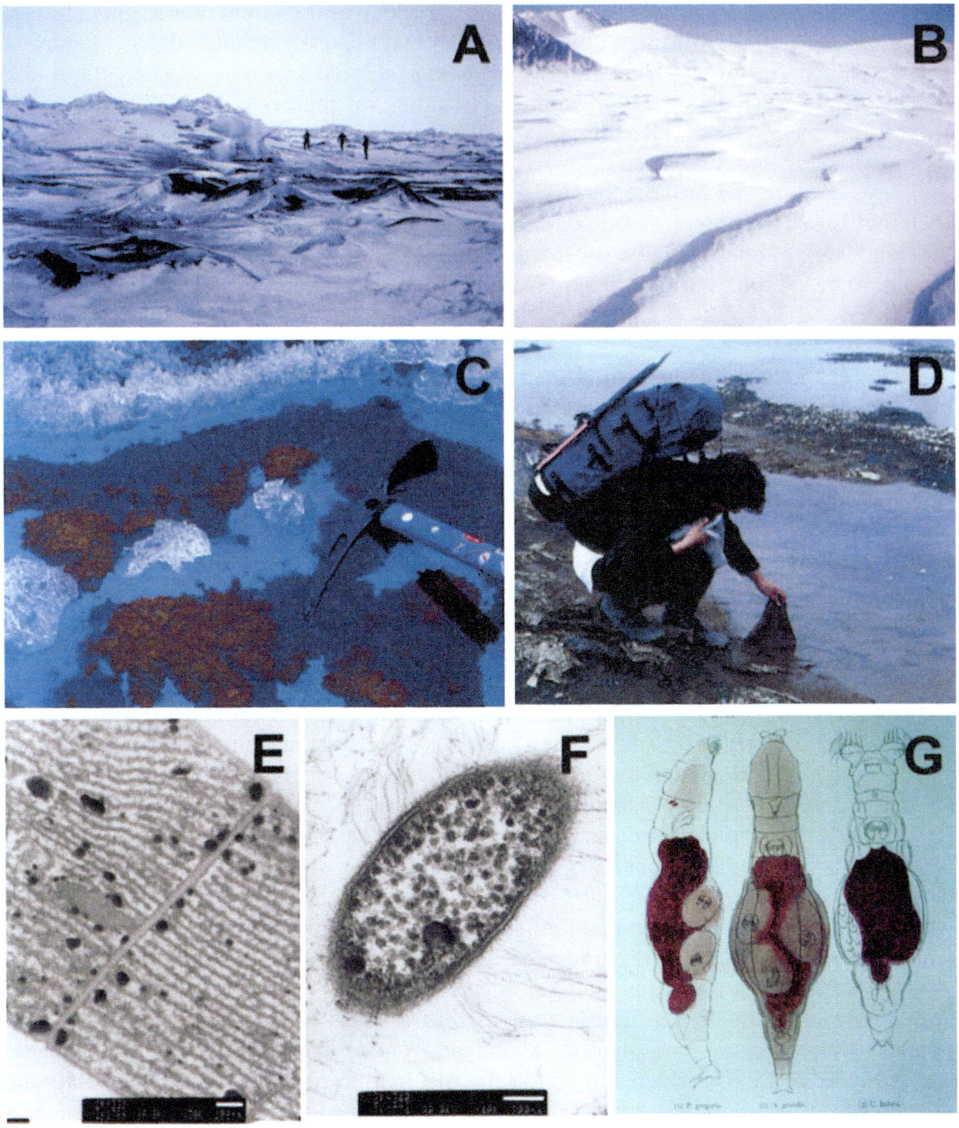

Figure 3. Ice shelf cryo-ecosystems. A. The McMurdo Ice Shelf, Antarctica, with marine sediments over the undulating, deep-frozen ice in late October. B. The undulating ice of the Ward Hunt Ice Shelf, Arctic Canada, with snow cover in early summer (June 1999). C. Microbial ice-mats in a meltwater pool on the Ward Hunt Ice Shelf. D. Microbial mats in a pond on the Koettlitz Glacier, at the edge of the McMurdo Ice Shelf. E, F. Transmission electron micrographs of a microbial mat from the McMurdo Ice Shelf showing thylakoid membranes of an oscillatorian, and a bacterium full of phage-like particles (Valérie Villeneuve et al., Centre d'Études Nordiques, Université Laval; unpublished); G. The rotifer *Philodina gregaria* rich in carotenoids derived from feeding on Antarctic cyanobacterial mats (Murray 1910).

active blue-green layer grow under highly shaded, orange-light conditions. Infrared radiation penetrates through the cyanobacterial layers to the deep stratum of photosynthetic sulfur bacteria below (Vincent et al. 1993).

There have been many ecophysiological studies on the both the intact mats of the McMurdo Ice Shelf as well as on taxa that have been isolated from these ice-mats and investigated in laboratory culture. All of the mats have proven to show nitrogenase activity, even those dominated by oscillatorians, but with some presence of the heterocystous genera *Nostoc*, *Anabaena* and *Nodularia*. These oscillatorian mats averaged a fixation rate that was estimated as 1 g m^{-2} a^{-1}, which would account for about one third of the total nitrogen requirement of the mat community (Fernández-Valiente et al. 2001).

Nitrogen fixation as well as other physiological processes continue over a broad range of salinities indicating substantial tolerance to the variable conditions during freeze-up and melting. For example, Hawes et al. (1999) found that photosynthesis and respiration by McMurdo Ice Shelf microbial mats continued at conductivities > 20 mS cm^{-1}, and it was only at 80 mS cm^{-1} (almost 2x seawater) that respiration declined. The microbial mats are also tolerant of prolonged desiccation, particularly those dominated by *Nostoc* (Hawes et al. 1992). In terms of temperature responses, most of the cyanobacterial isolates from the ice shelf mats have proved to be psychrotolerant rather than psychrophilic (Tang et al. 1997), although with some exceptions (Nadeau and Castenholz 2000). Growth optima are above 15 °C, well above the warmest ambient conditions of most of the ponds.

The cyanobacteria of the mats seem to be highly tolerant of ambient UV radiation and *in situ* screening of solar UVA and UVB radiation produced no enhancement of their physiological activities (Quesada et al. 2001). This pronounced UV-resistance is the result of four lines of defense against the damaging effects of bright solar radiation (Vincent and Quesada 1997): escape through migration (Nadeau et al. 1999); screening by UV-absorbing pigments; quenching of reactive oxygen species, for example by carotenoids (Roos and Vincent 1998); and damage-repair mechanisms.

A variety of interesting bacteria are now known to occur in the anoxic underlayers of these mats. This community is fuelled by autotrophically derived carbon from the overlying cyanobacteria. The carbon flux is primarily to CO_2 rather than to methane, and sulfate reduction is the main terminal anaerobic process (Mountfort et al. 1998). One isolate from below the cyanobacterial mat of a freshwater pond on the ice shelf, *Clostridium vincentii*, was found to be obligately anaerobic and psychrophilic. Optimal growth was at 12 °C, which is high relative to marine psychrophiles (e.g., relative to bacteria from the Southern Ocean). This isolate survived freeze-thaw cycles and grew well under salinities up to 30 ‰. It utilized a variety of mono- and disaccharides for growth including the cyanobacterial cell wall constituent N-acetyl glucosamine (Mountfort et al. 1997). Another novel psychrophile, *Psychromonas antarcticus*, was obtained from below the mat of a saline pond on the ice shelf. This isolate grew over the range 2 – 17 °C (optimal growth at 12 °C), tolerated high salt concentrations (10-40 ‰), and like *C. vincentii,* fermented N-acetyl glucosamine (Mountfort et al. 1998).

Studies on below-mat sediments have shown that freeze-thaw cycles influence the balance between methanogenesis and sulfate reduction (Mountfort et al. 2003). In low salinity sediments from ponds on the McMurdo Ice Shelf, both of these processes declined after freezing, but the carbon flow from acetate to methane increased relative to sulfate reduction. In sediments from high salinity ponds, sulfate production always dominated but became uncoupled from acetate oxidation after freezing.

Although these mat communities are dominated by photosynthetic and heterotrophic prokaryotes, they also contain a variety of other organisms. Virus-like particles have been detected throughout the mat profiles (up to 5 x 10^9 particles ml^{-1} of interstitial water, Valérie Villeneuve et al., unpublished), and TEM analysis of the mats has indicated the presence of bacteriophage (Fig. 3). Various eukaryotic algae occur within the mats, particularly small diatoms and green algae. Other protists occur within the mats such as ciliates and flagellates, and several micro-invertebrates occur including nematodes and abundant populations of rotifers (Suren 1990, Fig. 3).

6.2. WARD HUNT ICE SHELF

The Ward Hunt Ice Shelf extends northwards from the coast of Ellesmere Island in the Canada high Arctic and is one of several remnants of a much larger ice shelf system that extended 500 km along the northern coastline at the beginning of the 20th century (Vincent et al. 2001). Its upper ice surface of about 450 km^2 is a parallel system of wave-like troughs and ridges (Fig. 3) that contain elongate meltwater lakes and streams, typically 10 m wide, up to 3 m deep and 5 - 20 km long. Cylindrical holes up to 40 cm wide and 30 cm deep occur at the base of some of the elongate lakes, and have a higher conductivity and pH than the rest of the lake water. Similar holes but of more variable width (1-70 cm) also occur as discrete meltwater-filled pools at higher elevations on the ice ridges. The meltwaters of the lakes and pools have mostly low conductivities, however like the McMurdo Ice Shelf there is evidence that the ice and meltwater solutes are derived from the sea. The ratio of potassium to chloride relative to the same ratio for seawater is often around 1.0 in both systems, while the same ratio for sulfate is much higher (up to 67) consistent with the redissolution of mirabilite precipitated during the freezing of seawater. In parts of the Ward Hunt Ice Shelf, the surface contains marine sediments as well as sponges, molluscs and benthic crustaceans, further attesting to the marine origins of this glacial system. Similar marine fossils are found over the surface of the McMurdo Ice Shelf.

As in Antarctica, microbial mats are a feature of the Ward Hunt Ice Shelf, and also the nearby Markham Ice Shelf. These occur over the marine sediments that are periodically submerged by melt waters as well as in the cylindrical depressions in the ridge ice and at the base of the elongate lakes. The mats consist of a 2-10 mm layer of sediment that is loosely bound together by oscillatorian cyanobacteria and sometimes highly pigmented by a thin surface layer of cyanobacteria and chlorophytes containing orange carotenoids (Fig. 3). In an initial survey, chlorophyll *a* concentrations in these mats ranged from 0.4 to 15.3 µg Chl *a* cm^{-2} which overlaps with the McMurdo values (Howard-Williams et al. 1990). However, the mats cover a small area (less than 1 %) of the total Ward Hunt Ice Shelf and the standing stock of microbial biomass is thus much lower than on the Antarctic ice shelf.

Nitrogen-fixing cyanobacteria appear to be sparse in the Ward Hunt communities, perhaps reflecting low phosphorus concentrations (and low N:P ratios) by comparison with those at McMurdo, and there is a high biomass of chlorophytes, in particular

Palmellopsis, *Chlorosarcinopsis*, *Pleurastrum*, *Chlamydomonas*, *Chlamydocapsa*, *Chlorella*, *Bracteococcus*, *Chlorococcum* and *Klebsormidium*. Like the McMurdo mats, these communities also contain abundant populations of small benthic and aerophilic diatoms. The most common diatom species is *Chamaepinnularia* (*Navicula*) *begeri*, with subdominance by species of *Nitzschia*, *Navicula*, *Luticola*, *Achnanthes* and *Pinnularia*. The diatom flora of the McMurdo mat is similarly dominated by the genera *Navicula*, *Nitzschia*, *Pinnularia* and *Achnanthes* (Howard-Williams et al. 1990). The microbial mats of the Ward Hunt Ice Shelf, like those in Antarctica also harbor a variety of other organisms including viruses, ciliates, flagellates, heterotrophic bacteria and microinvertebrates (Vincent et al. 2000). The latter group includes rotifers, tardigrades, nematodes and turbellaria (flatworms).

7. Evolutionary Implications

What would be the evolutionary consequences of one or more snowball Earth cycles as depicted in Figure 2? The initial phase of freeze-up (steps 1 to 3) would likely be accompanied by mass extinction and a reduction in genetic diversity. Thermophilic species would only survive in isolated refugia such as geothermal springs, volcanic vents and (for non-phototrophs) deep-sea hydrothermal systems. There would be increasing selection for cold-tolerant genotypes that could gradually adapt to decreased temperatures and freeze-up conditions. The strategies found today in the ice shelf ecosystems would seem well suited to such conditions, with psychotrophic species adapted to prolonged freezing, intermittent thaw cycles and brief periods of slow growth during melt-out. As today, the microbial mats would allow a broad range of taxa to thrive under relatively protected conditions. The cyanobacterial UV–screens and quenching agents for reactive oxygen species would greatly reduce the impact of UV and bright PAR exposure, and the matrix of exopolymeric substances (EPS) produced by oscillatorians would protect all organisms within the fabric of the mat from freeze-thaw damage, as in annual sea-ice ecosystems (Krembs et al. 2003).

How long could such communities persist in the face of ongoing cooling (steps 3 to 4 in Fig. 2)? The global freeze-up scenarios to date have tended to focus on the low mean annual temperatures and the need for continuous liquid water conditions for the persistence of life. Kirschvink et al. (2000), in considering a hyperthermal ancestry for the three kingdoms of life, note that 'during a long snowball state, hydrothermal springs may have been one of the few places on Earth where liquid water was continuously maintained in the presence of sunlight'. However, the modern day ice shelf ecosystems experience only intermittent liquid water. Furthermore, high concentrations of solutes depress the freezing point of water to temperatures well below 0 °C, and in some Antarctic microbial mat environments (Ross Island ponds) the freeze-concentration of solutes results in liquid water conditions that persist well into winter despite air temperatures below – 40 °C (Schmidt et al. 1991). The presence of large quantities of marine derived salts and sediments on the modern day Ward Hunt and McMurdo ice shelves also suggests mechanisms that would tend to favor the persistence of microbial habitats. These solutes allow liquid water at low subzero temperatures, while black sediments derived from the sea floor enhance local radiative heating within or at the surface of the ice. The low albedo due to sediments would also affect sea ice dynamics and thickness (Warren et al. 2002).

Microbial mats have long been considered ideal environments for evolutionary processes to operate, although the focus to date has been on hot water systems. The microbial consortia of such mats consist of highly concentrated populations from diverse functional groups in contact with each other. Physical and chemical interactions are likely to be particularly strong in such communities and could lead to mutualism, symbiosis, even eukaryogenesis. For example, Nisbet and Fowler (1999) suggest that eukaryotes could have formed by fusion of symbiotic partners living across the redox boundary in thermophilic mats. Ice-mats of the type that exist today and perhaps during Precambrian glaciations, have surface layers that are charged in oxygen because of its high solubility at low temperatures, and many also have anoxic bottom layers, thereby resulting in strong gradients in redox conditions and associated microbial community structure. Extracellular DNA would be relatively stable in these ice-cool conditions, perhaps further stabilized by the cyanobacterial EPS, increasing the opportunities for genetic transformation. As seen in today's mats, bacteriophage (Fig. 3) and virus-like particles are common, and transduction could thus also facilitate lateral gene transfer.

Under the most extreme cold of the snowball Earth scenario (step 3 in Fig. 2) near-surface liquid water environments could have been rare or non-existent. However, as seen in today's ice shelf environments, prolonged dormancy is an important feature of ice-mat ecology, and allows the communities to maintain a large, perennial biomass despite the prevalence of non-growth conditions in their surrounding habitat. The transition to hot-house conditions (step 4 to 5, Fig. 2) may have been the most severe constraint on microbial growth and survival. Such a regime would have largely extirpated truly psychrophilic species from surface environments, although the deep ocean would have offered a vast refuge for cold-loving chemotrophs. Phototrophic species such as cyanobacteria and eukaryotes with broad thermal tolerances, as seen in today's ice-mats, would have been especially successful. The final return to warm conditions could have been the trigger for accelerated evolution and genetic exchange from isolated refugia, and the emergence of diverse life forms as observed in the Cambrian fossil record (Hoffman and Schrag 2000).

For some evolutionary biologists and Earth scientists, only prokaryotes could have survived the global glaciations of the Proterozoic. Thus Cavalier-Smith (2002) postulates that 'there is no problem, as plastids evolved just after the Varangerian [final Neoproterozoic glaciation] snowball Earth melted. If this is true, only bacterial photosynthesizers need have survived the near global glaciations and eukaryotic algae could have originated and radiated immediately after the climate rewarmed, with animals following hard on their heels in the Vendian and Cambrian.' Yet as shown here, today's ice shelf ecosystems contain rich microbial mats that contain many eukaryotes including diatoms, flagellates, ciliates, colonial green algae, and also metazoa such as rotifers (Fig. 3) and flatworms. These organisms thrive despite prolonged freeze-up and dormancy, and offer insights into how complex life may have persisted and evolved, even during the glacial upheavals of the Precambrian.

8. Acknowledgements

We thank the Natural Sciences and Engineering Research Council of Canada; les Fonds québecois de la recherche sur la nature et les technologies; Antarctica New Zealand; Polar Continental Shelf Project (this is PCSP publication no. 00903); and P. Hoffman, J. Seckbach and two anonymous reviewers for their critical comments on the manuscript.

9. References

Arnaud, E., and Eyles, C.H. (2002) Glacial influence on Neoproterozoic sedimentation: the Smalfjord Formation, northern Norway. Sedimentology **49**:765-788.

Bada, J.L., Bigham, C., and Miller, S.L. (1994) Impact melting of frozen oceans on the early Earth: Implications for the origin of life. Proceedings of the National Academy of Sciences USA **91**:1248-1250.

Bada, J.L., and Lazcano, A. (2002) Some like it hot, but not the first biomolecules. Science **296**:1982-1983.

Bowman, J. (2003) Distribution, ecology and taxonomy of psychrophilic bacteria. Aquatic Microbial Ecology In Press.

Brochier, C., and Philippe, H. (2002) A non-hyperthermophilic ancestor for Bacteria. Nature **417**:244.

Cavalier-Smith, T. (2002) The phagotrophic origin of eukaryotes and phylogenetic classification of Protozoa. International Journal of Systematic and Evolutionary Microbiology **52**:297-354.

Condon, D.J., Prave, A.R., and Benn, D.I. (2002) Neoproterozoic glacial-rainout intervals: Observations and implications. Geology **30**:35-38.

Debenham, F. (1920) A new mode of transportation by ice: the raised marine muds of South Victoria Land. Quarterly Journal of the Geological Society **75**:51-76.

Deming, J.W., and Huston, A.L. (2000) An oceanographic perspective on microbial life at low temperatures with implications for polar ecology, biotechnology and astrobiology, In: J. Seckbach, (ed.) *Journey to diverse microbial worlds: adaptation to exotic environments*. Kluwer Academic Publishers, Dordrecht, pp.149-160.

Doolittle, W.F. (1999) Phylogenetic classification and the universal tree. Science **284**:2124-2128.

Doran, P.T., Fritsen, C.H., McKay, C.P., Priscu, J.C., and Adams, E.E. (2002) Formation and character of an ancient 19-m ice cover and underlying trapped brine in an "ice-sealed" east Antarctic lake. Proceedings of the National Academy of Sciences USA **100**:26-31.

Fernández-Valiente, E., Quesada, A., Howard-Willams, C., and Hawes, I. (2001) N_2-fixation in cyanobacterial mats from ponds on the McMurdo Ice Shelf, Antarctica. Microbial Ecology **42**:338-349.

Fisher, C.R., MacDonald, I.R., Sassen, R., Young, C.M., Macko, S.A., Hourdez, S., Carney, R.S., Joye, S., and McMullin, E. (2000) Methane ice worms: *Hesiocaeca methanicola* colonizing fossil fuel reserves. Naturwissenschaften **87**:184-197.

Forterre, P. (2002) The origin of DNA genomes and DNA replication proteins. Current Opinion in Microbiology **5**:525-532.

Galtier, N., Tourasse, N., and Gouy, M. (1999) A nonhyperthermophilic common ancestor to exant life forms. Science **283**:220-221.

Gribaldo, S., and Philippe, H. (2002) Ancient phylogenetic relationships. Theoretical Population Biology **61**:391-408.

Hawes, I., Howard-Williams, C., and Pridmore, R.D. (1993) Environmental control of microbial biomass in the ponds of the McMurdo Ice Shelf, Antarctica. Archiv für Hydrobiologie **127**:27-287.

Hawes, I., Howard-Williams, C., and Vincent, W.F. (1992) Desiccation and recovery of cyanobacterial mats. Polar Biology **12**:587-594.

Hawes, I., Smith, R., Howard-Williams, C., and Schwarz, A.M. (1999) Environmental conditions during freezing, and response of microbial mats in ponds of the McMurdo Ice Shelf, Antarctica. Antarctic Science **11**:198-208.

Hoffman, P.F., Kaufman, A.J., Halverson, G.P., and Schrag, D.P. (1998) A Neoproterozoic snowball Earth. Science **281**:1342-1346.

Hoffman, P.F., and Schrag, D.P. (2000) Snowball Earth. Scientific American **282**:68-75.

Hoffman, P.F., and Schrag, D.P. (2002) The snowball Earth hypothesis: testing the limits of global change. Terra Nova **14**:129-155.

Hoham, R.W., and Ling, H.U. (2000) Snow algae: the effects of chemical and physical factors on their life cycles and populations, In: J. Seckbach, (ed.) *Journey to Diverse Microbial Worlds: Adaptation to Exotic Environments*. Kluwer Academic Publishers, Dordrecht, pp. 131-145.

Howard-Williams, C., Pridmore, R.D., Broady, P.A., and Vincent, W.F. (1990) Environmental and biological variability in the McMurdo Ice Shelf ecosystem, In: K.R. Kerry and G. Hempel, (eds.) *Antarctic Ecosystems: Ecological Change and Conservation*. Springer-Verlag, Berlin, pp. 23-31.

Huber, R., Huber, H., and Stetter, K.O. (2000) Towards the ecology of hyperthermophiles: biotopes, new isolation strategies and novel metabolic properties. FEMS Microbiology Reviews **24**:615-623.

Hyde, W.T., Crowley, T.J., Baum, S.K., and Peltier, W.R. (2000) Neoproterozoic snowball Earth simulations with a coupled climate ice-sheet model. Nature **405**:425-429.

Jenkins, G.S., and Smith, S.R. (1999) GCM simulations of Snowball Earth conditions during the late Proterozoic. Geophysical Research Letters **26**:2263-2266.

Jones, H.G., Pomeroy, J.W., Walker, D.A., and Hoham, R.W., (eds.) (2001) *Snow Ecology: An interdisciplinary examination of snow-covered ecosystems*. Cambridge University Press, Cambridge.

Kanavarioti, A., Monnard, P.A., and Deamer, D.W. (2001) Eutectic phases in ice facilitate nonenzymatic nucleic acid synthesis. Astrobiology **1**:271-282.

Karner, M.B., DeLong, E.F., and Karl, D.M. (2001) Archaeal dominance in the mesopelagic zone of the Pacific Ocean. Nature **409**:507-510.

Kennedy, M.J., Christie-Blick, N., and Prave, A.R. (2001) Carbon isotopic composition of Neoproterozoic glacial carbonates as a test of paleoceanographic models for snowball Earth phenomena. Geology **29**:1135 - 1138.

Kirschvink, J.L. (1992) Late Proterozoic low-latitude global glaciation: the snowball earth, In: J.W. Schopf and C. Klein, (eds.) *The Proterozoic Biosphere*. Cambridge University Press, Cambridge, pp. 51-52.

Kirschvink, J.L. (2002) Quand tous les océans étaient gelées. La Recherche **355**:26-30.

Kirschvink, J.L., Gaidos, E.J., Bertani, L.E., Beukes, N.J., Gutzmer, J., Maepa, L.N., and Steinberger, R.E. (2000) Paleoproterozoic snowball Earth: Extreme climatic and geochemical global change and its biological consequences. Proceedings of the National Academy of Sciences USA **97**:1400-1405.

Krembs, C., Eicken, H., Junge, K., and Deming, J.W. (2003) High concentrations of exopolymeric substances in Arctic winter sea ice: Implications for the polar ocean carbon cycle and cryoprotection of diatoms. Deep Sea Research Part I **49**:2163-5181.

Leather, J., Allen, P.A., Brasier, M.D., and Cozzi, A. (2002) Neoproterozoic snowball Earth under scrutiny: Evidence from the Fiq glaciation of Oman. Geology **30**:891-894.

Levy, M., and Miller, S.L. (1998) The stability of RNA bases: Implications for the origin of life. Proceedings of the National Academy of Sciences USA **95**:7933-7938.

Lovejoy, C., Carmack, E., Price, N., and Legendre, L. (2002) Water column interleaving: a new mechanism for protist biodiversity and bacterial states. Limnology and Oceanography **47**:1819-1831.

Massana, R., Taylor, L.T., Murray, A.E., Wu, K.Y., Jeffrey, W.H., and DeLong, E.F. (1998) Vertical disruption and temporal variation of marine planktonic archaea in the Gerlache Strait, Antarctica, during early spring. Limnology and Oceanography **43**:607-617.

McKay, C.P. (2000) Thickness of tropical ice and photosynthesis on a snowball Earth. Geophysical Research Letters **27**:2153-2156.

Miyakawa, S., Yamanashi, H., Kobayashi, K., Cleaves, H.J., and Miller, S.L. (2002) Prebiotic synthesis from CO atmospheres: implications for the origins of life. Proceedings of the National Academy of Sciences USA **99**:14628-14631.

Moulton, V., Gardner, P.P., Pointon, R.F., Creamer, L.K., Jameson, G.B., and Penny, D. (2000) RNA folding argues against a hot-start origin of life. Journal of Molecular Evolution **51**:416-421.

Mountfort, D., Kaspar, H., Asher, R., and Sutherland, D. (2003) Influences of pond geochemistry, temperature, and freeze-thaw on terminal anaerobic processes occurring in sediments of six ponds of the McMurdo Ice Shelf, near Bratina Island, Antarctica. Applied and Environmental Micorbiology **69**:583 - 592.

Mountfort, D.O., Rainey, F.A., Burghardt, J., Kaspar, H.F., and Stackebrandt, E. (1997) *Clostridium vincentii* sp. nov., a new obligately anaerobic saccharolytic, psychrophilic bacterium isolated from low-salinity pond sediment of the McMurdo Ice Shelf, Antarctica. Archives of Microbiology **167**:54-61.

Mountfort, D.O., Rainey, F.A., Burghardt, J., Kaspar, H.F., and Stackebrandt, E. (1998) *Psychromonas antarcticus* gen. nov., sp. nov., a new aerotolerant anaerobic, halophilic psychrophile isolated from pond sediment of the McMurdo Ice Shelf, Antarctica. Archives of Microbiology **169**:231-238.

Mueller, D.R., Vincent, W.F., Pollard, W.H., and Fritsen, C.H. (2001) Glacial cryoconite ecosystems: A bipolar comparison of algal communities and habitats. Nova Hedwigia Beiheft **123**:173-197.

Mukhin, L.M., Gerasimov, M., and Safonova, E. (1989) Origin of precursors of organic molecules during evaporation of meteorites and mafic terrestrial rock. Nature **340**:46-48.

Mukhin, L.M., and Pimenov, K.Y. (2002) Impact hot spots on the cold surface of the early Earth. Planetary and Space Science **50**:41-43.

Murray, J. (1910) On collecting at Cape Royds, In: J. Murray, (ed.) *British Antarctic Expedition 1907-9: Reports on the Scientific Investigations Vol. 1 Biology*. William Heinemann, London, pp. 1-15.

Nadeau, T., Howard-Williams, C., and Castenholz, R. (1999) Effects of solar UV and visible irradiance on photosynthesis and vertical migration of *Oscillatoria* sp. in microbial mats. Aquatic Microbial Ecology **20**:231-243.

Nadeau, T.L., and Castenholz, R.W. (2000) Characterization of psychrophilic oscillatorians (Cyanobacteria) from Antarctic meltwater ponds. Journal of Phycology **36**:914-923.

Nelson, K.E., Clayton, R.A., Gill, S.R., Gwinn, M.L., Dodson, R.J., Haft, D.H., Hickey, E.K., Peterson, J.D., Nelson, W.C., Ketchum, K.A., McDonald, L., Utterback, T.R., Malek, J.A., Linher, K.D., Garrett, M.M., Stewart, A.M., Cotton, M.D., Pratt, M.S., Phillips, C.A., Richardson, D., Heidelberg, J., Sutton, G.G., Fleischmann, R.D., Eisen, J.A., White, O., Saltzberg, S.L., Smith, H.O., Venter, J.C., and Fraser, C.M. (1999) Evidence for lateral gene transfer between Archea and Bacteria from genome sequence *Thermotoga maritima*. Nature **399**:323-328.

Nisbet, E.G., and Fowler, C.M.R. (1999) Archaean metabolic evolution of microbial mats. Proceedings of the Royal Society of London Biological Sciences Series B **266**:2375-2382.

Pedersen, K. (1997) Micrbial life in deep granitic rock. FEMS Microbiology Reviews **20**:399-414.

Poulsen, C.J., Jacob, R.L., Pierrehumbert, R.T., and Huynh, T.T. (2002) Testing paleogeographic controls on a Neoproterozoic snowball Earth. Geophysical Research Letters **29**:10.1029/2001GL014352.

Price, P.B. (2000) A habitat for psychrophiles in deep Antarctic ice. Proceedings of the National Academy of Sciences USA **97**:1247-1251.

Priscu, J.C., and Christner, B.C. (2003) Earth's icy biosphere, In: A. Bull, (ed.) *Microbial Diversity and Bioprospecting*. American Society for Microbiology, Washington D.C. In Press.

Priscu, J.C., and Fritsen, C.H. (1998) Antarctic lake-ice microbial consortia: Origin, distribution, and growth physiology. Antarctic Journal of the United States **31**:223-224.

Quesada, A., Sánchez-Contreras, M., and Fernández-Valiente, E. (2001) Tolerance of Antarctic cyanobacterial microbial mats to natural UV radiation. Nova Hedwigia Beiheft **123**:275-290.

Rivkina, E.M., Friedmann, E.I., McKay, C.P., and Gilichinsky, D.A. (2000) Metabolic activity of permafrost bacteria below the freezing point. Applied and Environmental Microbiology **66**:3230-3233.

Roos, J.C., and Vincent, W.F. (1998) Temperature dependance of UV radiation effects on Antarctic cyanobacteria. Journal of Phycology **34**:118-125.

Rothschild, L.J., and Mancinelli, R.L. (2001) Life in extreme environments. Nature **409**:1092-1101.

Sagan, C., and Mullen, G. (1972) Earth and Mars: Evolution of atmospheres and surface temperatures. Science **177**:52-56.

Sattler, B., Puxbaum, H., and Psenner, R. (2001) Bacterial growth in supercooled cloud droplets. Geophysical Research Letters **28**:239-242.

Schmidt, S., Moskal, W., De Mora, S.J., Howard-Williams, C., and Vincent, W.F. (1991) Limnological properties of Antarctic ponds during winter freezing. Antarctic Science **3**:379-388.

Schopf, J.W., and Klein, C. (1992) *The Proterozoic biosphere : a multidisciplinary study*. Cambridge University Press, Cambridge.

Schrag, D.P., Berner, R.A., Hoffman, P.F., and Havlerson, G.P. (2002) On the initiation of a snowball Earth. Geochemistry Geophysics Geosystems **3**:10.1029/2001GC000219.

Schwartzman, D. (1999) *Life, temperature and the Earth: The self-organizing biosphere*. Columbia University Press, New York.

Shain, D.H., Mason, T.A., Farrell, A.H., and Michalewicz, L.A. (2001) Distribution and behaviour of ice worms (*Mesenchytraeus solifugus*) in south-central Alaska. Canadian Journal of Zoology **79**:1-9.

Skidmore, M.L., Foght, J.M., and Sharp, M.J. (2000) Microbial life beneath a High Arctic glacier. Applied and Environmental Microbiology **66**:3214-3220.

Staley, J.T., and Gosink, J.J. (1999) Poles apart: Biodiversity and biogeography of sea ice bacteria. Annual Review of Microbiology **53**:189-215.

Suren, A. (1990) Microfauna associated with algal mats in melt ponds of the Ross Ice Shelf. Polar Biology **10**:329-335.

Tang, E.P.Y., Tremblay, R., and Vincent, W.F. (1997) Cyanobacteria dominance of polar freshwater ecosystems: are high-latitude mat-formers adapted to low temperature? Journal of Phycology **33**:171-181.

Thomas, D.N., and Dieckmann, G.S. (2002) Antarctic sea ice - a habitat for extremophiles. Science **295**:641-644.

Valley, J.W., Peck, W.H., King, E.M., and Wilde, S.A. (2002) A cool early Earth. Geology **30**:351-354.

Vincent, W.F. (1988) *Microbial ecosystems of Antarctica*. Cambridge University Press, Cambridge.

Vincent, W.F. (2000) Cyanobacterial dominance in the polar regions, In: B.A. Whitton and M. Potts, (eds.) *The Ecology of Cyanobacteria*. Kluwer Academic Publishers, Dordrecht, pp. 321-340.

Vincent, W.F., Castenholz, R.W., Downes, M.T., and Howard-Williams, C. (1993) Antarctic cyanobacteria: Light, nutrients, and photosynthesis in the microbial mat environment. Journal of Phycology **29**:745-755.

Vincent, W.F., Gibson, J.A., Pienitz, R., Villeneuve, V., Broady, P.A., Hamilton, P.B., and Howard-Williams, C. (2000) Ice shelf microbial ecosystems in the High Arctic and implications for life on snowball earth. Naturwissenschaften **87**:137-141.

Vincent, W.F., Gibson, J.A.E., and Jeffries, M.O. (2001) Ice shelf collapse, climate change, and habitat loss in the Canadian high Arctic. Polar Record **37**:133-142.

Vincent, W.F., and Quesada, A. (1997) Microbial niches in the polar environment and the escape from UV radiation in non-marine habitats, In: B. Battaglia, J. Valencia, and D. Walton, (eds.) *Antarctic Communities: Species, Structure and Survival*. Cambridge University Press, Cambridge, pp. 388-395.

Warren, S.G., Brandt, R.E., Grenfell, T.C., and McKay, C.P. (2002) Snowball Earth: Ice thickness on the tropical ocean. Journal of Geophysical Research **107**:3167.

Wilde, S.A., Valley, J.W., Peck, W.H., and Graham, C.M. (2001) Evidence from detrital zircons for the existence of continental crust and oceans on the Earth 4.4 Gyr ago. Nature **409**:175-178.

Williams, D.M., Kasting, J.F., and Frakes, L.A. (1998) Low-latitude glaciation and rapid changes in the Earth's obliquity explained by obliquity-oblateness feedback. Nature **396**:453-455.

Williams, G.E. (1993) History of Earth's obliquity. Earth Science Review **34**:1-45.

Woese, C.R. (1987) Bacterial evolution. Microbiological Reviews **51**:221-271.

Wolf, Y.I., Rogozin, I.B., Grishin, N.V., and Koonin, E.V. (2002) Genome trees and the tree of life. Trends in Genetics **18**:472-479.

Biodata of **Dan Cohen** author of ***"The Evolutionary Ecology Of Species Diversity In Stressed And Extreme Environments"***

Dr. Dan Cohen is a Professor Emeritus in the Department of Evolution, Systematic, and Ecology of The Hebrew University of Jerusalem. He obtained his Ph.D. in 1960 from Oxford University (UK) and served in The Hebrew University since then. Dr. Cohen had served, among various functions and activities in the University, as Chairman of the Environmental Biology Curriculum and the Committee for Graduate Students. He had served as the Chairman of the Life Science Institute and the Director of The Arid Systems Research Center.

E-mail: **dancohen@vms.huji.ac.il**

J. Seckbach (ed.), Origins, 503–514.

THE EVOLUTIONARY ECOLOGY OF SPECIES DIVERSITY IN STRESSED AND EXTREME ENVIRONMENTS.

DAN COHEN
Dept. of Evolution, Systematics and Ecology.
The Silberman Institute of Life Sciences. and
The Center for Rationality and Interactive Decisions
The Hebrew University of Jerusalem. Israel.

1. Introduction: The Definitions of Stress and Extreme Environments

1.1. THE ECO-PHYSIOLOGICAL DEFINITIONS OF STRESS IN PHYSIOLOGICAL AND ECOLOGICAL TIME SCALES.

A wide range of different definitions of stress conditions has been suggested. Some definitions deal mainly with the effects of extreme abiotic conditions (Oren, 1999). Other definitions are partly or completely circular, because the stress conditions are defined by the effect of low species diversity (Schink, 1999). Since most natural populations are limited in the long run by some combinations of limiting environmental factors, most species should therefore be considered to be stressed to about the same extent by the different limiting factors. (Chave et al. 2002).

In this short essay I suggest that all stresses or extreme conditions should be defined as extreme levels of abiotic and biotic limiting factors relative to the distribution of the tolerance or utilization range of the dimensions or variables that define the habitats or the niche resources of the species. The ecological and physiological processes and conditions that determine the diversity of different types of coexisting species in different habitats and areas, for any existing species' pools, are reasonably well understood in principle. These processes and patterns will be discussed therefore only very briefly in this presentation.

1.2. THE DEFINITIONS OF STRESS OR EXTREME CONDITIONS IN EVOLUTIONARY TIME SCALE.

Such definitions are very problematic and difficult to realize. This is because the 'reference' normal utilization range distribution of the habitat resources or conditions utilized by the species, and/or the tolerance range of the niche dimensions of the species, can and do change in evolutionary time scales. The normal range may change by evolution in shifting environments, and/or by slow or fast evolutionary changes in many important features of the species.

However, the mechanisms and constraints on the evolutionary changes of the normal range distribution of exploited habitat conditions or of the tolerance range within a species or lineage are still incompletely understood.

The normal range of habitat conditions or of the niche dimensions of species or lineages can be defined therefore only for relatively short-term evolutionary changes in the existing types of organisms, and may not exist for long-term evolutionary time scales. A proper definition of stress and extreme conditions relative to their normal range in evolutionary time scales is essential however for modeling and explaining the wide range of species diversity of different groups in different areas and habitats. (Rosenzweig 1995).

The evolutionary and ecological processes that determine the normal range of the habitat conditions exploited or tolerated by any particular lineage of species, usually change at very different rates for different types of organisms at different environmental conditions. These different time scales have to be taken into account in order to model and explain the existing and changing patterns of species diversity.

1.3. THE PATTERNS AND RATES OF SPECIATION

The patterns and rates of speciation of different types of species in different lineages and in different conditions and areas have a major influence on the numbers and types of species in the species pools of different ecological types. The conditions and processes that determine or influence the rates and directions of the formation of new species are only partly understood however, but they have a major influence on the different levels of diversity of different types of species at different habitats and conditions. (Rosenzweig 1995).

2. Models of Species Diversity

A number of different ecological, genetical, and evolutionary models and explanations have been suggested for the very wide range of species diversity of different types of organisms in different areas and environments. (Simberloff and Dayan 1991, Zobel 1992, Tilman and Pacala 1993, Iwasa et al 1993, Bengtsson et al 1994, Rosenzweig 1995, Chesson 2000, Chave et al 2002,). As noted above, it is important to distinguish between processes that operate at ecological time scales and processes that operate at evolutionary time scales, and between local and geographical spatial scales.

2.1. ECOLOGICAL EQUILIBRIUM MODELS OF SPECIES DIVERSITY IN COMMUNITY ECOLOGY.

Ecological models of the conditions for the stable equilibrium coexistence of competing species that utilize an overlapping range of limiting resources have been extensively developed and analysed for a wide range of variation of stable and varying environmental conditions. (Pacala and Tilman 1994, Chesson 2000, Huntley and Chesson 1997).

Some form of constant or shifting resource partitioning appears to be an essential component in all the models and experiments of the conditions that allow a long term stable coexistence of competing species, while also taking into account all the relevant spatial and temporal scales of the variations in the environment. (Cohen 1994, Huntley and Chesson 1997, Kinzig et al 1999, Chesson 2000).

According to such ecological models, the stable equilibrium of coexisting species from any given pool of species that compete for the same distribution of limiting resources, depends to a large extent on the degree of partition and specialization of the utilization of the available range of resources between the species. According to such models, the existing patterns of diversity and coexistence of competing species in different environments and ecosystems are maintained in approximate equilibrium in ecological time scales by the interactions between the ecological properties of the environment and the partition of resources by all the potentially available competing species. For example, a higher spatial heterogeneity of the environment in physical or resource space provides a larger number of opportunities for resource partitioning and for the coexistence of a higher diversity of specialized species. Higher predictable temporal variability of resources or conditions can also provide more opportunities for partitioning and specialization. On the other hand, unpredictable variability in the environment favours the coexistence of a smaller number of generalist strategies of utilization of a wide range of resources under a wide range of conditions. Specific developmental and physiological constraints also restrict very strongly the possibilities for adaptive responses to extreme or unpredictable temporal changes. (Oren 1999).

The temporal and spatial scales of the variation and heterogeneity of the environment in an ecosystem are expected to influence the coexistence conditions of different species differently according to their ecology, mobility, and life history characteristics. (Chave et al 2002). For example, long lived continuously active animals must feed all year around, and must therefore utilize a wide range of different resources that are available at different seasons and places. On the other hand, species that can become dormant and inactive for long periods may be able to maintain themselves by utilizing relatively rare specific temporary resources. Note however that the ecological models typically deal only with the conditions for coexistence of species from a given species pool. Note also that the evolution of the ecological characteristics of competing species that lead to higher or lower levels of specialization and diversity is still only very partly understood and ignored by most of these ecological models. (Zobel 1992, Tilman and Pacala 1993, Rosenzweig 1995).

2.2. NON-EQUILIBRIUM MODELS OF SPECIES DIVERSITY IN ECOLOGICAL TIME SCALES.

The coexisting competing species in any local ecosystem at any one time may very likely represent a ecological non-equilibrium. Clearly, the number and composition of species in any ecosystem at any one time represent the balance between the input of new species by immigration of new species from the existing species' pool in accessible areas, and by the formation of new species by speciation. The input of species is balanced by extinction of existing species at some probability, which decreases the number of species. (Hubbell 1979, Hubbel and Foster 1986, Chesson and Case 1986, Chave et al. 2002).

Ecological models of this type make explicit testable predictions about the effects of the spatial patterns of heterogeneity and of species mobility on the species diversity of different types of organisms in a wide range of different ecosystems in ecological time scales, for any given species' pool. The predictions of such island models have been tested and verified in a wide range of conditions. (Chesson and Case 1986).

The qualitative properties of non-equilibrium steady states of species diversity are similar however to some aspects of the equilibrium models. The probabilities or rates of extinction of species are influenced to a similar extent by the conditions that decrease the probability of coexistence in the equilibrium models. Similarly, the probabilities of establishment of immigrating or speciating new species are influenced to the same extent by the same conditions that increase the probability of coexistence in the equilibrium models.

However, such ecological non-equilibrium models also ignore the important long-term adaptive and non-adaptive genetic and evolutionary changes of the ecological properties and adaptations of all the species, which influence the probabilities for their coexistence. No less important is the lack of explicit modeling of the processes of genetic diversification within species and of the evolution of new species by speciation from existing species.

2.3. THE EVOLUTIONARY ECOLOGY OF SPECIATION

The genetical and ecological processes and conditions that determine the rates and levels of speciation have been investigated and modeled very extensively, but no consensus exists yet. (Rosenzweig 1995, Futuyma 1997) Many investigators have emphasized the genetical processes of speciation by the evolution of reproduction barriers between and within populations of the same species. One approach emphasizes the evolution of isolating mechanisms caused by the developmental or ecological low fitness of hybrids between differently adapted ecotypes, as in host specialized herbivorous insects.

A different approach emphasizes the evolution of isolating mechanisms by random drift or the random founder effect, or as a by-product of random differential selection in isolated populations of the species. Diversifying mate recognition behaviors and chromosomal rearrangements have also been suggested as important components in such processes of speciation. (Futuyma 1997).

2.4. SPECIATION IS PROMOTED BY SPECIFIC ECOLOGICAL ADAPTATIONS

Different local or specific ecological adaptations are expected to increase the probability of the formation and maintenance of ecologically and genetically distinct new species with different specifically adapted ecological characteristics. This is because hybrids between differently adapted parents are expected to have lower fitness than of each of the parents. On the other hand, strong unpredictable fluctuations of environmental conditions are expected to have the opposite effect of strongly selecting against any local or habitat specific ecological specialization and genetic isolation. Frequent local random extinctions and re-colonization cause a similar effect of selection against the formation of specifically adapted new ecological or genetic types. Strong fluctuations of the selection regime that act on one or few major components of fitness, also decrease the fitness and select against the formation and establishment of different specific epistatic genotypic combinations. Such selection would reduce the stability of different adaptive peaks, and the probability for coexistence of a large ecological diversity of species.

Strong selection for a single major fitness component, even in a constant environment, is expected to decrease the number of distinct adaptive peaks, and thus acts to decrease the stable or steady state diversity of competing species in the ecosystem or community. The low diversity of some types of species in some types of stressed and extreme environments and habitats may represent therefore the combined effects of a number of different ecological and evolutionary processes on species diversity in under such conditions.

3. Models and Mechanisms

I suggest mechanisms that explain some of the general patterns in the processes and factors that decrease species diversity under some types of stress and extreme conditions.

3.1. BIOTIC STRESSES REDUCE SPECIES DIVERSITY LESS THAN ABIOTIC STRESSES IN BOTH ECOLOGICAL AND EVOLUTIONARY TIME SCALES.

This is because most biotic stresses, such as competition, predation, parasites, and diseases, cause mortality and other losses of fitness that are strongly increasing functions of the densities of the species in both ecological and evolutionary scales. Rare species are very much less likely to be a major component of the diet of any consumer species. Specific parasites and diseases are much less likely to cause high mortality in rare species. There is no or only very weak selective advantage for a common species to expand its niche to exploit the small range of partly negatively correlated resources exploited by any other coexisting rare species. The same arguments apply to diet choice by foraging animals, which tend to consume less of the rarer food organisms.

Both theoretical models and field data also strongly indicate that the co-evolution of predators, herbivores, parasites, and diseases with their biotically stressed food or host species, leads in many cases to specialization and speciation of a higher diversity and coexistence of both stressed and stressor species (Rosenzweig 1995). Such diversifying co-evolution may lead to high levels of species diversity in remarkably short evolutionary time of a few thousand generations. The species diversity may then increase more slowly or reach a steady state.

Biotic stresses such as new predators, diseases, or strong competitors have caused the extinction of large numbers of species on many occasions during evolutionary history. However, most such biotic extinctions did not eliminate whole lineages, which may therefore recover and diversify quickly by new speciation under the new biotic stresses.

3.2. IN CONTRAST, ABIOTIC STRESSES STRONGLY DECREASE SPECIES DIVERSITY IN ECOLOGICAL AND EVOLUTIONARY TIME SCALES.

Abiotic stresses cause in most cases density independent mortality and other losses of fitness, which do not therefore reduce the probability of extinction as species become rare. In addition, rarity as such increases the probability of extinction of species by stochastic fluctuations in their population numbers or habitat conditions.

3.2.1. The Effects of Strong Constant Abiotic Stresses.

Strong selection for adaptation to a major abiotic stress necessarily reduces the force of selection and the probability of existence of a large number of diverse genotypes with different adaptive specializations. This is because the total strength of selection on all fitness components in mutation-selection equilibrium is approximately constant. Thus, a very strong selection in any one major stress component necessarily reduces the strength of selection and the adaptive level of all other weaker components. (Kassen 2002). Also, even a weak negative trade-off between the fitness of specific specializations and the fitness of the major stress component would be selected against. The rare different types would not benefit by being rare, because the abiotic stress is also frequency independent. The number of different abiotic stresses is also very much smaller than the number of biotic stresses, so that the probability of evolving a large diversity of different specialized adaptations is necessarily also much smaller, and so is the expected realized diversity.

Metabolic, physiological, or ecological adaptations to extreme abiotic stresses may also require higher and increasing levels of energy expenditure for physiological and metabolic homeostasis and maintenance. This necessarily reduces the possible number and diversity of species with different viable less efficient specialized adaptations that can coexist or can evolve even under constant stresses in long term evolutionary time. Such a process has been documented in halophilic bacteria occurring in increasing salinity stresses. (Oren 2001).

3.4. A GENERAL MODEL OF THE OPPOSING EFFECTS OF BIOTIC AND ABIOTIC STRESSES ON SPECIES DIVERSITY.

I suggest therefore that a major fraction of the variance of the large scale global, geographical, historical, evolutionary, ecological, and physiological patterns of species diversity can be explained by the relative strengths of the opposing effects of biotic and abiotic stresses and limiting factors.

3.4.1. Stronger *Complex Biotic Stresses Increase Species Diversity*

The commonly occurring biotic stresses of predation, competition, parasites, and diseases, promote the coexistence in ecological time scales of a higher diversity of species with different specialized adaptations. Biotic stresses also promote the co-evolution of a higher diversity of species with different specialized adaptations in evolutionary time scales. This would be the typical situation in species rich ecological communities, such as tropical rainforests and coral reefs, where biotic stresses dominate over abiotic stresses.

3.4.2. Stronger Abiotic Stresses Decrease Species Diversity.

Increasing levels of different types of abiotic stresses, such as extreme temperatures, salinity, or drought, and of local extinctions, reduce the diversity and number of coexisting species from any potential species' pool in ecological time scale. Abiotic stresses also select against the evolution of a large diversity of specialized ecological or physiological adaptations and species in evolutionary time scales. This would be the typical situation in strongly stressed species-poor ecological communities, such as hot springs, high salinity waters, extreme cold, extreme drought, etc. The biotic stress of extreme shade under forest canopies acts like an abiotic stress, typically causing a very low diversity of photosynthetic plant species.

3.5. THE OPPOSING EFFECTS OF CONSTANT VS. CHANGING STRESSES.

The relative strengths of the two opposing processes of constant vs. changing stresses have already been identified as major causes for the different levels and patterns of species diversity. (Chesson and Huntly 1997, Rosenzweig 1995, Tilman and Pacala 1993). Briefly, the arguments are:

3.5.1. Constant or Predictable Heterogeneous Abiotic and Biotic Stresses increase Diversity

Constant or predictable heterogeneous abiotic and biotic stresses at any environmental conditions and ecological communities allow the stable coexistence of a large diversity of species that are differentially adapted to different stresses and conditions. The same conditions of constant stresses in heterogeneous or predictable environments also select for the evolution of an increased diversity of high-fitness specialized adaptations that are better adapted for the particular stresses and conditions in the community.

3.5.2. *Changing Unpredictable Biotic and Abiotic Stresses Decrease Diversity*

In contrast, changing unpredictable biotic and abiotic stresses favor a low diversity of high-fitness generalist adaptations for a wide range of resources, habitats, and conditions in ecological time scale. Changing unpredictable stresses also provide a selective advantage in evolutionary time scales for a low diversity of species with high-fitness generalist adaptations for a wide range of resources, habitats, and conditions, in such changing unpredictable selection regimes.

3.6. DISCUSSION

In this short essay I suggest a very general model for the effects of two major processes that determine and explain the major trends of the effects of stresses on the patterns of species diversity:

1. The newly proposed concept of the differential effects of a limited number of severe abiotic stresses that strongly decrease the diversity, as opposed to the effects of diverse biotic stresses that strongly increase the diversity.
2. The documented effects of the predictability and constancy of the ecological conditions and of the selection regime that increase the diversity vs. the effects of change and unpredictability that decrease the diversity. This general model that includes the combined effects of these two factors is compatible with the observed general patterns of species diversity in the world, and could explain a large fraction of the total variance of the distribution of the global patterns of species diversity. Additional appropriate scaling and normalization would probably be necessary to explain the details of the variations of species diversity in different types of environments, functional ecological groups, or evolutionary lineages. Many of these differential effects could probably be incorporated as modifying components of the two main factors.

3.6.1 *Long term evolutionary and ecological changes in species diversity in different habitats and lineages during the history of the Earth.*

The long-term evolutionary expansion of the range of utilized resources and habitats during the history of the Earth by very many evolving lineages, probably began by an initial colonization of previously unexploited or unoccupied habitats or resources by very few colonizing species. Such new habitats were by definition stressful initially for these colonizing species.

Subsequent adaptive radiation by evolutionary and ecological diversification that occurred at later stages by the evolution of additional functionally adaptive specializations caused the existing rich patterns of species diversity.

3.7. CONCLUSIONS

Formulating and constructing a unifying framework for all the major processes and causes of species diversity, and especially the effects of stress and extreme conditions, is a major still unfinished task. The models in this essay, which emphasize the opposing effects of biotic vs. abiotic stresses and of constant vs. changing stresses, may provide some useful insights and ideas and some testable predictions. A large uncertainty still concerns the relative magnitude and appropriate scaling of the effects of different types of both biotic and abiotic stresses and of constant vs changing stresses on different types of organisms, and on the same organisms in different environments. Functional correlations and trade-offs between different types of adaptations are also difficult to include explicitly in the model without some specific assumptions.

The proposed models necessarily leave out many important issues and details. One major unexplained issue concerns the evolutionary constraints and costs of the evolution of different types and ranges of adaptations. The long-term evolutionary history of many different types of adaptations may provide us with some clues. Many types of adaptations have occurred independently many times in different lineages, suggesting that the evolution of such adaptations had to overcome only weak constraints. Other types of adaptations have occurred very infrequently or only once, suggesting that strong constraints have reduced the probability of their occurrence to extremely low levels. Large scale drastic adaptations in evolutionary history have occurred most often under conditions of very strong stresses and directional selection and reduced competition, which could have provided the necessary conditions to overcome even strong constraints.

4. References

Bengtsson, J., Fagerstrom, T., and Rydin, H. (1994) Competition and coexistence in plant communities. Trends Ecol. Evol. **9**: 246–50.

Chave, J., Muller-Landau, H.C. and Levin, S. (2002) Comparing Classical Community Models: Theoretical Consequences for Patterns of Diversity. Am. Nat. **159**, 1-23.

Chesson, P.L. (1994) Multispecies competition in variable environments. *Theor. Pop. Biol.* **45**: 227–76.

Chesson, P.L. (2000) Mechanisms of Maintenance of Species Diversity. Ann. Rev. Ecol. Syst. **31**: 343-66.

Chesson, P. L. and Huntly, N. (1997) The roles of harsh and fluctuating conditions in the dynamics of ecological communities. *Am. Nat.* **150**: 519–53.

Chesson, P.L. and Case, T.J. (1986) Overview: Non-equilibrium community theories: chance, variability, history, and coexistence in *Community Ecology*: In: J. Diamond and T. J. Case (eds.) *Community Ecology*, New York: Harper and Row. pp. 229–239.

Cohen, D. (1994) Modeling the coexistence of annual and perennial plants in temporally varying environments. Plant Species Biol. **9**:1-10.
Futuyma, D.J. (1997) Evolutionary Biology. 3-rd Ed. Sinauer Associates.
Hubbell, S.P. (1979) Tree dispersion, abundance, and diversity in a tropical dry forest. *Science* **203**: 1299–309.
Hubbell, S.P. and Foster, R.B. (1986) Biology, Chance, and History, and the Structure of Tropical Rain Forest Tree Communities: In: J. Diamond and T. J. Case (eds.) *Community Ecology*, New York: Harper and Row. pp. 314-330.
Hubbel, S.P. (1997) A unified theory of biogeography and relative species abundance, and its application to tropical rain forests and coral reefs. *Coral Reefs* **16**: S9–21.
Iwasa, Y., Sato, K., Kakita, M. and Kubo, T. (1993) Modeling biodiversity: latitudinal gradient of forest species diversity. In: E.D. Schulze and H.A. Mooney (eds.) *Biodiversity and Ecosystem Function*, Springer-Verlag, Berlin, pp. 433–51.
Kassen, R. (2002) The experimental evolution of specialists, generalists, and the maintenance of diversity, J. Evol. Biol. **15**(2): 173-190.
Kinzig, A.P., Levin, S.A., Dushoff, J. and Pacala, S. (1999) Limiting similarity, species packing, and system stability for hierarchical competition-colonization models. Am. Nat. **153**: 371–83.
Oren, A. (1999) Bioenergetic aspects of halophilism. Microbiol. Mol. Biol. Rev. **63**: 334– 348.
Oren, A. (2001) The bioenergetic basis for the decrease in metabolic diversity at increasing salt concentrations: implications for the functioning of salt lake ecosystems. Hydrobiol. **466**: 61–72.
Pacala, S.W. and Tilman, D. (1994) Limiting similarity in mechanistic and spatial models of plant competition in heterogeneous environments Am. Nat. **143**: 222–57.
Rosenzweig M. (1995) *Species Diversity in Space and Time.* Cambridge Univ. Press.
Schink, B. (1999) Habitats of Prokaryotes. In: J. W. Lengeler, G. Drews and H. G. Schlegel (eds.), *Biology of the Prokaryotes.* Thieme, Stuttgart, pp. 763–803
Shmida, A. and Ellner, S. (1984) Coexistence of plant species with similar niches. *Vegetatio* **58**: 29–55.
Simberloff, D. and Dayan, T. (1991) The guild concept and the structure of ecological communities. Ann. Rev. Ecol. Syst. **22**:115–43.
Tilman, D. and Pacala, S. (1993) The maintenance of species richness in plant communities, In: D. Schluter and R.E. Ricklefs (eds.) *Species Diversity in Communities: Historical Ecological and Geographical Perspectives*. Univ. Chicago Press, pp. 13–25.
Zobel, M. (1992) Plant species coexistence: The role of historical, evolutionary and ecological factors. *Oikos* **65**: 314–20

Biodata of **Vinod C. Tewari** author of ***"Microbial Diversity in Meso-Neoproterozoic Formations, with particular reference to the Himalaya"***

Dr. Vinod C. Tewari is a senior scientist at Wadia Institute of Himalayan Geology, Dehra Dun, and a regular associate of International Center for Theoretical Physics, Trieste, Italy. He received his Ph.D. in Geology in 1986 from the University of Lucknow, India. Dr. Tewari taught Geology at Kumaon University, Nainital, India as Professor of Geology. He has done research on early evolution and diversification of life in the Proterozoic sequences of the Lesser Himalayan stromatolitic carbonate belts. He has published more that sixty research papers and edited several volumes of ***Himalayan Geology Journal*** *and* ***Journal of Nepal Geological Society,*** Kathmandu, Nepal. Dr. Tewari has organized the first Indo-Soviet Symposium on Stromatolites and Stromatolitic Deposits, and other International Geological Correlation Program conferences in India.

E-mail: **vctewari@rediffmail.com**

J. Seckbach (ed.), Origins, 515–528.

MICROBIAL DIVERSITY IN MESO-NEOPROTEROZOIC FORMATIONS, WITH PARTICULAR REFERENCE TO THE HIMALAYA

VINOD CHANDRA TEWARI
Wadia Institute of Himalayan Geology, Dehra Dun 248 001, Uttaranchal, India and International Center for Theoretical Physics, Miramare 34100 Trieste, Italy

1. Introduction

Proterozoic microbial life is recorded from over 150 localities in the world. The microbial assemblage represent considerable biodiversity and ecological variability. Microfossiliferous Proterozoic cherts from Australia, Canada, Africa, Greenland, USA, Russia, China and India (Himalaya) have yielded authentic biota (Schopf, 1983; Hofmann and Schopf, 1992; Schopf and Klein, 1992; Knoll, 1985; Westall, 1999; Tewari, 1989, 2001, a,b,c, 2002). The chert biota is three dimensionally preserved in Deoban and other cherty stromatolitic limestones of the Lesser Himalaya ranging in age from Meso to Neoproterozoic (Fig. 1a). The diversified microbiota includes filament,coccoids, vase-shaped microfossils, acritarchs, isolated hexactinellid and monoaxon sponge spicules. The stromatolite assemblage is dominantly *Conophyton garganicum*, *Kussiella kussiensis*, *Colonnella columnaris* and *Baicalia nova*.

The Terminal Proterozoic succession of the Lesser Himalaya in India is represented by the Blaini-Krol - Tal sediments. The base of the Terminal Proterozoic is defined in the microbial pink limestone of the Blaini Formation and the Precambrian-Cambrian boundary is established in the Lower Tal Formation (Fig. 1b) based on small shelly fossils and carbon isotopic signatures. The middle part of the sequence, Krol Formation has yielded large acanthomorphic acritarchs typical of the Vendian age, eukaryotic multicellular algae Vendotaenids, *Krolotaenia* and Ediacaran metazoans *Cyclomedusa davidi, Charniodiscus* sp., *Kimberella* cf. *quadrata, Zolotytsia biserialis* and *Conomedusites lobatus*. This type section has been proposed as a Global Stratotype and Section Point (GSSP) for the Terminal Proterozoic (Tewari, 1999, 2001a). The Ediacaran metazoans are soft-bodied animals that occur above the highest Varangian glaciation (=Blaini diamictites) and below the lowest Cambrian deposits (=Tal Formation) on all continents. The present chapter deals with the important aspect of the Proterozoic diversity in microbial life of the Himalaya and its global significance.

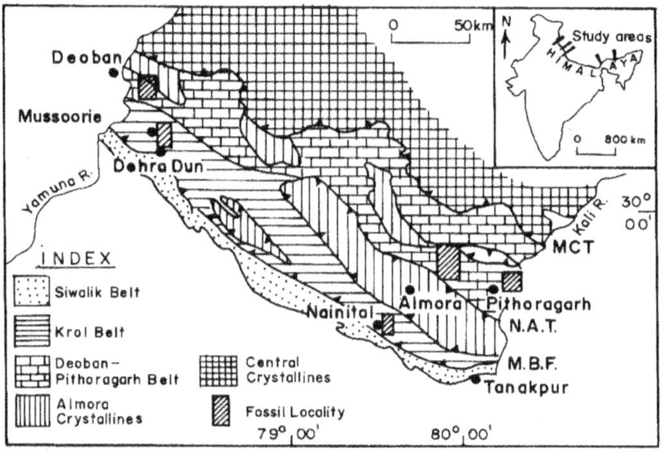

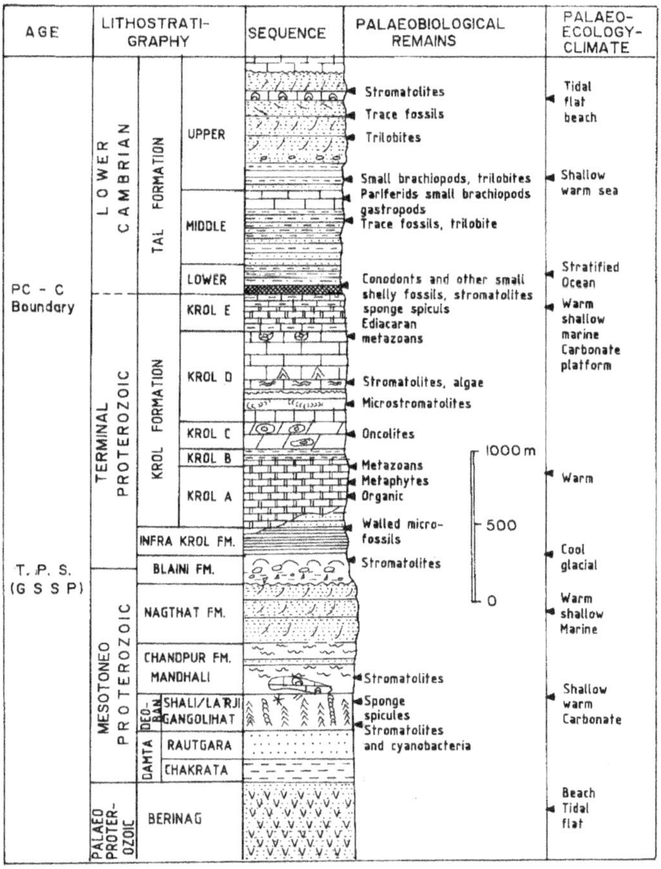

Figure 1a. Geological map of the Lesser Himalaya, India showing occurrences of the fossiliferous formations.

Figure 1b. Lithostratigraphic column of the Lesser Himalaya showing microbial diversification (Tewari, 2001).

2. Meso-Neoproterozoic Microbial Diversification

The oldest known record of life on Earth is bacteria described by Walsh (1992) and Westall et al. (2001) from Early Archean Onverwacht Group (3.3-3.5 Ga), South Africa. The Meso-Neoproterozoic microbial diversification has been recorded from all continents and the microfossils are preserved in early diagenetic silicified carbonates Table 1). The Deoban Limestone is a well developed carbonate buildup in the Lesser Himalaya of northern India (Fig.1a, b). It has been assigned Meso-Neoproterozoic age on the basis of Riphean stromatolite assemblage *Conophyton garganicum, Kussiella kussiensis, Colonnella columnaris* and *Baicalia nova* (Tewari, 1989, 1993a, 2002, Fig. 2). Deoban cherts show highly diversified microbial assemblage in the Meso-Neoproterozoic of the Lesser Himalaya in India and its global distribution is shown in Table 1. Twenty species belonging to eleven genera of coccoid forms (Fig. 3) are *Myxococcoides minor, M. inornata, Huronispora psilata, H. microreticulata, Eoentophysalis belcherensis, E. magna, E. cumulus, Glenobotrydion aenigmatis, G. majorinum, Tetraphycus major, T. conjunctum, Melasmatosphaera media, M. parva, Gloeodiniopsis lamellosa, G. gregaria, G.* sp., *Globophycus rugosum, Sphaerophycus parvum, Eosynechococcus isolatus* and *Caryosphaeroides pristina* (Shukla et al., 1986; Tewari, 1989, 2001b, 2002; Kumar and Srivastava, 1992). Five genera and eight species of filamentous forms have been recorded by Shukla et al., 1986; Tewari, 1989, 2001b, 2002 and Kumar and Srivastava, 1992 (Fig. 4). These are *Eomycetopsis robusta, E. filiformis, E. siberiensis, Gunflintia minuta, G. grandis, Biocatenoides* sp., *Oscillatoriopsis* sp. and *Siphonophycus kestron.* The above named twenty eight species are recorded in the Deoban assemblage, whereas the maximum number of common species is known from Bitter Springs Formation of Australia (900 Ma, Table 1). According to Hofmann (1976) *Glenobotrydion, Myxococcoides, Globophycus, Caryospheroides*, and *Melasmatosphaera* possibly represent the degradational variants. Tiwari et al. (2000) and Tewari (2002) have reported isolated hexactinellid and monaxon sponge spicule and calcified algae *Epiphyton* sp. and *Renalcis* sp. from the Gangolihat Dolomite (eastern extension of Deoban Limestone in the Uttaranchal Lesser Himalaya, Fig. 1a) which suggest Neoproterozoic to early Cambrian age for the Gangolihat Dolomite.

3. Terminal Proterozoic Microbial Diversification

The reviews of fossil bacteria in Schopf (1983), Knoll (1985) and Westall (1999) have dealt with Terminal Proterozoic microbial diversification in general. A significant increase in diversity of acritarchs took place during the Neoproterozoic (late Riphean and early Vendian) followed by a decrease in diversity in middle to late Vendian and a subsequent increase throughout the Cambrian (Knoll, 1985; Knoll and Walter, 1992; Tiwari and Knoll, 1994). Early Vendian large acanthomorphic acritarchs occur in the Infra-Krol (chert) and Lower Krol Formations of the Lesser Himalaya.

Vendotaenids are the oldest known multicellular brown algae and diversified around 650 Ma Vendian rocks (Gnilovskaya, 1988).

Deoban microbial diversity	1	2	3	4	5	6	7	8	9	10	11	12	13
Myxococcoides minor	G	G		G	S	S	G	S	G	G	G	S	G
Myxococcoides inornata		G	G	G	S	G	G	G	G	G	G	S	G
Eosynechococcus isolatus		G				S	G(?)				G		G
Glenobotrydion majorinum				G	S	G		S		cf.G		S	G
Melasmatosphaera media						G		G		cf.S		S	G
Melasmatosphaera parva						G		G		cf.G		S	
Tetraphycus major		G				G			S	cf.S			
Tetraphycus conjunctum		S				G			G	cf.G			
Globophycus rugosum					S					G		S	
Huroniospora psilata	G	G								G			S
Huroniospora microreticulata	cf. S	G								G	S		S
Caryosphaeroides pristina		G			G					cf.G	S	G	
Eoentophysalis magna		G		G			G	G	G		G	G	
Eoentophysalis cumulus		G		G			G	G	G		G	G	
Eoentophysalis belcherensis		G		G		S	G	S	S	cf.S	S	S	
Gloeodiniopsis gregaria			G		S		G						
Gloeodiniopsis lamellosa			S		S		G						
Gloeodiniopsis sp.				cf. S									
Sphaerophycus parvum	cf. G	G(?)	G		S	S	S		S	S	S	S	G
Gunflintia grandis						G							S
Gunflintia minuta						S							S
Eomycetopsis robusta	S	G	S	S	S	G		G	G	G	S	G	
Eomycetopsis filiformis	G	G	G	G	S	G		G	G	G	S	S	
Eomycetopsis siberiensis	G	S	G	G	G	S		G	G	G	G	G	
Siphonophycus kestron			G	G	S								
Oscillatoriopsis sp.	S		G		G		G			G			
Biocatenoides sp.			G		G		G(?)		G	G		G	

1= Jinning Yunnan, China, 610 Ma. 2= Yudoma Suite, 650 Ma. 3= Minyar Formation, 680-790 Ma. 4=Hunnberg Formation 750 - 800 Ma. 5. Bitter Springs Formation, 900 Ma. 6= Kheinjua Formation, 1100-1300 Ma. 7= Dismal Lakes Groups Ma. 8= Bungle Bungle Dolomite, 1600 Ma. 9= Balbirini Dolomite, 1600 Ma. 10= HYC Pyritic Shale, 1600 Ma. Amelia Dolomite, 1700 Ma. 12=Mcleary Formation, 1760-2370 Ma 13=Gunflinit Iron Formation, 1800-2000 Ma. G=Genus; S=Species present

Table 1 : Correlation of Deoban microbiota with other Proterozoic microbial assemblages of the world (Shukla et al., 1986; Tewari, 1989, 2001, Kumar and Srivastava, 1992)

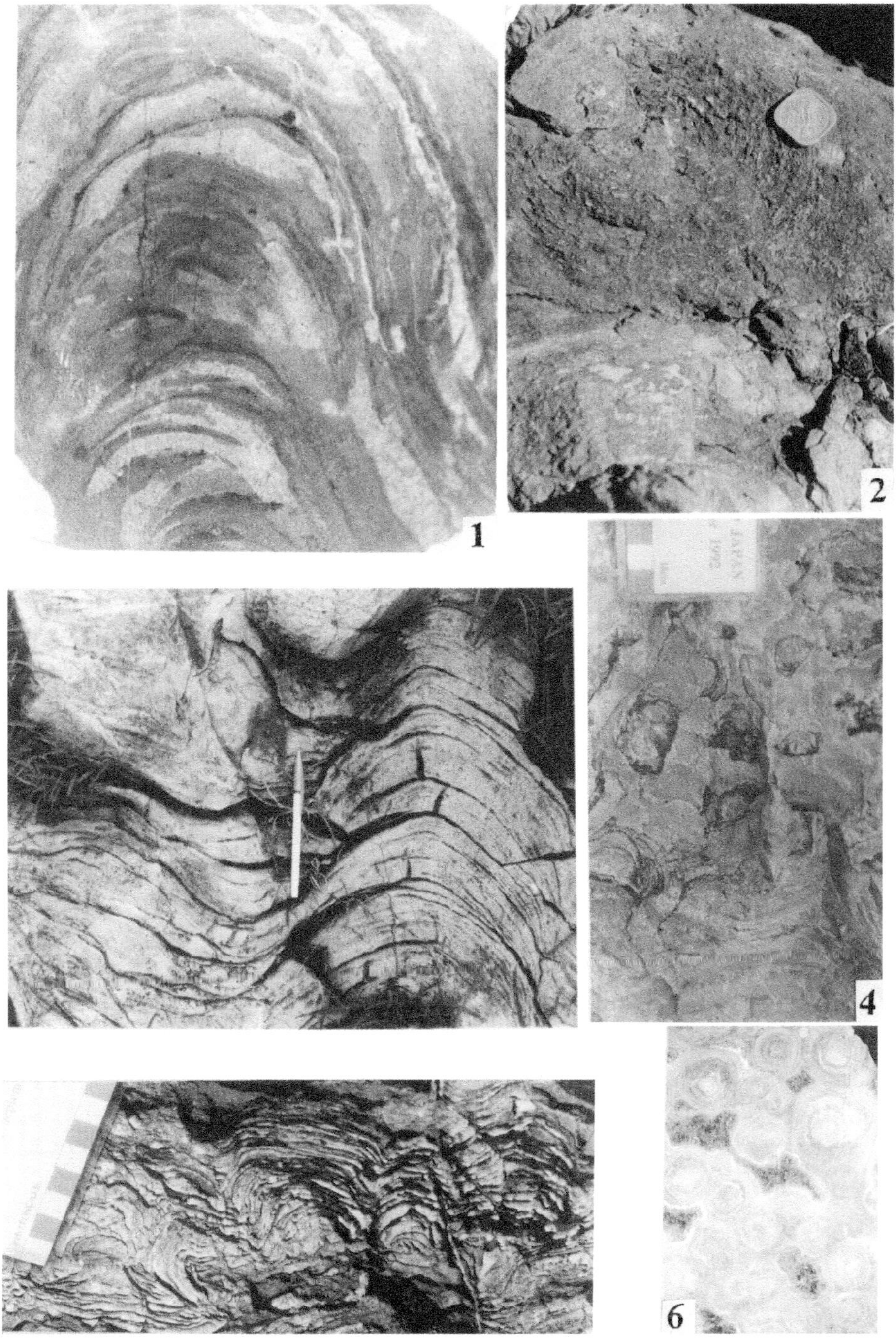

Figure 2. Meso-Neoproterozoic stromatolite diversification in the Deoban-Gangolihat Dolomite, Lesser Himalaya 1,2. *Conophyton garganicus*, 3. *Kussiella kussiensis*, 4. *Baicalia* sp. 5. Domal form 6. Oncolites (Krol)

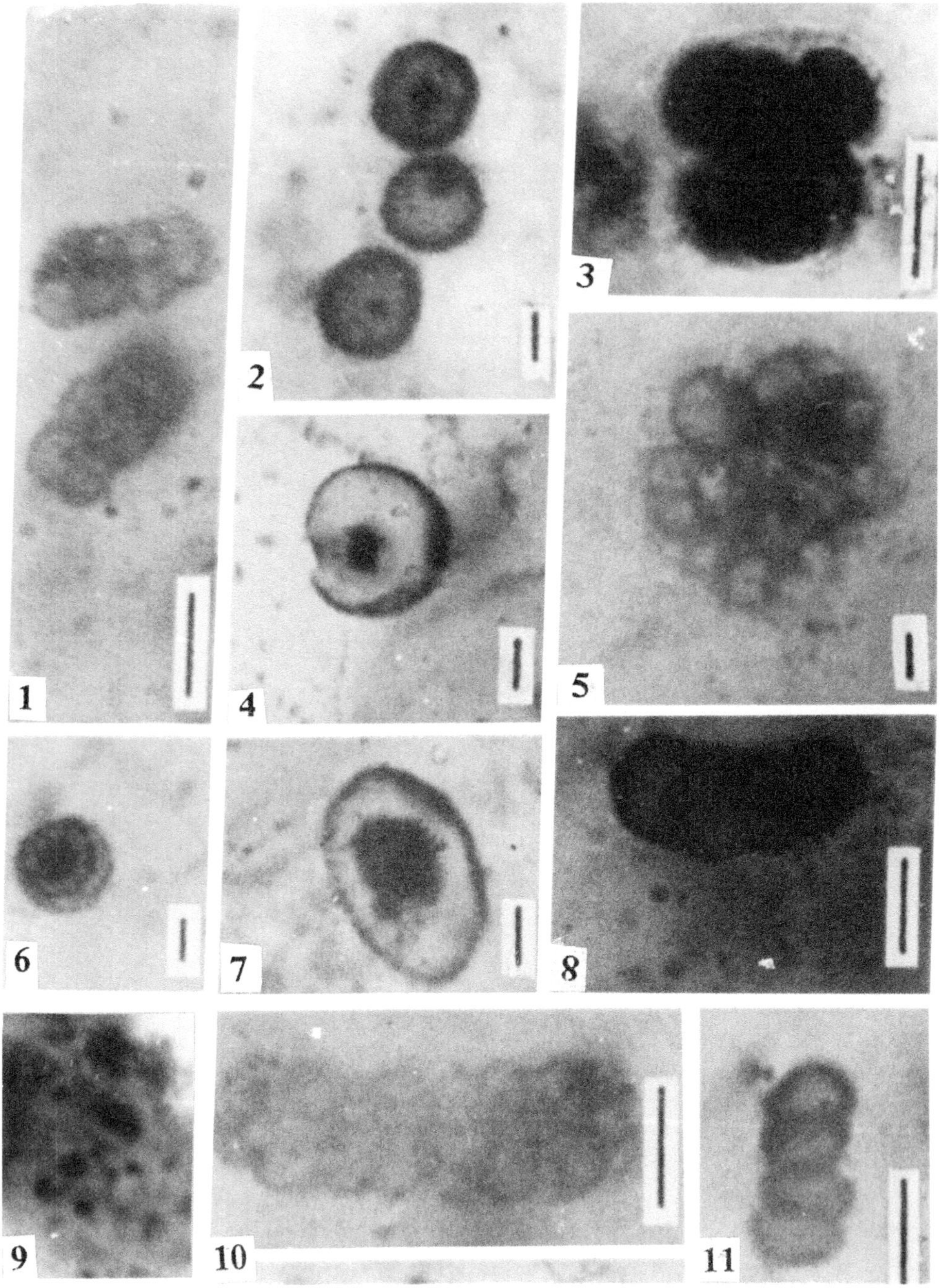

Figure 3. Coccoid assemblage of the Deoban Limestone, Lesser Himalaya

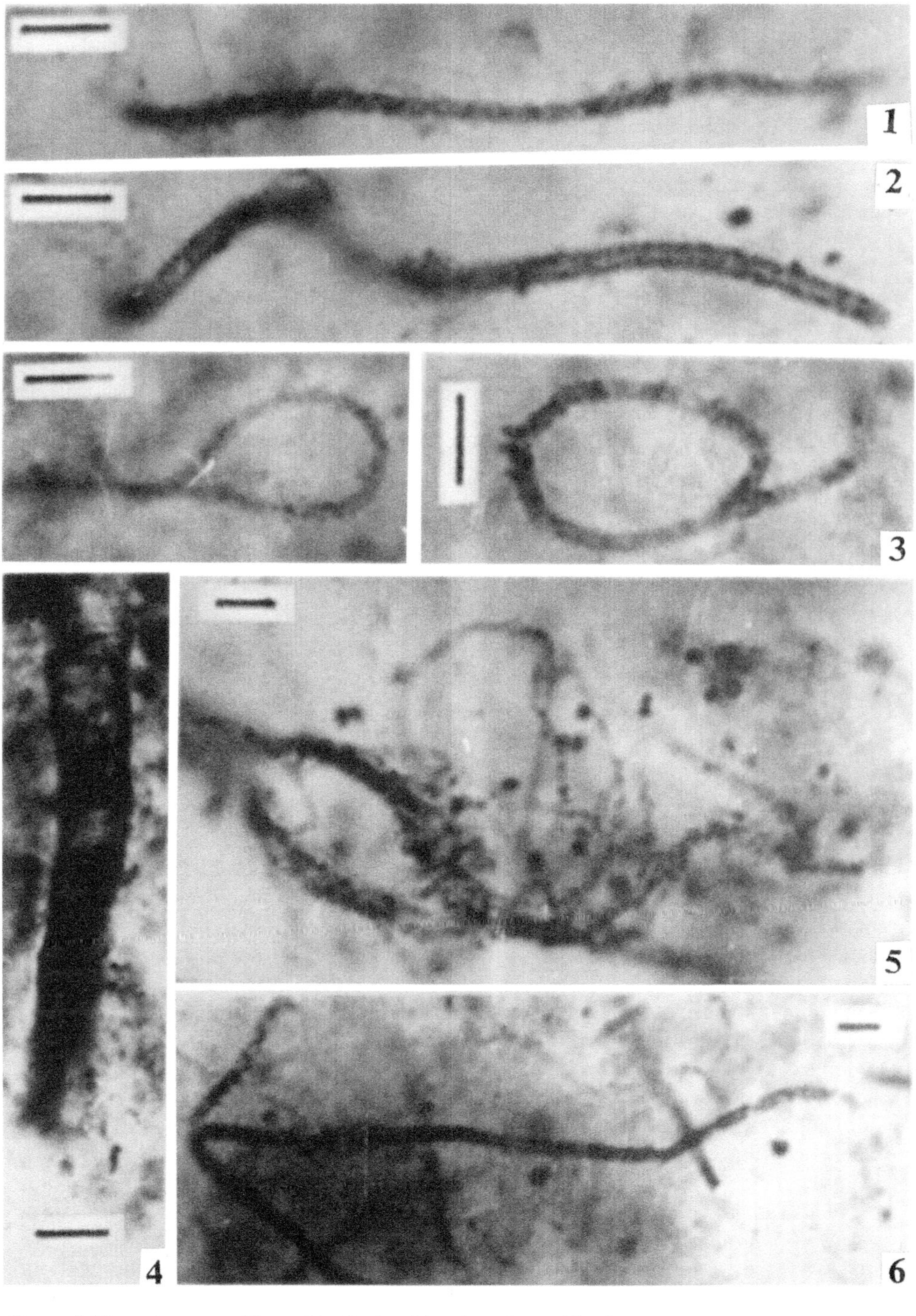

Figure 4. Filamentous assemblage of the Deoban Limestone, Lesser Himalaya

Tewari (1988) discovered Vendotaenid algae from the Lower Krol Formation of the Lesser Himalaya. A new genus *Krolotaenia gnilovskayi* was established and subsequently the genus *Tyrasotaenia* and *Vendotaenia* (Fig. 5; 3 and 4) have been recorded from the Nainital Syncline (Tewari, 1993b, 1999b). Hofmann (1992) regarded some carbonaceous megascopic ribbon shaped remains as algae (metaphyte) affinities of Neoproterozoic age. Schopf et al. (1973) described *Vendotaenia* as a multicellular macroscopic benthic metaphyte. The Vendotaenids are also recorded from the Sinian System (Late Proterozoic) of China and this eukaryotic algae has a global distribution. Ediacaran life diversified in the Terminal Proterozoic and includes soft bodied metazoans (coelenterates-medusoid, frondoid forms) and trace fossils. These evolved after the Varangian glaciation and are well known from the Ediacara type locality in Australia. These have been recorded from China, Eurasia, India and Wernecke and Mackenzie mountains of Canada (Glaessener, 1984; Narbonne and Hofmann, 1987; Hofmann, 1992; Walter, 1989; Mathur and Shankar 1989; Shankar and Mathur, 1992; Shankar et al., 1997; Tewari, 1992, 1996, 2001,b). The oldest pre Ediacaran fauna has been recorded from the intertillite beds of Windermere Supergroup, Mackenzie Mountains, Canada (Hofmann et al. 1990).

In India, the Ediacaran assemblage has been recorded from the Upper Krol Formation of the Lesser Himalaya (Figs. 1a,b). The assemblage includes the soft bodied metazoans *Cyclomedusa davidi, Charniodiscus* sp. fronds and disc, *Kimberella* cf. *quadrata, Zolotytsia biserialis* Fedonkin and *Conomedusites lobatus* Glaessener and Wade (Fig. 6, 1 to 6; Shanker and Mathur, 1992; Tewari, 1992).

4. Discussion and Conclusions

A major event in the diversity of fossil algae and unicellular eukarya is recorded in 1.2-1.0 Ga old rocks (Knoll, 1985). The Meso-Neoproterozoic and Terminal Proterozoic succession of the Lesser Himalaya in the northern India shows excellent preservation of the highly diversified microbial assemblages (Fig. 7). The microbiota of the Deoban cherts and the Bitter Springs Formation of Central Australia (Schopf, 1968) are remarkably similar. The well preserved filamentous and spheroidal unicells of the Deoban cherts have been compared with other known localities in the world (Table 1). The stromatolites and carbon isotopes suggest that the depositional environment for the Deoban-Gangolihat Dolomite and Krol-Buxa was a shallow marine tidal flat (Tewari, 1997, 2002). The hexactinellid and monoaxon sponge spicules from Gangolihat Dolomite provide evidence of metazoan silica biomineralization and Vendian age, whereas calcified algae *Epiphyton* and *Renalcis* are indicative of early Cambrian calcification event.

The occurrence of large acanthomorphic acritarchs from the Infra Krol Formation of the Lesser Himalaya strongly suggest that they have survived the Varanger glaciation (=*Blainian* glaciation, Tewari, 2001 a,b). However, there has been a decrease in size of the acanthomorphic acritarchs from the Vendian to the Cambrian in the Lesser Himalaya as also recorded in China and Europe. It is quite interesting that the Vendotaenid assemblage coincides with Ediacaran metazoan biozones in Canada

(Narbonne and Hofmann, 1987) and Lesser Himalayan Krol belt carbonates (Tewari, 1993b, 1999b, Figs. 5, 6 and 7). This supports the simultaneous eukaryotic evolution of

Figure 5. *Vendotaenids* (3, 4) calcified algae (1, 2, 5) ? *Epiphyton, dendrites* and *Renalcis* and *microstromatolites* from the NE Lesser Himalaya (6)

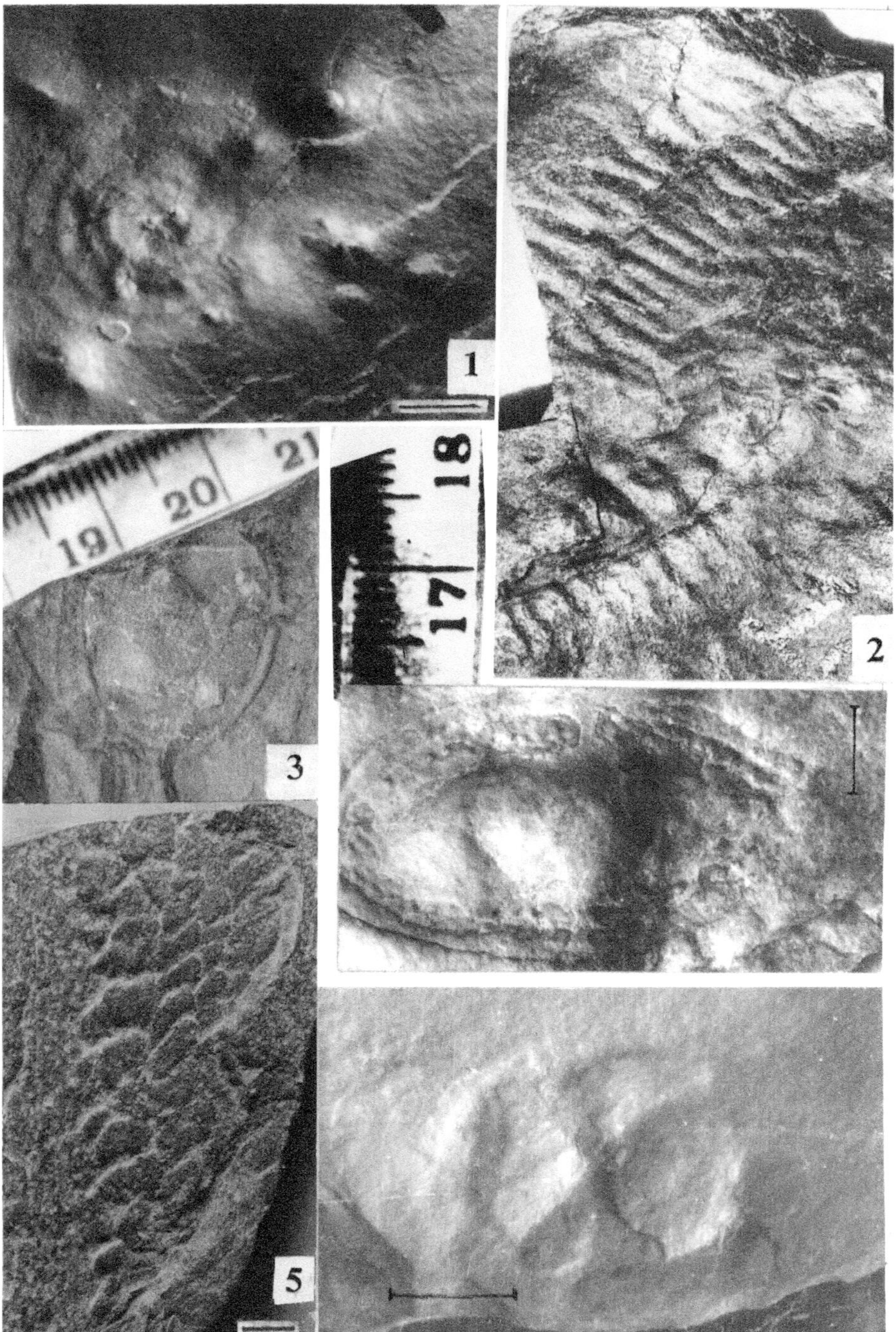

Figure 6. Ediacaran metazoans from the Upper Krol Formation of the Lesser Himalaya. 1. *Cyclomedusa davidi,* bar = 5 mm 2. *Charniodiscus* sp. bar=1 cm., 3. Base of *Charniodiscus*, bar = 1 cm. 4. *Kimberella* sp. (*K. quadrata*), 5. *Zolotytsia biserialis*, 6. *Conomedusites lobatus*, bar = 5 mm.

Figure 7. Proterozoic microbial diversity in the Deoban - Blaini-Infra Krol -Krol- Tal succession of the Lesser Himalaya

metaphytes and metazoans during the Terminal Proterozoic. The Krol Group of the Lesser Himalaya, India is correlated with Dengying Formation of China on the basis of similar occurrence of Neoproterozoic stromatolites, Vendotaenids, Ediacaran metazoans and acanthomorphic arcitarchs (Tewari, 1999 a,b). Gnilovskaya (1988) considered Vendotaenia the oldest multicellular phaeophyte (Brown algae) or a rhodophyte (Red algae). Seckbach (1994) studied the red algae *Cyanidium caldarium* in detail and interpreted that cyanidiophyceae is a transitional algal group between cyanobacteria and Rhodophyta. The eukaryotic algae was abundant and diversified in open marine, coastal tidal flat depositional environment (Tewari, 1999b). The eukaryotes occupied the Neoproterozoic planktonic realm while the prokaryotes occupied the benthonic realm (Shukla et al. 1986; Kumar and Srivastava, 1992). The stromatolites developed on these microbial mats and formed large microbial buildups in Upper Proterozoic time (Tewari, 1989, 1993a, 2001c, 2002). The size of microbial buildups decreased in Terminal Proterozoic (Fig. 7).

The Terminal Proterozoic diversification of life that led to the radiation of animals and plants occurred between 0.59 and 0.53 billion years ago on Earth. The prokaryotic to eukaryotic evolution and diversification of life on Earth, palaeoclimatic events of Neoproterozoic snowball Earth and the extinction and further emergence of highly organized life after Varanger (*Blainian*) glaciation can also be used as a possible model for the search of extraterrestrial life (astrobiological research). The stromatolites can also be used in the search for past life on Mars and elsewhere in the universe (Tewari, 1998, 2001b).

5. Acknowledgements

The author is grateful to Professor H.J. Hofmann, Mc Gill University, Canada and Dr. F. Westall, CNRS, Orleans, France for kindly reviewing the article and valuable suggestions. Prof. Joseph Seckbach, Hebrew University Jerusalam, Israel is thanked for his constant encourgement. Director, Wadia Institute of Himalayan Geology, Dehra Dun is thanked for facilities and permission to publish the paper. Mrs. Anita Chaudhary is thanked for ably typing the article.

6. References

Glassener, M.F. (1984) *The dawn of animal life, A historical study*, **224**, Cambridge London, New York, New Rochelle, Melbourne, Sydney.

Gnilovskaya, M.B. (ed.) (1988) *Vendotaenids of the East European platform* **143**, Nauka, USSR (in Russian).

Hofmann, H.J. (1976) Precambrian microflora, Belcher Islands Canada: significance and systematics. J. Paleontol. **50**, 1040-1073.

Hofmann, H.J. (1992) Proterozoic carbonaceous films, in: J.W. Schopf and C. Klein (eds.) *The Proterozoic Biosphere*. Cambridge University Press, U.K., pp. 349-357.

Hofmann, H.J. and Schopf, J.W. (1992) The Proterozoic Biosphere In: J.W. Schopf and C. Klein (eds.) Combridge University Press, New York, pp. 321-360.

Hofmann, H.J., Narbonne, G.M. and Aitken, J.D. (1990) Ediacaran fossils from intertillite beds in northwestern Canada. Geology, **18**, 1199-1202.

Knoll, A.H. (1985) The distribution and evolution of microbial life in the late Proterozoic Era. A Rev. Microbiol. **39**, 391-417.

Knoll, A.H. and Walter, M.R. (1992) Latest Proterozoic stratigraphy and earth history. Nature **356**, 673-678.

Kumar, S. and Srivastava, P. (1992) Middle to Late Proterozoic microbiota from the Deoban Limestone, Garhwal Himalaya, India. Precamb.Res. **56**, 291-318.

Mathur, V.K. and Shankar, Ravi (1989) First record of Ediacaran fossils from the Krol Formation, Nainital Synlcine. J. Geol. Soc. India, **34**, 245-254.

Narbonne, G.M.and Hofmann, H.J. (1987) Ediacaran biota of the Wernecke Mountains, Yukon Canada, Paleontol. **30**, 647-676.

Schopf, J.W. (1968) Microflora of the Bitter Springs Formation, Late Precambrian, Central Australia. J. Paleont. **42**, 651-688.

Schopf, J.W. (1983) *Earth's Earliest Biosphere.* Princeton University Press, New Jersey.

Schopf, J.W. and Klein, C. (1992) *The Proterozoic Biosphere* Cambridge University Press, Cambridge.1348p.

Schopf, J.W. (1993) Microfossils of the Early Archaean Apex Chert : New evidence of the Antiquity of life. Sci. **260**, 640-646.

Schopf, J.W. Haugh, B.H., Molnar, R.E. and Satterthwait, O.F. (1973) On the development of the metaphytes and metazoans. J. Paleontol. **47 (1)**, 1-9.

Seckbach, J. (1994). The natural history of *Cyanidium* (Geitler 1933): past and present perspectives. In: J. Seckbach (ed.) *Evolutionary pathways and enigmatic algae : Cyanidium caldarium, (Rhodophyta) and related cells*, Kluwer Academi Publishers, Dordrecht, The Netherlands. pp.99-112.

Shankar Ravi and Mathur, V.K. (1992) The Precambrian - Cambrian sequence in Krol belt and additional Ediacaran fossils. In: B.S. Venkatachala, K.P. Jain and N. Awasthy (eds.). *Geophytology* **22**, pp. 27-39.

Shanker Ravi, Mathur, V.K., Kumar, Gopendra and Srivastava, M.C. (1997) Additional Ediacaran biota from the Krol Group, Lesser Himalaya, India and their significance Geosci. J. **28(1)**, 79-94.

Shukla, M., Tewari, V.C. and Yadav, V.K. (1986) Late Precambrian microfossils from Deoban Limestone Formation, Lesser Himlaya, India. The Palaeobotanist **35(3)**, 347-356.

Tewari, V.C. (1988) Discovery of Vendotaenids from India. Indo - Soviet Symposium on Stromatolites and Stromatolitic Deposits, Wadia Institute of Himalaya Geology, Dehra Dun, 25-28.

Tewari, V.C. (1989) Upper Proterozoic - Lower Cambrian stromatolites and Indian Stratigraphy. Him. Geol., **13**, 143-180.

Tewari, V.C. (1992) Global decline of Pre-Ediacaran (Riphean) stromatolites and the emergence of Ediacaran biota : palaeobiological and stable isotope evidences from the Lesser Himalaya. Geol. Soc. Ind. **39**, 260-261.

Tewari, V.C. (1993a) Precambrian and Lower Cambrian stromatolites of the Lesser Himalaya, India. Geophytology, **23(1)**, 19-39.

Tewari, V.C. (1993b) Ediacaran metaphytes from the Lower Krol Formation, Lesser Himalaya, India. Geosci. J. **14(1,2)**, 143-148.

Tewari, V.C. (1996) Discovery of Pre Ediacaran acritarch *Chuaria cicularis* from the Deoban mountains, Lesser Himalaya, India. Geosci. J. **17(1)**, 25-39.

Tewari, V.C. (1997) Carbon and Oxygen isotope stratigraphy of the Deoban Group (Mesoproterozoic), Garhwal Lesser Himalaya. Geosci. J. **18(1)**, 95-101.

Tewari, V.C. (1998) Earliest microbes on Earth and possible occurrence of stromatolites on Mars. In: J. Chela Flores and F. Raulin (eds.) *Exobiology : Matter, Energy and Information in the Origin and Evolution of Life in the Universe,* Kluwer Academic Publihsers, Netherlands. pp. 261-265.

Tewari, V.C. (1999a) Candidate GSSP for the initial boundary of the Terminal Proterozoic Blaini Formation (Baliana Group) Mussoorie Syncline, Lesser Himalaya, India, In : A.H. Knoll (ed.) *Terminal Proterozoic System* 12th Circular, May, 1999. IUGS Sub commission on the Terminal Proterozoic System, pp. 17-24.

Tewari, V.C.(1999b) Vendotaenids : earliest megascopic multicellular algae on Earth. Geosci. J. **20(1)**, 77-85.

Tewari, V.C. (2001a) Neoproterozoic glaciation in the Uttaranchal Lesser Himalaya and the global palaeoclimate change. Geol. Surv. India. Spl. Publ. **65(3)**, 49-56.

Tewari, V.C. (2001b) Origins of life in the universe and earliest prokaryotic microorganisms on Earth. In: J. Chela Flores *et al.* (eds.) *First Steps in the Origin of Life in the Universe*, Kluwer Academic Publishers, the Netherlands. pp. 251-254.

Tewari, V.C. (2001c) Discovery and sedimentology of microstromatolites from Menga Limestone (Neoproterozoic/Vendian), Upper Subansiri district, Arunachal Pradesh, Northeastern Himalaya, India. Curr. Sci. **80(1)**, 1440-1444.

Tewari, V.C. (2002) Lesser Himalayan stratigraphy, sedimentation and correlation from Uttaranchal to Arunachal. Gyanodaya Prakashan, Nainital, 63-88.

Tiwari, M. and Knoll, A.H. (1994) Large acanthomorphic acritarchs from the Infra Krol Formation of the Lesser Himalaya and their stratigraphic significance. J. Him. Geol. **5(2),** 193-201.

Tiwari, M., Pant, C.C. and Tewari, V.C. (2000) Neoproterozoic sponge spicules and organic walled microfossils from the Gangolihat Dolomite, Lesser Himalaya, India.Curr. Sci., **79(5),** 651-654.

Walsh, M.M. (1992) Microfossils and possible microfossils from the Early Archean Onverwacht Group, Barberton Mountain Land, South Africa. Precamb. Res. **54**, 271-292.

Walter, M.R. (1989) The timing of major evolutionary innovation from the origin of life to the origins of the Metaphyta and Metazoa; the geological evidence. In: K.S.W. Campbell and M.F. Day (eds.) *Rates of Evolution*, Allan and Unwin, London. pp. 15-38

Westall, Frances (1999) Fossil Bacteria, In: J. Seckbach (ed.) *Enigmatic Micro-organisms and Life in Extreme Environments*, Kluwer Academic Publishers, Netherlands. pp. 73-88.

Westall, Frances, Maarten J. de wit, Jesse Dann, Sjerry Van der Gaast, Cornel E.J. de Ronde, Dane Gerneke (2001) Early Archean fossil bacteria and biofilms in hydrothermally influenced sediments from the Barberton from the Barberton greenstone belt, South Africa.

Biodata of **Michele K. Nishiguchi** and **Bryan W. Jones** authors of ***"Microbial Biodiversity within the Vibrionaceae"***

Dr. Michele K. Nishiguchi is affiliated with the Department of Biology at New Mexico State University (Las Cruces). She obtained her Ph.D. from the University of California, Santa Cruz in 1994. Her interests are in marine animal-bacterial symbiosis, coevolution of symbiotic relationships, evolutionary and ecological basis of speciation, and the population genetics of host-symbiont assemblages. The model system that her laboratory studies is the sepiolid squid-*Vibrio* association.

E-mail: **nish@nmsu.edu**

Bryan W. Jones is a pre-doctoral student with Dr. Nishiguchi at New Mexico State University. He received his B.S. from the University of Texas at Austin in ecology, conservation, and evolutionary biology in 2000. His current project addresses the impact of the sepiolid squid-*Vibrio* mutualism on bacterioplankton community structure.

E-mail: **bryjones@nmsu.edu**

M.K. Nishiguchi and B.W. Jones

J. Seckbach (ed.), Origins, 531–548.

MICROBIAL BIODIVERSITY WITHIN THE VIBRIONACEAE

MICHELE K. NISHIGUCHI AND BRYAN W. JONES
New Mexico State University, Department of Biology
Box 30001, MSC 3AF, Las Cruces, NM, 88003-8001 USA

1. The Vibrionaceae

1.1. A GENERAL DESCRIPTION

Vibrio takes its name from the Latin word *Vibrare*, meaning 'to wave'. Otto Müller first used the word *Vibrio* as a descriptor in the 18th century to describe bacteria with an elongated shape observed in culture (Rossello-Mora and Amann 2001). The family Vibrionaceae, first described by Véron (1965), resides within the γ-proteobacteria, one of the five subdivisions of the phylum Proteobacteria within the domain Bacteria.

All Proteobacteria are gram negative. This extremely metabolically diverse group of bacteria of medical and industrial importance has been divided into five subdivisions (α, β, δ, γ, and ε) on the basis of 16S rRNA sequence data. The γ-Proteobacteria are divided into three major subgroups, denoted γ-1, γ-2, and γ-3. Members of the γ-1 and γ-2 subgroups contain sulfur producing photosynthetic bacteria and members of the family *Legionella*, respectively. The γ-3 subgroup consist of a "*potpourri*" of organisms, including oceanospirilla, many pseudomonads, the enterics, and the Vibrionaceae (Woese *et al.* 1985).

Vibrionaceae is one of 22 families within the 14 orders of the γ-Proteobacteria. Members of this family were first described as oxidase-positive, motile bacteria with polar flagella (Krieg and Holt 1984). More recent data demonstrate that all Vibrionaceae species are either straight or curved rods, with some species not having oxidase-positive phenotypes (*Vibrio metschnikovii, V. gasogenes, Photobacterium phosphoreum, P. angustum,* and some strains of *P. leiognathi* are oxidase-negative), and some being non-motile. Vibrionaceae currently contains six genera: *Vibrio*, *Allomonas*, *Enhydrobacter*, *Listonella*, *Photobacterium* and *Salinivibrio*. Each genera has its own distinguishing morphological and physiological characteristics (Krieg and Holt 1984). Phenotypic data used to define each genus are G/C content, the presence of sheathed polar flagella, requirement of sodium for growth, lipase activity, D-mannitol utilization, and sensitivity to vibriostatic compound 0/129 (Holt *et al.* 1994). The most recent online version (July, 2002) of Bergey's Taxonomic Outline lists 72 species of *Vibrio*, 7 of *Photobacterium*, 3 of *Listonella*, and one of the other three genera (Table 1). These facultative anaerobes are found almost exclusively in aquatic environments, and have G/C contents ranging from 38-66% (Krieg and Holt 1984).

The genus *Enhydrobacter* is very unique among the Vibrionaceae. The only identified species in this genus, *E. aerosaccus*, forms gas-vacuoles, lacks luminescence, has a very high G/C content (66% compared to 38-51% for *Vibrio* and 40-44% for *Photobacterium* species). Furthermore, this species is non-motile and resistant to vibriostatic compound 0/129, making it quite an anomaly within the Vibrionaceae. Due to its very slow growth on unusual media (30-60 days), this species has not been studied to the extent of the other members of Vibrionaceae. In fact, a search in the Biological Abstracts and Science Citation Index only yields the paper in which *Enhydrobacter* is first described (Staley *et al.* 1987). Furthermore, the 16S rRNA has not been sequenced, and more work is necessary to determine the proper phylogenetic position of this distinct genus.

1.2. TAXANOMIC HISTORY

The advent of the polymerase chain reaction (PCR) and the resultant ease of sequencing DNA in the mid 1980's brought about a series of dramatic re-organizations within the Vibrionaceae. A complete review of all recent changes in species names within the Vibrionaceae is beyond the scope of this review; however, when examining the changes in genera within the past 16 years, it is interesting that only *Vibrio* and *Photobacterium* have not been deleted from the Vibrionaceae (Table 1).

TABLE 1. Changes in Vibrionaceae genera since 1984. Numbers in parentheses indicate the total number of described species in that genus.

1984[1]	Current Family	July 2002[2]	Date Added
Vibrio (20)	Unchanged	*Vibrio* (72)	-------
Photobacterium (3)	Unchanged	*Photobacterium* (7)	-------
Aeromonas (4)	Aeromonadaceae	*Allomonas* (1)	Kalina *et al.*(1984)
Pleisomonas (1)	Enterobacteriaceae	*Enhydrobacter* (1)	Staley *et al.* (1987)
		Listonella (3)	MacDonell and Colwell (1985)
		Salinovibrio (1)	Mellado *et al.* (1996)

[1]From Krieg and Holt (1984)
[2]From Garrity *et al.* (2002)

Between 1984 and the present, two families have been proposed and subsequently moved from Vibrionaceae. The first, *Shewanella*, was added together with *Listonella* by MacDonell and Colwell (1985). *Shewanella* was soon moved to the Alteromonadaceae based on 16S rRNA sequence analysis. *Colwellia* was then added to the Vibrionaceae (Demming *et al.* 1988) but subsequently relocated to the Alteromonadaceae. As more species are added to different families within the Vibrionaceae in the near future, the taxonomy will undoubtedly undergo many more significant changes.

2. Evolution of the Vibrionaceae

2.1. EVERCHANGING PHYLOGENY OF THE VIBRIOS

The family Vibrionaceae is one of the most well studied heterotrophic bacterial groups in marine ecosystems. They are important players in the upper surface waters, and inhabit a number of ecological niches, including fish intestinal tracts (Simidu *et al.* 1977), squid and fish light organs (Ruby and Nealson 1976; McFall-Ngai 2000), and are found as crustacean ectosymbionts (Roszak and Colwell 1987). Because of their broad ecological distribution and importance in marine community structure, their taxonomic identification has been extensively studied. Initially, the first phylogenetic accounts of vibrios were completed using phenotypic and genotypic analyses. These studies relied on DNA-DNA hybridzation, DNA-rRNA hybridization, enzymatic activity, restriction fragment length polymorphism analysis (RFLP), physiological growth at various temperatures, carbon source utilization, protein fingerprinting, and substituted amino acids to name a few (Baumann and Baumann 1977, Baumann *et al.* 1980a, 1983, 1984; Baumann and Baumann 1981; Colwell 1984; Bryant *et al.* 1986a, 1986b; Martin-Kearley and Gow 1994; Montilla *et al.* 1995; Urakawa *et al.* 1997, 1999; Lunder *et al.* 2000). Although these methods produced initial taxonomic and phylogenetic information, the relationships among related taxa, particularly those that are now renamed, was still vague. According to Bergey's manual of determinative bacteriology, species of *Vibrio* increased from 5 in 1974 to 72 in 2002 (Shewn *et al.* 1974; Garrity *et al.* 2002). This increase was due to the fact that the family has undergone some drastic taxonomic revisions (with some of the species being placed in an entirely different genus) as well as the new discovery of psychrophilic and psychrotrophic species. In 1987, Carl Woese (Woese 1987) proposed the 3 kingdom tree of life, using 16S rRNA as a molecular chronometer for determining phylogenetic relationships. With the advancement of techniques for rapidly sequencing DNA and RNA, the revision of prokaryotic taxonomy (and in particular, the Archaea and Bacteria), has been a fundamental instigator for the revision of the Vibrionaceae family. Initial studies using DNA sequences primarily focused on the nuclear ribosomal RNAs (5S and 16S rRNAs) to delineate species among Bacteria and Archaea. From this, investigators interested in the identification and classification of *Vibrio* species used the small ribosomal subunit in a number of studies to confer previous proposed phylogenies. Some of these phylogenetic analyses conferred known relationships (*Photobacterium* and *Vibrio* as monophyletic; Baumann *et al.* 1980a; Urakawa *et al.* 1997) but also changed the relationships of others. The genus *Aeromonas* originally was within the Vibrionaceae, but after the work of Colwell and others (Bryant *et al.* 1986a; Colwell *et al.* 1986; Kita-Tsukamoto *et al.* 1993; Ruimy *et al.* 1994) using 5S and 16S rRNA sequence analyses, they confirmed that this genus was within the family Aeromonadaceae.

As for inter-species relationships within the genus *Vibrio*, several phylogenetic relationships have been elucidated with molecular data. *V. marinus* has been classified as significantly different from other species of vibrios analyzed (MacDonell and Colwell 1984; Steven 1990), and recently has been proposed to be renamed *Moritella marinus* based on 5S rRNA sequence and DNA-DNA hybridization studies. Also, the relationship between *V. cholerae* and *V. mimicus* has been ambiguous; the two species are found to be similar based on 16S DNA sequence analyses (Davis *et al.* 1981) as well as a variety of phenotypic characters. Another example of incongruency is that of the relationship between *Vibrio anguillarum* and other related taxa. This species was previously placed as sister to *V. splendidus*, *Photobacterium angustum* and *V. metschnikovii* (Alsina and Blanch 1994), but has now been placed as sister to *V. damsela* and *V. ordalii* (Wiik *et al.* 1995). Again using the 16S rRNA as well as a combined analyses approach (genes and morphology) allows better resolution among taxa that have unclear associations (Figure 1). The 16S sequence data base for *Vibrio* species has also allowed the design of rRNA oligonucleotide probes to identify a number of pathogenic and free-living vibrios in the water column (Aznar *et al.* 1994). This technique can detect *Vibrio* cells from filtered water samples, and has initiated many studies examining the population and community structure that vibrios have on the total bacterioplankton ecology (Rehnstam *et al.* 1989). New, innovative approaches using additional morphlogical, DNA and RNA sequence data in combination will further the investigation of species divergence within the Vibrionaceae, and will help determine whether the taxonomy of many *Vibrio* species should be re-examined to provide a clearer picture of the diversity that this group represents (Figure 1).

2.2. THE EVOLUTION OF VIRULENCE IN MUTUALISTIC ASSOCIATIONS

Symbiosis among the Vibrionaceae occurs with many marine host species. Generally, vibrios colonize either crustacean (Bowser *et al.* 1981), mollusc (McFall-Ngai 2002), or fish (Wiik *et al.* 1995) hosts, either causing severe disease or death (Schiewe *et al.* 1981; Wiik *et al.* 1989; Toranzo and Barja 1990). Although a number of these pathogenic vibrios have common physiological attributes, it has always been a question of whether virulence or virulence factors (i.e., pathogencity islands) were common among these types of symbionts. Investigations assaying biochemical features (Lunder *et al.* 2000), iron sequestration (Tolmasky *et al.* 1985), and plasmid profiling (Sorum *et al.* 1990) have grouped many of the pathogens together, according to their specific hosts that they infect. Although this may provide a "common ground" for all species studied, 5S and 16S rRNA molecular data provide evidence that most of these alliances are not robust (Nishiguchi and Nair in press, Figure 1; Wiik *et al.* 1995) and the pathogenic species of *Vibrio* are not monophyletic. This is probably due to the fact that most phenotypic characters are more likely to place species or species groups according to their ecological niches; that is, phenotypic characters tend to reflect the type of habitat and the abiotic factors that influence the phenotype of that particular species or strain (Cohan 2002). One example of such behavior is found in the mutualistic association between vibrios and their sepiolid squid hosts (Nishiguchi 2001). Although this association is specific for 2 known *Vibrio* species (*V. logei* and *V. fischeri*; Fidopiastis *et al.* 1998), this particular symbiosis can be host specific (as seen in Indo-west Pacific populations of *Euprymna*; Nishiguchi *et al.* 1998, Nishiguchi 2002), or can be influenced by abiotic factors such as temperature (Nishiguchi 2000). Thus, virulence or virulence factors may be similar in gene homology (see section 4.2) but are not congruent with the phylogenic data based on 16S rRNA sequence (Nishiguchi, unpublished data, Figure 1).

Figure 1. Strict consensus tree for 75 taxa of representative *Vibrio* strains and related species. Phylogenetic relationships are based on the complete 16S rRNA sequence from a number of Vibrionaceae representatives. Molecular data were analyzed using the direct optimization method (Wheeler 1996) as implemented in the computer program POY (Wheeler *et al.* 2002). A parameter space of two variables (gap/change ratio and transversion/transition ratio) was explored, as in Wheeler (1995). Transversion/transition ratios of 1 (equal weights) and 2 (transversions receive twice as much weight than transitions) were explored for these analyses. All species are designated with strain identification in the figure. For key, see http://mvar.nmsu.edu/nish/index.htm.

3. Ecology of the Vibrionaceae

3.1. DIVERSE HABITATS AND ASSOCIATIONS

As previously mentioned, members of Vibrionaceae are ubiquitous in the marine environment, with species found in hydrothermal vents (Raguenes *et al.* 1997), deep sea (Maruyama *et al.* 2000), open water (Eilers *et al.* 2000), estuaries, and marine sediments (Lee and Ruby 1994). Traditionally, the Vibrionaceae were thought to comprise a very large portion of bacterioplankton communities (ZoBell 1946); however, modern molecular techniques (such as fluorescent *in situ* hybridization) have shown these estimates to be wildly overstated (Eilers *et al.* 2000). More recent estimates range from 1 (Eilers *et al.* 2000) to 10 (Nishimura *et al.* 1995) percent of total cells, depending on the habitat. Although they comprise only a small percentage of total free-living bacteria in the marine environment, the importance of Vibrionaceae should not be underestimated. Recently, many studies have suggested that vibrios degrade some ecologically hazardous compounds, such as polycyclic aromatic hydrocarbons (Ramaiah *et al.* 2000), and are major decomposers of chitin in the ocean (Nagasawa and Terazaki 1987; Hedlung and Staley 2001). Vibrios have also been found in rivers (Kenzaka *et al.* 1998), and *V. cholerae* is known to inhabit freshwater environments (Baumann *et al.* 1980b). One species, *V. navarrensis* has even been found in sewage outfalls (Urdaci *et al.* 1991).

Free-living vibrios have not been taken directly from soil, but have been found in the gut of soil invertebrates (Byzov *et al.* 1996; Tret'yakova *et al.* 1996). Assuming the gut of these animals was colonized environmentally from soil (as opposed to vertical or horizontal colonization), there should be free-living vibrios in these habitats. Further evidence that vibrios are found in the free-living state in soil was determined by Hallmann *et al.* (1999) who found both *V. cholerae* and *V. fluvialis* in soil samples following the addition of chitin to a final concentration of 1% (w/w). Since many vibrios exist in a viable but non-culturable state (Colwell *et al.* 1985), they may not be detected in experiments which rely on traditional culture based techniques for data collection.

Perhaps ecologically and economically more important than the free-living species is the variety of parasitic and mutualistic relationships existing within the Vibrionaceae. The diversity of parasitic associations exhibited by vibrios is truly remarkable. The most common parasitic association is the attachment to the surface of fish and other animals as saprophytes. In fact, one of the traditional methods used to isolate strains of luminescent bacteria was to incubate a dead fish and inspect it for glowing colonies (Nealson and Hastings 1979). *V. cholerae*, the bacterium whose virulent form causes the disease cholera, attaches to chitinaceous zooplankton, such as copepods (Colwell 1996), using them as a vector and transport mechanism, causing spread of this deadly disease. Members of Vibrionaceae have also been shown to cause potentially lethal diseases in humans and fish (Farmer *et al.* 1985; Brayton *et al.* 1986; Kusuda and Kawai 1998; McCarter 1999). More recently, studies have shown *V. shiloi* to be a coral pathogen, producing toxins that inhibit photosynthesis and lyse zooxanthellae resulting in bleaching (Banin *et al.* 2000a, 2000b).

Mutualistic interactions with vibrios have also been widely studied. One of the most common places to find vibrios is the gut of marine animals. Since many Vibrionaceae species contain chitinase (Ramaiah *et al.* 2000), they most likely aide their host in digestion. Vibrios are also the symbionts in teleost fish light organs. In these mutualistic associations, the bacteria receive nutrients from the light organ, and light produced has a variety of functions, including prey attraction (the angler fish), predator evasion via light bursts or counterillumination (the pony fish), and interspecies communication (flashlight and pinecone fishes; Nealson and Hastings 1979). More recently, the sepiolid squid/*Vibrio* symbiosis has become a model system for the study of mutualistic interactions (Nishiguchi *et al.* 1998; McFall-Ngai 1999). In this association, the *Vibrio* symbiont inhabits a complex, bilobed light organ inside the squid mantle cavity. Light produced by the *Vibrio* symbiont aides the squid host as camouflage in the form of counterillumination (Jones and Nishiguchi, unpublished data). In return, the squid provides nutrients for the symbiotic *Vibrio*, resulting in a generation time of under 30 minutes (Boettcher and Ruby 1990; McFall-Ngai and Ruby 1998), which is much faster than the free-living state.

3.2. BIOLUMINESCENCE AND THE VIBRIONACEAE

Light production is a common characteristic of many members from the genera *Vibrio* and *Photobacterium*. Although species of other genera (including *Shewanella* and *Photorhabdus*) are also luminescent, *Photobacterium* and *Vibrio* species have been the most extensively studied (Meighen 1994; Bassler and Silverman 1995; Bourgois *et al.* 2001). In its purest form, the light emitting reaction involves the catalysis by the luciferase enzyme of a number of substrates, including reduced flavin mononucleotide ($FMNH_2$), a long chain fatty aldehyde (RCHO), and molecular oxygen (Meighen 1991).

Luminescence is created and regulated by the *lux* genes, of which only five are common to all light-producing bacteria. The genes *luxA* and *luxB* encode for the α and β subunits of luciferase, respectively. The *luxC*, *D*, and *E* genes encode proteins involved in the ultimate construction of the long chain fatty aldehyde from tetradecanoyl-ACP (Meighen 1994). Other species-specific genes include various regulatory genes (*luxR*, *I*, *L*, *M*, *N*, and *O*; see section 3.3 for discussion of quorum sensing), genes responsible for the modification of light wavelength or emission efficiency (*luxL* and *Y*), and genes whose function have not been clearly defined (*luxG* and *H*; Bassler and Silverman 1995). In addition to the many species-specific genes in the operon, gene order also appears to be species-specific, with the exception of *luxA* and *B*, which are always situated adjacent to each other in the operon. The *lux*-regulated light producing reactions are metabolically expensive. It has been estimated that up to 17-20% of total respiration in *V. fischeri* cells is used in luminescent reactions (Makemson 1986; Bourgois *et al.* 2001). This begs the question of why bacteria should produce light and what types of selective pressures have led to the development of such a genetically complex system.

Nealson and Hastings (1979) summarized hypotheses as to why this phenomenon occurs in some bacteria. The first hypothesis relates bioluminescence to its biological and ecological functions. Since most luminescent species are enterics, light production on a substrate would be advantageous if it attracted a predator whose gut would be colonized by the luminescent colony. Although some theoretical work has been presented on the amount of bacteria required for a fish to detect a colony for consumption (Nealson and Hastings 1979), no known empirical tests have been conducted to assess this seemingly plausible argument.

The second set of hypotheses states that bioluminescence has a biochemical function, such as allowing the cell to benefit from photochemical reactions. Recently, Czyz *et al.* (2000) proposed that bioluminescence functions to repair damaged DNA. In this study, the authors noticed that UV irradiation-sensitive mutants of *V. harveyi* tended to lack a functional luminescence system. When these mutants were incubated in the dark following UV irradiation, their survival was significantly depressed compared to non-mutants incubated under the same conditions. Another biochemical possibility would be the use of the luminescence system to aide in the re-oxidation of excess reducing equivalents in the cell when the respiratory system nears saturation in near anaerobic environmental conditions (Makemson and Hastings 1986; Bourgois *et al.* 2001).

McElroy and Seliger (1962) proposed a separate hypothesis associated with the evolution of bacterial luminescence that is similar to many of the biochemical function hypotheses. This hypothesis states that bioluminescence first evolved as a detoxification mechanism via the reduction of molecular oxygen. Watanabe *et al.* (1993) suggested that prior to the evolution of the *lux* genes responsible for the formation of the long-chain aldehyde, luciferase may have acted to detoxify H_2O_2. This hypothesis is discussed in detail by Timmins *et al.* (2001). Although the hypotheses dealing with a biochemical function of light-production are intriguing, most (with the exception of Czyz *et al.* 2000) fail to provide a functional role of the light produced. More research into the biological and ecological advantages of bioluminescence will help to illuminate answers as to why these complex genes have evolved.

3.3. QUORUM SENSING

Light production in many members of the Vibrionaceae is controlled by a phenomenon termed quorum sensing, or autoinduction. This density-dependent control of transcription is a unique form of bacterial communication using extracellular signaling molecules known as autoinducers (*N*-acyl-L-homoserine lactones; Dunlap 1997). As the number of cells in a population grows, autoinducers diffuse from the cell into the local environment. Once a threshold autoinducer concentration is reached, the autoinducer diffuses back into the cell, activating transcription.

Quorum sensing was first described in *V. fischeri* and has since been found to control a variety of different activities in a diverse number of microorganisms (see Dunlap 1997 for a complete review). In *V. fischeri*, the autoinducer created by LuxI interacts with the regulator protein (LuxR) to activate transcription of the *luxICDEABEG* operon, resulting in light production (Engebrecht and Silverman 1984). A second autoinduction system controlling luminescence in *V. fischeri* has also been identified (Gilson *et al.* 1995). This autoinducer, whose production is directed by *ainS* appears to interfere with the binding of the LuxI produced autoinducer to LuxR, effectively inhibiting luminescence at low densities of bacteria (Kuo *et al.* 1996). A transcriptional activator, LitR also appears to play a role in quorum sensing through regulation of *luxR* (Fidopiastis *et al.* 2002).

V. harveyi has proven to be a valuable organism for the comparative study of autoinduction of the *lux* genes. The complex *V. harveyi* regulation system lacks genes homologous to *luxR* and *I* and autoinduction is controlled by a two component phosphorelay circuit (Bassler 1999). Briefly, the two autoinducers AI-1 and AI-2 interact with LuxQ and LuxN, respectively. The AI-1/LuxQ and AI-2/LuxN interaction is relayed to LuxO via the phosphotransferase, LuxU. LuxO indirectly inhibits transcription of the *luxCDABEGH* operon through a yet undiscovered repressor protein (Lilley and Bassler 2000).

Recent experiments have shown that virulence gene expression in *V. cholerae* is controlled by quorum sensing. Although *V. cholerae* lacks *luxCDABE* (genes for light production), genes analogous to those required for response to AI-2 in *V. harveyi* are present. In this system, LuxO inhibits transcription of *hapR*, which in turn inhibits *tcpP* transcription, an essential requirement for the expression of the toxin-coregulated pilus (TCP) virulence factor (Zhu *et al.* 2002).

4. Pathogenesis in the Vibrionaceae

4.1. DISEASES IN THE VIBRIONACEAE

Most diseases caused by vibrios are usually wound infections that have been in contact with saltwater or shellfish. These infections usually lead to "vibriosis" a terminal hemorrhagic septicemia in marine and freshwater fishes (Thune *et al.* 1993; Okuda *et al.* 2001), as well as in humans (Stelma *et al.*. 1992). Other *Vibrio* species, like *V. cholerae*, have been extensively studied due to their broad range of host virulence (Mooi and Bik 1997). Evolution of virulent strains has been of importance to human health and control of epidemics; although only a few serotypes have been fully characterized, there are still continual outbreaks of disease due to newly formed serotypes that are closely related to well studied strains. One example is that of the O1 serotype of *V. cholerae*; the main subdivisions of this strain are the classical and El Tor biotypes. These strains were linked to several outbreaks between 1881 and 1961, and yet a third serotype (O139 or Bengal) was discovered to be the culprit of a subsequent outbreak in India in 1992 (Mooi and Bik 1997). DNA fingerprinting confirms that the O139 strain is similar to El Tor and classic strains, however there are major differences that exist in several virulence factors that cause differences in virulence phenotypes. Thus, although strains are closely related phylogenetically, other factors have changed the mechanism as to which serotype invades or causes disease in its host.

Although *V. cholerae* is the most well-studied pathogenic *Vibrio* among the Vibrionaceae, other species of *Vibrio* have also been shown to contain similar virulence pathways that provide a means for infection and pathogenicity. Species such as *Vibrio (Listonella) anguilarum, V. parahaemolyticus, V. vulnificus*, and *V. mimicus* have been shown to express virulence-related properties such as production of the *toxR* gene (which regulates expression of the toxin-coregulated pilus (TCP) and cholera toxin; Lin *et al.* 1993; Okuda *et al.* 2001), production of phenolate siderophore (for iron sequestration; Stelma *et al.* 1992), as well as cell-mediated agglutination and bacterial adherence (Alam *et al.* 1996). Other benign mutualistic species, such as *V. fischeri*, have been shown to possess homologs to the transmembrane transcriptional activator *toxR* (Reich and Schoolnik 1994), and 2 types of halovibrin, which are both ADP-ribosyltransferases that have similar enzymatic activities to cholera toxin (Reich and

Schoolnik 1996; Reich *et al.* 1997). Since the infection mechanism is similar to *V. cholerae*, it is not surprising that *V. fischeri* possesses genes that have similar roles for infection and colonization of their respective eukaryotic partners (McFall-Ngai 2002; Nishiguchi 2002). Thus, proteins like cholera toxin and halovibrin are the "communication signals" that initiate interactions for bacteria to recognize and colonize their specific hosts.

4.2. HORIZONTAL GENE TRANSFER OF VIRULENCE FACTORS

As stated previously, genomic data has revealed that there is no strict congruence between phylogenetic relatedness and the physiological mechanisms that cause virulence. Several genes or gene families have been found that are responsible for transfer; cell-wall polysaccharide genes of several *V. cholerae* strains (El Tor, O139, O69, and O141) are associated with a mobile element (IS1358) that allows homologous recombination between lipopolysaccharide (LPS) gene clusters between different strains (Mooi and Bik 1997; Stroeher *et al.* 1995) Homologous gene clusters are also found in a number of other bacteria (not all pathogens), and may also play a role in mobilizing DNA directly by transposition to a phage (or plasmid) as well as in recombination. Both clinical and environmental isolates of *V. cholerae* have been found to contain the CTX phage, which encode the genes for cholera toxin (CT; Dalsgaard *et al.* 2001).

CTXΦis a lysogenic filamentous bacteriophage, and various isolates have been found in different strains of *V. cholera* (Davis *et al.* 1999). Analyses of these phage types have shown that some similarity is shared between phage repressor genes (*rstR*), which allow the phages to become infective or not (Davis *et al.* 1999). These phage are also known to use toxin co-regulated pili (TCP) as its receptor, and it has been hypothesized that this operon has also been introduced into strains of *V. cholerae* and related species through a bacteriophage (Dalsgaard *et al.* 2001; Mooi and Bik 1997; Nakayama *et al.* 1999; Nandi *et al.* 2000). Similarly, there are other virulence related genes that are clustered around the TCP operon as well as a putative integrase and bacteriophage attachment site (Mooi and Bik 1997). Thus, most of the current evidence points to the evolution of virulence through horizontal gene transfer via bacteriophage.

Although CTXΦ demonstrates the acquisition of pathogenicity islands among virulent strains of *V. cholerae*, mutational studies have demonstrated that TCP production and regulation is not entirely responsible for colonization and adhesion to the host epithelia. The *rfb* genes in *V. cholerae* are responsible for the production of enzymes necessary for lipopolysaccharide biosynthesis, which was originally thought to be linked to the expression and assembly of TCP (Iredell and Manning 1997). Recent studies have shown that mutations in the *rfb* gene do not affect TCP production, but do cause inhibition of intestinal colonization (Chiang and Mekalanos 1999). Because this mutant did not affect TCP production, changes in bacterial LPS may cause more susceptibility to host immune responses (such as complement), or antibiotics. Several recent studies have shown that LPS from symbiotic strains cause host responsiveness and virulence (Foster *et al.* 2000; Zhang *et al.* 1997), and therefore play an important role along with TCP production in the infection of host tissues.

5. Conclusions

The family Vibrionaceae is one of the most diverse and widely distributed groups of prokaryotes that have radiated into hundreds of existing niches in the environment. Our present understanding of the evolution, ecology, and virulence of this important group of Proteobacteria has been greatly enhanced in the past decade. With future work on a number of specific *Vibrio* species in their free-living and symbiotic state, we can hope to uncover mechanisms of virulence, pathogenesis, and speciation within this dynamic family of microorganisms.

6. Acknowledgements

The authors would like to thanks members of the Nishiguchi lab (J. Browne-Silva, A. Lindgren, J. Lopez, V. Nair, T. Powers, S. Saenz, W. Soto, S. Stevenson, and A. Weaver) for all the research support and help with reviewing the manuscript. Thanks to G. Giribet at Harvard University for help with preliminary analyses of the 16S rRNA phylogeny. This research was supported by NSF DBI-0079820 and NIH S06-GM08136-26 to M.K.N. and NSF ADVANCE SBE-0123690 at NMSU.

7. References

Alam, M., Miyoshi, S-I., Yamamoto, S., K-I Tomochika and Shinoda, S. (1996) Expression of virulence-related properties by, and intestinal adhesiveness of, *Vibrio mimicus* strains isolated from aquatic environments. Appl. Environ. Microbiol. **62**(10): 3871-3874.

Alsina, M., and Blanch, A. R. (1994) Improvement and update of a set of keys for biochemical identification of *Vibrio* species. J. Appl. Bacteriol. **77**: 719-721.

Aznar, R., Ludwig, W., Amann, R.I., and Schleifer, K. H. (1994) Siderophores and related outer membrane proteins produced by pseudomonads isolated from eels and freshwater. Int. J. Syst. Bacteriol. **44**(2): 330-337.

Banin, E., Ben-Haim,Y., Israely, T., Loya, Y. and Rosenberg, E. (2000a) Effect of the environment on the bacterial bleaching of corals. Water, Air, and Soil Poll. **123**: 337-352.

Banin, E., Israely, T., Kushmaro, A., Loya, Y., Orr, E. and Rosenberg E. (2000b) Penetration of the coral-bleaching bacterium bacterium *Vibrio shiloi* into *Oculina patagonica.* Appl. Environ. Microbiol. **66**(7): 3031-3036.

Bassler, B. L. (1999) *Cell-Cell Signaling in Bacteria.* S. C. Winans, ed. Washington D.C., American Society for Microbiology Press: 259-273.

Bassler, B. L. and M. R. Silverman (1995) Intercellular communication in marine *Vibrio* species: Density-dependent regulation of the expression of bioluminescence. In: T. J. Silhavy. (ed.) *Two-Component Signal Transduction.* Washington, D.C., American Society for Microbiology.

Baumann, L., and Baumann, P. (1977) Biology of the marine enterobacteria: genera *Benekea* and *Photobacterium.* Ann. Rev. Microbiol. **31**: 39-61.

Baumann, L., Bang, S. S. and Baumann, P. (1980a) Study of relationship among species of *Vibrio, Photobacterium,* and terrestrial enterobacteria by an immunological comparison of glutamine synthestase and superoxide dismutase. Curr. Microbiol. **4**: 133-138.

Baumann, P., and Schubert, R. H. W. (1984) *Bergy's manual of systematic bacteriology.* N. R. Krieg and J. G. Holt, eds. Baltimore, The Williams and Wilkins Co. pp 516-517.

Baumann, P., Baumann, L., Bang, S. S. and Woolkalis, M. J. (1980b) Reevaluation of the taxonomy of *Vibrio, Beneckea,* and *Photobacterium:* Abolition of the genus *Beneckea.* Curr. Microbiol. **4**: 127-132.

Baumann, P., Baumann, L.. Woolkalis, M. J. and Bang, S. S. (1983) Evolutionary relationships in *Vibrio* and *Photobacterium*: a basis for natural classification. Ann. Rev. Microbiol. 37: 369-398.

Baumnann, P. and Baumann, L. (1981) The marine Gram-negative eubacteria. In: H. S. M. P. Starr, H. G. Truper, A. Balows, and H. G. Schlegel (eds.) *The Prokaryotes, a handbook on habiats, isolation, and identification of bacteria.*. New York, Springer Verlag pp 1352-1394.

Boettcher, K. J. and Ruby, E. G. (1990) Depressed light emission by symbiotic *Vibrio fischeri* of the sepiolid squid *Euprymna scolopes*. J. Bacteriol. 172: 3071-3076.

Bourgois, J. J., Sluse, F. E. Baguet, F. and. Mallefet, J (2001) Kinetics of light emission and oxygen consumption by bioluminescent bacteria. J. Bioenerg. Biomem. **33**(4): 353-363.

Bowser, P. R., Rosemark, R. and Reiner, C. R. (1981) A prelminary report of vibriosis in cultured American lobsters, *Homarus americanus*. J. Invertebr. Pathol. **37**: 80-85.

Brayton, P. R., Bode, R. B., Colwell, R. R., MacDonell, M. T., Hall, H. L., Grimes, D. J., West, P. A. and Bryant, T. N. (1986) *Vibrio cincinnatiensis* sp. nov., a new human pathogen. J. Clin. Microbiol. **23**(1): 104-108.

Bryant, T. N., Lee, J. V., West, P. A. and R. R . Colwell (1986a) Numerical classification of species of *Vibrio* and related genera. J. Appl. Bacteriol. **61**: 437-467.

Bryant, T. N., Lee, J. V., West, P. A. and Colwell, R. R. (1986b) A probability matrix for the identification of species of *Vibrio* and related genera. J. Appl. Bacteriol. **61**: 469-480.

Byzov, B. A., Chernjakovskaya, T. F. , Zenova, G. M. and Dobrovol'skaya, T. G. (1996) Bacterial communities associated with soil diplopods. Pedobiologia **40**: 67-79.

Chiang, S. L. and Mekalanos, J. J. (1999) *rfb* mutations in *Vibrio cholerae* do not affect surface production of toxin-coregulated pili but still inhibit intestinal colonization. Infect. Immun. **67**(2): 976-980.

Cohan, F. M. (2002) What are bacterial species? Ann. Rev. Microbiol. **56**: 457-87.

Colwell, R. R. (1984) *Vibrios in the environment*. New York, John Wiley and Sons.

Colwell, R. R. (1996) Global climate and infectious disease: The cholera paradigm. Science **274**: 2025-2031.

Colwell, R. R., Brayton, P. R., Grimes, D. J., Roszak, D. B., Huq, S. A., and Palmer, L. M. (1985) Viable but non-culturable *Vibrio cholerae* and related pathogens in the environment: Implications for release of genetically-engineered microorganisms. Biotechnology **3**: 807-820.

Colwell, R. R., MacDonell, M. T., and De Ley, J. (1986) Proposal to recognize the family Aeromonadaceae fam. nov. Int. J. Syst. Bacteriol. **36**: 473-477.

Czyz, A., Wrobel, B. and Wegrzyn, G. (2000) *Vibrio harveyi* bioluminescence plays a role in stimulation of DNA repair. Microbiology **146**: 283-288.

Dalsgaard, A., Serichantalergs, O., Forslund, A., Lin, W., Mekalanos, J., Mintz, E., Shimida, T. and Wells, J. G. (2001) Clinical and environmental isolates of *Vibrio cholerae* serogroup O141 carry the CTX phage and the genes encoding the toxin-coregulated pili. J. Clin. Microbiol. **39**: 4086-4092.

Davis, B. R., Fanning, G. R., Madden, J. M., Steigerwalt, A. G., Bradford Jr., H. B., Smith Jr., H. L. and Brenner, D. J. (1981) Characterization of biochemically aptypical *Vibrio cholerae* strains and designation of a new pathogenic species, *Vibrio mimicus*. J. Clin. Microbiol. **14**: 631-639.

Davis, B. M., Kimse, H. H., Change, W., and Waldor, M. K. (1999) The *Vibrio cholerae* O139 Calucutta bacteriophage CTXΦ is infectious and encodes a novel repressor. J. Bacteriol. **181**(21): 6779-6787.

Demming, J. W., Somers, L. K., Straube, W. L., Swartz D. G. and MacDonell, M. T. (1988) Isolation of an obligately barophilic bacterium and description of a new genus, *Colwellia* gen. nov. Syst. Appl. Microbiol. **10**: 152-160.

Dunlap, P. V. (1997) *N*-acyl-L-homoserine latone autoinducers in bacteria: Unity and diversity. In: *Bacteria as Multicellular Organisms*. J. A. Shapiro and M. Dworkin, eds. New York, Oxford University Press.

Eilers, H., Pernthaler, J., Glockner, F. O. and Amann, R. (2000) Culturability and *in situ* abundance of pelagic bacteria from the North Sea. Appl. Environ. Microbiol. **66**(7): 3044-3051.

Engebrecht, J. and Silverman, M. R. (1984) Identification of genes and gene products necessary for bacterial bioluminescence. Proc. Natl. Acad. Sci. USA **81**: 4154-4158.

Farmer, J. J. III, Hickman-Brenner, F. W. and Kelly, M. T. (1985) *Vibrio*. In: *Manual of clinical microbiology*. H. J. Shadomy, ed. Washington, D. C., American Society for Microbiology.

Fidopiastis, P. M., Boletzky, S. V., and Ruby, E. G. (1998) A new niche for *Vibrio logei,* the predominant light organ symbiont of squids in the genus *Sepiola*. J. Bacteriol. **180**: 59-64.

Fidopiastis, P. M., Miyamoto, C. M., Jobling, M. G., Meighen, E. A. and Ruby, E. G. (2002) LitR, a new transcriptional activator in *Vibrio fischeri*, regulated luminescence and symbiotic light organ colonization. Mol. Microbiol. **45**(1): 131-143.

Foster, J. S., Apicella, M. A. and McFall-Ngai, M. J. (2000) *Vibrio fischeri* lipopolysaccharide induces developmental apoptosis, but not complete morphogenesis, of the *Euprymna scolopes* symbiotic light organ. Dev. Biol. **226**: 242-254.

Garrity, G. M., Johnson, K. L., Bell, J. and Searles, D. B. (2002) Bergey's Manual of Systematic Bacteriology, Second Edition, Release 3.0. New York, Springer-Verlag. http://dx.doi.org/10.1007/bergeysoutline200210.

Gilson, L., Kuo, A. and Dunlap, P. V. (1995) AinS and a new family of autoinducer synthesis proteins. J. Bacteriol. **177**(23): 6946-6951.

Hallman, J., Rodriguez-Kabana, R. and Kloepper, J. W. (1999) Chitin-mediated changes in bacterial communities of the soil, rhizosphere and within roots of cotton in relation to nematode control. Soil Biol. Biochem. **31**: 551-560.

Hedlung, B. P. and Staley, J. T. (2001) *Vibrio cyclotrophicus* sp. nov., a polycyclic aromatic hydrocarbon (PAH)-degrading marine bacterium. Int. J. Syst. Evol. Microbiol. **51**: 61-66.

Holt, J. G., Krieg, N. R., Sneath, P. H. A., Staley, J. T. and Williams, S. T. (1994) *Bergey's manual on determinative bacteriology*. Baltimore, Williams and Wilkins.

Iredell, J.R. and Manning, P. A. (1997) Outer membrane translocation arrest of the TcpA pilin subunit in *rfb* mutants of *Vibrio cholerae* 01 strain 569B. J. Bacteriol. **179**: 2038-2046.

Kalina, G. P., Antonov, A. S., Turova, T. P. and Grafova, T. I. (1984) *Allomonas enterica* gen. nov., sp. nov.: deoxyribonucleic acid homology between *Allomonas* and some other members of the Vibrionaceae. Int. J. Syst. Bacteriol. **34**: 150-154.

Kenzaka, T., Yamaguchi, N., Tani, K. and Nasu, M. (1998) rRNA-targeted fluorescent *in situ* hybridization analysis of bacterial community structure in river water. Microbiology-UK **144**(8): 2085-2093.

Kita-Tsukamoto, K., Oyaizu, H., Nanba, K. and Simidu, U. (1993) Phylogenetic relationships of marine bacteria, mainly members of the family Vibrionaceae, determined on the basis of 16S rRNA sequences. J. Syst. Bacteriol. **43**(1): 8-19.

Krieg, N. R. and Holt, J. G. (1984) *Bergy's manual of determinative bacteriology*, volume 1, Baltimore, The Williams and Wilkins Co.

Kuo, A., Callahan, S. M. and Dunlap, P. V. (1996) Multiple *N*-acyl-L-homoserine lactone autoinducers of luminescence in the marine symbiotic bacterium *Vibrio fischeri*. J. Bacteriol. **178**: 971-976.

Kusuda, R. and Kawai, K. (1998) Bacterial diseases of cultured marine fish in Japan. Fish Pathol. **33**(4): 221-227.

Lee, K-H. and Ruby, E. G. (1994) Effect of the squid host on the abundance and distribution of symbiotic *Vibrio fischeri* in nature. Appl. Environ. Microbiol. **60**: 1565-1571.

Lilley, B. N. and Bassler, B. L. (2000) Regulation of quorum sensing in *Vibrio harveyi* by LuxO and Sigma-54. Mol. Microbiol. **36**: 940-954.

Lin, A., Kumagai, K., Baba, K., Mekalanos, J.J. and Nishibuchi, M. (1993) *Vibrio parahaemolyticus* has a homolog of the *Vibrio choleare toxRS* operon that mediates environmentally induced regulation of the thermostable direct hemolysin gene. J. Bacteriol. **175**: 3844-3855.

Lunder, T., Sorum, H., Holstad, G., Steigerwalt, A. G., Mowinckel, P. and Brenner, D. J. (2000) Phenotypic and genotypic characterization of *Vibrio viscosus* sp. nov. and *Vibrio wodanis* sp. nov. isolated from Atlantic salmon (*Salmo salar*) with 'winter ulcer'. Int. J. Syst. Evol. Microbiol. **50**: 427-450.

MacDonell, M. T. and Colwell, R. R. (1984) The nucleotide sequence of 5S ribosomal RNA from *Vibrio marinus*. Microbiol. Sci. **1**: 229-231.

MacDonell, M. T. and Colwell, R. R. (1985) Phylogeny of the Vibrionaceae and recommendations for two new genera, *Listonella* and *Shewanella*. Syst. Appl. Microbiol. **6**: 171-182.

Makemson, J. C. (1986) Luciferace-dependent oxygen consumption by bioluminescent vibrios. J. Bacteriol. **104**: 313-322.

Makemson, J. C. and Hastings, J. W. (1986) Luciferase-dependent growth of cytochrome-deficient *Vibrio harveyi*. FEMS Microbiol. Ecol. **38**: 79-85.

Martin-Kearley, J. and Gow, J. A. (1994) Numerical taxonomy of Vibrionaceae from Newfoundland coastal waters. Can. J. Microbiol. **40**: 355-361.

Maruyama, A., Honda, D., Yamamoto, H., Kitamura, K. and Higashihara T. (2000) Phylogenetic analysis of psychrophilic bacteria isolated from the Japan Trench, including a description of the deep-sea species *Psychrobacter pacificensis* sp. nov. Int. J. Syst. Evol. Microbiol. **50**(2): 835-846.

McCarter, L. (1999) The multiple identities of *Vibrio parahaemolyticus*. J. Mol. Microbiol. Biotechnol. **1**(1): 51-57.

McElroy, H. H. and Seliger, H. H. (1962) Origin and evolution of bioluminescence. In: *Horizons in Biochemistry*. B. Pullman, ed. New York, Academic Press pp**:** 91-101.

McFall-Ngai, M. J. (1999) Consequences of evolving with bacterial symbionts: Lessons from the squid-vibrio association. Ann. Rev. Ecol. Syst. **30**: 235-256.

McFall-Ngai, M. J. (2000) Negotiations between animals and bacteria: the 'diplomacy' of the squid-vibrio symbiosis. Comp. Biochem. Physiol. A **126**: 471-480.

McFall-Ngai, M. J. (2002) Unseen Forces: The influence of bacteria on animal development. Dev. Biol. **242**: 1-14.

McFall-Ngai, M. J. and Ruby, E. G. (1998) Sepiolids and Vibrios: When first they meet. BioScience **48**(4): 257-265.

Meighen, E. A. (1991) Molecular biology of bacterial bioluminescence. Microbiol. Rev. **55**: 123-142.

Meighen, E. A. (1994) Genetics of Bacterial Bioluminescence. Ann. Rev. Gen. **28**: 117-139.

Mellado, E., Moore, E. R. B., Mieto, J. J. and Ventosa, A. (1996) Analysis of 16S rRNA gene sequences of *Vibrio costicola* strains: description of *Salinivibrio costicola* gen. nov., comb. nov. Int. J. Syst. Bacteriol. **46**: 817-821.

Montilla, R., Palomar, J., Fuste, M. C. and Vinas, M. (1995) Taxonomy and protein fingerprinting of halophilic *Vibrio* isolates from bivalves of the Ebre delta. Can. J. Microbiol. **41**: 61-69.

Mooi, F. R. and Bik, E. M. (1997) The evolution of epidemic *Vibrio cholerae* strains. Trends Microbiol. **5**(4): 161-165.

Nagasawa, S. and Terazaki, M. (1987) Bacterial epibionts of the deep-water copepod *Calanus cristatus* Kroyer. Oceanologica Acta **10**: 475-479.

Nakayama, K., Kanaya, S., Ohnishi, M. Terawaki, Y. and Hayashi, T. (1999) The complete nucleotide sequence of ΦCTX, a cytotoxin-converting phage of *Pseudomonas aeruginosa*: implications for phage evolution and horizontal gene transfer via bacteriophages. Mol. Microbiol. **31**(2): 399-419.

Nandi, B., Nandy, R. K., Vicente, A. C. P. and Ghose, A. C. (2000) Molecular characterization of a new variant of toxin-coregulated pilus protein (TcpA) in a toxigenic Non-01/Non-0139 strain of *Vibrio cholerae.* Infect. Immun. **68**(2): 948-952.

Nealson, K. H. and Hastings, J. W. (1979) Bacterial bioluminescence: its control and ecological significance. Microbiol. Rev. **43**: 496-518.

Nishiguchi, M. K. (2000) Temperature affects species distribution in symbiotic populations of *Vibrio.* Appl. Environ. Microbiol. **66**(8): 3550-3555.

Nishiguchi, M. K. (2001) Co-evolution of symbionts and hosts: The sepiolid-*Vibrio* model. In: *Symbiosis: Mechanisms and Model Systems*. J. Seckbach, ed. Dordrecht, The Netherlands, Cole-Kluwer Academic pp**:** 757-774.

Nishiguchi, M. K. (2002) Host recognition is responsible for symbiont composition in environmentally transmitted symbiosis. Microb. Ecol. **44**(1): 10-18.

Nishiguchi, M. K. and V. S. Nair (2003) Evolution of symbiosis in the Vibrionaceae: A combined approach using molecules and physiology. Int. J. Syst. Evol. Biol. In press.

Nishiguchi, M. K., Ruby, E. G. and McFall-Ngai, M. J. (1998) Competitive dominance among strains of luminous bacteria provides an unusual form of evidence for parallel evolution in sepiolid squid-*Vibrio* symbioses. Appl. Environ. Microbiol. **64**: 3209-3213.

Nishimura, M., Kogure, K., Kita-Tsukamoto, K. and Ohwada, K. (1995) Detection and direct count of specific bacteria in natural seawater using 16S rRNA oligonucleotide probes. Bull. J. Soc. Microb. Ecol. **10**(3): 109-113.

Okuda, J., Nakai, T., Chang, P. S. E., Oh, T., Nishino, T., Koitabashi, T. and Nishibuchi, M. (2001) The *toxR* gene of *Vibrio* (*Listonella*) *anguillarum* controls expression of the major outer membrane proteins but not virulence in a natural host model. Infect. Immun. **69**: 6091-6101.

Raguenes, G., Christen, R., Guezennec, J., Pignet, P. and Barbier, G. (1997) *Vibrio diabolicus* sp. nov., a new polysaccharide-secreting organism isolated from a deep-sea hydrothermal vent polychaete Annelid, *Alvinella pompejana.* Int. J. Syst. Bacteriol. **47**(4): 989-995.

Ramaiah, N., Hill, R. T., Chun, J., Ravel, J., Matte, M. H., Straube, W. L. and Colwell, R. R. (2000) Use of a *chiA* probe for detection of chitinase genes in bacteria from the Chesapeake Bay. FEMS Microb. Ecol. **34**: 63-71.

Rehnstam, A-S., Norquist, A., Wolf-Watz, H. and Hagstrom, A. (1989) Identification of *Vibrio anguillarum* in fish by using partial 16s rRNA sequences and a specific 16S rRNA oligonucleotide probe. Appl. Environ. Microbiol. **55**: 1907-1910.

Reich, K.A. and Schoolnik, G. K. (1994) The light organ symbiont *Vibrio fischeri* possess a homolog of the *Vibrio cholerae* transmembrane transcriptional activator *ToxR*. J. Bacteriol. **176**(10): 3085-3088.

Reich, K.A. and Schoolnik, G. K. (1996) Halovibrin, secreted from the light organ symbiont *Vibrio fishceri,* is a member of a new class of ADP-Ribosyltransferases. J. Bacteriol. **178**(1): 209-215.

Reich, K.A., Biegel, T. and Schoolnik, G. K. (1997) The light organ symbiont *Vibrio fischeri* possesses two distinct secreted ADP-Ribosyltransferases. J. Bacteriol. **179**(5): 1591-1597.

Rossello-Mora, R. and Amann, R. (2001) The species concept for prokaryotes. FEMS Microbiol. Rev. **25**: 39-67.

Roszak, D. B. and Colwell, R. R. (1987) Survival strategies of bacteria in the natural environment. Microbiol. Rev. **51**(3): 365-379.

Ruby, E. G. and Nealson, K. H. (1976) Symbiotic associations of *Photobacterium fischeri* with the marine luminous fish *Monocentris japonica*: a model of symbiosis based on bacterial studies. Biol. Bull. **151**: 574-586.

Ruimy, R., Breittmayer, V., Elbaze, P., Lafay, B., Boussemart, O., Gauthier, M. and Christen, R. (1994) Phylogenetic analysis and assessment of the genera *Vibrio*, *Photobacterium*, *Aeromonas*, and *Plesiomonas* deduced from small-subunit rRNA sequences. Int. J. Syst. Bacteriol. **44**: 416-426.

Schiewe, M. H., Trust, T. J. and Crosa, J. H. (1981) *Vibrio ordalii* sp. nov.: causative agent of vibriosis in fish. Curr. Microbiol. **6**: 343-348.

Shewn, J. M., and Veron, M. (1974) *Bergy's manual of determinative bacteriology*. R.E. Buchanan and N.E. Gibbons, eds. Baltimore, The Williams and Wilkins Co.

Simidu, U., Kaneko, E. and Taga, N. (1977) Microbiological studies of Tokyo Bay. Microb. Ecol. **3**: 173-191.

Sorum, H., Hvaal, A. B., Heum, M., Daae, F. L. and Wiik, R. (1990) Plasmic profiling of *Vibrio salmonicida* for epidemiological studies of cold-water vibriosis in Atlantic salmon (*Salmo salar*) and cod (*Gadus morhua*). Appl. Environ. Microbiol. **56**: 1033-1037.

Staley, J. T., Irgens, R. L. and Brenner, D. J. (1987) *Enhydrobacter aerosaccus* gen. nov., a gas-vacuolated, facultatively anaerobic, heterotrophic rod. Int. J. Syst. Evol. Microbiol. **37**(3): 289-291.

Stelma Jr., G. N., Reyes, A. L., Peeler, J. T., Johnson, C. H. and Spaulding, P. L. (1992) Virulence characteristics of clinical and environmental isolates of *Vibrio vulnificus*. Appl. Environ. Microbiol. **52**: 2776-2782.

Steven, S. E. (1990) Molecular systematics of *Vibrio* and *Photobacterium*. Ph.D. thesis, College Park, University of Maryland.

Stroeher, U. H., Jedani, K. E., Dredge, B. K., Morona, R., Brown, L. Karageorgos, E., Albert, M. J. and Manning, P. A. (1995) Genetic rearrangements in the *rfb* regions of *Vibrio cholerae* O1 and O139. Proc. Natl. Acad. Sci. USA **92**: 10374-10378.

Thune, R. L., Stanley, L. A. and Cooper, K. (1993) Pathogenesis of gram-negative bacterial infections in warm water fish. Ann. Rev. Fish Dis. **3**: 37-68.

Timmins, G. S., Jackson, S. K. and Swartz, H. M. (2001) The evolution of bioluminescent oxygen consumption as an ancient oxygen detoxification mechanism. J. Mol. Evol. **52**: 321-332.

Tolmasky, M. E., Actis, L. A., Toranzo, A. E., Barja, J. L. and Crosa, J. H. (1985) Plasmids mediating iron uptake in *Vibrio anguillarum* strains isolated from turbot in Spain. J. Gen. Microbiol. **131**: 1989-1997.

Toranzo, A. E., and Barja, J. L. (1990) A review of the taxonomy and seroepizootiology of *Vibrio anguillarum*, with special reference to aquaculture in the northwest of Spain. Dis. Aquat. Org. **9**: 73-82.

Tret'yakova, E. B., Dobrovol'skaya, T. G., Byzov, B. A. and Zvyagintsev, D. G. (1996) Bacterial communities associated with soil invertebrates. Microbiology **65**(1): 102-110.

Urakawa, H., Kita-Tsukamoto, K. and Ohwada, K. (1997) 16S rDNA genotyping using PCR/RFLP (restriction fragment length polymorphism) analysis among the family Vibrionaceae. FEMS Microbiol. Lett. **152**: 125-132.

Urakawa, H., Kita-Tsukamoto, K. and Ohwada, K. (1999) Restriction fragment length polymorphism analysis of psychrophilic and psychrotrophic *Vibrio* and *Photobacterium* from the north-western Pacific Ocean and Otsuchi Bay, Japan. Can. J. Microbiol. **45**: 67-76.

Urdaci, M. C., Marchand, M., Ageron, E. and Arcos, J. M. (1991) *Vibrio navarrensis* sp. nov., a species from sewage. Int. J. Syst. Bacteriol. **41**(2): 290-294.

Veron, M. M. (1965). La position taxonomique des *Vibrio* et de certaines bacteries comparables. C. R. Acad. Sci. Paris **261**: 5243-5246.

Watanabe, H., Nagoshi, T. and Inaba, H. (1993) Luminescence of a bacterial luciferase intermediate by reaction with H_2O_2: the evolutionary origin of luciferase and the source of endogenous light emission. Biochim. Biophys. Acta **1141**: 397-302.

Wheeler, W.C. (1995) Sequence alignment, parameter sensitivity, and the phylogenetic analysis of molecular data. System. Biol. **44**:321-331.

Wheeler, W.C. (1996) Optimization alignment: the end of multiple sequence alignment in phylogenetics? Cladistics **12**:1-9.

Wheeler, W.C., Gladstein, D. and DeLaet, J. (2002) POY version 3.0. Ver. Program and documentation available at ftp.amnh.org/pub/molecular. American Museum of Natural History.

Wiik, R., Stackebrandt, E., Valle, O., Daae, F. L., Rodseth, M. and Andersen, K. (1995) Classification of fish-pathogenic vibrios based on comparative 16S rRNA analysis. Internatonal J. Syst. Bacteriol. **45**: 421-428.

Wiik, R., Andersen, K., Daae, F. L. and Hoff, K.A. (1989) Virulence studies based on plasmid profiles of the fish pathogen *Vibrio salmonicida*. Appl. Environ. Microbiol. **55**: 819-825.

Woese, C. R. (1987) Bacterial evolution. Microbiol. Rev. **51**(2): 221-271.

Woese, C. R., Weisburg, W. G., Hahn, C. M., Paster, B. J., Zablen, L. B., Lewis, B. J., Macke, T. J., Ludwig, W. and Stackebrandt E. (1985) The phylogeny of purple bacteria: The gamma subdivision. Syst. Appl. Microbiol. **6**: 25-33.

Zhang, L., Radziejewska-Lebrecht, J., Krajewska-Pietrasik, D., Toivanen, P. and Skurnik, M. (1997) Molecular and chemical characterization of the lipopolysaccharide O-antigen and its role in the virulenced of *Yersinia enterocolitica* serotype O:8. Mol. Microbiol. **23**: 63-76.

Zhu, J., Miller, M. B., Vance, R. E., Dziejman, M., Bassler, B. L. and Mekalanos, J. J. (2002) Quorum sensing regulators control virulence gene expression in *Vibrio cholerae*. Proc. Natl. Acad. Sci. USA **99**(5): 3129-3134.

ZoBell, C. E., ed. (1946) *Marine Microbiology*. Waltham, Mass., Chronica Botanica Company.

VII. Distribution & Destiny of Life in the Universe

Biodata of **Vinod C. Tewari** author of ***"Microbial Diversity in Meso-Neoproterozoic Formations, with particular reference to the Himalaya"***

Dr. Vinod C. Tewari is a senior scientist at Wadia Institute of Himalayan Geology, Dehra Dun, and a regular associate of International Center for Theoretical Physics, Trieste, Italy. He received his Ph.D. in Geology in 1986 from the University of Lucknow, India. Dr. Tewari taught Geology at Kumaon University, Nainital, India as Professor of Geology. He has done research on early evolution and diversification of life in the Proterozoic sequences of the Lesser Himalayan stromatolitic carbonate belts. He has published more that sixty research papers and edited several volumes of ***Himalayan Geology Journal*** *and* ***Journal of Nepal Geological Society,*** Kathmandu, Nepal. Dr. Tewari has organized the first Indo-Soviet Symposium on Stromatolites and Stromatolitic Deposits, and other International Geological Correlation Program conferences in India.

E-mail: **vctewari@rediffmail.com**

J. Seckbach (ed.), Origins, 515–528.

COMETS AND THE ORIGIN OF LIFE ON THE PRIMITIVE EARTH

JOAN ORÓ
Department of Biology & Biochemistry, University of Houston
Houston, TX 77204-5934, USA[*]

1. Introduction

This review on the origin of life on the primitive Earth is based on previous experimental work concerning the prebiological synthesis of molecules such as adenine from hydrogen cyanide, a cometary molecule (Oró, 1960) and the subsequent hypothesis developed by the author on the role that cometary matter may have had on the formation of biochemical molecules on the primitive Earth (Oró, 1961), a hypothesis supported by other authors and further developed by Delsemme (2000). It is also well known that the biogenic elements (H, C, N, O, S, P) and organic molecules are some of the major constituents of the universe. Indeed, more than three fourths of the individual chemical species identified in the interstellar medium are simple compounds of carbon and/or of the biogenic elements (e.g., NH_3, H_2O, HCN, HCHO, HC_3N, etc.).

When the protosolar nebula condensed to form the Solar System, the organic compounds became part of the outer planetary bodies, as well as comets, dark asteroids and carbonaceous chondrites. On the other hand, because of their proximity to the Sun, the terrestrial planets became significantly depleted of water, and other volatile compounds. The acquisition of water and organic compounds presumably took place in good measure by late accretion from comets and other planetesimals. Upon capture of comets by the Earth, the synthesis of biochemical compounds such as amino acids, purines and pyrimidines could take place from simple cometary molecules. This, together with the fact that Earth's orbit lies within a circumstellar habitable zone is considered to have made possible the emergence of life on our planet some four billion years ago (Schopf, 1999).

In summary, this brief review discusses relevant aspects of: (a) The formation of biogenic elements and the organic matter in the interstellar medium. (b) The formation of the Earth-Moon system and the role of comets and other planetesimals in contributing organic matter to the primitive Earth. (c) The prebiotic synthesis of biochemical compounds and the emergence of life on our planet. (d) The discoveries of protoplanetary disks around other stars, which suggest that the processes which occurred in our Solar System are probably occurring now in extrasolar protoplanetary systems. In order to place into proper perspective the role of comets in the origin of life it will be useful to discuss it in the sequential order of the above evolutionary phases.

[*] Present address: Travessera de les Corts, 272, esc. B, 2-2 08014-Barcelona (Spain) E-mail: fjo@fjo.es

2. Formation of Biogenic Elements

Hydrogen and helium make about 98% of the chemical species of the Universe. Of the remaining 2% of matter, approximately 1% is made by the five biogenic elements (C, N, O, S, P) in addition to the large amount of the primordial hydrogen also a biogenic element. From an organic chemist point of view the nuclear synthesis of the nuclide of carbon and of the other biogenic elements are some of the most fundamental processes in the cosmic evolution of the Universe towards life. These elements are needed not only to make the organic compounds that are present in the circumstellar and interstellar medium, in comets and other cosmic bodies, but most importantly they are necessary for the formation of the biochemical compounds of living systems. All of this is limited to carbon based life using a chemistry very similar to that of terrestrial life. Other chemistries may be based on quite different histories.

The nuclear syntheses of these elements are reasonably well understood and are described elsewhere (Macià et al., 1997). Only a few words about the first two nuclear reactions need to be mentioned here. Starting with the hydrogen nuclide, formed during the primordial nucleosynthesis, which took place soon after the Big Bang, stars produce helium during the major part of their existence. Stellar nuclear synthesis of the helium nuclide takes place by the so-called proton-proton chain. This occurs in the core of many ordinary stars and specifically in our Sun at a temperature of ~ 15 million degrees. The small deficiency of mass which results from the condensation of four protons into one nuclide of helium is converted into energy in accordance with Einstein's equation, $E=mc^2$. The resulting large amount of energy in the form of heat and radiation has been essential for the appearance of life on our planet and for its evolution during the past 4 Gy.

3. The Nucleosynthesis of Carbon

The next most important nuclear reaction is the one that involves the condensation of three helium nuclides into one ^{12}C nuclide. This nuclear reaction is known as the triple-alpha process. It occurs inside all super-giant stars, some of them refereed to as carbon stars, at temperatures of 10^8 degrees. Obviously without the triple-alpha process, we would not be able to talk about life in the Universe. In fact we would not be here. Once ^{12}C is made by the collision of three alpha particles, subsequent alpha-capture processes give rise to the oxygen and sulfur nuclides. On the other hand, the formation of the phosphorus nuclide requires the participation of many complex nuclear reactions. This explains the relatively lower cosmic abundance of this element, which as we know, is one of the most important for the bioenergetics of living systems and therefore is indispensable for the existence of life (Macià et al., 1997), for biochemistries using RNA and/or DNA and ATP.

4. The Interstellar Medium

Carbon stars are a rich source of carbon compounds and organic molecules. From these and other stars the biogenic elements (H, C, N, O, S, P) are ejected and migrate to the circumstellar and cooler regions of the stars. There, at temperatures of a few thousand degrees, ordinary chemical reactions give rise to the formation of diatomic and triatomic species such as C_2, CN, CO, CH, NH, OH and H_2O, as can be seen, for instance, in the atmosphere of our Sun. Also it is of interest for prebiological chemistry that the first five neutral molecules detected in interstellar space were: ammonia, water, formaldehyde, hydrogen cyanide and cyanacetylene. So far, more than one hundred chemical species have been identified in the interstellar medium (ISM), by their gas-phase molecular spectra primarily by microwave spectrometry. All these molecules, ions and radicals are relatively simple. The ones specifically identified are made from two, three, etc., up to 13 atoms ($HC_{11}N$). About 75% of them are organic and contain carbon. Therefore, by the composition of the interstellar medium, one could say, that the universe is essentially organic.

In addition to the relatively low molecular weight species detected by their microwave spectra, we should also consider the high molecular weight compounds like the polycyclic aromatic hydrocarbons (PAHs) detected by their infrared spectra. According to L.J. Allamandola et al. (1989) to date they have measured the infrared spectra of 23 neutral and 19 cationic polycyclic aromatic hydrocarbons (PAHs) ranging from naphthalene ($C_{10}H_8$) to dicoronylene ($C_{48}H_{20}$). They conclude that the infrared emission band spectrum associated with many different interstellar objects can be modeled successfully by using combined laboratory spectra of neutral and positively charged polycyclic aromatic hydrocarbons (PAHs). The formation of cosmic PAHs probably results from the condensation of C_2 of C_2H_2 in the atmospheres or circumstellar envelopes of carbon stars. More than one hundred fifty interstellar and circumstellar molecules radicals and ions have been detected by their microwave or infrared spectra (see Oró, 2000).

5. The Solar System

Approximately 4.6 billion years ago a dense interstellar cloud gravitationally collapsed to form the Sun and the surrounding, flattened solar nebula, within which the planets, comets and asteroids formed. The initiation of the collapse was probably triggered by the shock wave of a supernova explosion. Hoppe et al. (1997) obtained evidence for the presence of SiC grains in the Murchison meteorite where several isotopic ratios indicate that they are matter from a type II supernova. Subsequently, during the first several million years, the evolving Solar System was in a state of great upheaval where the norm was the continuous collision of planetesimals and comets with other major bodies. The primary precursors of the terrestrial planets were rocky planetesimals partially devoid, or deficient, of light molecules due to the relatively high temperature prevailing around the orbit of the Sun within the space of the terrestrial planets up to the inner part of the asteroid belt. On the other hand light molecules and carbonaceous material were rather abundant in the outer Solar System as well as in the dark matter of the outer belt asteroids and comets.

6. The Earth-Moon System

The impact record on the Moon provides evidence about the turbulent state of flux during the early stages of formation of the Solar System. The proto-Earth was obviously subject to many collisions from large and small planetesimals. A model proposed by Cameron and other investigators, e.g., Cameron and Benz (1991), has led to a new theory that avoids the deficiencies of earlier theories. According to this theory a planetesimal named Theia with a mass approaching that of Mars collided with the proto-Earth and caused the melting and ejection into orbit of part of the Earth's mantle. Portions of the mantle eventually coalesced to form the Earth's only natural satellite, the largest moon of the terrestrial planets. Most of the volatiles were evaporated and ejected into interplanetary space. This impact origin explains most of the similarities and slight differences between the chemical composition of the Earth's mantle and the Moon. For instance, the iron-poor nature of our satellite. In the dynamic aspects, it also explains the angular momentum of the Earth-Moon system. Recent measurements of the oxygen isotopes from the Earth and 31 lunar samples are in agreement with the giant impact model of formation of the Moon by the collision of the planetesimal Theia with the proto-Earth (Wiechert et al., 2001). In spite of the massive collision, traces of noble gases and water must have remained in the internal or rocky reservoir of the core of the proto-Earth (Owen and Bar-Nun, 1993).

7. Cometary Matter Captured by Earth

Aside from the small amounts of water, noble gases, and other volatiles which may have been retained by the EARTH's, we may ask the question: If the Earth-Moon system was significantly depleted of volatiles as the result of the above mentioned catastrophic collision, what is the source of most water and organic matter present on the Earth? The author suggested some time ago (Oró, 1961) that comets were supposedly the major source of water and simple organic compounds captured from comets by the primitive Earth. In the presence of liquid water, the simple organic and other cometary compounds could react to form the more complex biochemical molecules which eventually led to the emergence of life on our planet. A significant amount of the organic compounds in comets were captured as such. Some were decomposed by the impact of the collision as pointed out by Chyba and others (e.g., Pierazzo and Chyba, 1999). However it should be considered that their decomposition products would have been converted again in to amino acids (Miller, 1953) and other biochemical molecules (Oró, 1965) by means of electrical discharges, heat, ultraviolet light and other processes occurring on the primitive Earth.

In the first calculations, in part, based on H. Urey's probability estimates of cometary collisions with the Earth, I arrived at the amount of up to 10^{18} gm from the cometary matter captured by the Earth. This is equivalent to the total mass of our biosphere. This was apparently too low since later calculations (Oró et al., 1992) gave 10^{23} gm which has been confirmed by other authors and extended to values ranging from 10^{21} – 10^{26} gm. According to Delsemme (1998, 1992) this would allow all the terrestrial water to be of cometary origin. Also, according to Delsemme (1998, 2000) the comets contributing this water and cometary matter to Earth had to be comets from the Jupiter region during the early phases of formation of the Solar System. They could not be comets from the Oort cloud, like Halley, Hyakutake and Hale-Bopp, since their D/H ratios are too high. Estimates of cometary matter captured by Earth and other terrestrial bodies are shown in Table 1.

TABLE 1. Cometary matter trapped by solar system bodies.

	Cometary matter (g)	Time-span	References*
Venus	$4.0x10^{20}$	$2x10^{9}$ years	Lewis 1974
Moon	$2.0x10^{20}$	Late-accretion	Wetherill 1975
Earth	$2.0x10^{14-18}$	$2x10^{9}$ years	Oró 1961
	$1.0x10^{25-26}$	Late-accretion	Whipple 1976
	$3.5x10^{21}$	Late-accretion	Sill & Wilkening 1978
	$7.0x10^{23}$	$4.5x10^{9}$ years	Chang 1979
	$2.0x10^{22}$	$4.5x10^{9}$ years	Pollack & Yung 1980
	$1.0x10^{23}$	$2.0x10^{9}$ years	Oró et al. 1980
	$1.0x10^{24-25}$	$1.0x10^{9}$ years	Delsemme 1984, 1991
	$6.0x10^{24-25}$	$1.0x10^{9}$ years	Ip & Fernandez 1988
	$1.0x10^{23-26}$	$4.5x10^{9}$ years	Chyba et al. 1990

* (For references see Oró et al., 1992)

8. Comets

Comets are aggregates of interstellar matter at the very low temperatures of interstellar space (Oort cloud) or slightly less cold from the outer regions of the Solar System (Kuiper belt). The Fred Whipple dirty-ice model has prevailed with minor modifications. The composition and many aspects of the relation of comets to terrestrial planets has been studied in detail by Delsemme (1992) and other authors, and the model for the condensation of interstellar grains to larger aggregates to generate cometesimals and planetesimals has been developed by Greenberg and Hage (1990). Several interesting studies have been reported on Halley's comet by Kissel and Krueger (1987). Some of the organic molecules detected in the tail dust of Halley's comet by the Giotto spacecraft mass-spectrometer by Kissel and Krueger (1987) are shown in Table 2.

It is also of interest that the first space polymer was also identified in Comet Halley (Huebner, 1987). It is the linear polymer of formaldehyde, polyoxymethylene (POM in short) also known ordinarily as paraformaldehyde.

In table 2 one may notice that a large number of the small molecules are cyclic compounds. Kissel and Krueger also found adenine and two other purines as could be expected from the large abundance of hydrogen cyanide. The authors were surprised by not finding aminoacids. However it should not be surprising since their formation would require the presence of liquid water to allow first the condensation of aldehydes with HCN and then the hydrolysis to the corresponding amino acids. This is essentially the same reason that life as well a viruses can not exist in comets as Fred Hoyle has suggested in his book "The Life Cloud". The water present in comets is frozen, at very low temperatures and not available for hydrolysis of amino nitriles or for the existence of life either.

TABLE 2. Organic molecules detected in Comet Halley dust grains

C-H	C-N-H	C-O-H
Pentyne	Hydrocyanic acid	Formaldehyde
Hexyne	Acetonitrile	Acetaldehyde
Butadiene	Propanenitrille	Formic acid
Pentadiene	Iminomehane	Acetic Acid
Cyclopentene	Iminoethane	Isocyanic acid
Cyclohexene	Iminopropene	Methanol imine
Cyclohexadiene	Imidazole	Oximidazole
Benzene	Pyridine, Pyrimidine	Oxypyrimidine
Toluene	Purine, Adenine	Xanthine

More recently other interesting studies have been made of comet Shoemaker-Levy-9 by Gautier and other investigators; and of Hyakutake and Hale-Bopp by others. Mumma (1997) applied high resolution infrared spectroscopy for the first time to Hyakutake. He detected strong emissions from H_2O, HDO, CO, CH_4, C_2H_2, C_2H_6, CH_3OH, H_2CO, OCS, HCN, OH and other chemical species. Of particular interest are the large amounts of methane and ethane in Hyakutake's nucleus. It is also of particular interest the large number of species detected by different methods in comet Hale-Bopp as reported by Bachiller and Planesas (1997). These volatile molecules are listed in Table 3.

TABLE 3. Molecules, radicals and ions detected in comet Hale-Bopp *

C,H	C,H,O	C,H,O	C,H,N	C,H,N	C,H,O,N	C,H,O,S
C_2	CO	H_2O+	CN	NH_3	HNCO	SO
C_3	CO+	H_3O+	HCN	CH_3CN	NH_2CHO	SO_2
CH_4	CO_2	HCO+	HC_3N			H_2S
C_2H_2	HDO	H_2CO	HNC			OCS
C_2H_6	OH	CH_3OH	NH			CS
	H_2O	HCOOH	NH_2			H_2CS

* Personal communication Bachiller and Planesas, 1997.

Even though the cosmic abundance of phosphorus is lower than that of the other biogenic elements, Kissel and Krueger probably detected it in the grains of the Halley's comet. Also, it is known that anhydrous interplanetary dust particles (IDPs) of probable cometary origin contain the element phosphorus at an abundance of about 0.3% (Macià et al., 1997). Additional support for the presence of phosphorus in the form of phosphates in comets is provided by the finding of PO_2 and PO_3 anions in some of the IDPs of probable cometary origin (see Macià et al., 1997). We assume that interstellar molecules must have all the ingredients for the generation of the biochemical compounds necessary for life, and since comets are samples of interstellar clouds they should have the same ingredients. Among the phosphorus compounds PN and PC have been detected in the interstellar medium. We have suggested that PO and PO_2 should also be found in the envelopes of oxygen-rich stars (Macià et al., 1997). At any rate, the fact that phosphates are found in most meteorites and in particular in carbonaceous chondrites and IDPs assures us that they were also present as one of the essential ingredients captured by the primitive Earth.

9. Carbonaceous Chondrites

In addition to comets, the Earth was and is bombarded by asteroids and meteorites. Thousands of meteorites have been recovered from Antartica and many parts of the world. Some of them are carbonaceous chondrites, which contain organic compounds. The Alais carbonaceous chondrite, which fell in France in 1806, was analysed by Berzelius. In the past few years a large number of meteorites haven been found in Antartica. So far, one of the most interesting carbonaceous chondrites is the one that fell in Murchison, Australia, in 1969. Following the initial findings, Cronin and his collaborators have systematically analyzed this meteorite for organic compounds, and have recently reviewed this information (Cronin & Chang, 1993). The organic matter is largely macromolecular, and possibly related to the refractory organic mantle or interstellar dust grains of Greenberg's model (Greenberg & Hage, 1990).

The complex mixture of monomeric organic compounds in the Murchison meteorite includes carboxylic acids, amino acids, hydroxy acids, sulfonic acids, phosphonic acids, amides, purines (adenine and guanine), and a pyrimidine (uracil), alcohols, carbonyl compounds, aliphatic, aromatic, and polar hydrocarbons. Some seventy-five amino acids have been characterized. Of these, eight are common constituents of proteins, such as glycine, alanine and aspartic acid. A few others are of metabolic interest such as γ–aminobutyric acid (GABA), but many of them are not found in biological systems. With relation to the chirality of the α–carbon, all the amino acids are racemic (equal mixtures of D- and L- isomers). According to Cronin there are some exceptions on the chirality of some amino acids. This, together with other properties such as the higher D/H and $^{13}C/^{12}C$ ratios, suggest that these amino acids were synthesized from extraterrestrial precursors (Cronin & Chang, 1993) predating the formation of the Solar System. Alkyl phosphonates, with the alkyl group from C_1 to C_4, attached to the phosphorus atom of the phosphonate, are also present in the meteorite. This finding suggests a possible derivation of phosphonic acids from the interstellar molecule PC. Phosphates are presumably derived from interstellar PO_2 as indicated by the IDPs of cometary origin (Macià et al.,1997). An extensive list of the organic compounds found in the Murchison and other common carbonaceous chondrites is shown in Table 4. In addition to the complex organic polymer, more than 615 individual organic compounds have been detected in the meteorites.

10. Chemical Evolution and the Origin of Life

One of the earlier hypothesis on the origin of life was that of Darwin who in a letter to Hooker (Calvin, 1961) wrote in a very condensed manner: "But if (and what a big if) we would conceive of some warm little pond with all salts of ammonia, and phosphoric salts, light, heat, electricity, etc. present, that a protein compound was chemically formed, ready to undergo still more complex changes…". A. I. Oparin was the first to propose in 1924 a reasonable hypothesis centered about chemical evolution. It was based on the prebiotic chemical synthesis of biochemical molecules of increasing complexity, which he judged, was a necessary precondition for the appearance of the first life forms, out of which descended all other organisms according to Darwin's theory of biological evolution. It is possible that Oparin (1936) was inspired by Darwin's ideas and also by Mendeleev's theory for the origin of petroleum.

TABLE 4. Organic compounds identified in carbonaceous chondrites (Llorca, 1999)

Compound type	Conc. (ppm)	Identified compds.	Carbon Length
Aliphatic Hydrocarbons	>35	210	C_1-C_{30}
Aromatic Hydrocarbons	15-28	87	C_6-C_{20}
Alcohols and polyhydroxy compounds	11	>8	C_1-C_4
Aldehydes and ketones	27	9	C_1-C_5
Carboxylic acids	>300	42	C_1-C_{12}
Dicarboxylic acids	>30	67	C_2-C_9
Hydroxycarboxylic acids	15	51	C_2-C_8
Amines	8	10	C_1-C_4
Amides	55-70	4	C_1-C_3
Amino acids	60	78	C_2-C_9
Purines and pyrimidines	1-2	5	-
Other heterocycles	7	32	-
Sulfonic and fosfonic acids	5	12	C_1-C_4
Total	>570	>615	

J. Llorca: 1999 Tribuna de Astronomia y Universo I, **3**, 20-24.

According to Mendeleev (Oparin, personal communication), petroleum was formed when superheated water from Earth's interior proceeded through geological formations containing iron carbides which would be hydrolyzed to the hydrocarbons which constitute petroleum. We know today that petroleum is generated from living organisms upon their death and transformation within sediments into hydrocarbons. At any rate, a general theory of chemical evolution which would include the author's cometary theory (Oró, 1969) could be summarized as follows. Once the Earth was formed and acquired water and the cometary precursors for the synthesis of biochemical compounds, there had to be the appropiate conditions (liquid water, reasonable temperatures, evaporation processes, etc.) that could allow the synthesis of biochemical monomers and polymers (polynucleotides, polypeptides, and lipids) that upon self-organization eventually gave rise to the appearance of the first self-reproductive living system (Oró et al., 1994; Lazcano, 1994).

11. Prebiotic Synthesis of Aminoacids

As we know the first experiment that successfully tested Oparin's hypothesis was performed by S.L. Miller (1953) with the synthesis of amino acids from methane, ammonia and water, by means of electrical discharges. The mixture of amino acids (glycine, alanine, aspartic, etc) is similar but less complex than that of the Murchison meteorite. The presence of α-alkyl, α-amino acids (isovaline, α-aminoisobutyric) as essentially racemic mixtures in both cases demostrates (a) the extraterrestrial nature of the meteorite aminoacids and (b) that the synthesized amino acids in the laboratory are not biological contamination products. It should also be pointed out that glycine, alanine and aspartic acid, and other amino acids and biochemical compounds, can also be obtained from hydrogen cyanide a cometary molecule in a aqueous-amonia environment (Oró and Kamat, 1961).

12. Prebiotic Synthesis of Purines and Pyrimidines

The next most important question relating to the formation of biomonomers on the primitive Earth was how purines and pyrimidines could be synthesized prebiotically. Having my own ideas about the role of comets on the origin of life on Earth, the solution to this problem came to me in a intuitive way. I knew that hydrogen cyanide was a major constituent of comets and also a very reactive compound. So the reaction of hydrogen cyanide in a aqueous ammonia mixture produced a world of biochemical compounds the most significant of which was adenine (Oró, 1960). The amino acids glycine, alanine, aspartic acid, together with glycinamide, formamidine, formic acid and other biochemical compounds were also found among the reaction products.

The mechanism of synthesis of adenine involves at least 6 steps but it can occur from hydrogen cyanide and ammonia (Wakamatsu et al., 1966), and even from hydrogen cyanide alone, at very low temperatures. This may be the explanation of why adenine and other purines were found, by mass spectrometry, in the dust grains of Halley's comet. The mechanism of adenine synthesis in an aqueous-ammonia solution involves the formation of the intermediates 4-amino-5-imidazolecarboxamide and its analog amidine which can lead in the presence of urea or guanidine to the formation of the other 3 purines, namely guanine, hypoxanthine and xanthine. The pyrimidines cytosine and uracil can also be synthesized from a cometary compound cyanacetylene or its hydration product cyanacetaldehyde (Robertson and Miller, 1995). In fact practically all of the bases for RNA or DNA can be generated readily from cometary molecules (eg. Oró, 1965; Levy et al., 1999).

13. Formation of Monosaccharides, Fatty Acids and Lipids

From the time of Butlerow and Loew and other investigators, we know that formaldehyde easily condenses by base catalysis to form many monosaccharides including ribose among all the trioses, tetroses, pentoses and hexoses. The synthesis of 2-deoxyribose can also be accomplished by condensation of glyceraldehyde and acetaldehyde (Oró, 1965). We also know that these two simple aldehydes formaldehyde and acetaldehyde are present in comets and therefore should have been present on the primitive Earth. Another abundant cometary molecule, carbon monoxide, in the atmospheric presence of some hydrogen and minute amounts of nickel-iron grains of meteoritic origin could generate aliphatic hydrocarbons and fatty acids by Fischer-Tropsch processes. The latter compounds can lead to the synthesis of phospholipids by condensing them with glycerol phosphate as well as with choline or serine as has been done in our laboratory (Oró et al., 1990). Such a prebiological synthesis of phospholipids and the formation of liposomes for the encapsulation of replicating and catalytic biomolecules, prior to the formation of a protocell is discussed in a paper from our laboratory (Oró, 1994). The simplified version of the role of cometary compounds in the formation of amphiphilic lipids necessary to generate a protocell is schematically represented in Fig 1.

Figure 1. Prebiotic synthesis from cometary molecules

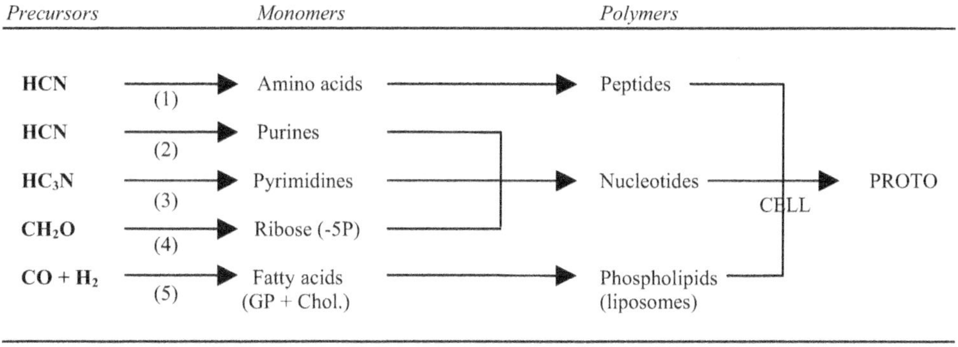

(1,2) + H_2O + NH_3 (3) + H_2O + Urea (4) + PO_4^{3-} (5) FTT Catalysis + GP + Chol.

14. Formation of Nucleosides and Nucleotides

The synthesis of purine nucleosides in small yields is not difficult. Thus, when a mixture of purine bases are heated with ribose and salts of sea water (which contain magnesium chloride, a catalyst for this reaction) small but reasonable yields of different purine nucleosides are obtained, namely, 4,5 % β-inosine, 2% β-adenosine, 3,2% β-guanosine and 1,8% β-xanthosine (Fuller et al., 1972). This is what would be expected during the evaporation of coastal lagoons of the primitive Earth's oceanic cometary waters. The synthesis of pyrimidine nucleosides and nucleotides is more difficult, because it requires that once the α-anomers are formed they be rearranged to the natural isomers (β-anomers) with the help of irradiation with ultraviolet light (Sanchez and Orgel, 1970). The phosphorylation of nucleosides can be carried out with ortophosphates by dry heating at relatively low temperatures (below 100°C) in the presence of urea and ammonium chloride, as shown by Lohrmann and Orgel (1971). Minor modifications of this method produce reasonable yields of 5'-nucleosidemonophosphates minimizing the formation of 2':3' cyclic phosphate derivatives, as shown recently by Reimann and Zubay (1999). This was done with the four nucleosides, uridine, cytidine, inosine and adenosine. In summary, the phosphorylation of nucleosides involve reactions using moderate temperatures and ambient humidity simulatng the conditions resulting from the evaporation of primordial lakes from cometary waters.

15. High Energy Phosphates and Coenzymes

The synthesis of pyrophospate under possible primitive Earth conditions has been shown by Miller and Parris (1964). It occurs by reaction of orthophosphate with cyanate on the surface of hyroxylapatite, presumably forming carbamyl phosphate which reacts with another phosphate molecule to generate pyrophosphate. As is known, phosphate has been detected in fluffy interplanetary dust particles (IDP'S) of probable cometary origin (Macià et al., 1997). Once inorganic pyrophosphate is available, transphosphorylations can take place. In this way AMP has been further transphosphorylated to ADP and ATP (Miller and Parris, 1964). Furthermore the synthesis of all deoxyribomonucleosides mono-, di- and triphosphates has been accomplished in our laboratory by simple evaporation in aqueous solutions in the presence of inorganic phosphate and cyanamide. Under similar prebiotic conditions a number of simple phosphate containing coenzymes UDPG, CDP-choline, CDP-ethanolamine and other coenzymes have also been obtained in our laboratory (Oró, 1994).

16. Synthesis of Polynucleotides and Polypeptides

The formation of biochemical monomers presumably took place during the first phase of prebiotic evolution. The conditions for the formation of biopolymers had to be somewhat milder and at any rate more specific. In the presence of condensing agents, or activating compounds, it has been possible to synthesize oligopeptides and oligonuccleotides. In our laboratory we have been able to synthesize oligodeoxyribonucleotides up to 8 nucleotides long and polyglycines of up to 15 monomers long by means of the condensing agent cyanamide (Stephen-Sherwood et al., 1974). The synthesis of polynucleotides has been studied particularly in the laboratories of Orgel and of Ferris, using activated nucleotides. For instance, Ferris and coworkers (Ferris, 1998) have shown that a common clay mineral, montmorillonite, is an excellent catalyst for the polymerization of 5'–phosphorimidazolides of the nucleoside bases. Polymers up to 40 to 50 units long have been obtained in this way.

17. Replication of Polynucleotides

In Orgel's laboratory they have synthesized polymers of GMP, (PolyG) starting with a template of PolyC (Inoue and Orgel, 1982). The monomers of GMP were activated with 2-Me Imidazolide. In this way PolyG is obtained as the template of PolyC. However, the reaction only works with the isomeric D-mononucleotides, not with racemic mixtures of D-and L-mononucleotides.

18. The RNA world

The well known work on ribozymes by Cech and Altmann, and the Gilbert notion of an RNA only world elevated the importance of RNA over DNA. The idea of an RNA world first is possibly a breakthrough on the theory of the origin of life that may solve some of the problems in prebiotic evolution. The fact that a single molecule may embody the three major characteristics of life: encoding information, catalysis and replication, appears to be a simplifying principle. An optimist view is that of "The molecular biologist dream" of Joyce and Orgel (1999). Simply stated this says that the primitive Earth was loaded with a pool of activated ribonucleotides waiting to be polymerized into RNA. But, as we pointed out, there are problems with the replication of RNA. An alternative possible solution to the RNA world problem is that of chemical coevolution of ribonucleotide oligomers and catalytic peptide protoenzymes as discussed by Lahav (1993). Another important suggestion is that of peptide nucleic acid (PNA) as a model structure for the primordial genetic material. This would be a pre-RNA world as proposed by Nielsen (1993) and more recently elaborated by Miller and coworkers (2000). See also Eschenmoser's views in Orgel (2001).

19. Protoplanetary Systems

We are beginning to obtain pertinent evidence concerning the generalization of the concept of the role of comets in the origin of life to other planetary systems: starting with the large cloud around the star β-Pictoris discovered a long time ago by Smith and Terrile, several protoplanetary disks have been discovered surrounded by protoplanetary nebulae or cometary clouds. They include, aside from β-Pictoris, HR4796A, Fomalhaut, Vega, Epsilon Eridani and others. Moreover, a large number of these protoplanetary systems have been discovered by O'Dell with the Hubble Space Telescope in the Orion nebula. All these protoplanetary systems appear to be large clouds of dust and gas around a central star, usually very young, of a few million years in age. It has been shown in the case of β-Pictoris there appear to be comets that continuously collide with the central star and with some yet undiscovered planets that may be forming in the inner regions of the protoplanetary cloud or disk. It is quite probable that their stage of evolution is similar to that of the protoplanetary cloud of the Solar System 4.5 Gy ago. The observation of all these protoplanetary disks gives us a reasonable picture, in action, of the formation of our Solar System and how the creation of other planetary systems continues in the Universe.

20. Extrasolar Planetary Systems

Our lack of knowledge about the existence of other planetary systems has changed since 1995. As we know in that year Michel Mayor and Didier Queloz discovered a major giant planet orbiting the star 51 Pegasi. Since then more than eighty additional planets orbiting other stars have been discovered by the group at the University of California, Berkeley, led by Geoff Marcy and Paul Butler (1999) and other astronomers. Most of the planets are very close to the star and only in one case, 47 Ursae Majoris, there are three planets that have orbital characteristics approaching those of our Solar System.

The new science of astrobiology has been initiated with these findings that will be further developed during the XXI century. An orbital satellite with four telescopes, using the technique of infrared interferometry is planned to be constructed and placed in an orbit beyond the terrestrial planets in order to determine the possible presence of an exo-planet or planets with water vapor, oxygen and ozone in its atmosphere. This would be an indication of the existence of an extrasolar system planet (or several) which may harbor life. So life on Earth may not be a unique reality in our Milky Way or the universe at large. Without doubt this would be the greatest discovery of humankind, only surpassed by the possible existence and discovery of extraterrestrial intelligent life.

21. Acknowledgments

I appreciate the dedication of Maite Latorre (FJO) for helping me in the preparation of this paper. The author is indebted (1) to the Astronomical Society of the Pacific for permission to reproduce in revised and updated form part of what was published by myself under the title "Organic Matter and the Origin of Life in the Solar System" in *ASP conference series vol 213. Bioastronomy 99. A new Era in Bioastronomy © 2000.* pp 285-314, and (2) to *Kluwer Academic Press, The Netherlands* for permission to reproduce in revised and updated form part of what was published by myself under the title "Cometary Molecules and Life's Origin In: J. Chela-Flores et al. (eds.), First steps in the Origin of Life in the Universe" © 2001 Kluwer Academic Press, The Netherlands.

22. References

Allamandola, L. J., Tielens, A. G.G.M. and Barker, J.R. (1989) Interstellar polycyclic aromatic hydrocarbons: the infrared emission bands, the excitation-emission mechanism and the astrophysical implications, *Astrophys. J. Suppl.* **71**, 733-755.

Calvin, M. (1969) Chemical Evolution. Oxford Univ. Press, New York, 278 pp.

Cameron, A.G.W. and Benz, W. (1991) The origin of the Moon and the single impact hypothesis IV, *Icarus*, **92**, 204-216.

Cronin, J.A. and Chang, S. (1993) Organic Matter in meteorites: Molecular and isotopic analyses of the Murchison meteorite, In: J. M. Greenberg et al. (eds.) *The Chemistry of Life's Origins*, Kluwer Academic Publishers, Dordrecht pp. 209-258.

Delsemme, A. (1992) Cometary origin of carbon, nitrogen and water on the Earth. In *Proc. Eau Claire Conference on Comets and the Origin and Evolution of Life. Origins of Life*, **21,** 279-296.

Delsemme, A. (1998) The deuterium enrichment observed in recent comets is consistent with the cometary origen of sea water, *Planetary and Space Science*, **47**, 125-131.

Delsemme, A. (2000) Cometary Origin of the Biosphere, *Icarus*, **146**, 313-325.

Ferris, J.P. (1998) Catalyzed RNA synthesis for the RNA world. In Brack, A. (Ed.) *Molecular origins of life: Assembling pieces of the puzzle,* Cambridge Univ. Press, pp 255-268.

Ferris, J.P., Sanchez, R.A. and Orgel, L.E. (1968) Synthesis of pyrimidines from cyanoacetylene and cyanate. *J. Mol. Bio.* **33**, 693.

Fuller, W.D., Sanchez, R.A. and Orgel, L.E. (1972) Solid state synthesis of purine nucleosides. *J. Mol. Evol.* **1**, 249.

Greenberg, J.M. and Hage, J.I. (1990) From interstellar dust to comets: A unification of observational constraints, *Ap. J.,* **361**, 260-274.

Hoppe, et al. (1997) Type II supernova matter in a silicon carbide grain from the Murchison meteorite, *Science* **272**, 1314-1317.

Huebner, W.E. (1987) Fist polymer in space identified in Comet Halley. *Science,* **237**, 628-630.

Inoue, G.F., and Orgel, L.E. (1982) Oligomerization of (guanosine 5' phosphor) -2-methylimidazolide on polyC: and RNA polymerase model. *J. Mol. Biol.*, **162**, 201-207.

Joyce, G.F. and Orgel, L.E. (1999) Prospects for understanding the origin of the RNA world. In Gesteland R.F., Cech T.R. and Atkins, J.F. (Eds.). *The RNA world. The Nature of Modern RNA suggests a prebiotic RNA.* Cold Spring Harbor Laboratory Press, pp 49-77.

Kissel, J. and Krueger, F.R. (1987) The organic component in dust from comet Halley as measured by the PUMA mass spectrometer on board Vega I, *Nature*, **326**, 755-760.

Lahav, N. (1993) The RNA world and co-evolution hypotheses and the origin of life: Implications research, strategies and perspectives. *Origins Life Evol. Biosph.*, **23**, 329-344.

Lazcano, A. (1994) The RNA world, its predecessors an descendants, in Bengston 5 (ed.) Early Life on Earth: Nobel Symposium **84**, Columbia University Press, pp. 70-80.

Levy, M., Miller, S.L. and Oró, J. (1999) Production of guanine from NH_4CN polymerizations. *J. Mol. Evol.* **49**,165-168.

Llorca, J. (1999) Meteoritos, cometas y el origen de la Vida (I) in: *Tribuna de Astronomia y Universo*, 3, pp. 20-24, spain.

Lohrmann, R. and Orgel, L.E. (1971) Urea-inorganic phosphate mixtures as prebiotic phosphorylating agents. *Science*, **171**, 490-494.

Macià, E., Hernandez, V. and Oró, J. (1997) Primary sources of phosphorus and phosphates in chemical evolution. *Origins Life Evol. Biosph.,* **27**, 459-480.

Marcy, G.W. and Butler, R.P. (2000) Extrasolar planets: Techniques, results, and the future, In: C.J. Lada and N.D. Kylafis (eds.) *The Origin of Stars and Planetary Systems*, pp. 681-708.

Miller, S.L. (1953) Production of aminoacids under possible primitive earth conditions. *Science*, **117**, 528-529.

Miller, S.L. and Parris, M. (1964) Synthesis of pyrophosphates under primitive Earth conditions, *Nature*, **204**, 1248-1250.

Miller, S.L. et al., (2000) Peptide Nucleic Acid rather than RNA may have been the first genetic molecule *(Presented, sixth Trieste Conference on Chemical Evolution, september 18-20).*

Mumma, M.J. (1997) Organics in comets, in Cosmovici, C.B., Bowyer, S. And Werthimer, D. (eds.), *Astronomical and Biochemical Origins and the Search for Life in the Universe,* Editrice Compositori, Bologna, pp. 121-142.

Nielsen, P.E. (1993) Peptide Nucleic Acid (PNA): A model structure for the primordial genetic material? *Origins Life Evol. Biosph.*, **23**,323-327.

Oparin, A. (1936) *The Origin of Life.* (English Transl. by S. Margolis). Republished 1953 by Dover, New York.

Orgel, L. (2001) The origin of biological information. In: J. W. Shcopf (ed.) *The Emergence of Life, Proceedings of the Gold Medal Simposium* (*In press*)

Oró, J. (1960) Synthesis of adenine from ammonium cyanide, *Biochem. Biophys. Res. Commun.* **2**, 407-412.

Oró, J. (1961) Comets and the formation of biochemical compounds on the primitive Earth, *Nature*, **190**, 389-390.

Oró, J. (1965) Stages and mechanisms of prebiological organic synthesis. In *The Origins of Prebiological Systems* (S.W.Fox, ed.), Academic Press, New York, New York, pp 137-171.

Oró, J. (1994) Chemical synthesis of lipids and the origin of life. *J. Biol. Physics,* **20**, 135-147.

Oró, J. (2000) Organic Matter and the Origin of Life in the Solar System, In: G. Lemarchand, K. Meech (eds.) *A new era in the Search for Life in the Universe*. ASP Conference Series, Vol. 213, 2000, pp. 285-299.

Oró, J. and Kamat, S.S. (1961) Amino acid synthesis from hydrogen cyanide under possible primitive earth conditions. *Nature*, **190**, pp. 442-443.

Oró, J., Miller, S.L. and Lazcano, A. (1990) The origin and early evolution of life on Earth. *Ann. Rev. Earth and Planet. Sci.*, **18**, 317-356.

Oró, J., Mills, T. and Lazcano, A. (1992) Comets and the formation of biochemical compounds on the primitive Earth-A review, *Origins Life Evol. Biosph,* **21**, 267-277.

Owen, T.A., and Bar-Nun, A.(1993) Noble gases in atmospheres. *Nature*, **361**, 693-694.

Pierazzo, E. and Chyba C.F. (1999) Amino acid survival in large cometary impacts. *Meteoritics and Planetary Science,* **34**, 909-918.

Reimann, R. and Zubay, G. (1999) Nucleoside phosphorylation: A feasible step in the prebiotic pathway to RNA. *Origins Life Evol. Biosph.*, **29**, 229-247.

Robertson, M.P. and Miller, S.L. (1995) An efficient prebiotic synthesis of cytosine and uracil, *Nature*, **375**, 772-774.

Sanchez, R.A. and Orgel, L.E. (1970) Synthesis and photoanomerization of pyrimidine nucleosides. *J. Mol. Biol.*, **47**, 531.

Schopf, J.W. (1999) The Cradle of Life: The Discovery of Earth's Earliest Fossils, *Princeton University Press*, Princeton, New Jersey, pp. 367

Stephen-Sherwood, E., Odom, D.G. and Oró, J. (1974) The prebiotic synthesis of deoxythymidine oligonucleotides, *J. Mol. Evol.*, **3**, 323-330.

Wakamatsu, et al. (1966) Synthesis of adenine by oligomerization of hydrogen cyanide, *J. Org. Chem.*, **31**, 2035-2036.

Wiechert, U. et al. (2001) Oxygen Isotopes and the Moon Forming Giant Impact, *Science*, **294**, 345-348.

Biodata of **Humberto Campins** author (with co-authors T.D. Swindle and D.A. Kring) of ***"Evaluating Comets as a Source of Earth's water"***

Humberto Campins is currently Professor of Physics and Astronomy at the University of Central Florida in Orlando and adjunct faculty at the Lunar and Planetary Laboratory at the University of Arizona. He received his Ph.D. in Planetary Sciences from the University of Arizona in 1982. His main research interest is the study of comets, asteroids and other small bodies in the solar system.

E-mail: **campins@phy.physics.ucf.edu**

Timothy D. Swindle is currently an Associate Professor of Planetary Sciences and Geosciences at the University of Arizona. He received his Ph.D. in Physics at Washington University (St. Louis) in 1986.

David A. Kring is currently an Associate Professor of Planetary Sciences and Geosciences at the University of Arizona. He received his Ph.D. in geological sciences, Harvard University in1989.

J. Seckbach (ed.), Origins, 569–591

EVALUATING COMETS AS A SOURCE OF EARTH'S WATER

HUMBERTO CAMPINS[1], TIMOTHY D. SWINDLE[2] AND DAVID A. KRING[2]
[1]Physics Department, University of Central Florida, Orlando, FL 32816, USA. [2]Lunar and Planetary Laboratory, University of Arizona, Tucson, AZ 85721, USA

1. Introduction

Comets are ice-rich primitive bodies that may have contributed to Earth's hydrosphere and atmosphere. However, there are other potential sources of terrestrial water. Water may have been part of the original accreting solid material or it may have been provided later by hydrous asteroids. It should be possible to determine if comets were an important source of Earth's water and the relative proportions of other sources, because different sources should have distinct chemical and isotopic signatures.

Comets have also been implicated in the origin and early evolution of life, as they may be a source of biogenic and life-sustaining substances (*e.g.*, Oró 1961, Chyba and Sagan 1992, Delsemme 2000). This issue is also complex since there are other potential sources of water and biogenic materials, because water and organic molecules in comets can be destroyed during impacts (although the individual atoms survive), and also because some of the water may be blown off the planet by explosive impacts (Thomas *et al.* 1997 and references therein). The possible role that cometary organic material may have played in the origin of life on Earth is discussed elsewhere in this volume by Oró.

New studies of the chemical and isotopic composition of comets allow a reevaluation of the possible sources of Earth's water. We begin by outlining the chronology of the delivery of water to Earth. We then describe the chemical and isotopic signatures of comets and compare them with the signatures of other sources. We discuss the implications and outline the uncertainties that future observations need to address.

2. Water in Early Earth

The geologic record indicates that water was available on the surface of the Earth ~3.85 Ga[1], when aqueously deposited sediments were produced (*e.g.*, Nutman *et al.* 1996). These same sediments may also contain the earliest isotopic remnants of life (*e.g.*, Mojzsis *et al.* 1996; Mojzsis and Harrison 2000). This period followed what may have been a particularly intense epoch of bombardment on Earth, which was first gleaned from

[1] Ga is Giga anna and Ma is Mega anna, which equal 10^9 and 10^6 years, respectively. In geology, it is used to indicate how many years ago something occurred.

studies of the Moon and commonly known as the lunar cataclysm (Turner *et al.* 1973, Tera *et al.* 1974). We now understand that the Earth suffered more impacts than the Moon (by factors of 13 to 500; Zahnle and Sleep 1997, Hartmann *et al.*, 2000). It is also possible that the event affected the entire inner solar system (Bogard 1995, Kring and Cohen 2002). If life evolved before the cataclysm and was extinguished one or more times, there may have been a period of "impact frustration" (Maher and Stevenson 1988). If life evolved before the cataclysm and only a few species with specific capabilities were able to survive, then there may have been an "impact bottleneck" in the evolution of life (Chyba 1993).

There are hints that a hydrosphere had developed on Earth prior to the impact cataclysm. Recent studies of early Archean sediments have uncovered isolated zircon crystals that were eroded from older, 4.3 to 4.4 Ga, igneous rocks (Mojzsis *et al.* 2001, Wilde *et al.* 2001). Zircons imply a highly differentiated source, like granite, which, in turn, implies water was involved in the petrogenesis of the source. In addition, oxygen isotopes in those zircon grains reveal they formed in a crust that was interacting with an evolved hydrosphere (Mojzsis *et al.* 2001, Wilde *et al.* 2001). It is also clear that the cataclysm, as violent as it was when over 20,000 impact craters with diameters ranging from 20 km to ~5000 km were produced on Earth, was not the principal source of Earth's water (Swindle and Kring 2001). Thus, the bulk of the water had to arrive earlier than 3.9 to 4.0 Ga, and possibly earlier than 4.4 Ga.

The impact flux in the interval between Earth's accretion 4.5 Ga and the cataclysm 500 to 600 Ma later is still being debated and may be crucial to our understanding of the delivery of water to Earth and the origin and early evolution of life. In one model, the impact rate is considered to be rather high throughout the entire 600 Ma period, followed by a sharp decline in the impact rate around 3.85 Ga. In another model, the cataclysm is a sharp increase (spike) in the impact rate, which was preceded by a relatively low impact rate between the time of accretion and the cataclysm (~4.5 to 4.0 Ga). The lack of impact melt ages older than 4.0 Ga in lunar samples (Ryder 1990; Dalrymple and Ryder 1993, 1996, Cohen *et al.* 2000) and the presence of 4.2 to 4.4 Ga zircons (Valley *et al.* 2002) imply that the latter model may be more applicable. It is also possible conditions were intermediate between the two models.

The terminology that has developed about the first 500 to 600 Ma of Earth history is sometimes confusing. In some cases, the term "late heavy bombardment" has been used to describe the entire period, while in other cases it has been used to describe only the impact events 3.9 to 4.0 Ga. For that reason, a distinctly different term, cataclysm, has been used recently to describe the spike in impact activity that may have occurred 3.9 to 4.0 Ga. The impactor flux in either case was a mixture of comet-like (icy) and asteroid-like (rocky) objects. Comparisons between present day asteroids, comets, and meteorites and the objects that impacted Earth during this early period can be useful. However, significant differences are also possible and could limit the validity of such comparisons, particularly in the case of asteroids and meteorites (see below).

Comets have been a very attractive source of biogenenic materials because they contain ices and organic molecules. Some asteroids also contain water, in the form of hydrated silicates, and organic material including amino acids, so objects like them may have also been a source of biogenic material. While comets contain more biogenic materials on a per mass basis, the flux of impacting comets relative to impacting asteroids is uncertain and the relative survivability of key biogenic material following an impact is unclear. The evidence thus far suggests that the flux of asteroid-like bodies (both anhydrous and hydrous) was far greater (approximately ≥ 80%) than that of comet-like bodies (approximately ≤ 20%; Chyba 1987, Swindle and Kring 2001), including the objects involved in the cataclysm (Kring and Cohen, 2002). However, it has also been suggested that relatively small comets (less than a few km diameter) may have been the principal source of terrestrial amino acids (Pierazzo and Chyba, 1999). Another term often used in these discussions is "late veneer," which some people mistakenly believe refers to material added only to the surface of the Earth. As Drake and Righter (2002) point out, this term refers to the hypothesized late arrival of a siderophile (metal-seeking) element source that was mixed into the Earth's mantle, not just a coating of the planet's surface.

The debate is also being shaped by discussions about the zone within the nascent solar system that fed material to Earth during its accretion. In some cases, comparisons to modern meteorites are made, implying that material currently between 2 and 5 AU may have been an important constituent of the material that accreted to form Earth. The validity of such comparisons is unclear because meteorites also indicate that the amount of radial mixing in their formation zones was small. Analyses of meteorites indicate there were at least 70 to 80 planetesimals between the orbits of Mars and Jupiter (Wasson 1990) with distinct chemistry. This is *prima facie* evidence that there was not enough radial mixing to homogenize the material in that small region of the solar system, much less over larger regions that encompassed Earth. Consistent with the meteorite evidence are distinctions in asteroid spectral classes with heliocentric distance, which also suggest there were limits to the amount of radial mixing in the asteroid belt (*e.g.*, Gaffey et al. 2003). Recent comparisons between the chemistry of meteorites and Earth also indicate that Earth accreted from material unlike any individual group of meteorites (Drake and Righter 2002). Similar comparisons between Earth and comets can also be informative and are the topic of the next section.

3. Comets and their Composition

Comets are among the most primitive members of our solar system. Because of their orbits and small sizes, comets have undergone relatively little processing, unlike larger bodies such as the Moon and the Earth that have been modified considerably since they formed. This primitive nature of comets is evidenced in their high abundance of volatile compounds, which are easily lost under processing. The composition of comets contains a wealth of information relevant to the origin and evolution of the entire solar system.

3.1. SOURCES AND FORMATION SITES

There are two known sources of comets. The Oort cloud is a quasi-spherical shell about 50,000-100,000 Astronomical Units (AU) from the Sun. The Kuiper belt is disk-like distribution that starts just beyond Neptune's orbit (30 AU) and extends outward to an unknown distance, possibly 48 AU (*e.g.*, Trujillo and Brown 2001). Comets in the Oort cloud are believed to have formed in the same region as the giant planets (Jupiter, Saturn, Uranus and Neptune) and were scattered out during the formation of these planets. On the other hand, Kuiper belt objects (also known as transneptunian objects or TNOs) formed primarily in situ. Therefore, Oort cloud comets formed closer to the Sun, even though the Oort cloud is much more distant than the Kuiper belt. The Kuiper belt is the main source of Jupiter-family comets, also know as short-period comets. The nucleus of a comet is made of a mixture of ices and dust, which is heterogeneous and unequilibrated down to very small (sub-micron) scales. The nature of this dirty ice provides insights into cometary formation and evolution processes. The rest of this section presents a summary of the composition of cometary ice and dust. Because the direct observation of cometary nuclei is difficult, most of our information about the composition of comets comes from studies of the gases and dust particles in the coma.

3.2. GAS COMPOSITION

In the coma of a comet, parent molecules are those that sublimate directly from ices in the comet's nucleus. Daughter molecules are photodissociation fragments of parent molecules. The most prominent daughter fragments in the visible spectra of comets are OH, CN, C_2, C_3, and NH. Water is the most abundant volatile species in comets; however, because of absorption by H_2O in Earth's atmosphere it is difficult to study cometary H_2O. The first direct detection of H_2O vapor in a comet was achieved in 1985, in comet Halley. In the solar radiation field H_2O will photodissociate mainly into OH plus H. Hydroxyl (OH) is relatively easy to detect from Earth's surface, and is used as a tracer of the H_2O in comets when H_2O can not be observed directly.

3.2.1. Chemical Classification

Emission bands from OH, CN, C_2, C_3, and NH dominate the visible and near ultraviolet spectrum of most comets. Filter photometry has been used to study the absolute and relative abundances of these species in approximately 100 comets (A'Hearn *et al.* 1995). This has resulted in the first chemical classification of comets into two main groups, those with "typical" abundances and those depleted in C_2 and C_3 with respect to OH and CN. All the comets likely to have originated in the Oort cloud show typical abundances. Comets likely to have originated in the Kuiper belt (Jupiter-family) can show either typical abundances or C_2 and C_3 depletions. The origin and significance of these two compositional groups are difficult to establish at this time because we do not yet know the parent molecules of these fragments. However, it is clear that the icy components in comets can have distinctly different chemical compositions.

3.2.2. Parent Molecules

All the chemical species discussed above, except for H_2O, are photodissociation fragments of parent molecules. It is preferable to study parent molecules whenever possible because they are more indicative of the composition of the ices. The development of new instrumentation for spectroscopy at infrared and microwave wavelengths, coupled with the apparition of two bright comets, Hyakutake in 1996 and Hale-Bopp in 1997, resulted in an explosion in the identification and study of new parent molecules. Recent reviews of this subject are given by Bockelée-Morvan and Crovisier (2002), Campins (2000), Despois *et al.* (1999), and Irvine *et al.* (2000). One of the conclusions from the molecular abundances in comets is that cometary ices have remained sufficiently cold and unprocessed to preserve at least partial signatures of the interstellar molecular cloud environment that preceded the formation of the solar nebula. Nevertheless, processing of cometary ices in the solar nebula has been discussed by a number of authors (*e.g.*, Irvine *et al.* 2000). Most recently, Iro *et al.* (2003) present interpretations of the nitrogen deficiency in comets in terms of processing of interstellar material in the solar nebula during the formation of precometary grains.

3.2.3. The Composition of Comet 1999 S4.

This comet deserves special attention because of its unusual composition. It was an Oort cloud comet, however, its composition was significantly different from that of other well-studied Oort cloud comets (including Hale-Bopp, Halley and Hyakutake). Comet 1999 S4 was depleted in CO, CH_4, C_2H_6, C_2H_2, CH_3OH, with respect to H_2O, but not depleted in HCN, H_2S, and H_2CO. Is the different composition due to origin or processing? Mumma *et al.* (2001) argue that it is due to processing in the Jupiter region before ejection to the Oort cloud, and they speculate that this may be the first Jovian-class comet observed. According to Mumma *et al.* (2001) Jovian class comets are those that spent sufficient time in a Jovian sub-nebula to have their original chemistry altered by the higher temperatures and the unique mixing of material that occurred in this region. In fact, Iro *et al.* (2003) state that comets formed near Jupiter could exhibit deuterium to hydrogen (D/H) ratios significantly lower than the values measured for Comets Halley, Hyakutake, and Hale-Bopp (Section 3.2.4). This is the type of comet proposed by Delsemme (2000) to have been the main source Earth's water. If Comet 1999 S4 is one these Jovian-class comets, could it have a D/H ratio and noble gas abundances compatible with Earth? (Section 4). Unfortunately, we will never know because this comet suffered a catastrophic disruption and disintegrated. However, future observations of similar comets may clarify this tantalizing question. Recent measurements of the composition of Jupiter's atmosphere are relevant to the possible contribution of comets to the formation of this planet (Section 4.2.2).

3.2.4. Isotopic Ratios

Isotopic abundances can provide clues on the formation conditions and subsequent processing of cometary material. The measurement of isotopic ratios in comets using remote sensing is difficult because it requires high resolution spectroscopy and comets bright enough for this type of study are rare. Consequently, isotopic ratios had been measured for few elements and in few comets (Vanysek 1991). Observations of recent bright comets have contributed greatly in this area.

So far, the deuterium-to-hydrogen, or D/H, ratio is the most diagnostic isotopic ratio in comets. All of the deuterium is believed to have formed in the early universe and nucleosynthesis converts deuterium into helium, thus lowering the universe's D/H ratio with time. However, fractionation processes can produce local enhancements in the D/H ratio. For example, low-temperature ion-molecule reactions in the cores of molecular clouds can enhance the D/H ratio in icy grains as much as two orders of magnitude above that observed in the interstellar medium (*e.g.*, Gensheimer *et al.* 1996). In our solar system, there is evidence for more than one, and possibly three reservoirs of hydrogen (Drouart et al. 1999, Mousis et al. 2000, Robert 2001, Hersant *et al.* 2001 and references therein). The solar nebula gas D/H ratio was low and estimated at $2.1 \pm 0.4 \times 10^{-5}$ (Lellouch *et al.* 2001) from observations of CH_4 in Jupiter and Saturn, which likely obtained most of their hydrogen directly from the solar nebula gas (this estimate is also consistent with protosolar D/H value inferred from the solar wind implanted into lunar soils; Geiss and Gloecker, 1998). A second reservoir, enriched in deuterium compared with the solar nebula gas, contributed to bodies that accreted from solid grains, including comets and meteorites. A possible third source of gaseous deuterated molecules (*e.g.*, HDO, CH_3D, NH_2D, DCN, etc.) would be multiple transitory concentrations of these molecules that may have resulted from (deuterium-rich) ices vaporizing as they fell into the solar nebula from the presolar cloud.

We now have D/H ratios from H_2O in three comets (all from the Oort cloud), Halley ($3.2 \pm 0.1 \times 10^{-4}$; Eberhardt *et al.* 1995), Hyakutake ($2.9 \pm 1.0 \times 10^{-4}$; Bockelée-Morvan *et al.* 1998), and Hale-Bopp ($3.3 \pm 0.8 \times 10^{-4}$; Meier *et al.*1998). These are all about twice the value for terrestrial water (1.49×10^{-4}; Lecuyer *et al.* 1998), about fifteen times the value for the solar nebula gas ($2.1 \pm 0.4 \times 10^{-5}$, Lellouch *et al.* 2001), and consistent with the range of values for "hot cores" of dense molecular clouds (2 to 6×10^{-4}; Gensheimer *et al.* 1996). Carbonaceous chondrites have the highest water abundance of all meteorites (up to 16%) (Jarosewich 1990) and their D/H ratio ranges from 1.20×10^{-4} to 3.2×10^{-4} (Lecuyer 1998). The largest deuterium enrichment in a water-bearing mineral in a meteorite was measured at $7.3 \pm 1.2 \times 10^{-4}$ in the LL3 chondrite Semarkona (Deloule et al. 1998). These results are illustrated in Figure 1, and one of the implications is that cometary ices formed in dense molecular clouds and retain some chemical memory of that environment (in agreement with observed molecular abundances in comets). These measurements have also contributed to the debate on the origin of Earth's water (Section 4).

A D/H ratio of $2.3 \pm 0.4 \times 10^{-3}$ was measured for HCN in comet Hale-Bopp (Meier *et al.* 1998b). The difference from the value for H_2O is similar to that observed in molecular clouds and is considered by some as another interstellar signature in the composition of comets (Meier *et al.* 1998b). However, others argue that some nebular reprocessing of the interstellar ices may have taken place and modified the D/H ratio in H_2O as well as in HCN; in that case the presolar chemical memory in cometary ices would only be partial (Drouart *et al.* 1999, Mousis *et al.* 2000, Hersant *et al.* 2001).

Isotopic ratios for heavier elements in comet Halley were all consistent with solar system values (*e.g.*, Vanysek 1991). Isotopic ratios measured more recently include $^{12}C/^{13}C$ (90 ± 15, Lis *et al.* 1999), $^{14}N/^{15}N$ (299 ± 30 Matthews *et al.* 1997) and $^{32}S/^{34}S$ (27 ± 3, Matthews *et al.* 1997) measured in comet Hale-Bopp, and $^{16}O/^{18}O$ measured in comet Ikeya-Zhang (Lecacheux and Biver 2002). These four observed isotopic ratios are consistent with solar system values (89, 270, 22.6, and 490 respectively). Isotopic fractionation in the heavier elements is more difficult to achieve due to the smaller difference in mass between the isotopes. Hence, this agreement in the isotopic composition of the heavier elements does not contradict the implications of the measured D/H ratios. A detailed review of D/H ratios in the solar system is provided in Robert *et al.* (2000).

3.2.5. Noble Gases

We do not know much about noble gases in comets, and some of the measurements we do have appear contradictory. A neon upper limit in comet Hale-Bopp of 0.5% of the solar Ne/O ratio was reported by Krasnopolsky *et al.* (1997). A tentative argon detection

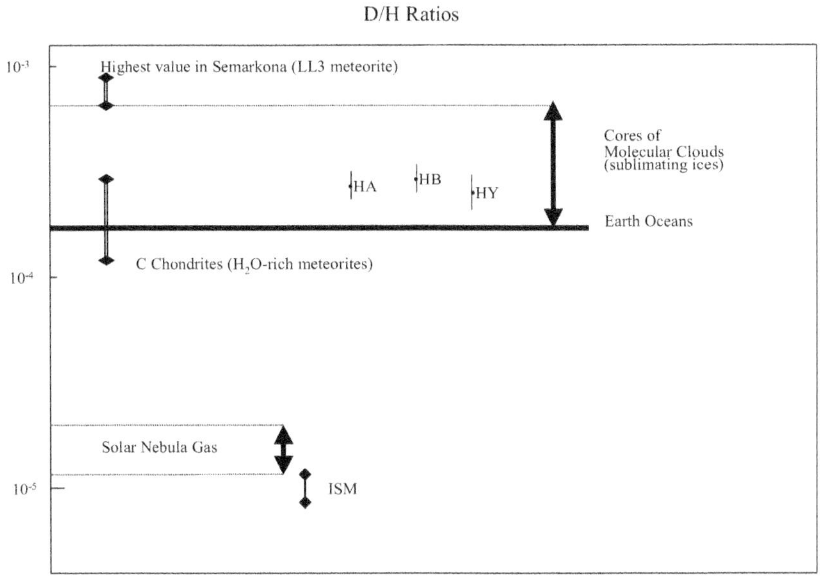

Figure 1. The D/H ratios are plotted for Earth's oceans, the three comets observed so far, the cores of molecular clouds, carbonaceous meteorites, the solar nebula gas, and the current value for the interstellar medium (ISM). The values for the cores of molecular clouds represent the composition of ices formed in the coldest regions of these clouds, and the similarity with cometary values supports the view that comets retain a chemical signature of the environment that preceded the solar nebula. If the all the solar nebula ices had equilibrated with the gas, the D/H ratios would all be near 2.6 x 10^{-5}. Other types of chondritic meteorites (not shown in this figure), such as enstatite and unequilibrated ordinary chondrites have a range of D/H values that also include the value for Earth's oceans. The value for the ISM is lower than that for the solar nebula gas most likely because nucleosynthesis has decreased the D/H in the 4.6 Ga since the formation of the solar system.

in comet Hale-Bopp with a roughly solar abundance of argon to oxygen was reported by Stern *et al.* (2000). More recently, upper limits for Ar/O of <10% and <8% the solar value were reported in comets LINEAR 2001 A2 and LINEAR 2000 WM1, respectively by Weaver, *et al.* (2002). All these observations diagnostic of Ar have been made in comets from the Oort cloud. The more sensitive upper limits for Ar in other comets are not consistent with the detection reported by Stern *et al.* (2000) in Comet Hale-Bopp. ** At this point it is not clear if Comet Hale-Bopp was unusually rich in Ar or if the tentative detection is somehow flawed.** Additional information relevant to comets comes from laboratory experiments of noble gases trapped in ices, and is discussed in Section 4.2.

3.3. DUST COMPOSITION

Ices are not the only carrier of water and organics in comets, cometary dust can also contain these two components. Hence, the composition of cometary dust is also relevant to the origin of Earth's volatiles. So far, remote sensing and *in situ* mass spectrometers have identified two major components of cometary dust, carbonaceous material and silicates. Recent reviews of this topic can be found in Hanner and Bradley (2003) and in Wooden (2003); here, we summarize the main characteristics of cometary dust.

3.3.1. Carbonaceous Material

Delsemme (1982) had predicted that a significant fraction of the carbon in comets was tied up in dust (*i.e.*, material less volatile than CO or CO_2 ice). The carbonaceous component in cometary dust was first detected by in situ spacecraft measurements of the dust coma of comet Halley. The mass spectra obtained by the Giotto and VEGA spacecraft sampled just a few nanograms of the smallest dust grains in the coma of this comet (*e.g.*, Fomenkova 1997 and references therein). Three major types of particles were measured. The first type, about one quarter of the total, were silicate-free carbonaceous particles. Of this 25%, one-third are elemental carbon (probably amorphous carbon) and the other two-thirds are made almost exclusively of the elements H, C, N, and O. This led to the designation of "CHON" particles to the grains that contained mainly these four light elements (*e.g.*, Gruen and Jessberger 1990). The molecular composition of the CHON material is not well defined, because the mass spectra yielded only atomic compositions. The likely molecular composition of these CHON particles includes kerogen-like materials and refractory organics, and is discussed by several authors including Khare *et al.* (1993), Huebner and Boice (1997) and Kissel *et al.* (1997).

The second type of particle detected by the Giotto and VEGA mass spectrometers was also about one-quarter of the total number measured. These particles were silicates (see Section 3.3.2) and showed mass spectra where the rock-forming elements Mg, Si, Fe and O were dominant. The third type of particle was the most abundant, about one-half of the total, and was composed of a mixture of carbon-bearing materials and silicates. The spatial distribution of the different particles measured in the coma of comet Halley support the idea that the mixed grains are held together by the organic refractory "glue" that sublimates and releases smaller silicate particles in the coma at larger distances from the nucleus (Boehnardt *et al.* 1990, Fomenkova 1997, Mumma 1997).

3.3.2. Silicates

In the past five years there has been considerable progress in our understanding of cometary silicates. To date, crystalline and amorphous olivine $[Mg,Fe]_2SiO_4$ and pyroxene $[Mg,Fe]SiO_3$ have been identified in cometary dust (Wooden 2003 and references therein). Crystalline silicates have also been observed in some shells around oxygen-rich evolved stars (Waters *et al.* 1996), and in disks around young stars such as β Pictoris (Knacke *et al.* 1993, Wooden 2003), but not in interstellar dust clouds. These results raise an important issue involving the origin of crystalline silicates in comets: if all interstellar silicates are amorphous, how and when did the silicates in comets crystallize? How does this crystallization happen without heating the ices in comets and erasing the observed chemical signatures indicative of a low formation temperature? Although the specific mechanism for crystallization of amorphous silicates has not been established (*e.g.*, Nuth *et al.* 2000; Wooden 2003), it has been suggested that mixing of material from different regions in the solar nebula is a way to incorporate high temperature crystalline silicates with low temperature ices (Bockelée-Morvan *et al.* 2002).

3.3.3. Interplanetary Dust Particles

Interplanetary dust particles (IDPs) are microscopic (typically 10-50 μm) samples of extraterrestrial material collected primarily in the Earth's stratosphere (*e.g.*, Bradley *et al.* 1988). Some of these IDPs, in particular the chondritic porous (CP) ones, are believed to be cometary. Laboratory analyses of CP IDPs provide valuable information relevant to cometary dust and are nicely complementary to remote sensing results. These CP IDPs are mixtures of an aggregate matrix with one or more 0.1–1.0 μm sized components, including anhydrous silicate crystals (olivine and pyroxene), hydrous layer lattice silicate crystals (mainly smectite and some serpentine), and refractory iron sulfide or oxides. The matrix is composed of amorphous and poorly graphitized carbon, and an organic "glue" that holds together other phases such as iron sulfide, and "GEMS". These GEMS (Glass with Embeded Metal Sulfide) are 0.1 μm amorphous silicate spherules thought to be produced in the envelopes of oxygen-rich stars (Bradley 1994). The aggregate nature of CP IDPs, with an "organic glue" holding together different mineral components, is in excellent agreement with the in situ measurements of the dust in comet Halley.

Deuterium enrichments up 50 times greater than the D/H in Earth's oceans have been measured in a number of CP IDPs (Messenger *et al.* 1996 and Messenger 2000) and identified with the carbonaceous phase of these IDPs (Aléon *et al.* 2001). This deuterium enrichment is thought to have been produced by the same low-temperature ion-molecule reactions in molecular clouds that enriched the deuterium in cometary water ice. The isotopic composition of the carbonaceous component in CP IDPs implies a similar deuterium enrichment for the cometary carbonaceous material. If so, cometary carbonaceous material may be (in addition to water) another significant carrier of deuterium in comets.

There are few observational constraints on the existence and abundance of hydrated silicates in comets; hence, estimates of the abundance and measurements of their D/H ratios are highly desirable. Because of the unequilibrated nature of cometary material one cannot assume that the D/H ratio in cometary hydrated silicates and in cometary water ice is the same. Deuterium enrichments have been measured in hydrated (also know as chondritic smooth or CS) IDPs (*e.g.*, Bradley *et al.* 1988). There are arguments suggesting an asteroidal origin for CS IDPs and it is reasonable to assume that hydrated silicates in comets would also be enriched in deuterium, but by how much? If hydrated silicates are a significant fraction of the mass in cometary nuclei, these minerals could be an important source of terrestrial water and deuterium not previously considered. A quantitative estimate of the contribution of terrestrial water from cometary hydrated silicates would require better constraints on the mineralogy of cometary solids. One interesting possibility is that the capture efficiency of water of hydration may be higher than that of water ice in large cometary impacts.

4. Origin of Earth's Water

In this section we review the evidence relevant to the origin of terrestrial water and consider possible sources. We divide these sources into four categories a) primordial (gas captured from the solar nebula), b) early accretion from hydrated planetesimals in the Earth's accretion zone, c) early accretion from comet-like and asteroid-like bodies, and d) a late bombardment of comet-like and asteroid-like bodies. In order to test the merits of each of the proposed mechanisms, we consider the most pertinent evidence and uncertainties. Based principally on Earth's D/H ratio in water and noble gas abundances (see below), one can rule out the first source, namely a primordial atmosphere captured from the solar nebula.

4.1. ISOTOPIC RATIOS

The determination of the D/H isotopic ratios for water in three comets is among the most significant developments in this field; curiously, it is interpreted very differently by different authors. Some (*e.g.*, Dauphas *et al.* 2000; Morbidelli *et al.* 2000; Drake and Righter 2002, Robert 2001), consider the high D/H ratio in these comets as evidence against a cometary origin of most of the terrestrial water. Others (*e.g.*, Delsemme 2000, Owen and Bar-Nun 2001), argue that comets are the main reservoir of deuterium-rich H_2O that raised the terrestrial D/H a factor of six above the protosolar value. In carbonaceous chondrite meteorites, the D/H ratios for H_2O overlap the range observed in terrestrial water (Lecuyer *et al.* 1998). Can the sources of water-rich meteorites also be a major source of Earth's water? We discuss this point in Section 4.4.

4.2. NOBLE GASES

Noble gases are chemically inert and extremely volatile. Hence they probably arrive at a planet along with other volatiles, quickly move to the planet's atmosphere, and avoid the chemical complications of planetary evolution. This means that the noble gas characteristics of a planetary atmosphere can be extremely useful as tracers of the source of the planet's volatiles.

4.2.1. Noble Gases Trapped in Ices

As discussed in section 3.2.5, we do not know much about noble gases in comets; there is a tentative detection of argon in comet Hale-Bopp and upper limits for neon and argon in comets Hale-Bopp and LINEAR 2001 A3. However, laboratory measurements of noble gases trapped in amorphous H_2O ices formed at low temperatures (simulating comets) suggest that the relative abundance of three noble gases (argon, krypton and xenon) in comets is similar to that observed on Earth and in Mars (Figure 2; Owen and Bar-Nun 1995, 2001). In particular, the amounts of various noble gases trapped in amorphous ice depends on the temperature at which trapping occurs, with the Ar/Xe ratio increasing by more than two orders of magnitude as the temperature drops from 60K to 30K, and Ne beginning to be trapped at even lower temperatures (see review in Owen and Bar-Nun, 2001). The noble gas measurements in comets so far are ambiguous about the applicability of these laboratory simulations. Furthermore, there are alternate models for the trapping of noble gasses by water ice that involve trapping in the form of clathrate hydrates (*e.g.*, Iro *et al.* 2003).

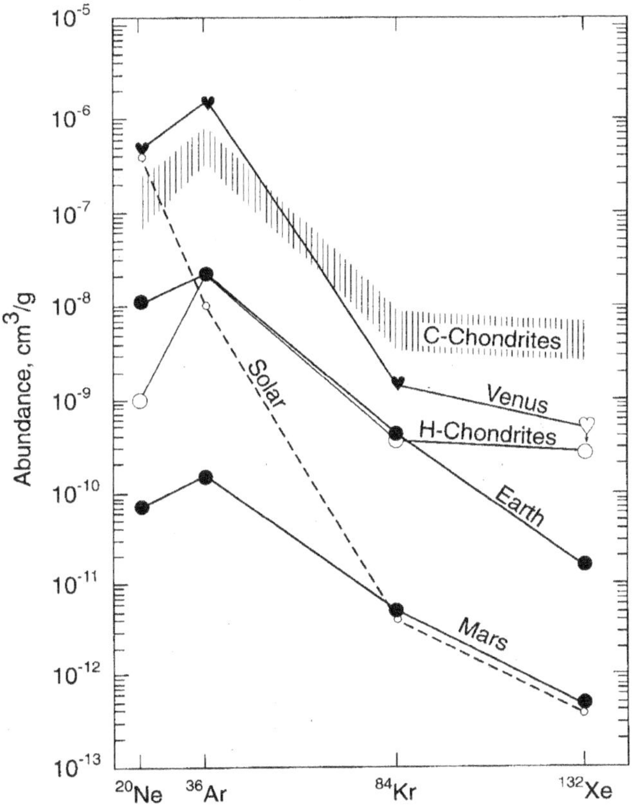

Figure 2. Abundances of noble gases in the atmospheres of Venus, Earth, and Mars compared with solar abundances and those in two kinds of meteorites (from Owen and Bar-Nun 1995). Note that the noble gas abundances in the three planets differ significantly from solar abundances, indicating that these atmospheres were not derived directly from the solar nebula. The large argon-to-xenon and krypton-to-xenon ratios on Earth relative to those in meteorites suggest that meteorites were not an important source of Earth's noble gases.

4.2.2. Noble Gases in Jupiter's Atmosphere

Recent observations by the atmospheric probe of the Galileo spacecraft of the Ar, Kr and Xe abundances in Jupiter's atmosphere (Owen *et al.* 1999), can constrain the nature and composition of the material accreted by Jupiter. The noble gas abundances in Jupiter are significantly different from those in the Sun (and the solar nebula). Based on these abundances, Owen *et al.* (1999) argue that the icy planetesimals (comets) that contributed to Jupiter's mass were formed at temperatures lower than those predicted by present models of giant-planet formation. They suggest that the enrichment in Ar, Kr, and Xe is the result of a significant fraction of Jupiter's inventory of elements heavier than hydrogen and helium coming from objects that formed at temperatures below 30 K. As Owen *et al.* (1999) point out, this is inconsistent with Jupiter's location just outside the expected "snow line," where H_2O condenses at relatively high temperatures. A different interpretation is offered by Gautier *et al.* (2001) and by Hersant *et al.* (2003). These authors argue that volatiles must have been trapped in the form of clathrate hydrates in the cooling feeding zone of Jupiter until the epoch when all hydrogen and matter contained in this zone collapsed onto the core of the planet. This model reproduces the enrichment measured in Jupiter in Ar, Kr, Xe, C, N and S, and is consistent with reasonable evolutionary models of the solar nebula.

4.2.3. Noble Gases in Earth and Mars

Figure 2 (from Owen and Bar-Nun 1995) shows the abundances of noble gases in Venus, Earth and Mars compared with solar abundances and those in two kinds of meteorites. Note that the elemental abundance patterns of the planets are distinct from the solar pattern, ruling out direct capture from the solar nebula. Pepin and Porcelli (2002) have recently reviewed the origin of these noble gases – we will concentrate on the possibility of a cometary contribution. Many researchers have concluded that the match in Ne/Ar/Kr ratios of Earth, Mars and meteorites implies that meteorites were the major source of the atmospheres of terrestrial planets. However, the difference in Kr/Xe ratios has never been explained satisfactorily. Owen and Bar-Nun (1995, 2001) have argued that the large argon-to-xenon and krypton-to-xenon ratios on Earth and Mars, which are not matched by meteorites, are consistent with laboratory simulations of cometary ices, and hence suggest that comets were an important source. However, the similarity in Ne/Ar ratios between terrestrial planets and some meteorites then becomes a problem. In other words, neither comets nor meteorites are a perfect fit to the abundances of noble gases in the atmospheres of terrestrial planets.

4.2.4. Xenon Isotopes.

The relative abundances of the xenon isotopes (124, 126, 128, 129, 130, 131, 132, 134, and 136) measured for CI carbonaceous chondrites, Earth's atmosphere, Mars' atmosphere, and the solar wind, show that no two sources are exactly alike. However, there seem to be two "families" with Earth and Mars in one, and the Sun and the CI chondrites in the other. This difference between the xenon isotopic abundances in CI chondrites and the atmospheres of Earth and Mars argues that some fractionation process must have occurred to the gases that became the atmospheres of the terrestrial planets. Pepin has argued for fractionation during an episode of hydrodynamic escape from the young planet (Pepin and Porcelli, 2002). Notescu *et al.* (1999) have investigated the possibility of fractionation during trapping in low-temperature ices, but their simulations have not yielded the amount required to explain the terrestrial planet atmospheres. Hence the current leading theory would have the Xe isotopic pattern dominated by a loss process that occurred on the planets. This is not what would be expected if the noble gas elemental patterns reflect cometary precursors.

4.2.5. The Ar/H_2O Ratio in Comet Hale-Bopp

As mentioned above, there is considerable uncertainty in the abundance of argon in comet Hale-Bopp. Swindle and Kring (2001) pointed out that comets with a solar ratio of Ar/H_2O could not contribute significantly to Earth's water, because the Earth's water would be accompanied by more than four orders of magnitude more Ar than Earth possesses. In fact, the oversupply of Ar is such that it is not sensitive to their use of the controversial high Ar measurement in comet Hale-Bopp. Even if the Ar/O ratio is two orders of magnitude lower than solar, as in the simulations by Owen and Bar-Nun (1995), supply of the Earth's water would be accompanied by more than 100 times as much Ar as on Earth. A way out of this contradiction would be a scenario in which much of the Ar could then be lost while much or all of the water is retained. If loss of noble gases is going to be invoked, then it must be non-fractionating if the similarity of the noble gas ratios between the terrestrial planets and the comet simulations (Owen and Bar-Nun, 2001) is significant. This argument can be phrased another way. Some observations support the case for comets providing part or all of the atmospheres of the terrestrial planets. However, the amounts of noble gases in those planetary atmospheres are so small that unless the comets that brought those volatiles were extremely poor in noble gases, they would not be expected to carry enough water to be a significant fraction of Earth's budget.

Recent results support a very low Ar abundance in at least some comets. A model for the trapping of volatiles in the nebula as clathrates made by Iro *et al.* (2003) argues that unless the local amount of ice (O/H) was higher than solar by a factor of 2.5 in the region of cometary grain formation, the probability of detecting Ar in Oort cloud comets is quite small. The situation may be different for comets formed at larger heliocentric distances (in the Kuiper belt), in which Ar trapped in the form of amorphous ice in the presolar cloud could be present if the nebula was tenuous enough to avoid the vaporization of ices falling from the presolar cloud.

4.2.6. Noble Gases and Water in Meteorites

The ratios of water to noble gases in CI chondrite meteorites are comparable to those in models by Iro *et al.* (2003), and five or six orders of magnitude higher than the H_2O/Ar would be in a comet with a solar ratio (Stern *et al.*, 2000). Noble gas and water contents of many carbonaceous chondrites have been reported by Mazor *et al.* (1970) and Jarosewich (1990), respectively. Water to ^{36}Ar ratios for these meteorites are typically about 10^7 to 10^8 by weight, while the solar ratio of O to ^{36}Ar is about 120 (Anders and Grevesse, 1989). Although carbonaceous chondrite meteorites can be considered a plausible source for some of Earth's water, their noble gas abundances are so low that they cannot be the sole source of Earth's noble gases.

4.3. NEON AND NITROGEN

Because neon and nitrogen have approximately the same cosmic abundance with respect to hydrogen, any atmosphere that was captured from the protosolar nebula should have a ratio of Ne/N_2 of approximately 2. On Earth, Mars and Venus, Ne$\ll N_2$; furthermore, the isotopes of neon have been highly fractionated in these atmospheres. Owen and Bar-Nun (2001) argue that cometary delivery of nitrogen and other volatiles, but not neon, can yield this result. Since Ne will condense with ice only at very low temperatures, a lack of Ne in a comet that supplies N_2 is quite plausible. However, the Ne/N ratio in meteorites tends to be much lower than in terrestrial planet atmospheres (cf. Table II, Pepin, 1991). If the terrestrial planet atmospheres were supplied by Ne-poor comets (as suggested by Owen and Bar-Nun 2001), the source of the Ne is a problem because it can be provided by neither comets nor meteorites.

4.4. WATER IN METEORITES FROM DIFFERENT HELIOCENTRIC DISTANCES

Most meteorites are believed to come from the asteroid belt; however, only one asteroid (4 Vesta) has been identified as the parent body of a specific group of meteorites, the eucrites-diogenites-howardites. Within our limited understanding of the connections between specific asteroids and meteorite types, the contents of H_2O in meteorites indicates a decrease in water abundance in the asteroid belt with decreasing heliocentric distance. Water in CI carbonaceous chondrites can be as high as 16% in mass (Jarosewich 1990), while the driest meteorites are 0.05-0.1% H_2O in mass. It has been argued (*e.g.*, Morbidelli *et al.* 2000 and references therein) that the planetesimals that formed in Earth's zone should have had even lower water content. Estimates for the mass of water in Earth's oceans, crust and mantle are given in Table 1 (Lecuyer *et al.*1998), and they range from 0.03% to about 0.8% of Earth's mass. Because the amount of water in Earth's mantle is so uncertain, models for the mass of water delivered to Earth by different sources are poorly constrained. If one were to assume Earth formed entirely out of material similar to the driest meteorites, this material would have been sufficiently wet to be the only source of Earth's oceans. If the mantle contains much more water than the oceans (as much as 50 times the Earth's oceans have been proposed by Abe *et al.* 2000 and more recent results by Murakami *et al.* 2002, suggest 11 times Earth's oceans), the material that formed Earth had to be wetter or additional sources are required.

TABLE 1. Mass of Water in Earth's Oceans, Crust and Mantle

Reservoir	H_2O Mass in grams	% of Earth's Mass in H_2O
Oceans and Ice Sheets	1.4×10^{24}	0.02%
Crust (not including oceans)	2.7×10^{23}	0.005%
Mantle	$5\text{-}50 \times 10^{24}$	0.08-0.8%

Delsemme (1997 and 2000) argues that based on the estimated formation temperatures of carbonaceous and ordinary chondrite meteorites, the temperature for Earth's accretion region (which he defines as 0.8 to 1.3 AU) was between 800 and 1500 K. This argument favors the condensation (from the solar nebula) of anhydrous solid particles in this region, and the accretion of a dry Earth. While Delsemme's idea has merit, the implications on the composition of planetesimals in Earth's region are model dependent, as the composition of condensates could have changed significantly if the temperature dropped before the dissipation of the gas in the solar nebula. Given the miniscule percentage of water needed in the planetesimals to provide Earth's water (Table 1) we do not consider Delsemme's argument a strong constraint on a possible source of water in the material from Earth's accretion region.

4.5. COMPOSITION OF EARTH'S CRUST AND MANTLE

The presence and relative abundances of some siderophile (metal-seeking) elements on Earth's crust and upper mantle suggest that these elements arrived after the formation of Earth's core (*e.g.*, Drake and Righter (2002). The relative abundances of these elements on Earth are consistent with a "chondritic" source, *i.e.*, chondritic meteorites and/or comets. Drake and Righter (2002) argue that Earth-building materials shared some but not all the properties with extant meteorites, *i.e.*, no primitive material similar to Earth's mantle is currently in our meteorite collection. This argument is based on the observed similarities between the a) $^{187}Os/^{188}Os$ ratio in Earth's upper mantle with that of ordinary chondrite meteorites, b) the similarity of the oxygen isotopic composition in Earth's upper mantle with enstatite meteorites, c) the differences in the Al-Mg-Si ratios on Earth, Mars, and most primitive meteorites, and d) the differece in D/H ratios in Earth's water and three comets.

4.6. RATIO OF ASTEROIDAL TO COMETARY IMPACTS ON THE MOON

The impact record on the Moon suggests that the flux of asteroids was far greater than that of comets (Chyba 1987, Swindle and Kring 2001), including the objects involved in the cataclysm (Kring and Cohen, 2002). The approximate percentages were greater than or equal to 80% asteroids and less than or equal to 20% comets. Although it may be a coincidence, these are close to the percentages of asteroids (~90%) and extinct comets (~10%) estimated, by several independent methods, in today's population of near-Earth objects (*e.g.*, Bottke *et al.* 2002, Fernández *et al.* 2002, Whiteley 2001).

5. Discussion and Conclusions

Can one make sense of these diverse and sometimes contradictory results? We believe the answer is yes, particularly if we keep in mind that different constraints may be diagnostic of contributions from different bodies and probably at different times. As discussed in Section 4, the D/H ratios and the noble gases rule out a primordial atmosphere captured from the solar nebula. Beyond this point matters are less clear and we consider three options. The first option is that a single source brought all the water to Earth. The second option would be a combination of water sources, some of which may still exist. The third option is that we may not recognize the source(s) yet, because post-accretional fractionation processes altered its chemical and isotopic signatures.

5.1. SINGLE AND MULTIPLE SOURCES OF WATER

If only one type of material delivered all the water, the main constraint would be a D/H ratio like that in Earth's water. If this material came early from planetesimals in Earth's accretion zone, it was likely in the form of hydrated silicates because that zone was too hot for ices to condense or survive long. According to Drake and Righter (2002), such material is not currently represented in either meteorites or comets.

Although a single source of terrestrial water cannot be ruled out, there may be evidence in favor of multiple sources. More specifically, the zircon evidence indicates some of Earth's water arrived early, probably by 4.4 Ga, before the lunar cataclysm that occurred about 3.9 to 4.0 Ga. The population that produced the lunar cataclysm included asteroids and possibly comets, which also impacted Earth. It is not clear if the cataclysm population contributed significantly to Earth's water (*e.g.*, Kring and Cohen 2002); if it did, then multiple sources are likely. Water is not necessarily linked to the material that produced the bulk of Earth's mass, and there can be a distinction between the water that accreted on or before 4.4 Ga and water that was delivered later. The composition of the source and the timing of the delivery, determine the constraints that apply to each potential source.

a) If the source or sources were mostly water (*i.e.*, comets and water-rich asteroids) they must have had the D/H and possibly the noble gas composition (both elemental and isotopic) of Earth.
b) If water was only a trace component of the material, then the source(s) must meet (in addition to the D/H ratio) the other constraints discussed in Section 4.5, namely osmium isotopes, siderophile elements, Mg-Al-Si ratios, and oxygen isotopes. The order of these constraints is roughly correlated with decreasing abundance of water in the source. In other words, the oxygen isotope constraint would apply only if the source of Earth's water had a high rock-to-water ratio, because the oxygen isotopic signature on Earth is likely due to oxygen in silicates rather than in water (note that about 60% of a silicate is oxygen).

Since members of potential sources still exist, their study may establish if they did contribute to Earth's water and in what proportions. Some models (*e.g.*, Morbidelli *et al.* 2000) propose that asteroids and comets were the first water deliverers, when the Earth was half its present mass. D/H ratios for carbonaceous chondrite meteorites are believed to be indicative of those for primitive asteroids; however, no measurement of the D/H ratio in an asteroid has been made. Comets might have also been among the bodies that produced the lunar cataclysm (less than or equal to 20%), and their contribution to the water delivered late to Earth may have been significant. A cometary contribution higher than 20% of the terrestrial water delivered during the cataclysm is possible because water is a higher fraction of the mass in comets than in hydrous asteroids. The compositional diversity among comets is now well established and must be considered when evaluating comets as sources of terrestrial water. Three specific populations of comets have been invoked to deliver water to Earth, a) comets from the Jupiter region that formed near the "snow-line" (Delsemme 2000), b) comets from the Jupiter-Saturn region (Morbidelli *et al.* 2000), and comets from the Kuiper belt (Owen and Bar-Nun 2001). Some of the comets in populations a and b might still be observable as subset of the Oort cloud, while the Kuiper belt objects are the main source of what we call Jupiter-family or short-period comets. Jupiter-Family comets are relatively faint but predictable and frequently observable. The chemical and isotopic composition of comets is consistent with a cometary origin of some of Earth's water and organic molecules, but the magnitude of that contribution remains highly uncertain. Hence, the continued study of comets is likely to reveal new clues about the origin of Earth's water.

5.2. FRACTIONATED WATER SOURCES

The processes involved in planetary accretion, degassing, and the evolution of a hydrosphere and atmosphere are complicated and may have fractionated the chemical and isotopic signatures of water's source(s), complicating its identification. Hydrogen, for example, may be an important constituent in the outer core and possibly inner core of Earth (*e.g.*, Okuchi 1997, 1998). If hydrogen and deuterium were fractionated in that process, the residual D/H in the hydrosphere may not reflect that of the original source. Following accretion, the surface of the Earth continued to be modified by large impacting asteroids and comets (Section 2). Large impact events have the capacity to completely volatilize any oceans (Zahnle and Sleep 1997) and blow off portions of Earth's atmosphere (Melosh and Vickery, 1989). Fractionation of D/H and noble gas/water ratios may occur as a consequence of impact processes, which may also mask the signatures of the original source material.

5.3. FUTURE WORK

Observations and calculations most diagnostic of the sources of Earth's water include the following:

a) measurements of the D/H ratio in several sites, including water in a statistically significant number of Oort cloud and Jupiter-family comets, in cometary organics, in water and/or hydrated silicates in primitive asteroids, in hydrated silicates and organics in IDPs and meteorites, in rocks from the Earth's deep mantle (which can help identify early and late sources of terrestrial water)
b) measurements of noble gas abundances in a statistically significant number of comets
c) determining if the abundance of hydrated silicates in comets is significant, and if so, measuring the D/H ratios in those hydrated silicates
d) experimental and theoretical evaluations of the degree to which planetary evolution and impact processes may fractionate D/H and noble gases on Earth, and
e) measurements of the quantity and D/H ratios of water reservoirs on Mars.

The Stardust spacecraft is on its way to encounter comet Wild 2, collect samples of its dust coma and return them to Earth in 2006. Measurements in “a” and “c” above may be achieved by analyses of dust samples delivered by the Stardust mission.

6. Summary

New studies of the chemical and isotopic composition of comets allow a reevaluation of possible sources of terrestrial water. We review the evidence and consider four sources: a) primordial (gas captured from the solar nebula), b) early accretion from hydrated planetesimals in the Earth's accretion zone, c) early accretion from asteroids and comets, and d) a late bombardment of asteroids and comets. The first source can be ruled out based on the D/H ratio in Earth's water and the terrestrial noble gases. Zircon crystals eroded from very old igneous rocks suggest some of Earth's water arrived early, probably by 4.4 Ga. The lunar cataclysm that occurred about 3.9 to 4.0 Ga may have contributed some of Earth's water, but was probably not the principal source. If only one type of material delivered all the water, the main constraint on such material would be a D/H ratio like that in Earth's water. If this material came early from planetesimals in Earth's accretion zone, it was likely in the form of hydrated silicates, and according to Drake and Righter (2002), such material is not currently represented in either meteorites or comets. Although a single source of terrestrial water cannot be ruled out, there are arguments in favor of multiple sources. One possibility is that we may not recognize the source(s) yet, because post-accretional fractionation processes altered its chemical and isotopic signatures. The chemical and isotopic composition of comets is consistent with a cometary origin of some of Earth's water and organic molecules, but the magnitude of that contribution remains highly uncertain.

7. Acknowledgments

H. Campins acknowledges the support of grants from NASA and NSF, and helpful comments from J. Ward. D. Kring was supported by the NASA Astrobiology Program through a subcontract from Arizona State University to the University of Arizona. T. Swindle was supported by the NASA Cosmochemistry Program.

8. References

Abe, Y., Ohtani, E., Okuchi, T., Righter, K., Drake, M. (2000) Water in the early Earth. In: R.M. Canup and K. Righter (eds) *Origin of the Earth and Moon*. University of Arizona Press, Tucson, pp. 413-433.

A'Hearn, M.F., Millis, R.L., Schleicher, D.G., Osip, D.J., and Birch, P.V. (1995) The ensemble properties of comets: Results from narrowband photometry of 85 comets, 1976-1992. Icarus **118**, 223-270.

Aléon, J.; Engrand, C.; Robert, F.; Chaussidon, M. (2001) Clues to the origin of interplanetary dust particles from the isotopic study of their hydrogen-bearing phases. Geochim. Cosmochim. Acta, **65**, 4399-4412.

Anders, E. and Grevesse, N. (1989) Abundances of the elements: Meteoric and solar. Geochim. Cosmochim Acta **53**, 197-214.

Bockelée-Morvan, D., Gautier, D., Lis, D.C., Young, K., Keene, J., Phillips, T., Owen, T., Crovisier, J., Goldsmith, P.E., Bergin, E.A., Despois D. and Wooten, A. (1998) Deuterated water in comet C/1996 B2 (Hyakutake) and its implications for the origin of comets, Icarus **133**, 147-162.

Bockelée-Morvan, D. and Crovisier, J. (2002) Lessons of comet Hale-Bopp for coma chemistry: Observations and theory. Earth, Moon and Planets **89**, 53-71.

Bockelée-Morvan, D., Gautier, D., Hersant, F., Huré, J-M., Robert, F. (2002) Turbulent radial mixing in the solar nebula as the source of crystalline silicates in comets Astron. Astrophys. **384**, 1107-1118.

Boehnhardt, H., Fechtig, H. and Vanysek, V. (1990) The possible role of organic polymers in the structure and fragmentation of dust in the coma of comet P/Halley. Astron. Astrophys. **231**, 543-547.

Bogard, D.D. (1995) Impact ages of meteorites: A synthesis. Meteoritics **30**, 244-268.

Bottke, W.F., Morbidelli, A., Jedicke, R., Petit, J-M., Levison, H.F., Michel, P. and Metcalfe, T.S. (2002) Debiased orbital and absolute magnitude distribution of the near-Earth objects. Icarus **156**, 399-433.

Bradley, J.P., Sandford, S.A., and Walker, R.M. (1988) Interplanetary dust particles. In: J.F. Kerridge and M.S. Matthews (eds.) *Meteorites and the Early Solar System*. University of Arizona Press, Tucson, pp. 861-895.

Bradley, J.P. (1994) Chemically anomalous, preaccretionally irradiated grains in interplanetary dust from comets. Science **265**, 925-929.

Campins, H. (2000) The chemical composition of comets. In: J. Chela-Flores, G. A. Lemarchand and J. Oró (eds.) *Astrobiology: Origins from the Big Bang to Civilisation*. Kluwer, Dordrecht, pp.163-176.

Chyba, C. and Sagan, C. (1992) Endogenous production, exogenous delivery and impact-shock synthesis of organic molecules: an inventory for the origins of life, Nature **355**, 125-132.

Chyba, C.F. (1987) The cometary contribution to the oceans of primitive Earth, Nature **330**, 632-635.

Chyba, C.F. (1993) The violent environment of the origin of life: Progress and uncertainties, Geochim. Cosmochim. Acta **57**, 3351-3358.

Chyba, C.F., Thomas, P.J., Brookshaw, L. and Sagan, C. (1990) Cometary delivery of organic molecules to the early Earth, Science **249**, 366-373.

Cohen, B.A., Swindle, T.D. and Kring, D.A. (2000) Lunar meteorites support the lunar cataclysm hypothesis, Science **290**, 1754-1756.

Dalrymple, G.B. and Ryder, G. (1993) $^{40}Ar/^{39}Ar$ age spectra of Apollo 15 impact melt rocks by laser step-heating and their bearing on the history of lunar basin formation, J. Geophys. Res. **98**, 13085-13095.

Dalrymple, G.B. and Ryder, G. (1996) $^{40}Ar/^{39}Ar$ age spectra of Apollo 17 highlands breccia samples by laser step heating and the age of the Serenitatis basin, J. Geophys. Res. **101**, 26069-26084.

Dauphas, N., Robert, F. and Marty, B. (2000) The late asteroidal and cometary bombardment of Earth as recorded in water deuterium to protium ratio, Icarus **148**, 508-512.

Delsemme, A.H. (1982) Chemical composition of cometary nuclei In: L.L. Wilkening (ed.) *Comets*, Univ. of Arizona Press, Tucson, pp. 85-163.

Delsemme, A.H. (1997) The origin of the atmosphere and of the oceans. In: P. J. Thomas, C.F. Chyba and C.P. McKay (eds.) *Comets and the Origin and Evolution of Life*. Springer-Verlag, New York, pp. 29-67.

Delsemme, A.H. (2000) 1999 Kuiper prize lecture cometary origin of the biosphere. Icarus **146**, 313-325.

Despois, D. (1999) Radio Line Observations of Molecular and Isotopic Species in Comet C/1995 O1 (Hale-Bopp). Earth, Moon, and Planets **79**, 103-124.

Drake, M.J. and Righter, K. (2002) Determining the composition of the Earth. Nature **416**, 39-44.

Drouart, A., Dubrulle, B., Gautier, D. and Robert, F. (1999) Structure and transport in the solar nebula from constraints on deuterium enrichment and giant planets formation. Icarus **140**, 129-155.

Eberhardt, P., Reber, M., Krankowsky, D. and Hodges, R.R. (1995) The D/H and $^{18}O/^{16}O$ ratios in water from comet P/Halley. Astron. & Astrophys. **302**, 301.

Fernández, Y.R., Jewitt, D. C. and Sheppard, S. S. (2002) Low albedos among extinct comet candidates. Astrohys. J. **553**, L197-L200.

Fomenkova, M.N. (1997) Organic components of cometary dust. In: Y.J. Pendelton and A.G.G.M. Tielens (eds.) *From Stardust to Planetesimals*, Asron. Soc. Pac. Conf. Ser. 122, Astron. Soc. Pac., Provo, pp. 415-421.

Gaffey, M.J., Cloutis, E.A., Kelley, M.S. and Reed, K.L (2003) Mineralogy of Asteroids. In: W.F. Bottke Jr., A. Cellino, P. Paolicchi, and R.P. Binzel (eds.) *Asteroids III*. University of Arizona Press, Tucson, pp. 183-204.

Gautier, D., Hersant, F., Mousis, O. and Lunine, J.I. (2001) Enrichments in volatiles in Jupiter: A new interpretation of the Galileo measurements. Astrohys. J. **550**, L227-230. Erratum, Astrohys. J.Lett. 559, L183.

Gensheimer, P.D., Mauersberger, R., and Wilson, T.L. (1996) Water in galactic hot cores. Astron. & Astrophys. **314**, 281-294.

Gruen, E. and Jessberger, E.K. (1990) Dust. In: W.F. Huebner (ed.) *Physics and Chemistry of Comets*. Springer-Verlag, New York, pp. 113-176.

Hagemann, R., Nief, G., and Roth, E. (1970) Absolute isotopic scale for deuterium analysis natural waters: Absolute D/H ratio for SMOW. Tellus **20**, 712.

Hanner M.S. and Bradley, J. (2003) Composition and mineralogy of cometary dust. In: M. Festou, H.U. Keller and H.A. Weaver (eds) *Comets II*, University of Arizona Press, Tucson, in press.

Hartmann, W.K., Ryder, G., Dones, L. and Grinspoon, D. (2000) The time-dependent intense bombardment of the primordial Earth/Moon system. In: R.M. Canup and K. Righter (eds) *Origin of the Earth and Moon*. University of Arizona Press, Tucson, pp. 493-512.

Hersant, F., Gautier, D. and Lunine, J.I. (2003) Enrichment in the giant planets of the solar system. Planet. Space Sci., submitted.

Hersant, F., Gautier, D. and Hure, J.-M. (2001). A two-dimensional model for the primordial nebula constrained by D/H measurements in the solar system: Implications for the formation of giant planets. Astrohys. J. **554**, 391-407.

Huebner, W.F. and Boice, D.C. (1997) Polymers and other macromolecules in comets. In: P. J. Thomas, C.F. Chyba and C. P. McKay (eds.) *Comets and the Origin and Evolution of Life*. Springer-Verlag, New York, pp. 111-129.

Iro, N., Gautier, D., Hersant, F., Bockelee-Morvan, D. and Lunine, J.I. (2003) An interpretation of the nitrogen deficiency in comets. Icarus **161**, 511-532.

Irvine, W.M., Schloerb, F.P., Crovisier, J., Fegley, B. and Mumma, M. (2000) Comets: a link between interstellar and nebular chemistry. In: V. Mannings, A. Boss and S. S. Russell (eds) *Protostars and Planets IV*. Univ. Arizona Press, Tucson, p. 1159-1200.

Jarosewich E. (1990) Chemical analyses of meteorites: A compilation of stony and iron meteorite analyses. Meteoritics **25**, 323-337.

Khare, B.N., Thompson, W.R., Cheng, L., Chyba, C., Sagan, C., Arakawa, E.T., Meisse, C. and Tuminello, P.S. (1993) Production and optical constraints of ice tholin from charged particle irradiation of (1:6) C_2H_6/H_2O at 77 K. Icarus **103**, 290-300.

Kissel, J., Krueger, F.R. and Roessler, K. (1997) Organic chemistry in comets from remote and in situ observations. In: P.J. Thomas, C.F. Chyba and C.P. McKay (eds.) *Comets and the Origin and Evolution of Life*. Springer-Verlag, New York, pp. 70-109.

Knacke, R. F., Fajardo-Acosta, S. B., Telesco, C. M., Hackwell, J. A., Lynch, D. K. and Russell, R. W. (1993) The silicates in the disk of beta Pictoris. Astrophys. J. **418**, 440-450.

Krasnopolsky, V.A., Mumma, M.J., Abbott, M., Flynn, B.C., Meech, K.J., Yeomans, D.K., Feldman, P.D. and Cosmovici, C.B. (1997) Detection of soft x-rays and a sensitive search for noble gases in comet Hale-Bopp (C/1995 01). Science **277**, 1488-1491.

Kring, D.A. and Cohen, B.A. (2002) Cataclysmic bombardment throughout the inner solar system 3.9-4.0 Ga, J. Geophys. Res. **107**, pp. 4-1 through 4-6.

Lecacheux A. and Biver, N. (2002) Comet C/2002 C1 (Ikeya-Zhang), Internat. Astron. Union Cirular 7910.

Lecuyer, C., Gillet, P. and Robert, F. (1998) The hydrogen isotope composition of seawater and the global water cycle. Chem. Geol. **145**, 249-261.

Lellouch, E., Bezard, B., Fouchet, T., Feuchtgruber, H., Encrenaz, T. and de Graauw, T. (2001) The deuterium abundance in Jupiter and Saturn from ISO-SWS observations. Astron. & Astrophys. **370**, 610-622.

Lis, D.C., Mehringer, D.M., Benford, D., Gardner, M., Phillips, T.G., Bockelée-Morvan, D., Biver, N., Colom, P., Crovisier, J., Despois, D. and Rauer, H. (1999) New molecular species in comet C/1995 O1 (Hale-Bopp) observed with the caltech submillimeter observatory. Earth, Moon and Planets, **78**, 13-20.

Maher, K.A. and Stevenson, D.J. (1988) Impact frustration of the origin of life. Nature **331**, 612-614.

Matthews, H. E., Biver, N., Senay, M., Davies, J. K., Dent, W. R. F., Bockelée-Morvan, D., Jewitt, D., Owen, T., Crovisier, J., Rauer, H., Gautier, D., Sarmecanic, J., Lavezzi, T., Fomenkova, M. and Jones, B. (1997) Comet C/1996 B2 (Hyakutake). Internat. Astron. Union Circular 6353.

Mazor E., Heymann D. and Anders, E. (1970) Noble gases in carbonaceous chondrites. Geochim. Cosmochim. Acta **34**, 781-824.

Meier, R., Owen, T.C., Matthews, H.E., Jewitt, D., Bockelée-Morvan, D., Biver, N., Crovisier, J. and Gautier, D. (1998) A determination of the HDO/H20 ratio in Comet C/1995 01 (Hale Bopp). Science **279**, 842-844.

Meier, R., Owen, T.C., Jewitt, D.C., Matthews, H.E., Senay, M., Biver, N., Bockelée-Morvan, D., Crovisier J. and Gautier, D. (1998b) Deuterium in Comet C/1995 O1 (Hale-Bopp): Detection of DCN. Science **279**, 1707-1710.

Melosh, H.J. and Vickery, A.M. (1989) Impact erosion of the primordial atmosphere of Mars. Nature **338**, 487-489.

Messenger, S. (2000) Identification of molecular-cloud material in interplanetary dust particles. Nature **404**, 968-971.

Messenger, S., Walker, R.M., Clemett, S. J. and Zare, R.N. (1996) Deuterium enrichments in cluster IDPs Lunar Planet. Sci. **27**, 867-868.

Mojzsis, S.J. and Harrison, T.M. (2000) Vestiges of a beginning: Clues to the emergent biosphere recorded in the oldest known sedimentary rocks. GSA Today **10**, 1-6.

Mojzsis, S.J., Arrhenius, G., McKeegan, K.D., Harrison, T.M., Nutman, A.P. and Friend, C.R.L. (1996) Evidence for life on Earth before 3,800 million years ago. Nature **384**, 55-59.

Mojzsis, S.J., Harrison, T.M. and Pidgeon, R.T. (2001) Oxygen-isotope evidence from ancient zircons for liquid water at the Earth's surface 4,300 Myr ago. Nature **409**, 178-181.

Morbidelli, A., Chambers, J., Lunine, J.I., Petit, J.M., Robert, F., Valsecchi, G.B., and Cyr, K.E. (2000) Source regions and timescales for the delivery of water to the Earth. Meteoritics **35**, 1309-1320.

Mousis, O., Gautier, D., Bockelee-Morvan, D., Robert, F., Dubrulle, B. and Drouart, A. (2000) Constraints on the formation of comets from D/H ratios measured in H_2O and HCN. Icarus **148**, 513-525.

Mumma, M.J., Dello Russo, N., DiSanti, M.A., Magee-Sauer, K., Novak, R.E., Brittain, S., Rettig, T., McLean, I.S., Reuter, D.C. and Xu, L.H. (2001) Organic composition of C/1999S4: a comet formed near Jupiter? Science **292**, 1334-1339.

Mumma, M. J. (1997) Organic volatiles in comets:Their relation to interstellar ices and solar nebula material. In: Y.J. Pendelton and A.G.G.M. Tielens (eds.) *From Stardust to Planetesimals*, Asron. Soc. Pac. Conf. Ser. 122, Astron. Soc. Pac., Provo, pp. 369-396.

Murakami, M. Key, H., Yurimoto H., Nakashima, S. and Takafuji, N. (2002) Water in Earth's lower mantle. Science, **295**, 1885-1887.

Notescu, G., Laufer, D., Bar-Nun, A. and Owen T. (1999) An experimental study of isotopic enrichments in Ar, Kr, and Xe when trapped in water ice. Icarus **142**, 298-300.

Nuth, J.A. III, Hallenbeck, S.L. and Rietmeijer, F.J.M. (2000) Laboratory studies of silicate smokes: Analog studies of circumstellar materials. J. Geophys. Res. **105**, 10387-10396.

Nutman, A.P., McGrego, V.R., Friend, C.R.L., Bennett, V.C. and Kinny, P.D. (1996) The Itsaq Gneiss Complex of southern West Greenland: The world's most extensive record of early crustal evolution (3900-3600 Ma. Precambrian Research **78**, 1-39.

Okuchi, T. (1997) Hydrogen partitioning into molten iron at high pressure: implications for Earth's core. Science **278**, 1781-1784.

Okuchi, T. (1998) Melting temperature of iron hydride at high pressures and its implications for tempurature of the Earth's core. J. Phys. Condensed Matter **10**, 11595-11598.

Oro, J. (1961) Comets and the formation of biochemical compounds on the primitive Earth. Nature **190**, 389-390.

Owen, T. and Bar-Nun, A. (1995) Comets, impacts and atmospheres. Icarus **116**, 215-226.

Owen, T. and Bar-Nun, A. (2001) Contributions of icy planetesimals to Earth's early atmosphere. Origins Life Evol. Biosphere **31**, 435-458.

Owen, T., Lutz, B.L. and De Bergh, C. (1986) Deuterium in the outer solar system - Evidence for two distinct reservoirs. Nature **320**, 244-246.

Owen, T., Mahaffy, P., Niemann, H.B., Atreya, S., Donahue, T., Bar-Nun A. and de Pater, I. (1999) A low-temperature origin for the planetesimals that formed Jupiter. Nature **402**, 269-270.

Pepin R.O. and Porcelli D. (2002) Origin of noble gases in the terrestrial planets. Reviews in Mineralogy and Geochemistry **47**, 191-246.

Pepin R.O. (1991) On the origin and early evolution of terrestrial planet atmospheres and meteoritic volatiles. Icarus **92**, 2-79.

Pierazzo, E. and Chyba, C.F. (1999) Amino acid survival in large cometary impacts. Meteoritics. **34**, 909

Robert, F. (2001) Isotope geochemistry: the origin of water on Earth. Science **293**, 1056-1058.

Robert, F., Gautier, D. and Dubrulle, B. (2000) The solar system D/H Ratio: Observations and theories. Space Sci. Rev. **92**, 201-224.

Ryder, G. (1990) Lunar samples, lunar accretion and the early bombardment of the Moon. Eos Trans. Am. Geophys. Union **313**, 322-323.

Stern, S.A., Slater, D.C., Festou, M.C., Parker, J.W., Gladstone, G.R., A'Hearn, M.F. and Wilkinson, E. (2000) Discovery of argon in comet C/1995 O1 (Hale-Bopp). Astrophys. J. **544**, L169-L172.

Swindle, T.D. and Kring, D.A. (2001) Cataclysm + cold comets = lots of asteroid impacts, Lunar Planet. Sci. XXXII, Abstract #1466, Lunar and Planetary Institute, Houston (CD-ROM).

Tera, F., Papanastassiou, D.A. and Wasserburg, G.J. (1974) Isotopic evidence for a terminal lunar cataclysm. Earth Planet. Sci. Lett. **22**, 1-21.

Thomas, P.J., Chyba, C.F. and McKay, C.P. (1997) Comets and The Origin and Evolution of Life, Springer-Verlag, New York.

Trujillo, C.A. and Brown M.E. (2001) The Radial distribution of the Kuiper belt, Astrophys. J. **554**, L95-L98.

Turner G., Cadogan, P.H. and Yonge, C.J. (1973) Argon selenochronlogy. Proc. Lunar Sci. Conf. 4, 1889-1914.

Valley, J.W., Peck, W.H., King, E.M. and Wilde, S.A. (2002) A cool early Earth. Geology **30**, 351-354.

Vanysek, V. (1991) Isotopic ratios in comets. In: Newburn, R.L. Neugebauer, M. and Rahe, J. (eds) *Comets in the Post-Halley Era,* vol. 2. Kluwer, Dordrecht, pp. 879-895.

Wasson, J.T. (1990) Ungrouped iron meteorites in Antarctica: Origin of anomalously high abundance. Science **249**, 900-902.

Waters, L.B.F.M., Molster, F.J., de Jong, T., Beintema, D.A., Waelkens, C., Boogert, A.C.A., Boxhoorn, D.R., de Graauw, T., Drapatz, S., Feuchtgruber, H. and 27 coauthors (1996) Mineralogy of oxygen-rich dust shells. Astron. & Astrophys. **315**, L361-364.

Weaver, H.A., Feldman, P.D., Combi, M.R., Krasnopolsky, V., Lisse, C.M. and Shemansky, D.E. (2002) A search for argon and O VI in three comets using the far ultraviolet spectroscopic explorer. Astrophys. J. **576**, L95-L98.

Whiteley, R.J. (2001) A compositional and dynamical survey of near-Earth asteroids. P.D. dissertation, University of Hawaii, Honolulu.

Wilde, S.A., Valley, J.W., Peck, W.H. and Graham, C.M. (2001) Evidence from detrital zircons for the existence of continental crust and oceans on Earth 4.4 Gyr ago. Nature **409**, 175-178.

Wooden, D.H. (2003) Comet grains: Their IR emission and their relation to ISM grains. Earth, Moon and Planets, **89**, 247-287

Zahnle K.J. and Sleep, N.H. (1997) Impacts and the early evolution of life. In: P. J. Thomas, C.F. Chyba and C. P. McKay (eds.) *Comets and the Origin and Evolution of Life*. Springer-Verlag, New York, pp. 175-208.

Biodata of **Jan K. W. Toporski** author of ***"From Microbial Fossils to Astrobiology"***

Dr. Jan Toporski is a research Fellow in Geo- and Astrobiology at the Carnegie Institute of Washington, Geophysical Laboratory, Washington D.C. He obtained his Ph.D. (2001) from the University of Portsmouth, School of Earth and Environmental Sciences, Portsmouth, UK. His current research focuses on the mineralization, preservation and detection of bacterial biomarkers. He obtained the Gerry Soffen Award (2001) by Mary Ann Liebert Publishing Inc for outstanding young investigators in Astrobiology.

E-mail: **j.toporski@gl.ciw.edu**

J. Seckbach (ed.), Origins, 593–605.

FROM MICROBIAL FOSSILS TO ASTROBIOLOGY

JAN TOPORSKI AND ANDREW STEELE
Geophysical Laboratory
Carnegie Institution of Washington
5251 Broad Branch Road NW
Washington DC, 20015 USA

1. Introduction

Crucial programs in Astrobiology seek to understand the distribution of life in our solar system and ultimately the Universe (as defined in the NASA Astrobiology Road Map; http://astrobiology.arc.nasa.gov/roadmap) t*o be able to recognize the signature of life in rocks from Earth's fossil record as well as in extraterrestrial materials*" is stated as Goal 7 of this road map. Understanding the processes involved in microbial preservation and recognizing the markers left behind in rocks as a result of their interaction with solid and fluid phases are key issues in astrobiology and geo(micro)biology research. To identify true microfossils, a number of criteria have been devised and applied in the past (Schopf and Walter, 1983). They must be authentic constituents of the host rock, occur in multiples, be of carbonaceous composition, display morphological microbial characteristics and be demonstrably biogenic in origin. The difficulty associated with the unambiguous recognition of evidence of microbial activity in rocks has lately become exposed in the scientific and public domain due to eminent scientific controversies over the earliest evidence of life on Earth and the continuing discussions over possible evidence of biogenic activities in Martian meteorite ALH84001 (see below). The common denominator in these debates is the underlying difficulty or inability to demonstrate conclusively the biogenicity of the respective evidence. In either of the above cases, this would be evidence of fossil microbial life (both Bacteria and Archaea). A further consensus that has emerged from these discussions, and is now seen as a critical requirement, is the demand for chemical/molecular evidence in addition to the morphological evidence to support such claims. Carbon isotopes have successfully been correlated with individual Proterozoic microfossils (House et al., 2000) and Raman spectra were obtained on supposed early Archaean microfossils (Schopf et al., 2002), although this evidence is far from being conclusive (Brasier et al., 2002; Pasteris and Wopenka, 2002). Notably, Toporski et al. (2002b) demonstrated the possibility to correlate more comprehensive molecular information with fossil bacterial biofilms using Time of Flight – Secondary Ion Mass Spectroscopy (ToF-SIMS), a technique that has been shown to be capable of detecting unambiguous biomarkers, e.g. of the hopane class and others, in complex organic mixtures (Steele et al., 2001; Toporski, 2001). Such efforts contribute to a comprehensive understanding of Earth's fossil record and may eventually help to unambiguously identify traces of microbial activities in rocks, regardless of whether terrestrial or extraterrestrial.

2. Microbial Fossils in Earth's Fossil Record

2.1 MORPHOLOGICAL EVIDENCE

In an attempt to identify morphological evidence of microbial fossils in rocks, the same criteria used to describe and characterize living microorganisms apply. These include characteristics such as size and shape of individual cells, their organization and distribution e.g. in a biofilm, cellular structures such as cell walls and membranes, or ancillary features such as flagella (Westall, 1999). A further morphological consideration is the presence of extracellular polymeric substances (EPS). Microorganisms, especially bacteria, may secrete highly hydrated EPS, in both sessile and planktonic state (Westall et al., 2000). The composition and structural state of EPS can vary greatly depending on environmental conditions with diverse compositional ranges possible, even within a single species. EPS play multiple roles in microbial cells and biofilms including protection, storage, biofilm stabilization and support of attachment to surfaces (Wimpenny, 1995). Regardless of the exact role and composition of EPS in a biofilm, it was shown that EPS has a high fossilization, and thus preservation, potential (Westall et al., 2000), due to its high cation binding capacity resulting from a net-negative charge of the organic compounds (*e.g.* Beveridge, 1989; Schultze-Lam et al., 1993). Fossilized polymeric substances (FPS) in bacterial biofilms appear to have the potential to retain the same morphological characteristics as the unfossilized counterpart, as compellingly demonstrated by Toporski et al. (2002a) on fossil bacterial biofilms from the Upper Oligocene Enspel formation, Germany, as well as by studies on cyanobacterial sheaths (Westall et al., 2000). The presence of FPS has been described from numerous fossil environments ranging up to 3.5 Ga in age (Westall et al., 2001). Through observations from the fossil record, it appears in fact that FPS may be more common than cellular microbial fossils, especially in the older part of Earth's fossil record (Walsh, 1992; Westall, 2000).

Microbial fossils have been described from most of Earth's fossil record, from as early as 3.5 Ga ago to recent ongoing microbial fossilization. Among the earliest morphological indicators of life on Earth are permineralized microorganisms found in the ca. 3.5 Ga old Barberton formation, South Africa (Westall et al., 2001). Electron microscopy methods have revealed the presence of sub-micron to micron sized fossil-like structures in Early Archaean cherts from the Barberton greenstone belt in South Africa (Westall, 1998; Westall and Gerneke, 1998; Westall et al., 2000). These structures, mostly coccoid and sometimes rod-shaped, are interpreted as the remains of ancient non-cyanophytae microorganisms and their associated polymeric substances (Westall et al., 2001). The shapes and size-ranges of these structures suggest that the ancient bacteria were morphologically identical to modern eubacteria and Archaea, although their biogenicity is still in doubt due to the lack of definitive chemical and/or molecular evidence. Equally old are filamentous structures discovered in the Early Archaean Apex Cherts of Australia (Schopf, 1993). Initially interpreted as fossilized cyanobacterial filaments, this interpretation has recently been challenged (Brasier et al., 2002) and the initial interpretation has been revoked (Kerr, 2002). Although morphological evidence has been supported by in situ Raman spectroscopy and imagery supplying corroborative data proving the biogenicity of the filaments (Schopf et al., 2002), Pasteris and Wopenka (2002) contend the applicability of the provided evidence, as the Raman spectra do not *per se* allow the inference that the analyzed features are of biogenic origin. Regardless of the exact nature of the Apex Chert filaments, however, there is ample evidence that (microbial) life was present in the early Archaean, with early fossil life forms consisting mostly of filamentous structures in 3.2 to 3.5 Ga old Early Archaean rocks (e.g. Brooks et al., 1973; Schopf, 1974; Muir and Grant, 1976; Dunlop et al., 1978; Schopf and Walter, 1983; Walsh and Lowe, 1985; Walsh, 1992), although the same rigorous standards developed from the Apex fossils and ALH84001 debates (see below) have yet to be applied to these examples. In later periods microorganisms preserved in siliceous rocks have been reported to occur in Proterozoic-age rocks such as the Gunflint Formation (e.g. Barghoorn and Tyler, 1965; Cloud, 1965, Westall et al., 2000 and references therein) and in Phanerozoic - age rocks (e.g. Knoll, 1985; Carson, 1991; Monty et al., 1991; Westall, 1994; Toporski et al., 2002b and references therein).

Research on the preservation of microbes in recent environments has concentrated on *in situ* silicification of microorganisms in a variety of hot spring milieus (e.g. Schultze-Lam et al., 1995; Cady and Farmer, 1996; Jones and Renaut, 1996; Renaut et al., 1998; Jones et al., 1999), indicating sometimes rapidly progressing fossilization. This rapidity of preservation may, in fact, be the critical factor as to why microorganisms can be preserved in such excellent quality. Rapidly proceeding fossilization leaves little time for cellular degradation, and with the simultaneously occurring precipitation and/or polymerization of a mineralizing solid phase, cells would largely be preserved in their authentic shape. Indeed, there is experimental evidence that fossilization of bacteria by means of preservation in silica may commence after only 24 hours in a silicifying solution (Toporski et al. 2002a). This study clearly shows that the preservation of cellular detail is possible under such circumstances, by not allowing cell lysis and other degenerative processes to occur. Extrapolating this information to natural settings would help explain the high quality of preservation of some microbial fossils.

Therefore, through the study of young, well-preserved microbial fossils from the fossil record in combination with fossilization experiments, we may gain detailed insight into the mechanisms involve in microbial preservation. This would contribute significantly to the understanding of formation of microbial fossils, even for those buried in ancient rocks. It is only through this knowledge and expertise, that we may eventually be successful in the hunt for evidence of extraterrestrial life.

2.2 MOLECULAR EVIDENCE

Clearly, there are microbial fossils preserved in recognizable quality, and able to be identified by means of light or electron microscopy. If diagenetic conditions do not favor preservation of morphological fossils, there may be traces left behind in the form of molecular marker molecules, or biomarkers (Eglinton and Calvin, 1967). Biomarkers are defined as organic molecules found in the fossil record with their precursor compounds being produced by living systems. In the search for evidence of microbial life biomarkers derived from Bacteria and Archaea are of particular interest. Both groups represent candidate groups of organisms adapted to environments potentially relevant in the origin of life on Earth or elsewhere in the solar system. Evidence of previous life on Mars or Jupiter's moon Europa may still be present in the form of molecular biomarkers and could be targeted in solar system exploration. Biomarkers are an important source of information in the search for evidence of life in geological samples, possibly including extraterrestrial materials. Organic geochemists have made considerable progress in the isolation, identification, and taxonomic attribution of biologically informative organic compounds in sediments. This bacterial compound fossil record has been mainly uncovered by research within the petroleum industry for mechanisms to classify the formation and composition of oil (Engel and Macko, 1993; Peters and Moldowan, 1993). Over the years a number of chemicals have been discovered, which point to the presence of life millions of years ago. Specific classes of compounds, for which we have a reliable information base, include pentacyclic triterpane hydrocarbons, so called hopanes, which are the fossil remnants of hopanoids biosynthesized by numerous eubacteria. Further classes include steranes (derived from sterols) from eukaryotes, acyclic isoprenoids from various precursors but particularly from lipids of Archaea, and certain hydrocarbons derived from carotenoid precursors (Summons and Walter, 1990). Other classes of compounds include porphyrin-based molecules from the diagenesis of chlorophyll and cytochromes and the derived isoprenoids from phototrophic microorganisms. There are various other organic marker molecules that may originate from bacteria. However, these are non-specific and thus are not of diagnostic character without ample additional information (e.g. alkanes and polycyclic aromatic hydrocarbons, PAHs). Hopanes, in contrast, are generally regarded as unambiguous bacterial biomarkers (Rohmer et al., 1984; Engel and Macko, 1993). The hopanes are ubiquitous on Earth due to their resistant carbon skeleton (Innes et al., 1997). Indeed recent research has shown the presence of a specific cyanobacterial hopane (2-methyl hopane) in 2.7 billion year old sediments (Brocks et al., 1999). This research shows that the diagenetic products of bacteria can be detected on Earth dating from a time when Mars had more favorable environmental conditions for life, rendering this class of bacterial biomarkers an interesting target molecule for planetary exploration. It is generally important to realize that the duration a biomarker can survive in a natural environment not only depends on its refractory properties but also on the diagenetic environment and the burial conditions (e.g. P/T-condition, pH, redox potential, etc.). Thus it is possible that theoretically less refractory compounds such as squalene can be preserved for up 1690 million years (Summons and Walters, 1990). A brief synopsis on bacterial biomarkers can be found in Toporski and Steele (2003).

Indication of biological activities in rocks may further be present in the form of carbon-isotopic evidence. Kinetic isotope effects during microbial catabolism and anabolism impart an isotope fractionation, discriminating against the heavy ^{13}C isotope over ^{12}C, which is largely retained by organic matter in rocks (Schidlowski, 1988, 2002). This biological isotopic effect is propagated from the biosphere to the geosphere almost unaltered, which allows tracing the isotopic signatures of life back into the geological past. These systematics have been drawn on in an attempt to identify the oldest evidence for early life on Earth. Until recently, the earliest evidence was seen in carbonaceous inclusions within grains of apatite (Mojzsis et al., 1996; Mojzsis and Harrison, 2000) and microparticles (Rosing, 1999) from the oldest known sediments of the Isua supracrustal belt, West Greenland, based on stable carbon isotopic ratios approximating those of biotic abundance. These studies suggest that life was abundant on Earth as early as 3.8 Ga ago (Schidlowski, 1988). As with the oldest morphological evidence of life, however, recent reassessment of these materials shed doubt on this evidence, as the carbon-containing phases have been reinterpreted as results of metasomatic formation, clearly lacking biological relevance (Fedo and Whitehouse, 2002; van Zuilen et al., 2002). Furthermore, it is uncertain whether abiotic processes may also lead to C-isotope fractionation (Brasier et al., 2002). These examples indicate the difficulties connected with unambiguously identifying evidence of traces of life in rocks, further emphasizing that knowledge of geological and chemical context is extremely important in this endeavor.

Although laden with intriguing complexity, Earth's fossil record provides a wealth of sources of information, helping us understand the distribution of life on Earth over space and time. With the information and experience gained through studies of Earth materials, we are constructing a baseline for reasonable evaluation of evidence of traces of life, if encountered in extraterrestrial materials. This is true for morphological evidence, as well as molecular and isotopic traces. All these categories of evidence being apt for preservation on planetary bodies and indeed required for the identification of traces of life.

3. The Debate on Evidence of Life in Martian Meteorite ALH84001

Is there a record of life in extraterrestrial materials? The discovery of evidence of life in meteorites has been claimed several times in the past, mostly without compelling success. Thus far, the evidence provided by McKay et al. (1996) on "possible relic biogenic activity in Martian meteorite ALH84001" arguably represents the most comprehensive suite of evidence for the possible detection of traces of life in extraterrestrial materials. Multiple arguments were provided, each of which taken on its own would not be proof, but taken together, were interpreted as possibly indicating evidence of life. Each of these arguments has been the subject of contentious, still ongoing, debate (e.g. Frankel and Buseck, 2000; Thomas-Keprta et al., 2000, 2001; Treiman, 2001; Golden et al., 2000, 2002; Thomas-Keprta et al., 2002).
McKay et al. (1996) argued that the association of Fe-Mg-Ca-carbonates and sulphide and oxide minerals in discrete zoned carbonate globules of Martian origin (Mittlefehldt, 1994) could be related to biogenic processes. This association, however, has been argued not to be proof of life (Schopf, 1999). Further debate emerged over the origin of the carbonates as both high- and low-temperature formation has been proposed (e.g. Harvey and McSween, 1996; Kirshvink et al., 1997; Valley et al., 1997; Warren, 1998). It has further been suggested that the zoning of the globules could possibly be a result of microbial activity (McKay et al., 1996) and indeed can be produced under biologic influence (Barrat et al., 1999; Zhang et al., 1999). Chemical zoning in Fe-Mg-Ca-carbonates has also been produced abiotically at elevated temperatures in the laboratory (Golden et al., 2000), although zoning and oxygen isotope non-equilibrium argue against prolonged heating of carbonates (Treiman and Romanek, 1998). Further complications may have arisen from impact shock melting of the carbonates (Scott et al., 1997; Scott, 1999). Recent research further suggests that the rims may be a product of weathering processes and formed through dissolution and reprecipitation of a primary Martian carbonate while the meteorite was exposed at the surface in Antarctica (Kopp, 2002). Clearly, the presence of zoned carbonates cannot be seen as unambiguous indicator of biogenicity.

McKay at al. (1996) described the presence of minute iron sulphide (pyrrhotite and greigite) and magnetite crystals associated with the Fe- and Mg-rich globule rims. Pyrrhotite has not been established as a biogenic product (Posfai et al., 1998) and has been suggested as a possible derivative of pyrite. Greigite, in contrast, is a product known from certain marine magnetotactic and sulphate reducing bacteria (Posfai et al., 1998). However, the presence of greigite in the ALH84001 carbonates has not been confirmed (Frankel and Buseck, 2000), rendering both types of minerals inadequate for life detection in this case. Some (ca. 27%) of the elongated prismatic (truncated hexa-octahedral) magnetite crystals present in the carbonate globules show similarities to those found in some magnetotactic bacteria from Earth (Thomas-Keprta et al., 2000, 2001, 2002). These are being interpreted as biogenically produced and are now considered to represent the strongest line of evidence (Thomas-Keprta et al., 2002). It is argued that this subset of magnetite is virtually identical in size, elongation, crystal shape, chemical purity and the absence of structural defects to that produced by a marine strain of magnetotactic bacteria (MV-1). Friedman et al. (2001) reported chains of the magnetite crystallites in the carbonate globules, suggesting the possibility of intact magnetosome preservation. However, magnetite grains with some of these characteristics form from thermal decomposition of siderite (Golden et al., 2000, 2002), providing a suitable model for the abiogenic formation of the magnetites and leaving the discussion on the magnetites as biomarkers devoid of consensus. Furthermore, work conducted by Taylor et al. (2000) showed that the criteria chosen by Thomas-Keprta et al. (2000) do not necessarily apply to all strains of biogenic magnetite on Earth, which casts doubt on their suitability for a biogenic marker in extraterrestrial materials.

The presence of organic molecules (PAHs) associated with the carbonates was described by McKay et al. (1996), and was interpreted as having formed through diagenetic activity of microorganisms. Other researchers, however, argue that PAHs can also form abiotically by Fischer-Tropsch type reactions (Anders, 1996; Zolotov and Shock, 2000), as long as molecular hydrogen is available. Decarbonatization from siderite to form magnetite releases CO and CO_2, which can react with H_2 below 300°C to form PAHs (Zolotov and Shock, 2000). A further point of discussion is the initial source of the PAHs. PAHs can readily form abiotically in space (Léger et al., 1987), and thus be introduced to the meteorite during its journey to Earth, and possibly on the surface of Mars without any biogenic intervention (Zolotov and Shock, 1999). Becker et al. (1997) suggested that the PAHs may have been introduced by melted Antarctic ice and were preferentially scavenged by the carbonates within the meteorite. The ^{14}C values of the organics in the meteorite support this argument as 80 % of the organic material in ALH84001 has a probable terrestrial source (Jull et al., 1998). Clemett et al. (1998), however, made a case against the contamination of ALH84001 with terrestrial PAHs by showing that the concentration of PAHs to the carbonate globules is not an effect of selective deposition. The PAHs present in the ice differ from those found in the meteorite, which makes the ice an unlikely source. Other meteorites collected in the Allan Hills (ALH) region did not show evidence of PAH contamination and finally, the PAH concentration is greater inside the meteorite than on its exterior. That Stefan et al. (1999) found the PAH broadly distributed in the meteorite and not concentrated in the carbonate globules shows that the argument over PAH in ALH84001 as biomarkers is far from settled and may be due to which sample distribution has been analyzed.

Finally, McKay et al. (1996) presented images of "features resembling terrestrial microorganisms" as possible fossils of Martian microorganisms. Other researchers interpreted these structures as irregularities in the surfaces of minerals, which were accentuated by the metal coating for electron microscope examination (Bradley et al., 1997). McKay et al. (1997) replied that they found such surface irregularities, but they were unrelated to the possible Martian nanofossils. It was shown that the metal coating would not cause artifacts in the size-range of the supposed Martian nanofossils (Steele et al., 1998). As Steele et al. (2000) reported the presence of recent contaminants in ALH84001, it cannot be ruled out that some of the described microbe-shaped objects may be derived from terrestrial biota. Furthermore, the size of these features was considered too small to represent fossil microorganisms (Schopf, 1999). However, several researchers have presented evidence for the existence of nano-scale organisms (Çiftçioglu et al., 1997; Kajander et al., 1997, 1998; Hubet et al., 2002). More importantly, Gillet et al. (2000) described bacterium-shaped objects from within the Tatahouine meteorite that matched the size of the structures in ALH84001. Bacteria isolated from the soil around this meteorite revealed similar morphologies. Transmission electron micrographs of these cultures showed the presence of cell walls and 16S rDNA analysis confirmed a terrestrial bacterial origin. These findings render the size-argument used to refute McKay et al.'s hypothesis incongruous, as it is conceivable that nanobes (if present) can become fossilized just as well as microbes can. Putative fossil nanobes have been reported from a number of terrestrial palaeo-environments (Folk, 1993) but these claims are subject to considerable debate, and alternative abiogenic explanations have been suggested (Kirkland et al., 1999; Vecht and Ireland, 2000; Vali et al., 2001). In fact, Folk and Taylor (2002) recently reported nanobacterial alteration from the pyroxene matrix of ALH84001. The evidence is based on morphological observations alone and, as with the supposed Martian nanofossils, their identity as true nanofossils is being understandably disputed, as once again there is a complete lack of relevant chemical and/or molecular information.

4. Perspectives

Clearly, the vestiges of microbes, particularly in ancient rocks from Earth and in extraterrestrial materials, are difficult to recognize. Over decades scientists have painted a comprehensive picture of the distribution of (microbial) life on Earth over space and time, although some landmark data are currently being debated. Methods have been established that in many cases allow the unambiguous identification of traces of life and the use of new technologies is currently being pioneered. However, due to contamination and sample size, attempting to identify those traces in extraterrestrial materials, regardless of whether meteoritic samples studied on Earth or remotely on planetary surfaces is extremely difficult and arguably a field in its infancy. Although there are promising efforts, it is clear that a great deal of further systematic investigation and experimentation must be undertaken and new technologies explored (e.g. Toporski et al., 2002b, Steele and Toporski, 2003), in order to unambiguously identify the traces of microbial life in rocks. We feel that this can only be achieved by coupling morphology with chemical and/or molecular information as reported by House et al. (2000) and Toporski et al. (2002b). It is only through this kind of investigation that new technologies, methodologies and philosophies for the detection of extraterrestrial life can be honed.

5. Acknowledgements

James Hall (GL, CIW, Washington DC, USA) for useful comments on the manuscript, Frances Westall (CNRS, Orleans, France), D.S. McKay (NASA JSC, Houston, Texas), R. Avci (MSU, Bozeman, Montana) for discussions on the subject matter, and B. Kessler for general support.

6. Reference

Anders E. (1996) Evaluating the evidence for past life on Mars (letter). *Science,* **274**, 2119-2121.

Bada J.L., Glavin D.P., McDonald G.D., and Becker L. (1998) A search for endogenous amino acids in Martian meteorite ALH84001. *Science,* **279,** 362–365.

Barghoorn E.S. and Tyler S.A. (1965) Microfossils from the Gun Flint Chert. *Science*, **147**, 563-577.

Barrat J.A., Gillet Ph., Lesourd M., Blichert-Toft J., and Popeau G.R. (1999) The Tatahouine diogenite: mineralogical and chemical effects of sixty-three years of terrestrial residence. *Meteorit. Planet. Sci.*, **34**, 91-97.

Becker L., Glavin D. P., and Bada J. L. (1997) Polycyclic aromatic hydrocarbons (PAH) in Antarctic Martian meteorites, carbonaceous chondrites, and polar ice. *Geochim. Cosmochim. Acta,* **61**, 475-481.

Beveridge T.J. (1989) Metal ions and bacteria. In: T.J. Beveridge and D.J. Doyle (eds.) *Metal Ions and Bacteria*, Wiley, New York. pp. 1-30,

Bradley J.P., Harvey R.P., and McSween H.Y. (1997) No 'nanofossils' in Martian meteorite. *Nature*, **390**, 454.

Brasier, M.D., Green O.R., Jephcoat A.P., Kleppe A.K., van Kranendonk M.J., Lindsay J.F., and Steele A. (2002) and Grassineau N.V. Earth's oldest fossils: questioning the evidence. *Nature*, **416**, 76-81.

Brooks J., Muir M.D., and Shaw G. (1973) Chemistry and morphology of Precambrian microorganisms. *Nature*, **244**, 215-217.

Brocks J.J., Logan G.A., Buick R. and Summons R.E. (1999) Archaean Molecular fossils and the early rise of eukaryotes. *Science*, **285**, 1033-1036.

Cady S.L. and Farmer J.D. (1996) Fossilization processes in siliceous thermal springs: trends in preservation along thermal gradients. In: G.R. Bock and J.A. Goode (eds.). Evolution of hydrothermal ecosystems on Earth (and Mars?). Wiley, Chichester (Ciba Foundation Symposium). pp. 150-173.

Carson G.A. (1991) Silicification of fossils. In: P.A. Allison and D. E. G. Briggs (eds.), *Taphonomy, releasing the data locked in the fossil record.* Plenum Press, New York, London. Pp. 455-499.

Çiftçioglu N., Pelttari A., and Kajander E.O. (1997) Extraordinary growth phases of nanobacteria isolated from mammalian blood. *Proceedings of SPIE*, **3111**, 429-435.

Clemett S.J., Dulay M.T., Gilette J.S., Chillier X.D.F., Mahajan T.B., and Zare R.N. (1998) Evidence for the extraterrestrial origin of polycyclic aromatic hydrocarbons (PAHs) in the Martian meteorite ALH 84001. *Faraday Discussions (Royal Soc. Chem.),* **109**, 417-436.

Cloud P.E. (1965) Significance of the Gun Flint (Precambrian) microflora. *Science*, **148**, 27-35.

Dunlop J.S.R, Muir M.D., Milne V.A., and Groves D.I. (1978) A new microfossil assemblage from the Archaean of Western Australia. *Nature*, **274**, 676-678.

Eglinton G. and Calvin M. (1967) Chemical fossils. *Sci. Am.*, **261**, 32-43.

Engel M.E. and Macko S.A. (eds.) (1993) *Organic Geochemistry - Principles and Applications.* Plenum Press, New York, London.

Fedo C.M. and Whitehouse M.J. (2002) Metasomatic origin of quartz-pyroxene rock, Akilia, Greenland, and implications for Earth's earliest life. *Science*, **296**, 1448-1452.

Folk R.L. (1993) SEM imaging of bacteria and nannobacteria in carbonate sediments and rocks. *J. Sed. Petrol.*, **63**, 990-999.

Folk R.L. and Taylor L.A. (2002) Nannobacterial alteration of pyroxenes in Martian meteorite Allan Hills 84001, *Met. Plan. Sci.*, **37**, 1057-1070.

Frankel R.B. and Buseck P.R. (2000) Magnetite biomineralization and ancient life on Mars. *Current Opinion on Chemical Biology*, **4**, 171-176.

Friedmann E.I., Wierzchos J., Ascaso C., and Winklhofer M. (2001) Special feature: chains of magnetite crystals in the meteorite ALH84001: evidence of biological origin, *Proc. Natl. Acad. Sci.*, **98**, 2176-2181.

Gillet Ph., Barrat J.A., Heulin Th., Achouak W., Lesourd M., Guyot F., and Benzerara K. (2000) Bacteria in the Tatahouine meteorite: nanometric-scale life in rocks. *Earth Plan. Sci. Let.*, **175**, 161-167.

Glavin D.P., Schubert M., Botta O., Kminek G., and Bada J.L. (2001) Detecting pyrolysis products from bacteria on Mars. *Earth Plan. Sci. Let.*, **185**, 1-5.

Golden D.C., Ming D.W., Schwandt C.S., Lauer H.V., Socki R., Morris R., Lofgren G.E., and McKay G.A. (2000) Inorganic formation of zoned Fe-Mg-Ca carbonate globules with magnetite and sulphide rims similar to those in Martian meteorite ALH84001. *Lunar Planet. Sci. XXXI*, Lun. Plan. Inst., Houston TX, #1799 (abstr.).

Golden D.C., Ming D.W., Lauer H.V. Jr., Schwandt C.S., Morris R., Lofgren G.E., and McKay G.A. (2002) Inorganic formation of truncated hexa-octahedral magnetite: implications for inorganic processes in Martian meteorite ALH84001. *Lunar Planet. Sci. XXXIII*, Lun. Plan. Inst., Houston TX, #1839 (abstr.).

Harvey R.P. and McSween H.Y. Jr. (1996) A possible high-temperature origin for the carbonates in the Martian meteorite ALH84001. *Nature,* **382**, 49-51.

Hoover R.B., Rozanov A.Yu., Zhmur S.I., and Gorlenko, V.M. (1998) Further evidence of microfossils in carbonaceous chondrites. *Proc. of SPIE*, **3441**, 203-216.

House C.H., Schopf J.W., McKeegan K.D., Coath C.D., Harrison T.M., and Stetter K.O. (2000) Carbon isotopic composition of individual Precambrian microfossils. *Geology*, **28**, 707-710.

Huber H., Hohn N.J., Rachel R., Fuchs T., Wimmer V.C., Stetter K.-O. (2002) A new phylum of Archaea represented by a nano-sized hypothermophilic symbiont. *Nature*, **417**, 63-67.

Innes H.E., Bishop A.N., Head I.M. and Farrimond P. (1997) Preservation and diagenesis of hopanoids in recent lacustrine sediments of Priest Pot, England. *Org. Geochem.*, **26**, 565-576.

Jones B. and Renaut R.W. (1995) Influence of thermophilic bacteria and calcite and silica precipitation in hot springs with water temperatures above 90°C: evidence from Kenya and New Zealand. *Can. J. Earth Sci.*, **32**, 72-83.

Jones B., Renault R.W., and Rosen M.R. (1999) Actively growing siliceous oncoids in the Waiotapu geothermal area, North Island, New Zealand. *J. Geol. Soc., Lond.*, **156**, 89-103.

Jull A. J. T., Courtney C., Jeffrey D. A., and Beck J. W. (1998) Isotopic evidence for a terrestrial source of organic compounds found in Martian meteorites Allan Hills 84001 and Elephant Moraine 79001. *Science,* **279**, 366-369.

Kajander O.E., Björkland M., and Çiftçioglu N. (1998) Mineralization by nanobacteria. *Proc. of SPIE*, **3441**, 86-94.

Kajander E.O., Kuronen I., Åkerman K., Pelttari A., and Çiftçioglu N. (1997) Nanobacteria from blood, the smallest culturable, autonomously replicating agent on Earth. *Proc. of SPIE*, **3111**, 420-428.

Kerr R.A. (2002) Earliest signs of life just oddly shape crud? *Science*, **295**, 1812-1813.

Kirkland B.L., Lynch F.L., Rahnis M.A., Folk R.L., Molineux I.J., and McLean R.J. (1999) Alternative origins for nanno-bacteria-like objects in calcite. *Geology*, **27**, 247-250.

Kirshvink J.L., Maine A.T., and Vali H. (1997) Paleomagnetic evidence of a low-temperature origin of carbonates in the Martian meteorite ALH84001. *Science*, **275**, 1629-1633.

Knoll A.H. (1985) Exceptional preservation of photosynthetic organisms in silicified carbonates and silicified peats. *Phil. Trans. R. Soc. Lond.*, **311**, 11-122.

Kopp R.E. (2002) An Antarctic origin for proposed signs of past Martian life? MSc (hons.) Thesis, University of Chicago, USA.

Léger A., d'Hendcourt L., and Boccara N. (1987) *Polycyclic aromatic hydrocarbons and astrophysics*. Reidel, Boston, USA.

McKay D.S., Gibson E.K., Thomas-Keprta K., Vali, H., Romanek C., Clemett S., Chillier X.D.F., Maechling C.R. & Zare R.N. (1996) Search for past life on Mars: possible relic biogenic activity in Martian meteorite ALH84001. *Science,* **273**, 924-930.

McKay D. S. Gibson E. K. Jr., Thomas-Keprta K. L., and Vali H. (1997) No 'nanofossils' in Martian meteorite: reply. *Nature,* **390**, 455-456.

Mittlefehldt D. W. (1994) ALH84001, a cumulate orthopyroxenite member of the SNC meteorite group. *Meteoritics,* **29**, 214-221

Mojzsis S.J., Arrheinius G., MvKeegan K.D., Harrison T.M., Nutman A.P., and Friend R.L. (1996) Evidence for life on Earth before 3.800 million years ago. *Nature*, **384**, 55-59.

Mojzsis S.J. and Harrison T.M. (2000) Vestiges of a beginning: clues to the emergent biosphere recorded in the oldest known sedimentary rocks. *GSA Today*, **10**, 1-7.

Monty C.L.V., Westall F., and Van der Gaast S. (1991) Diagenesis of siliceous particles in sub-Antarctic sediments, ODP Leg 114, Hole 699A: possible microbial mediation. In: P.F. Ciesielski, and 18 others. (eds.). *Proceedings of the ODP scientific Research*, **114**. Ocean Drilling Program, College Station, Texas, pp. 685-710.

Muir M.D. and Grant P.R. (1976) Micropalaeontological evidence from the Onverwacht Group, South Africa. *In*: B.F Windley (ed.), *The Early History of Earth*. John Wiley and Sons, London, pp. 595-604.

Pasteris J.D. and Wopenka B. (2002) Images of the Earth's earliest fossils? *Nature*, **420**, 476-477.

Peters K.E. and Moldowan M.J. (1993) *The biomarker guide*. Prentice Hall, New Jersey, USA. 271 – 273.

Posfai M., Buseck P.R., Bazylinski D.A., and Frankel R.B. (1998) Reaction sequence of iron sulphide minerals in bacteria and their use as biomarkers. *Science*, **280**, 880-883.

Renaut R.W., Jones B., and Tiercelin J.-J. (1998) Rapid *in situ* silicification at Loburu hot springs, Lake Bogria, Kenya Rift Valley. *Sedimentology*, **45**, 1083-1103.

Rohmer M., Bouvier-Nave P. and Ourisson G. (1984) Distribution of hopanoids triterpenes in prokaryotes. *J. Gen. Microbio.*, **130**, 1137-1150.

Rosing M.T. (1999) ^{13}C-depleted carbon microparticles in >3700 Ma sea-floor sedimentary rocks from West Greenland. *Science*, **283**, 674-676.

Schidlowski M. (1988) A 3.800 million year isotopic record of life from carbon in sedimentary rocks. *Nature*, **333**, 313-318.

Schidlowski M. (2002) Search for morphological and biogeochemical vestiges of fossil life in extraterrestrial settings: utility of terrestrial evidence. In: G. Horneck and C. Baumstark-Khan (eds.) Astrobiology – *the quest for the conditions of life*.). Springer, Berlin, Heidelberg. pp. 373-386.

Schopf J.W. (1974) The development and diversification of Precambrian life. *Origins of Life*, **5**, 119-135.

Schopf J.W. and Walter M.R. (1983) Archaean microfossils: new evidence of ancient microbes. In: J.W. Schopf (ed.) *Earth's earliest biosphere, it's origin and evolution*. Princeton University Press, pp. 214-239.

Schopf J.W. (1993) Microfossils of the Early Archaean Apex Chert: New evidence of the Antiquity of Life. *Science*, **260**, 640-646.

Schopf, J.W. (1999) Fossils and pseudofossils: lessons from the hunt for early life on Earth. In: *Size Limits of very small Microorganisms – Proceedings of a Workshop*. National Academic Press, Washington D.C., USA, pp. 88-93.

Schopf J.W., Kudryavtsev A.B., Agresti D.G., Wdowiak T.J. and Czaja A.D. (2002) Laser-Raman imagery of Earth's earliest fossils. *Nature*, **416**, 73-76.

Schultze-Lam S., Thompson J.L., and Beveridge T.J. (1993) Metal ion immobilization by bacterial surfaces in fresh water environments. *Water Poll. Res. J. Canada*, **28**, 51-81.

Schultze-Lam S., Ferris F.G., Konhauser K.O., and Wiese R.G. (1995) *In situ* silicification of an Iceland hot spring microbial mat: implications for microfossil formation. *Can. J. Earth Sci.*, **32**, 2021-2026.

Scott E.R.D. (1999) Origin of carbonate-magnetite-sulphide assemblages in Martian meteorite ALH84001. *Jour. Geophys. Res.*, ***104***, 3803-3813.

Scott E. R. D., Yamaguchi A., and Krot A. N. (1997) Petrological evidence for shock melting of carbonates in the Martian meteorite ALH 84001. *Nature*, **387**, 377-379.

Steele A., Goddard D.T., Beech I.B., Tapper R.C., Stapleton D., and Smith. J.R. (1998) Atomic force microscopy imaging of fragments from the Martian meteorite ALH84001. *J. Microsc.*, **189**, 2-7.

Steele A., Goddard D.T., Stapleton D., Toporski J.K.W., Peters V., Bassinger, V., Sharples G., Wynn-Williams D.D., and McKay D.S. (2000) Imaging of an unknown organism on the Martian Meteorite ALH84001. *Met. Plan. Sci.*, **35**, 237-241.

Steele A., Toporski J.K.W., Avci R., Guidry S.A., and McKay D.S. (2001) Time of Flight – Secondary Ion Mass Spectrometry (ToF-SIMS) of a number of bacterial hopanoids. *Org. Geochem.*, **32**, 905-911.

Steele A. and Toporski J. (2003) Astrobiotechnology. *2nd European Exo- and Astrobiology Workshop, Graz, Austria, 16th – 19th September 2002*, ESA Publication Division Noordwijk, The Netherlands, **ESA SP-518**, in press.

Stephan T., Heiss C.H., Rost D., and Jessberger E.K. (1999) Polycyclic aromatic hydrocarbons in meteorites: Allan Hills 84001, Murchison, and Orgueil. *Lunar Planet. Sci. XXX*, Lunar and Planetary Institute, Houston, #1569 (Abstr.).

Summons R.E. and Walter M.R (1990) Molecular fossils and microfossils of prokaryotes and protests from Proterozoic sediments. *Am. J. Sci.*, **290**, 212-244.

Taylor A.P., Barry J.C., and Webb R.I. (2000) Structural and morphological anomalies in magnetosomes: possible biogenic origin for magnetite in ALH84001. *J. Microsc.*, **201**, 84-106.

Thomas-Keprta K.L., Bazylinski D.A., Kirschvink J.L., Clemett S.J., McKay D.S., Wentworth S.J., Vali H., Gibson E.K. Jr. and Romanek C.S. (2000) Elongated prismatic magnetite crystals in ALH84001 carbonate globules: Potential Martian magnetofossils. *Geochim. Cosmochim. Acta*, **64**, 4049-4081.

Thomas-Keprta K.L., Clemett S.J., Bazylinski D.A., Kirschvink J.L., McKay D.S., Wentworth S.J., Vali H., Gibson E.K. Jr., McKay M.F., and Romanek C.S. (2001) Special Feature: truncated hexa-octahedral magnetite crystals in ALH84001: Presumptive biosignatures, *Proc. Natl. Acad. Sci.*, **98**, 2164-2169.

Thomas-Keprta K.L., Clemett S.J., Bazylinski D.A., Kirschvink J.L., McKay D.S., Wentworth S.J., Vali H., Gibson E.K. Jr. and Romanek C.S. (2002) Magnetofossils from ancient Mars: a robust biosignature in the Martian meteorite ALH84001. *Appl. Env. Micro.*, **68**, 3663-3672.

Toporski J. (2001) The preservation and detection of morphological and molecular bacterial biomarkers and their implications for astrobiological research. PhD Thesis, University of Portsmouth, Portsmouth, UK.

Toporski J., Steele A., Westall F., Thomas-Keprta K., and McKay D.S. (2002a) The simulated silicification of bacteria - new clues to the modes and timing of bacterial preservation and implications for the search for extraterrestrial microfossils. *Astrobiology*, **2**, 1-26.

Toporski J., Steele A., Westall F., Avci R., Martill M., and McKay D.S. (2002b) Morphological and spectral investigation of exceptionally well preserved bacterial biofilms from the Oligocene Enspel formation, Germany. *Geochim. Cosmochim. Acta*, **66**, 1773-1791.

Toporski J. and Steele A. (2003) The relevance of bacterial biomarkers in astrobiological research. *2nd European Exo- and Astrobiology Workshop, Graz, Austria, 16th – 19th September 2002*, ESA Publication Division Noordwijk, The Netherlands, **ESA SP-518**, in press.

Treiman A.H. and Romanek C.S. (1998) Bulk and stable isotopic compositions of carbonate minerals in Allan Hills 84001: no proof of high formation temperature. *Meteorit. Planet. Sci.* **33**, 737-742.

Vali H., McKee M.D., Ciftcioglu N., Seras S.K., Plows F.L., Chevet E., Ghiabi P., Plavsic M., Kajander E.O., and Zare R.N. (2001) Nanoform: a new type of protein-associated mineralisation. *Geochim. Cosmochim Acta*, **65**, 63-74.

Valley J.W., Eiler J.M., Graham C.M., Gibson E.K.Jr., Romanek C.S., and Stolper E.M. (1997) Low-temperature carbonate concretions in the Martian meteorites ALH 84001: evidence from stable isotopes and mineralogy. *Science,* **275**, 1633-1638.

Walsh M.M. (1992) Microfossils and possible microfossils from the early Archaean Onverwacht Group, Barberton Mountain Land, South Africa. *Precamb. Res.*, **54**, 271-293.

Walsh M.M. and Lowe D.R. (1985) Filamentous microfossils from the 3,500 Myr. old Onverwacht Group, Barberton Mountain Land, South Africa. *Nature*, **314**, 530-532.

Warren P.H. (1998) Petrologic evidence for low-temperature, possibly flood evaporitic origin of carbonates in the ALH84001 meteorite. *Jour. Geophys. Res.*, **103**, 16759-16773.

Westall F. (1994) Silicified bacteria and associated biofilm from deep-sea sedimentary environment. *Kaupia - Darmstädter Beiträge zur Naturgeschichte*, **4**, 29-43.

Westall F. (1998) The oldest fossil mineral bacteria from the Early Archaean of South Africa and Australia. In: J. Chela-Flores and F. Raulin (eds). *Exobiology: Matter, Energy, and Information in the Origin and Evolution of Life in the Universe.* Kluwer Academic Publishers, Amsterdam. pp.181-186.

Westall F. and Gerneke D. (1998) Electron microscope methods in the search for the earliest life forms on Earth (in 3.5-3.3 Ga old cherts from the Barberton greenstone Belt, South Africa): applications for extraterrestrial life studies. *Proc. SPIE Int. Soc. Opt. Eng.*, **3441**, 158-169.

Westall F. (1999) The nature of fossil bacteria: a guide to the search for extraterrestrial life. *J. Geophys. Res. - Planets*, **104**, 16.437-16,451.

Westall F., Steele A., Toporski J., Walsh M.M., Allen C., Guidry S., McKay D.S., Gibson E., and Chafetz H. (2000) Polymeric substances and biofilms as biomarkers in terrestrial and extraterrestrial materials. *J. Geophys. Res - Planets*, **105**, 24,511-24,527.

Westall F., de Wit M.J., Dann J., van der Gaast S., de Ronde C.E.J., and Gerneke D. (2001) Early Archaean fossil bacteria and biofilms in hydrothermally-influenced, shallow water sediments, Barberton greenstone belt, South Africa, *Precamb. Res.*, **106**, 93-116.

Wimpenny J.W.T. (1995) The spatial organization of biofilms. In: The life and death of biofilm; edited by Wimpenny et al., pp. 1-6, Bioline, Cardiff, Wales.

Zhang C., Vali H., Romaneck C.S., Roh Y., Sears S.K., and Phelps T.J (1999) Chemical and morphological characterisation of siderite formed by iron reducing bacteria. *Lunar Planet. Sci. XXX*, Lunar and Planetary Institute, Houston, #1855 (Abstr.).

Zhmur S.I. and Gerasimenko L.M. (1999) Biomorphic forms in carbonaceous meteorite Allende and possible ecological system - producer of organic matter chondrites. In: R.B. Hoover (ed.) *Instruments, Methods and Missions for Astrobiology II, Proc. SPIE*, **3755**, 48-58.

Zhmur S.I., Rozanov A.Yu., and Gorlenko V.M. (1997) Lithified remnants of microorganisms in carbonaceous chondrites. *Geochem. Internat.*, **35**, 58-60.

Zolotov M. and Shock E. (1999) Abiotic synthesis of polycyclic aromatic hydrocarbons on Mars. *J. Geophys. Res.*, **104 E6**, 14.033-14,050.

Zolotov M.Y. and Shock E.L. (2000) An abiotic origin for hydrocarbons in the Allan Hills 84001 Martian meteorite through cooling of magmatic and impact-generated gases. *Met. Planet. Sci.*, **35**, 629-638.

van Zuilen M., Lepland A. and Arrhenius G. (2002) Reassessing the evidence for the earliest traces of life. *Nature*, **418**, 627-630.

Biodata of **David Wharton** author of ***"Organisms in Extreme Environments: from the Origin and Early Evolution of Life on Earth to Astrobiology"***

Dr. David A. Wharton is an Associate Professor in the Department of Zoology of the University of Otago, New Zealand. The University of Bristol, England awarded him a Ph.D. in 1977 and a D.Sc. in 1997. He is the author of two books: ***A Functional Biology of Nematodes*** (Croom Helm/John Hopkins University Press, 1986) and ***Life at the Limits: Organisms in Extreme Environments*** (Cambridge University Press, 2002). His research focuses upon the mechanisms of cold tolerance and anhydrobiosis.

E-mail: **david.wharton@stonebow.otago.ac.nz**

J. Seckbach (ed.), Origins,607–620.

ORGANISMS IN EXTREME ENVIRONMENTS: FROM THE ORIGIN AND EARLY EVOLUTION OF LIFE ON EARTH TO ASTROBIOLOGY

DAVID A. WHARTON
Department of Zoology, University of Otago, P.O. Box 56, Dunedin, New Zealand.

1. Introduction

Organisms in extreme environments live or survive in conditions that are outside the range experienced and/or tolerated by the majority of organisms (Wharton, 2002). There are two distinct groups of such organisms. Extremophiles grow and reproduce under extreme environmental conditions whilst cryptobiotes survive in a dormant state; growing and reproducing when normal conditions return. These represent contrasting adaptive strategies to extreme conditions. Extremophiles display capacity adaptation whilst cryptobiotes display resistance adaptation (Wharton, 2002). Capacity adaptation enables the organism to grow and reproduce under extreme conditions and is favoured where the environment is constantly extreme. Resistance adaptation allows the organism to survive extreme conditions in a dormant state and is favoured where the environment is periodically extreme.

A number of organisms display resistance adaptation and enter a dormant state in response to environmental stress, for example winter hibernation in mammals and birds (Lyman *et al.*, 1982). Cryptobiosis is distinguished from the other forms of dormancy displayed by tolerant organisms by the cessation of metabolism in the cryptobiotic state. Tolerant organisms do not show the extreme resistance of cryptobiotes. They will survive an environmental stress with a lowered metabolism but there is a limit as to how far metabolism can be depressed before death ensues. Tolerant organisms may grow slowly under an environmental stress or they may cease growing. In contrast, extremophiles require the extreme conditions for growth and reproduction. The three types of organisms can thus be distinguished in terms of their metabolic response to extreme conditions. For extremophiles, extreme conditions are optimal for metabolism. For tolerant organisms, metabolism is optimal under normal conditions but they will survive stresses with a reduced metabolism. For cryptobiotes, metabolism is also optimal under normal conditions but they will survive more extreme stresses than do tolerant organisms by ceasing metabolism. These responses are compared in Fig. 1.

It is difficult to define what is normal and extreme for organisms. Each species will survive in a range of conditions and grow and reproduce under a somewhat narrower

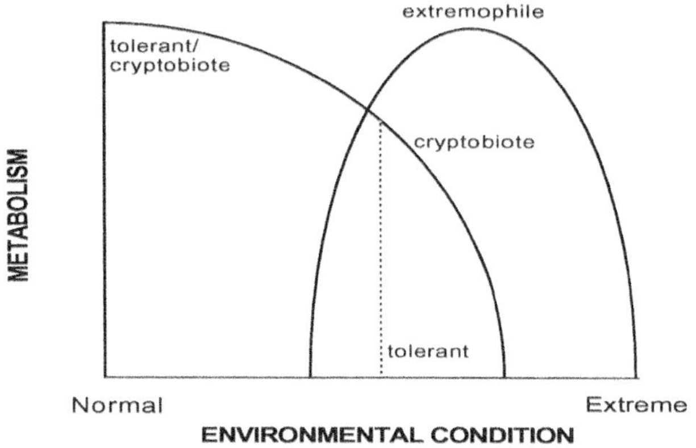

Fig. 1 Changes in metabolism in response to shifts in environmental conditions in extremophiles, cryptobiotes and tolerant organisms. Extremophiles have an optimum rate of metabolism under extreme conditions. For tolerant and cryptobiotic organisms the optimum rate of metabolism occurs under normal conditions. As conditions shift from normal to extreme the rate of metabolism declines. For a tolerant organism if conditions become too extreme metabolism ceases and the organism dies (dotted line). For a cryptobiote metabolism continues to decline to zero as conditions become more extreme and the organism survives.

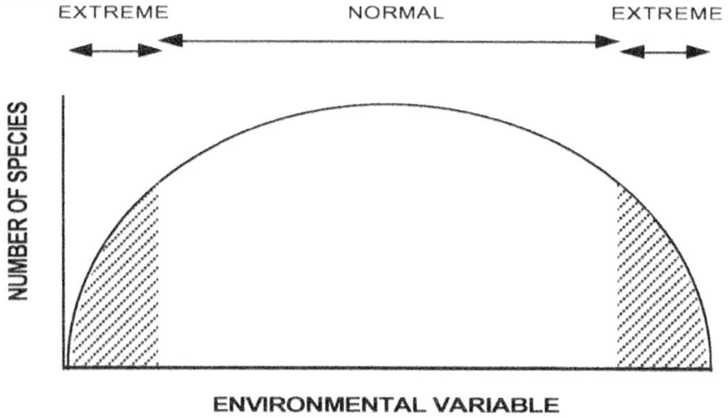

Fig. 2 Normal and extreme (shaded area) defined in terms of the conditions experienced by the majority of organisms.

range. A more objective measure of what is extreme and normal for life on Earth in general might be to compare the numbers of species found in various parts of the range for a particular environmental variable (Fig. 2). For temperature, for example, the majority of organisms survive between -10°C and 45°C: the normal range of surface temperatures on Earth (Cossins and Bowler, 1987). Those surviving outside this range may be considered to be surviving extreme conditions.

There are different physical stresses (high and low temperatures, pressure, high salt concentrations, high and low pH, low oxygen, desiccation) with extremophiles and cryptobiotes inhabiting environments where these stresses are experienced. The classification and terminology of organisms from extreme environments is shown in Table 1.

In the last 30 years or so our knowledge of organisms from extreme environments has increased enormously (Gross, 1998; Horikoshi and Grant, 1998; Postgate, 1994; Rothschild and Mancinelli, 2001; Seckbach, 1999; Wharton, 2002). In this review I will consider how this may inform our speculations on the origin of life on Earth and our search for life elsewhere in the Universe.

TABLE 1. Classification of organisms from extreme environments

Stress	Required for growth and metabolism *extremophilic*	No or slow growth, metabolism slows *tolerant*	No growth, metabolism ceases *cryptobiotic*
Cold			
psychro	psychrophilic	psychrotolerant	
cryo		cold tolerant	cryobiotic
Heat			
thermo	thermophilic	thermotolerant	thermobiotic
Pressure			
piezo	piezophilic	piezotolerant	-
baro	barophilic	barotolerant	
Osmotic			
halo	halophilic	halotolerant	
osmo	osmophilic	osmotolerant	osmobiotic
Acid/Alkali			-
acido	acidophilic	acidotolerant	
alkali	alkaliphilic	alkalitolerant	
No oxygen			
anaero	obligate	facultative	
anoxy	anaerobes	anaerobes	anoxybiotic
Desiccation			
anhydro			anhydrobiotic
xero	xerophilic	xerotolerant	

2. Organisms in Extreme Environments: the Origin and Early Evolution of Life

2.1 TEMPERATURE EXTREMES AND THE ORIGIN OF LIFE

Thermophilic Bacteria and Archaea have optimum growth temperatures of 50-80°C, whilst that of hyperthermophiles is 80-106°C and they grow at up to 113°C (Stetter,

1989). Some molecular phylogenies have placed these organisms close to the root of the tree of life (Di Giulio, 2001; Nisbet and Sleep, 2001), suggesting that the last universal common ancestor was a hyperthermophile and that life arose under hot conditions – in association with hydrothermal vents or after the heating of the Earth following a major meteorite impact (Nisbet and Sleep, 2001).

However, a hot origin for life has been challenged. Evidence has been presented that the last common universal ancestor was likely to be a non hyperthermophile (Brochier and Philippe, 2002; Galtier *et al.*, 1999). Low temperatures favour the long-term survival of organic compounds, particularly the nucleic acids that carry the genetic information (Bada and Lazcano, 2002). The high temperatures of hydrothermal vent systems are likely to destroy, and not create, organic compounds (Miller, 1998). Hot conditions strongly reduce RNA folding (Moulton *et al.*, 2000) and stability (Levy and Miller, 1998; Lindahl, 1967). This suggests a cold origin for life under mesophilic or psychrophilic conditions. A cold origin does not, however, preclude the possibility that organisms may have had to survive periods of high temperatures on Earth, where the survival of thermophiles would be favoured (Gogarten-Boekels *et al.*, 1995).

Psychrophiles have an optimal growth temperature of 15°C or lower, whilst psychrotolerant organisms grow at low temperatures but have an optimal growth temperature at 20-40°C (Madigan *et al.*, 2000). Psychrophiles are quite common, inhabiting the deep sea (-1°C to 4°C) and a variety of polar habitats (Deming, 2002). There are even metabolically-active bacteria at the South Pole (Carpenter *et al.*, 2000). There is evidence that the Earth has experienced periods of intense, and perhaps total, glaciation: as envisaged by the 'Snowball Earth' theory (Hoffman and Schrag, 2002; Kirschvink, 1992). Psychrophilic and psychrotolerant organisms would be in a better position than mesophilic organisms to survive these conditions. Organisms could also survive in a state of cryptobiosis to emerge when milder conditions return. Viable bacteria isolated from permafrost may be millions of years old (Kochkina *et al.*, 2001; Vishnivetskaya *et al.*, 2000).

2.2 ANHYDROBIOSIS AND THE ORIGIN OF LIFE

Anhydrobiosis ('life without water') refers to the phenomenon of cryptobiotic organisms surviving the cessation of metabolism due to desiccation. It is, of course, not possible to demonstrate that metabolism ceases completely during anhydrobiosis, since any technique for measuring metabolism has a limited sensitivity and could not detect a level of metabolism that was occurring in an anhydrobiotic organism at a rate below this limit. However, metabolism, if it is occurring at all, must be at one tenthousandth of normal levels in some anhydrobiotic organisms (Barrett, 1982). It may not be necessary for all water to be lost for metabolism to cease. Metabolism, in the sense of controlled integrated pathways of enzyme-mediated reactions, ceases below a water content of 0.1 g water g^{-1} dry weight (Clegg, 1986; Clegg, 2001).

Some ability to survive desiccation appears to be almost universal amongst prokaryotes (both Bacteria and Archaea) (Potts, 1994). Prokaryotes do, however, vary in their desiccation-survival abilities and it is difficult to make comparisons between species, since different studies have used different desiccation and storage conditions. Survival is better if samples are dried slowly, rather than quickly, and some prokaryotes produce resting stages, cysts or spores that may be more tolerant. For those that were dried slowly (in soil or plant material) survival times are in the range 1 – 300 years

(Potts, 1994). Even longer survival times are claimed for bacteria isolated from permafrost and salt inclusions (Kennedy *et al.*, 1994; Potts, 1994; Vreeland *et al.*, 2000) but this is controversial (Potts, 2001). A wide range of eukaryotic organisms are capable of anhydrobiosis, including: small invertebrate animals (some rotifers, nematodes, tardigrades and arthropod larvae), plants (seeds, pollen, moss, resurrection plants), protists, lichens and fungal spores (Clegg, 2001; Wharton, 2002).

The widespread occurrence of anhydrobiosis, particularly amongst the simpler organisms, led the entomologists Hinton and Blum to suggest that it may be a primitive property and that anhydrobiosis played an important role in the origin and early evolution of life (Hinton and Blum, 1965). Hinton and Blum (1965) thought that life may have arisen in a terrestrial environment, rather than in the sea. Their proposal was inspired by Hinton's studies on *Polypedilium vanderplanki*, one of the few insects and the largest animal, capable of anhydrobiosis (Hinton, 1960). Since life can exist in a purely morphological state (with no metabolism) when free of water, it may have originated by the chance arrangement of dried constituents in rock crevices or pools on land. The alternate flooding and drying of such sites would allow the concentration of chemicals washed out from rocks and the atmosphere. The enormous number of such sites and the variability in their physical and chemical characteristics could have led to the conditions necessary for the origin of life.

Hinton and Blum's proposal finds some echoes in the 'prebiotic beach' (or drying beach or drying lagoon) hypothesis of the origin of life. This envisages that dilute prebiotic organic compounds in the sea became concentrated by evaporation on beaches, or in tidal pools or lagoons, and that this concentration effect allowed more complex products to form (Bernal, 1951; Miller and Orgel, 1974). Cyanoacetaldehyde reacts with urea at high concentrations to produce the pyrimidines cytosine and uracil (Robertson and Miller, 1995), and with guanidine hydrochloride to give diaminopyrimidine, which hydrolyzes to cytosine and uracil, and thiocytosine (Robertson *et al.*, 1996). Modified purines can be produced under similar conditions from adenine and guanine (Levy and Miller, 1999). Other authors have suggested that cycles of desiccation and rehydration could play a role in prebiotic reactions by providing an energy source (Muller, 1995). The Moon was much closer to the early Earth than it is now and tides were up to 30 times larger (Delsemme, 1998). This would provide numerous beaches and tidal pools where concentration by evaporation could play a role in prebiotic reactions, unless the Earth was entirely covered by ocean (McClendon, 1999).

As well as playing a role in prebiotic syntheses, desiccation could also have been important in the functioning of early cells. The cells of all organisms are surrounded by a membrane that separates them from their environment. A lipid membrane is impermeable and organisms have complex transport systems that enable nutrients and metabolites to be carried across the lipid membrane barrier. How could the passage of these materials occur in the first organisms, which presumably lacked transporter systems? When a membrane is dried and then rehydrated it becomes transiently leaky. The removal of water results in a phase change in the membrane phospholipids from a liquid crystalline to a gel state, in which the phospholipid molecules are packed more closely together. Rehydration of the membrane allows the molecules to move apart again and to revert to the liquid crystalline state. The resulting loss of material from the cell, as the membrane becomes leaky, could be fatal for an anhydrobiotic organism. The 'water replacement hypothesis' suggests that the synthesis of trehalose by anhydrobiotes prevents these potentially fatal phase changes by replacing the water molecules associated with their membranes (Crowe *et al.*, 1992). However, in the first organisms, that were surrounded by a soup of prebiotic organic compounds, the phase changes in their membrane lipids would have allowed them to take up materials from their surroundings. Molecules as large as proteins and nucleic acids are captured by liposomes under such conditions (Deamer, 1998). Cycles of desiccation and rehydration could thus have been crucial in the formation and functioning of the first organisms and this is reflected in the widespread occurrence of anhydrobiosis amongst the simpler organisms present today.

2.3 ANAEROBIC AND AEROBIC ORGANISMS

Conditions on Earth were predominantly anaerobic up to about 2.2 billion years ago. Oxygen began to accumulate in the atmosphere following the evolution of photosynthetic organisms and after the supply of oxidizable minerals had been exhausted (Holland, 1994). This does not preclude the possibility that pockets of aerobic conditions occurred much earlier. The first photosynthetic organisms, similar to cyanobacteria, are thought to have evolved earlier than 2 billion years ago (Schopf, 1998). The earliest claimed bacterial fossil assemblage, from the Apex chert, may indicate the evolution of cyanobacterial-like organisms as early as 3.5 billion years ago. The bacterial nature of these fossils is, however, hotly debated (Brasier *et al.*, 2002; Schopf *et al.*, 2002; van Zullen *et al.*, 2002).

Aerobic catabolism is of considerable advantage to organisms since it yields 13 to 38 times as much energy for a given amount of food than does the anaerobic catabolism of sugars (Dawes, 1986). However, oxygen metabolism produces highly reactive molecules and ions, such as hydrogen peroxide, hydroxyl radicals and superoxide ions (Cadenas, 1995). These need to be detoxified by aerobic organisms. It could thus be argued that all aerobic organisms are extremophiles (Rothschild and Mancinelli, 2001). However, I have suggested that 'normal' and 'extreme' conditions should be defined in terms of the conditions experienced by the majority of species of organisms (Wharton, 2002). The vast majority of species of organisms today are aerobic, since oxygen is pervasive in the atmosphere and oceans of the world. To be aerobic is thus clearly normal, under present-day Earth conditions, and most obligate aerobic organisms die if deprived of oxygen (a few are capable of cryptobiosis e.g. Clegg, 1997). Obligate anaerobic organisms are restricted to the few remaining environments that provide the conditions they need to survive – the absence of oxygen in deep muds and sediments, the intestines of vertebrate animals and deep underground. It could be argued that such habitats, although not as common as aerobic environments, are fairly common and so perhaps we should consider neither aerobes nor anaerobes to be organisms from extreme environments. If we define extreme in terms of the conditions experienced by the majority of organisms (Fig. 2), however, we should consider anaerobic conditions extreme – at least on the present-day Earth.

On the early Earth, however, conditions were predominantly anaerobic. The first aerobic organisms must have evolved in proximity to the oxygen-producing photosynthetic organisms (which were, of course, aerobic themselves) but before oxygen became widespread. What is normal and extreme has thus changed through Earth history as conditions and organisms have changed.

3. Organisms in Extreme Environments and Astrobiology

The habitable zone around a star is the region in which an Earth-like planet may have surface temperatures capable of supporting life (Franck *et al.*, 2001). Our increasing knowledge of the extreme conditions under which some organisms can grow and survive expands our estimates of the sizes of habitable zones. The study of organisms from extreme environments, observations on planets and their moons by space probes, and improvements in the instrumentation and techniques available to astronomers have all contributed to a renewed interest in the search for life in our own solar system. The two most likely candidates are Mars and Europa.

3.1 MARS

Mars is a cold, dry desert world. Its average surface temperature is -53°C but can rise above 0°C near the equator. The polar ice caps consist mainly of solid carbon dioxide, which melts seasonally, but this covers a permanent ice cap of water ice. The atmosphere is thin, consisting mainly of carbon dioxide and with an atmospheric pressure about 1/100 that of Earth. Water is present in the atmosphere, as ice crystals and vapour. Liquid water is unlikely to persist at the surface given the low vapour pressure in the atmosphere (Jakosky, 1998).

Gamma-ray spectroscopy has been used to map the distribution of hydrogen on Mars and is thought to indicate the presence of widespread subsurface water ice (Boynton *et al.*, 2002). There is abundant evidence for the activity of liquid water at the Martian surface at some time in its history (Baker, 2001) and images from the Mars Global Surveyor's Mars Orbital Camera indicate the presence of recent (and perhaps current) water flows at the surface (Malin and Edgett, 2000). The subsurface water reserves are likely to be largely frozen. There is, however, evidence of past volcanism on Mars (Jakosky, 1998) and if hydrothermal activity is present today these sites may contain life. The most likely scenario for life is in subsurface areas where the Martian permafrost is melted by the internal heat of the planet (Walter, 1999). Microbes on Earth have been found at a depth of 5278 m (Pedersen, 2000). Methanogenic microbes that utilize the oxidation of hydrogen and the reduction of carbon dioxide as an energy source are a possibility. Such communities of organisms have been found on Earth (Chapelle *et al.*, 2002).

The soils of the Dry Valleys of Antarctica provide the closest analogue on the surface of Earth to conditions on Mars (Wynn-Williams and Edwards, 2000). The Dry Valleys have an annual temperature average of -20°C and precipitation of less than 10 cm water equivalent. Soil surface temperatures are above 0°C during summer but evaporation is so rapid that the permafrost, which is at 10-30 cm depth, cannot supply sufficient moisture to support the microbial growth that would stabilize the surface (Treonis *et al.*, 1999; Virginia and Wall, 1999). Despite this, Dry Valley soils contain a community of organisms; consisting of bacteria, fungi, protists, nematodes, rotifers and tardigrades. The abundance of these organisms seems to be driven by factors such as salinity and the legacy of carbon derived from the productivity of ancient lakes, rather than the moisture content of the soils (Virginia and Wall, 1999). The survival of these organisms is critically dependant upon their anhydrobiotic abilities (Treonis *et al.*, 2000).

3.2 EUROPA

Europa is one of the moons of Jupiter. Its surface is dominated by water ice and has features that indicate fracturing and the activity of liquid water. Tidal heating, from the gravitational pull of Jupiter, may be sufficient to produce liquid water and there is evidence, such as magnetic field data, that suggests the presence of a subsurface ocean (Showman and Malhotra, 1999). Analysis of areas of impact craters indicate that the ice is 3 - 4 km thick in those locations (Turtle and Pierazzo, 2001). There is also evidence for the presence of water on Callisto and Ganymede (Showman and Malhotra, 1999).

If liquid water is present on Europa so may be life. Pockets of sea ice on Earth can remain liquid down to -35°C and metabolic activity in psychrophilic bacteria has been demonstrated to -5°C in glacial and lake ice, to -17°C in Antarctic snow-ice and to -20°C in Siberian tundra (Deming, 2002). Lake Vostok, beneath the Antarctic Vostok Station, is the largest of over 80 subglacial lakes that have been discovered lying between the Antarctic ice sheet and the substrate. It is covered by up to 4,000 m of ice (Siegert, 2000). Lake Vostok provides the closest analogue on Earth to the situation on Europa (Wynn-Williams and Edwards, 2000). Ice from 3590 m below Vostok Station is thought to be accreted from the liquid water associated with Lake Vostok. It contains microbes, which may indicate microbial populations in the Lake (Karl *et al.*, 1999; Priscu *et al.*, 1999). However, microbes from the accreted ice are similar to present-day microbes. They may thus be derived from deep glacial ice and are not representative of the microbiota of Lake Vostok, which has yet to be sampled (Siegert, 2000). The microbes of Lake Vostok, if they exist, are remote from the solar energy that powers photosynthesis and they must use chemical energy sources.

3.3 PANSPERMIA AND INTERPLANETARY TRANSFER

In the desiccated state anhydrobiotic organisms are capable of surviving extreme environmental stresses that would be fatal to them if they were hydrated (Keilin, 1959). Tardigrades, microscopic invertebrate animals, will survive exposure to within a few degrees of absolute zero, temperatures up to 125°C, high salinities, organic solvents, high and low pressure and high levels of ionizing radiation (Copley, 1999; Wright, 2001). These survival abilities have led some to suggest that they could survive exposure to space (Copley, 1999).

Although the ability of anhydrobiotic animals to survive in space has yet to be tested, some microbes have survived such exposure. *Bacillus subtilis* spores are one of the most resistant of organisms, surviving a variety of stresses in a state of cryptobiosis. They will survive wet heat (100°C), dry heat, extended periods of desiccation, high vacuum, toxic chemicals, high pressure and they are significantly more resistant to UV and gamma radiation than are actively growing cells (Nicholson *et al.*, 2000). These spores are the most popular model for space exposure where they have survived for nearly six years (Horneck *et al.*, 1994). The vegetative cells of the halophilic cyanobacterium *Synechococcus* sp. and the halophilic archaeon *Haloarcula* sp. have also survived exposure to space (Mancinelli *et al.*, 1998).

Although *B. subtilis* spores survive the vacuum of space well, they are rapidly killed if exposed to the full spectrum of solar UV radiation (Horneck *et al.*, 1984). They survive, however, if shielded from UV radiation during space exposure by clay, rock or meteorite material (Horneck *et al.*, 2001). Cosmic radiation (HZE particles) is fatal to spores but a large meteorite may provide sufficient shielding to allow spores to survive for a million years (Mileikowsky *et al.*, 2000).

The theory of Panspermia suggests that life in the form of spores can be dispersed from one planet, or solar system, to another (Arrhenius, 1903). Given the vulnerability of spores to radiation the transfer of isolated spores from planet to planet seems unlikely. Large impacts may eject rocks that eventually fall as meteorites on other planets. The ability of bacterial spores to survive exposure to space, if protected from radiation, suggests such meteorites as a potential mode of interplanetary transfer of organisms (Clark, 2001; Mileikowsky *et al.*, 2000).

4. Conclusions

Extremophiles grow and function in extreme environments whilst cryptobiotes survive periodic exposure to extreme conditions in an ametabolic dormant state. Several scenarios for the origin of life suggest the involvement of organisms from extreme environments; thermophiles, psychrophiles and/or anhydrobiotes; in the earliest stages of the evolution of life on Earth. Studies on the biology of organisms from extreme environments have dramatically expanded our understanding of the conditions that organisms can survive and of the environments where life is possible. This makes life seem possible not just under relatively mild conditions on Earth but also under the much harsher conditions likely to be found on Mars and Europa. The survival abilities of cryptobiotes suggest that organisms could even survive the rigors of interplanetary transfer.

5. References

Arrhenius, S. (1903) Die Verbreitung des Lebens im Weltenraum. Die Umschau **7**, 481-485.

Bada, J.L. and Lazcano, A. (2002) Origin of life - Some like it hot, but not the first biomolecules. Science **296**, 1982-1983.

Baker, V.R. (2001) Water and the martian landscape. Nature **412**, 228-236.

Barrett, J. (1982) Metabolic responses to anabiosis in the fourth stage juveniles of *Ditylenchus dipsaci* (Nematoda). Proceedings of the Royal Society of London **B216**, 159-177.

Bernal, J.D. (1951) *The Physical Basis of Life.* Routledge and Kegan Paul, London.

Boynton, W.V., Feldman, W.C., Squyres, S.W., Prettyman, T.H., Bruckner, J., Evans, L.G., Reedy, R.C., Starr, R., Arnold, J.R., Drake, D.M., Englert, P.A.J., Metzger, A.E., Mitrofanov, I., Trombka, J.I., d'Uston, C., Wanke, H., Gasnault, O., Hamara, D.K., Janes, D.M., Marcialis, R.L., Maurice, S.,

Mikheeva, I., Taylor, G.J., Tokar, R. and Shinohara, C. (2002) Distribution of hydrogen in the near surface of Mars: Evidence for subsurface ice deposits. Science **297**, 81-85.

Brasier, M.D., Green, O.R., Jephcoat, A.P., Kleppe, A.K., Van Kranendonk, M.J., Lindsay, J.F., Steele, A. and Grassineau, N.V. (2002) Questioning the evidence for Earth's oldest fossils. Nature **416**, 76-81.

Brochier, C. and Philippe, H. (2002) A non-hyperthermophilic ancestor for Bacteria. Nature **417**, 244.

Cadenas, E. (1995) Mechanisms of oxygen activation and reactive oxygen species detoxification, in: S. Ahmed (ed.) *Oxidative Stress and Antioxidant Defences in Biology*, Chapman & Hall, New York, pp. 1-61.

Carpenter, E.J., Lin, S.J. and Capone, D.G. (2000) Bacterial activity in South Pole snow. Applied and Environmental Microbiology **66**, 4514-4517.

Chapelle, F.H., O'Neill, K., Bradley, P.M., Methé, B.A., Ciufo, S.A., Knobel, L.L. and Lovley, D.R. (2002) A hydrogen-based subsurface microbial community dominated by methanogens. Nature **415**, 312-314.

Clark, B. (2001) Planetary interchange of bioactive material: Probability factors and implications. Origins of Life and Evolution of the Biosphere **31**, 185-197.

Clegg, J.S. (1986) The physical properties and metabolic status of *Artemia* cysts at low water contents: the water replacement hypothesis, in: A.C. Leopold (ed.) *Membranes, Metabolism and Dry Organisms*, Cornell University Press, New York, pp. 169-187.

Clegg, J.S. (1997) Embryos of *Artemia franciscana* survive four years of continuous anoxia: the case for complete metabolic rate depression. Journal of Experimental Biology **200**, 467-475.

Clegg, J.S. (2001) Cryptobiosis - a peculiar state of biological organization. Comparative Biochemistry and Physiology B-Biochemistry and Molecular Biology **128**, 613-624.

Copley, J. (1999) Indestructible. New Scientist **23 October 1999**, 44-46.

Cossins, A.R. and Bowler, K. (1987) *Temperature Biology of Animals.* Chapman & Hall, London.

Crowe, J.H., Hoekstra, F. and Crowe, L.M. (1992) Anhydrobiosis. *Annual Review of Physiology* **54**, 579-599.

Dawes, E.A. (1986) *Microbial Energetics.* Blackie, Glasgow.

Deamer, D.W. (1998) Membrane compartments in prebiotic evolution, in: A. Brack (ed.) *The Molecular Origins of Life*, Cambridge University Press, Cambridge, pp. 189-205.

Delsemme, A.H. (1998) Cosmic origin of the biosphere, in: A. Brack (ed.) *The Molecular Origins of Life*, Cambridge University Press, Cambridge, pp. 100-118.

Deming, J.W. (2002) Psychrophiles and polar regions. Current Opinion in Microbiology **5**, 301-309.

Di Giulio, M. (2001) The universal ancestor was a thermophile or a hyperthermophile. Gene **281**, 11-17.

Franck, S., Block, A., von Bloh, W., Bounama, C., Garrido, I. and Schellnhuber, H.J. (2001) Planetary habitability: is Earth commonplace in the Milky Way? Naturwissenschaften **88**, 416-426.

Galtier, Z., Tourasse, N. and Gouy, M. (1999) A non-hyperthermophilic common ancestor to extant life forms. Science **283**, 220-221.

Gogarten-Boekels, M., Hilario, E. and Gogarten, J.P. (1995) The effects of heavy meteorite bombardment on the early evolution - the emergence of the three domains of life. Origins of Life and Evolution of the Biosphere **25**, 251-264.

Gross, M. (1998) *Life on the Edge.* Plenum Press, New York.

Hinton, H.E. (1960) Cryptobiosis in the larva of *Polypedilium vanderplanki* Hint. Journal of Insect Physiology **5**, 286-300.

Hinton, H.E. and Blum, M.S. (1965) Suspended animation and the origin of life. New Scientist **28 October 1965**, 270-271.

Hoffman, P.F. and Schrag, D.P. (2002) The snowball Earth hypothesis: testing the limits of global change. Terra Nova **14**, 129-155.

Holland, H.D. (1994) Early proterozoic atmospheric change, in: S. Bengston (ed.) *Early Life on Earth, Nobel Symposium No. 84*, Columbia University Press, New York, pp. 237-244.

Horikoshi, K. and Grant, W.D. (1998). *Microbial Life in Extreme Environments.* Wiley-Liss, New York.

Horneck, G., Bücker, H. and Reitz, G. (1994) Long-term survival of bacterial spores in space. Advances in Space Research **14**, (10)41-(10)45.

Horneck, G., Bücker, H., Reitz, G., Requardt, H., Dose, K., Martens, K.D., Mennigmann, H.D. and Weber, P. (1984) Microorganisms in the space environment. Science **225**, 226-228.

Horneck, G., Rettberg, P., Reitz, G., Wehner, J., Eschweiler, U., Strauch, K., Panitz, C., Starke, V. and Baumstark-Khan, C. (2001) Protection of bacterial spores in space, a contribution to the discussion on Panspermia. Origins of Life and Evolution of the Biosphere **31**, 527-547.

Jakosky, B. (1998) *The Search for Life on Other Planets.* Cambridge University Press, Cambridge.

Karl, D.M., Bird, D.F., Bjorkman, K., Houlihan, T., Shackelford, R. and Tupas, L. (1999) Microorganisms in the accreted ice of Lake Vostok, Antarctica. Science **286**, 2144-2147.

Keilin, D. (1959) The Leeuwenhoek Lecture. The problem of anabiosis or latent life : history and current concept. Proceedings of the Royal Society of London **B150**, 149-191.

Kennedy, M.J., Leader, S.L. and Swierczynski, L.M. (1994) Preservation records of micro-organisms: Evidence of the tenacity of life. Microbiology **140**, 2513-2529.

Kirschvink, J.L. (1992) Late Proterozoic low-latitude global glaciation: the snowball earth, in: J.W. Schopf and C. Klein (eds.) *The Proterozoic Biosphere*, Cambridge University Press, Cambridge, pp. 51-52.

Kochkina, G.A., Ivanushkina, N.E., Karasev, S.G., Gavrish, E.Y., Gurina, L.V., Evtushenko, L.I., Spirina, E.V., Vorob'eva, E.A., Gilichinskii, D.A. and Ozerskaya, S.M. (2001) Survival of micromycetes and actinobacteria under conditions of long-term natural cryopreservation. Microbiology **70**, 356-364.

Levy, M. and Miller, S.L. (1998) The stability of the RNA bases: implications for the origin of life. Proceedings of the National Academy of Science USA **95**, 7933-7938.

Levy, M. and Miller, S.L. (1999) The prebiotic synthesis of modified purines and their potential role in the RNA world. Journal of Molecular Evolution **48**, 631-637.

Lindahl, T. (1967) Irreversible heat inactivation of transfer ribonucleic acids. Journal of Biological Chemistry **242**, 1970-1973.

Lyman, C.P., Willis, J.P., Malan, E. and Wang, L.C.H. (1982) *Hibernation and Torpor in Mammals and Birds*. Academic Press, London.

Madigan, M.T., Martinko, J.M. and Parker, J. (2000) *Brock Biology of Microorganisms*. 9th ed., Prentice Hall, Upper Saddle River, NJ.

Malin, M.C. and Edgett, K.S. (2000) Evidence for recent groundwater seepage and surface runoff on Mars. Science **288**, 2330-2335.

Mancinelli, R.L., White, M.R. and Rothschild, L.J. (1998) Biopan-survival I. Exposure of the osmophiles *Synechococcus* sp. (Nageli) and *Haloarcula* sp. to the space environment. Advances in Space Research **22**, 327-334.

McClendon, J.H. (1999) The origin of life. Earth-Science Reviews **47**, 71-93.

Mileikowsky, C., Cucinotta, F.A., Wilson, J.W., Gladman, B., Horneck, G., Lindegren, L., Melosh, J., Rickman, H., Valtonen, M. and Zheng, J.Q. (2000) Risks threatening viable transfer of microbes between bodies in our solar system. Planetary and Space Science **48**, 1107-1115.

Miller, S.L. (1998) The endogenous synthesis of organic compounds, in: A. Brack (ed.) *The Molecular Origins of Life*, Cambridge University Press, Cambridge, pp. 59-85.

Miller, S.L. and Orgel, L.E. (1974) *The Origin of Life on Earth*. Prentice-Hall, New Jersey.

Moulton, V., Gardner, P.P., Pointon, R.F., Creamer, L.K., Jameson, G.B. and Penny, D. (2000) RNA folding argues against a hot-start origin of life. Journal of Molecular Evolution **51**, 416-421.

Muller, A.W.J. (1995) Were the first organisms heat engines - a new model for biogenesis and the early evolution of biological energy-conversion. Progress in Biophysics and Molecular Biology **63**, 193-231.

Nicholson, W.L., Munakata, N., Horneck, G., Melosh, H.J. and Setlow, P. (2000) Resistance of *Bacillus* endospores to extreme terrestrial and extraterrestrial environments. Microbiology and Molecular Biology Reviews **64**, 548-572.

Nisbet, E.G. and Sleep, N.H. (2001) The habitat and nature of early life. Nature **409**, 1083-1091.

Pedersen, K. (2000) Exploration of deep intraterrestrial microbial life: current perspectives. FEMS Microbiology Letters **185**, 9-16.

Postgate, J.R. (1994) *The Outer Reaches of Life*. Cambridge University Press, Cambridge.

Potts, M. (1994) Desiccation tolerance of prokaryotes. Microbiological Reviews **58**, 755-805.

Potts, M. (2001) Desiccation tolerance: a simple process? Trends in Microbiology **9**, 553-559.

Priscu, J.C., Adams, E.E., Lyons, W.B., Voytek, M.A., Mogk, D.W., Brown, R.L., McKay, C.P., Takacs, C.D., Welch, K.A., Wolf, C.F., Kirshtein, J.D. and Avci, R. (1999) Geomicrobiology of subglacial ice above Lake Vostok, Antarctica. Science **286**, 2141-2144.

Robertson, M.P., Levy, M. and Miller, S.L. (1996) Prebiotic synthesis of diaminopyrimidine and thiocytosine. Journal of Molecular Evolution **43**, 543-550.

Robertson, M.P. and Miller, S.L. (1995) An efficient prebiotic synthesis of cytosine and uracil. Nature **375**, 772-774.

Rothschild, L.J. and Mancinelli, R.L. (2001) Life in extreme environments. Nature **409**, 1092-1101.

Schopf, J.W. (1998) Tracing the roots of the universal tree of life, in: A. Brack (ed.) *The Molecular Origins of Life*, Cambridge University Press, Cambridge, pp. 336-362.

Schopf, J.W., Kudryavtsev, A.B., Agresti, D.G., Wdowiak, T.J. and Czaja, A.D. (2002) Laser-Raman imagery of Earth's earliest fossils. Nature **416**, 73-76.

Seckbach, J. (1999). *Enigmatic Microorganisms and Life in Extreme Environments*. Dordrecht: Kluwer Academic Publishers.

Showman, A.P. and Malhotra, R. (1999) The Galilean satellites. Science **286**, 77-84.

Siegert, M.J. (2000) Antarctic subglacial lakes. Earth-Science Reviews **50**, 29-50.
Stetter, K.O. (1989) Extremely thermophilic chemolithoautotrophic archaebacteria., in: H.G. Schlegel and B. Bowien (eds.) *Autotrophic Bacteria*, Science Tech Publishers and Springer-Verlag, Berlin, pp. 167-171.
Treonis, A.M., Wall, D.H. and Virginia, R.A. (1999) Invertebrate biodiversity in Antarctic Dry Valley soils and sediments. Ecosystems **2**, 482-492.
Treonis, A.M., Wall, D.H. and Virginia, R.A. (2000) The use of anhydrobiosis by soil nematodes in the Antarctic Dry Valleys. Functional Ecology **14**, 460-467.
Turtle, E.P. and Pierazzo, E. (2001) Thickness of a European ice shell from impact crater simulations. Science **294**, 1326-1328.
van Zullen, M.A., Lepland, A. and Arrhenius, G. (2002) Reassessing the evidence for the earliest traces of life. Nature **418**, 627-629.
Virginia, R.A. and Wall, D.H. (1999) How soils structure communities in the Antarctic Dry Valleys. BioScience **49**, 973-983.
Vishnivetskaya, T., Kathariou, S., McGrath, J., Gilichinsky, D. and Tiedje, J.M. (2000) Low-temperature recovery strategies for the isolation of bacteria from ancient permafrost sediments. Extremophiles **4**, 165-173.
Vreeland, R.H., Rosenzweig, W.D. and Powers, D.W. (2000) Isolation of a 250 million-year-old halotolerant bacterium from a primary salt crystal. Nature **407**, 897-900.
Walter, M. (1999) *The Search for Life on Mars.* Perseus Books, Cambridge, Massachusetts.
Wharton, D.A. (2002) *Life at the Limits: Organisms in Extreme Environments.* Cambridge University Press, Cambridge.
Wright, J.C. (2001) Cryptobiosis 300 years on from van Leeuwenhoek: What have we learned about tardigrades? Zoologischer Anzeiger **240**, 563-582.
Wynn-Williams, D.D. and Edwards, H.G.M. (2000) Antarctic ecosystems as models for extraterrestrial surface habitats. Planetary and Space Science **48**, 1065-1075.

Biodata of **Victor R. Baker** author of ***"Geological History of Water on Mars"***

Dr. Victor R. Baker is Regents Professor and Head of Department, Hydrology and Water Resources, and also Professor of Planetary Sciences, The University of Arizona. He earned his Ph.D. from the University of Colorado (1971) and subsequently rose to the rank of Professor of Geological Sciences, The University of Texas at Austin (1971-1981). At the University of Arizona since 1981, Dr. Baker has also been a Visiting Research Fellow (Fulbright-Hays Senior Scholar) at The Australian National University (1979-80), a Visiting Research Fellow at National Institute of Hydrology, India (1987-88), a Visiting Fellow at the University of Adelaide, Australia (1988), and Adjunct Professor, Anna University, India (2002). He was the 1998 President of the Geological Society of America. He is a Fellow of the American Geophysical Union, and Honorary Fellow of the European Union of Geosciences, and a Foreign Member of the Polish Academy of Sciences.

Dr. Baker is an author or editor of 14 books or monographs, and he has generated over 250 scientific papers/articles and 350 abstracts/presentations concerned with planetary geology, flood geomorphology, paleoflood hydrology, Quaternary geology, and history/philosophy of the Earth and planetary sciences. He has been honored with the Don J. Easterbrook Distinguished Scientist Award of The Geological Society of America (2001), the Grand Prix International de Cannes de l'Eau et de la Science (2001), and the David Linton Award of the British Geomorphological Research Group (1995).

E-mail: **baker@hwr.arizona.edu**

J. Seckbach (ed.), Origins, 621–631.

A BRIEF GEOLOGICAL HISTORY OF WATER ON MARS

VICTOR R. BAKER
Department of Hydrology and Water Resources
The University of Arizona
Tucson, Arizona 85721-0011 USA

1. Introduction

The history of water on Mars has proven to be one of the most controversial topics in all science. Past theories have ranged from (1) speculations that a water-rich ancient history afforded opportunities for the evolution of life to the level of inhabitants capable of planetary-scale engineering (Lowell 1908), to (2) physics-based theories that Mars has always been both cold and dry up to the present day (Leighton and Murray 1966). The discovery of channels and valleys on Mars in the early 1970s led to considerable theoretical work to explain their existence, including theories that deny any role for water in their formation (Hoffman 2000). The theoretical controversy continues, with various atmospheric models for early Mars proposing (1) warm, wet conditions generated by an intense CO_2 greenhouse (Pollack et al. 1987), (2) denial that such a greenhouse is possible (Kasting 1991), so that geothermal heat is necessary to produce temporary water flow for valley formation (Squyres and Kasting 1994), or (3) that high impact rates early in Mars history explain the release of water for valley formation (Segura et al. 2002).

In contrast to the theoretical debates, the geomorphological and geological interpretations of space mission imagery have long indicated that Mars, like Earth, is a profoundly water-rich planet, on which aqueous processes have been active, perhaps sporadically, throughout its history (Baker 1982, 2001; Masson et al. 2001). The landforms that lead to this conclusion have been interpreted to have fluvial, lacustrine, littoral, glacial, wet mass-movement, ice-rich permafrost, and volcano-ice-water origins. Of course, these interpretations have all been intensely debated (e.g. Carr 1996), and alternative, nonaqueous explanations have been proposed for nearly every land form to which aqueous origins have been ascribed. While all these explanations are certainly reasonable for isolated *ad hoc* instances, they do not provide any sense of a unifying theme as to how the Martian landscape works. They are united only by their denial of a significant role for liquid water and/or greatly changed past Martian climates. This view may be partly motivated by a Mars-specific version of the uniformitarian principle, which holds that the present-day, cold-dry Mars conditions (7 millibar atmospheric pressure, mean annual temperature of about -60^0 C and colder) are better postulated for the Martian past than are any speculations about warmer, wetter conditions. Though logically flawed, this mode of reasoning has a long history of influence in geological thinking (Baker 1998). A potentially more fruitful approach for interpreting the Martian landscape is somewhat provocatively stated as follows: if this entire assemblage of (apparently) water-related landforms, as manifested both in time and spatial association (these being demonstrated by geological mapping), were observed in some newly discovered region of Earth, there would be absolutely no question in regard to its aqueous origins. It is in this spirit that this brief review is pursued.

In outlining a geological history of water on Mars it is essential to distinguish an "early Mars" epoch during which impacting rates by meteors and comets were much higher than afterward. This epoch, also known as the Noachian, is currently considered to have begun just after accretion of the planet, extending from 4.5 to about 3.8 billion years ago (Hartmann and Neukum 2001). The early Mars (Noachian) epoch is best recorded in the heavily cratered highlands of the planet, mostly in equatorial and southern latitudes. There is also evidence that the low-lying northern plains of Mars are underlain by large impact basins that were emplaced during this early heavy bombardment period (Frey et al. 2002), but these are now extensively buried by younger lava flows and/or sediments. Based on a selective interpretation of the Mars fluvial history, a common view, especially among theoreticians, is that nearly all the aqueous activity on Mars was concentrated into this early epoch. For convenience I will label this the MIDDEN hypothesis (**M**ars **I**s continuously **D**ead and **D**ry, **E**xcept during the **N**oachian).

2. Fluvial Activity

The two main varieties of fluvial landforms on Mars are valley networks and outflow channels, morphological attributes of which are reviewed by Baker et al. (1992). A great many of the valley networks occur in the old cratered highlands of Mars, leading to the view that nearly all of them formed during the heavy bombardment, as presumed by the MIDDEN hypothesis. The outflow channels, in contrast, involve the immense upwelling of cataclysmic flood flows from subsurface sources (Baker and Milton, 1974), mostly during post-Noachian periods of Martian history. The transition from a more aqueous phase in the Noachian, with a progressively thickening ice-rich permafrost zone in post-Noachian time, is the basis for theories that explain the outflow channels as products of subsurface water confined by this process (Carr 1979, 1998; Clifford and Parker 2001). Certainly, there is strong evidence, notably from impact crater morphologies (Carr 1996), that much of the Martian surface is underlain by a thick ice-rich permafrost zone, a "cryolithosphere" (Kuzmin et al. 1988). Nevertheless, the geological record shows that highland valley formation extended into the period after the heavy bombardment (Baker and Partridge 1986, Grant 2000), and much younger valley networks are extensively developed on some Martian volcanoes (Gulick and Baker 1989, 1990). Very young post-Noachian fluvial erosion is also indicated in the extensive mid-latitude area that is mantled by the Medusa Fossae Formation, which is itself the likely product of explosive volcanism (Bradley et al. 2002). Mars has many other areas of post-Noachian volcano-ice-water interactions, some of which are associated with channels and valleys.

An alternative to the linear model of cryosphere thickening is that outflow channel activity is related to episodic heat flow and volcanism (Baker et al. 1991). This hypothesis is now known by the acronym MEGAOUTFLO: **M**ars **E**pisodic **G**lacial **A**tmospheric **O**ceanic **U**pwelling by **T**hermotectonic **F**lood **O**utburst. It envisions long periods (perhaps on the order of 10^8 years) in which Mars has a stable atmosphere that is cold and dry like that of today, with nearly all its water trapped as ground ice and underlying ground water. The stable state is punctuated by relatively short-duration (perhaps 10^4 or 10^5 years) episodes of quasi-stable conditions that are warmer and wetter than those at present. The motivation for MEGAOUTFLO is the observed reality of water-related landforms briefly described in this review, especially those of post-Noachian age. Extensive criticism of the hypothesis by Carr (1996) questioned the significance of the water-related landforms and denied that significant epochs of climate change occurred after the heavy bombardment, as presumed by the MIDDEN hypothesis.

The MIDDEN hypothesis is itself under attack for its presumption of warmer atmospheric conditions during the Noachian. The criticisms arise from the role of internal geothermal heating for valley formation (e.g., Gulick 2001). However, it is clear that extensive geothermal heating will also impact the atmosphere by its injection of water vapor and other gases. Perhaps the strongest argument that early Mars cannot have been continuously cold and dry is that highland craters and basins are extensively eroded, most likely by processes involving rainfall and surface runoff (Craddock and Maxwell 1993; Craddock and Howard 2002). Prolonged, intense fluvial erosion occurred, with cratering competing with drainage basin development, such that the latter was restricted to localized areas (Irwin and Howard 2002). Relatively high denudation rates are inferred for the Noachian, which are much greater than those of later periods (Golombek and Bridges 2000). These observations are consistent with the discovery that the ancient Martian crust of the highlands is layered to considerable depths, probably because sedimentary rocks were emplaced during the intense denudation phase (Malin and Edgett 2000a). Imagery from the Mars Orbiter Camera (MOC) of the Mars Global Surveyor (MGS) Mission shows that the Martian highlands do not consist of an initial lunar-like surface, underlain by an impact-generated megaregolith, as presumed in previous hydrogeological models (Clifford 1993; Carr 1996). Instead, cratering, fluvial erosion, and deposition of layered materials probably all occurred contemporaneously, leading to a complex interbedding of lava flows, igneous intrusions, sediments, buried crater forms, and erosional unconformities (Malin and Edgett 2001).

3. Lakes, Seas, and an "Ocean"

Evidence for persistent standing bodies of water on Mars is abundant, but also more controversial than that for fluvial activity. On Earth such bodies of water include (1) lakes, in which the water is surrounded by extensive land areas, (2) seas, for which saline waters cover the greater part of the planetary surface, and (3) the ocean, which is the vast, interconnected body of water that covers about 70% of Earth's surface. For Mars there is no direct geomophological evidence that the majority of its surface was ever covered by standing water, though the term "ocean" has been applied to temporary ancient inundations of the northern plains, which did not persist through the whole history of the planet. Although initially inferred from sedimentary landforms on the northern plains (e.g., Lucchitta et al. 1986), inundation of the northern plains has been most controversially tied to identifications of "shorelines" made by Parker et al. (1989, 1993). Failure to confirm some of the shoreline landforms on the newer MOC imagery led some to reject the ocean hypothesis (Malin and Edgett 1999, 2001), though it is only various shoreline interpretations that can be rejected in this manner, not the hypothesis of plains indundation. Nevertheless, the MGS data confirm the initial observations of a regionally mantling layer of sediment, now called the Vastitas Borealis Formation, covering perhaps 3 x 10^7 km^2 of the northern plains (Head et al. 2002). This sediment is contemporaneous with the post-Noachian outflow channels, and it was likely emplaced as the sediment-laden outflow channel discharges became hyperpycnal flows upon entering water ponded water on the plains (Ivanov and Head 2001). In another scenario, Clifford and Parker (2001) envision a Noachian "ocean," contemporaneous with the highlands valley networks, and fed by a great fluvial system extending from the south polar cap, through Argyre and the Chryse Trough, to the northern plains.

Though the debate over the Martian "ocean" has received much attention, even more compelling evidence supports the existence of numerous lakes, which were temporarily extant on the surface of Mars at various times in the planet's history (Cabrol and Grin 1999). The more ancient lakes occupied highland craters during the heavy bombardment epoch, spilling over to feed valleys such as Ma'adim Vallis (Irwin et al. 2002). Even more abundant crater paleolakes seem to have developed just after the heavy bombardment, and expecially large lakes occupied the floors of the impact basins, Hellas (Moore and Wilhelms 2001) and Argyre. Even younger lacustrine activity is indicated by the finely layered deposits of the Valles Marineris (Nedell et al. 1987). These are up 8 km thick, which could indicate a very prolonged period of deep water inundation of this immense tectonic trough.

Considerations of lake mass balances (Grin and Cabrol 1997) and of likely formation times for observed deltas and wave-eroded terraces (Ori et al. 2000) suggest that the crater lakes had lifetimes on the order of about 1000 years. Calculations with a general circulation model demonstrate that ice-covered lakes of this duration might be possible in a quasi-stable state for some portions of the Martian surface even under present-day conditions (Haberle et al. 2001). Of course, the water would first need to be mobilized to liquid form, suggesting that in its present frozen state Mars is merely hydrologically dormant (Cabrol and Grin 2001).

4. Glacial and Periglacial Landforms

Evidence for glacial activity though Martian history is also abundant (Kargel and Strom 1992) and controversial (Carr 1996). Resistance to the idea of ancient glaciers on Mars is especially curious, given that there is a general scientific consensus that Mars displays an immense variety of periglacial landforms, most of which require the activity of ground ice (Lucchitta 1985, Squyres et al. 1992). The periglacial landforms include debris flows, polygonally patterned ground, thermokarst, frost mounds, pingos, and rock glaciers. On Earth most of these landforms develop under climate conditions that are both warmer and wetter than the conditions for cold-based glacial landforms (Baker 2001). Because the Mars glacial landforms are all post-Noachian in age, and some are very young, they are completely inconsistent with the MIDDEN hypothesis. Glaciers require substantial transport of atmospheric water vapor to sustain the snow accumulation that generates the positive mass balance needed for glacial growth. There are no known Earth glaciers that develop from water supplied by the melting of ground ice, though this mechanism has been proposed for ancient glaciers on Mars that are hypothesized to have occupied the outflow channels (Lucchitta 1982).

The glacial landforms of Mars are erosional (grooves, streamlined/sculpted hills, drumlins, horns, cirques, and tunnel valleys), depositional (eskers, moraines, and kames), and ice-marginal (outwash plains, kettles, and glaciolacustrine plains). Of course, the landform names are all genetic designations, and *ad hoc* alternatives have been suggested for many. What is not *ad hoc*, however, is that all the glacial landforms occur in spatial associations, proximal-to-distal in regard to past ice margins, that would be obvious in a terrestrial setting. Areas of past glaciation on Mars include the Tharsis volcanoes, uplands surrounding Argyre and Hellas (Kargel and Strom 1992), and the polar regions, where the ice caps were much more extensive during portions of post-Noachian time (Fishbaugh and Head 2000, Head and Pratt 2001).

5. Very Recent Fluvial and Volcano-Ice-Water Activity

One of the most striking recent discoveries is that many water-related landforms on Mars are exceptionally young in age. This fact was prominently demonstrated by MOC images from the MGS orbiter showing numerous small gullies generated by surface runoff on hillslopes (Malin and Edgett 2000b). The gullies are most likely formed by debris-flow processes and the melting of near-surface ground ice (Costard et al. 2002). Melting can be induced at the appropriate latitude by changes in the solar insolation that would be induced by the immense shifts in Martian obliquity that are retrodicted to have occurred during the past few million years (Laskar and Robutel 1993, Touma and Wisdom 1993). The gullies are uncratered, and their associated debris-flow fan deposits are superimposed on both on eolian bedforms (dunes or wind ripples) and on polygonally patterned ground, all of which cover extensive areas that are also uncratered (Malin and Edgett 2000b). The patterned ground is itself a very strong indicator of near-surface, ice-related processes in the active (seasonally thawed) layer above the Martian permafrost zone (Siebert and Kargel 2001).

Exceptionally young outflow channels and associated volcanism occur in both the Cerberus Plains and the Tharsis regions of Mars (Hartmann and Berman 2000, Mouginis-Mark 1990). Data from MGS show that localized water releases, interspersed with lava flows, occurred approximately within the last 10 million years (Berman and Hartmann 2002, Burr et al. 2002). The huge discharges associated with these floods and the temporally related volcanism should have introduced considerable water into active hydrological circulation on Mars. It is tempting to hypothesize that the young outflow processes and volcanism are genetically related to other very young water-related landforms. The latter include the gullies, a thin ice-rich mantling layer covering about 23% of Mars (Mustard et al. 2001, Kreslavsky and Head 2002), and perhaps some of the dark slope streaks on Martian hillslopes, at which shallow water is possibly being mobilized (Ferris et al. 2002). The genetic connection for all these phenomena might well be climate change, induced by the water vapor and gases introduced to the atmosphere by both flooding and volcanism, as proposed in the MEGAOUTFLO hypothesis. This would explain the recent detection by neutron and gamma-ray instruments of a near-surface zone of water-ice abundance at high Mars latitudes (Boynton et al. 2002). That water may reflect very recent emplacement during the same aqueous episode.

6. Discussion

Recent discoveries from MOC images show that Mars displays a diverse suite of exceptionally young, globally distributed landforms that are water-related. If observed on Earth these landforms would all be well understood to have aqueous origins that were capable of generating them on relative short time scales (100s to 1000s of years) in a much warmer, wetter, and denser atmosphere than occurs on Mars today. Likewise, in contrast to the MIDDEN hypothesis, the surface of Mars displays older post-Noachian landforms of fluvial, glacial, periglacial, and hydrovolcanic origins. These phenomena are all consitent with the episodic climatic changes envisioned by the MEGSAOUTFLO hypothesis.

A much-discussed "conundrum" of Mars science is the problem of the "missing carbonate deposits." The argument made is that in a warm, thick atmosphere (e.g., Pollack et al. 1987) reactions of CO_2 gas and water would lead inevitably to weathering of surface rocks and the deposition of extensive carbonate deposits, as occur on Earth (Kahn 1985). The spectral observations of Mars, however, have failed to detect any carbonates (Blaney and McCord 1989, Christensen et al. 2001). Moreover, recent data from MGS show that the Martian surface extensively exposes unaltered feldspar and pyroxene in essentially unwearthered basalt outcrops (Christensen et al. 2000). This lack of weathering is actually to be expected if the post-Noachian history of Mars has nearly always been extremely cold and dry. The MEGAOUTFLO hypothesis envisions only very brief wet episodes, no longer than those of hyperarid and cold-desert regions of Earth, which also preserve essentially unweathered rock outcrops. Weathering in a Noachian wet period might by obscured by burial, and the lack of carbonate spectral signatures could result from suppression of those signatures by processes likely to have occurred on Mars (Craddock and Howard in press). Alternatively, the carbonates could indeed be absent. They may have accumulated on the floor of a very ancient ocean (Schaefer 1993), but this ocean floor could have been subducted during a plate-tectonic regime in the early Noachian (Baker et al. 2002).

A very early phase of plate tectonics could have generated the Martian highland crust by continental accretion (Fairen et al. 2002). By concentrating volatiles in a local region of the Martian mantle, the early plate-tectonic phase of Mars would have led to a superplume at Tharsis. The resulting immense concentration of volcanism at Tharsis would itself have a great influence on climate change (e.g., Phillips et al. 2001). The persistence of this volcanism episodically through later Martian history (e.g., Anderson et al. 2001) would provide a mechanism for the episodic, short-duration aqueous phases that produced the above-described landforms.

7. References

Anderson, R.C., Dohm, J.M, Golombek, M.P., Haldemann, A.F.C., Franklin, B.J., Tanaka, K.L., Lias, J. and Peer, B. (2001) Primary centers and secondary Concentrations of tectonic activity through time for the western hemisphere of Mars. *J. Geophys. Res.*, **106**, 20,563-20,585.

Baker, V.R. (1982) *The Channels of Mars*, University of Texas Press, Austin, Texas. Baker, V.R. (1998) Catastrophism and uniformitarianism: logical roots and current relevance. In: D.J. Blundell and A.C. Scott (eds.) *Lyell: The Past is the Key tothe Present*, Geological Society Special Publication 143, London, pp. 171-182.

Baker, V.R. (2001) Water and the martian landscape. *Nature*, **412**, 228-236.

Baker, V.R. and Milton, D.J. (1974) Erosion by catastrophic floods on Mars and Earth. *Icarus*, **23**, 27-41.

Baker, V.R. and Partridge, J. (1986) Small martian valleys: pristine and degraded morphology. *J. Geophys. Res.*, **91**, 3561-3572.

Baker, V.R., Maruyama, S. and Dohm, J.M. (2002) A theory of early plate tectonics and subsequent long-term superplume activity on Mars. *Electronic Geosciences*, **7** (http://lin.springer.de/service/journals/10069/free/conferen/superplu/)

Baker, V.R., Carr, M.H., Gulick, V.C., Williams, C.R. and Marley, M.S. (1992) Channels and valleys. In: H.H. Kieffer, B.Jakosky and C. Snyder (eds.) *Mars*, University of Arizona Press, Tucson, pp. 493-522.

Baker, V.R., Strom, R.G., Gulick, V.C., Kargel, J.S., Komatsu, G. and Kale, V.S. (1991) Ancient oceans, ice sheets and hydrological cycle on Mars. *Nature*, **352**, 589-594.

Berman, D.C. and Hartmann, W.K. (2002) Recent fluvial, volcanic, and tectonic activity on the Cerberus Plains of Mars. *Icarus*, **159**, 1-17.

Boynton, W.V. and 24 others (2002) Distribution of hydrogen in the near surface of Mars: evidence for subsurface ice deposits. *Science*, **297**, 81-85.

Bradley, B.A., Sakimoto, S.E.H., Frey, H. and Zimbelman, J.R. (2002) Medusa Fossae Formation: new perspectives from Mars Global Surveyor. *J. Geophys. Res.*, **107,** doi: 10.1029/2001JE001537.

Burr, D.M., Grier, J.A., McEwen, A.S. and Keszthelyi, L.P. (2002) Repeated aqueous flooding from the Cerberus Fossae: evidence for very recently extant, deep ground- water on Mars. *Icarus*, **159**, 53-73.

Blaney, D.L. and McCord, T.B. (1989) An observational search for carbonates on Mars. *J. Geophys. Res.*, **94**, 10,159-10,166.

Cabrol, N.A. and Grin, E.A. (1999) Distribution, classification, and ages of martian impact crater lakes. *Icarus*, **142**, 160-172.

Cabrol, N.A. and Grin, E.A. (2001) The evolution of lacustrine environments on Mars: Is Mars only hydrologically dormant? *Icarus*, **149**, 291-328.

Carr, M.J. (1979) Formation of Martian flood features by release of water from confined aquifers. *J. Geophys. Res.*, **84**, 2995-3007.

Carr, M.J. (1996) *Water on Mars*, Oxford University Press, Oxford.

Carr, M.J. (1998) Martian oceans, valleys and climate. *Astronomy and Geophysics*, **41**, 3.20-3.26.

Christensen, P.R., Bandfield, J.L., Smith, M.D., Hamilton, V.E. and Clark, R.N. (2000) Identification of a basaltic component on the martian surface from Thermal Emission Spectrometer data. *J. Geophys. Res.*, **105**, 9609-9621.

Christensen, P.R. and 25 others (2001) Mars Global Surveyor Thermal Emission Spectrometer experiment: investigation description and surface science results. *J. Geophys. Res.*, **106**, 23,823-23,871.

Clifford, S.M. (1993) A model for the hydrologic and climatic behavior of water on Mars. *J. Geophys. Res.*, **98**, 10,973-11,016.

Clifford, S.M. and Parker, T.J. (2001) The evolution of the martian hydrosphere: implications for the fate of a primordial ocean and the current state of the northern plains. *Icarus*, **154**, 40-79.

Costard, F., Forget, F., Mangold, N. and Peulvast, J.P. (2002) Formation of recent martian debris flows by melting of near-surface ground ice at high obliquity. *Science*, **295**, 110-113.

Craddock, R.A. and Howard, A.D. (2003) The case for rainfall on a warm, wet early Mars. *J. Geophys. Res.*, **107**, doi: 10.1029/2001JE001505.

Craddock, R.A. and Maxwell, T.A. (1993) Geomorphic evolution of the martian highlands through ancient fluvial processes. *J. Geophys. Res.*, **98**, 3453-3468.
Fairen, A.G., Ruiz, J. and Angula, F. (2002) An origin for the linear magnetic anomalies on Mars through accretion of terranes: implications for dynamo timing. *Icarus*, **160**, 220-223.
Ferris, J.C., Dohm, J.M., Baker, V.R. and Maddock, T. (2002) Dark slope streaks on Mars: are aqueous processes involved? *Geophys. Res. Letters*, **29,** doi: 10.1029/2002GL014936.
Fishbaugh, K.E. and Head, J.W., III (2000) North polar region of Mars: Topography of circumpolar deposits from Mars Orbiter Laser Altimeter (MOLA) data and evidence for asymmetric retreat of the polar cap. *J. Geophys. Res.*, **105**, 22,455-22,486.
Frey, H.V., Roark, J.H., Shockey, K.M., Frey, E.L. and Sakimoto, S.E.H. (2002) Ancient lowlands on Mars. *Geophys. Res. Letters*, **29,** noi: 10.1029/2001GL013832.
Golombek, M.P. and Bridges, N.T. (2000) Erosion rates on Mars and implications for climate change: constraints from the Pathfinder landing site. *J. Geophys. Res.*, **105**, 1841-1853.
Grant, J.A. (2000) Valley formation in Margaritifer Sinus, Mars, by precipitation-recharged ground-water sapping. *Geology*, **28**, 223-226.
Grin, E.A. and Cabrol, N.A. (1997) Limnologic analysis of Gusev crater paleolake. *Icarus*, **130**, 461-474.
Gulick, V.C. (2001) Origin of the valley networks on Mars: a hydrological perspective. *Geomorphology*, **37**, 241-268.
Gulick, V.C. and Baker, V.R. (1989) Fluvial valleys and Martian paleoclimates. *Nature*, **341**, 514-516.
Gulick, V.C. and Baker, V.R. (1990) Origin and evolution of valleys on Martian volcanoes. *J. Geophys. Res.*, **95**, 14,325-14,344.
Haberle, R.M., McKay, C.P., Schaeffer, J., Cabrol, N.A., Grin, E.A., Zent, A.P. and Quinn, R. (2001) On the possibility of liquid water on present-day Mars. *J. Geophys. Res.*, **106**, 23,317-23,326.
Hartmann, W.K. and Berman, D.C. (2000) Elysium Planitia lava flows: crater count chronology and geological implications. *J. Geophys. Res.*, **105**, 15,011-15,025.
Hartmann, W.K. and Neukum, G. (2001) Cratering chronology and the evolution of Mars. *Space Sci. Reviews*, **96**, 165-194.
Head, J.W., III, and Pratt, S. (2000) Extensive Hesperian-aged south polar ice sheet on Mars: Evidence for massive melting and retreat, and lateral flow and ponding of melt-water. *J. Geophys. Res.*, **106**, 12,275-12,299.
Head, J.W., Kreslavsky, M.A. and Pratt, S. (2002) Northern lowlands of Mars: evidence for widespread volcanic flooding and tectonic deformation of the Hesperian Period. *J. Geophys. Res.*, **107** doi: 10.1029/2000JE001445.
Hoffman, N. (2000) White Mars: A new model for Mars' surface and atmosphere based on CO_2. *Icarus,* **146**, 326-342.
Irwin, R.P., III, and Howard, A.D. (2002) Drainage basin evolution in Noachian Terra Chimmeria, Mars. *J. Geophys. Res.*, **107** doi: 10.1029/2001 JE001818.
Irwin, R.P., III, Maxwell, T.A., Howard, A.D., Craddock, R.A. and Leverington, D.W. (2002) A large paleolake basin at the head of Ma'adim Vallis, Mars. *Science*, **296**, 2209-2212.
Ivanov, M.A., and Head, J.W., III (2001) Chryse Planitia, Mars: topographic configuration, outflow channel continuity and sequence, and tests for hypothesized ancient bodies of water using Mars Orbiter Laser Altimeter (MOLA) data. *J. Geophys. Res.*, **106**, 3275-3295.
Kahn, R. (1985) The evolution of CO_2 on Mars. *Icarus*, **62**, 175-190.
Kargel, J.S. and Strom, R.G. (1992) Ancient glaciation on Mars. *Geology*, **20**, 3-7.
Kasting, J.F. (1991) CO_2 condensation and the climate of early Mars. *Icarus*, **94**, 1-13.
Kreslavsky, M.A. and Head, J.W., III (2002) Mars: Nature and evolution of young, latitude-dependent water-ice-rich mantle. *Geophys. Res. Lettrs.*, **15** doi: 10.1029/2002GL015292.
Kuzmin, R.O., Bobina, N.N., Zabalueva, E.V. and Shashkina, V.P. (1988) Structural inhomogeneities of the Martian cryolithosphere. *Solar SystemResearch*, **22**, 195-212.
Laskar, J. and Robutel, P. (1993) The chaotic obliquity of the planets. *Nature*, **361**, 608-612.
Leighton, R.B. and Murray, B.C. (1966) Behavior of carbon dioxide and other volatiles on Mars. *Science,***153**, 136-144.
Lowell, P, (1908) *Mars as an Abode of Life*, Macmillan, London.
Lucchitta, B.K. (1982) Ice sculpture in the Martian outflow channels. *J. Geophys. Res.*, **87**, 9951-9973.
Lucchitta, B.K. (1985) Geomorphic evidence for ground ice on Mars, in: J. Klinger, D. Benest, A. Dollfus and R. Smoluchowski (eds.)*Ices in the Solar System*, Reidel, N.Y., pp. 583-604.
Lucchitta, B.K., Ferguson, H.M. and Summers, C. (1986) Sedimentary deposits in the northern lowland plains, Mars. *J. Geophys. Res.*, **91**, E166-E174.

Malin, M.C. and Edgett, K.S. (1999) Oceans or seas in the Martian northern lowlands: high-resolution imaging tests of proposed coastlines. *Geophys. Res. Lettrs.*, **26**, 3049-3052.
Malin, M.C. and Edgett, K.S. (2000a) Sedimentary rocks of early Mars. *Science*, **290**, 1927-1937.
Malin, M.C. and Edgett, K.S. (2000b) Evidence for recent groundwater seepage and surface runoff on Mars. *Science*, **288**, 2330-2335.
Malin, M.C. and Edgett, K.S. (2001) Mars Global Surveyor Orbiter Camera: Inter-planetary cruise through primary mission. *J. Geophys. Res.*, **106**, 23,429-23,570.
Masson, P., Carr, M.H., Costard, F., Greeley, R., Hauber, E. and Jauman, R. (2001) Geomorphologic evidence for liquid water. *Space Science Reviews*, **96**, 333-364.
Moore, J.M. and Wilhelms, D.E. (2001) Hellas as a possible site of ancient ice-covered lakes on Mars. *Icarus*, **154**, 258-276.
Mouginis-Mark, P.J. (1990) Recent water release in the Tharsis region of Mars. *Icarus*, **84**, 362-373.
Mustard, J.F., Cooper, C.D. and Rifkin, M.K. (2001) Evidence for recent climate change on Mars from the identification of youthful near-surface ground ice. *Nature*, **412**, 411- 414.
Nedell, S.S., Squyres, S.W. and Anderson, D.W. (1987) Origin and evolution of the layered deposits in the Valles Marineris, Mars. *Icarus*, **70**, 409-441.
Ori, G.G., Marinangeli, L. and Baliva, A. (2000) Terraces and Gilbert-type deltas in crater lakes in Ismenius Lacus and Memnonia (Mars). *J. Geophys. Res.*, **105**, 17,629-17,641.
Parker, T.J., Saunders, R.S. and Schneeberger, D.M. (1989) Transitional morphology in the west Deuteronilus Mensae region of Mars: implications for modification of the lowland/upland boundary. *Icarus*, **82**, 111-145.
Parker, T.J., Gorsline, D.S., Saunders, R.S., Pieri, D. and Schneeberger, D.M. (1993) Coastal geomorphology of the Martian northern plains. *J. Geophys. Res.*, **98**, 11,061-11,078.
Phillips, R.J. and 9 others (2001) Ancient geodynamics and global change hydrology on Mars. *Science*, **291**, 2587-2591.
Pollack, J.B., Kasting, J.F., Richardson, S.M. and Poliakoff, K. (1987) The case for a wet, warm climate on early Mars. *Icarus*, **71**, 203-224.
Schaefer, M.W. (1993) Aqueous geochemistry on early Mars. *Geochim. Csmochim. Acta*, **57**, 4619-4625.
Segura, T.L., Toon, O.B., Coaprete, A. and Zahnle, K. (2002) Environmental effects of large impacts on Mars. *Science*, **298**, 1977-1980.
Siebert, N.M. and Kargel, J.S. (2001) Small-scale Martian polygons liquid surface water. *Geophys. Res. Lett.*, **28**, 899-903.
Squyres, S.W. and Kasting, J.F. (1994) Early Mars: How warm and how wet? *Science*, **265**, 744-749.
Squyres, S.W., Clifford, S.M., Kuzmin, R.O., Zimbelman, J.R. and Costard, F.M. (1992) Ice in the Martian regolith. In: H.H. Kieffer, B.Jakosky and C. Snyder (eds.) *Mars*, University of Arizona Press, Tucson, pp. 523-554.
Touma, J. and Wisdom, J. (1993) The chaotic obliquity of Mars. *Science*, **259**, 1294-1297.

Biodata of **Helmut Lammer** author (with co-authors H.L.M. Lichtenegger and A. Hosseinmardi) of ***"Evolution of The Martin Water Inventory"***

Dr. **Helmut Lammer** is a scientist at the *Austrian Space Research Institute*, *Department for Extraterrestrial Physics* of the *Austrian Academy of Sciences* in Graz, Austria (http://www.iwf.oeaw.ac.at). He participated as a scientist on various international space missions. He serves as the Head of the Planetary Atmosphere Research Group of the Department for Extraterrestrial Physics. RCL-Representative on Mars Express Atmosphere-Surface-Interaction studies (ESA). Dr. H. Lammer obtained his PhD. in 1992 in Geophysics from the Karl Franzens University in Graz Austria. His main fields of research are Atmospheric evolution of Venus, Mars and Titan, and several other planetary studies. Dr. Lammer is on the Editorial Board of Astrobiology journal (Mary Ann Libert, Inc. Publishers, USA) and on International Journal of Astrobiology (Cambridge University Press, UK). He has been among the organizers of conference organizations and served in the organizing committee at the Water in the Upper Martian Surface (April 17-19, 2002, Potsdam, Germany) and served as Main local organizer of the 2nd European Workshop on Exo / Astrobiology (September 16-19, 2002, Graz, Austria).

E-mail: **helmut.lammer@oeaw.ac.at**

J. Seckbach (ed.), Origins, 633–643.

EVOLUTION OF THE MARTIN WATER INVENTORY

HELMUT LAMMER[1], HERBERT I. M. LICHTENEGGER[2] AND AZARNOUSH HOSSEINMARDI[3]

[1]*Department of Extraterrestrial Physics Space Research Institute, Austrian Academy of Sciences, Schmiedlstr. 6, A-8042 Graz, Austria*

[2]*Department of Experimental Space Physics, Space Research Institute, Austrian Academy of Sciences, A-8042, Graz, Austria*

[3]*Institute of Chemical Engineering Fundamentals and Plant Design, Graz University of Technology Infeldgasse 25, A-8010, Graz, Austria*

1. Introduction

The discovery of high concentrations of water-ice just below the Martian surface polar areas by NASA's Mars Odyssey spacecraft has strengthened the debate about the search for life on Mars. Generally it is believed that life on Earth emerged in liquid water from the processing of organic molecules. Since the possible origin of life on early Mars should have been related to the evolution of the planetary water inventory it is important to estimate the amount of the water-ice below the planetary surface.

High resolution altimetric data from the Mars Orbiter Laser Altimeter (MOLA) instrument on board of the Mars Global Surveyor (MGS) spacecraft defined a detailed topography of the northern Martian lowlands (Head III et al. 1999). A wide range of data are now consistent with the hypothesis that a lowland-encircling geologic contact represents the ancient shoreline of a large standing body of H_2O in the Martian past.

Large outflow channels empty into the northern lowlands (Baker 2001). Some gullies on the Martian surface have been attributed to recent H_2O seepage and runoffs suggesting H_2O or ice close to the surface. Actual observations by the High Energy Neutron Detector (HEND) on board of Mars Odyssey give strong evidence that H_2O-ice is concentrated in the subsurface in the northern and southern hemisphere (Mitrofanov et al. 2002).

There are also indications that large standing bodies of H_2O ranging from lakes to an ocean may have existed there in the past history of Mars (Carr 1987). The MOLA data support this hypothesis since the flatness and smoothness of these areas have shown that they are part of the largest watershed on Mars (e.g., Head III et al. 1999).

Since Mars does not have a strong intrinsic magnetic field today (Acuña et al. 1998) there are only two possibilities for the fate of early H_2O: it is either sequestered in the Martian subsurface somewhere on the planet or it was lost to space during the Martian history. The ancient fluvial networks on the Martian surface suggest that the atmosphere was also warmer, denser and wetter at least 3.5 - 4.5 Gyr ago.

2. Loss of Water from Mars

When water evaporates from the Martian surface and subsurface ice reservoirs, its vapor reaches the atmosphere, where solar UV radiation can break up the molecules into H, H_2 and O atoms (e.g., Hunten and McElroy 1970) resulting to escape to space and surface oxidation and hydration reactions. The production of reactive radical atmospheric HO_x species is initiated by photolysis with:

$$H_2O + O(^1D) \rightarrow OH + OH, \qquad (1)$$

where the excited oxygen atom is derived from photolysis of ozone. The HO_x radicals participate in catalytic cycles for the oxidation of CO to CO_2 molecules. Further, the HO_x species are destroyed by the following chemical reaction:

$$OH + HO_2 \rightarrow H_2O + O_2, \qquad (2)$$

which transforms them back to H_2O. This reaction limits the amount of HO_x radicals in the Martian atmosphere. Another reaction removes HO_x with a minor yield by:

$$H + HO_2 \rightarrow H_2 + O_2, \qquad (3)$$

resulting in the formation of H_2 and O_2 molecules. The majority of the H_2 produced via reaction (3) is removed by:

$$H_2 + OH \rightarrow H_2O + H \qquad (4)$$

$$H_2 + O(^1D) \rightarrow OH + H \qquad (5)$$

A small part of the H_2 produced in reaction (5) escapes chemical via HO_x radicals in the lower Martian atmosphere. They get transported to the upper atmosphere, where they can participate in ionospheric reactions with molecular CO_2^+ ions in:

$$H_2 + CO_2^+ \rightarrow CO_2H^+ + H, \qquad (6)$$

$$CO_2H^+ \rightarrow CO_2 + H, \qquad (7)$$

where the H_2 molecules are transformed to atomic H. H atoms produced in the lower Martian atmosphere have a short chemical lifetime and cannot diffuse to the upper atmosphere where they can escape (e.g., Hunten and McElroy 1970). The H_2 molecules produced via reaction (5) are very stable, diffuse to the upper atmosphere where a fraction of them react with CO_2^+ ions and produce H which can escape thermally from the upper atmosphere.

The thermal escape rate of hydrogen from Mars was estimated by Anderson and Hord 1972 by fitting exosphere temperatures which satisfied the Mariner 6 and 7 Lyman α dayglow observations. The UV spectrometers that flew by Mars in 1969 observed the Lyman α dayglow of neutral atomic H in an altitude range between 200 km and 24,000 km. They found that an exospheric temperature of about 350 K ± 100 K and a number density for H at the exobase level around 240 km altitude of about $3.0 \times 10^4 cm^{-3}$ fits their data well.

By using these data one gets an escape flux for H atoms of about $1.8 \times 10^8 cm^{-2}s^{-1}$. However, an average exosphere temperature of about 350 K on Mars might be to large since very low energetic H atoms (VLENA) and energetic H atoms should have influenced the Lyman α observations (Lichtenegger et al. 2002). Since no better data are available before ESA's Mars Express and the Japanese Nozomi spacecraft we estimate the thermal escape rate from Mars of about 1.5×10^{26} H atoms s^{-1}.

Recently, Krasnopolsky and Feldman 2001 observed four H_2 lines in a spectrum of Mars with the Far Ultraviolet Spectroscopic Explorer (FUSE) satellite. The line intensities correspond to a column abundance of molecular hydrogen of 1.17×10^{13} cm^{-2}, 140 km above the Martian surface, resulting in a H_2 mixing ratio of 15 ± 5 parts per million in the lower Martian atmosphere. By using this value we get a thermal H_2 escape rate of about 3.3×10^{24} s^{-1}. One can see that the thermal molecular H_2 loss rate is negligibly small compared to the atomic hydrogen loss rate. Therefore, we derive a total thermal loss rate of neutral H atoms of about $1.5 \times 10^{26} s^{-1}$.

Measurements carried out on the Phobos 2 spacecraft have shown that the plasma tail of Mars consists mainly of ions which have their origin in the Martian atmosphere (Lundin et al. 1989, 1990; Rosenbauer et al. 1989). Cold H^+ ions of planetary origin with densities of about 1-2 cm^{-3} were observed by the ASPERA (Automatic Space Plasma Experiment with a Rotating Analyzer) instrument inside the Martian magnetic boundary.

Rosenbauer et al. 1989 have recorded two proton populations which were clearly distinguished from the energy spectra. The flux of cold H^+ ions, which was attributed to the planetary source, increased abruptly at the bow shock. In the magnetosheath, a gradual thermalization of both components provided their mixing and a separation of both populations becomes less evident.

For studying the nonthermal hydrogen loss rates from Mars we used a test particle model which involves the motion in the environmental electric and magnetic fields of Mars based on the *Spreiter-Stahara* model (Spreiter and Stahara 1993; Lichtenegger and Dubinin 1998; Lichtenegger et al. 2002). This test particle model was successfully used to explain several characteristic features obtained by the Pioneer Venus Orbiter at Venus (Luhmann 1993) and by the Phobos 2 plasma measurements at Mars (Lichtenegger et al. 1995, 1998).

The calculation of the ion fluxes has been performed for three different stages in the history of Mars where the evolution of the EUV flux was obtained from solar proxies studied inside the *Sun in Time* program (Guinan and Ribas 2002, Lammer et al. 2003): 1 EUV represents the present Martian conditions, 3 EUV represents Mars at 2 Gyr in the past while 6 EUV refers to 3.5 Gyr in the past.

The simulated total planetary ion escape rates for H^+, H_2^+ and O^+ are given in Table 1. We note that these model results are in close agreement with the estimation of the ion escape rates based on the Phobos-2 measurements which is of the order of a few times 10^{25} s^{-1} (Lundin et al. 1989, Rosenbauer et al. 1989). Since these measurements were obtained at high solar wind conditions, they correspond to simulation results in between 1 EUV and 3 EUV.

We get for 2 Gyr ago thermal loss rates for H atoms of about 2.4×10^{26} s^{-1} and about $5.7 \times 10^{26} s^{-1}$, 3.5 Gyr ago. The escape flux of H today is about half the maximum possible given the amount of H present near the homopause (Donahue 1995). Another limitation to the H escape flux is set by the amount of H available in the mixed atmosphere below the homopause and by the rate at which H reaches the exosphere where escape can occur (Hunten 1973; Hunten and Strobel 1974).

If the escape flux was once much higher than the mixing ratio of the presently dominant hydrogen compound near the Martian homopause H_2 must then have been also higher than today. Such an enhancement could have been caused by an increase in the H_2O vapor concentration or in an enhanced loss of O corresponding to a mechanism postulated by McElroy and Donahue 1972 and Liu and Donahue 1976 where a change in O loss stabilize the H escape flux at *twice* the O escape flux.

This mechanism is known as *self-regulation*. The production rate of H_2 is proportional to H and HO_2 which is the most abundant HO_x radical in the Martian atmosphere. Its concentration is larger than OH and H and this reaction determines the concentration of H atoms in the Martian atmosphere. At high O_2 concentrations H and the production of H_2 are small and vice versa. If the total escape flux ϕ_O into space and into the Martian surface is < 1/2 ϕ_H O_2 will build up and the rate of the production of H_2 would fall as a result of the increase of O_2. A new steady state may be reached in which the escape flux ϕ_H would be smaller so that $\phi_O = 1/2\ \phi_H$ is satisfied again. McElroy and Donahue 1972 argued that the same could be applied to the case in which $\phi_O > 1/2\ \phi_H$, a condition which should have happened during Mars' past. In this case the O_2 concentration will decrease due to the higher O loss rate, resulting in an increase in the production of H_2 and in a greater H escape flux. A new steady state would be reached that would satisfy the stoichiometric ratio of 2:1 for H and O again.

TABLE 1. Model results for all possible atmospheric loss Processes (Th: thermal, Pu: ion pick-up, Dr: dissociative recombination, Sp: sputtering) on Mars corresponding to three different times in units of [s^{-1}] . Present (1 EUV), 2 Gyr in the past (3 EUV), 3.5 Gyr ago (6 EUV)

particle	present	2 Gyr ago	3.5 Gyr ago
Th: H	1.5×10^{26}	2.0×10^{26}	6.0×10^{26}
Pu: H^+	1.2×10^{25}	1.9×10^{25}	1.5×10^{26}
Pu: H_2^+	8.6×10^{24}	2.4×10^{25}	7.9×10^{25}
Pu: O^+	3.2×10^{24}	4.0×10^{25}	8.3×10^{26}
Dr: O	3.0×10^{24}	2.0×10^{25}	8.0×10^{25}
Sp: O	3.0×10^{23}	2.0×10^{26}	1.5×10^{27}
Su: CO_2	4.0×10^{22}	3.0×10^{24}	2.0×10^{25}

Table 1 shows all thermal and nonthermal loss processes where hydrogen and oxygen atoms and ions are involved from the present to 3.5 Gyr ago. The hydrogen and oxygen ion loss rates are based on the ion pick up model of Lammer et al. 2003. The thermal H escape rates are related to the solar EUV input obtained from the *Sun in Time* program

(Lammer et al. 2003, Guinan and Ribas 2002). The sputter loss rates are calculated from the sputter yields of Table I of Leblanc and Johnson 2002 and our incident particle pick up fluxes. The present dissociative recombination based O loss rate is taken from Lammer and Bauer 1991 and Fox 1993. The 3 EUV and 6 EUV O are taken from Luhmann 1997. Table 2 shows the total loss of H_2 and O from present to 3.5 Gyr ago.

TABLE 2. Total escape of H_2 and O from Mars corresponding To three different times in units of [s^{-1}]. Present (1 EUV), 2 Gyr in the past (3 EUV), 3.5 Gyr in the past (6 EUV).

particle	present	2 Gyr ago	3.5 Gyr ago
H_2	8.0×10^{25}	1.0×10^{26}	4.0×10^{26}
O	6.0×10^{24}	2.0×10^{26}	2.8×10^{27}

One can see that there is a lower total O escape rate to space compared to a total H loss rate expressed as H_2 loss to space back to about 2.0 Gyr ago. We suggest that since this time oxygen reacted very efficient with the surface material by chemical weathering (Lammer et al. 2003)

3. Total Loss of Water from Mars over 3.5 Gyr ago

By integrating the escape rate of H_2O from Mars over the past 3.5 Gyr one finds that a mass ΔM_a of about 2.5×10^{18} kg H_2O was lost. By using:

$$d=\Delta M_a / (4 \pi R_{Mars}^2 \rho_{H2O}), \tag{8}$$

with R_{Mars} the planetary radius of 3393 km and ρ_{H2O} the density of H_2O, one finds that Mars had lost a global H_2O ocean to space with the depth d of 12 m to 17 m depending on the efficiency of ion erosion processes (Lammer et al. 2003). Interestingly, our estimation lies between former estimations by Lammer et al. 1996 of about 5 m and Krasnopolsky and Feldman 2001 of about 30 m but is significantly lower than the estimations based on the high loss rate models of Kass and Yung 1995, 1996, 1999.

One should also note that the estimated 17 m depth is based on a full active self regulation coupling mechanism between O and H as proposed by McElroy and Donahue 1972 and more realistic values for the evolution of the solar activity during the past based on data obtained inside the *Sun in Time* program.

4. Estimation of the Past and Present Martian Water Reservoir

Measurements of D/H isotope ratios in the Martian atmosphere indicate an enrichment of D compared to H of about 5.5 times that of the Earth ocean value (Owen et al. 1988). The D and H can escape over the Martian past, where the atmosphere will be enriched in the heavier D isotope. We assume that the loss process of D and H produced from H_2O vapor is a simple Rayleigh distillation, following the relationship (Kass and Yung 1999):

$$R(t)=R(t=0)\ [(c+d)/c]^{(1-f)}, \qquad (9)$$

where R(t) is the D/H ratio at a time t, c is the size of the residual mixed reservoir at time t, d is the total water amount lost of about 17 m and f is a fractionation factor between 0 and 0.02 (Cheng et al. 1999). Therefore, the water-ice reservoir can be written as:

$$c=d\ /\ [(R(t)\ /\ R(t=0))^{1/(1-f)}-1]. \qquad (10)$$

Within the last decade a number of investigations have been made which brought almost conclusive evidence that among the thousands of known meteorites there is a small group, which originates from Mars. Table 3 shows the D/H ratio and the age of Martian SNC meteorites.

Ion microprobe studies of hydrous amphibole, biotite, and apatite in SNC meteorites, except Chassigny and Governador contain water with a significantly higher D/H ratio relative to terrestrial sea water value (Leshin-Watson et al. 1994, Mathew and Marti, 2001), but in good agreement with measured D/H ratios in comets.

TABLE 3. D/H ratio and age of Martian SNC meteorites.

meteorite	D/H ratio	age [Myr]
Chassigny	0.000164	9.9
Shergotty	0.00048843	165 ± 11
EETA79001A	0.00038464	173 ± 10
Zagami	0.00025119	180 ± 4
Lafayette	0.00029502	400
Governador	0.00019668	1360
Nakhla	0.00027121	1370
ALH84001	0.00027743	3850

Table 4 shows observed D/H ratios from H_2O from three comets (Eberhardt et al. Bockelée-Morvan et al. 1998, Meier et al. 1998) and the terrestrial seawater value. These rates are all about twice the value for terrestrial seawater, but are consistent with rates in *hot cores* of molecular clouds. It is believed that ion-molecule reactions in dense molecular clouds at temperature close to 35 K can produce these D/H enrichments.

TABLE 4. Cometary D/H ratios compared to the terrestrial seawater level

source	Halley	Hyakutake	Hale-Bopp	seawater
D/H	0.000316	0.00029	0.00033	0.000156

Our study implies a present H_2O-ice reservoir below the Martian surface idealized to a spherical shell with a thickness of about 10 m if the water originated from cometary impacts and about 3.5 m if one uses the terrestrial seawater D/H isotope ratio.

By using the D/H measurements and our atmospheric escape values one can also estimate the reservoir 3.5 Gyr ago, idealized as described above, with a depth of about 27 m if the water originates from comets and about 20.5 m by using the terrestrial seawater standard. Since the D/H ratio determined in old Martian meteorites like ALH84001 show an enrichment comparable with cometary D/H ratios one finds strong indications that the majority of the Martian water may have its origin from cometary impacts and a global H_2O-ice layer with a depth of about 10 m may represent the current Martian subsurface ice reservoir. Correlated to the HEND data of Mars Odyssey the observed polar ice caps may be about 50 m deep.

If the observed D/H ratio may not be related to the primordial water but to cometary impacts this ratio should be coupled in a complex relationship to the average D/H, the time-integrated cometary flux, the actual mass and D/H of any large comets which may have impacted on Mars in the last several 100 Myr. In this context it would be very important to study in-situ the D/H isotope ratio in the Martian soil and/or ice by a future lander of rover.

5. Conclusion

It is suggested that hydrogen atoms in the Martian atmosphere originate from the planetary water inventory hidden in subsurface ice and permafrost. When water evaporates from these reservoirs, its vapor reaches the upper atmosphere solar UV radiation can break up the molecules into hydrogen and oxygen. After this photochemical process the evolution of the Martian water inventory is influenced by thermal escape of H, H_2 and nonthermal escape of H^+, H_2^+, O, O^+, CO_2, and O_2^+ and chemical weathering of oxygen with the surface soil. By including the evolution of the solar radiation and particle environment, we estimate the total loss of water from Mars since 3.5 Gyr ago of an amount which is equivalent to a depth of a global Martian water ocean of about 17 m.

We use the observed D/H isotope fractionation in the Martian atmosphere, measured D/H ratios in SNC meteorites and the terrestrial seawater and our loss estimation for the calculation of the present and early water-ice reservoir. By comparing the D/H ratio of Martian meteorites one find a strong indication that the majority of the Martian H_2O has its origin from cometary impacts and an idealized global water-ice layer with a depth of about 10 m may represent the current Martian subsurface H_2O-ice reservoir.

Further, we suggest that hydrodynamic escape of hydrogen related to a more active early Sun may have been responsible for removing more H_2O which was out-gassed from the planetary interior before 3.5 Gyr. Future studies related to research corresponding to the origin of life should concentrate on the possibility how long water may have been liquid on the Martian surface.

6. References

Acuña, M. H., Connerney, J. E. P., Wasilewski, P., Lin, R. P., Anderson, K. A., Carlson, C. W., McFadden, J., Curtis, D. W., Mitchell, D., Reme, H., Mazelle C., Sauvaud, J. A., d'Uston, C., Cros, A., Medale, J. L., Bauer, S. J., Cloutier, P., Meyhew, M., Winterhalter, D. and Ness, N.F. 1998) Magnetic field and plasma observations at Mars: Initial results of the Mars Global Surveyor mission. Science, **279**, 1676-1680.

Anderson Jr., D. E., and Hord, C.W. (1972) Mariner 6 and 7 Ultraviolet spectrometer experiment: Analysis of hydrogen Lyman-Alpha data. J. Geophys. Res., **76**, 6666-6673.

Baker, V. R. (2001) Water and the Martian landscape. Nature, **412**, 228-236.

Bockelee-Morvan, D., Goutier, D., D. C., Lis, K., Young, J., Keene, T., Phillips, T. C., Owen, J. Crovisier, P., Goldsmith, E. A., Bergin, Despois, D., and Wootten, A. 1998) Deuterated water in comet C/1996 B2 (Hyakutake) and its implications for the origin of comets. Icarus, **133**, 147-162.

Carr, M. H. (1987) Water on Mars. Nature, **326**, 30-34.

Cheng, B.-M., Chew, E. P., Liu, C.-P., Bahou, M., Lee, Y-P., Yung, Y. L., and Gerstell, M.F. (1999) Photo-induced fractionation of water isotopomers in the Martian atmosphere. Geophys. Res. Lett., **26**, 3657-3660.

Donahue, T. M. (1995) Evolution of water reservoirs on Mars from D/H ratios in the atmosphere and crust. Nature, 374, 432-434.

Eberhardt, P., Reber, M., Krankowsky, D., and Hodges, R. R. (1995) The D/H and 18O/16O ratios in water from comet P/Halley. Astron. Astrophys., 302, 301-316.

Fox, J. L., and Dalgarno, A. (1983) Nitrogen escape from Mars. Geophys. Res., 88, 9027-9032.

Guinan, E. F., and Ribas, I. (2002) Our changing Sun: the role of solar nuclear evolution and magnetic activity on Earth's atmosphere and climate. In: B. Montesinos, A. Giménz, and E. F. Guinan, (eds.) The Evolving Sun and its Influence on Planetary Environments, ASP, San Francisco, 269, 85-107.

Head III, J. W., Hiesinger, H., Ivanov, M. A., Kreslavsky, M. A., Pratt, S., and Thomson, B. J. (1999) Possible ancient oceans on Mars: evidence from Mars Orbiter Laser Altimeter Data. Science, 286, 2134-2137.

Hunten, D. M., and McElroy, M. B. (1970) Production and escape of hydrogen on Mars. J. Geophys. Res., 75, 5989-6001.

Hunten, D. M. (1973) The escape of light gases from planetary atmospheres. J. Atmos. Sci., 30, 1481-1494.

Hunten, D. M., and Strobel, D. F. (1974) Production and escape of terrestrial hydrogen. J. Atmos. Sci., 31, 305-317.

Kass, D. M., and Yung, Y. L. (1995) Loss of atmosphere from Mars due to solar wind induced sputtering. Science, 268, 697-699.

Kass, D. M., and Yung, Y. L. (1996) Loss of atmosphere from Mars: response to Johnson and Liu (1995). Science, 268, 697-699.

Kass, D. M., and Yung, Y. L. (1999) Water on Mars: isotopic constraints on exchange between the atmosphere and surface. Geophys. Res. Lett., 26, 3653-3656.

Krasnopolsky, V. A., and Feldman, P. D. (2001) Detection of molecular hydrogen in the atmosphere of Mars. Science, 294, 1914-1917.

Lammer, H., and Bauer, S. J. (1991) Nonthermal atmospheric escape from Mars and Titan. J. Geophys. Res., 96, 1819-1825.

Lammer, H., Stumptner, W., and Bauer, S. J. (1996) Loss of H and O from Mars: implications for the planetary water inventory. Geophys. Res. Lett., 23, 3353-3356.

Lammer, H., Lichtenegger, H. I. M., Kolb, C., Ribas, I., Guinan, E. F., Abart, R., and Bauer, S. J. (2003) Loss of water from Mars: Implications for the oxidation of the soil. Icarus, in press, 2003.

Leblanc, F., and Johnson, R. E. (2002) Role of molecular species in pick up ion sputtering of the Martian atmosphere. J. Geophys. Res., 107, E2, (5), 1-6.

Leshin-Watson, L., Hutcheon, I. D., Epstein, S., and Stolper, E. M. (1994) Water on Mars: clues from deuterium/hydrogen and water contents of hydrous phases in SNC meteorites. Science, 265, 86-90.

Lichtenegger, H. I. M., Schwingenschuh, K., Dubinin, E. M., and Lundin, R. (1995) Particle simulation in the Martian magnetotail. J. Geophys. Res., 100, 21,659-21,667.

Lichtenegger, H. I. M., and Dubinin, E. M. (1998) Model calculations of the planetary ion distribution in the Martian tail. Earth Planets Space, 50, 445-452.

Lichtenegger, H. I. M., Lammer, H., and Stumptner, W. (2002) Energetic neutral atoms at Mars III: Flux and energy distributions of planetary energetic H atoms, J. Geophys. Res., 107 (A10), 1279, doi:10. 1029/2001JA000322, 2002.

Liu, S. C., and Donahue, T. M. (1976) The regulation of hydrogen and oxygen escape from Mars. Icarus, 28, 231-246.

Luhmann, J. G. (1993) A model of the ionospheric tail rays of Venus. J. Geophys. Res., 98, 17,615-17,621.

Luhmann, J. G. (1997) Correction to "The ancient oxygen exosphere of Mars: implications for atmospheric evolution". J. Geophys. Res., 102, 1637.

Lundin, R., Zakharov, A., Pellinen, R. , Hultquist, B., Borg, H., Dubinin, E. M., Barabash, S. W., Pissarenko, N., Koskinnen, H., and Liede, I. (1989) First results of the ionospheric plasma escape from Mars. Nature, 341, 609-612.

Lundin, R., Zakharov, A., Pellinen, R., Barabash, S. W., Borg, H., Dubinin, E. M., Hultquist, B., Koskinen, H., Liede, I., and Pissarenko, N. (1990) ASPERA/Phobos measurements of the ion outflow from the Martian ionosphere. Geophys. Res. Lett., 17, 873-876.

Mathew, K. J., and Marti, K. (2001) Early evolution of Martian volatiles: Nitrogen and noble gas components in ALH84001 and Chassigny. J. Geophys. Res., 106, 1401-1422.

McElroy, M. B., and Donahue, T. M. (1972) Stability of the Martian atmosphere. Science, 177, 986-988.

Meier, R., Owen, T. M., Matthews, H. E., Jewitt, D. C., Bockelee-Morvan, D., Biver, N., Crovisier, J., and Gautier, D. (1998) A determination of the HDO/H2O ratio in comet C/1995 O1 (Hale-Bopp). Science, 279, 842-844.

Mitrofanov, I., Anfimov, D., Kozyrev, A., Litvak, M., Sanin, S., Tret'yakov, V., Krylov, A., Shvetsov, V., Boynton, W., Shinohara, C., Hamara, D., and Saunders, R. S. (2002) Maps of subsurface hydrogen from the high energy neutron detector, Mars Odyssey. Science, 297, 78-81.

Owen, T. C., Maillard, J. P., deBergh, C., and Lutz, B. L. (1988) Deuterium on Mars: the abundance of HDO and the value of D/H. Science, 240, 1767-1770.

Rosenbauer, H., Shutte, N., Apathy, I., et al. (1989) Ions of Martian origin and plasma sheet in the Martian magnetosphere: initial results of the TAUS experiment. Nature, 341, 612-614.

Spreiter, J. R., and Stahara, S. S. (1993) Solar wind flow past Venus: Theory and comparisons, J. Geophys. Res., 98, 17,251-17,262.

Biodata of **Michael Simakov** author of ***"Possible Biogeochemical Cycles on Titan"***

Dr. Michael Simakov is a scientific researcher in the Group of Exobiology, Institute of Cytology, Russian Academy of Sciences, St. Petersburg, Russia. He obtained his PhD (1991) in Organic chemistry at St. Petersburg State University. His current research interests include (1) the origins of life, chemical evolution, abiogenic synthesis of biologically important compounds; (2) the role of energy sources of open space (ultraviolet radiation of different wavelengths, high charged particles) in abiogenic synthesis of nucleotides and oligopeptides; (3) exobiology — possible sites of exobiological activity on Titan. He designed and conducted the exobiological scientific experiments on board of space satellites "Bion" and Russian module of International Space Station. Dr. Simakov is a member of Scientific Council of Russian Astrobiology Center as well as a member of other societies.

E-mail: **exobio@mail.cytspb.rssi.ru**

J. Seckbach (ed.), Origins, 645–665.

POSSIBLE BIOGEOCHEMICAL CYCLES ON TITAN

M.B. SIMAKOV
Group of Exobiology, Institute of Cytology RAS,
Tikhoretsky Av., 4, St.Petersburg, 194064, RUSSIA

1. Introduction

Titan, the largest satellite of Saturn, has a great exobiological significance. Its dense atmosphere is composed primarily of N_2 with about 8% CH_4 and a large number of minor constituents such as carbon monoxide, carbon dioxide, ethane, ethylene, acetylene, cyanoacetylene, hydrogen cyanide, and many others. The existence of such substantial atmosphere is an unresolved question today. The main attention from exobiological point of view devotes now to a very complex atmospheric chemistry. The interaction of UV-light and cosmic rays with atmospheric constituents produces a complex radical chemistry which leads to formation of a high variety of organic molecules. The main products of these reactions are different hydrocarbons both saturated and unsaturated, nitriles and other N-organic (Yung, 1987; Lara et al., 1996). Dissociation of N_2, coupled with photolysis of CH_4, form the basis for many further chemical processes (Clarke and Ferris, 1997). The complex gas-phase organic chemistry induced by corona discharges (Ramirez et al., 2001) or lightning (Lammer et al., 2001) also is expected into dense nitrogen-methane atmosphere.

The atmosphere has at least three haze layers which are very rich in organic (McKay et al., 2001). Titan's organic haze is a potentially important sink of the photochemically produced carbon and nitrogen compounds. The complex organic mixture of simple alkanes, aromatic compounds, heteropolymers and amino acids precursors forms tholins — the solid organic product with very poorly known molecular composition. Analysis of the experimental tholins revealed more than 75 constituent compounds. The production of aerosols in large amounts implies that carbon is being rained out and the source of methane or other condensed hydrocarbons has to be on surface of the satellite in form of lakes or seas. The products of photolysis could cover the surface with 100-200 meters and more (Lunine and McKay, 1995). The O-bearing compounds (CO, H_2O and CO_2) have recently been discovered on Titan and these compounds can induce the production of several oxygenated compounds through energetic action.

2. Brief History of Titan

Accretion models of the Saturnian satellite suggest that heating released during late stages of its formation was sufficient to create a warm, dense atmosphere with mass at least 30 times greater then the present value (Lunine and Stevenson, 1983) and large open water ocean on its surface. Such juvenile Titan's ocean could exist during period of 10^8 years. As the great part of the primordial Titan's atmosphere could be supplied by comets during or after accretion (Zahnle et al., 1992), the composition of such atmosphere would have consisted of mostly H_2O, N_2, CO and CO_2, since the cometary carbon appears concentrated in the form of CO (ranging from a few to 45% relative to water), CO_2 (~15%) and heavy organic (Jessberger et al., 1989; Whittet et al., 1996). The mass of volatile acquired by Titan from comets would be expected to be ~10^{20}-10^{22} g for CO and 10^{20}-10^{21} g for N (Griffith and Zahnle, 1995). So we can see that the Titan's primordial atmosphere could be warm, dense and consist of CO_2-(CO)-N_2.

Complex organic compounds also may have been acquired from cometary and chondritic material (English et al., 1996). The first stages of chemical evolution would have took place in these atmosphere and ocean under action of such energy sources as ultraviolet radiation, solar wind, galactic cosmic rays, magnetospheric plasma ion bombardment, electrical discharges and radiogenic heat. Recent attempts to establish a lower limit for the time required for emergence of life suggest that 10-100 million years was enough in case of Earth (Orgel, 1998). The time of existence of the Titan's juvenile ocean was enough for arising of the first protoliving objects.

As the planet developed through time several energetic processes (irradiation, lightnings, meteoritic and comet impacts) could produce different forms of fixed nitrogen from the atmospheric one. Theoretical and experimental studies give information regarding N_2 fixation rate for the early Earth's CO_2-N_2 atmosphere (Yung and McElroy, 1979; Kasting and Walker, 1981). Yung and McElroy (1979) have shown that up to 10^9 kg N/yr would be produced by lightnings and deposited in the ocean primarily as HNO_3 with some HNO_2. Kasting and Walker (1981) have postulated the formation of HNO in the atmosphere and its deposition into the ocean. The future fate of HNO in the aqueous solution lead to form of N_2O, NO_2^- and NO_3^-. All nitrogen could have been in the fixed form at the short after the accretional period. Such scenario has been supposed by Mancinelli and McKay (1988) for the evolution of prebiotic nitrogen cycling on Earth, and the similar processes could be proposed in the case of the Saturn's satellite. Hence, in the absence of a recycling mechanism dissolved NO_2^- and NO_3^- would accumulate in the ocean or on the surface.

During the phase of cooling, Titan's ocean was roofed over with icy crust. If life had originated by then, it could survive in some places up to the present (Fortes, 2000). On Earth microbial life exists in all locations where microbes can survive. In other case the variety of prebiotic processes can take place inside Titan at present time. Many volatiles and inorganic salts were probably presented in the primordial liquid layer from leaching of the mantle components. The composition of the rich atmosphere which is host to extensive organic photochemistry and internal liquid layer must be very complex and Titan's putative ocean might harbor life or complex prebiotic structures.

The most recent models of the Titan's interior lead to the conclusion that a substantial liquid layer exists today under relatively thin ice cover inside Titan (Lunine and Stevenson, 1987; Grasset and Sotin, 1996; Grasset et al., 2000). Lorenz (2001) has found that the internal oceans are mandated for the large icy satellites. Thermal evolution models also predict the existence of thick (~300 km) liquid layer with relatively thin (~80 km) ice cover (Grasset et al., 2000). Spohn and Schubert (2003) have shown that even radiogenic heating in a chondritic core alone may suffice to keep a water ocean inside large icy satellites. Taking into account non-Newtonian viscosity of the water ice in planetary condition, the water ocean on Titan might have survived to date due to only radioactive heat source (Ruiz and Fairen, 2002). *Galileo* spacecraft has given indications, primarily from magnetometer and gravity data, of the possibility that three of Jupiter's four large moons, Europa, Ganymede and Callisto have such oceans (Anderson et al., 1998; Zimmer et al., 2000). So, the existing of liquid water ocean within icy world can be a consenquence of the physical properties of water ice, and they not require the addition of antifreeze substabces nor any other special conditions.

3. The Possible Biogeochemical Cycles in the Water Layer

3.1. UPPER WATER LAYER

The present composition of the putative liquid layers of the ice satellites is probably very complex. Heavily hydrated salt minerals, such as magnesium and sodium sulfates, sodium carbonate and their mixtures have been detected on Jupiter's satellites by near infrared mapping spectrometer (NIMS) on *Galileo* (McCord *et al.*, 1999; 2001), along with organic material like tholins bearing different functional groups, including C-H, S-H, SO_2, CO_2, and C=N (McCord et al., 1998).

Mass balance calculations modeled an extraction of the elements into the aqueous phase from chondritic material (Zolotov and Shock, 2000a) show that Titan's extensive subsurface ocean likely also contains dissolved salts from endogenic materials resembling to carbonaceous chondrite rocks incorporated into the satellite during its formation and released at the time of planetary differentiation. The low and high-temperature alteration of primitive accreted material leads to form of a complex water solution of such cations as K, Na, Mg, Ca, Mn, Fe and anions as SO_4^{2-}, Cl^-, Br^-, CO_3^{2-}, HCO_3^- and others (Fanale et al., 2001) along with nitrogen compounds. Phosphorus, sulfur, micro- and macronutrients have to be abundant inside bottom Titan's rocks. Aqueous weathering of these basalts would release nutrients to fluid where they would be available to microorganisms.

The comets, meteorites and IDPs fallen into the early ocean would be subjected to conditions producing high temperatures and pressures due to possible volcanism and hydrothermal vents (Mautner et al., 1995). Such treatment could release a large quantities of prebiotic organic also from insoluble kerogen-like polymer containing up to 70% of organic material in meteorites (Levy et al., 1973). Up to 50% of the organic is released in aqueous environments which simulate submarine volcanic, hydrothermal or impact-induced conditions (Mautner et al., 1995). This organic becomes available for the future chemical evolution. It was very important the release of amphiphilic compounds which could form compartments - the starting point on the way of a protobiological organizing, for example, the extracts of the Murchison meteorite can form stable bilayers and vesicles (Deamer and Pashley, 1989). Mautner (2002) has shown, that even a very gentle (natural) extraction of a sample of the meteorite (4 days at 20 °C) yields a large essential inorganic components, such as PO_4^{3-}, SO_4^{2-}, Cl^-, Ca^{2+}, Mg^{2+}, Na^+, K^+ as well as organic matter.

Since the light energy in the form of the solar radiation is not accessible in such conditions (the solar flux to the top of Titan' atmosphere is ~1.1% from the Earth's one, and only a tenth of that reaches its surface) the chemical energy has to be the main source which drives the life and other disequilibrium processes. The initial components, such as NO_3^-, SO_4^{2-}, CO_3^{2-} for the origin of lithoautotrophic processes could exist in the Titan's putative ocean from the earlier stages of the satellite's evolution and provide biologically useful electron donor-acceptor pairs in the upper layer where the temperature and pressure are not very hostile. Nitrate accumulated in the ocean at the first stage of atmosphere's evolution would have allowed the first protobiosystems to use it as the primary source of energy (Simakov, 2000). Recently it was a common view that the first living organism was one with an anaerobic fermentation, the simplest form of energy conversation on a substrate-level phosphorylation. The energetic reactions can ever proceed in a homogenous solution and in the absence of any supramolecular elements of cells, such as membranes. These organisms had to grow on prebiotically formed organic compounds, but the sources of these compounds were probably limited on early Titan in comparison with the abundance of nitrogen and sulfur inorganic compounds. We would like to propose the idea that the first protoliving systems in Titan's ocean could had internal energy source, namely, the chemical potential of an inorganic reaction. The all energetic metabolism should be constituted on the base of some kind of inorganic redox reaction — «Basic Reaction» (BR).

There are some candidates on the role of the BR. Electron acceptors such as NO_3^-, SO_4^{2-}, Fe^{3+}, Mn^{4+}, or CO_2 have to be coupled with the electron donors. Electron donors that may be important in such process include H_2, CO, CH_4, Fe^{2+}, Mn^{2+}, pyrite, sulfur compounds and organic material (Ottley et al., 1997). Some of these molecules could be generated abiotically on the bottom of the internal ocean by the reaction of water with rocks of the silicate mantle and by the reaction of water with meteoritic materials, others could be synthesized under the action of radiation. A radiation-driven ecosystem has been proposed for Europa (Chyba, 2000; Chyba and Phillips, 2001; Chyba and Hand, 2001) and can work in the case of Titan, at least on the first stage of satellite's evolution.

Four energetic full operative biogeochemical cycles are possible inside Titan's ocean, namely nitrogen (N-cycle), sulfur (S-cycle), iron (Fe-cycle) and carbon (C-cycle) and all of them could be connected each with other.

The BR of nitrate reduction to dinitrogen is a more thermodynamically favorable in the row of different inorganic substrates (Gaidos et al., 1999). All gaseous nitrogen in the contemporary Titan's atmosphere can be the product of this reaction (Simakov, 2000). Experiments of Hart with co-workers (2000) have shown that nitrate would be rapidly metabolized to N_2 in a CO_2 atmosphere by some microorganisms if liquid water and organic materials were present.

A very interesting bacteria have been discovered recently which use ammonium as an inorganic electron donor for denitrification (Jetten et al., 1999):

$$NH_4^+ + NO_3^- \rightarrow N_2 + 2H_2O$$

This reaction has a very favorable energetic (-357 kJ/mol) and presents a good source of free energy for the metabolism. Hydroxylamine (NH_2OH) and hydrazine (N_2H_4) are formed as intermediates and bicarbonate is the sole carbon source. This is the first case when hydrazine, a rocket fuel, is a free intermediate in any biological system. Both these components could be widespread in the Titan's environments and be used by microorganisms for energy transduction and the buildup of an electrochemical gradient. It is possible to hypothesize a start reaction as:

$$N_2H_4 \rightarrow N_2$$

which can evolve through:

$$NH_2OH \rightarrow N_2H_4 \rightarrow N_2$$

to

$$NO_3^- \rightarrow NH_2OH \rightarrow N_2H_4 \rightarrow N_2$$

at the rout of microbial evolution.

Another example we could see in using of Fe as an electron donor. Although reduction of nitrate by Fe(II) is thermodinamically feasible, this reaction is not occurring spontaneously, but it is ready catalyzed. Catalysis is usually defined as a process that accelerates a specific chemical reaction toward equilibrium through the use of a substance (i.e., catalysis) that does not undergo any net chemical change during the conversion of reactant to product. Iron can rapidly reduce nitrate at presence of Cu^{2+}, Sn^{2+} or Ag^+ (Jones, 1984). This reaction may be catalyzed via homogeneous, heterogeneous or microbial catalysis. The transition between these types of catalysis could enhance the efficiency of the Basic Reaction.

Briefly we can suppose a few evolutionary steps of the first protobiological metabolism on base of $NO_3^- \rightarrow N_2$ reaction: (1) the homogeneous catalysis of the BR with inorganic ions; (2) the origin of the simplest inorganic (for example, sulfur) membrane on base of Fe-containing clusters and others colloidal systems; (3) the evolution of inorganic membranes toward a complex organic system by scavenging surrounding organic molecules; (4) the increasing of a summary surface of the membrane which would serve to raise the efficiency of the BR (the origin of protomitochondrion); (5) the origin of the fermentation as an essential step for supplying the redox equivalents for the BR; (6) the origin of genetic information; (7) the transfer of all serving processes for the BR (respiration) out of protomitochondrion (it is the point of the division of life between prokariotes and eukariotes).

Dissimilatory ferric iron-reducing and ferrous iron-oxidizing organisms also can form the basis for a closed ecosystem which gains energy through cyclic reduction and oxidation of iron minerals, sometime by NO_3^--dependent way. Fe(III) oxyhydroxides are readily reduced with H_2S, inorganic sulfides (Pizzik and Sommer, 1981), elemental sulfur and various organic acids (Stumm and Morgan, 1981). On Earth microbial Fe(III) reduction is the major way of organic carbon oxidation in anaerobic environment (Lowe et al., 2000). Microorganisms utilizing Fe(III) as an electron acceptor were discovered into mesobiotic marine and freshwater anoxic sediments and submerged soils (Lovley, 1995). The denitrifying bacterium was isolated from the mud of Mariana Trench (Tamegai et al., 1997). It show greater tolerance to low temperature and high hydrostatic pressure (50 MPa). Thermobiotic ecosystems also contain bacteria able to reduce Fe(III) with formate, lactate or molecular hydrogen. The production of reduced end products, e.g. Fe(II), FeS by Fe-reduction and H_2S with such processes could resupply the basic reaction with reagents. Ferrous iron is oxidized chemically by a number of inorganic compounds, most notably molecular oxygen, manganese oxide (MnO_2) and nitrate. The Fe(III)-reducers could couple oxidizing of H_2 by ferric iron with chemolithoheterotrophic growth (Lovley, 1995). But the pure cultures of thermophilic microorganisms able to reduce Fe(III) coupled to the oxidation of H_2 also have been described (Slobodkin and Wiegel, 1997).

The main product of amorphic Fe(III) oxide reduction are magnetite (Fe_3O_4) and siderite ($FeCO_3$) which in some cases can be considered as biomarkers. At the other hand the biogenic solid-phase Fe(II) compounds could be subjects of microbially catalyzed NO_3^- dependent oxidation. The magnetite formation by biooxidation of Fe(II) coupled to denitrification could occur simultaneously (Chaudhuri et al., 2001). Gaidos with co-workers (1999) stated that an isolated subsurface ocean will approach chemical equilibrium and annihilate any ecosystems dependent on redox gradients unless there is a substantial alternative energy source. They missed a great potential of the biosystems to adapt at the environment and a great flexibility of the metabolic processes. The origin of the first living systems could reverse a process of the production of the entropy and disequilibrium concentrations of redox reactants could be driven by microorganisms themselves. The closed ecosystems on Earth are good examples. Besides, there are some mechanisms which could prevent the establishment of the chemical equilibrium into the ocean, such as volcanic and hydrothermal supplying of the reagents, radiolitic production of O_2 and H_2O_2 species caused by a decay of ^{40}K in the ice and in the ocean (Chyba and Hand, 2001), atmospheric supplying of highly reactive species, for example, molecules with triple C≡C bonds, and some kinds of prebiological chemical cycles connected with each other.

Reduction and oxidation of iron may proceed through a wide variety of biogeochemical pathways, including homogeneous, mineral surface-controlled, photochemical (unfortunately impossible inside Titan), and catalytically (including enzymatically) mediated electron-transfer reactions. There can be microorganisms which reduce of local ferric iron of the environment to the lesser oxidizer magnetite, using the hydrocarbons as the reducing agent. This cycle would demand only a few flow of external energy and reagent sources.

The reaction of nitrate/nitrite reduction could have served not only as an energy source but also as a source of reduced (or «fixed») nitrogen compounds, which are essential for the synthesis of biologically important ones. For example, NO_2^- and NO_3^- are quickly and efficiently reduced in contact with FeS into NH_4^+ at higher temperature and pressure (Brandes et al., 1998) which could exist on the bottom of Titan's ocean, whereas magnetite and basalt converted NO_3^- and NO_2^- to gaseous species, in particular, NO, NO_2 and N_2. Moreover, the Strecker synthesis and the Fe(II) reduction of nitrate to ammonia can be combined to form of amino acids from nitrite, Fe(II), cyanide, and formaldehyde (Summers, 1999). The lithotrophic iron oxidizers that use Fe^{2+} as the only energy source for their metabolism and use nitrate as an electron acceptor have been described (Straub et al., 1996). This bacterial process may significantly contribute to ferric iron formation in the zone of sediments. Others metal ions, such as Mn and U also could be used in the microbial respiration. One Mn^{2+} oxidizing bacterium has the genes for the key enzyme for fixing CO_2

(Caspi et al., 1996). There are many redox couples that can be used advantageously by microbes in the course of the evolution.

It is important to pay attention to a possible inhomogeneity of the ocean. Given its great depth, considerable ionic and osmotic gradients could extend from the ocean floor upward. Titan's internal ocean also could have different spatial gradients in salinity, temperature, density, ion ratios, part of which is driven by redox cyclic reactions. Thermodynamic models show that different temperatures are favorable to different chemosynthetic reactions, for example, an oxidation of H_2S, CH_4, Fe^{2+} and Mn^{2+} more readily occurs only below ~38 °C and methanogenesis and reduction of sulfate or S^0 favored at higher temperature (McCollom and Shock, 1997). Such gradients are very common in terrestrial see environments. For example, redox cycling of Fe and Mn occurs when opposite transport fluxes move the reduced and oxidized species in the opposite direction (van Cappellen et al., 1998). Large changes in the concentrations of dissolved cations and their phase conditions give an additional driving force for biogeochemical (or protobiogeochemical) cycle. The layers of water with different temperatures are very probable into the global ocean. Liquid water is densest at a temperature a few degrees above the freezing point, so water at the base of the icy shell must be more buoyant than warmer water inside ocean. Depending of convective moving the layers could mixing forming flows of different components. The hot plumes could rise directly to the ice and periodically melt a part of them. Several processes could cause persistent horizontal pressure gradients that induce long-lived flow patterns in the bulk of water. Long-lived thermal upwellings in the deep ocean may breach the stable stratified layer and permit the voluminous warm water to directly impinge on the ice shell. This would result in rapid melting so long as the supply of warm water is continued.

The putative life inside Titan does not depend on solar energy and photosynthesis for its primary energy supply and it is essentially independent of the surface circumstances. There could be microorganisms having a great similarity with the last common ancestor (LCA) on Earth. We can see that all Earth's thermophiles which are near the root of the universal phylogenetic tree show very interesting features (Stetter, 2001): First, all of them rely on respiration and there are no phototrophs among them, and, second, the majority of them is capable of anaerobic respiration. Both observations fit the «respiration early» scenario. This scenario states that the respiration metabolism evolved earlier than photosynthesis. Phylogenic analysis of the protein sequences from archaebacterial respiratory chains gives support to the view that several of the most important respiratory chains were present already in the LCA (Castresana and Moreira, 1999). A number of proteins involved in respiratory processes are homologous in arhaeal and bacterial species. At the other hand, there are no molecular data supporting the view that the LCA could perform oxygenic photosynthesis.

It has been proposed in several works that cytochrome oxidase, a membrane protein complexes that contain haem and copper, reducing oxygen to water as the terminal step in aerobic respiration, evolved from denitrification enzymes (Saraste and Castresana, 1994). The contemporary denitrification is an anaerobic respiratory process which reduces nitrate to dinitrogen and it involves some different steps. The evolution of enzymes from denitrification may have been through a change in the object of catalytic reaction from N_2O to O_2. Nitric oxide reductase shows a great similarity in primary structure to *cb*-type cytochrome-*c*-oxidase (Saraste and Castresana, 1994). The evolution never develops new things *de novo*, but it always uses previous structures and functions to create novelities. The gap between aerobic and anaerobic respiratory is not as large as previously thought. For example, the «strictly anaerobic» sulfate-reducing bacterium *Desulfovibrio gigas* was recently shown to have an aerobic respiratory chain, including a terminal oxidase of the cytochrome *bd*-type (Lemos et al., 2001).

The electron transfer chains could be very adaptable and have a few different terminal oxidases (Richardson, 2000). So, depending on availability of energetic sources, one of the metabolic way may predominate. Many present day Earth organisms maintain several types of respiratory chains that are differentially expressed in response to the availability of electron donors and acceptors in the medium. For example, *Shewanella putrefaciens* displays remarkable respiratory versatility and can utilize a variety of liquid and solid phases as sole terminal electron acceptor, including O_2, NO_3^-, Mn(IV), Fe(III), U(VI), SO_3^{2-}, $S_2O_3^{2-}$, trimethylamine-N-oxide, and fumarate (Blakeney et al., 2000). And it can utilize a variety of organic substances and H_2 as electron donors during anaerobic respiration.

The possible alkalinity of the putative ocean is not a serious problem for biological activity. On the contrary, the alkalinity of the ocean could be favorable to biogenesis by diminishing of Ca^{2+} and Mg^{2+} concentrations needed for biochemical function of proteins and origins of living things (Kempe and Kazmierczak, 2002). Alkaline oceans have been proposed for early Earth (Kempe and Degens, 1985) and Mars (Kempe and Kazmierczak, 1997) on base of mass balance consideration of the weathering of the early planetary crust. Primordial Titan's ocean could be alkaline as well. High carbonate concentration have originated through weathering of the satellite's silicate core. One of the most basic extremes on Earth known to allow growth is pH 12.5 (Duckworth et al., 1996). It has been shown that species of bacteria are living in the extremely alkaline (pH>12) and metal-rich conditions in the Lake Calumet region of southeast Chicago. One of these species, *Leptothrix mobilis* is associated with iron and manganese oxidation. These metals are deposited as oxides in sheaths that surround a chain of cell. After cell's death such sheaths could serve as source of nutrients for others species.

A rich chemosynthetic ecosystems could be associated with methane clathrate areas on the icy bed. Clathrates are cage structures formed by H_2O ice having 8 cavities for each 46 water molecules that can trap large «guest» molecules of nonpolar methane molecules. These clathrates could supply methane into Titan's atmosphere. A free or dissolved methane represents a major microbial source of energy. The cold methane vents induced by liquid methane could serve as a source for forming of the chemosynthetic ecosystems. These processes could involve a transfer of electrons from methane to sulfate or others electrons acceptors. The examples of such systems could be found around methane clathrate on the Earth's sea bed (Boetius et al., 2000). Some organisms are capable of disproportioning of methanol, methylamines or methylsulphides to methane and carbon dioxide, oxidizing $-CH_3$ groups to CO_2 anaerobically.

Microbial consortia based on anaerobic oxidation of methane coupled to sulfate reduction can support other microbial communities by generating of substantial biomass accumulation derived from methane. Massive microbial mats covering up to

4-meter-high carbonate buildups prosper at methane seeps in anoxic waters of the northwestern Black Sea shelf has been recently discovered (Michaelis et al., 2002). Methane is oxidized with equimolar amounts of sulfate, yielding carbonate and sulfide, respectively (Nauhaus et al., 2002). Generation of alkalinity favors the precipitation of methane-derived bicarbonate according to the following net reaction:

$$CH_4 + SO_4^{2-} + Ca^{2+} \rightarrow CaCO_3 + H_2S + H_2O$$

There is increasing evidence that the capacity for anaerobic oxidation of methane is present in several deep-branching clades of archaea (Orphan et al., 2001), which indicates an early evolutionary origin, just as for methanogenesis. Anaerobic oxidation of methane may have influenced the carbon isotope record. Moreover, recent studies suggest that on Earth sulfate reduction has already been prolific 3.5 Ga ago (Shen et al., 2001), close to the first occurrence of microfossils (3.6 Ga ago) and the first isotopic traces of bioorganic carbon cycling (3.8 Ga ago) (Mojszis et al., 1996). Hence, anaerobic oxidation of methane may have represented an important link in the biological cycling of carbon in an anoxic biosphere.

Moreover, it was shown experimentally that there are sulfate-reducing bacteria using aliphatic and aromatic hydrocarbons or fatty acids as electron donors (Rueter et al., 1994). A moderately thermophilic pure culture utilizes *n*-alkanes in oil for sulfate reduction to sulfide on the reaction:

$$C_{10}H_{22} + 7.75SO_4^{2-} + 5.5H^+ \rightarrow 10HCO_3^- + 7.75H_2S + H_2O$$

A mesophilic sulfate-reducing enrichment culture is shown to oxidize alkylbenzenes in oil to CO_2 also, and the upper water layer of Titan's ocean covered with film from hydrocarbons sunken from the surface could be a very good niche for such microorganisms.

Several alkylbenzenes, alkanes or alkenes are anaerobically utilized as carbon and electron sources by several species of denitrifying and ferric iron-reducing bacteria, also (Heider et al., 1999). The complex organic photochemistry in the Titan's atmosphere is a great source of these initial compounds. Hydrocarbon oxidation at the expense of nitrate or ferric iron yields a high amount of free energy. The conversion of hydrocarbons to CO_2, H_2 and acetate also gives a great source of H_2 for methanogens community. Hydrogen sulfide, the vast product of sulfate-reducing hydrogen consuming bacteria could be utilized by different H_2S-oxidizing organisms. Hydrogen sulfide itself is a strong reductant because of the stability of pyrite and elemental sulfur.

$$Fe^{2+} + 2H_2S \rightarrow FeS_2 + 2H^+ + H_2$$

A part of the H_2S could be oxidized to $S_2O_3^{2-}$ which is a key intermediate of the sulfur cycle. The $S_2O_3^{2-}$ might be reduced again to H_2S, oxidized to SO_4^{2-}, or disproportionated to H_2S and SO_4^{2-}. Sulfide oxidizing microorganisms gain metabolic energy by converting H_2S to sulfate and oxidation of H_2S supplies the greatest amount of energy for living community.

A coupling of H_2S oxidation to the reduction of either N_2 and CO_2 is energetically favorable and could provide a mechanism for nitrogen fixation. It has been shown that N_2 can be reduced by hydrogen sulfide to ammonia in the freshly precipitated FeS. A part of the H_2S reacts with iron compounds and precipitates as FeS or FeS_2. The FeS/FeS_2 redox couple could be used as a reductant for synthesis of C-H-O-N compounds — organic and amino acids. Chemosynthetic communities based on the consumption of gas-bearing (hydrogen, methane, carbon dioxide, hydrogen sulfide) cold fluids are widespread on the Earth (Barbieri et al., 2001). On Earth an anaerobic FeS oxidation with NO_3^- as electron acceptor may be catalyzed by *Thiobacillus denitrificans* or by anaerobic Fe(II) oxidizing and NO_3^- reducing bacteria (Benz et al., 1998). This is a good example of interaction between N-, Fe- and S-cycles. Iron sulfides may also be oxidized within the anoxic sediment where NO_3^-, Fe(III)-oxides, or MnO_2 are available as potential electron acceptors. A variety of sulfur compounds such as elemental sulfur, sulfate, thiosulfate, tritonate, tetrathionate and pentathionate are the products of this oxidation. The processes are summarized by equations (Schippers and Sand 1999):

$$FeS_2 + 7.5MnO_2 + 11H^+ \rightarrow Fe(OH)_3 + 2SO_4^{2-} + 7.5Mn^{2+} + 4H_2O$$
$$FeS + 1.5MnO_2 + 3H^+ \rightarrow Fe(OH) + S^0 + 1.5Mn^{2+}$$

An anaerobic regeneration of sulfate is possible via oxidation by Fe(III) which is a well-known oxidant for FeS_2 even at neutral pH (Bottrell et al., 2000):

$$FeS_2 + 14Fe^{3+} + 8H_2O \rightarrow 15Fe^{2+} + 2H_2SO_4 + 12H^+$$

Fe(III) is assumed also to be the oxidant for FeS and FeS_2 in the presence of MnO_2. On the iron sulfide surface, Fe(III) is reduced to Fe(II) which is reoxidized to Fe(III) by MnO_2. Thus, an Fe(II)/Fe(III) shuttle could transport electrons between the surfaces of the two solid compounds (Schippers and Jorgensen, 2001).

Fe^{3+} are widely considered as an important early respiratory substrate, for example, several organisms closely related to the LCA are able to use it as an electron acceptor, and even *Thermogota maritima*, which was for a long time considered to rely exclusively on fermentation, can respire using Fe^{3+} (Vargas et al., 1998). Iron-sulfur clusters are often found in bioenergetic proteins (Beinert, 2000) and they could be considered as relicts from earlier evolution steps when the clusters along, without proteins, catalyzed bioenergetic reactions (Cammack, 1996).

Along with the upper layer of the internal water ocean where a temperature and pressure are suitable for living processes there are some additional appropriate sites for biological and/or prebiological activity at present day (Simakov, 2001): (1) water pockets and liquid veins inside icy layer; (2) the places of cryogenic volcanism; (3) macro-, mini- and microcaves in icy layer connecting with cryovolcanic processes; (4) the brine-filled cracks in icy crust caused by tidal forces; (5) liquid water pools on the surface originated from meteoritic strikes; (6) the sites of hydrothermal activity on the bottom of ocean; two of them will be considered further.

3.2. WATER POCKETS INSIDE A LOWEST PART OF THE ICE LAYER

The formation of ice during freezing leads to concentrate of the impurities both organic and inorganic that have a lower melting temperature than water (e.g. brines) in the water ice crystal interfaces. Thus the small portion of the initial volume likely remained unfrozen and consisted of highly concentrated brines of different compounds. For example, complete freezing of H_2O-salts system occurs at -45 °C (Zolotov and Shock, 2000b). In clay permafrost soils in Siberia a liquid water was analyzed at temperatures up to -60 °C. Approximately 7% of Earth's surface is covered with sea ice and its evolution provides an useful analog to understand the possible behavior of the briny ice on Titan. The resulting Titan's ice could consist of pure water ice with brine inclusions which could have different properties depending on many conditions. The precise details of the formation of the satellite's ice layer are unknown, but we can conclude that a continuous network of aqueous veins could be formed in the lowest layer of the ice. The ions (SO_4^{2-}, HSO_4^-, NO_3^-) from solution and organic admixtures are concentrated in these veins (Fukazawa et al., 1998). Eutectic freezing also is the most promising concentration mechanism and a potential prebiological chemistry could exist in such environment (Levy et al., 2000; Miyakava et al., 2002). It has been shown that the ice matrice may accelerate certain prebiotic polymerization reactions. At freezing the increased concentration offsets the effect of the lower temperature and accelerates the reaction of polymerization of phosphoimidazolide-activated uridine, cytidine, adenine and guanine forming oligomers up to 11 bases long with high yields (Kanavarioti et al., 2001). The network of the channels and pores could be a good microbial habitat since a fluid flow is possible between these veins allowing nutrient exchange to occur. The extensive community of psychrophilic microorganisms containing algae, protozoa and bacteria colonized the sea ice on Earth (Staley and Gosink, 1999). Some of the gramm-negative sea ice bacteria have among the lowest maximum temperatures for growth known, <10°C for some strains. The liquid water, main electron acceptors and carbon source are in these environments. The lowest ice layer could be highly porous and extremely permeable for oceanic water containing electron acceptor and donors to create an energy flow. The Earth's brines tend to migrate within solid sea ice due to temperature and concentration variations, density/buoyancy characteristics, and the resulting gravity gradients. There are a great spatial heterogeneity to any sea-ice zone. The brines can be distributed throughout sea ice, making up volume percentages typically 5-40%. They vary in size from a few micrometers to millimeters and form labyrinth that the sea-ice organisms live. So, the brines could transport substances and heat inside the ice layer. A wide range of conditions can lead to brine mobilization, their vertical and lateral migration. Such migration could rapidly change the physical state of the ice. The most important feature of this migration is the possible transfer of ions and nutrients. The heat generation might be associated with ice I crystal structure changes and hydration reactions in the brine components. Such hypersaline environments are good habitats for Earth halotolerant and halophile microorganisms (Javor, 1989). Extremely halotolerant microbes have been found in low-temperature Antarctic hypersaline lakes (Franzmann et al., 1988). These salt saturated lakes remain unfrozen year-round despite on minimal winter temperatures about -60 °C, and all of them have complex microbial community. The family of «salt-loving» bacteria consisted of such microorganisms as *Halobacterium halobium*, along with species from such genera as *Oscillatoria*, *Phormidium, Nostoe*, and some others is well adapted to surviving in saturated salt solution. One of the most intriguing properties of these organisms is an ability to survive after dessicated into solid salt crystals and reviving after dissolution of these crystals (Grant et al., 1998). From exobiological point of view the members of halophilic family have been proposed as a possible life form on present Mars (Landis, 2001) or sometime in the past of the Red planet (Litchfield, 1998). Recently long-lived colonies of bacteria and fungi have been discovered in the salty (with 3.000 ppm salt concentration) soils of Antarctic's Dry Valleys (Mahaney et al., 2001). The hypersaline environment is a good candidate for a place of life's origin since a very diminished water activity could enhance the formation of macromolecules from biological monomers (Dundas, 1998). Salt-saturated brines may provide also isohaline environment for primitive vesicle structures.

A minimum temperature for enzyme activity was reported from -25 °C to -100 °C (Bragger et al., 2000). Microbes, some of which may be viable, have been found in ice cores drilled at Vostok Station at depth down to ~3.600 m (Abysov et al., 1999). The lower part of the Titan's ice crust could provide good habitable niches both for many forms of the Earth's life and for any domestic forms.

The capacity for survive at very low temperature is a prerequisite for life in the ice. To survive in a liquid habitat for extended periods of time, microorganisms must have exploitable energy sources. The lowest layer of the icy crust has permeable structure that permit the mass transfer of liquid water and copious nutrients for supporting viable microorganisms. The whole system of different channels, pores and brine pockets could serve for these purposes. The existence of bacterial communities on microscopic scale influences on surrounding environments also. On Earth it has been shown that cyanobacterial mats in Antarctic lakes are extremely resistant to dessication and freezing (Hawes et al., 1992) and they capable to colonize and grow within the extreme in-situ conditions of the permanent ice covers (e.g. long periods of freezing, freeze-thaw cycles, cold temperature).

Microbial community presented in the brine network have to base on chemolitho-autotrophic organisms gaining energy by the oxidation of inorganic substrates. Lithoautotrophic growth is an important presumption for long-term survival of microbes in extreme environments (Morita, 1999). Methanogenesis is a process on base of reduction of carbon dioxide by molecular hydrogen $4H_2 + CO_2 \rightarrow CH_4 + 2H_2O$ and carbon dioxide can be used as the sole carbon source. Instead CO_2 methanogens could also use a few simple organic compounds, such as acetate, formate, methanol and methylamines. A lot of Earth's methanogens are able to adapt to high salinity and they are found in millions of years old permafrost sediments (Rivkina et al., 1998). An unique hydrogen-based methanogenic subsurface community has been described recently (Chapelle et al., 2002) in which hydrogen-consuming methanogenic microorganisms form the base of a food chain. To survive in Titan's briny network lithotrophic microbial communities require a sustainable energy source such as hydrogen. A volumetric electrolysis of dirty ice due to an action of electric currents excited by strong magnetic field of Saturn could supply the methanogens with a large quantity of H_2 (Drobyshevski, 2002). The electrolysis products H_2 and O_2 could form a solid solution (clathrate hydrates) with concentration up to ~10-15% and redistributed by solid-state thermal convection. Hydrogen will diffuse out from the icy-matrice into brine drainage network and it could be consumed by the microorganisms. At high contents of the H_2 and O_2 (more then 15 wt%) ice becomes capable of detonation. Such global explosions inside icy satellites could serve as driving force of the chemical evolution on these objects (Drobyshevski et al., 1995) and form a temporal habitats inside icy layer. Others sources of hydrogen could be: 1) vented gas from hydrothermal vents. Berndt with co-workers (1996) found that both H_2 and CH_4 were generated by the reaction of CO_2-bearing, NaCl aqueous fluid with common igneous mineral olivine, (Mg, Fe)$_2$SiO$_4$ at 300 °C and 500 bars. This mineral is abundant in the chondrites, the most common type of meteorites. Other electron donors (Fe^{2+}, Mn^{2+}, and S) are constituents of both volcanic rocks and meteorites also and could be released by weathering; 2) water reactions with fresh mineral surfaces and ferrous iron; or 3) a process of H_2 production at time of rocks crushing (Freund et al., 2002).

Some scientists consider methanogenic Archea as one of the initial organisms from the origin of life on Earth. The gas formed by methanogens could have been recycled, for example via anaerobic oxidation of methane discussed earlier, leading to substrates (inorganic carbon and sulfide) for chemosynthesis of new biomass.

3.3. HYDROTHERMAL VENTS

It has been suggested that the satellite may has a metallic core and silicate mantle and this core can be assumed to provide internal heat through radioactive decay, subjecting the ocean floor to volcanic eruption. Tidal dissipation and radiogenic heat could cause in substantial heating in the rocky mantle, generating volcanism at the base of the water layer, similar to sub-oceanic one on Earth. The release of heat from Titan's core also could be accompanied by magma degassing and hydrothermal activity. Hydrothermal activity level could be estimated on calculations of internal heating. Up to 10^{10} liters per year of hydrothermal fluid may be generated at Titan's hydrothermal vents (Jakosky and Shock, 1998). The hydrothermal fluid is a major source of reductants such as H_2S, H_2, CH_4 and Fe^{2+}. Iron-sulfur chemistry in combination with geothermal settings has been implicated in the prebiotic production of organic molecules (Clark et al., 1998). Several abiotic processes could lead to synthesis of organic compounds and supply them to the putative ocean: (1) Fischer-Tropsch type (FTT) synthesis from cooling volcanic gases (Zolotov and Shock, 2000c); (2) aqueous synthesis in mixing zones of hydrothermal fluids and with oceanic water (Shock and Schulte, 1998); (3) aqueous FTT synthesis during initial stages of chemical interaction of oceanic water with basic/ultrabasic igneous rocks (Berndt et al., 1996). Hydrothermal reactions between the ocean and underlying rock layer should have reprocessed a portion of the ocean, along with cometary and chondritic material, into more complex organic compounds. Hydrothermal vents provide a great source of disequilibrium and metabolic energy for chemolithotrophic microorganisms. A mixing of hydrothermal fluids with seawater gives a metabolic energy potentially available from chemosynthetic reactions involving S, C, Fe, N and Mn compounds (McCollom and Shock, 1997). Theoretically, hydrothermal vents present very diverse environments for microbial life, such as black smoker chimney walls and orifices, warm vents and seeps, surface-attached microbial mats and floating particles in vent plumes. They could maintain the carbon cycle (Shock, 1997) and supply heat and biology useful substances for possible suboceanic habitats. The hydrothermal systems could support a wide spectrum of chemolithoautotrophic processes and they have been proposed as a possible site for the origin of the first life forms on any planet. Molecular hydrogen along with other substances such as sulfide, sulfur, and ferrous serve as important electron donors for different types of anaerobic respiration (nitrate-, sulfate-, sulfur-, and carbon dioxide respiration). Organic matter produced by chemolithoautotrophs serves as nutrients for heterotrophs. For Earth, the total potential for chemosynthetic primary production at deep see hydrothermal vents is estimated to be 10^{13} g biomass per year (McCollom and Shock, 1997). An organism with the highest growth temperature on Earth is *Pyrolobus fumarii* exhibiting an upper temperature border above 113 °C (Blochl et al., 1997) and it gains energy by chemolithoautotrophic nitrate reduction, forming ammonia ($NO_3^- \rightarrow NH_4^+$). The members of other genus *Pyrodictium* gain energy by reduction of S^0 ($S^0 \rightarrow H_2S$). The capacity to reduce of an elemental sulfur to H_2S is widespread among *Bacteria* and *Archaea* as an energy-conserving reaction (Stetter, 1996). Others chemolithoautotrophs able to grow by reduction of SO_4^{2-}, SO_3^{2-}, $S_2O_3^{2-}$ with H_2 (Stetter, 2001). Some microorganisms could gain energy by reduction of NO_3^- ($NO_3^- \rightarrow NO_2^-$) with Fe^{2+} (*Ferroglobus*), S^0 (*Aquifex*) and H_2S (*Ferroglobus*) as electron donors.

Carbon monoxide is regarded as the common component of volcanic gases in deep-sea environments, and this gas could serve as a sole source of carbon and energy to a variety of diverse bacteria on Earth (Sokolova et al., 2002). In anoxic environment CO could be oxidized with production of acetate, methane or CO_2 and H_2 and these wast products could be used as nutrients by the others microorganisms. For example, molecular hydrogen could be used in processes of methano- or acetogenesis, sulfate, sulfur, iron and nitrate reduction.

4. Conclusions

The environments mentioned above indicate that all conditions capable of supporting life are possible on Titan. All requirements needed for exobiology — liquid water which exists within long geological period, complex organic and inorganic chemistry and energy sources for support of biological processes are on Saturnian moon. On Earth life exists in all niches where water is in liquid form for at least a portion of the year. Subglacial life may be widespread among such planetary bodies as Jovanian satellites, Titan, and satellites of others giant planets, detected in our Galaxy at last decade (Perryman, 2000). The low temperature hypersaline brines has been proposed as habitat for microbial communities on Mars (Wynn-Williams et al., 2001). The existence of rich atmosphere is the main difference of Titan from the Jupiter's moons. This atmosphere could supply the large quantity of different organic compounds to putative ocean. There are some possible mechanisms for extensive, intimate interaction of a liquid water ocean with the surface of the ice crust. Titan provides also insights regarding the geological and biological evolution of early Earth during ice-covered phase. There is a huge deficiency of carbon in the contemporary environment and this missing carbon could be contained as biomass and dissolved organic carbon in the putative ocean. Possible metabolic processes, such as nitrate/nitrite reduction, sulfate reduction and methanogenesis could be suggested for Titan. Nitrate and sulfate could be predominant forms of N and S in the ocean and nitrate and/or sulfate reduction would have been potential sources of energy for primitive life forms. Given the possibility that organic compounds may be widespread in the ocean from synthesis within hydrothermal systems (Shock, 1992), derived from atmospheric chemistry and delivered by comets and meteorites these putative nitrate and sulfate reducers may have been either heteritrophic or autotrophic. Furthermore, at the presence of substantial amount of methane, the methanogenesis along with methanotrops also have been energetically favorable. Excreted products of the primary chemoautotrophic organisms could serve as a source for other types of microorganisms (heterotrophes) as it has been proposed for Europa (Gaidos et al., 1999; Chyba, 2000) and Mars (Boston et al., 1992).

5. References

Abyzov, S., Mitskevich, I.N., Poglazova, M.N., Barkov, N.I., Lipenkov, V.Y., Bobin, N.E., Kudryashov, B.B., and Pashkevich, V.M. (1999) Antarctic ice sheet as an object for solving some methodological problems of exobiology. *Adv.Space Res.* **23**, 371-76.

Anderson, J.D., Schubert, G., Jacobson, R.A., Lau, E.L., Moore, W.B., and Sjogren, W.L. (1998) Europa's differentiated internal structure: Inferences from four Galileo encounters. *Science* **281**, 2019-2022.

Barbieri, R., Ori, G.G., and Taviani, M. (2001) Phanerozoic submarine cold vent biota and its exobiological potential. In: P. Ehrenfreund, O. Angerer and B. Battrick (eds.) *Proc. First European Workshop on Exo/Astro-Biology,* Frascati, *21-23, May 2001*, ESA SP-496, pp. 295-298.

Beinert, H. (2000) Iron-sulfur proteins: ancient structures, still full of surprises. *J.Biol.Inorg.Chem.* **5**, 2-15.

Benz, M., Brune, A., and Schink, B. (1998) Anaerobic and aerobic oxidation of ferrous iron at neutral pH by chemolithotrophic nitrate-reducing bacteria. *Arch.Microbiol.* **169**, 159-165.

Berndt, M.E., Allen, D.E., and Seyfried, W.E. Jr. (1996) Reduction of CO_2 during serpentinization of olivine at 300 degrees C and 500 bar. *Geology* **24**, 351-354.

Blakeney, M.D., Moulaei, T., and DiChristina, T.J. (2000) Fe (III) reduction activity and cytochrome content of Shewanella putrefaciens grown on ten compounds as sole terminal electron acceptor. *Microbiol.Res.* **155**, 87-94.

Blochl, E., Rachel, R., Burggraf, S., Hafenbradl, D., Jannasch, H. W., and Stetter, K.O. (1997) *Pyrolobus fumarii*, gen. and sp. nov., represents a novel group of Archaea, extending the upper temperature limit for life to 113 degrees C. *Extremophiles* **1**, 14-21.

Boetius, A., et al. (2000) A marine microbial consortium apparently mediating anaerobic oxidation of methane. *Nature* **407**, 623-626.

Borucki, J.G. and Khare, B.N. (2001) Synthesis of organic molecules in the fracture zone of meteorite impacts on Europa. [abstract]. 1st Workshop of the Europa Focus Group, 2-3.

Boston, P.J., Ivanov, M.V., and McKay, C.P. (1992) On the possibility of chemosynthetic ecosystems in subsurface on Mars. *Icarus* **95**, 300-308.

Bottrell, S.H., Parkes, R.J., Cragg, B.A., and Raiswell, R. (2000) Isotopic evidence for anoxic pyrite oxidation and stimulation of bacterial sulphate reduction in marine sediments. *J.Geolog.Soc.* **157**, 711-714.

Bragger, J.M., Dunn, R.V., and Daniel, R.M. (2000) Enzyme activity down to -100 degrees C. *BBA-Protein Structure* **1480**, 278-282.

Brandes, J.A., et al. (1998) Abiotic nitrogen reduction on the early Earth. *Nature* **395**, 365-367.

Cammack, R. (1996) Iron and sulfur in the origin and evolution of biological energy conversion systems. In: H. Baltscheffsky (ed.) *Origin and evolution of biological energy conversion.* VCH, pp. 43-70.

Caspi, R., Haygood, B.M., and Tebo, B.M. (1996) Unusual ribulose-1,5-bisphosphate carboxylase/oxygenase genes from marine manganese-oxidizing bacterium. *Microbiology* **142**, 2548-2559.

Castresana, J. and Moreira, D. (1999) Respiratory chains in the Last Common Ancestor of living organisms. *J.Mol.Evol.* **49**, 453-460.

Chapelle, F., et al. (2002) A hydrogen-based subsurface microbial community dominated by methanogens. *Nature* **415**, 312-315.

Chaudhuri, S.K., Lack, J.G., and Coates, J.D. (2001) Biogenic magnetite formation through anaerobic biooxidation of Fe(II). *Appl.Envir.Microbiol.* **67,** 2844-2848.
Chyba, C.F. (2000) Energy for microbial life on Europa. *Nature* **403**, 381.
Chyba, C.F. and Hand, K.P. (2001) Life without photosynthesis. *Science* **292**, 2026-2027.
Chyba, C.F. and Phillips, C.B. (2001) Possible ecosystems and the search for life on Europa. *PNAS* **98**, 801-804.
Clark, P.D., Dowling, N.I., and Huang, M. (1998) Comments of the role of H_2S in the chemistry of Earth's early atmosphere and in prebiotic synthesis. *J.Mol.Evol.* **47**, 127-132.
Clarke, D.W. and Ferris, J.P. (1997) Chemical evolution on Titan: comparisons to the prebiotic Earth. *Orig.Life Evol.Biosphere* **27**, 225-248.
Deamer, D.W. and Pashley, R.M. (1989) Amphiphilic components of the Murchison carbonaceous chondrite: surface properties and membrane formation. *Orig.Life Evol.Biosphere* **19**, 21-28.
Drobyshevski, E.M., Chesnakov, V.A., and Sinitsyn, V.V. (1995) The electrolytical processes in dirty ices: Implications for origin and chemistry of minor bodies and related objects. *Adv.Space Res.* **16**, 73-84.
Drobyshevski, E.M. (2002) Galilean satellites as sites for incipient life, and the Earth as its shelter. In: M.B. Simakov and A. K. Pavlov (eds.) *Astrobiology in Russia*, Proc. Int. Workshop, St.Petersburg, Russia, 25-28, March 2002, pp. 47-62.
Duckworth, A.W., Grant, W.D., Jones, B.E., and Steenbergen, R.V. (1996) Phylogenetic diversity of soda lake alkaliphiles. *FEMS Microbiol.Ecol.* **19**, 181-191.
Dundas, I. (1998) Was the environment for primordial life hypersaline? *Extremophiles* **2,** 375-377.
English, M., Lara, L.M., Lorenz, R., Ratcliff, P.M., and Rodrigo, R. (1996) Ablation and chemistry of meteoritic materials in the atmosphere of Titan. *Adv.Space Res.* **17**, 157-160.
Fanale, F.P., Li, Y.-H., De Carlo, E., Farley, C., Sharma, S.K., Horton, K., and Granahan, J.C. (2001) An experimental estimate of Europa's "ocean" composition-independent of Galileo orbital remote sensing. *J.Geophys.Res.* **106**, 14595-14600.
Fortes, A.D. (2000) Exobiological implications of a possible ammonia-water ocean inside Titan. *Icarus* **146**, 444-452.
Franzmann, P.D., Stackebrandt, E., Sanderson, K., Volkman, J.K., Cameron, D.E., Stevenson, P.L., McMeekin, T.A., and Burton, H.R. (1988) *Halobacterium lacusprofundi* sp. nov., a halophilic bacterium isolated from Deep Lake, Antarctica. *System.Appl.Microbiol.* **11**, 20-27.
Freund, F., Dickinson, J.T., and Cash, M. (2002) Hydrogen in rocks: An energy source for deep microbial communities. *Astrobiology* **2**, 83-92.
Fukazawa, H., Sugiyama, K., Mae, S., Narita, H., and Hondoh, T. (1998) Acid ions at triple junctions of Antarctic ice observed by Raman scattering. *Geophys.Res.Lett.* **25**, 2845-2848.
Gaidos, E.J., Nealson, K.H., and Kirschvink, J.L. (1999) Life in ice-covered oceans. *Science* **284**, 1631-1633.
Grant, W.D., Gemmell, R.T., and McGenity, T.J. (1998) Halobacteria: the evidence for longevity. *Extremophiles* **2**, 279-287.
Grasset, O. and Sotin, C. (1996) The cooling rate of a liquid shell in Titan's interior. *Icarus* **123**, 101-112.
Grasset, O., Sotin, C., and Deschamps, F. (2000) On the internal structure and dynamics of Titan. *Planet. Space Sci.* **48**, 617-636.
Griffith, A.C. and Zahnle, K. (1995) Influx of cometary volatiles to planetary moons: The atmospheres of 1000 possible Titans. *J.Geophys.Res.* **100**, 16907-16922.
Hart, S.D., Currier, P.A., and Thomas, D.J. (2000) Denitrification by *Pseudomonas aeruginosa* under simulated engineered Martian conditions. *J.Brit.Interplanet.Sci.* **53**, 357-359.
Hawes, I., Howard-Williams, C., and Vincent, W. (1992) Dessication and recovery of Antarctic cyanobacterial mats. *Polar.Biol.* **12**, 587-594.
Heider, J., Spormann, A.M., Beller, H.R., and Widdel, F. (1999) Anaerobic bacterial metabolism of hydrocarbons. *FEMS Microbiol.Rew.* **22**, 459-473.
Jakosky, B.M. and Shock, E.L. (1998) The biological potential of Mars, the early Earth, and Europa. *J.Geophys.Res.* **103**, 19359-19364.
Javor, B. (1989) *Hypersaline Environments: Microbiology and Biogeochemistry*. Springer-Verlag.
Jessberger, E.K., Kissel, J., and Rahe, J. (1989) The composition of comets. In: *Origin and evolution of planetary and satellite atmospheres*. University of Arizona Press, pp. 167-191.
Jetten, M.S.M., et al. (1999) The anaerobic oxidation of ammonium. *FEMS Microbiol.Rev.* **22**, 421-437.
Jones, M.N. (1984) Nitrate reduction by shaking with cadmium. *Water Res.* **18**, 643-646.
Kanavariotti, A., Monnard, P.-A., and Deamer, D.W. (2001) Eutectic phases in ice facilitate nonenzymatic nucleic acid synthesis. *Astrobiology* **1**, 271-281.
Kasting, J.F. and Walker, J.C.G. (1981) Limits on oxygen concentration in the prebiological atmosphere and the rate of abiotic fixation of nitrogen. *J.Geophys.Res.* **86**, 1147-1158.
Kempe, S. and Degens, E.T. (1985) An early soda ocean? *Chem.Geol.* **53**, 95-108.

Kempe, S. and Kazmierczak, J. (1997) A terrestrial model for an alkaline martian hydrosphere. *Planet.Space Sci.* **45**, 1493-1499.

Kempe, S. and Kazmierczak, J. (2002) Biogenesis and early life on Earth and Europa: Favored by an alkaline ocean? *Astrobiology* **2**, 123-130.

Lammer, H., et al. (2001) Lightning activity on Titan: can Cassini detect it? *Planet.Space Sci.* **49**, 561-574.

Landis, G.A. (2001) Martian water: Are there extant halobacteria on Mars? *Astrobiology* **1**, 161-164.

Lara, L.M., Lellouch, E., Lopez-Moreno, J.J., and Rodrigo, R. (1996) Vertical distribution of Titan's atmospheric neutral constituents. *J.Geophys.Res.* **101**, 23262-23283.

Lemos, R.S., Gomes, C.M., Santana, M., LeGall, J., Xavier, A.V., and Teixeira, M. (2001) The 'strict' anaerobe *Desulfovibrio gigas* contains a membrane-bound oxygen-reducing respiratory chain. *FEBS Lett.* **469**, 40-43.

Levy, M., Miller, S.L., Brinton, K., and Bada, J.L. (2000) Prebiotic synthesis of adenine and amino acids under Europa-like conditions. *Icarus* **145**, 609-613.

Levy, R.L., Grayson, M.A., and Wolf, C.J. (1973) The organic analysis of the Murchison meteorite. *Geochim. Cosmochim.Acta* **37**, 467.

Litchfield, C. (1998) Survival strategies for microorganisms in hypersaline environments and their relevance to life on early Mars. *Meteor.Planet.Sci.* **33**, 813-819.

Lorenz, R.D. (2001) Of course Ganymede and Callisto have oceans: Application of a principle of maximum entropy production to icy satellite convection [abstract 1160]. *32nd Lunar and Planetary Science Conference.*

Lovley, D.R. (1995) Microbial reduction of iron, manganese, and other metals. *Adv.Agron.* **54**, 175-231.

Lowe, K.L., DiChristina, T.J., Roychoudhury, A.N., and van Cappellen, P. (2000) Microbiological and geochemical characterization of microbial Fe (III) reduction in salt marsh sediments. *Geomicrobiol.J.* **17**, 163-178.

Lunine, J.I. and McKay, C.P. (1995) Surface-atmosphere interactions on Titan compared with those on the pre-biotic Earth. *Adv.Space.Res.* **15**, 303-311.

Lunine, J.I. and Stevenson, D.J. (1983) Formation of the Galilean satellites in a gaseous nebula. *Icarus* **52**, 14-38.

Lunine, J.I. and Stevenson, D.J. (1987) Clathrate and ammonia hydrate at high pressure: Application to the origin of methane on Titan. *Icarus* **70**, 61-77.

Mahaney, W.C., et al. (2001) Morphogenesis of antarctic paleosols: Martian analogue. *Icarus* **154**, 113-130.

Mancinelli, R.L. and McKay, C.P. (1988) The evolution of nitrogen cycling. *Orig.Life Evol.Biosphere* **18**, 311-325.

Mautner, M.N. (2002) Planetary resources and astroecology. planetary microcosm models of asteroid and meteorite Interiors: Electrolyte solutions and microbial growth — implications for space populations and panspermia. *Astrobiology* **2**, 59-76.

Mautner, M.N., Leonard, R.L., and Deamer, D.W. (1995) Meteorite organics in planetary environments: hydrothermal release, surface activity, and microbial utilization. *Planet.Space Sci.* **43**, 139-147.

McCollom, T.M. and Shock, E.L. (1997) Geochemical constraints on chemolithoautotrophic metabolism by microorganisms in seafloor hydrothermal systems. *Geochim.Cosmochim.Acta* **61**, 4375-4391.

McCord, T.B., et al. (1998) Non-water-ice constituents in the surface material of the icy Galilean satellites for the Galileo near-infrared mapping spectrometer investigation. *J.Geophys.Res.* **103**, 8603-8626.

McCord, T.B., et al. (1999) Hydrated salt minerals on Europa's surface from the Galileo near-infrared mapping spectrometer (NIMS) investigation. *J.Geophys.Res.* **104**, 11827-11851.

McCord, T.B., Hansen, G.B., and Hibbitts, C.A. (2001) Hydrated salt minerals on Ganymede's surface: Evidence of an ocean below. *Science* **292**, 1523-1525.

McKay, C.P., et al. (2001) Physical properties of the organic aerosols and clouds on Titan. *Planet.Space Sci.* **49**, 79-99.

Melosh, H.J., Ekholm, A.G., Showman A.P. and Lorenz, R.D. (2002) Is Europa's subsurface water ocean warm? [abstract 1824]. *33rd Lunar and Planetary Science Conference.*

Michaelis, W., et al. (2002) Microbial reefs in the Black sea fueled by anaerobic oxidation of methane. *Science* **297**, 1013-1015.

Miyakawa, S., Cleaves, H.J., and Miller, S.L. (2002) The cold origin of life: An implications based on the hydrolytic stabilities of hydrogen cyanide and formamide. *Orig.Life Evol.Biosphere* **32**, 195-208.

Mojzsis, S.J., Arrhenius, G., McCleesan, K.D., Harrison, T.M., Nutman, A.P., and Friend, C.R.L. (1996) Evidence for life on Earth before 3.8 billion years ago. *Nature* **384**, 55-59.

Morita, R.Y. (1999) Is H_2 the universal energy source for long-term survival? *Microb.Ecol.* **38** 307-320.

Nauhaus, K., Boetius, A., Kruger, M., and Widdel, F. (2002) In vitro demonstration of anaerobic oxidation of methane coupled to sulphate reduction in sediment from a marine gas hydrate area. *Environm.Microbiol.* **4**, 296-305.

Orgel, L.E. (1998) The origin of life — How long did it take? *Orig.Life Evol.Biosphere* **28**, 91-96.

Orphan, V.J., et al. (2001) Comparative analysis of methane-oxidizing archaea and sulfate-reducing bacteria in anoxic marine sediments. *Appl.Envirron.Microbiol.* **67**, 1922-1934.

Ottley, C.J., Davison, W., and Edmunds, W.M. (1997) Chemical catalysis of nitrate reduction by iron (II). *Geochim.Cosmochim.Acta* **61**, 1819-1828.

Perryman, M.A.C. (2000) Extra-solar planets. *Rep.Progr.Phys.* **63**, 1209-1272.

Pizzik, A.J. and Sommer, S.E. (1981) Sedimentary iron monosulfides: Kinetics and mechanism of formation. *Geochim.Cosmochim.Acta* **45**, 687-689.

Ramirez, S.I., Navarro-Gonzalez, R., Coll, P., and Raulin, F. (2001) Possible contribution of different energy sources to the production of organics in Titan's atmosphere. *Adv.Space Res.* **27**, 261-270.

Richardson, D.J. (2000) Bacterial respiration: a flexible process for a changing environment. *Microbiology-UK* **146**, 551-571.

Rivkina, E., Gilichinsky, D., Wagener, S., Tiedje, J., and McGrath, J. (1998) Biogeochemical activity of anaerobic microorganisms from buried permafrost sediments. *Geomicrobiol.J.* **15**, 187-193.

Rueter, P., et al. (1994) Anaerobic oxidation of hydrocarbons in crude-oil by new types of sulfate-reducing bacteria. *Nature* **372,** 455-458.

Ruiz, J. and Fairen, A.G. Seas under ice: Stability of water oceans whithin icy worlds. In: H.Lacoste (ed.) *Proc. 2nd European Workshop on Exo/Astrobiology,* Graz, Austria, 16-19 September 2002, ESA SP-518, pp. 551-552.

Saraste, M. and Castresana, J. (1994) Cytochrome oxidase evolved by tinkering with denitrification enzymes. *FEBS Lett.* **341**, 1-4.

Schippers, A. And Jorgensen, B.B. (2001) Oxidation of pyrite and iron sulfide by manganese dioxide in marine sediments. *Geochim.Cosmochim.Acta* **65**, 915-922.

Schippers, A. And Sand, W. (1999) Bacterial leaching of metal sulfides proceeds by two indirect mechanisms via thiosulfate or via polysulfides and sulfur. *Appl.Environ.Microbiol.* **65**, 319-321.

Shen, Y.A., Buick, R., and Canfield, D.E. (2001) Isotopic evidence for microbial sulphate reduction in the early Archaean era. *Nature* **410**, 77-81.

Shock, E.L. 1992 Chemical environments of submarine hydrothermal systems. *Orig.Life Evol.Biosphere* **22**, 67-107.

Shock, E.L. 1997 High-temperature life without photosynthesis as a model for Mars. *J.Geophys.Res.* **102,** 23687-23694.

Shock, E.L. and Schulte, M.D. (1998) Organic synthesis during fluid mixing in hydrothermal systems. *J.Geophys.Res.* **103**, 28513-28527.

Simakov, M.B. (2000) Dinitrogen as a possible biomarker for exobiology: The case of Titan. In: G. A. Lemarchand and K. J. Meechs.) *Bioastronomy'99: A new era in Bioastronomy*, Sheridan Books, pp. 333-338.

Simakov, M.B. (2001) The possible sites for exobiological activity on Titan In: P.Ehrenfreund, O.Angerer and B. Battrick (eds.) *Proc.First European Workshop on Exo/Astro-Biology,* Frascati, *21-23, May 2001,* ESA SP-496, pp. 211-214.

Slobodkin, A. and Wiegel, J. (1997) Fe (III) as an electron acceptor for H_2 oxidation in thermophilic anaerobic enrichment cultures from geothermal areas. *Extremophiles* **1**, 106-109.

Sokolova, T.G., et al. (2002) Anaerobic CO-oxidizing, H_2-producing prokariotes from volcanic habitats. In: : M.B. Simakov and A .K Pavlovs.) *Astrobiology in Russia*, Proc. Int. Workshop, St.Petersburg, Russia, 25-28, March 2002, pp. 156-163.

Spohn, T. and Schubert, G. (2003) Oceans in the icy Galilean satellites of Jupiter? *Icarus*, **161**, 456-467.

Staley, J.T. and Gosink J.J. (1999) Poles apart: Biodiversity and biogeography of sea ice bacteria. *Ann.Rev.Microbiol.* **53**, 189-215.

Stetter, K.O. (1996) Hyperthermophilic procariotes. *FEMS Microbiol.Rev.* **18**, 149-158.

Stetter, K.O. (2001) Hyperthermophilic microorganisms. In: G. Horneck (ed.) *Astrobiology. The Quest for the Conditions of Life*. Springer, pp. 169-184.

Straub, K.L., Benz, M., Schink, B., and Widdel, F. (1996) Anaerobic, nitrate-dependent microbial oxidation of ferrous iron. *Appl.Environm.Microbiol.* **62**, 1458-1460.

Stumm, W. and Morgan, J.J. (1981) *Aquatic chemistry*. John and Wiley and Sons.

Summers, D.P. (1999) Sources and sinks for ammonia and nitrite on the early Earth and the reaction of nitrite with ammonia. *Orig.Life Evol.Biosphere* **29**, 33-46.

Tamegai, H., Li, L., Masui, N., and Kato, C. (1997) A denitrifying bacterium from the deep sea at 11000-m depth. *Extremophiles* **1**, 207-211.

Thompson, R.W. and Sagan, C. (1992) Organic chemistry on Titan — Surface interactions. In: *Proceedings of the Symposium on Titan*, Toulouse, September 1991, ESA SP-338, pp. 167-182.

van Cappellen, C.P., Viollier, E., and Roychoudhury, A. (1998) Biogeochemical cycles of manganese and iron at the oxic-anoxic transition of a stratified marine basin (Orca Basin, Gulf of Mixico). *Envirn. Sci. Technol.* **32**, 2931-2939.

Vargas, M., Kashefi, K., Blunt-Harris, E.L., and Lovley, D.R. (1998) Microbial evidence for Fe(III) reduction on early Earth. *Nature* **395**, 65-67.

Whittet, D.C.B., et al. (1996) An ISO SWS view of interstellar ices: first results. *Astron.Astrophys.* **315**, L357-L360.

Wynn-Williams, D.D., Cabrol, N.A., Grin, E.A., Haberle, R.M., and Stoker, C.R. (2001) Brines in seepage channels as eluants for subsurface relict biomolecules on Mars? *Astrobiology* **1,** 165-184.

Yung, Y.L. (1987) An update of nitrile photochemistry on Titan. *Icarus* **72,** 468-472.

Yung, Y.L. and McElroy, M.B. (1979) Fixation of nitrogen in the prebiotic atmosphere. *Science* **203**, 102-104.

Zahnle, K., Pollack, J.B., Grinspoon, D., and Dones, L. (1992) Impact-generated atmospheres over Titan, Ganymede, and Callisto. *Icarus* **95**, 1-23.

Zimmer, C., Khurana, K.K., and Kivelson, M.G. (2000) Subsurface oceans on Europa and Callisto: Constraints from Galileo magnetometer observations. *Icarus* **147**, 329-347.

Zolotov, M.Y. and Shock, E.L. (2000a) Mass balance constraints on the elemental composition of the ocean on Europa [abstract 1580]. *31st Lunar and Planetary Science Conference.*

Zolotov, M.Y. and Shock, E.L. (2000b) Freezing of oceanic water on Europa: Theoretical modelling [abstract 1726]. *31st Lunar and Planetary Science Conference.*

Zolotov, M.Y. and Shock, E.L. (2000c) A thermodynamic assessment of the potential synthesis of condensed hydrocarbons during cooling and dilution of volcanic gases. *J.Geophys.Res.* **105**, 539-560.

Biodata of **Julian Chela-Flores** author of ***"Astrobiology's Last Frontiers: Distribution and Destiny of Life in the Universe"***

Dr. Julian Chela-Flores current positions are: Staff Associate of the Abdus Salam International Center for Theoretical Physics (ICTP), Trieste and Professor, Institute for Advanced Studies, Caracas. His particular area of expertise is Astrobiology. He is a Fellow of: The Latin American Academy of Sciences, The Third World Academy of Sciences, the Academy of Creative Endeavors (Moscow) and a Corresponding Member of the Academia de Fisica Matematicas and Ciencias Naturales (Caracas). Dr. Chela-Flores obtained his Ph.D. (1969) in quantum mechanics in the University of London, England. He was a researcher at the Venezuelan Institute for Scientific Research and Professor at Simon Bolivar University (Caracas) until his retirement in 1990. Dr. Chela-Flores has been the organizer of the Trieste Conferences on ***Chemical Evolution and the Origin of Life***, every three years since 1991 (and is the co-director of ICTP since 1995). He is also the editor of all Proceedings from the Trieste conferences. In 1999 he co-directed the ***Iberoamerican School of Astrobiology*** in Caracas, Venezuela. His book ***The Science of*** **Astrobiolog [http://www.wkap.nl/prod/b/0-7923-7125-9]**, has been published in 2001 within the frame of the **C**ellular **O**rigin and **L**ife in **E**xtreme Habitats book series [http://www.wkap.nl/prod/s/COLE].

E-mail: **chelaf@ictp.trieste.it**

J. Seckbach (ed.), Origins, 667–679.

ASTROBIOLOGY'S LAST FRONTIERS:
Distribution and Destiny of Life in the Universe.

JULIAN CHELA-FLORES
The Abdus Salam, International Centre for Theoretical Physics
P.O.Box 586; Strada Costiera 11; 34136 Trieste, Italy and
Instituto de Estudios Avanzados, Caracas 1015A, Venezuela.

1. Introduction

The four areas that define the new science of astrobiology are the origin, evolution, distribution and destiny of life in the universe. It is undoubtedly the fourth one, which is most likely to encourage interdisciplinary dialogue. This is not only in relation with the more evident physically- and biologically-oriented disciplines, but also with other not so close, but yet significant branches of culture. Indeed, the destiny of life in the universe will depend on the cosmological model that will eventually be best supported by experiments in the high-energy physics domain; it will also depend on observation of the largest red-shifts of receding galaxies.

The case of an ever-expanding universe is an evident example of great interest to several areas of culture; for if there would not be a reversal of the big bang expansion (which seems to have some data in its favor), the possibility of the phenomenon of life surviving on a large enough scale of time seems unlikely (Dyson, 1979; Krauss and Starkman, 2000). The question of eternal life provides a frontier between astrobiology and the main monotheistic religions of the world.

Natural theology is the body of knowledge about divine action, which may be obtained by human reason alone without the aid of revelation, in other words, independent of the process by means of which truth contained in the Holy Books has been communicated to humans. Philosophers have distinguished between 'truths' of reason and truths of revelation. Leaving aside those truths of revelation, the subject matter of natural theology requires to rationalize the origin and evolution of man, a phenomenon that until very recently was a question that had been strictly framed within the context of this planet. Now with the SETI project and the existence of an ever increasing number of extra solar planets being brought to our attention, those questions lead us to a fruitful dialogue between natural theology and astrobiology - neither of the two areas, however, overlap; but astrobiology is needed as the obvious scientific background against which to rationalize the place of man in the universe, a topic of perennial interest to other branches of culture.

In this context, it is opportune to recall that the belief that mankind has a special position in the universe is central to the mosaic tradition. It is therefore relevant for natural theologians to be aware of the recent developments in the frontier of astrobiology.

The community at large should be aware of the consequences of discovering a "second genesis', which astrobiology is actively searching in our solar system as well as other solar systems. While the realms of natural theology and natural science are distinct areas of culture, since the Renaissance Newtonian mechanics has influenced natural theologians. The destiny of life in the universe provides a cornerstone not only for discussions of science and religion, but it is also a subject of interest for theoretical philosophy: To put the question of "What is the position of the human being in the cosmos?" in its proper perspective, we should carry the argument a little further:

The question of what is our place in cosmic evolution has been raised not only across the cultural boundaries, but also the question has emerged frequently at various times in the past: almost a century and a half separate us from the concern expressed by Sir Charles Lyell regarding the position of man amongst all living organisms. It seems that the underlying difficulty is still related to inserting Lyell's "ugly facts" of Darwinian evolution into our culture (Desmond and Moore, 1991). Darwin was prudent enough to avoid ideological issues. Darwin concentrated on the narrow, but transcendental problem of establishing the theory of evolution of life on Earth; he prudently postponed the wider issue of the position of man in the universe.

On the other hand, the distribution of life in the universe, in spite of lacking solid theoretical, or observational bases, has in its favor that it is an aspect of astrobiology that can be probed in terms of a much wider range of projects and space missions especially dedicated to explore the question of whether or not we are alone in the universe. In the companion paper (Chela-Flores, 2002) we have persevered with the hypothesis that was first formulated during the commemoration of the centenary of Charles Darwin's death (Dawkins, 1982). The points made on that occasion are relevant in the context of this paper: Darwin's theory of evolution is assumed to be the only theory that can adequately account for the phenomena that we associate with life anywhere in the universe. In the present paper we examine the question whether evolution of intelligent behavior is just a matter of time and preservation of steady planetary conditions, and hence ubiquitous in the universe. This question is motivated by the problem of understanding the bases on which we can get significant insights into the question of the distribution of life in the universe. Such information would also have deep implications on the other frontier of astrobiology mentioned above, the destiny of life in the universe.

In Secs. 2-4 arguments are presented within the theory of evolution that lead us to a rationale, within the life sciences, for testing the universality of Darwin's theory.

2. To What Extent is Natural Selection Stronger than Contingency?

2.1. NATURAL SELECTION VERSUS CONTINGENCY

For simplicity, in Sec. 4 we shall focus on various degrees of convergence across all scientific disciplines that are relevant to astrobiology. Features that become more, rather than less similar through independent evolution, will be called 'convergent'. In fact, convergence in biology is often associated with similarity of function, as in the evolution of wings in birds and bats. An example is provided by New World cacti and the African spurge family, such as some euphorbs (for example, the *Euphorbia stapfii*), and even some members of the Madagascar Didieraceae *(Didiera madagascariensis).* These plants are similar in appearance, being succulent, spiny, water-storing, and adapted to desert conditions (Tudge, 1991; Nigel-Hepper, 1982). However, they are classified in separate and distinct families, sharing characteristics that have evolved independently in response to similar environmental challenges, and hence this is a typical case of convergence.

On the other hand, adaptive radiation is a second Darwinian concept that will be necessary for understanding the examples in this section that will argue in favor of restricted predictability in astrobiology. In fact, adaptive radiation simply means evolution of an animal, or plant group into a wide variety of types adapted to specialized modes of life. In other words, adaptive radiation signifies evolutionary diversification of a single lineage into a variety of species with different adaptive properties. Darwin's finches provide the classical example of adaptive radiation. Thirteen species of Darwin's finches live in the Galapagos Islands. They differ in the shape of their beaks. It is remarkable how versatile their beaks can be: keratin, the substance form which they are made, can be easily molded by evolutionary pressures, thereby facilitating the origin of all the species now inhabiting these islands. (Besides, there is an additional species inhabiting Cocos Island in the Costa Rican territorial waters in the Pacific Ocean, north of the Galapagos Islands.) We shall consider below, some less familiar examples of adaptive radiation that will argue in favor of certain degree of predictability in biology (and clearly in all its branches, particularly in astrobiolgy, in view of the assumed universality of Darwinism).

Before we approach the central question of convergent evolution, however, we should recall, firstly that so far factors influencing the relative degree to which the Earth biota has been shaped are still a debatable topic. According to the hypothesis of universal Darwinism (cf., Sec. 1), life on Earth, and elsewhere, may have been shaped mainly by:

(i) Contingency, or by

(ii) The gradual action of evolution's main driving force (i.e., natural selection).

However, it may be possible to ascertain experimentally whether, independent of historical contingency, natural selection is powerful enough for organisms living in similar environments to be shaped to similar ends. For this reason, it is important to highlight the following three recent examples, which suggest that, to a certain extent and in certain conditions, natural selection may be stronger than chance:

• Sticklebacks *(Gasterosteus spp.)* are members of the fish Gasterosteide Family; they are, in fact, small northern hemisphere fishes with spines in front of their dorsal and ventral fins. Empirical evidence has been presented (Rundle *et al.*, 2000), which demonstrates that natural selection plays a fundamental role in the early stages of the evolution of new species. The case in question concerns a striking example of convergent evolution: sticklebacks were originally of marine origin *(G. aculeatus),* but were trapped in three different lakes on the Canadian Pacific coast (Lakes Priest, Enos and Paxton) at the end of the last ice age, as the glaciers retreated. In spite of the fact that the lakes have remained isolated, the same two species have formed in all three lakes. Two non-interbreeding varieties have evolved in each lake; firstly, a bulky bottom-dwelling (benthic) type and, secondly, streamlined actively swimming individuals (limnetic), which feed in the open water. The mating preferences of the fish were tested. It was found that benthic mated with benthic, both from their own lake and form the other two. In addition, it was found that limnetic mated with limnetic. This remark demonstrates that natural selection has been the driving force in the evolution that has taken place in the above three Canadian lakes.

• Black European fruit flies *(Drosophila subobscura)* were transported to California over twenty years ago. This event has provided the possibility of testing the role of natural selection in two different continental environments. Pacific coast *D. subobscura* (Santa Barbara to Vancouver) was compared in wing-length with European ones (from Southern Spain to the middle of Denmark). After half a dozen generations were observed in similar conditions, the increase in wing length was almost identical (4%). This is a compelling case in favor of the key role played by natural selection in evolution (Huey *et al.*, 2000).

• Anole lizards from some Caribbean islands (*Anolis spp.)* provide another example of evolutionary convergence. The islands are Cuba, Hispaniola (shared by Haiti and the Dominican Republic), Jamaica and Puerto Rico (the so called Greater Antilles). The observed phenomenon suggests that in similar environments adaptive radiation can overcome historical contingencies in order to produce strikingly similar evolutionary outcomes. We could even say that there has been *replicated adaptive radiation* in the various islands. In fact, what has been shown (Losos *et al.,* 1998) is that although it had been known for some time that dozens of species thrive on these islands, some groups of lizards from different islands living in similar environments also look similar. Genetic analysis has shown that similar traits have evolved in distantly related species for coping with similar environments (such as tree-tops or ground-dwelling): anoles that live on the ground have long, strong hind legs, while those living at tree tops have large toe-pads and short legs. Repeated evolution of similar groups of species (both morphologically and ecologically) suggests that adaptation is responsible for the predictable evolutionary responses of the anole lizards of the Caribbean. Indeed, we can speak in this case of evolutionary history repeating itself (Vogel, 1998).

These arguments support the following views:

(a) It is necessary to argue that within certain limits the outcome of evolutionary processes might be rather predictable (Conway Morris, 1998).

(b) Certain directions of evolution may carry such decisive selective advantages as to have high probability of occurring elsewhere [in the universe] as well (De Duve, 1995).

(c) The ubiquity of evolutionary convergence argues against the view that biological diversity on this planet is unique to Earth (Conway Morris, 2002).

2.2. CONSTRAINTS ON CHANCE

It is instructive to appreciate that several constraints on chance are relevant to the question of whether life elsewhere might follow pathways analogous to the transition terrestrial ones. Various examples of constraints on chance have been enumerated elsewhere (De Duve, 1995). We shall comment on them and later we shall provide an additional example:

• Not all genes are equally significant targets for evolution. The genes involved in significant evolutionary steps are few in number; they are the so-called regulatory genes. In these cases mutations may deleterious and are, therefore, not fixed.

• Once a given evolutionary change has been retained by natural selection, future changes are severely constrained; for example, once a multicellular body plan has been introduced, future changes are not totally random, as the viability of the organisms narrows down the possibilities. For instance, once the body plan of mammals has been adopted, mutations such as those that are observed in *Drosophila*, which exchange major parts of their body, are excluded. Such fruit-fly mutations are impossible in the more advanced mammalian body plan.

• Not every genetic change retained by natural selection is equally decisive. They may lead more to increasing biodiversity, rather than contributing to a significant change in the course of evolution. This may be illustrated within the Solanaceae family (Brown, 1966): one tomato chromosome has a region between its center and end ('centromere' and 'telomere'), which consists of a row of segments in which DNA is compacted into tight masses, largely inactive in transcription ('chromomeres') in Petunia, in spite of being another genus of the same family, the abundance of chromomeres is not preserved, since larger blocks of heterochromatin are observed. A tiny bit of heterochromatin may be superficially indistinguishable from a eukaryotic 'chromomere'. These two genera illustrate how quickly evolution can induce rearrangements of heterochromatin, while preserving general chromosome structure; this mutation has not contributed any significant change in the course of evolution.

Besides the constraints on chance mentioned in the last section, we should recall the eternal confrontation deep in the fabric of evolutionary theory, brought to popular attention by Jacques Monod in his book *Chance and Necessity* (Monod, 1972).

Indeed, implicit in Darwin's work we have chance represented by the randomness of mutations in the genetic patrimony, and their necessary filtering by natural selection. However, the novel point of view that astrobiology forces upon us accept that randomness is built into the fabric of the living process. Yet, contingency, represented by the large number of possibilities for evolutionary pathways, is limited by a series of constraints, as mentioned before. What we feel is even more significant for astrobiology, is to recognize that natural selection necessarily seeks solutions for the adaptation of evolving organisms to a relatively limited number of possible environments. We will see in cosmochemistry (cf., Sec. 4.1) that the elements used by the macromolecules of life are ubiquitous in the cosmos.

To sum up, the finite number of environments forces upon natural selection a limited number of options for the evolution of organisms. From these remarks we expect convergent evolution to occur repeatedly, wherever life arises. It will make sense, therefore, to search for the analogues of the attributes that we have learnt to recognize in our own particular planet.

3. If Local Environments can Shape how Organisms Change with Time Through Natural Selection, How Frequent are Cosmic Environments Favorable to Life's Origin?

In the original theory of Darwin the possibility had been raised that local environments shape how organisms change with time through natural selection. In view of the evidence presented in Secs. 1 and 2, we assume that natural selection is the main driving force of evolution in the universe. For these reasons it is relevant to question whether local environments that were favorable for the emergence of life on the early Earth, were at all unique, occurring exclusively in our own solar system.

Alternatively, we may question, as we do in the present paper, whether other environments fulfill conditions favorable to life's origin, either within our solar system, or in any of the planets, or satellites, in the multiple examples of solar systems known at present. In addition, we suppose that on such bodies steady conditions are preserved. By steady conditions it should be understood that the planet where life may evolve is bound to a star that lasts long enough: in other words, the time available for the origin and evolution of life should be sufficient to allow life itself to evolve before the solar system of the host planet, or satellite, reaches the final stages of stellar evolution, such as the red-giant and supernova phases. It is also assumed that major collisions of large meteorites with the world supporting life are infrequent after the solar system has passed through its early period of formation.

In such steady conditions, the gradual action of natural selection will be expected to be the dominant mechanism in evolution, in view of the assumed universality of Darwinism (cf., Sec. 1). Besides, convergence is expected to occur at various levels: in prebiotic evolution (cf., Sec. 4.1), chemical evolution (cf., Sec. 4.7), and finally, in biological evolution (cf., Sec. 4.4 and 4.5). Fortunately, the existence of steady Earth-like planetary conditions is an empirical question for which we will be able to give partial answers in the foreseeable future; observational means will be provided by progress in space interferometry. One possibility for achieving this objective will be the Darwin Project (Fridlund, 2001).

4. Evolutionary Convergence Beyond the Framework of Biology

The above examples suggest that natural selection is powerful enough to shape organisms to similar ends, independent of contingency. We shall enumerate examples of general aspects of convergence at different levels of evolution: cosmic, planetary, molecular, biochemical and, once again, biological.

4.1. COSMIC CONVERGENCE

Hydrogen and helium make up almost the totality of the chemical species of the Universe. Only 2% of matter is of a different nature, of which approximately one half is made by the five additional biogenic elements (C, N, O, S, P). From organic chemistry we know that nuclear synthesis is relevant for the generation of the elements of the Periodic Table beyond hydrogen and helium and, eventually, for the first appearance of life in solar systems. The elements synthesized in stellar interiors are needed for making the organic compounds that have been observed in the circumstellar, as well as in interstellar medium, in comets, and other small bodies. The same biogenic elements are also needed for the synthesis of biomolecules of life (cf., Sec. 4.7, "Convergence in Biochemistry"). Besides, the spontaneous generation of amino acids in the interstellar medium is suggested by general arguments based on biochemistry: the detection of amino acids in the room-temperature residue of an interstellar ice analogue that was ultraviolet-irradiated in a high vacuum has yielded 16 amino acids, some of which are also found in meteorites (Muñoz Caro *et al.*, 2002; cf., also Bernstein *et al.*, 2002).

4.2. CONVERGENCE IN THE FORMATION OF SOLAR SYSTEMS

Our solar system formed in the midst of a dense interstellar cloud of dust and gas. This event may have been triggered by the shock-wave of a supernova explosion. Indeed, there is some evidence for the presence of silicon carbide (carborundum, SiC) grains in the Murchison meteorite, where isotopic ratios demonstrate that they are matter from a type II supernova (Hoppe *et al.*, 1997). In the case of our solar system this occurred 4.6

billion years before the present (Gyr BP). We may be observing such a circumstellar disk around a young sun-like star (3 million years, Myr, of age) in the constellation of Monoceros:

A spinning cloud around the young star is having the brightness of its light regularly faded (it goes dim for 18 days every 48.3 days) by the interference of stellar photons and the cloud itself (Kerr, 2002). Several earlier examples of circumstellar disks are known, including a significantly narrow one around an 8 Myr old star. (The narrowness suggests the presence of planets constraining the disk.) The observation was by means of a spectrometer on board of the Hubble Space Telescope (Schneider *et al.,* 1999). The matter of the original collapsing interstellar cloud that does not coalesce into the star collapses into a spinning circumstellar disk, where planets are thought to be formed in a process of accretion (some planetesimals collide and stay together, due to the gravitational force). In addition, a variety of small bodies are formed in the disk, prominent amongst which are comets, asteroids and meteorites.

4.3. CONVERGENT ORIGINS OF HYDROSPHERES AND ATMOSPHERES

Collisions of comets are thought to have played a role in the formation of the hydrosphere and atmosphere of habitable planets, such as the Earth. (This question may be settled in the next few years by the fleet of space missions that will interact with a variety of comets.) An alternative scenario supposes that volatiles emerged form the planet's interior through volcanic vents. The source of comets is the Oort cloud and Kuiper belt. These two components of the outer solar system seem to be common for other solar systems.

Hence we can also recognize evolutionary convergence in this cosmic sense. But there are additional factors, which contribute to the formation of habitable planets. We have already mentioned meteorites in the context of the formation of solar systems. In fact, the Murchison meteorite may even play a role in the origin of life: According to chemical analyses, some amino acids have been found in several meteorites: in Murchison we find basic molecules for the origin of life such as lipids, nucleotides, and over 70 amino acids (Cronin and Chang, 1993). Most of the amino acids are not relevant to life on Earth and may be unique to meteorites. This remark demonstrates that those amino acids present in the meteorite, which also play the role of protein monomers, are indeed of extraterrestrial origin.). In addition, chemical analysis has exposed the presence of a variety of amino acids in the Ivuna and Orgueil meteorites (Ehrenfreund et al., 2001). If the presence of biomolecules on the early Earth is due in part to the bombardment of interplanetary dust particles, or comets and meteorites, then the same phenomenon could have taken place in any of other solar systems.

4.4. CONVERGENCE AT THE LEVEL OF INVERTEBRATES

The evolutionary biology of the Bivalvia, both at the level of zoology and paleontology, provide multiple examples of convergence and parallel evolution, a fact that makes difficult the interpretation of their evolutionary history (Harper *et al.,* 2000). Specific examples of convergence in mollusks have been pointed out earlier in the case of the camaenid, helminthoglyptid and helcid snails (Chela-Flores, 2001).

4.5. CONVERGENCE AT THE LEVEL OF VERTEBRATES

The examples of sticklebacks and anole lizards (cf., Sec. 2) provide two additional examples of evolutionary convergence in the vertebrates. Earlier we pointed out further examples amongst the vertebrates: Passeriformes is a group of birds that is often confused with Apodiformes, but is not related to them. Since this example of swallows and swifts constitutes a classical example of evolutionary convergence, we need not repeat it here (Chela-Flores, 2001).

4.6. CONVERGENCE AT THE MOLECULAR BIOLOGY LEVEL

Convergent evolution is manifest at the active sites of enzymes, in whole proteins, as well as in the genome itself, as we proceed to show:

- The lipases and serine proteases. They have identical active sites (histidine, serine, aspartate form a catalytic triad). On the other hand, their folding is completely different (Tramontano, 2002).
- The northern sea cod (*Boreogadus saida*, Svalbard Norway) is an economically important marine fish of the family Gadidae. It is found on both sides of the North Atlantic. The distantly related order Perciformes with its suborder Percoidei, contains the sea basses, sunfishes, perches, and, more relevant to our interest, the notothenioid fishes from the Antarctic (*Dissotichus mawsoni*, McMurdo Sound, Antarctica). In spite of their distant relationship with cods, they have evolved the same type of antifreeze proteins, in which there are repeats of the amino acids threonine, alanine and proline (Chen *et al.*, 1997.) These proteins are active in the fish's blood and avoid freezing by preventing the ice crystals from growing larger. The Antarctic fish protein arose over 7 million years BP, while the Arctic cod first appeared about 3 million years BP (both species arose in different episodes of genetic shuffling).
- The blind cave fish *Astyanax fasciatus* are sensitive to two long wavelength visual pigments. In humans the long wavelength green and red visual pigments diverged about 30 Myr BP. The mammalian lineage diverges form fishes about 400 Myr BP, but a recent episode in evolution has granted fish multiple wavelength-sensitive green and red pigments. Genetic analysis demonstrates that the red pigment in humans and fish evolved independently from the green pigment by a few identical amino acid substitutions (Yokoyama and Yokoyama, 1990), a clear case of evolutionary convergence at the molecular level.

4.7. CONVERGENCE IN BIOCHEMISTRY

The universal nature of biochemistry has been discussed form the point of view of the basic building blocks (Pace, 2001). One of the main points made in that paper is that it seems likely that the basic building blocks of life anywhere will be similar to our own. Amino acids are formed readily from simple organic compounds and occur in extraterrestrial bodies such as the above example of meteorites (cf., the Murchison meteorite in the section on "Convergent origins of hydrospheres and atmospheres"). Themes that are suggested to be common to life elsewhere in the cosmos are the capture of adequate energy from physical and chemical processes to conduct the chemical transformations that are necessary for life: lithotropy, photosynthesis and chemosynthesis. Other factors that militate in favor of the universality of biochemistry are physical constraints (temperature, pressure and volume), as well as genetic constraints.

5. Discussion

The arguments presented in this paper militate in favor of planning experiments based on standard biology in solar system search for microorganisms, in view of both evolutionary convergence and universal Darwinism.

Ever since the publication of *The origin of species*, it has been argued that the possible course of evolution may be dominated by either contingency or the gradual action of natural selection. Random gene changes accumulating over time may imply that the course of evolution is generally unpredictable over time. But some care is needed in this assertion: What is certainly unpredictable is the future of a given lineage. This is due to the strong role in shaping life's evolutionary pathways played by contingent factors, such as extinction of species due to asteroid collisions with a given inhabited world, or other calamities. However, such uncertainties are of lesser interest to the larger issues that are relevant to astrobiology, namely the inevitability of the appearance of biological features, such as vision, locomotion, nervous systems, brains and, consequently intelligent behavior.

We have argued that contingency does not contradict a certain degree of predictability of the eventual biological properties that are likely to evolve. We should underline "biological property", as opposed to a "lineage", which is clearly a strongly dependent on contingency. In the companion paper we have sketched preliminary suggestions that would test whether the early stages of evolution of intelligent behavior are being driven mainly by natural selection at the cellular level in the context of planetary science (Chela-Flores, 2002).

Strictly speaking, a certain degree of predictability in astrobiology is not on the same footing as predictability in either physics (for instance, the exact time for an eclipse), or chemistry (for example, the result of a chemical reaction).

It is relevant to keep in mind that from neuroscience other aspects of the limited predictability of biology have been pointed out for some time (Manger et al, 1998). In fact, the tremendous expansion of the cetacean neocortex, while preserving the basic modular subdivisions (analogous to small-brained mammals such as the mouse) has led to a conjecture: There is a limited set of underlying mechanisms that are accessed to building brains.

Thus in the course of evolutionary history repetition of similar structures does occur. This also suggests that module size is evolutionary stable *across species*. Indeed with Krubitzer we can even question whether species differences are really so different (Krubitzer, 1995). In other words, in order to understand the evolution of the cortex and how organisms increase in behavioral complexity, it may be necessary to assume that anatomical alterations are generated from similar mechanisms.

Further, while the future of a given lineage cannot be predicted with certainty, the conjectured common mechanisms of cortex evolution, give us some certainty of the types of modification that are likely to occur in brain evolution. This aspect of brain evolution gives us some confidence that to a certain extent evolutionary history repeats itself in the context of the neurosciences; as in the earlier examples from the evolution of multicellular organisms (cf., Sec. 2). However, from what we have explained above, such limited predictability is nevertheless relevant to aspects of astrobiology's last frontiers: the question of life in the solar system, as well as evolution of intelligent behavior in the universe.

6. References

Bernstein, M.P., Dworkin, J.P., Sandford, S.A., Cooper, G.W. and Allamandola, L.J. (2002) Racemic amino acids from the ultraviolet photolysis of interstellar ice analogues. Nature **416**, 401-403.

Brown, S.W. (1966) Heterochromatin. Science **151**, 417-425.

Chela-Flores, J. (2001) The New Science of Astrobiology From Genesis of the Living Cell to Evolution of Intelligent Behavior in the Universe. Kluwer Academic Publishers: Dordrecht, The Netherlands, p. 161.

Chela-Flores, J. (2002) Can evolutionary convergence be tested on Europa? European Space Agency Special Publication ESA SP **518**, 337-340.

Chen, L., DeVries, A.L. and Cheng, C-H. C. (1997) Convergent evolution of antifreeze glycoproteins in Antarctic notothenioid fish and Arctic cod. Proc. Natl. Acad. Sci. USA **94,** 3817-3822.

Conway Morris, S. (1998) *The Crucible of Creation. The Burgess Shale and the Rise of Animals,* London, Oxford University Press, p. 202.

Conway Morris, S. (2002) First Steps towards defining galactic niches. Paper presented at the IAU Symposium 213 Bioastronomy 2002 *Life Among the Stars*. Summary of Proceedings, Hamilton Island, Great Barrier Reef, Australia July 8-12, Australian Centre for Astrobiology, Sydney, p. 12.

Cronin, J. R. and Chang, S. (1993) Organic matter in meteorites: Molecular and isotopic analyses of the Murchison meteorite, In: J. M., Mendoza-Gomez, C. X. and Pirronello, V. (eds.) *The chemistry of life's origins*. Greenberg, Dordrecht: Kluwer Academic Publishers, pp. 209-258.

Dawkins, R. (1983) Universal Darwinism, In: D.S. Bendall (ed.) *Evolution from molecules to men*, London, Cambridge University Press, pp. 403-425.

De Duve, C. (1995) *Vital Dust. Life as a cosmic imperative,* New York, Basic Books, A Division of HarperCollins Publishers, pp. 296-297.

Desmond, A. and Moore, J. (1991) *Darwin.* Michael Joseph, London, pp. 412-413.

Dyson, F.J. 1979 Time without End: Physics and Biology in an Open Universe. Reviews of Modern Physics **51,** 447-460.

Ehrenfreund, P., Glavin, D.P., Botta, O., Cooper, G. and Bada, J.L. (2001) Extraterrestrial amino acids in Orgueil and Ivuna: Tracing the parent body of CI type carbonaceous chondrites. Proc. Natl. Acad. Sci. USA, **98**, 2138-2141.

Fridlund, M. (2001) The Darwin Mission and Exo-/Astrobiology. ESA SP **496**, 267-271.

Harper, E.M. Taylor, J.D. and Crame, J.A. (2000) *The evolutionary biology of the Bivalvia,* Geological Society Special Publication **177,** London. www.palaeo-electronica.org/2001_2/books/evo.pdf

Hoppe, P., Strebel, R., Eberhadt, P., Amari, S. and Lewis, R.S. (1997) Type II supernova matter in a silicon carbide grain from the Murchison meteorite. Science **272**, 1314-1317.

Huey, R., Gilchrist, G., Carlson, M., Berrigan, D. and Serra, L. (2000) Rapid evolution of a geographic cline in size in an introduced fly. Science **287**, 308-309.

Kerr, R. A. (2002) Winking star unveils planetary birthplace. Science **296**, 2312-2313.

Krauss, L.M. and Starkman, G.D. (2000) Life, the Universe and Nothing: Life and Death in an Ever-Expanding Universe. *Astrophysical Journal* **531**, pp. 22-30.

Krubitzer, L. (1995) The organization of neocortex in mammals: are species differences so different? Trends in Neurosci., **18**, 408-417.

Losos, J. B., Jackman, T.R., Larson, A., de Queiroz, K. and Rodriguez-Schettino, L. (1998) Contingency and determinism in replicated adaptive radiations of island lizards. Science **279**, 2115-2118.

Manger, P., Sum, M., Szymanski, M., Ridgway, S. and Krubitzer, L. (1998) Modular subdivisions of dolphin insular cortex: Does evolutionary history repeats itself? J. Cognitive Neurosci., **10**, 153-166.

Monod, J. (1972) *Chance and Necessity An Essay on the Natural Philosophy of Modern Biology,* Collins, London, p. 136.

Muñoz Caro, G.M. , Meierhenrich, U.J. , Schutte, W.A., Barbier, B., Arcones Segovia, A., Rosenbauer, H.,Thiemann, W.H.P., Brack, A., Greenberg, J.M. (2002) Amino acids from ultraviolet irradiation of interstellar ice analogues. Nature **416**, 403-406.

Nigel-Hepper, F. (1982) *Kew: Gardens for Science and Pleasure*, London, Her Majesty's Stationary Office, p. 81.

Pace, N. R. (2001) The universal nature of biochemistry. Proc. Natl. Acad. Sci. USA **98**, 805-808.

Rundle H. D., Nagel, L., Wernick Boughman, J. and Schluter, D. 2000, Natural selection and parallel speciation insympatric sticklebacks. Science **287**, 306-308.

Schneider, G., Smith, B. A., Becklin, E. E., Koerner, D. W., Meier, R., Hines, D. C., Lawrence, P. J., Terrile, R.J., Thomson, R. I. and Rieke, M. 1999, NICMOS Imaging of the HR 4796A Circumstellar Disk. Ap. J. **513,** L1217-L130.

Tramontano, A. (2002) Private communication.

Tudge, C. (1991) *Global Ecology,* London: Natural History Museum Publications. p. 67.

Vogel, G. (1998) For island lizards, history repeats itself. Science, **279**, 2043.

Yokoyama, R., and Yokoyama, S. (1990) Convergent evolution of the red- and green-like visual pigment genes in fish, *Astynax fasciatus,* and human. Proc. Natl. Acad. Sci. USA**, 87**, 9315-9318.

ORGANISM INDEX

SUBJECT INDEX

B

D

F

G

H

I

U

V

W

Cellular Origin and Life in Extreme Habitats

1. J. Seckbach (ed.): *Enigmatic Microorganisms and Life in Extreme Environments.* 1999 ISBN 0-7923-0-7923-5492-3
2. J. Seckbach (ed.): *Journey to Diverse Microbial Worlds: Adaptation to Exotic Environments.* 2000 ISBN 0-7923-602060
3. J. Chela-Flores: *The New Science of Astrobiology: From Genesis of the Living Cell to Evolution of Intelligent Behavior in the Universe.* 2001 ISBN 0-7923-7125-9
4. J. Seckbach (ed.): *Symbiosis: Mechanisms and Model Systems.* 2001 ISBN 1-4020-0189-4
5. A. Oren: *Halophilic Microorganisms and their Environments.* 2002 ISBN 1-4020-0829-5
6. J. Seckbach (ed.): *Origins.* Genesis, Evolution and Diversity of Life. 2004 ISBN 1-4020-1813-4

KLUWER ACADEMIC PUBLISHERS – DORDRECHT / BOSTON / LONDON

GPSR Compliance

The European Union's (EU) General Product Safety Regulation (GPSR) is a set of rules that requires consumer products to be safe and our obligations to ensure this.

If you have any concerns about our products, you can contact us on ProductSafety@springernature.com

In case Publisher is established outside the EU, the EU authorized representative is:

Springer Nature Customer Service Center GmbH
Europaplatz 3
69115 Heidelberg, Germany

Batch number: 10755458

Printed by Printforce, the Netherlands